Essentials of Fluidization Technology

Essentials of Fluidization Technology

Edited by

John R. Grace

Xiaotao Bi

Naoko Ellis

WILEY-VCH

Editors

Prof. John R. Grace
University of British Columbia
Chemical and Biological Engineering
Vancouver Campus
2360 East Mall
Canada
V6T 1Z3 NK

Prof. Xiaotao Bi
University of British Columbia
Chemical and Biological Engineering
Vancouver Campus
2360 East Mall
Canada
V6T 1Z3 NK

Prof. Naoko Ellis
University of British Columbia
Chemical and Biological Engineering
Vancouver Campus
2360 East Mall
Canada
V6T 1Z3 NK

Cover Credit: iStock # 1153898634/ DamienGeso.

Library of Congress Card No.:
applied for

British Library Cataloguing-in-Publication Data
A catalogue record for this book is available from the British Library.

Bibliographic information published by the Deutsche Nationalbibliothek
The Deutsche Nationalbibliothek lists this publication in the Deutsche Nationalbibliografie; detailed bibliographic data are available on the Internet at <http://dnb.d-nb.de>.

Print ISBN: 978-3-527-34064-4
ePDF ISBN: 978-3-527-69947-6
ePub ISBN: 978-3-527-69949-0
oBook ISBN: 978-3-527-69948-3

Cover Design: Adam-Design, Weinheim, Germany
Typesetting SPi Global, Chennai, India
Printing and Binding CPI books GmbH, Leck

Printed on acid-free paper

10 9 8 7 6 5 4 3 2 1

Contents

Preface *xix*
Acknowledgement *xxi*

1 **Introduction, History, and Applications** *1*
John R. Grace
1.1 Definition and Origins *1*
1.2 Terminology *2*
1.3 Applications *3*
1.4 Other Reasons for Studying Fluidized Beds *4*
1.5 Sources of Information on Fluidization *8*
References *8*
Problems *9*

2 **Properties, Minimum Fluidization, and Geldart Groups** *11*
John R. Grace
2.1 Introduction *11*
2.2 Fluid Properties *11*
2.2.1 Gas Properties *11*
2.2.2 Liquid Properties *12*
2.3 Individual Particle Properties *12*
2.3.1 Particle Diameter *12*
2.3.2 Particle Shape *12*
2.3.3 Density and Internal Porosity *13*
2.3.4 Surface Roughness *14*
2.3.5 Terminal Settling Velocity *14*
2.3.6 Coefficients of Restitution (Particle–Particle and Particle–Wall) *15*
2.3.7 Dielectric Constant and Electrical Conductivity *16*
2.3.8 Thermal Properties *16*
2.4 Bulk Particle Properties *16*
2.4.1 Mean Particle Diameter and Particle Size Distribution *16*
2.4.2 Bulk Density, Voidage, and "Flowability" *16*
2.5 Minimum Fluidization Velocity *18*
2.5.1 Measuring U_{mf} Experimentally *18*

2.5.1.1 Pressure Drop vs. Superficial Velocity Method *18*
2.5.1.2 Other Experimental Methods of Determining U_{mf} Experimentally *20*
2.5.2 Predicting U_{mf} Based on Particle and Fluid Properties *20*
2.5.3 Other Factors Influencing the Minimum Fluidization Velocity *23*
2.6 Geldart Powder Classification for Gas Fluidization *24*
2.7 Voidage at Minimum Fluidization *27*
Solved Problem *28*
Notations *28*
References *29*
Problems *31*

3 Liquid Fluidization *33*
Renzo Di Felice and Alberto Di Renzo
3.1 Introduction *33*
3.2 Field of Existence *33*
3.3 Overall Behaviour *35*
3.4 Superficial Velocity–Voidage Relationship *37*
3.5 Particle Segregation and Mixing *40*
3.6 Layer Inversion Phenomena *41*
3.6.1 Predicting the Layer Inversion Voidage (via the Particle Segregation Model) *43*
3.6.2 Layer Inversion Velocity *46*
3.7 Heat and Mass Transfer *46*
3.7.1 Interphase Transfer *46*
3.7.2 Bed-to-Surface Transfer *47*
3.8 Distributor Design *48*
Solved Problems *48*
Notations *51*
References *52*
Problems *53*

4 Gas Fluidization Flow Regimes *55*
Xiaotao Bi
4.1 Onset of Fluidization *55*
4.2 Onset of Bubbling Fluidization *55*
4.3 Onset of Slugging Fluidization *57*
4.4 Onset of Turbulent Fluidization *58*
4.5 Termination of Turbulent Fluidization *62*
4.6 Fast Fluidization and Circulating Fluidized Bed *62*
4.7 Flow Regime Diagram for Gas–Solid Fluidized Beds *64*
4.8 Generalized Flow Diagram for Gas–Solid Vertical Transport *65*
4.9 Effect of Pressure and Temperature on Flow Regime Transitions *68*
Solved Problems *70*
Notations *71*
References *72*
Problems *74*

5 **Experimental Investigation of Fluidized Bed Systems** *75*
Naoko Ellis
5.1 Introduction *75*
5.1.1 Design Goals *75*
5.1.2 Purpose of Experiments *76*
5.2 Configuration and Design *76*
5.2.1 Column Material *78*
5.2.2 Distributor Plate *79*
5.2.3 Plenum Chamber *81*
5.2.4 Feeding System (See Also Chapter 11) *81*
5.2.4.1 Hopper Design *82*
5.2.5 Auxiliary Components *83*
5.2.5.1 Secondary Injection of Fluid *84*
5.3 Fluidizability and Quality of Fluidization *84*
5.3.1 Characterization of Flowability of Particles *84*
5.3.2 Particle Properties and Fluidizability *86*
5.3.3 Quality of Fluidization *86*
5.4 Instrumentation and Measurements *87*
5.4.1 Pressure Measurements *87*
5.4.2 Thermal Sensors *89*
5.4.3 Optical Probes *89*
5.4.4 Non-invasive Measurements *90*
5.4.5 Visualization Measurements *91*
5.4.6 Acoustic Emission Measurements *92*
5.4.7 Solids Circulation Flux *92*
5.4.8 Gas or Solids Sampling *92*
5.4.9 Mixing and Residence Time Distribution *93*
5.5 Operation of Fluidized Beds *93*
5.5.1 Start-Up and Shutdown *94*
5.5.2 Steady-State Operation *94*
5.6 Data Analysis *95*
5.6.1 Frequency Analysis *95*
5.6.2 Bivariate Time Series *96*
5.6.2.1 Joint Probability Density Function *96*
5.6.2.2 Cross-Correlation Function *96*
5.6.2.3 Cross-Spectral Density Function *97*
5.6.3 Other Signal Analyses *98*
Solved Problem *98*
Notations *98*
References *100*
Problems *104*

6 **Computational Fluid Dynamics and Its Application to Fluidization** *109*
Tingwen Li and Yupeng Xu
6.1 Two-Fluid Model *110*
6.1.1 Governing Equations *110*

6.1.2 Kinetic Granular Theory *112*
6.1.3 Frictional Model *114*
6.1.4 Limitations of TFM *115*
6.2 Discrete Particle Method *115*
6.2.1 Governing Equations *116*
6.2.2 Limitations of CFD–DPM *119*
6.3 Gas–Solid Interaction *119*
6.3.1 Gas–Solid Drag *119*
6.3.2 Gas–Solid Heat Transfer *121*
6.4 Boundary Conditions *122*
6.5 Example and Discussion *123*
6.5.1 TFM Simulation of a Bubbling Fluidized Bed with Tube Bundle *123*
6.5.2 CFD–DPM Simulation of a Small-Scale Circulating Fluidized Bed *124*
6.5.3 Discussion *125*
6.6 Conclusion and Perspective *126*
Solved Problem *126*
Notations *127*
References *128*

7 Hydrodynamics of Bubbling Fluidization *131*
John R. Grace
7.1 Introduction *131*
7.2 Why Bubbles Form *133*
7.3 Analogy Between Bubbles in Fluidized Beds and Bubbles in Liquids *134*
7.4 Hydrodynamic Properties of Individual Bubbles *135*
7.4.1 Rising Velocity of Single Bubbles *135*
7.4.2 Bubble Wakes *135*
7.4.3 Bubble Breakup and Maximum Stable Size *137*
7.4.4 Interphase Mass Transfer and Cloud Formation *137*
7.5 Bubble Interactions and Coalescence *139*
7.6 Freely Bubbling Beds *139*
7.6.1 Flow of Gas by Translation of Bubbles *139*
7.6.2 Mean Bubble Diameter as a Function of Height and Gas Velocity *140*
7.6.3 Rising Velocity of Bubbles in Freely Bubbling Bed *142*
7.6.4 Bubble Volume Fraction (Holdup) *142*
7.6.5 Bed Expansion *142*
7.6.6 Radial Nonuniformity of Bubbles and Its Effect on Mixing *143*
7.6.7 Turnover Time, Solids Mixing, and Particle Segregation *144*
7.6.8 Gas Mixing *145*
7.7 Other Factors Influencing Bubbles in Gas-Fluidized Beds *146*
Solved Problem *147*
Notations *147*
References *148*
Problems *152*

8 Slug Flow *153*
John R. Grace
8.1 Introduction *153*
8.2 Types of Slug Flow *153*
8.3 Analogy Between Slugs in Fluidized Beds and Slugs in Liquids *155*
8.4 Experimental Identification of the Slug Flow Regime *155*
8.5 Transition to Slug Flow *156*
8.6 Properties of Single Slugs *156*
8.7 Hydrodynamics of Continuous Slug Flow *158*
8.7.1 Slug Rising Velocity *158*
8.7.2 Slug Spacing and Length *158*
8.7.3 Time Between Successive Slugs and Slug Frequency *159*
8.7.4 Bed Expansion *159*
8.7.5 Uniformity and Symmetry of Flow *159*
8.8 Mixing of Solids and Gas in Slugging Beds *159*
8.9 Slugging Beds as Chemical Reactors *160*
Solved Problem *160*
Notations *161*
References *161*

9 Turbulent Fluidization *163*
Xiaotao Bi
9.1 Introduction *163*
9.2 Flow Structure *165*
9.2.1 Axial and Radial Voidage Distribution *165*
9.2.2 Local Void Size and Rise Velocity *166*
9.2.3 Void Phase Volume Fraction, Void Phase, and Dense Phase Solids Holdup *167*
9.3 Gas and Solids Mixing *168*
9.3.1 Gas Mixing *168*
9.3.2 Solids Mixing *171*
9.4 Effect of Column Diameter *172*
9.5 Effect of Fines Content *173*
Solved Problem *173*
Notations *175*
References *176*
Problems *180*

10 Entrainment from Bubbling and Turbulent Beds *181*
Farzam Fotovat
10.1 Introduction *181*
10.2 Definitions *182*
10.2.1 Transport Disengagement Height (TDH) *182*
10.2.2 Elutriation *184*
10.3 Ejection of Particles into the Freeboard *184*
10.4 Entrainment Beyond the Transport Disengagement Height *185*
10.5 Entrainment from Turbulent Fluidized Beds *190*

10.6 Parameters Affecting Entrainment of Solid Particles from Fluidized Beds *191*
10.6.1 Properties of Particles *191*
10.6.2 Geometry and Shape of Freeboard *192*
10.6.3 Dense Bed Height *192*
10.6.4 Internals *192*
10.6.5 Pressure and Temperature *193*
10.6.6 Electrostatic Charges *194*
10.7 Possible Means of Reducing Entrainment *195*
Solved Problem *195*
Notations *196*
References *197*
Problems *201*

11 Standpipes and Return Systems, Separation Devices, and Feeders *203*
Ted M. Knowlton and Surya B. Reddy Karri
11.1 Standpipes and Solids Return Systems *203*
11.1.1 Packed Bed Flow *206*
11.1.2 Fluidized Bed Flow *206*
11.1.3 Types of Standpipes *206*
11.1.3.1 Overflow Fluidized Standpipe *207*
11.1.3.2 Underflow Packed Bed Standpipe *208*
11.1.3.3 Underflow Fluidized Standpipes *210*
11.2 Standpipes in Recirculating Solids Systems *212*
11.2.1 Automatic Solids Recirculation Systems *212*
11.2.2 Controlled Solids Recirculation Systems *213*
11.2.3 Function of a Standpipe *215*
11.3 Standpipes Used with Nonmechanical Solids Flow Devices *216*
11.3.1 Nonmechanical Solids Control Mode Operation *216*
11.3.2 Automatic Solids Flow Devices *219*
11.3.2.1 Cyclone Diplegs *219*
11.4 Solids Separation Devices *222*
11.4.1 Cyclones *222*
11.4.1.1 Cyclone Types *223*
11.4.1.2 Flow Patterns in Cyclones *225*
11.4.1.3 Cyclones in Series *226*
11.4.1.4 Cyclones in Parallel *226*
11.4.1.5 Internal vs. External Cyclones *227*
11.4.1.6 Cyclone Inlet Design *227*
11.4.1.7 Effect of Solids Loading *227*
11.4.1.8 Gas Outlet Tube *228*
11.4.1.9 Inlet Gas Velocity *228*
11.4.1.10 Cyclone Dimensions and Design *228*
11.4.2 Other Separation Devices *229*
11.4.2.1 Particulate Scrubbers *229*
11.4.2.2 Fabric Filters *229*

11.4.2.3 Granular Bed Filters *230*
11.4.2.4 Electrostatic Precipitators *230*
11.4.2.5 U-Beams *230*
11.5 Solids Flow Control Devices/Feeders *230*
Solved Problem *232*
Notations *233*
References *235*
Problems *237*

12 Circulating Fluidized Beds *239*
Chengxiu Wang and Jesse Zhu
12.1 Introduction *239*
12.1.1 What Is a Circulating Fluidized Bed? *239*
12.1.2 Key Characteristics of Circulating Fluidized Beds *240*
12.2 Basic Parameters *241*
12.3 Axial Profiles of Solids Holdup/Voidage *243*
12.4 Radial Profiles of Solids Distribution *246*
12.5 The Circulating Turbulent Fluidized Bed *249*
12.6 Micro-flow Structure *250*
12.7 Gas and Solids Mixing *256*
12.8 Reactor Performance of Circulating Fluidized Beds *258*
12.9 Effect of Reactor Diameter on CFB Hydrodynamics *261*
Notations *262*
References *263*
Problems *268*

13 Operating Challenges *269*
Poupak Mehrani and Andrew Sowinski
13.1 Electrostatics *269*
13.1.1 Measurement and Prediction of Electrostatic Charge *271*
13.1.2 Mitigation Techniques *272*
13.1.3 Summary *273*
13.2 Agglomeration *273*
13.3 Attrition *274*
13.3.1 Modelling Attrition *276*
13.4 Wear *278*
Solved Problems *280*
Notations *286*
References *287*
Problem *290*

14 Heat and Mass Transfer *291*
Dening Eric Jia
14.1 Heat Transfer in Fluidized Beds *291*
14.1.1 Interphase Heat Transfer *293*
14.1.2 Bed-to-Surface Heat Transfer *294*
14.1.2.1 General Considerations *294*

14.1.2.2 Particle Convective Component *294*
14.1.2.3 Gas Convection Component *298*
14.1.2.4 Maximum Heat Transfer Coefficient *299*
14.1.3 Heat Transfer Correlations for Fluidized Beds *302*
14.1.3.1 Correlations for Bed-to-Surface Heat Transfer *303*
14.1.4 Heat Transfer Between Fluidized Bed and Immersed Surfaces *304*
14.1.4.1 Correlations for Vertical Tubes *306*
14.1.4.2 Martin's Correlations for Heat Transfer to Immersed Surfaces *309*
14.1.4.3 Finned Tubes and Non-cylindrical Tubes *312*
14.1.4.4 Tubes in Freeboard Region *313*
14.1.4.5 Methods of Augmenting Bed-to-Surface Heat Transfer *314*
14.1.5 Radiative Heat Transfer *314*
14.1.6 Heat Transfer in Fast and Circulating Fluidized Beds *316*
14.2 Mass Transfer in Fluidized Beds *318*
14.2.1 Particle and Fluid Mass Transfer in the Dense Phase *318*
14.2.2 Bubble to Dense-Phase Interphase Mass Transfer *320*
Solved Problem *320*
Notations *323*
References *325*
Problem *329*

15 Catalytic Fluidized Bed Reactors *333*
Andrés Mahecha-Botero
15.1 Introduction *333*
15.2 Reactor Design Considerations *334*
15.2.1 Suitability of Fluidized Beds for Catalytic Processes *334*
15.2.2 Reactor Types by Flow Regime and Phase *334*
15.3 Reactor Modelling *334*
15.3.1 Model Development *337*
15.3.2 Model Structure and Reaction Considerations *338*
15.3.2.1 Flow Regime Considerations *338*
15.3.2.2 Reaction Equilibrium Considerations *339*
15.3.2.3 Reaction Kinetics Considerations *340*
15.4 Fluidized Bed Catalytic Reactor Models *342*
15.4.1 Mass/Mole and Energy Balances *345*
15.4.2 Reaction Rate Expressions *345*
15.4.2.1 Single-Phase Models *346*
15.4.3 Models Based on Standard Two-Phase Theory *349*
15.4.3.1 Division of Flow and Calculation of Fluidized Bed Parameters *349*
15.4.4 More Sophisticated Models *352*
15.4.4.1 Comprehensive Reactor Modelling *352*
15.4.4.2 Computational Fluid Dynamics (CFD) Models for Fluidized Bed Catalytic Reactors *353*
15.4.5 Model Verification and Validation *353*
15.4.6 Recommendations for Programming and Numerical Solution of Reactor Models *356*

15.5 Conclusions *356*
Notations *357*
References *358*
Problems *361*

16 Fluidized Beds for Gas–Solid Reactions *363*
Jaber Shabanian and Jamal Chaouki
16.1 Introduction *363*
16.2 Gas–Solid Reactions for a Single Particle *364*
16.2.1 Reaction Models for Non-porous Particles *365*
16.2.1.1 Shrinking Particle *366*
16.2.1.2 Shrinking Unreacted Core Model *369*
16.2.2 Reaction Models for Porous Particles *373*
16.2.2.1 Reactions of Complete Consumption of the Particle *373*
16.2.2.2 Reactions for Porous Particles of Unchanging Overall Size *374*
16.2.3 Reaction Models for Solid–Solid Reactions Proceeding Through Gaseous Intermediates *377*
16.3 Reactions of Solid Particles Alone *377*
16.4 Conversion of Particles Bathed by Uniform Gas Composition in a Dense Gas–Solid Fluidized Bed *378*
16.5 Conversion of Both Solids and Gas *381*
16.5.1 Reactor Performance Calculation for a Bed of Fine Particles (Case I) *382*
16.5.2 Reactor Performance Calculation for a Bed of Coarse Particles (Case II) *385*
16.6 Thermal Conversion of Solid Fuels in Fluidized Bed Reactors *386*
16.7 Final Remarks *390*
Solved Problems *391*
Acknowledgments *398*
Notations *398*
References *401*
Problems *403*

17 Scale-Up of Fluidized Beds *405*
Naoko Ellis and Andrés Mahecha-Botero
17.1 Challenges of Scale *405*
17.2 Historical Lessons *407*
17.3 Influence of Scale on Hydrodynamics *408*
17.3.1 Bubbling Fluidization *408*
17.3.2 Turbulent Fluidization Flow Regime *410*
17.3.3 Fast Fluidization *411*
17.4 Approaches to Scale-Up *412*
17.4.1 Framing Questions *412*
17.4.2 General Approaches *412*
17.4.3 Dimensional Similitude (Scaling Models) *413*
17.4.4 Other Models *414*

17.5 Practical Considerations *415*
17.5.1 Purpose of Pilot-Scale Units *415*
17.5.2 Pilot-Scale Units *416*
17.5.2.1 Biomass Combined Heat and Power (CHP) Güssing Case *416*
17.5.2.2 Dual Gasifier with CO_2 Capture *417*
17.5.2.3 Calcium Looping Technologies *419*
17.6 Scale-Up and Industrial Considerations of Fluidized Bed Catalytic Reactors *419*
17.6.1 Challenges of Scale-Up of Fluidized Bed Catalytic Reactors *419*
17.6.2 Practical Recommendations for Industrial Implementation of Reactor Systems *422*
Solved Problems *424*
Notations *426*
References *426*
Problems *429*

18 Baffles and Aids to Fluidization *431*
Yongmin Zhang
18.1 Industrial Motivation *431*
18.2 Baffles in Fluidized Beds *432*
18.2.1 Clarification of Baffles in Low-Velocity Dense Fluidized Beds *432*
18.2.2 Geometric Characteristics of Baffles *432*
18.2.2.1 Horizontal Baffles *432*
18.2.2.2 Vertical Baffles *435*
18.2.2.3 Fixed Packings *436*
18.2.3 Baffles in Low-Velocity Dense Fluidized Beds *439*
18.2.3.1 Effect of Baffles on Bed Hydrodynamics *439*
18.2.3.2 Performance of Baffles in Industrial Fluidized Bed Reactors *445*
18.2.3.3 Other Findings and Applications *446*
18.2.4 Baffles or Inserts in High-Velocity Fast Fluidized Beds *446*
18.2.5 Design of Baffles for Industrial Fluidized Beds *447*
18.3 Other Aids to Fluidization *449*
18.3.1 Brief Introduction *449*
18.3.2 Electrical Fields *450*
18.3.3 Magnetic Fields *450*
18.3.4 Pulsations and Vibrations *451*
18.3.5 Glidants and Antistatic Agents *451*
18.4 Final Remarks *452*
Notations *452*
References *452*
Problem *455*

19 Jets in Fluidized Beds *457*
Cedric Briens and Jennifer McMillan
19.1 Introduction *457*
19.2 Jets at Gas Distributors *457*
19.2.1 Criterion for Uniform Gas Distribution *459*

19.2.1.1 "Dry" Distributor Pressure Drop *461*
19.2.1.2 Actual Distributor Pressure Drop *461*
19.2.2 Defluidized Zones *462*
19.2.3 Erosion of Internals *464*
19.2.3.1 Penetration of Upward Vertical Jets *465*
19.2.3.2 Penetration of Horizontal, Inclined, and Downward Vertical Jets *466*
19.2.3.3 Angle of Upward Jets *466*
19.2.3.4 Merging and Coalescence *466*
19.3 Mass Transfer, Heat Transfer, and Reaction in Distributor Jets *467*
19.4 Particle Attrition and Tribocharging at Distributor Holes *467*
19.5 Jets Formed in Fluidized Bed Grinding *469*
19.5.1 Mechanisms *469*
19.6 Applications *471*
19.7 Jet Penetration *471*
19.8 Solids Entrainment into Jets *471*
19.9 Nozzle Design *472*
19.9.1 Nozzle Inclination *472*
19.9.2 Impact of Bed Hydrodynamics *473*
19.9.3 Opposing Jets *473*
19.10 Jet-Target Attrition *473*
19.10.1 Prediction of Attrition Rates *474*
19.11 Jets Formed When Solids Are Fed into a Fluidized Bed *475*
19.11.1 Mechanisms *475*
19.11.2 Applications *475*
19.11.3 Jet Penetration *476*
19.11.4 Solids Entrainment *476*
19.11.5 Injection System Design *476*
19.11.6 Nozzle Inclination *476*
19.11.7 Impact of Bed Hydrodynamics *476*
19.12 Jets Formed When Liquid Is Sprayed into a Gas-Fluidized Bed *477*
19.12.1 Pure Liquid Jets *477*
19.12.2 Mechanism for Gas–Liquid Jets *477*
19.12.3 Applications *478*
19.13 Jet Penetration *478*
19.13.1 Solids Entrainment *478*
19.13.2 Injection System Design *479*
19.13.2.1 Upstream Piping Design *479*
19.13.2.2 Spray Nozzle Design *480*
19.13.2.3 Laboratory Nozzles *480*
19.13.2.4 Non-rodable Commercial Nozzles *480*
19.13.2.5 Rodable Commercial Nozzles *480*
19.13.2.6 Downstream Attachments *481*
19.13.2.7 Impact Attachments *481*
19.13.2.8 Shrouds *481*
19.13.2.9 Gas Jets *482*

19.13.2.10 Draft Tubes *482*
19.13.3 Nozzle Inclination *482*
19.13.4 Interactions Between Spray Jets *483*
19.13.5 Impact of Bed Hydrodynamics *483*
Solved Problems *483*
Notations *487*
References *488*
Problem *497*

20 Downer Reactors *499*
Changning Wu and Yi Cheng
20.1 Downer Reactor: Conception and Characteristics *499*
20.2 Hydrodynamics, Mixing, and Heat Transfer of Gas–Solid Flow in Downers *501*
20.2.1 Basic Hydrodynamic Behaviour *501*
20.2.2 Mixing Behaviour of Solids in Downers *503*
20.2.3 Heat Transfer in Downers *506*
20.3 Modelling of Hydrodynamics and Reacting Flows in Downers *508*
20.3.1 Reaction Engineering Model *509*
20.3.2 Eulerian–Eulerian Model *509*
20.3.3 Eulerian–Lagrangian Model *511*
20.4 Design and Applications of Downer Reactors *514*
20.4.1 Inlet Design *514*
20.4.2 Fast Separation of Gas and Solids at Downer Exit *517*
20.4.3 High-Density Downer *518*
20.4.4 Downer–Riser Coupled Reactors *518*
20.4.5 Application Case 1: FCC *519*
20.4.6 Application Case 2: Gasification *521*
20.4.7 Application Case 3: Coal Pyrolysis in Plasma *521*
20.5 Conclusions and Outlook *523*
Solved Problem *523*
Notations *525*
References *526*
Problems *528*

21 Spouted (and Spout-Fluid) Beds *531*
Norman Epstein
21.1 Introduction *531*
21.2 Hydrodynamics *532*
21.2.1 Constraints on Fluid Inlet Diameter and Cone Angle *532*
21.2.2 Minimum Spouting Velocity *533*
21.2.3 Maximum Spoutable Bed Height *534*
21.2.4 Fluid Flow in Annulus *535*
21.2.5 Fluid Flow in Spout *536*
21.2.6 Pressure Drop *536*
21.2.7 Behaviour of Solid Particles *537*
21.3 Heat and Mass Transfer *538*

21.4 Chemical Reaction *538*
21.5 Spouting vs. Fluidization *539*
21.6 Spout-Fluid Beds *540*
21.7 Non-conventional Spouted Beds *543*
21.8 Applications *546*
21.9 Multiphase Computational Fluid Dynamics *547*
Solved Problem *547*
Notations *548*
References *549*

22 Three-Phase (Gas–Liquid–Solid) Fluidization *553*
Dominic Pjontek, Adam Donaldson, and Arturo Macchi
22.1 Introduction *553*
22.1.1 General Description and Classification *553*
22.1.2 Applications *554*
22.2 Reactor Design and Scale-up *556*
22.2.1 Reactor Design *556*
22.2.2 Reactor Scale-up *558*
22.3 Compartmental Flow Models *558*
22.3.1 Plenum and Fluid Distributor *560*
22.3.2 Fluidized Bed *561*
22.3.3 Freeboard *562*
22.4 Fluid Dynamics in Three-Phase Fluidized Beds *562*
22.4.1 Flow Regimes *562*
22.4.1.1 Minimum Fluidization *562*
22.4.1.2 Bubbling Regimes *564*
22.4.2 Phase Holdups *565*
22.4.2.1 Modelling: Global (Bed Volume Averaged) *565*
22.4.2.2 Bed Contraction vs Expansion *568*
22.4.2.3 Particle Entrainment *568*
22.5 Phase Mixing, Mass Transfer, and Heat Transfer *569*
22.5.1 Phase Mixing *569*
22.5.2 Surface-to-Bed Heat Transfer *570*
22.5.3 Interphase Gas–Liquid and Liquid–Solid Mass Transfer *571*
22.5.3.1 Gas–Liquid Mass Transfer *572*
22.5.3.2 Liquid–Solid Mass Transfer *572*
22.6 Summary *574*
Solved Problems *574*
Notations *582*
References *585*
Problems *587*

Index *591*

Preface

We are pleased to present the first comprehensive teaching book on *fluidized beds* to be published in nearly three decades since the second edition of the Kunii and Levenspiel, *Fluidization Engineering* book in 1991 and the *Gas Fluidization Technology* book edited by Geldart, published in 1986. During the intervening period, there has been considerable progress, leading to new understanding in such areas as multiphase computational fluid dynamics (CFD), interparticle forces, electrostatics, jets, downers, and advanced experimental methodologies (such as particle tracking, MRI, and various types of tomography). These new areas are, to a degree, covered in this book, while we have also drawn heavily on the "more classical" fluidization literature. We have also included chapters on liquid and three-phase fluidization, spouted beds, CFD, and downers, topics not included in previous fluidization books intended as educational texts.

There have also been a number of new fluidized bed applications and processes in recent times, most notably in chemical looping, processing of silicon-containing materials for solar applications, extraction of advanced materials, thermochemical conversion of biomass residues to energy and biofuels, and efforts to produce or utilize nanoparticles. While these new processes are not dealt with explicitly in depth in the book, they have influenced the fluidization research community and topics of research articles, hence affecting the knowledge reflected in this book. The authors who have contributed to the book combine some who have been engaged in this field for many decades with a new generation of fluidization experts, eager to advance the understanding and applicability of fluidized beds.

In choosing the material to be included in the book, we have been guided by the word "Essentials" in the title. Thus we have had to leave out material, which, while interesting, is not essential for most beginners and general readers. However, readers should, after close reading of the chapters, be able to delve into the extensive specialized research literature with a good general background. Our book will have served its purpose if it helps readers, whether these be young engineers working in industry or graduate students undertaking research projects related to fluidization, become familiar with the broad areas of fundamental and practical knowledge underlying the field. Incorporation of a small number of solved problems and unsolved problem exercises is intended to further the understanding of the topics covered. In addition to single-reader usage, we intend that this book be available as a textbook for courses related to fluidization and multiphase systems.

Vancouver
31 October 2019

John Grace
Naoko Ellis
Xiaotao Bi

Acknowledgement

We thank the authors for responding with enthusiasm to our proposal to write chapters of the book and for their help in preparing and revising the material. We thank Zezhong John Li for assistance with figures, logistics, and administrative details. We are grateful to the Natural Sciences and Engineering Research Council of Canada for funding some of the expenses related to the preparation of this book, as well as for covering the costs of a number of studies that have contributed to our experience and expertise in fluidization and related areas.

1

Introduction, History, and Applications

John R. Grace

University of British Columbia, Department of Chemical and Biological Engineering, 2360 East Mall, Vancouver, Canada V6T 1Z3

1.1 Definition and Origins

Fluidization occurs when solid particles are supported and allowed to move relative to each other as a result of vertical motion of a fluid (gas or liquid) in a defined and contained volume. Most commonly, the fluid is a gas blown upwards by a blower or compressor through a perforated flat plate or a series of orifices, but many other configurations are possible. Once an assembly ("bed") of particles has been actuated in this manner, it is said to be a "fluidized bed."

The origin of fluidized beds is unclear, but liquid-fluidized beds likely preceded gas-fluidized beds. For example, early fluidization has been attributed to Agricola [1] when he described and illustrated hand jigging for ore dressing. The first industrial applications of fluidized beds were likely beds of ore particles fluidized by liquids in order to classify them by size or density in an operation known as "teetering" [2].

The first widespread application of gas-fluidized beds was in the 1920s in Germany when Winkler [3] patented a novel gasifier. However, the terms "fluidization" and "fluid bed" did not emerge until about 1940 when researchers in the United States developed gas-supported beds for catalytic cracking of heavy hydrocarbons [4, 5]. A plaque commemorating the development of the fluid bed reactor at a local oil refinery was erected at the Louisiana Art and Science Museum in Baton Rouge in 1998.

The term "circulating fluidized bed" (or "CFB") has been used since the 1980s to cover configurations where there is no upper bed surface, with particles supported by fluid contained in equipment that incorporates one or more gas–solid separator (usually cyclones), as well as recirculation piping as an integral part of the system. These have become popular, mostly for calcination, energy, and metallurgical operations [6].

Commercial fluidized bed reactors are now among the largest chemical reactors in the world. For example, in China fluidized bed combustors have reached a power capacity of 660 MW_e [7].

Essentials of Fluidization Technology, First Edition. Edited by John R. Grace, Xiaotao Bi, and Naoko Ellis.

1.2 Terminology

As in other fields, specialized terminology is used by the fluidization community. Definitions of the following terms may be helpful for those new to the field:

Agglomeration: Particles sticking together to form assemblies (agglomerates).
Attrition: Break-up of particles due to collisions or other interactions and stresses.
Bed expansion: Height of operating fluidized bed divided by static bed height or bed height at minimum fluidization.
Bubbles: Voids containing few, if any, particles, rising relative to the particles above them and behaving in a somewhat analogous manner to bubbles in liquids.
Choking: Collapse of dilute gas–solid suspension into dense phase flow when decreasing the gas velocity at constant solids flow. For different modes of choking, see [8].
Circulating fluidized bed: Fluid and particles in relative motion in a configuration where there is no distinct upper bed surface and entrained particles are continuously separated and returned to the base of a riser.
Cluster: Group of particles travelling together due to hydrodynamic factors.
Dense phase: Gas–solid region where the concentration of particles is sufficiently high that there are significant particle–particle interactions and contacts.
Dilute phase: Region where particle concentration is low enough that interparticle contacts are relatively rare.
Downer: Vessel in which particles are contacted with a fluid while they fall downwards.
Distributor: Horizontal plate with perforations, nozzles, or other openings or other means of introducing a fluidizing fluid to support the weight of particles and cause them to move while also supporting the dead weight of the particles when the flow of fluid is interrupted.
Elutriation: Progressive selective removal of finer particulates by entrainment.
Fines: Relatively small particles, typically those smaller than 37 or 44 μm in diameter.
Fluid: Either gas or liquid, usually the former in the context of fluidization.
Freeboard: Region extending from dense fluidized bed upper surface to top of vessel.
Geldart powder group: See Chapter 2.
Grid: Alternate name for gas distributor supporting the fluidized bed and assuring uniform entry of gas at its base.
Loop seal: Common configuration (see Chapter 11) for recirculating solids to the bottom of a fluidized bed or riser without reverse flow of gas.
Membrane walls: Containing wall consisting of vertical heat transfer tubes connected by parallel fins, commonly used in combustion applications (see Chapter 14).
Membrane reactor: Reactor containing solid surfaces ("membranes") that are selectively permeable to one or more component of the gas mixture.

Plenum chamber: Pressurized chamber below the distributor of a fluidization column from which fluidizing fluid is fed into the bed above the distributor.
Riser: Tall column in which particles are carried, on average, upwards by an ascending fluid.
Segregation: Tendency for particles of different physical characteristics (e.g. different size, density, and/or shape) to preferentially become more concentrated in different spatial regions.
Solids: Generic term referring to solid particles.
Superficial velocity: Volumetric flow rate of fluid divided by total column cross-sectional area.
Voidage: Fraction of bed volume or local volume occupied by fluid.
Windbox: Same as plenum chamber, but only when the fluidizing fluid is a gas.

Other terms are introduced and defined as needed in the text.

1.3 Applications

Gas-fluidized beds account for most of the commercial applications of fluidized beds. Relative to packed beds, gas-fluidized beds commonly offer the following advantages:

- Temperature uniformity (with variations seldom exceeding 10 °C in the dense bed and elimination of "hot spots.")
- Excellent bed-to-surface heat transfer coefficients (typically 1 order of magnitude better than in fixed beds and 2 orders of magnitude better than in empty columns.)
- Ability to add and remove particles continuously, facilitating catalyst regeneration and continuous operation.
- Relatively low pressure drops (essentially only enough to support the bed weight per unit cross-sectional area.)
- Scalable to very large sizes (e.g. there are commercial fluidized bed reactors hundreds of square metres in cross-sectional area.)
- Excellent catalyst effectiveness factors (i.e. very low intra-particle mass transfer resistances): With particles 1 order of magnitude smaller than in fixed beds, i.e. catalyst particles smaller than 100 μm, effectiveness factors usually approach 1.
- Good turndown capability: The gas flow rate can be varied over a wide range, typically by at least a factor of 2–3.
- Ability to tolerate some liquid: For example, in a number of processes, such as fluid catalytic cracking, liquids are sprayed into the column where they vaporize and then react.
- Wide particle size distributions (typically with a ratio of upper to lower decile particle diameter, d_{p90}/d_{p10}, of 10: 20).

These advantages must be significant enough to compensate for some significant disadvantages of gas-fluidized beds:

- ❖ **Substantial vertical (axial) mixing of gas**: Gas is dragged downwards by descending particles resulting in "backmixing" and large deviations from plug flow, with typical axial Peclet numbers of order 5–10.
- ❖ **Substantial axial dispersion of solids**: Vigorous motion of particles and their clusters results in substantial axial dispersion and backmixing of solids. As a result, in continuous processes, some particles spend very little time in the bed, while others spend much longer than the mean residence time.
- ❖ **Bypassing of gas**: Gas associated with a lower-density phase, e.g. rising as bubbles, passes through the bed more quickly and with less access to particles than gas associated with a denser phase in which there is better gas–solid contacting.
- ❖ **Limitations on particles that can be successfully fluidized**: Particles of extreme shapes (e.g. needle or flat disc shapes) or smaller than about 30 μm in mean diameter are difficult, or even impossible, to fluidize.
- ❖ **Entrainment**: Particles, especially fine ones, are carried upwards by the exhaust or product gas and leave the column through the exit. To minimize their losses, entrained particles must normally be continuously captured and returned to the bottom of the vessel.
- ❖ **Attrition**: Particles can break or be abraded when they collide/interact with each other and with fixed surfaces.
- ❖ **Wear on surfaces**: Particle motion tends causes erosion/wastage of fixed surfaces.
- ❖ **Complexity and risk**: Fluidized beds are more complex to design, operate, and model than comparable fixed bed reactors. As a result, there is greater risk of problems and less than desired performance.

The advantages identified above have been found to outweigh the disadvantages in a number of industrially significant processes. The most important of these processes are listed in Table 1.1. Useful reviews of the early years of these processes were provided by Geldart [9–11].

Practical information related to many of the processes listed in Table 1.1 was summarized by Yerushalmi [12]. For information on a recently commercialized process, see Tian et al. [13]. For applications related to food processing, see Smith [14]. The typical operating range for catalytic fluidized bed reactors are summarized in Table 1.2. Particles tend to be larger and gas superficial velocities to be higher in the case of physical operations and for gas–solid reactions than for catalytic processes.

Applications of liquid-fluidized beds, spouted beds, and gas–liquid–solid (i.e. three-phase) fluidized beds are covered in Chapters 3, 21, and 22, respectively.

1.4 Other Reasons for Studying Fluidized Beds

In addition to being useful in many commercial applications, as summarized above and as outlined in later chapters, there are other reasons for interest in the behaviour of fluidized beds:

Table 1.1 Industrial applications of gas-fluidized beds.

Physical operations	Solid-catalyzed reactions	Gas–solid reactions
Drying of particles	Fluid catalytic cracking	Combustion and incineration
Granulation	Acrylonitrile	Gasification
Coating of surfaces by Chemical Vapour Deposition	Ethylene dichloride	Pyrolysis
Particle mixing/blending	Catalytic combustion	Torrefaction
Preheating and heating	Ethanol dehydration	Roasting of ores
Steam raising	Ethylene synthesis	Reduction of iron oxide
Freezing	Maleic anhydride	Polyolefin production
Quenching/tempering	Fischer–Tropsch synthesis	Fluid coking and flexicoking
Carburizing, nitriding	Aniline	Calcination
Constant temperature baths	Methanol to olefins	Catalyst regeneration
Filtering of particles	Methanol to gasoline	Chlorination, fluoridation
Feeding of particles	Oxidative dehydrogenation	Hydrochlorination of silicon
Sorption of harmful gases	Phthalic anhydride	Silane decomposition → pure Si
Treatment of burn victims	Catalytic reforming	Carbon nanotubes via Chemical Vapour Deposition
	Tar cleaning	Gas–solid fermentation
	Steam reforming	Melamine production
	Methanation	Titanium dioxide pigment

Table 1.2 Usual operating ranges for solid-catalyzed gas-phase reactors.

Variable	Range and comments
Sauter mean particle diameter	50–100 μm
Particle size distribution	Broad, e.g. 0–200 μm
Reactor diameter	Up to ~7 m
Pressure	Up to ~80 bars
Temperature	Up to ~600 °C
Superficial gas velocity	~0.3–12 m/s
Static bed depth	1–10 m
Immersed surfaces	May contain horizontal or vertical heat transfer surfaces
Gas–solid separation	Heavily reliant on gas cyclones

Table 1.3 Summary of proceedings of major fluidization and CFB conferences.

Year	Designation	Conference location	Editor(s)	Publisher
1967	International Symposium on Fluidization	Eindhoven, Netherlands	Drinkenburg	Netherlands University Press
1975	Fluidization Technology	Asilomar, California	Keairns and Davidson	Hemisphere
1978	Fluidization	Cambridge, UK	Davidson and Keairns	Cambridge University Press
1980	Fluidization	Henniker, NH, USA	Grace and Matsen	Plenum Press
1983	Fluidization	Kashikojima, Japan	Kunii and Toei	Engineering Foundation
1985	CFB I	Halifax, Canada	Basu	Pergamon Press
1986	Fluidization V	Lyngby, Denmark	Ostergaard and Sorensen	Engineering Foundation
1988	CFB II	Compiègne, France	Basu and Large	Pergamon Press
1989	Fluidization VI	Banff, Canada	Grace, Shemilt, Bergougnou	Engineering Foundation
1990	CFB III	Nagoya, Japan	Basu, Horio, Hasatani	Pergamon Press
1992	Fluidization VII	Gold Coast, Australia	Potter and Nicklin	Engineering Foundation
1993	CFB IV	Hidden Valley, USA	Avidan	AIChE
1995	Fluidization VIII	Tours, France	Large and Laguérie	Engineering Foundation
1996	CFB V	Beijing, China	Kwauk and Li	Science Press, Beijing
1998	Fluidization IX	Durango, USA	Fan and Knowlton	Engineering Foundation
1999	CFB VI	Würzburg, Germany	Werther	DECHEMA
2001	Fluidization X	Beijing, China	Kwauk, Li, Yang	United Engineering Foundation

2002	CFB VII	Niagara Falls, Canada	Grace, Zhu, de Lasa	Canadian Society for Chemical Engineering
2004	Fluidization XI	Ischia, Italy	Arena, Chirone, Miccio, Salatino	Engineering Conferences International
2005	CFB VIII	Hangzhou, China	Cen	International Academic Publishers
2007	Fluidization XII	Harrison, Canada	Bi, Berruti, Pugsley	Engineering Conferences International
2008	CFB IX	Hamburg, Germany	Werther, Nowak, Wirth, Hartge	TuTech Innovation
2010	Fluidization XIII	Gyeong-ju, Korea	Kim, Kang, Lee, Seo	Engineering Conferences International
2011	CFB 10	Sunriver, Oregon, USA	Knowlton	Engineering Conferences International
2013	Fluidization XIV	Noordwijkerhout, Netherlands	Kuipers, Mudde, van Ommen, Deen	Engineering Conferences International
2014	CFB 11	Beijing, China	Li, Wei, Bao, Wang	Chemical Industry Press
2016	Fluidization XV	Montebello, Canada	Chaouki and Shabanian	Vol. 316 of Powder Technology, Elsevier
2017	CFB 12	Krakow, Poland	Nowak, Sciazko, Mirek	Journal of Power Technologies and Archivum Combustionis
2019	Fluidization XVI	Guilin, China	Wang and Ge	American Institute of Chemical Engineering
2020	CFB 13	Vancouver, Canada	Bi, Briens, Ellis, Wormsbecker	GLAB

- They are inherently fascinating to observe, even finding their way into kinetic art.
- Due to their complex flow patterns and the many factors involved, fluidized beds are challenging and difficult to model, with some surprising features.
- They may be related to some natural phenomena, in particular avalanches, pyroclastic flows associated with volcanic eruptions and atmospheric convection of water drops, snowflakes, and hailstones [15, 16]. There has even been speculation that some craters on the surface of the moon may be related to eruption of fluidization bubbles.

1.5 Sources of Information on Fluidization

Thousands of papers have been published in the scientific and engineering literature (journals and books) on fluidization fundamentals and applications. Due to length restrictions and its scope, this book cites only a small fraction of these articles. In addition to the many research articles that appear in journals like *Powder Technology, Particuology, Advanced Powder Technology,* and *the International Journal of Multiphase Flow,* many relevant papers appear in the major chemical engineering journals such as *Chemical Engineering Science, Industrial and Engineering Chemistry Research,* and *American Institute of Chemical Engineers (AIChE) Journal,* as well as a wide variety of other engineering- and physics-related journals. In addition, there are many published proceedings of conferences and symposia on fluidization. The most useful of these for those interested in fundamentals of fluidized beds have appeared in refereed proceedings of tri-annual Fluidization conferences, coordinated for many years by the Engineering Foundation and then by Engineering Conferences International, and tri-annual CFB conferences (recently renamed "International Conference on Fluidized Bed Technology.") Information on these proceedings is summarized in Table 1.3. Less rigorously refereed proceedings of fluidized bed combustion, originally coordinated and published by the American Society of Mechanical Engineers at two-year intervals, and more recently every three years, also contain many applied and fundamental fluidization articles. Periodic China–Japan Conferences on Fluidization have also led to a series of well-edited volumes.

References

1 Agricola, G. (1556). *De Re Metallica* (trans. H.C. Hoover and L.H. Hoover), 310–311. New York, 1950: Dover.

2 Epstein, N. (2005). Teetering. *Powder Technol.* 151: 2–14.

3 Winkler, F. (1922). Verfahren zum Herstellen Wassergas. German Patent 437,970.

4 Jahnig, C.E., Campbell, D.L., and Martin, H.Z. (1980). History of fluidized solids development at Exxon. In: *Fluidization* (eds. J.R. Grace and J.M. Matsen), 3–24. Plenum Press.

5 Squires, A.M. (1986). The story of fluid catalytic cracking: the first "circulating fluid bed.". In: *Circulating Fluidized Bed Technology* (ed. P. Basu), 1–19. New York: Pergamon Press.
6 Reh, L. (1971). Fluid bed processing. *Chem. Eng. Prog.* 67: 58–63.
7 Cai, R., Ke, X.W., Lyu, J.F. et al. (2017). Progress of circulating fluidized bed combustion technology in China: a review. *Clean Energy* 1 (1): 36–49. https://doi.org/10.1093/ce/zkx001.
8 Bi, H.T., Grace, J.R., and Zhu, J. (1993). Types of choking in vertical pneumatic systems. *Int. J. Multiph. Flow* 19: 1077–1092.
9 Geldart, D. (1969). Physical processing in gas fluidised beds. *Chem. Ind.* 33: 311–316.
10 Geldart, D. (1967). The fluidised bed as a chemical reactor: a critical review of the first 25 years. *Chem. Ind.* 31: 1474–1481.
11 Geldart, D. (1968). Gas-solid reactions in industrial fluidized beds. *Chem. Ind.* 32: 41–47.
12 Yerushalmi, J. (1982). Applications of fluidized beds, Chapter 8.5. In: *Handbook of Multiphase Systems* (ed. G. Hetsroni), 8-152–8-216. Washington, DC: Hemisphere Publishing.
13 Tian, P., Wei, Y., Ye, M., and Liu, Z. (2015). Methanol to olefins: From fundamentals to commercialization. *ACS Catal.* 5: 1922–1938.
14 Smith, P.G. (2007). *Applications of Fluidization to Food Processing*. Oxford, UK: Blackwell Science.
15 Wilson, C.J.N. (1984). The role of fluidization in the emplacement of pyroclastic flow: experimental results and their interpretation. *J. Volcanol. Geotherm. Res.* 20: 55–84.
16 Horio, M. (2017). Fluidization in natural phenomena, reference module. In: *Chemistry, Molecular Sciences and Chemical Engineering* (ed. J. Reedijk). Waltham, MA: Elsevier https://doi.org/10.1016/B978-0-12-409547-2.12185-7.
17 Gullichsen, J. and Harkonen, E. (1981). Medium consistency technology. *TAPPI J.* 64: 69–72. and 113–116.

Problems

1.1 Gullichsen and Harkonen [17] applied the term "fluidization" to the creation of a fluid-like state in pulp fibre aqueous suspensions due to rapid centrifugal mechanical mixing. Is this use of the term consistent with the definition of fluidization given in this chapter?

1.2 Imagine a reactor of cross-sectional area 100 m^2 containing catalyst particles of diameter 60 μm and density 1600 kg/m^3. The void fraction of the static material is 0.52. How many particles are needed to fill the reactor to a static bed depth of 6 m? What is the total mass of these particles?

2

Properties, Minimum Fluidization, and Geldart Groups

John R. Grace

University of British Columbia, Department of Chemical and Biological Engineering, 2360 East Mall, Vancouver, Canada V6T 1Z3

2.1 Introduction

This chapter identifies the key particle and fluid properties that affect the ability to fluidize particles and that play a major role in determining the properties of fluidized beds. It is important to characterize these properties, or at least to consider whether each could be relevant, when deciding whether or not a given process might benefit from fluidization, as well as when designing, operating, and modelling fluidized bed processes. The chapter also considers different methods of measuring and predicting both the minimum fluidization velocity and the bed voidage at minimum fluidization, two very important quantities affecting the properties of fluidized beds. Finally, we introduce the four Geldart powder groups for particles fluidized by gases. This classification is widely used in discussing, characterizing, and explaining gas fluidization.

2.2 Fluid Properties

2.2.1 Gas Properties

Gas properties that influence the properties of gas-fluidized beds are:

Density: Higher gas density leads to increased drag on particles and hence earlier and more vigorous fluidization. Gas density increases with increasing pressure and decreases with increasing temperature. Ideal gas behaviour can usually be assumed as a good approximation when assessing the roles of temperature and pressure on gas-fluidized beds.

Viscosity: Higher gas viscosity causes greater drag for small particles, but plays only a small role for larger (e.g. Geldart D) particles (see Section 2.6). Gas viscosity is almost independent of pressure, but increases with increasing temperature.

Essentials of Fluidization Technology, First Edition. Edited by John R. Grace, Xiaotao Bi, and Naoko Ellis.

Humidity: As discussed in Chapters 10 and 13, the humidity (water vapour content) of a gas can affect the electrostatic charges on particles, thereby affecting the properties of, and entrainment from, gas-fluidized beds.

Adsorptivity: The presence of gaseous components that adsorb on the surface of particles can affect van der Waals interparticle forces, thereby influencing the properties of fluidized beds, especially if the particles are relatively fine.

2.2.2 Liquid Properties

Liquid properties that affect the properties of liquid- and three-phase fluidized beds are:

Density: Higher liquid density leads to both higher buoyancy and increased drag on particles and hence promotes earlier and more vigorous fluidization. Liquid density is virtually independent of pressure and generally (except for water in the 0–4 °C range) decreases slightly with increasing temperature.

Viscosity: Higher viscosity causes greater drag on particles. Liquid viscosity is almost independent of pressure, but, contrary to the behaviour for gases, decreases with increasing temperature.

Contact angle: The contact angle of a liquid on the surface of particles determines the extent to which the liquid wets particles, thereby affecting capillary forces.

Surface tension: When gas is present in addition to the liquid, the surface tension of the liquid affects the formation and properties of drops and bubbles.

2.3 Individual Particle Properties

2.3.1 Particle Diameter

Particle size is profoundly important in fluidization processes. The size is expressed as a diameter, usually based on sieve (screening) analysis, as the mean of the opening sizes of the last screen through which the particle passed and the sieve through which it did not pass. For nonspherical particles, this corresponds approximately to the maximum dimension (or chord length) in the second of three principal (orthogonal) directions. Various sphere-equivalent diameters are also widely used, such as a volume-equivalent diameter (diameter of a sphere with the same volume as the particle). Of these equivalent diameters, the most appropriate average for fluidization is the Sauter mean:

$$d_{p32} = 6V_p/S_p \tag{2.1}$$

where V_p is the particle volume and S_p is its exterior surface area. Note that for a spherical particle, Eq. (2.1) gives the actual diameter.

2.3.2 Particle Shape

In practice, relatively few particles of practical interest are perfectly spherical. Engineers like to designate a single "shape factor" to characterize the degree of

Table 2.1 Typical sphericities for common materials.

Particulate material	Typical sphericity
Glass beads	0.99
Crushed coal	0.75
Rounded sand	0.90–0.95
Angular sand	0.6–0.8
Common salt	0.84
Mica flakes	0.28

deviation from spherical. The most common shape factor is the "sphericity," defined as

$$\text{Sphericity} = \phi = \frac{\text{Surface area of a sphere of the same volume as the particle}}{\text{Actual exterior surface area of the particle}} \tag{2.2}$$

The sphericity is 1 for a spherical particle and <1 for any other shape, decreasing in value as the shape deviates further from that of a sphere. Typical values of sphericity for common particulate materials are listed in Table 2.1. A much more extensive list has been provided by Ho [1].

In practice it is difficult to measure the exterior surface area of particles accurately. Hence the sphericity is sometimes estimated [2] by assuming that it is equal to its two-dimensional (2D) analogue:

$$\text{¢} = \text{Circularity} = \frac{\text{Perimeter of sphere of the same projected area}}{\text{Actual projected perimeter of the particle}} \tag{2.3}$$

based on (2D) viewing particles through a microscope and hence depending on orientation. It must be recognized that both the sphericity and the circularity are imperfect indicators of complex shapes. For example, very different shapes may lead to equal values of both of these shape factors. Nevertheless, the sphericity is widely used, and it provides a rough measure of the degree of deviation from the shape of a sphere.

2.3.3 Density and Internal Porosity

For nonporous particles, the particle density, ρ_p, is equal to the density of the material of which the particle is composed, e.g. silica (SiO_2) for silica sand particles. For porous particles, including most solid catalysts, the particle density is given by

$$\rho_p = \text{Material density} \times (1 - \text{particle porosity}) \tag{2.4}$$

where the particle porosity is the fraction of the particle volume occupied by internal pores. For practical materials, it is common to determine and utilize the bulk density (density of a mass of loose-packed particles). This is related to the particle density by

$$\rho_{bulk} = \rho_p \times (1 - \varepsilon_{bulk}) \tag{2.5}$$

where ε_{bulk} is the bulk voidage (sometimes called the void fraction) of the loose-packed bulk particulate material.

2.3.4 Surface Roughness

Roughness elements on the surface of particles augment frictional resistance to particles sliding past each other. In addition, they can influence drag (the force due to flow of gas over the particle). Asperities on the surface also affect how close particles can approach each other when in contact, hence playing a role in setting the van der Waals forces, frictional forces, bulk voidage, and electrostatic interparticle and particle–wall forces [3]. Asperities on the surface also affect attrition (covered in Chapter 13) and particle-to-wall and particle-to-particle heat transfer (see Chapter 14).

2.3.5 Terminal Settling Velocity

The terminal settling velocity of particles, v_t, is the velocity approached asymptotically by individual particles after falling a long distance through a large expanse of stagnant fluid (gas or liquid). It is relevant to entrainment of particles from fluidized beds and to the motion of particles in dilute flow conditions. At the terminal settling condition, the net gravity force (i.e. weight-minus-buoyancy) of a particle, acting in the downward direction, is exactly balanced by the upward drag imposed by the surrounding fluid due to the relative velocity between the fluid and the particle. The magnitude of the drag force is primarily a function of the particle size, its shape, the fluid density and viscosity, and the velocity of the fluid relative to that of the particle. Other factors such as surface roughness and particle rotation may also appreciably affect the drag force under some circumstances [2].

Many methods are available for predicting drag and terminal velocities when the particles are spherical [3]. It is helpful [4, 5] to define a dimensionless particle diameter and a dimensionless terminal velocity:

$$d_p^* = d_p \times \left\{ \frac{\rho_f g(\rho_p - \rho_f)}{\mu^2} \right\}^{1/3} \quad \text{and} \quad v_t^* = v_t \times \left\{ \frac{\rho_f^2}{\mu g(\rho_p - \rho_f)} \right\}^{1/3} \tag{2.6}$$

where v_t is the velocity of the particle relative to the fluid (equal to the terminal settling velocity for the case under consideration here). For spheres, v_t^*, and hence the terminal velocity, are then obtained from the relationships given in Table 2.2 [4] or, with less effort, but also less accuracy, from the empirical equation [5]

$$v_t^* = \left[\left(\frac{18}{d_p^{*2}} \right)^{0.824} + \left(\frac{0.321}{d_p^*} \right)^{0.412} \right]^{-1.214} \tag{2.7}$$

Table 2.2 Terminal settling velocity correlations for spheres [4].

d_p^* range	v_t^* range	Equation relating d_p^* and v_t^*, where $w = \log_{10} d_p^*$
≤3.8	≤0.624	$v_t^* = d_p^{*2}/18 - 3.1234 \times 10^{-4} d_p^{*5} + 1.6415 \times 10^{-6} d_p^{*8} - 7.278 \times 10^{-10} d_p^{*11}$
3.8–7.58	0.624–1.63	$\log_{10} v_t^* = -1.5446 + 2.9162w - 1.0432w^2$
7.58–227	1.63–28	$\log_{10} v_t^* = -1.64758 + 2.94786w - 1.0970w^2 + 0.17129w^3$
227–3350	28–91.7	$\log_{10} v_t^* = 5.1837 - 4.5103w + 1.687w^2 - 0.189\ 135w^3$

It is also often helpful to find the limiting terminal settling velocities for small and large spherical particles. The limit for small particles, called the Stokes terminal velocity, is

$$v_t = \frac{g(\rho_p - \rho_f)d_p^2}{18\mu} \quad (d_p^* \leq 1.2) \tag{2.8}$$

For large spherical particles, in what is often called the Newton's law range, the terminal velocity approaches the limit

$$v_t = 1.73\sqrt{\frac{g(\rho_p - \rho_f)d_p}{\rho_f}} \quad (57.3 \leq d_p^* \leq 3450) \tag{2.9}$$

For nonspherical particles, the sphericity, defined by Eq. (2.2), was used by Haider and Levenspiel [5], together with an equation of form similar to Eq. (2.7):

$$v_t^* = \left[\frac{18}{d_p^{*2}} + \frac{(2.3348 - 1.7439\phi)}{d_p^{*0.5}}\right]^{-1} \quad (0.5 \leq \phi \leq 1.0) \tag{2.10}$$

where the diameter used in Eq. (2.6) to find d_p^* is the volume-equivalent diameter, i.e. the diameter of a sphere having the same volume as the particle. A number of other empirical shape factors have also been utilized (e.g. [5–11]) to correlate drag coefficients and terminal velocities for nonspherical particles. These methods often require iterative solutions to estimate the terminal velocity, and the accuracy is commonly no better than ±30%, especially for shapes very different than spheres.

2.3.6 Coefficients of Restitution (Particle–Particle and Particle–Wall)

The coefficient of restitution is the ratio of particle relative speed immediately after a collision to that before the collision. Hence a restitution coefficient of 1 corresponds to elastic collisions (no energy loss), whereas a value of 0 would correspond to a perfectly inelastic collision with no "bounce." In practice, no collisions are perfectly elastic, so coefficients of restitution are always <1. Coefficients for particle–particle collisions differ, in general, from those for particle–wall collisions, both coefficients affecting the motion of fluidized particles by affecting the energy lost as particles collide with each other and with the walls of the containing vessel.

2.3.7 Dielectric Constant and Electrical Conductivity

The dielectric constant of particles and their electrical conductivity affect the tendency for particles to gain and retain electrostatic charges (see Chapters 10 and 13).

2.3.8 Thermal Properties

The specific heat and thermal conductivity of particles are important properties of particles affecting heat transfer as discussed in Chapter 14.

2.4 Bulk Particle Properties

2.4.1 Mean Particle Diameter and Particle Size Distribution

Particles of interest in fluidized beds are usually of a size that can be conveniently analyzed by sieve (sometimes called screen) analysis, although other techniques, such as Coulter counters and Malvern mastersizers, are often used to determine particle size distributions, particularly when a substantial fraction of particles are smaller than about 40 μm in diameter. For details of particle size analysis, see standard reference books on particle technology (e.g. [12–14]). The mean particle diameter of greatest interest is usually the Sauter (surface-to-volume) mean defined as

$$\overline{d}_p = 1/\sum(w_i/d_{pi}) \tag{2.11}$$

where w_i is the weight fraction of particles having a mean size of d_{pi}, often determined by sieve analysis. The particle size distribution can sometimes be fitted by well-known distributions, such as the log-normal distribution or Rosin-Rammler distribution [13], but experimentally measured particle size distributions are usually represented graphically in cumulative form (percent by mass smaller than particle diameter).

Fine particles, commonly referred to as "fines," are defined for convenience as the proportion of particles by mass passing through the finest standard sieves: 37 μm openings for the standard US ASTM sieve series and equivalent Tyler 400 mesh. These fines are of substantial importance in determining the behaviour of gas-fluidized beds. Hence this proportion should always be measured and reported.

2.4.2 Bulk Density, Voidage, and "Flowability"

When a significant volume of a particulate material is collected in a containing vessel with stationary air filling the interstices, the material is said to be a bulk material. The corresponding bulk density, ρ_{bulk}, is then simply the mass of the particles divided by the volume of the container occupied by the particulate material or powder. The corresponding void fraction, usually referred to as the bulk voidage, ε_{bulk}, is then obtained from Eq. (2.5) as simply

$$\varepsilon_{bulk} = 1 - \rho_{bulk}/\rho_p \tag{2.12}$$

In practice, the bulk volume of the material is affected by the manner in which the particulate material is placed in the vessel. For example, if very gently poured into the container, the volume will tend to be greater (and hence the corresponding bulk density lower) than if the bulk material were dumped at high speed and then tapped vigorously. The former is often referred to as “loose-packed” and leads to voidage values similar to those corresponding to minimum fluidization (ε_{mf}; see Section 2.7). A dimensionless group that describes the degree of compression of the material when vigorously tapped,

$$\text{Hausner index} = \frac{\text{Tapped bulk density}}{\text{Loose packed bulk density}} = \frac{\text{Loose packed bulk volume}}{\text{Tapped bulk volume}} \tag{2.13}$$

provides a simple means of characterizing the flow properties of particulate materials [15]. A particulate material with a Hausner index less than about 1.2 generally flows and fluidizes well, whereas a value of 1.5 or greater denotes poor flowability [16].

It is also important to gain a sense of the “flowability” of a powder (how well it flows) by tipping a beaker or inverting a closed container containing the particulate material to see how readily and smoothly the particles pour and flow. A material that pours and flows smoothly and continuously, without the need for shaking or tapping, will generally also fluidize well. On the other hand, particles that are “sticky” or interlock with each other when tested for pouring are likely to be poor candidates for fluidization.

Other properties that are generally conducive to particles being able to be fluidized successfully are listed in Table 2.3. Several of these properties are seen to be best within intermediate ranges of values, with extreme values on either side being undesirable (“Goldilocks-type choices”). Values that fall between the desirable and undesirable ranges in the table are often tolerable, especially if all of the other properties in the table fall within the desirable ranges. It is seen that particle shape, mean size and size distribution, surface roughness, rigidity, and robustness can all play a role. Relative humidity is important because excessive dryness

Table 2.3 Particle properties and gas relative humidities (RHs) that generally promote good “fluidizability.”

Property	Desirable	Undesirable
Shape	Rounded, sphericity $\gtrsim$0.7	Extreme, e.g. needles, discs
Mean diameter	Between ~50 μm and ~1.6 mm	$\lesssim$40 μm and $\gtrsim$3 mm
Size distribution	Unimodal, $2 \lesssim d_{p90}/d_{p10} \lesssim 20$	Bi- or tri-modal; excessively broad; completely uniform
Surface roughness	Smooth surfaces	Asperities higher than $d_p/20$
Rigidity/flexibility	Rigid	Flexible, deformable
Attrition resistance	Robust	Friable
Humidity	10% $\lesssim$ RH $\lesssim$ 60%	RH $\lesssim$ 2% and $\gtrsim$80%

tends to promote electrostatics, while excessive humidity leads to capillary forces between particles. Note that the simple indicators in Table 2.3 are less useful if the fluidized bed process under consideration is to take place at high temperature where particle and gas properties may differ greatly from the corresponding properties under atmospheric conditions, especially if the particle surfaces soften or melt at the elevated temperature.

More elaborate characterization methods, often used in conjunction with hopper flows, are also available for estimating the "flowability" of powders. For example, these can be based on the shear cell invented by Jenike [17] or on other empirical methods such as determination of the angle of repose or angle of internal friction of the particulate medium. However, these methods are primarily of qualitative value with respect to fluidization, with greater "flowability" generally corresponding to greater "fluidizability" and smoother fluidization.

2.5 Minimum Fluidization Velocity

The superficial gas or liquid velocity at which fluidization is initiated, called the minimum fluidization velocity and given the symbol U_{mf}, is an important property of particles and the fluidizing fluid. It is independent of bed depth. This velocity is important in setting the minimum flow rate required to achieve fluidization in a given column, and it also appears in many of the relationships summarized in this book for predicting other properties of fluidized beds. This section deals with methods for measuring and predicting the minimum fluidization velocity.

2.5.1 Measuring U_{mf} Experimentally

While it is possible to estimate U_{mf}, by calculation, as outlined in Section 2.5.2, it is always best if it can be determined experimentally, even if that can only be done at atmospheric pressure and temperature. There are several methods for measuring U_{mf}.

2.5.1.1 Pressure Drop vs. Superficial Velocity Method

In this method, certainly the most common one, and often regarded as the standard method, a sample of the particles of interest is placed in a vertical cylindrical column of inside diameter at least 20 times the mean particle diameter, to a depth of at least 50 times the mean particle diameter. The inner surface of the column must be smooth. The distributor at the bottom should be perfectly horizontal and must distribute the incoming gas or liquid uniformly over the cross section (see Chapter 5), with a pressure drop across it not less than 10% of the bed weight-minus-buoyancy at the minimum fluidization velocity. The top of the column should either be open, in the case of air fluidization, or be horizontal allowing uniform overflow, in the case of fluidization by water or another liquid. The column should be equipped with a port for pressure measurement just above the distributor with a filter or screen to prevent entry of particles into the port.

This port should be connected to one side of a differential pressure transducer or manometer. For gas fluidization, the other side of the transducer or manometer should be connected to a port opening in the freeboard region, above the bed surface, or may be left open to the atmosphere if the top of the vessel is itself open to atmosphere. In the case of fluidization by a liquid, the low-pressure end of the transducer or manometer should be connected to a port above the particulate bed surface.

Gently pour the particles into the column and make the surface level before beginning measurements by fluidizing the bed for at least two minutes, then reducing the gas or liquid flow stepwise, and recording the time-averaged pressure drop across the bed at each step, until the velocity is less than that where particles are in motion. The superficial velocity of the fluidizing fluid, U, can then be increased stepwise, with the time-mean pressure drop recorded at each step and then plotted against the superficial velocity. As indicated in Figure 2.1, U_{mf} is then identified by means of the intersection point of two extrapolated straight lines: (i) a line fitted to the decreasing packed bed portion of the pressure profile and (ii) a horizontal line corresponding to the flat portion of the trace beyond the velocity at which fluidization was initiated. Note that, as indicated in Figure 2.1, there is very likely to be hysteresis, with a greater peak pressure for the increasing pressure curve, depending on the particle packing, than for decreasing-velocity curve. The value of U_{mf} based on the decreasing-flow inclined line is therefore preferred when finding the intersection.

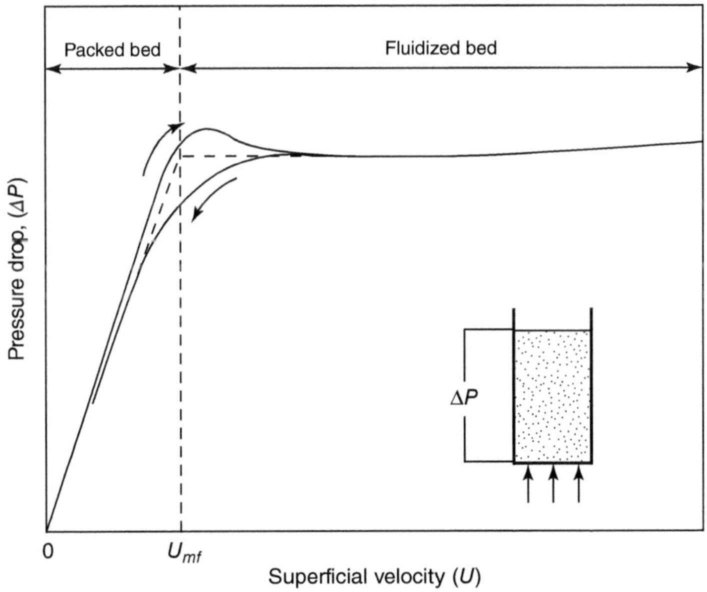

Figure 2.1 Graphical identification of the minimum fluidization velocity, U_{mf}, based on intersection of two straight lines as superficial velocity is decreased stepwise after briefly fluidizing the bed vigorously.

It is also important to check the magnitude of the horizontal line identified in (ii) above. If this ΔP value is significantly less than the bed weight-minus-buoyancy per unit cross-sectional column area, the bed is likely to be subject to substantial channelling.

2.5.1.2 Other Experimental Methods of Determining U_{mf} Experimentally

(a) **Visual:** If the column is transparent, visual observation can indicate the transition between a quiescent bed and a slightly expanded bed where the particles "lift off" and begin to show some movement. This works reasonably well if the particles are uniform in properties, but if there is a broad range of particle diameters or other properties, particle segregation will occur when the bed operates at low velocities. Therefore it is suggested that one should fluidize the bed vigorously before decreasing the flow rapidly to the range of interest near the incipient fluidization point.

(b) **Sinking of a weight at the surface:** In this method, a dense object, much larger and denser than the particles and attached to a string for retrieval, is placed gently on the upper surface of the static (unfluidized) bed. Then the flow through the bed is gradually increased to determine the superficial velocity at which the object sinks into the bed. This technique tends to under-estimate the true value of U_{mf}, since the object causes a higher local gas or liquid velocity around its periphery, therefore leading to its premature submersion.

(c) **Standard deviation of pressure fluctuations:** U_{mf} can also be determined quite accurately from where the standard deviation of pressure fluctuations, plotted against superficial velocity, is extrapolated back to 0 [18, 19]. Again it is best to use the results from decreasing the flow gradually, rather than from increasing it slowly.

2.5.2 Predicting U_{mf} Based on Particle and Fluid Properties

More than 100 empirical and semi-empirical equations have been suggested in engineering journals for the prediction of the minimum fluidization velocity, as detailed and tabulated in a recent review paper [20]. The best methods for predicting U_{mf} are based on the fact that minimum fluidization requires that the weight-minus-buoyancy of the particles be counterbalanced by drag exerted by the fluid at the transition from packed bed flow to fluidization. In a fixed bed, the drag is proportional to the pressure drop through the bed. Therefore, U_{mf} can be predicted from where the pressure drop across a fixed bed matches the buoyed weight of the bed per unit cross-sectional area of the bed.

The pressure drop through a fixed bed of height H, voidage ε_{mf}, and particles of diameter d_p and sphericity ϕ is given at $U = U_{mf}$ by the well-known Ergun equation [21]:

$$\frac{-\Delta P}{H} = \frac{150 U_{mf} \mu}{(\phi d_p)^2} \frac{(1-\varepsilon_{mf})^2}{\varepsilon_{mf}^3} + \frac{1.75 \rho_f U_{mf}^2}{\phi d_p} \frac{(1-\varepsilon_{mf})}{\varepsilon_{mf}^3} \tag{2.14}$$

The first term on the right-hand side, dominant at low Reynolds numbers, arises from viscous forces, whereas the final term, dominant at large Reynolds numbers, corresponds to inertial drag.

At minimum fluidization, the pressure drop equals the bed weight per unit area:

$$-\Delta P = (\rho_p - \rho_f)g(1 - \varepsilon_{mf})H \tag{2.15}$$

Equating the two expressions for $-\Delta P$ from Eqs. (2.14) and (2.15) leads to a quadratic equation for U_{mf}, which includes both ε_{mf} and ϕ. Wen and Yu [22] eliminated this dependence on voidage and sphericity by introducing the following approximations:

$$\phi\varepsilon_{mf}^3 \approx 1/14 \text{ and } \frac{(1 - \varepsilon_{mf})}{\phi^2\varepsilon_{mf}^3} \approx 11 \tag{2.16}$$

which can be solved to yield $\phi = 0.67$ and $\varepsilon_{mf} = 0.47$ and the following equation for U_{mf}:

$$U_{mf} = \frac{\mu}{\rho_f d_p}[\sqrt{\{33.7^2 + 0.0408Ar\}} - 33.7] \tag{2.17}$$

Here Ar is the Archimedes number (ratio of gravity to viscous forces), given by

$$Ar = \frac{\rho_f g(\rho_p - \rho_f)d_p^3}{\mu^2} \tag{2.18}$$

Many other numerical values of the two fitted numerical constants in Eq. (2.17) have been suggested in the literature [20, 23]. The Wen and Yu [22] values of 33.7 and 0.0408 above are the most widely applied ones, but, whatever constants are chosen from those suggested in the literature, the resulting minimum fluidization velocity can be over- or under-predicted, depending, among other factors, on particle sphericity and how one determines the average particle diameter [24]. For supercritical carbon dioxide [25] and supercritical water [26] as fluidizing media having properties well outside the range of common fluidizing fluids, the Wen and Yu correlation has been found to underpredict U_{mf} by ~20%.

Equations (2.17) and (2.18) can be applied to both gas-fluidized and liquid-fluidized beds. Predictions based on this relationship are usually accurate to within approximately ±20% for liquid–solid fluidization and for gas–solid fluidization of Geldart group B and D particles (see Section 2.7), but only to about ±40% for Geldart group A powders. The generally greater accuracy for Geldart group B and D particles than for group A powders [20] occurs because van der Waals and other interparticle forces, not included in the (solely hydrodynamic) variables used to derive Eq. (2.17), are significant for group A particles, but play only secondary roles for group B and group D solids.

As reviewed by Anantharaman et al. [20], a number of other forms of equations have also been fitted to experimental U_{mf} data, with limited success and wide scatter, especially for group A powders fluidized by gases. In some cases, relationships have been derived for specific types of particles, and, in such cases, it

Table 2.4 Effects of increasing temperature and pressure on minimum fluidization velocity, U_{mf}.

Fluidizing fluid	Directional change in fluid density and viscosity	For small particles, $Ar \leq 10^3$ $U_{mf} \propto (\rho_p - \rho_f)d_p^2/\mu$	For large particles, $Ar \geq 10^7$ $U_{mf} \propto \sqrt{[(\rho_p - \rho_f)d_p/\rho_f]}$
Liquid	As temperature ↑, ρ_f →, μ ↓	U_{mf} ↑	U_{mf} →
Liquid	As pressure ↑, ρ_f →, μ →	U_{mf} →	U_{mf} →
Gas	As temperature ↑, ρ_f ↓, μ ↑	U_{mf} ↓	U_{mf} ↑
Gas	As pressure ↑, ρ_f ↑, μ →	U_{mf} ↓ slightly	U_{mf} ↓

↑, Designates increase; ↓, signifies decrease; →, corresponds to almost no change.

may be useful to base predictions on these alternative equations. But, when experimental measurements for the particles of interest are not possible, Eq. (2.17) is recommended.

For small and large particles (low and high Archimedes numbers), Eq. (2.17) approaches the following limits:

$$U_{mf} = \frac{0.00061g(\rho_p - \rho_f)d_p^2}{\mu} \quad (Ar \leq 10^3) \tag{2.19}$$

$$U_{mf} = 0.202\sqrt{\frac{g(\rho_p - \rho_f)d_p}{\rho_f}} \quad (Ar \geq 10^7) \tag{2.20}$$

Except for the numerical constants, these two relationships are identical to the high and low Reynolds number limits for the terminal settling velocity of single particles in Section 2.3.5. This means that for small spherical particles where Stokes law applies to the terminal settling velocity, $v_t/U_{mf} \approx 1/(18 \times 0.00061) = 91$. Hence U/U_{mf} (sometimes called the "fluidization number") can reach ~90 before the superficial gas velocity reaches the particle terminal settling velocity.[1] The corresponding ratio for large particles (high Archimedes number) is much smaller, $v_t/U_{mf} \approx 1.73/0.202 = 8.5$.

Equations (2.19) and (2.20) also allow one to identify the directional effects (trends) of increasing temperature, pressure, and particle size on U_{mf}, as shown in Table 2.4 for both gases (where $\rho_f = \rho_g$) and for liquids (for which $\rho_f = \rho_L$). Note that these same directional effects apply to terminal settling velocities for small and large particles.

For convenience, Figure 2.2 plots the minimum fluidization velocities in air at 20 °C and 1 atm pressure and in water at room temperature (20 °C) based on the Wen and Yu [22] correlation for a broad range of mean particle sizes.

1 In early work on fluidization, it was often assumed that fluidization would be impossible once the superficial velocity reached the terminal settling velocity of individual particles due to all the particles then being entrained from the column. However, later work has demonstrated that fluidized beds can be operated in high-velocity flow regimes (see Chapters 9 and 12) due to particle clustering, saturated carrying capacity of the conveying fluid and efficient separation and return of particles carried over from the top of the column.

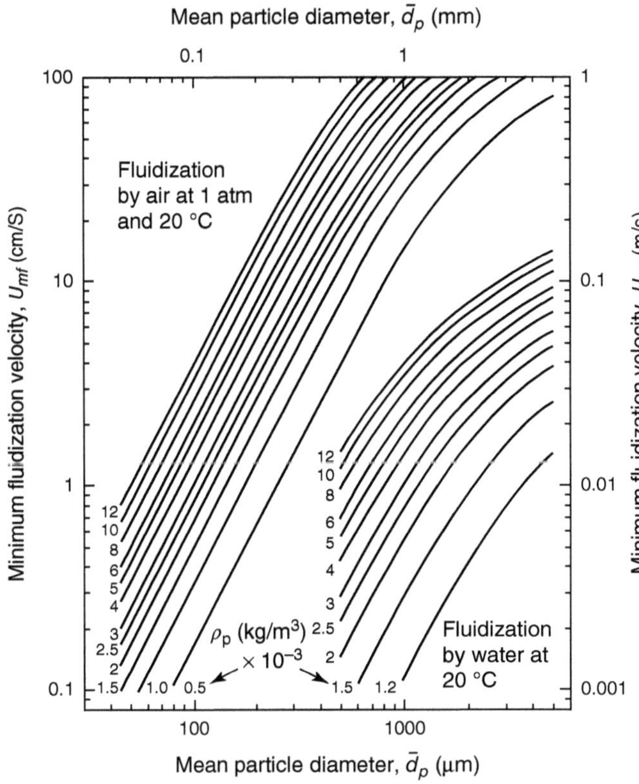

Figure 2.2 Minimum fluidization velocity, U_{mf}, for spherical particles of different densities and mean diameters in air at 20 °C and 1 atm pressure, and in water at 20 °C.

Note that minimum fluidization velocities can also be predicted based on computational fluid dynamics. For example, Ye et al. [27] applied a soft-sphere discrete particle model to find values of U_{mf}, which are consistent with values obtained from the empirical U_{mf} correlation of Abrahamsen and Geldart [28]

2.5.3 Other Factors Influencing the Minimum Fluidization Velocity

The above relationships are for standard conditions, i.e. unimodal particle size distributions of dry fluidizable powders. Some factors other than those included in Table 2.4 that affect the minimum fluidization velocity include:

(a) **Presence of moisture, causing capillary forces between the particles**: This can lead to an increase in U_{mf}, in extreme cases even making fluidization virtually impossible, i.e. transforming the powder into a Geldart type C material (see Section 2.6).
(b) **Mechanical and acoustic vibrations**: Introducing vibrations can assist in fluidizing materials, which would otherwise be difficult or even impossible to fluidize [29].

(c) **Binary and multicomponent mixtures**: When particles of substantially different properties (sizes, densities, shapes) are combined to make up a composite powder, segregation of components can occur, especially when fluidizing gently near the minimum fluidization condition. In extreme cases, the particles will separate into two or more layers, each of which will fluidize at its own U_{mf}, so there is an interval in which the smaller or less dense material is fluidized on top of the larger or denser species of particles.

Asif [30] provided a useful summary of the methods of predicting U_{mf} for multicomponent solid mixtures, distinguishing between systems that remain well mixed and those that segregate, at both small and large *Ar*, and considering cases where the components are of different densities, as well as where they differ in particle size. Contraction to a smaller ε_{mf} value when intermixing occurs tends to make U_{mf} less than would be expected on the basis of averaging either physical properties or component U_{mf} values.

For wide particle size distributions, segregation occurs when trying to measure U_{mf} experimentally. The superficial gas velocity at which the entire mixture is fluidized is then called the "final fluidization velocity." Various models are available for predicting this final velocity for binary mixtures (e.g. [31]).

2.6 Geldart Powder Classification for Gas Fluidization

Geldart [32, 33] proposed a popular and useful designation that assigns particulate materials to one of four powder groups in terms of their overall gas fluidization properties. In the earlier of the two papers [32], he proposed a division between the two most common types, which he called A and B, then added in the later paper [33] a smaller type C and a larger type D. This explains why the four Geldart powder groups, in order of increasing particle size, are C, A, B, and D, rather than being in alphabetical order. The first three of these powder groups have a convenient moniker starting with the appropriate letter:

Group C (cohesive): Group C powders are dominated by interparticle cohesive forces so that they do not fluidize readily, but instead tend to form fissures through which gas flows. The Hausner index [13] (Eq. (2.13)) tends to be >1.4. An example of a group C powder is flour. Valverde [34] divides the C region into two sub-regions: SFB, where there is transition with increasing gas velocity from solid-like to fluid-like to bubbling, and SFE, where, instead, there is transition with increasing superficial gas velocity from solid-like to fluid-like to elutriation, with visible bubbling suppressed and the gas viscosity playing a strong role. An extensive recent review of the fluidization of cohesive powders, including nanoparticles, has been provided by Raganati et al. [35].

Group A (aeratable): Interparticle forces play an appreciable, but not dominant, role for group A particles. When the upward gas flow through a bed of a group A material is sufficient to fluidize the particles, additional flow leads to bed expansion, initially without the formation of bubbles. Eventually a minimum bubbling velocity (U_{mb}) is reached (see Chapter 4). Group A materials generally fluidize smoothly. Fluid catalytic cracking (FCC) catalyst is an example of a Geldart group A material.

Group B (bubble readily): In the case of group B materials, interparticle forces play a negligible role. Bubbles form immediately when the gas superficial velocity exceeds the minimum fluidization velocity. Fluidization is generally of good quality, though less so than for group A powders. Most beach sands are excellent examples of group B particulate materials.

Group D: Group D materials are composed of coarser particles that can be fluidized, but less readily and less smoothly than for group A and B powders. Cereal grains and peas are examples of group D materials. These coarse particulate materials are often treated in spouted beds (see Chapter 21).

Geldart [33] proposed graphical boundaries between each successive pair of the four groups above for gas fluidization at room temperature and atmospheric pressure. His proposed boundaries are shown in Figure 2.3, a dimensional plot of $(\rho_p - \rho_g)$ vs. Sauter mean particle diameter. Various modifications of the Geldart boundaries have been proposed, for example, to account for shifting boundaries with nanoparticles subject to increased capillary interparticle forces [36].

It is important to understand that none of the transitions corresponding to the Geldart group boundaries is abrupt. Instead, the transitions are diffuse, with gradual changes as particle properties or operating pressure and temperature are adjusted. Because the CA and AB boundaries depend on properties other than those that are solely hydrodynamic, in particular on properties that affect interparticle forces such as Hamaker's constant, dielectric properties, particle compressibility and roughness, none of which is included in the variables that determine the coordinates on Geldart's density difference vs. mean particle diameter

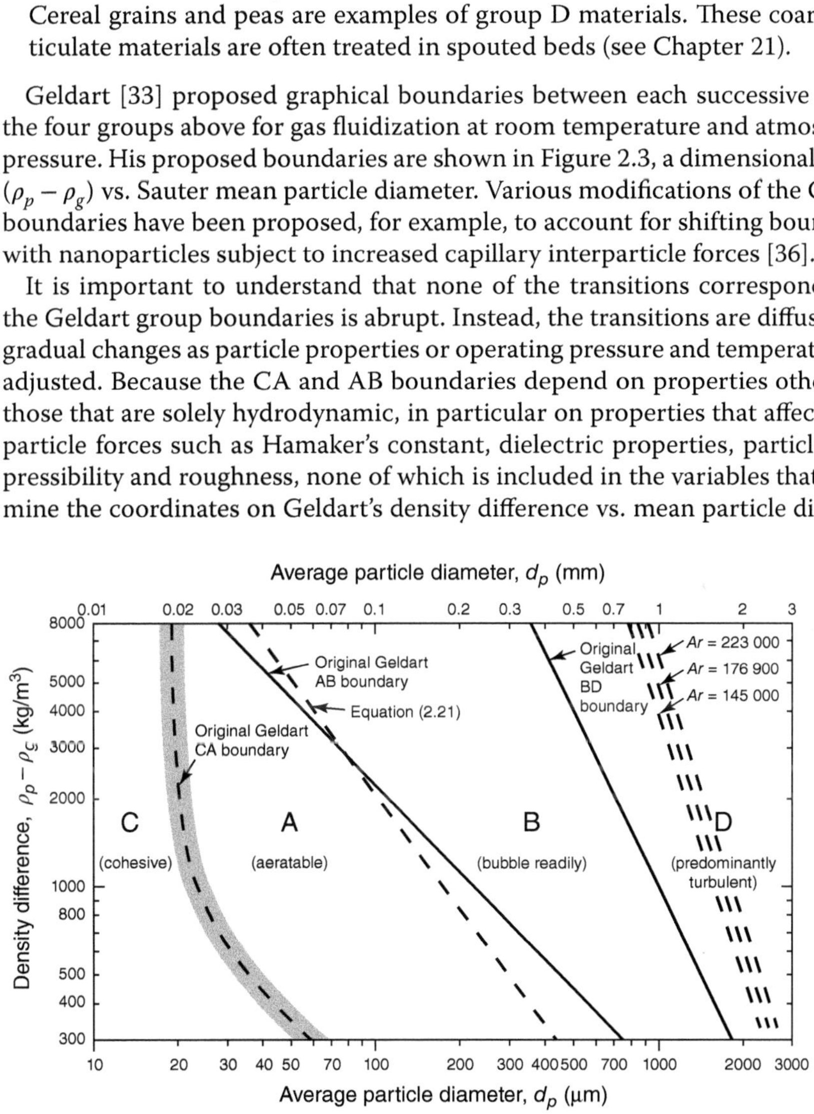

Figure 2.3 Geldart [33] powder classification group boundaries for fluidization in air at 20 °C and 1 atm pressure.

plot (Figure 2.3), the CA boundary is impossible to locate precisely on these coordinates. Hence the CA boundary shown on Figure 2.3 should be regarded as delineating a typical diffuse CA boundary range, rather than a sharp transition between group C and group A powders. In addition, adding small amounts (e.g. 0.01 wt%) of very fine dry particles (sometimes referred to as "fluidization aids") can, at least in some cases, coat the surfaces of C-type particles, making them fluidize like group A particles [37]. In practice, fine materials should always be tested experimentally, rather than relying on correlations or on figures like Figure 2.3.

Since interparticle forces also play some role with respect to the location of the AB boundary, it is also difficult to delineate this boundary sharply. Valverde [34] propose that the AB boundary should correspond to

$$\text{Granular bond number} = \frac{\text{Sum of interparticle attractive forces}}{\text{Particle weight}} = 1 \tag{2.21}$$

The interparticle forces can include van der Waals, capillary, electrostatic, electrical, and magnetic forces. By considering the equivalence or nonequivalence of published minimum fluidization and minimum bubbling velocity data for gases other than air and for temperatures and pressures higher than atmospheric, in addition to data for air under standard temperature and pressure conditions, Grace [4] showed that the AB boundary could be fitted by the empirical dimensional correlation:

$$d_{pAB} = \frac{101\mu^{2/3}\rho_g^{0.092}}{g^{1/3}(\rho_p - \rho_g)^{0.759}} \qquad \text{(SI units)} \tag{2.22}$$

The BD boundary is the only one of the three Geldart boundaries that is solely dependent on hydrodynamic considerations. By basing the BD transition on the condition where the pressure gradient in the dense phase of particles switches from being dominated by the viscous term to predominance of the inertia term, Grace [4] proposed a boundary at a constant Archimedes number of 145 000, corresponding to

$$d_{pBD} = 53\left\{\frac{\mu^2}{\rho_g g(\rho_p - \rho_g)}\right\}^{1/3} \tag{2.23}$$

This also corresponds to the particle size at which the maximum spoutable bed depth (see Chapter 21) reaches a maximum, providing a linkage to Geldart's relating this boundary to "spoutability." Goosens [38] also proposed a constant Archimedes number BD boundary, but with a somewhat higher value $Ar = 176\,900$, again based on turbulent effects dominating the hydrodynamics. If the Wen and Yu [22] constants are used, then $Ar = 223\,000$ [4]. As shown in Figure 2.3, all three values of Ar lead to similar boundaries, which, though not identical to the original boundary suggested by Geldart [33] based on sparse points related to bubble shape, wake size, and spouting, provide rational boundaries based on hydrodynamic principles. Note that the consequences of being on one side or the other of the BD boundary are not very consequential.

Fortunately, the boundaries between the Geldart powder groups for a given particulate material do not shift very much as the temperature, pressure, and gas composition change [39, 40] (see also Chapter 4). Thus if the group for a specific powder is assigned at room temperature and atmospheric pressure conditions (i.e. based on Figure 2.3), this assignment is likely to continue to be applicable under most other fluidization operating conditions. In Chapter 4, we have therefore been able to indicate the BD boundary and approximate CA and AB boundaries on plots of dimensionless superficial gas velocity vs. $Ar^{1/3}$ coordinates (e.g. see Figure 4.9).

2.7 Voidage at Minimum Fluidization

The bed voidage, ε_{mf}, corresponding to the minimum fluidization velocity, is also of considerable importance in determining the properties and performance of fluidized beds. Several methods are available for its determination:

(a) The voidage at minimum fluidization is similar to the loose-packed bed voidage of the particulate material, typically slightly ($\lesssim$0.01) higher. Hence, it is very simple to estimate ε_{mf} in most cases by utilizing the relationship based on Eqs. (2.12):

$$\varepsilon_{mf} \approx \varepsilon_o = 1 - \rho_{bulk}/\rho_p \tag{2.24}$$

(b) Gibson et al. [24] found that ε_{mf} and the particle shape could be related by

$$\frac{1-\varepsilon_{mf}}{¢^2 \varepsilon_{mf}^3} = 12.2 \pm 0.4 \tag{2.25}$$

where ¢ was the circularity measured optically, assumed to be equal to the sphericity.

(c) Measure ε_{mf} at the same time as measuring U_{mf} experimentally, as described in Section 2.5.1.

(d) If the material density is not available, or the particle density is unknown (e.g. because of internal porosity), then some tips may be helpful:
- A good first estimate for Geldart group B and D particles is $\varepsilon_{mf} = 0.41$.
- To account for mono-particle size distributions having higher void fractions than wide size distributions (e.g. [41]), increase the value by up to 0.02 if the particle size distribution is very narrow (e.g. from a single sieve when screening), or decrease by up to 0.02 if the size distribution is broad.
- Increase the estimated value by 0.01 if the particles are very angular.
- For Geldart group A materials, ε_{mf} is usually greater than for group B and group D particles due to interparticle forces hindering smaller and lighter particles from moving past each other. Follow (a), (b), or (c) above, and then add an increment as follows:

$$\varepsilon_{mf} \approx \varepsilon_{mf|\text{steps a to c}} + \frac{0.13(d_{p\text{AB}} - d_p)}{(d_{p\text{AB}} - d_{p\text{CA}})} \tag{2.26}$$

where $d_{p\text{AB}}$ is the particle diameter corresponding to the AB boundary from Eq. (2.22) and $d_{p\text{CA}}$ is from the CA boundary in Figure 2.3.

- Providing the temperature does not become high enough to cause sintering (particles sticking together due to surface melting), ε_{mf} can be assumed to be independent of temperature. However, counter-intuitively, the voidage at minimum fluidization and dense phase voidage for Geldart group A powders tend to be higher with increasing operating pressure [42, 43].

Solved Problem

2.1 Calculate the sphericity and circularity for a perfect cube.

Solution

$$\text{Volume} = L^3 = \pi d_v{}^3/6;\ d_v = L(6/\pi)^{1/3};$$
$$\phi = \pi(6/\pi)^{2/3}L^2/(6L^2) = (\pi/6)^{1/3} = 0.806.$$

Similarly,

$$\text{Projected area} = L^2 = (\pi/4)d_A{}^2;\ d_A = (4/\pi)^{0.5}L;$$
$$¢ = \pi d_A/4L = (\pi/4)^{0.5} = 0.886.$$

Notations

Ar	Archimedes number defined by Eq. (2.18) (–)
d_p	particle diameter (m)
d_p^*	dimensionless particle diameter defined by Eq. (2.6) (–)
d_{p32}	Sauter mean particle diameter (m)
d_{p90}	particle diameter corresponding to 90% by weight being smaller (m)
d_{p10}	particle diameter corresponding to 10% by weight being smaller (m)
g	acceleration of gravity (m/s^2)
H	expanded bed depth (m)
RH	relative humidity (–)
S_p	particle external surface area (m^2)
U	superficial gas velocity (m/s)
U_{mf}	superficial gas velocity at minimum fluidization (m/s)
V_p	particle volume (m^3)
v_t	terminal settling velocity (m/s)
v_t^*	dimensionless particle settling velocity defined by Eq. (2.6) (–)
w	dimensionless group defined in Table 2.2, $\log_{10} d_p^*$ (–)
w_i	weight fraction of particles having mean diameter = d_{pi}, in Eq. (2.11) (–)
ΔP	pressure drop across the bed of particles (Pa)

Greek letters

ϕ	particle sphericity defined by Eq. (2.2) (–)
ρ	density (kg/m^3)

ε	voidage (–)
μ	viscosity (Pa s)
¢	circularity (–)

Subscripts

AB	boundary between Geldart group A and B particles
BD	boundary between Geldart group B and D particles
bulk	bulk material
CA	boundary between Geldart group C and A particles
f	fluid
g	gas
L	liquid
mf	at minimum fluidization
p	particle
T	terminal settling value

References

1 Ho, T.C. (2003). Mass transfer, Chapter 11. In: *Handbook of Fluidization and Fluid-Particle Systems* (ed. W.-C. Yang), 287–307. New York: Marcel Dekker.
2 Clift, R., Grace, J.R., and Weber, M.E. (1978). *Bubbles, Drops and Particles.* New York: Academic Press. Republished by Dover in 2005.
3 Seville, J., Tüzün, U., and Clift, R. (1997). *Processing of Particulate Solids.* London: Blackie.
4 Grace, J.R. (1986). Contacting modes and behaviour classification of gas–solid and other two-phase suspensions. *Can. J. Chem. Eng.* 64: 353–363.
5 Haider, A. and Levenspiel, O. (1989). Drag coefficient and terminal velocity of spherical and non-spherical particles. *Powder Technol.* 58 (1): 63–70.
6 Ganzer, G.H. (1993). A rational approach to drag prediction of spherical and nonspherical particles. *Powder Technol.* 77: 143–152.
7 Hölzer, A. and Sommerfeld, M. (2008). New simple correlation formula for the drag coefficient of non-spherical particles. *Powder Technol.* 184: 361–365.
8 Dioguardi, F. and Mele, D. (2015). A new shape dependent drag correlation formula for non-spherical rough particles: experiments and results. *Powder Technol.* 277: 222–230.
9 Bagheri, G. and Bonadonna, C. (2016). On the drag of freely falling non-spherical particles. *Powder Technol.* 301: 526–544.
10 Breakey, D.E.S., Farid, V.G., Masliyah, J.H., and Sanders, R.S. (2018). Side-view only determination of drag coefficient and settling velocity for non-spherical particles. *Powder Technol.* 339: 182–191.
11 Wang, Y., Zhou, J., Wu, Y., and Yang, Q. (2018). New simple correlation formula for the drag coefficient of calcareous sand particles of highly irregular shape. *Powder Technol.* 326: 379–392.

12 Kaye, B.H. (1997). Particle size characterization, Chapter 1. In: *Handbook of Powder Science and Technology*, 2e (eds. M.F. Fayed and L. Otten), 1–34. London: Chapman & Hall.
13 Fan, L.-S. and Zhu, C. (1998). *Principles of Gas-Solid Flows*. Cambridge, UK: Cambridge University Press.
14 Rhodes, M. (1998). *Introduction to Particle Technology*. Chichester: Wiley.
15 Grey, R.O. and Beddow, J.K. (1969). On the Hausner ratio and its relationship to some properties of metal powders. *Powder Technol.* 2 (6): 323–326.
16 Hausner, H.H. (1967). Friction conditions in a mass of metal powders. *Int. J. Powder Metall.* 3 (7): 13.
17 Jenike, A.W. (1967). Quantitative design of mass-flow bins. *Powder Technol.* 1 (4): 237–244.
18 Puncochar, M., Drahos, J., Cermak, J., and Selucky, K. (1985). Evaluation of minimum fluidizing velocity in gas fluidized bed from pressure fluctuations. *Chem. Eng. Commun.* 35: 81–87.
19 Felipe, C.A.S. and Rocha, S.C.S. (2007). Prediction of minimum fluidization velocity of gas-solid fluidized beds by pressure fluctuation measurements. *Powder Technol.* 174: 104–113.
20 Anantharaman, A., Cocco, R.A., and Chew, J.W. (2017). Evaluation of correlations for minimum fluidization velocity (U_{mf}) in gas–solid fluidization. *Powder Technol.* 323: 454–485.
21 Ergun, S. (1952). Fluid flow through packed columns. *Chem. Eng. Prog.* 48: 89–94.
22 Wen, C.Y. and Yu, Y.H. (1966). A generalized method for predicting the minimum fluidization velocity. *AIChE J* 12: 610–612.
23 Grace, J.R. (1982). Fluidized-bed hydrodynamics, Chapter 8.1. In: *Handbook of Multiphase Systems* (ed. G. Hetsroni), 8-1–8-64. Washington: Hemisphere.
24 Gibson, H.A., Slim, C.J., Zheng, Y. et al. (2018). An examination of Wen and Yu's formula for predicting the onset of fluidization. *Chem. Eng. Res. Des.* 135: 103–111.
25 Rodriguez-Rojo, S. and Cocero, M. (2009). Supercritical fluidized bed modeling. *J. Supercrit. Fluids* 50: 54–60.
26 Lu, Y., Zhao, L.Z., Han, Q. et al. (2013). Minimum fluidization velocities for supercritical water fluidized bed within the range of 633–693 K and 23–27 MPa. *Int. J. Multiphase Flow* 49: 78–82.
27 Ye, M., van der Hoef, M.A., and Kuipers, J.A.M. (2005). The effects of particle and gas properties on the fluidization of Geldart A particles. *Chem. Eng. Sci.* 60: 4567–4580.
28 Abrahamsen, A.B. and Geldart, D. (1980). Behaviour of gas-fluidized beds of fine powders Part I. Homogeneous expansion. *Powder Technol.* 26: 35–46.
29 Escudero, D. and Heindel, T.J. (2013). Minimum fluidization velocity in a 3D fluidized bed modified with an acoustic field. *Powder Technol.* 231: 68–75.
30 Asif, M. (2013). Predicting minimum fluidization velocities of multi-component solid mixtures. *Particuology* 11: 309–316.
31 Vasconcelos, D., Batalha, G., Pereira, L.G.P., and Pires, C.A. (2018). Fluidization of binary mixtures of sisal residue and sand: a new model for deriving the final fluidization velocity. *Particuology* 40: 10–22.

32 Geldart, D. (1972). The effect of particle size and size distribution on the behaviour of gas-fluidized beds. *Powder Technol.* 6: 201–205.
33 Geldart, D. (1973). Types of gas fluidization. *Powder Technol.* 7: 285–292.
34 Valverde, J.M. (2013). *Fluidization of Fine Particles*. Dordrecht: Springer.
35 Raganati, F., Chironem, R., and Ammendola, P. (2018). Gas–solid fluidization of cohesive powders. *Chem. Eng. Res. Des.* 133: 347–387.
36 Valverde, J.M. and Castellanos, A. (2009). Bubbling suppression in fluidized beds of fine and ultrafine powders. *Part. Sci. Technol.* 26: 197–213.
37 Chen, Y., Yang, J., Dave, R.N., and Pfeffer, R. (2008). Fluidization of coated group C powders. *AIChE J* 54: 104–120.
38 Goosens, W.R.A. (1998). Classification of fluidized particles by Archimedes number. *Powder Technol.* 98: 48–53.
39 Geldart, D. and Abrahamsen, A.B. (1978). Homogeneous fluidization of fine powders using various gases and pressures. *Powder Technol.* 19: 133–136.
40 Bi, H.T. and Grace, J.R. (1996). Effects of pressure and temperature on flow regimes in gas–solid fluidization systems. *Can. J. Chem. Eng.* 74: 1025–1027.
41 Halvorsen, B.M. and Arvoh, B. (2009). Minimum fluidization velocity, bubble behaviour and pressure drop in fluidized beds with a range of particle sizes. In: *Computational Methods in Multiphase Flow V*, WIT Transactions on Engineering Sciences (eds. A. Mammoli and C.A. Brebbia), 227–238. Southampton, UK: WIT Press.
42 Chitester, D.C., Kornosky, R.M., Fan, L.S., and Danko, J.P. (1984). Characteristics of fluidization at high pressure. *Chem. Eng. Sci.* 39: 253–261.
43 Weimer, A.W. and Quarderer, G.J. (1985). On dense phase voidage and bubble size in high pressure fluidized beds of fine particles. *AIChE J* 31: 1019–1028.

Problems

2.1 Find the sphericity and the circularity for a cylindrical particle of length-to-diameter ratio 2 : 1, if, when viewed from above, (a) it stands on its end or (b) it lies on its side.

2.2 Estimate the terminal settling velocity in both air and water at 20 °C and atmospheric pressure for (a) a spherical glass bead of diameter 250 μm and density 2650 kg/m^3 and (b) a sand particle of the same volume and density, but with a sphericity of 0.75.

2.3 Alumina (Al_2O_3) has a density of 3690 kg/m^3. Alumina particles have an internal porosity of 0.30 and a bulk voidage of 0.41. Estimate the particle density and the bulk density. Which of the four Geldart powder groups (when fluidized by air at atmospheric temperature and pressure) would you expect the alumina powder to be in if the Sauter mean diameter is 100 μm?

2.4 Estimate the particle diameter at which glass beads (density 2650 kg/m^3) undergo transition from (a) Geldart group A to group B and (b) Geldart group B to group D.

3

Liquid Fluidization

Renzo Di Felice[1] and Alberto Di Renzo[2]

[1] *Università degli Studi di Genova, Dipartimento di Ingegneria Civile, Chimica e Ambientale, via Opera Pia, 15, Genova 16145, Italy*
[2] *Università della Calabria, Dipartimento di Ingegneria Informatica, Modellistica, Elettronica e Sistemistica, Via P. Bucci, Cubo 45A, Rende, Cosenza 87036, Italy*

3.1 Introduction

Liquid fluidization is an operation by which a liquid stream continuously passes through a batch of solid particles. It can be encountered in industrial practice in a variety of processes, as discussed by Epstein [1], such as:

- Particle classification and separation
- Backwashing of filter media
- Crystallization, e.g. for making struvite (magnesium ammonium phosphate)
- Leaching and washing
- Adsorption and ion exchange
- Flocculation
- Electrolysis
- Heat exchangers
- Thermal energy storage
- Bioreactors

In line with the objectives of this book, only essential basic information on liquid-fluidized beds is provided here. Readers can refer to other published works [2–4] for more extensive and detailed analysis. Unless specifically stated, we refer to "typical" liquid-fluidized beds, characterized by rigid, spherical solid particles of uniform size and density, Newtonian fluidizing liquid, and a vertical fluidizing column of constant cross-sectional area.

3.2 Field of Existence

As in any other fluidized bed system, under steady-state conditions, the weight of solid particles is fully supported by the fluid through fluid–solid dynamic drag and buoyancy, with particle–particle contacts, if present, playing only a

Essentials of Fluidization Technology, First Edition. Edited by John R. Grace, Xiaotao Bi, and Naoko Ellis.

secondary role. One of the consequences is that the basic equation of fluidization always applies over a bed height of H:

$$P(0) - P(H) = H(1 - \varepsilon)(\rho_p - \rho_f)g \tag{3.1}$$

where P is the piezometric pressure (pressure minus liquid elevation contribution).

A batch solid bed will therefore be in a fluidized state if the fluid velocity exceeds a minimum value sufficient for the continuous phase to fully support the weight of the solids and, at the same time, below a critical value where solids entrainment from the column would occur. Both boundaries can be estimated with a very good level of confidence.

The minimum fluidization velocity, the bottom boundary, is discussed in Chapter 2 where a commonly used empirical relationship linking Reynolds number, Re_{mf}, with Archimedes number, Ar, is reported (Eq. (2.17)).

At the other extreme, a liquid-fluidized bed ceases to exist when the superficial fluid velocity exceeds the single particle terminal velocity, causing solid particles to be carried out from the column outlet. For a given system, the single particle drag coefficient, C_D, is the only input needed to estimate the terminal conditions. If the simple empirical drag relationship suggested by Dallavalle [5] is used,

$$C_D = \left(0.63 + \frac{4.8}{Re^{0.5}}\right)^2 \tag{3.2}$$

then

$$Re_t = [-3.809 + (3.809^2 + 1.835Ar^{0.5})^{0.5}]^2 \tag{3.3}$$

The above expression does not take into account particle clustering commonly observed under extremely dilute conditions and therefore may actually underpredict the actual upper critical velocity by a factor of 10–20% [6].

Figure 3.1 depicts, as a function of Ar, the range of Reynolds number between which liquid-fluidized beds are expected to be found. Note that the two limiting Reynolds numbers, and therefore the two limiting superficial velocities (corresponding to minimum fluidization and terminal settling) differ by 2 orders of

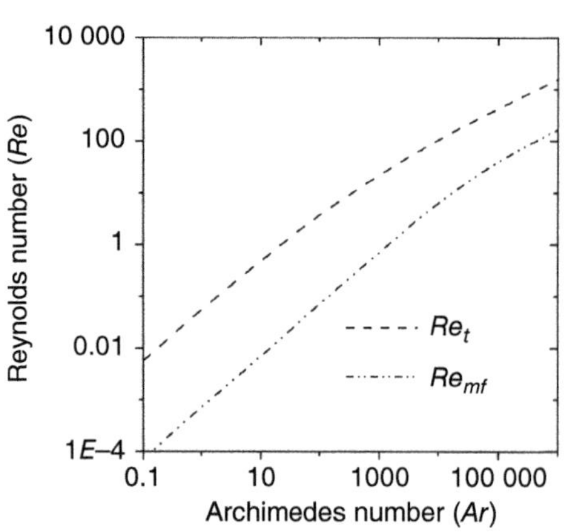

Figure 3.1 Limits of existence for a liquid-fluidized bed of spherical particles as a function of the Archimedes number of the system.

magnitude for low Ar systems, but only by 1 order of magnitude for high Ar systems.

In practice, the true upper velocity limit may be lower due to the fact that the real columns have a finite height, H. The bed material can expand only up to H before being elutriated, and therefore the maximum attainable voidage is given by

$$\varepsilon_{max} = 1 - \frac{H_0}{H}(1 - \varepsilon_0) \approx 1 - 0.6\frac{H_0}{H} \tag{3.4}$$

with this value being a function of the amount of material in the bed and the column height. Knowledge of how the bed voidage depends on fluid velocity allows the upper limit to be calculated.

3.3 Overall Behaviour

In any fluidized system overall bed height changes with fluid velocity; liquid-fluidized beds are no exception. However, before the correlation between voidage and fluid velocity is introduced, some observations on the apparent behaviour of liquid-fluidized beds must be reported and discussed.

It is generally believed that liquid-fluidized beds expand "homogeneously," implying that the particles are uniformly distributed spatially. This deceiving assumption owes its popularity to the fact that the large majority of experimental observation has been made with the naked eye, and, with this rather primitive device, liquid beds indeed appear to behave in a homogeneous manner. An example can be seen in Figure 3.2, where a photo of a "homogeneous" bed appears, although a numerical simulation shows a rather different picture of the system's behaviour. The only notable exceptions are liquid-fluidized systems of very dense solids, such as water fluidization of lead or copper, where large "bubbles" of liquid can be clearly seen rising through the bed, similar to bubbles in typical gas-fluidized beds, or fluidized systems of particles larger than 1 mm, where transverse waves of high voidage can be seen travelling upwards. The reality is quite different if bed behaviour is analyzed at a particle scale, as demonstrated by Kwauk [2], who fluidized glass particles of four different diameters (0.57, 1.1, 2, and 4 mm). Visual inspection of the 0.57 mm particle diameter bed led to the conclusion of "homogeneous" behaviour; however, for the 4 mm particles, waves travelling along the bed could be clearly observed. Kwauk photographed each bed and enlarged the negatives to obtain particle diameters, which appeared to be identical in all four cases. Inspection of the four photographs led to the conclusion that voidage inhomogeneities at a particle-scale level had the same relevance in each of the beds studied.

No liquid-fluidized bed behaves rigorously in a homogeneous manner, although the manner in which inhomogeneities appear (in the form of bubbles or travelling waves, so-called "parvoids" or particle clusters) depends on the system's physical parameters. No satisfactory prediction method exists on how a liquid-fluidized bed behaves and on what type of inhomogeneity in the spatial distribution of particles can be expected. Stability analyses have been presented starting from momentum balances for the solid and fluid phases, with

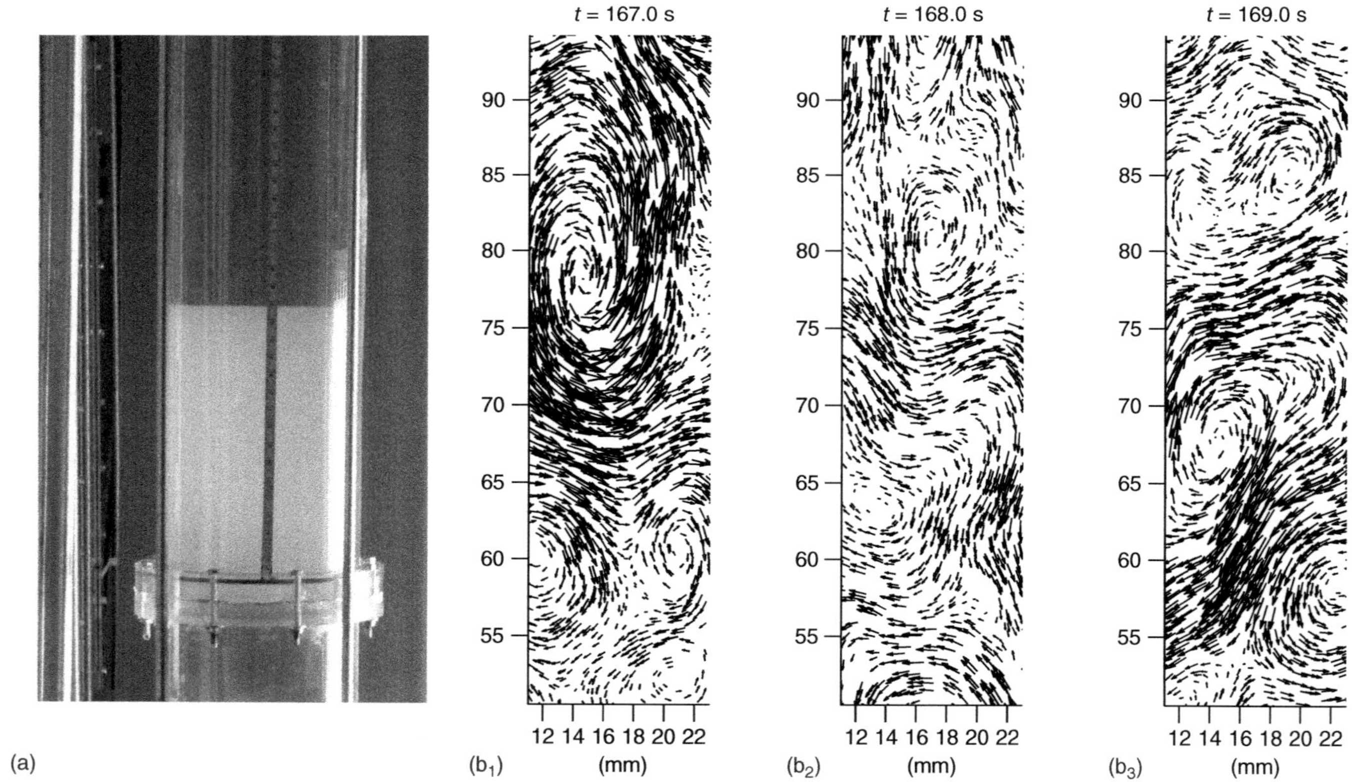

Figure 3.2 (a) Homogeneous expansion of a liquid-fluidized bed. (b_1–b_3) Snapshots of local motion of particles in liquid-fluidized beds computed by DEM–CFD simulations. Source: Reprinted with permission from Di Renzo et al. [7]. Copyright 2011, Elsevier.

satisfactory predictions generally claimed by the authors [8, 9], but, in reality, the results are dubious. Fundamental questions remain unanswered on how to write the governing equations, which terms should be included, and, for those included, such as the solid–fluid interaction force or the added mass effect, how to correctly quantify them.

3.4 Superficial Velocity–Voidage Relationship

Regardless of the specific behaviour, all liquid beds show clear expansion as the fluidizing velocity increases, and the relation between fluidizing velocity and overall bed voidage is reported here. Many data are available in the open literature, those for fluidized beds covering mostly voidages from that at minimum fluidization (i.e. ~0.4) to generally no more than 0.85, at which equipment limitations and top surface spreading make measurements less reliable. Collecting these data is a relatively simple task, which may explain their relative abundance. Overall bed voidages are estimated from knowledge of the particle mass, M, charged to the column by measuring the overall bed height, H, through

$$\varepsilon = 1 - \frac{M}{\rho_p A H} \tag{3.5}$$

whereas the velocity is expressed by the "superficial velocity":

$$U = \frac{Q}{A} \tag{3.6}$$

All published results appear to follow a common empirical relationship:

$$U = U_i \varepsilon^n \tag{3.7}$$

known as the Richardson–Zaki equation [10]. Equation (3.7) simply states that in a log–log plot, a linear relationship links velocity and voidage, and therefore, only two parameters, U_i and n, are needed to represent the observed behaviour, regardless of the system investigated. The simplicity of Eq. (3.7) is probably its most striking and surprising feature, where the complex influence of the fluid and particle physical characteristics on the particle–fluid interaction is characterized by only two parameters. Since its original appearance, many other equations have appeared in the literature, but the popularity of the Richardson–Zaki equation is unsurpassed. Empirical evidence has been used to quantify n and U_i. n has been found to be a function of the flow regime, as can be seen from Figure 3.3 where only the original Richardson and Zaki data are plotted. Values of n vary from ~4.8 to 5 at low Reynolds numbers (viscous flow regime) to just above 2 for the highest Reynolds numbers (inertial regime). In the original Richardson–Zaki work, an effect of the particle-to-column diameter ratio d_p/D on the value of the parameter n was suggested, but further investigations have failed to support that influence. The general dependency of n on the flow regime, on the other hand, has been confirmed and is universally accepted. Note, however, that the great majority of experimental work reported so far is relevant to suspensions in the

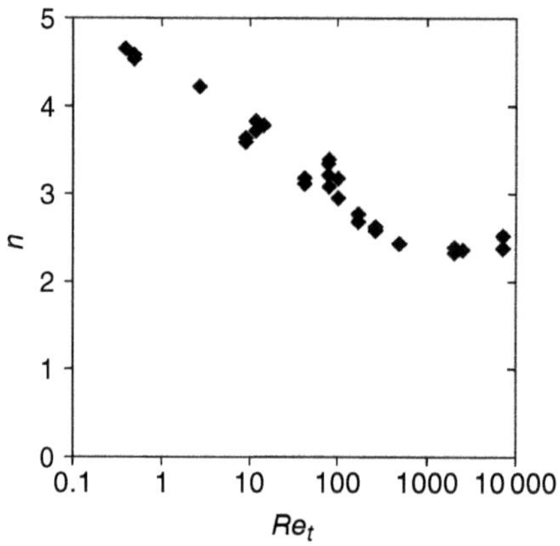

Figure 3.3 Parameter *n* as reported by Richardson–Zaki as a function of Re_t.

intermediate flow regime. Very few systems have been investigated in the inertial flow regime, where further experimental study is needed.

The second parameter, U_i, is automatically associated with the single particle terminal velocity, v_t, which is the value that it would assume if Eq. (3.7) were valid over the complete range of voidage up to 1. No definite experimental evidence has been reported to declare this assumption correct, with the very few studies of liquid-fluidized beds at voidages >0.85, not able to give a precise picture.

Figure 3.4 shows experimentally determined values of U_i for various systems relative to the single particle terminal velocity, v_t, as calculated from the Richardson–Zaki work. The U_i/v_t ratio seems to assume values of 0.8–1.0, independent of the experimental conditions. (If anything a weak dependency on the terminal Reynolds number could be detected.) Early finding of a dependency of U_i/v_t on the particle-to-column diameter ratio has not found subsequent support.

Nonspherical particles are used in many practical applications. From the numerous experimental investigations reported, it can be concluded that the expansion characteristics are again well represented on a log–log plot by straight lines, as is generally the case for spherical particles. A definite influence on the

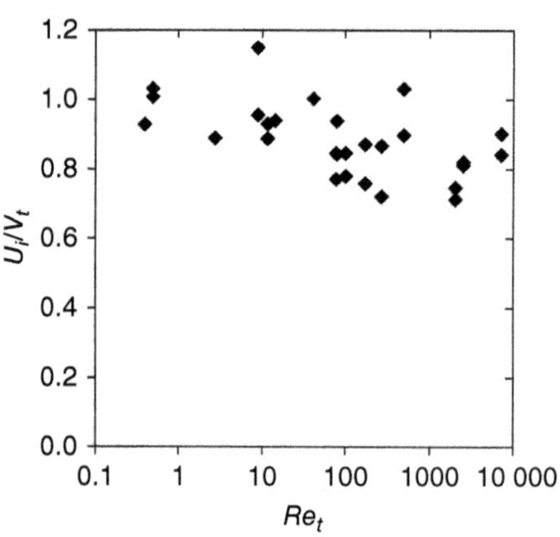

Figure 3.4 U_i/v_t ratio reported by Richardson–Zaki as a function of Re_t.

index, n, is reported, always being greater than for spheres (with values of n as high as 10 obtained). A robust quantitative correlation between particle shape factor and index n is lacking. One such correlation [11] is

$$n_{non-sphere} = n_{sphere}\phi^{b} \tag{3.8}$$

where

$$b = -2.937\phi^{0.884}Re_t^{-0.363} \tag{3.9}$$

Needless to say, extreme care should be used if the above equations are used.

Tapered columns possess characteristics that make them of great practical interest: the variation of fluid velocity with axial position allows for different types of fluidization at different levels in the column; for both monosized and polydisperse solids, the mixing can be strongly increased or reduced, depending on which direction the taper is set. Tapered beds can be quite easily designed to operate at fluid velocities exceeding the terminal settling velocity at one end, but less than the minimum fluidization velocity at the other. Kwauk [2] provided a thorough description of these phenomena.

The expansion law for a tapered column can be reasonably estimated by applying the Richardson–Zaki empirical equation to each layer of the bed, utilizing the local fluid superficial velocity. This claim has been supported by comparing overall tapered bed heights and axial voidage profiles with predicted values, for both diverging and converging columns, with reported deviations, for example lower bed expansions for diverging columns than those estimated by the simplified model described above, being of secondary relevance. Solids circulation is a main feature for tapered columns, and CFD may help in providing a coarse description.

Fluidized solids may not be rigid, as in the case of particles covered by a biofilm in the biological treatment of wastewater. Studies addressing this feature have shown that the fluid dynamic behaviour of such systems is not too dissimilar from that of rigid particle suspensions (so the Richardson–Zaki equation is still applicable). The main question in this case arises from the solid parameter quantification, as solid density and size are less clearly definable.

With the liquid density usually on the order of 1000 kg/m^3, the possibility of wishing to fluidize solids of lower density often arises. In this situation, solid particles could still be fluidized, but the liquid would have to be fed from the top of the column downwards, with the distributor plate located at the top of the column. This process is called "inverse fluidization." It is easy to recognize that this arrangement is identical with the classical upward-flow case, the difference consisting in the direction of the relevant quantities, as, for example, the pressure would decrease as one moves downwards in the bed. With this in mind, no deviant behaviour is expected, and the few studies carried out have confirmed this assumption. So the validity of the Richardson–Zaki equation has been fully confirmed, with the main doubt being the value of the coefficient U_i and its comparison with v_t. This is because, for solids of lower density than the fluid, the single particle terminal velocity is somewhat less definable, as the particle shows some unpredictable (zigzag) trajectory as it rises, rather than simply following a straight vertical path.

In summary (stressing again that the Richardson–Zaki equation is an empirical representation of experimental data and nothing more, so extrapolation outside the investigated boundaries should be avoided):

- The Richardson–Zaki equation gives an excellent representation of the expansion characteristics of concentrated fluidized suspensions ($\varepsilon < 0.85$); it works well both for spherical and nonspherical particles and for inverse fluidization.
- The parameter n is a function of the Reynolds number. For spherical, rigid particles, a relationship proposed by Rowe [12] covers the entire flow regime with a single equation:

$$\frac{4.7 - n}{n - 2.35} = 0.175 Re_t^{0.75} \tag{3.10}$$

  Even better, the relationship proposed by Khan and Richardson [13] relates n to the Archimedes number:

$$\frac{4.8 - n}{n - 2.4} = 0.043 Ar^{0.75} \tag{3.11}$$

  so n can be estimated directly once the physical parameters are known, without having to first calculate Re_t; for non-rigid particles or for inverse fluidization, the above relationship can still be used, but extra care is needed. For nonspherical particles the parameter n is somewhat higher.
- The parameter U_i is not a function of d_p/D, as originally stated [10], and its value is roughly 0.8–1.0 times that of the single particle terminal velocity, v_t.

3.5 Particle Segregation and Mixing

Liquid fluidization has long been used with multi-solids, in which different particle species can mix or segregate. For example, in binary solids sorting (classification by density), particles of similar sizes are separated, with the denser component settling to the bottom of the fluidized bed and the less dense one rising towards the surface. Alternatively, in sizing processes (classification by size), the difference in particle size is exploited, within a mixture composed of solid species differing little in density, with finer species accumulating at the top and coarser particles eventually found preferentially at the bottom.

The reason for this behaviour lays in the interplay between the action of the upward-flowing liquid, dragging the particles upwards, and gravity acting downwards. These two force contributions act differently on the various particle species, since gravity depends on the density and volume (cube of the diameter), while the drag force depends essentially on the particle surface area (diameter to the power of two or lower, depending on the flow regime), and is independent of the particle density.

Experiments in which a large solid spherical particle is immersed in a liquid-fluidized bed have been carried out to investigate the float/sink behaviour of the particle. The typical observation is that float or sink depends on whether the sphere density is lower or higher than the fluidized bed density, $\rho_b = \rho_p(1-\varepsilon) + \rho_f\varepsilon$. Oscillating behaviour is observed for neutrally buoyant (i.e. same density) spheres, much like what happens with simple liquids. This analogy of the fluid bed to a pseudo-fluid works well in cases where the immersed object is considerably bigger than the particles. Objects of comparable size and/or insignificant volume fraction may exhibit qualitatively and/or quantitatively different behaviour, as observed by Tatemoto et al. [14] in experiments with recycled plastic particles at volume fractions up to 50% in a fluidized bed of fine inerts.

Solid mixtures in which the particulate species differ only in size show a tendency to segregate, with the fine particles collecting at the surface and the coarser ones towards the bottom. Stratification can be counterbalanced by increasing the fluidization intensity (i.e. superficial velocity), as the solids recirculation may then be sufficient to redistribute the particle sizes throughout the bed. Stronger tendency towards the formation of separate segregated layers is observed if the solids differ in density. Without size difference, the denser solid tend to accumulate at the bed bottom, while less dense particles are pushed upwards. The formation of distinctly separated layers can be observed if the densities differ significantly and the size distributions are relatively sharp. The most common observations report two layers, one of which is a mixture and the other a pure component. In some cases, a neat interface between the fluidized layers cannot be traced as the proportion of the species varies gradually along the bed height.

It is of practical interest to know the overall expansion characteristics of a fluidized bed of solid mixtures. If the particle characteristics are similar, one can treat the solid phase as mono-component, utilizing some average particle size and density. On the other hand, when the solids are such that they fluidize in separate layers, the overall bed height can be readily obtained based on the equation

$$\frac{1}{1-\varepsilon} = \sum \frac{x_i}{1-\varepsilon_i} \tag{3.12}$$

where ε_i is the voidage when solid i is fluidized alone and x_i is its volume fraction on a dry basis in the mixture. The first approach relates to when the solids are completely mixed, the second to when they are completely segregated. In intermediate situations, neither approach is conceptually exact as the degree of mixing is not taken into account; however, the difference between calculated and experimental data is generally limited, and Eq. (3.12) can be used with sufficient confidence.

3.6 Layer Inversion Phenomena

The key actions determining the segregation tendency of multiparticle fluidized beds are gravity and fluid drag on the particles. Due to their different relationships with fluid and particle properties, the resulting expanded bed can give rise to peculiar phenomena, possessing intriguing characteristics, as well as practical implications. Combination of binary particle mixtures and fluid properties can generate expansion behaviour associated with species segregation or mixing, in which even the "segregation direction" can vary. One necessary, but not sufficient, requirement to observe such counter-intuitive behaviour is the – not-so-rare – combination of bigger particles of lower density fluidized with smaller, but denser, particles. This circumstance generates a complex interplay between fluid drag and gravity. The result may be that at lower liquid superficial velocities, the fluidized bed segregates, with the bigger solid accumulating at the top, while at higher velocities the segregation pattern and solid layers are reversed. The inversion of the segregation direction is gradual with increasing fluid velocity, and at some intermediate velocity, the expanded bed appears to be fully mixed. This peculiar behaviour is known as *layer inversion*, and it has been observed to occur upon changing the fluid velocity or the mixture composition. Figure 3.5 shows a computer-generated representation of the fluidization/segregation behaviour of a binary mixture. Photographs showing the experimental inverted layers at low and high fluid velocities are illustrated in Figure 3.6.

The unique combination of velocity and expansion degree (i.e. voidage) at which the intermediate homogeneously mixed bed is observed is known as the "layer inversion condition," characterized by the pair U_{li} and ε_{li}. Various methods have been proposed to compute the critical inversion conditions. Most

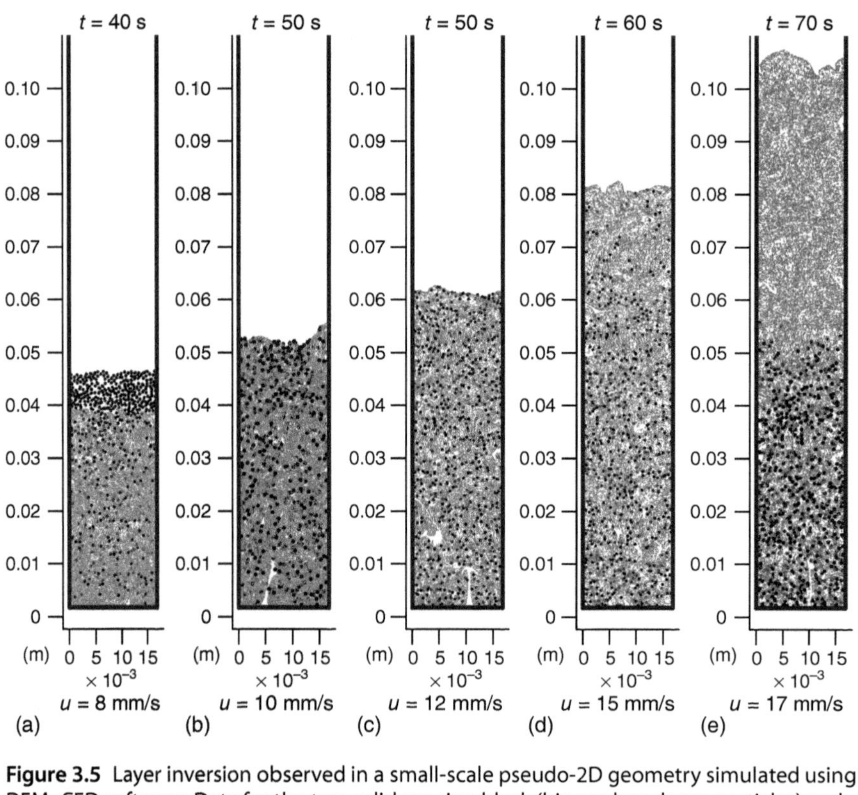

Figure 3.5 Layer inversion observed in a small-scale pseudo-2D geometry simulated using DEM–CFD software. Data for the two solid species, black (bigger, less dense particles) and grey (smaller, denser particles), and the fluid, and the range of velocities reflect those used in lab-scale experiments. The bed expansion and segregation pattern matched the measurements very well. Source: Reprinted with permission from Di Renzo et al. [7]. Copyright 2011, Elsevier.

(a) (b)

Figure 3.6 Layer inversion of lab-scale water-fluidized bed containing $x_1 = 58\%$ by volume of glass beads (average diameter = 875 μm, density = 2500 kg/m^3) and ABS/$BaSO_4$ particles (diameter = 5.95 mm, density = 1450 kg/m^3). Upon fluidization, starting from a mixed initial condition: (a) at low water velocity, glass beads sank to the bottom of the bed, while plastic spheres tended to float. (b) At higher velocities, the glass beads tended to segregate to the bed surface, while plastic particles sank.

of these require coupled calculations for both variables, sometimes via successive iterations. Below, we illustrate a simpler direct method to work out the inversion voidage, based on the mixture particle properties and composition, known as the particle segregation model [15]. A separate equation can be utilized to compute the corresponding inversion velocity, as discussed below.

3.6.1 Predicting the Layer Inversion Voidage (via the Particle Segregation Model)

In a fluidized system the total fluid drag force acting on each particle balances, on average, its net gravity (weight minus buoyancy) force. If we consider a uniform mixture of two species of homogeneously fluidized solids, we can consider a force balance equation on one fluidized particle in such a mixture, e.g. the larger, less dense species (species 2):

$$F_{Ab} + F_{pg} + F_d = F_g \tag{3.13}$$

Here F_{Ab} is the buoyancy, the pressure gradient F_{pg} and the drag force F_d are the forces exerted by the upward-flowing fluid, and F_g is the particle weight. Some of these forces depend on the average solid density, defined as

$$\rho_{av} = \rho_{p1}x_1 + (1 - x_1)\rho_{p2} \tag{3.14}$$

where x_1 is again the volume fraction of species 1 on a dry basis.

The most delicate ingredient of the force balance is the expression for the drag force acting on one particle in a fluidized mixture. It can be expressed by

$$F_d = y_2^2 \cdot \varepsilon(\rho_{av} - \rho_f)\frac{\pi}{6}d_{av}^3 g \tag{3.15}$$

where y_i is a species polydispersion index defined as $y_i = \frac{d_{p,i}}{d_{av}}$ and the average diameter d_{av} follows the Sauter mean definition.

By defining all terms of the force balance and rearranging, Eq. (3.15) simplifies to

$$\frac{\rho_{p2} - \rho_f}{\rho_{av} - \rho_f} = 1 - \varepsilon + \varepsilon \frac{d_{av}}{d_{p2}} \tag{3.16}$$

Equation (3.16) provides an important relationship between the variables. Conditions of particle properties, average diameter and density, and voidage fulfilling Eq. (3.16) denote a homogeneously fluidized system in which, in addition to the total equilibrium of forces typical of suspended mixtures, also one species (the big, less dense solid particles) individually satisfies the force equilibrium. Complementarily, also the smaller particles must experience a force equilibrium. Therefore, we have a tool to directly characterize the critically mixed state typical of the layer inversion condition, i.e. the layer inversion voidage for the mixture.

In practical terms, by introducing the species-to-average diameter ratio $\overline{d} = \frac{d_{av}}{d_{p2}}$ and density ratio $\overline{s} = \frac{\rho_{p2} - \rho_f}{\rho_{av} - \rho_f}$, a very compact and useful expression for the inversion voidage is obtained:

$$\varepsilon_{li} = \frac{1 - \overline{s}}{1 - \overline{d}} \tag{3.17}$$

An analogous result can be obtained in terms of the species-to-species diameter ratio, $d = \frac{d_{p1}}{d_{p2}}$, and density ratio, $s = \frac{\rho_{p2} - \rho_f}{\rho_{p1} - \rho_f}$. By making use of the Sauter mean diameter definition in $\overline{d}$ and the average solid density for $\overline{s}$, expansion of Eq. (3.17) yields

$$\varepsilon_{li} = \frac{1 - s}{1 - d} \cdot \frac{x_1 + (1 - x_1)d}{x_1 + (1 - x_1)s} \tag{3.18}$$

Similarly, a map of the layer inversion combinations of solids' properties and mixture composition can be plotted by rearranging Eq. (3.18) to

$$s = \frac{x_1}{\dfrac{1}{1 - \varepsilon + \varepsilon \left(\frac{d}{x_1 + (1 - x_1)d} \right)} - (1 - x_1)} \tag{3.19}$$

Representative lines are shown in Figure 3.7 as density ratios as a function of the size ratio, where inversion combinations appear in the shaded area. Note that due to the definitions of s and d and their ranges, both 0 to 1, all points in the plot represent combinations of small, denser particles with bigger, less dense ones. However, the existence region for layer inversion is located only within the grey area, according to the particle segregation model. Indeed, combinations outside the grey area would result in inversion voidages outside the admissible range of $\varepsilon = 0.4$–1, so the segregation tendency of the two species remains uniform for all velocities (or voidages). Indication of the segregation direction outside the inversion region, as illustrated by the inserts in Figure 3.7, is also possible, thanks to

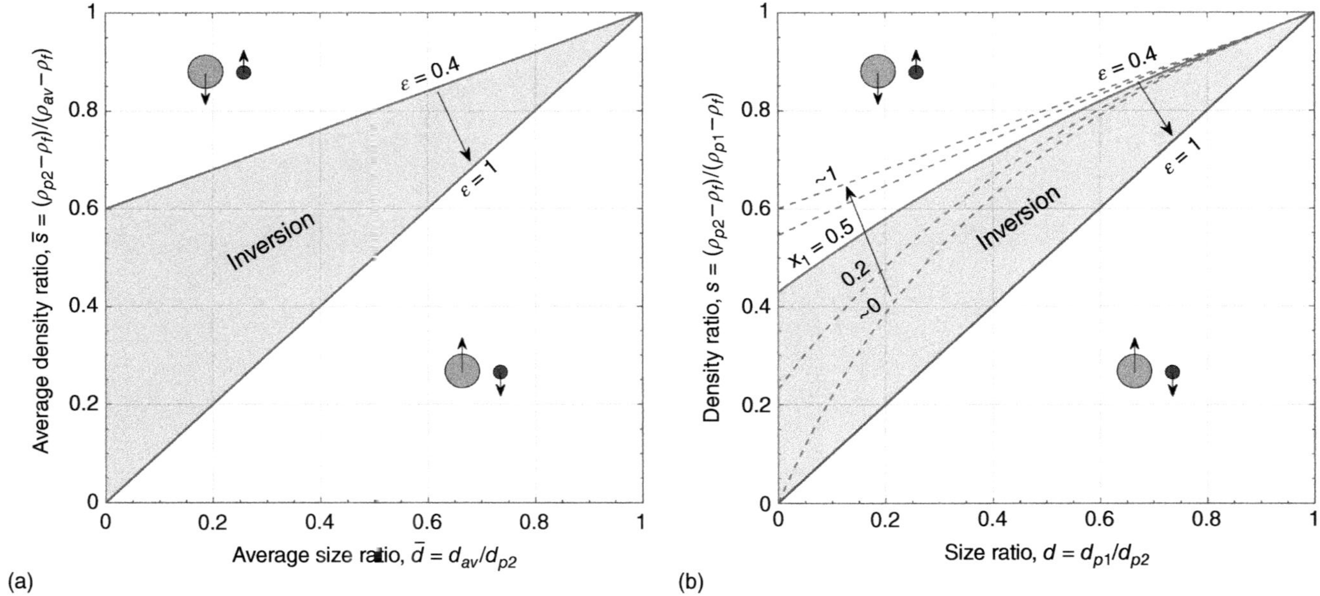

Figure 3.7 Region of possible occurrence of layer inversion characterized by a line of corresponding critical voidage, ε_{li}: (a) on an average density vs. average diameter ratio diagram. (b) On a density ratio vs. diameter ratio diagram. Inserts in the plots indicate the prevalence of the denser (black) and less dense solid (grey) along the bed height. In plot (b), the shaded area is bounded between a dilute limit straight line, independent of composition, and a dense limit curve that depends on the composition (see Eq. (3.19)).

the force balance nature of the results. More details on the derivation presented above are provided in Ref. [15].

3.6.2 Layer Inversion Velocity

A model, known as the complete segregation model, was developed by Gibilaro et al. [16], in which an overall momentum balance equation and a maximization criterion for the bulk density are used to find the velocity, voidage, and compositions at layer inversion conditions. Making use of the result for ε_{li} by Eq. (3.17) or (3.18), we illustrate here the simplified calculation procedure for the layer inversion velocity. Applied to the layer inversion condition, the overall momentum balance equation takes the form

$$\left[\left(\frac{17.3}{Re_{av,li}}\right)^{\alpha} + 0.336^{\alpha}\right]^{\frac{1}{\alpha}} u_{li}^{2} - \frac{d_{av}(\rho_{av} - \rho_f)g}{\rho_f \varepsilon_{li}^{-4.8}} = 0 \tag{3.20}$$

where the average Reynolds number is defined as

$$Re_{av,li} = \frac{\rho_f d_{av} U_{li}}{\mu} \tag{3.21}$$

The exponent α is given by

$$\alpha = 2.55 - 2.1[\tanh(20\varepsilon_{li} - 8)^{0.33}]^{3} \tag{3.22}$$

The layer inversion velocity appears non-explicitly in Eq. (3.20), but working it out can be achieved by a few iterations on U_{li} (see Solved Problem 3.2). Note that there is a very high sensitivity of the velocity prediction to the viscosity and hence on the system temperature (e.g. see Ref. [17]). Unfortunately, this operating parameter is often overlooked, being either unmonitored or measured with insufficient accuracy.

3.7 Heat and Mass Transfer

Transport of heat and mass in liquid-fluidized beds has been the subject of a large number of research studies. An account of some correlating result is introduced here for practical purposes, with additional details available in the references at the end of this chapter, particularly [4].

3.7.1 Interphase Transfer

Heat and mass transport in liquid-fluidized beds generally benefits from the agitation and homogeneous conditions throughout the system. In terms of the liquid-to-particle contact and exchange properties, in addition to solids mixing, the fluidized condition brings about substantially greater transfer coefficients than in comparable fixed beds. Mass transfer coefficients higher by a factor of 15 or more have been measured in fluidized bed bioreactors for bio-artificial organ applications compared with corresponding fixed beds [18].

However, expanding the bed causes a decrease in the efficiency due to smaller contact surface per unit reactor volume.

Under steady-state conditions, similar scalar transport differential equations describe the vertical temperature and solute concentration profiles, called energy and species transport balances for fluid flow under convection-dominated (i.e. plug flow) conditions:

$$\rho U c_p \frac{dT}{dz} = h_p a(T_p - T) \tag{3.23}$$

$$U c_p \frac{dC_A}{dz} = k_p a(C_{As} - C_A) \tag{3.24}$$

From many experimental observations, mass transfer is much slower than heat transfer, so correlations have been determined from solute transport measurements. For the sake of completeness, we suggest the relationship

$$Sh_p = K_0 Ar^{1/3} Sc^a \tag{3.25}$$

in which the constants, K_0 and a, within the ranges $K_0 = 0.23–0.31$, $a = 0.40–0.33$ (with extreme values used in the same order), have been shown to agree reasonably well with experimental values covering a broad range of Ar and Sc.

3.7.2 Bed-to-Surface Transfer

Heat flow between submerged surfaces, immersed ones or the column walls, and the fluidized suspension is characterized in terms of the heat transfer coefficient, h. Plots of its dependence on an increase of bed expansion (or liquid superficial velocity) show an increasing and then decreasing trend in the practical range between incipient fluidization and carryover. The maximum value has been experimentally observed to lie in the range $\varepsilon = 0.62–0.82$. A useful and simple correlation [19] is

$$Nu = 0.0734 Re^{0.75} Pr^{0.63} \frac{(1-\varepsilon)^{0.25}}{\varepsilon} \tag{3.26}$$

where $Nu = \frac{hd_p}{\lambda}$. Ranges of validity are as follows: $Re = 0.020–9400$, $Pr = 1.65–7700$, $\varepsilon = 0.40–0.99$, $Ar = 3.85–6.7 \times 10^7$, $d_p/D_h = 0.0013–0.210$ (D_h being the hydraulic diameter of the fluidization column). General trends reported are that the heat transfer coefficient increases with the particle size and particle density, and it has been argued to be relatively insensitive to the particle thermal properties.

Bed-to-surface mass transfer coefficients cannot be simply evaluated by adapting Eq. (3.26) by analogy with single-phase flow. However, the observed increase followed by decrease in the mass transfer coefficient with increasing voidage still applies, with a maximum value lying within the range $\varepsilon = 0.58 - 0.75$. Despite the practical importance of mass transfer, extensive measurement of mass transfer characteristics is generally complex, and correlations cannot claim good general validity. That said, the correlation proposed by Schmidt et al. [20] is suggested:

$$Sh_p = 0.14 Re^{\frac{1}{3}} Sc^{\frac{1}{3}} + 0.13(1-\varepsilon)(\varepsilon - \varepsilon_{mf})\left(\frac{Sc}{Re}\right)^{\frac{1}{3}} Ar^{\frac{2}{3}} \tag{3.27}$$

where $h = \frac{k_l d_p}{D}$, applicable within the following ranges: d_p = 0.325–3 mm, ρ_p = 2500–11 343 kg/m^3, $\frac{Re}{1-\varepsilon}$ = 0.9–1652, Sc = 151–7021, Ar = 32–4.8 × 10^5, and ε = 0.4–1.

3.8 Distributor Design

The aim of a good distributor is to provide as flat a velocity profile across the bed cross section as possible, thereby preventing liquid channelling and other inhomogeneities. Nonuniform distributors tend to reduce bed expansion and decrease the effectiveness of fluidized beds. Desirable behaviour is generally achieved by passing the liquid through a calming section located upstream, before flowing through the distributor. The calming section is often composed of a fixed bed of relatively large and dense spherical particles (e.g. lead shot). The calming section height is generally on the order of the column diameter or more; alternatively, flow straighteners can be used. The actual distributor is commonly a metal plate with regularly arranged holes, a sintered metal or glass porous plate or stacked metal gauze layers. The geometry of such elements is selected based on the key design property, namely, the liquid pressure drop due to the distributor compared with the pressure drop across the particle bed. Sufficient flow uniformity is achieved by requiring that

$$\frac{\Delta P_{dist}}{\Delta P_{bed}} > 0.2 \tag{3.28}$$

at the desired nominal flow conditions. The threshold 0.2 in Eq. (3.28) has been found to be suitable for nominal voidage < 0.48. From moderately to highly expanded beds (ε = 0.5–0.9), pressure drop ratios as high as 1 or higher are recommended. Moreover, the tendency towards instabilities has been found to decrease for expanded beds with large aspect (height-to-diameter) ratios, but such an effect is not easily quantifiable [4].

For a perforated plate distributor, variables such as the ratio of open area to total cross-sectional area and the hole diameter are based on a trade-off between small hole size, to prevent particles from falling through, and large holes, easier to machine.

Solved Problems

3.1 Estimate the field of existence when 2.5 kg of 0.5 mm sand particles (ρ_p = 2500 kg/m^3) are fluidized by ambient water (ρ_f = 1000 kg/m^3, μ = 0.0009 Pa s) in a 100 mm ID cylindrical column, 1 m tall.

Solution

The area of the column is

$$A = \frac{\pi \times D^2}{4} = \frac{3.14 \times 0.1^2}{4} = 0.0785\ \text{m}^2$$

First the Archimedes number is calculated:

$$Ar = \frac{d_p^3(\rho_p - \rho_f)\rho_f g}{\mu^2} = \frac{0.0005^3 \times 1500 \times 1000 \times 9.81}{0.0009^2} = 2271$$

Re_{mf} is obtained directly from Eq. (2.5):

$$Re_{mf} = (33.7^2 + 0.0408 \times 2271)^{0.5} - 33.7 = 1.57$$

from which

$$U_{mf} = \frac{Re_{mf}\mu}{d_p\rho_f} = \frac{1.57 \times 0.009}{0.0005 \times 1000} = 0.0028 \text{ m/s and } Q_{mf} = U_{mf}A$$
$$= 0.0028 \times 0.00785 = 0.000022 \text{ m}^3/\text{s} \equiv 80 \text{ l/h}$$

If the column was infinitively long, the maximum flow rate could be estimated from the terminal settling velocity, starting from Eq. (3.3):

$$Re_t = [-3.809 + (3.809^2 + 1.835 \times 2271^{0.5})^{0.5}]^2 = 39.5$$

from which

$$v_t = \frac{Re_t\mu}{d_p\rho_f} = \frac{39.5 \times 0.009}{0.0005 \times 1000} = 0.071 \text{ m/s and } Q_{max} = v_tA$$
$$= 0.071 \times 0.00785 = 0.00056 \text{ m}^3/\text{s} = 2000 \text{ l/h}$$

When the actual column height is taken into account, then, considering that we may want to expand the bed no more than 0.8 m and that the fixed bed voidage can be safely assumed to be equal to 0.4,

$$H_0 = \frac{M}{\rho_pA(1-\varepsilon_0)} = \frac{2.5}{2500 \times 0.00875 \times (1-0.4)} = 0.21 \text{ m}$$

and

$$\varepsilon_{max} = 1 - \frac{H_0}{H}(1-\varepsilon_0) = 1 - \frac{0.21}{0.8}(1-0.4) = 0.84$$

By using the Richardson–Zaki Eq. (3.7),

$$U_{max} = U_i\varepsilon_{max}^n$$

with $n = 2.56$ from Eq. (3.10) and assuming $U_i = 0.9 \times v_t$
$U_{max} = 0.9 \times 0.071 \times 0.84^{2.56} = 0.041$ m/s and $Q_{max} = U_{max}A = 0.041 \times 0.00785 = 0.00032$ m^3/s $= 1200$ l/h.

3.2 (*Data from an experiment in Ref.* [16]) A mixture composed by 100 g of 700 μm diameter zirconia (ρ_p = 3800 kg/m^3) and 560 g of 135 μm diameter copper (ρ_p = 8800 kg/m^3) particles is to be fluidized by water at ambient conditions. Based on the particle segregation model, determine the predicted inversion voidage. Assuming μ = 0.001 Pa s for the water viscosity, find also the inversion velocity.

Solution

The smaller particle solids are conventionally designated species 1. First, compute the volumes of solids and the volumetric fraction x_1:

$$V_1 = \frac{0.56}{8800} = 63.6 \text{ cm}^3$$

$$V_2 = \frac{0.1}{3800} = 26.3 \text{ cm}^3$$

$$x_1 = \frac{63.6}{63.6 + 26.3} = 0.707$$

Assuming a water density of $\rho_f = 1000\ \text{kg/m}^3$, calculate the size and net density ratios:

$$d = \frac{d_{p1}}{d_{p2}} = \frac{135}{700} = 0.193$$

$$s = \frac{\rho_{p2} - \rho_f}{\rho_{p1} - \rho_f} = \frac{3800 - 1000}{8800 - 1000} = 0.359$$

Hence, from Eq. (3.18),

$$\varepsilon_{li} = \frac{1-s}{1-d} \cdot \frac{x_1 + (1-x_1)d}{x_1 + (1-x_1)s} = 0.747$$

The accuracy of the predicted voidage can be assessed against the experimentally observed expansion voidage $\varepsilon_{exp} = 0.72$ [16].

To work out the inversion velocity, let us calculate the average solids density using Eq. (3.14):

$$\rho_{av} = \rho_{p1}x_1 + (1-x_1)\rho_{p2} = 7335\ \text{kg/m}^3$$

The Sauter average diameter is

$$d_{av} = \left(\frac{x_1}{d_{p1}} + \frac{1-x_1}{d_{p2}}\right)^{-1} = 176.8\ \mu\text{m}$$

So, the average Reynolds number at layer inversion is

$$Re_{av,li} = \frac{\rho_f d_{av} U_{li}}{\mu} = 176.8 \cdot U_{li}$$

Using the inversion voidage found above, i.e. $\varepsilon_{li} = 0.747$, the exponent α is

$$\alpha = 2.55 - 2.1[\tanh(20\varepsilon_{li} - 8)^{0.33}]^3 = 0.716$$

At this point, the velocity can be evaluated using Newton–Raphson iterations on Eq. (3.20), i.e.

$$U_{k+1} = U_k - \frac{f(U_k)}{f'(U_k)}$$

in which U is the velocity sought, k is the iteration number, the function $f(U)$ expresses the dependence of the LHS of Eq. (3.20) on the velocity, and $f'(U)$ is its derivative. The method starts from an initial estimate of the velocity and quickly proceeds to the converged value, as exemplified in the table below.

k	U_k (m/s)	$f(U_k)$	$f'(U_k)$	U_{k+1}
0	0.01 (initial estimate)	-1.606×10^{-3}	0.1193	0.023 46
1	0.023 46	1.310×10^{-4}	0.1382	0.022 51
2	0.022 51	5.777×10^{-7}	0.1369	0.022 50
3	0.022 50	1.151×10^{-11}	0.1369	0.022 50

The computed inversion velocity of 22.5 mm/s does not agree very well with the experimentally observed value of 13.8 mm/s [16]. However, such a discrepancy between predictions and measurements is not uncommon, owing to the high model sensitivity to variables and the somewhat difficult experimental measurement. Using the experimentally observed voidage, $\varepsilon_{li} = 0.72$, the above procedure would have yielded $U_{li} = 19.6$ mm/s.

Notations

a	area per unit volume (m^2/m^3)
A	column cross-sectional area (m^2)
Ar	Archimedes number $\left(= \frac{d_p^3(\rho_p-\rho)\rho g}{\mu^2}\right)$ (–)
b	numerical parameter (–)
C_D	particle drag coefficient (–)
c_p	specific heat (J/kg K)
d	particle-to-particle diameter ratio (–)
$\overline{d}$	particle-to-average diameter ratio (–)
d_p	particle diameter (m)
D	column diameter (m)
$\mathcal{D}$	diffusion coefficient (m^2/s)
D_h	hydraulic diameter (m)
F	force (N)
g	gravitational field strength (N/kg)
h	heat transfer coefficient (W/(m^2 K))
H	expanded bed height (m)
k	thermal conductivity (W/(m K))
k_l	mass transfer coefficient (m/s)
M	total mass of particles in the column (kg)
n	numerical parameter in Richardson–Zaki equation (–)
Nu	Nusselt number $\left(= \frac{hd_p}{k}\right)$ (–)
$\mathcal{P}$	piezometric pressure (Pa)
Pr	Prandtl number $\left(= \frac{\mu c_p}{\lambda}\right)$ (–)
Q	volumetric flow rate (m^3/s)
Re	particle Reynolds number $\left(= \frac{\rho d_p U}{\mu}\right)$ (–)
s	particle-to-particle net density ratio (–)
$\overline{s}$	average net particle density ratio (–)
Sc	Schmidt number $\left(= \frac{\mu}{\rho \mathcal{D}}\right)$ (–)
Sh_p	particle Sherwood number $\left(= \frac{k_l d_p}{\mathcal{D}}\right)$ (–)
T	temperature (K)
U	liquid superficial velocity (m/s)
U_i	parameter in Richardson–Zaki equation (m/s)
V	volume (m^3)
v	particle velocity (m/s)

x	solid volume fraction (–)
z	vertical coordinate (m)

Greek Letters

α	numerical parameter (–)
ε	voidage (–)
φ	sphericity of particles (–)
μ	viscosity (Pa s)
ρ	density (kg/m^3)

Subscripts

0	static bed
1,2	mixture component indices
av	average
b	bulk
bed	particle bed
dist	distributor
f	fluid
li	layer inversion
max	maximum
mf	minimum fluidization
p	particle
t	terminal settling condition

References

1 Epstein, N. (2003). Applications of liquid-solid fluidization. *Int. J. Chem. Reactor Eng.* 1 (R1): 1–16.
2 Kwauk, M. (1991). Particulate fluidization: an overview. *Adv. Chem. Eng.* 17: 207–360.
3 Di Felice, R. (1995). Hydrodynamics of liquid fluidisation. *Chem. Eng. Sci.* 50 (8): 1213–1245.
4 Epstein, N. (2003). Liquid-solids fluidization. In: *Handbook of Fluidization and Fluid-Particle Systems* (ed. W.-C. Yang). New York: Marcel Dekker Inc.
5 Dallavalle, J.M. (1948). *Micromeritics: The Technology of Fine Particles.* London, UK: Pitman Publishing Corporation.
6 Atta, A., Razzak, S.A., Nigam, K.D.P., and Zhu, J.-X. (2009). (Gas)–liquid–solid circulating fluidized bed reactors: characteristics and applications. *Ind. Eng. Chem. Res.* 48 (17): 7876–7892.
7 Di Renzo, A., Cello, F., and Di Maio, F.P. (2011). Simulation of the layer inversion phenomenon in binary liquid – fluidized beds by DEM–CFD with a drag law for polydisperse systems. *Chem. Eng. Sci.* 66 (13): 2945–2958.
8 Lettieri, P., Di Felice, R., Pacciani, R., and Oweyemi, O. (2006). CFD modelling of liquid fluidized beds in slugging mode. *Powder Technol.* 167: 94–103.

9 Peng, Z., Joshi, J.B., Moghtaderi, B. et al. (2016). Segregation and dispersion of binary solids in liquid fluidised beds: a CFD-DEM study. *Chem. Eng. Sci.* 152: 65–83.
10 Richardson, J.F. and Zaki, W.N. (1954). Sedimentation and fluidization. *Trans. Inst. Chem. Eng.* 32: 35–53.
11 Cleasby, J.L. and Fan, K. (1981). Predicting fluidization and expansion of filter media. *J. Environ. Eng. Div.* 107 (3): 455–471.
12 Rowe, P.N. (1987). A convenient empirical equation for the estimation of the Richardson–Zaki exponent. *Chem. Eng. Sci.* 43: 2795–2796.
13 Khan, A.R. and Richardson, J.F. (1989). Fluid–particle interactions and flow characteristics of fluidized beds and settling suspensions of spherical particles. *Chem. Eng. Commun.* 78: 111–130.
14 Tatemoto, Y., Ishii, A., Suzuki, Y., and Higashino, T. (2010). Effect of volume fraction of material on separation by density difference in a liquid-fluidized bed of inert particles. *Chem. Eng. Technol.* 33 (7): 1169–1176.
15 Di Maio, F.P. and Di Renzo, A. (2016). Direct modeling of voidage at layer inversion in binary liquid-fluidized bed. *Chem. Eng. J.* 284: 668–678.
16 Gibilaro, L.G., Di Felice, R., Waldram, S.P., and Foscolo, P.U. (1986). A predictive model for the equilibrium composition and inversion of binary-solid liquid fluidized beds. *Chem. Eng. Sci.* 41 (2): 379–387.
17 Escudié, R., Epstein, N., Grace, J.R., and Bi, H.T. (2006). Layer inversion phenomenon in binary-solid liquid-fluidized beds: prediction of the inversion velocity. *Chem. Eng. Sci.* 61 (20): 6667–6690.
18 Gautier, A., Carpentier, B., Dufresne, M. et al. (2011). Impact of alginate type and bead diameter on mass transfers and the metabolic activities of encapsulated C3a cells in bioartificial liver applications. *Eur. Cells Mater.* 21: 94–106.
19 Haid, M. (1997). Correlations for the prediction of heat transfer to liquid-solid fluidized beds. *Chem. Eng. Process.* 36: 143–147.
20 Schmidt, S., Büchs, J., Born, C., and Biselli, M. (1999). A new correlation for the wall-to-fluid mass transfer in liquid-solid fluidized beds. *Chem. Eng. Sci.* 54: 829–839.

Problems

3.1 Discuss the effect of temperature and pressure on the velocity–voidage relationship of a liquid-fluidized bed.

3.2 Rework Solved Problem 3.1 for a system with particle density of 800 kg/m^3 (i.e. an inverse fluidization system).

4

Gas Fluidization Flow Regimes

Xiaotao Bi

University of British Columbia, Department of Chemical and Biological Engineering, 2360 East Mall, Vancouver, Canada V6T 1Z3

Gas–solid fluidized beds can be operated in different hydrodynamic flow regimes ranging from fixed bed, delayed bubbling, bubbling, slugging, turbulent to fast fluidization, and pneumatic convey, as depicted in Figure 4.1. It is important to differentiate these flow regimes because hydrodynamics, mixing, heat and mass transfer behaviour, and reactor performance all vary from regime to regime.

4.1 Onset of Fluidization

The onset of fluidization occurs when the superficial gas velocity reaches the minimum fluidization velocity, U_{mf}, where the particles in the bed become fully suspended by the fluid. Experimental methods for measuring U_{mf} and equations for prediction of U_{mf} are provided in Chapter 2.

4.2 Onset of Bubbling Fluidization

Bubbling is defined to commence when the first bubble is observed in a gas-fluidized bed as the gas flow is increased very slowly. For Geldart group A powders, due to interparticle forces, bubbles do not form until the superficial gas velocity is significantly beyond the minimum fluidization velocity. The region between U_{mf} and the minimum bubbling velocity, U_{mb}, is then termed delayed bubbling or homogeneous fluidization.

The minimum bubbling velocity can be determined either by visual observation or by measuring pressure and/or voidage fluctuations [2]. Before bubbles form in the bed, both pressure fluctuations and voidage fluctuations are very small. Pressure and voidage fluctuations increase as soon as bubbles appear in the bed, as shown in Figure 4.2.

The minimum bubbling velocity, U_{mb}, is influenced significantly by interparticle forces, i.e. by van der Waals forces, capillary forces for wet particles, and electrostatic forces for electrically charged particles. For dry and rounded particles

Essentials of Fluidization Technology, First Edition. Edited by John R. Grace, Xiaotao Bi, and Naoko Ellis.

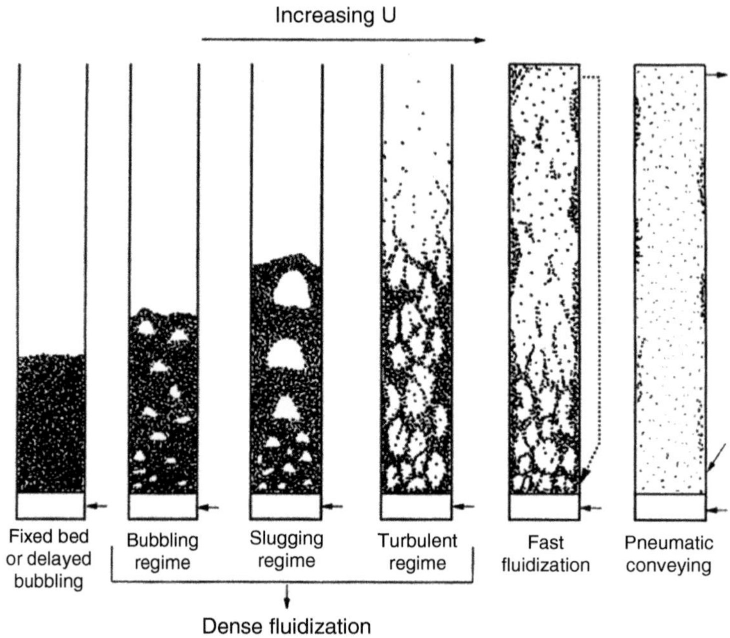

Figure 4.1 Flow patterns in gas–solid fluidization systems, modified from [1]. Source: Adapted from Grace and Bi 1997 [1].

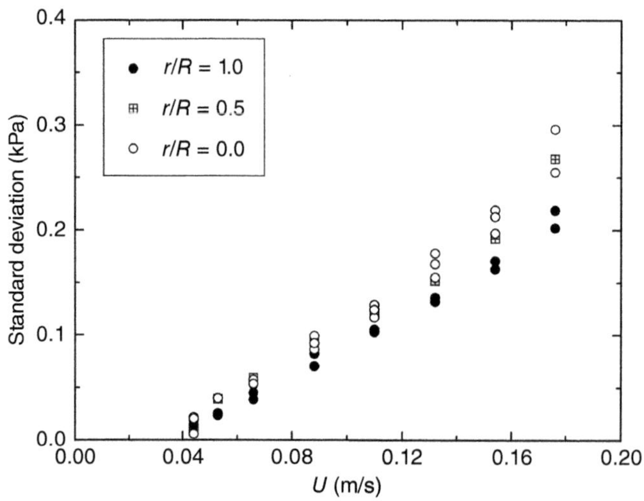

Figure 4.2 Standard deviation of absolute pressure fluctuations in a 0.1 m ID (Inner Diameter) column with 0.215 mm sand particles. Source: From Bi 1994 [3].

with negligible electrostatic charges, the Abrahamsen and Geldart [4] correlation can be used to predict U_{mb}:

$$\frac{U_{mb}}{U_{mf}} = \frac{2300\rho_g^{0.126}\mu_g^{0.523}\exp(0.716F_{45})}{d_p^{0.8}g^{0.923}(\rho_p-\rho_g)^{0.934}} \tag{4.1}$$

where F_{45} is the fraction of solids by mass smaller than 45 µm in diameter. As discussed in Chapter 2, the existence of delayed bubbling demarcates the boundary between group A and group B particles in Geldart's powder classification, with $U_{mb}/U_{mf} = 1$ for group B and D powders. Equation (4.1) predicts that the region of delayed bubbling or particulate fluidization is broader at high pressures and high temperatures. Therefore, some group B powders at ambient conditions show group A behaviour at high pressure and/or high temperature.

4.3 Onset of Slugging Fluidization

As the gas velocity increases in small-diameter fluidized beds, bubble coalescence during their rise leads to the formation of larger bubbles. The bubble shape and rise velocity are significantly affected by the column wall when the bubble volume-equivalent diameter, d_e, becomes larger than ~0.1D, where D is the column diameter. Wallis [5] showed that the rise velocity of an isolated gas bubble in a liquid influenced by the wall can be correlated by

$$u_b = 1.13u_{b0}\exp(-d_e/D) \tag{4.2}$$

where u_{b0} is the bubble rise velocity in isolation, i.e. in the absence of wall effects.

Bubbles may be elongated and developed into gas slugs when bubbles grow to a size comparable to the column diameter, e.g. $d_e/D > 0.6$, leading to a transition from bubbling to slugging fluidization. A correlation for predicting the superficial gas velocity at minimum slugging, U_{ms}, in gas–solid fluidized beds was derived by Stewart and Davidson [6]:

$$U_{ms} = U_{mf} + 0.07\sqrt{gD} \tag{4.3}$$

for axisymmetric round-nosed slugs.

Slugging cannot occur, even if the fluidization velocity is higher than U_{ms}, in some systems if the bed is too shallow, or if the bubble size is limited by splitting. To ensure the occurrence of slugging, the following criteria need to be satisfied:

- $U > U_{ms}$
- $H > H_{entry}$
- $d_{b,max} > 0.6D$
- $U < U_c$

The minimum bed height H_{entry} can be estimated [7] by

$$H_{entry} = 3.5D\{1 - 1/\sqrt{N_{or}}\} \quad (\text{with } N_{or} > 3) \tag{4.4}$$

where N_{or} is the number of orifices in the gas distributor plate.

For $N_{or} = 1, 2$, or 3,

$$H_{entry} = 1.5D \tag{4.5}$$

The maximum stable bubble size, which is determined by a balance between bubble coalescence and bubble splitting, is estimated [7] by

$$d_{b,\max} = 2.0v_t'^2/g \tag{4.6}$$

where v_t' is the terminal settling velocity of particles of diameter 2.7 times the average particle diameter. U_c is the superficial velocity for transition to turbulent fluidization, which is defined and discussed in Section 4.4.

4.4 Onset of Turbulent Fluidization

Turbulent fluidization occurs when bubble coalescence is balanced by bubble splitting, beyond which splitting becomes dominant with further increases in gas velocity and the mean bubble size decreases for group A powders or remains constant for group B and D powders. In bubbling fluidized beds, solids holdup inside gas bubbles tends to increase with increasing gas velocity [8]. In a turbulent fluidized bed of group A powders, large voids are distorted, no longer appearing as bubbles with smooth surfaces. Instead, distorted dilute transient voids are observed in the turbulent fluidization flow regime, as shown in Figure 4.3, with the mean void size decreasing with increasing superficial gas velocity.

In gas–solid fluidized beds, pressure fluctuations are caused mainly by the eruption, formation, and splitting of gas bubbles. As the gas velocity increases, pressure fluctuations increase as the average bubble size increases. On the other hand, as the void size decreases in the turbulent fluidization regime of group

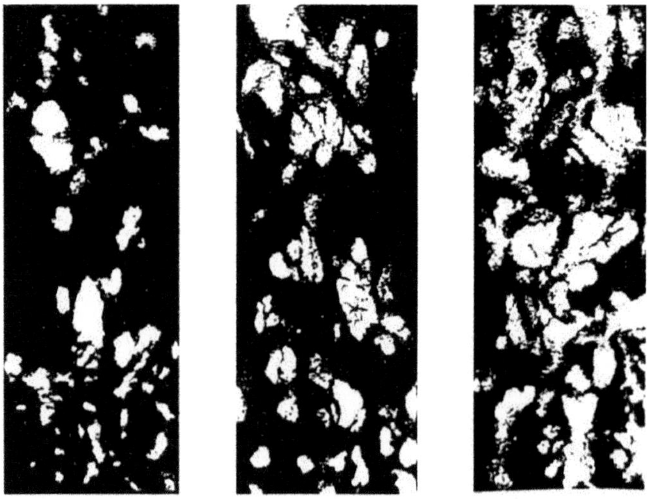

Figure 4.3 Instantaneous spatial distribution of voids in a gas–solid fluidized bed of fine particles ($U = 0.1$, 0.2, and 0.5 m/s for pictures from left to right). Source: Kai et al. 1995 [9]. Reproduced with permission of Elsevier.

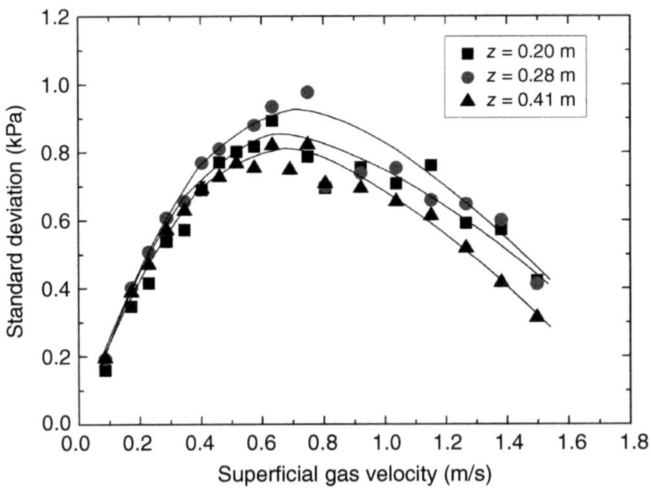

Figure 4.4 Standard deviation of pressure fluctuations at three heights in a gas–solid fluidized bed. Source: Bi and Grace 1995 [2]. Reproduced with permission of Elsevier.

A powders, the amplitude of pressure fluctuations decreases with increasing superficial gas velocity. The transition from bubbling to turbulent fluidization is said to occur when the standard deviation (or mean amplitude) of pressure fluctuations reaches a maximum with increasing U [10, 11]. As shown in Figure 4.4, the velocity at the maximum point is defined as the transition velocity U_c, denoting the onset of turbulent fluidization.

The transition velocity U_c can be predicted by assuming the existence of a maximum stable bubble size in fine particle fluidization systems. Splitting of trailing bubbles can be induced by the turbulent wake of leading bubbles. When a following bubble moves into the turbulent wake region of a leading bubble of maximum stable size, the trailing bubble breaks into small bubbles. The onset of massive breakdown of bubbles transforms the flow from bubbling to turbulent fluidization as bubble splitting becomes dominant. U_c can thus be predicted based on equations derived from the fluidized bed two-phase flow theory [11] or simulated pressure fluctuation signals [12].

Pressure fluctuations can be simulated by mechanistic models based on the pressure–bubble relationship of Davidson [13] or others. Bubbles can be assumed to be aligned in vertical chains in a deep fluidized bed with bubble growth dominant at low gas velocities. After reaching a maximum stable bubble size, the mean bubble size remains at its maximum stable size for group B/D powders, or decreases for group A powders, when the gas velocity is increased further, with the separation distance between adjacent bubbles starting to decrease. Differential pressure fluctuations across a vertical interval of the bed can be simulated using a bubble growth model. The simulated results in Figure 4.5 show that the standard deviation of pressure fluctuations reaches a maximum with increasing superficial gas velocity, regardless of whether the mean bubble size remains at a constant value or decreases with further increases in gas velocity after the maximum stable size has been reached. This implies that U_c corresponds to the

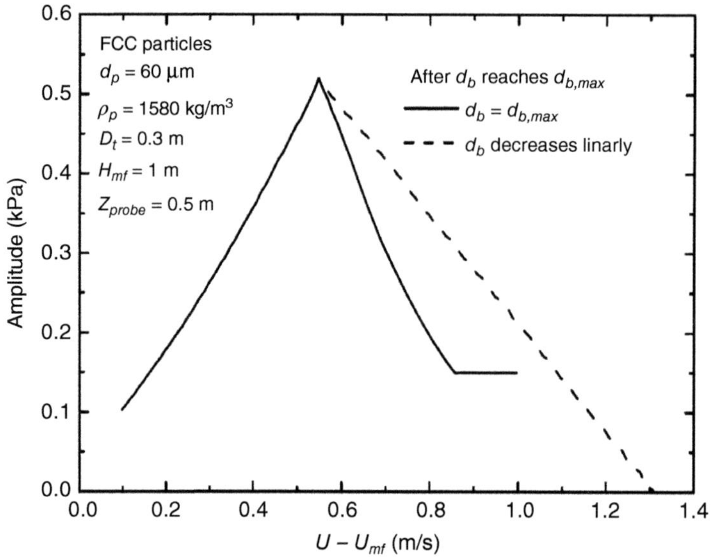

Figure 4.5 Comparison of simulated amplitude of local pressure fluctuations as a function of excess gas velocity for cases with linearly decreasing bubble size and with constant bubble size after maximum size is reached. Source: Chen and Bi 2003 [12]. Reproduced with permission of Elsevier.

breakdown of bubbles after they reach the maximum size for fine particles and the reduction of separation distance between consecutive bubbles when the separation distance between bubbles becomes comparable to the bubble diameter.

Pressure fluctuations can also be simulated using two-phase computational fluid dynamic models [14–16]. The standard deviation of pressure fluctuations simulated from the discrete particle/element method (DEM) in Figure 4.6 reaches a maximum at a gas velocity close to experimentally measured U_c values.

U_c has been extensively studied experimentally with different types of particles based on pressure fluctuations in fluidization columns of different sizes. In general, it has been found that U_c increases with increasing particle size, particle density, and reactor temperature and with decreasing reactor pressure. U_c determined from differential pressure fluctuations [2] also appears to vary axially, with lower values higher in the bed, because bubbles grow as they rise and large bubbles split near the bed surface at lower velocities. However, U_c measured from absolute or single-point pressure fluctuations of the bed is insensitive to the location, because the pressure at a single position is superimposed on pressure waves generated from other locations propagating inside the fluidized bed, representing the overall and average bed behaviour.

Several empirical correlations have been developed for the prediction of U_c [2, 10], mostly based on measurement of absolute pressure fluctuations. For the prediction of the average bed transition from bubbling to turbulent fluidization, Bi and Grace [1] recommended two equations:

$$Re_c = \frac{\rho_g U_c d_p}{\mu_g} = 0.565 Ar^{0.461} \tag{4.7}$$

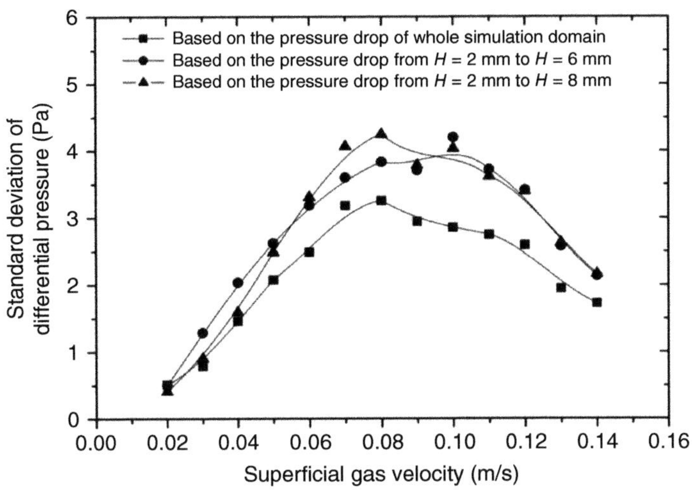

Figure 4.6 Simulated standard deviations of differential pressure fluctuations as a function of superficial gas velocity (d_p = 75 μm). Source: Wang et al. 2011 [15]. Reproduced with permission of Elsevier.

due to Bi and Grace [5] and

$$\frac{U_c}{\sqrt{gd_p}} = \left(\frac{0.211}{D^{0.27}} + \frac{0.00242}{D^{1.27}}\right)\left(\frac{\mu_{g20}}{\mu_g}\right)^{0.2}\left[\left(\frac{\rho_{g20}}{\rho_g}\right)\left(\frac{\rho_p - \rho_g}{\rho_g}\right)\left(\frac{D}{d_p}\right)\right]^{0.27} \tag{4.8}$$

proposed by Cai et al. [17].

In columns of small diameter with large particles where slugging is encountered as the gas velocity increases, a peak pressure fluctuation is also observed, although the peak region is much flatter than for non-slugging systems. This peak value is reached when the length of the gas slug and the vertical separation distance between consecutive slugs (occupied by dense plugs) become comparable. As the gas velocity increases further, the plugs become shorter than the slug length based on gas flow conservation, leading to a decrease in pressure fluctuations and deformation of trailing slugs due to the wakes of leading slugs. The flow pattern is then similar to "churn flow" in gas–liquid two-phase flow in small-diameter vertical pipes.

U_c also corresponds to "phase inversion" from bubbling or slugging fluidization, where voids are immersed in a dense emulsion phase, to a situation where the dilute void phase is dominant. If the emulsion phase retains the properties of a dense bed at minimum fluidization and the dilute phase contains no particles, the voidage, ε_c, at U_c is expected to be approximately 0.725 [$\varepsilon_c = 1 - (1 - \delta_c)(1 - \varepsilon_{mf})$] = [$1 - (1 - 0.5)(1 - 0.45)$]. The measured bed voidage at U_c is ~0.65 [11], somewhat less than 0.725, with the difference likely due to the nonuniform radial distribution of bubbles in laboratory fluidized beds, with higher void fraction near the axis.

4.5 Termination of Turbulent Fluidization

In a turbulent fluidized bed, the solids entrainment rate increases exponentially with increasing superficial gas velocity. Unless entrained particles are sufficiently captured and returned to the bottom of the bed, the bed may lose all its particles, making it impossible to maintain steady operation of a dense fluidized bed in the turbulent fluidization regime [18]. Transition of turbulent fluidization to higher gas velocity regimes is thus related to solids entrainment. Li and Kwauk [19] demarcated the transition from turbulent fluidization to fast fluidization, in which particles need to be continuously recycled to the bottom of the bed to maintain dense bed operation, by a critical velocity, above which dilute-phase upflow with continuous solids feed becomes possible. Similarly, a critical superficial gas velocity, U_{se}, was defined as the velocity at which significant particle entrainment occurs [18], determined by plotting the saturation carrying capacity vs. the superficial gas velocity, to quantify the transition from turbulent fluidization to fast fluidization. A correlation for predicting U_{se} was developed [18] based on published data:

$$Re_{se} = \frac{\rho_g U_{se} d_p}{\mu_g} = 1.53Ar^{0.5} \tag{4.9}$$

The critical velocity, U_{se}, is typically several times the terminal settling velocity, v_t, for Geldart group A and fine group B particles. It can be considered as a hindered or apparent terminal settling velocity of bed particles, reflecting the existence of clusters or agglomerates of particles in dense fluidized beds. For group D particles, U_{se} is very similar to the terminal settling velocity, v_t, of single particles. When U_{se} calculated from Eq. (4.9) exceeds the single particle terminal velocity, U_{se} should be taken as being equal to v_t.

4.6 Fast Fluidization and Circulating Fluidized Bed

At a given gas velocity in a vertical pipe, a single particle is likely to be carried upwards by the gas if the average gas velocity exceeds the terminal settling velocity of the particle (v_t). Each gas velocity corresponds to a constant elutriation flux of particles of given size, called the saturation carrying capacity, G_s^*. In low-velocity fluidized beds, the saturation carrying capacity corresponds to the elutriation rate constant, measured at an exit located well above the transport disengagement height (TDH) (see Chapter 10). In high-velocity gas–solid upflow systems, the saturation carrying capacity can be measured by gradually increasing the solids feed rate into the bottom of a transport line, with saturation reached when the dilute upward flow collapses, causing particles to begin to accumulate at the bottom of the transport line [18]. The saturation carrying capacity is a unique characteristic of pneumatic transport systems. It defines the maximum solids flow rate that an upward-flowing gas can carry while the entrance of the vertical line is not choked by a dense layer of particles. It also

Figure 4.7 Schematic of a bottom-restricted circulating fluidized bed system.

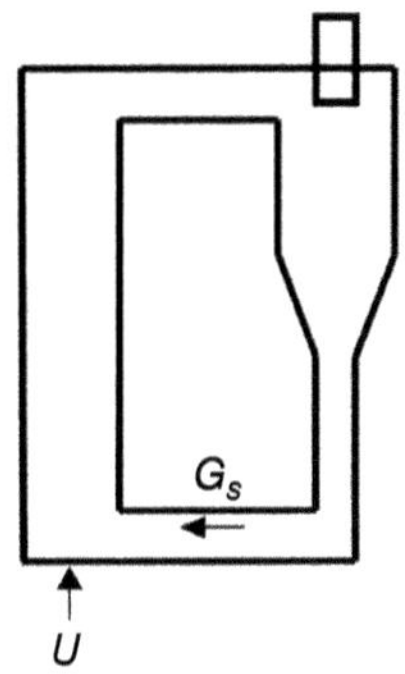

demarcates the termination of dilute uniform gas–solid upflow. The saturation carrying capacity, G_s^*, can be predicted [20] by

$$\frac{G_s^*}{\rho_g U} = 3.45 \times 10^{-3} Ar^{-0.194} \left(\frac{U}{\sqrt{g d_p}} \right)^{1.845} \tag{4.10}$$

If the bottom of the vertical transport line, also called a riser, is restrained at its bottom by a distributor to prevent particles from escaping from the bottom of the system, a dense fluidized bed forms at the bottom. With solids continuously carried up and out of the riser and returned to the bottom via a circulation loop, a circulating fluidized bed (CFB) is established, as shown in Figure 4.7 (see also Chapter 12). A CFB can be operated in either co-current upward dilute flow (i.e. pneumatic transport) or in a fast fluidization mode, depending on the gas velocity and solids circulation rate. When the solids feeding rate exceeds the saturation carrying capacity, a dense bed forms at the bottom of the riser and spreads upwards with further increase in the solids feeding rate for steady-state operation. The global flow pattern in the riser then resembles "fast fluidization," with a dense region in the lower section and a dilute region in the upper section of the riser.

The flow pattern in the CFB riser depends not only on the superficial gas velocity, U, but also the solids circulation flux, G_s, with the cyclone and standpipe being coupled to the riser to form a circulation loop. The maximum circulation rate in a CFB system is determined by the global pressure balance over the whole solids circulation loop [21]. Such a pressure balance is key to understanding "choking," defined as the critical condition when the CFB riser terminates its stable operation beyond a certain solids circulation rate at a given riser gas velocity. Choking occurs either because the gas blower is unable to provide enough pressure head to support dense flow in the riser or pressure head build-up in the standpipe is insufficient to push particles from the standpipe into the riser due to a lower solids inventory [22] (see Figure 4.8).

For a well-designed CFB system, the riser can operate at high solids circulation flux, e.g. up to ~1000 kg/m^{2}s or more in commercial fluid catalytic cracking (FCC) risers, when limitations from the gas blower and standpipe solids return

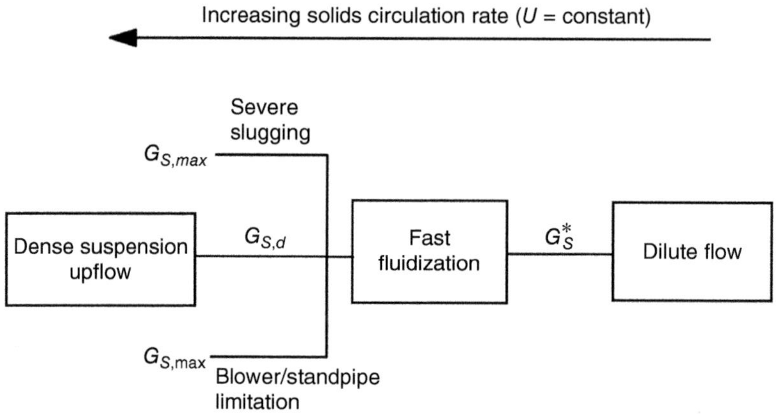

Figure 4.8 Flow pattern transitions in a circulating fluidized bed system.

line are eliminated. It has been found that beyond a certain solids flux and superficial gas velocity, the net downward flow of particles in the annulus region of the riser ceases, resulting in disappearance of the core–annulus flow structure of fast fluidization where particles have net upward flow in the dilute core region and net downward flow in a dense annular region. The riser can then be considered to operate in a "dense suspension upflow" flow regime [23]. The boundary between fast fluidization and dense suspension upflow regimes can be estimated [24] by

$$\frac{G_{s,d}}{\rho_p \varepsilon_s v_t} = 5 \times 10^{-3} Ar^{0.339} \left(\frac{U}{\varepsilon_s v_t} \right)^{0.339} \tag{4.11}$$

4.7 Flow Regime Diagram for Gas–Solid Fluidized Beds

Table 4.1 summarizes the key characteristics of the various flow regimes in gas–solid fluidized beds. A number of attempts have been made [25–29] to distinguish various flow regimes on phase diagrams. Each such diagram has its particular advantages. The one developed by Grace [28] seems to be most useful for engineering applications. With the dimensionless superficial gas velocity plotted vs. the dimensionless particle diameter, Grace [28] incorporated the typical operation regions of pneumatic transport lines and fast fluidized beds (see Figure 4.9). Typical operating regions for bubbling fluidized beds and spouted beds are also identified in this diagram. Bi and Grace [29] extended the Grace diagram by incorporating the U_c and U_{se} transition velocities based on Eqs. (4.8) and (4.9) to identify the turbulent and fast fluidization flow regimes.

It must be pointed out that severe slugging, blower, and standpipe limitations in CFB are not reflected in almost all of these phase diagrams, because slugging is restricted by the column diameter, while the maximum solids circulation rate is influenced by the blower pressure head and solids inventory in the standpipe of

Table 4.1 Major characteristics of gas–solid fluidization and adjacent flow regimes.

Operating range	Regime	Appearance and principal features
$0 < U < U_{mf}$	Fixed bed	Particles are stationary; gas flows through the interstices
$U_{mf} < U < U_{mb}$	Bubble-free fluidization	Bed expands smoothly and uniformly; top surface is well defined; some small-scale particle motion; little tendency for particles to aggregate; very little pressure fluctuation
$U_{mb} < U < U_{ms}$	Bubbling fluidization (Chapter 7)	Bubbles form near the distributor, grow mostly by coalescence, and rise to the surface; top surface is well defined, with bubbles breaking through periodically; irregular pressure fluctuations of appreciable amplitude. Bubble size increases as U increases
$U_{ms} < U < U_c$	Slugging fluidization (Chapter 8)	Gas slugs fill most of the column cross section; top surface rises and collapses periodically with a reasonably regular frequency; large and regular pressure fluctuations
$U_c < U < U_{se}$	Turbulent fluidization (Chapter 9)	Inversion of dense phase from continuous to dispersed phase; transition of well-defined bubbles to transient voids; voids may briefly connect to each other to develop a continuous long void; top surface is difficult to distinguish; small amplitude pressure fluctuations only
$U > U_{se}$ and $G_s > G_{s,d}$	Dense suspension upflow	Particles mostly move upwards across the whole cross section, with little or no upper dilute zone; limited axial and radial mixing of gas and particles
$U > U_{se}$ and $G_{s,d} > G_s > G_s^*$	Fast fluidization (Chapter 12)	No distinguishable upper bed surface; particles are transported out from the top and must be replaced by adding solids near the bottom. Clusters or strands of particles descend, mostly near the wall, while gas and dispersed particles move upwards in the interior
$U > U_{se}$ and $G_s < G_s^*$	Dilute-phase transport	No axial variation of solids concentration, except in the bottom acceleration section. Some particle strands may still be identifiable near the wall

the CFB system, none of which is accounted for in these diagrams. For a specific CFB system, these limits can be identified by using appropriate approaches [21, 22].

4.8 Generalized Flow Diagram for Gas–Solid Vertical Transport

If particles are fed from the middle section into a vertical tube with an open top and bottom in which gas is flowing upwards from bottom to top, as shown in

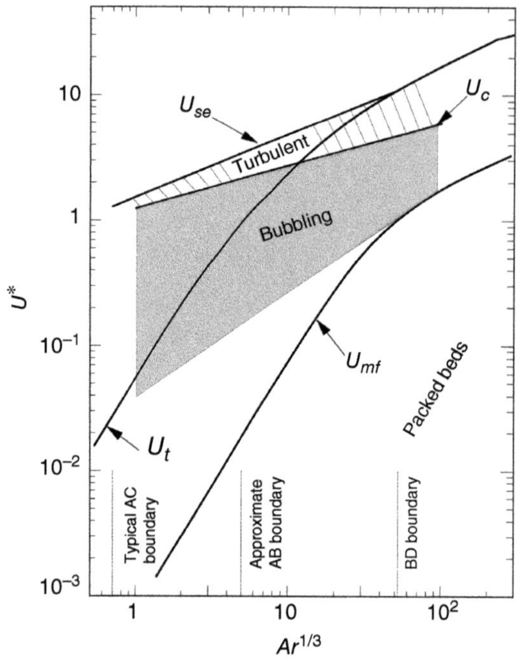

Figure 4.9 Fluidization flow regime diagram of Bi and Grace, extended from Grace [29]. Source: Bi and Grace 1995 [29]. Reproduced with permission of Elsevier.

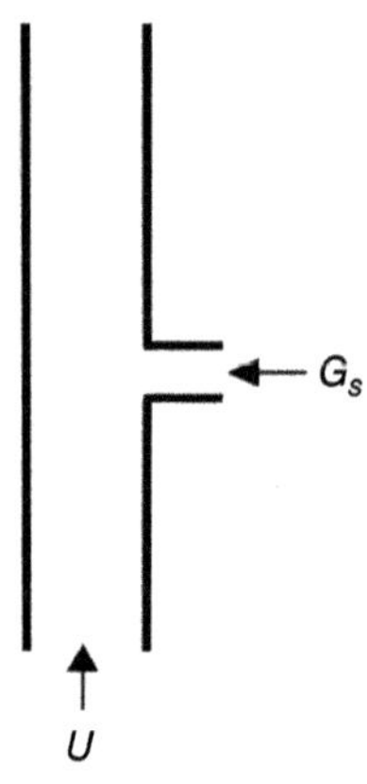

Figure 4.10 Schematic of nonrestricted vertical transport line with gas flow upwards.

Figure 4.10, both gas–solid co-current upward flow in the upper section above the feeding point and counter-current flow in the lower section below the feed level are possible, depending on the magnitudes of the gas velocity and solids feed rate. Counter-current flow ends when solids can no longer fall downwards (i.e. at the *flooding* point), and gas–solid co-current upflow ceases when the gas velocity is less than the particle terminal settling velocity. A fluidized bed reactor can be considered as a special case where the bottom of the vertical tube is restricted by a gas distributor to prevent solids from falling beneath the solids feed port.

A flow regime diagram for a given vertical tube, gas, and particle properties can be constructed based on the flooding velocity and the gas velocity corresponding

Table 4.2 Flow regimes and corresponding flow patterns in a vertical tube with open ends.

Regime	U	G_s	Upper section	Lower section
I	$<v_t$	$<G_{s,F}$	Single-phase gas flow	Dilute counter-current flow
II	$<v_t$	$>G_{s,F}$	Dense co-current flow	Dense counter-current flow
III	$>v_t$	$< G_s^*$	Dilute co-current flow	Single-phase gas flow
IV	$>v_t$	$< G_s^* < G_s < (G_s^* + G_{s,F})$	Dilute co-current flow	Dilute counter flow
V	$>v_t$	$> G_s^* + G_{s,F}$	Dense co-current flow	Dense counter flow

to the saturation carrying capacity. The saturation carrying capacity can be predicted by Eq. (4.10). The flooding velocity is predicted by Eq. (4.12) [30], modified from an equation originally developed to predict flooding in packed towers:

$$\left[\frac{U}{\sqrt{gD}}\left(\frac{\rho_g}{\rho_D}\right)^{1/2}\right]^{2/3} + \frac{1}{2^{1/3}}\left(\frac{G_{s,F}/\rho_p}{\sqrt{gD}}\right)^{2/3} = \left(\frac{1}{2\tan\theta}\right)^{1/3} \tag{4.12}$$

where U is the superficial gas velocity, $G_{s,F}$ is the solids flux at flooding, D is the column diameter, and θ is the angle of internal friction of particles (typically ~70° for round particles).

Figure 4.12 shows such a flow regime diagram for a vertical gas–solid transport line with upward gas flow. As summarized in Table 4.2, there are five unique flow regimes in the tube.

As shown in Figure 4.11, at a gas velocity less than the critical velocity, U_{se}, for fine group A particles or the particle terminal settling velocity for group B or D particles, most fed particles fall downwards at a low feed rate, leading to counter-current flow in the lower section of the tube and single-phase gas flow in the upper section, Regime I in Figure 4.11. However, with an increase in solids feed rate, flooding is reached with the particle discharge rate from the bottom end of the tube becoming smaller than the solids feed rate. As a result, solids start to accumulate and build up in the upper section, forming a dense fluidized bed in the upper section, corresponding to Regime II in Figure 4.11. This flooding phenomenon is analogous to flooding in gas–liquid counter-flow systems.

Consider cases where the gas velocity in the tube exceeds the critical velocity. At a low solids feed rate, all particles fed are carried upwards, giving rise to co-current upward flow in the upper section and to single-phase gas flow in the lower section, Regime III in Figure 4.11. When the solids feed rate increases to such an extent that the solids feed rate exceeds the saturation particle carrying capacity of the gas, excess particles fall downwards and leave the tube from the bottom, forming counter-current flow in the lower section, as well as co-current upward flow in the upper section, corresponding to Regime IV in Figure 4.11. If the solids feed rate is further increased to such an extent that the particle downflow exceeds the flooding rate, corresponding to the maximum discharge rate from the bottom of the column, dense suspension starts to build up above

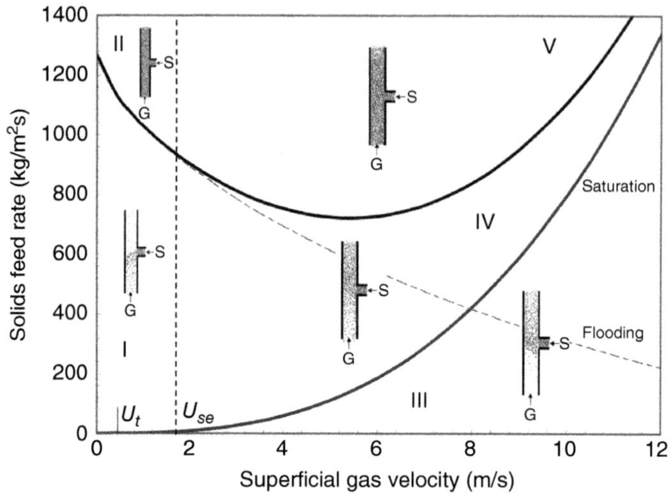

Figure 4.11 Flow regime diagram for nonrestricted vertical transport lines. FCC particles in ambient air: mean particle size = 60 μm; particle density = 1800 kg/m³. Source: Bi 2011 [32].

the solids feed point, forming co-current dense suspension upflow in the upper section and flooded counter-current flow in the lower section, i.e. Regime V in Figure 4.11. The flow pattern in Regime V resembles the flow pattern of fast fluidization encountered in bottom-restricted CFBs. However, instead of being restricted by a distributor at the bottom, as in CFBs, the downward solids flow is now restricted by flooding in the nonrestricted gas–solid vertical transport.

4.9 Effect of Pressure and Temperature on Flow Regime Transitions

As a first approximation, gas viscosity can be considered to be independent of system pressure, while, based on the ideal gas law, gas density is proportional to absolute pressure and inversely proportional to the absolute temperature. For many gases, the viscosity increases approximately with the square root of the absolute temperature, i.e. $\rho_g \propto P/T$ and $\mu_g \propto T^{0.5}$. Inserting these dependencies in the definitions of the dimensionless coordinates of Figure 4.12, it can be shown that

$$d_p^* \propto P^{1/3}/T^{2/3} \text{ and } U^* \propto P^{2/3}/T^{5/6} \quad (4.13)$$

To illustrate the influence of pressure and temperature, consider a particle with $d_p^* = 10$ operated at $U^* = 1$ and ambient temperature and pressure, a point that falls in the bubbling flow regime, as indicated in Figure 4.12. At a constant volumetric flow rate or superficial gas velocity, an increase in system pressure is predicted to shift the operation towards the turbulent fluidization regime. However,

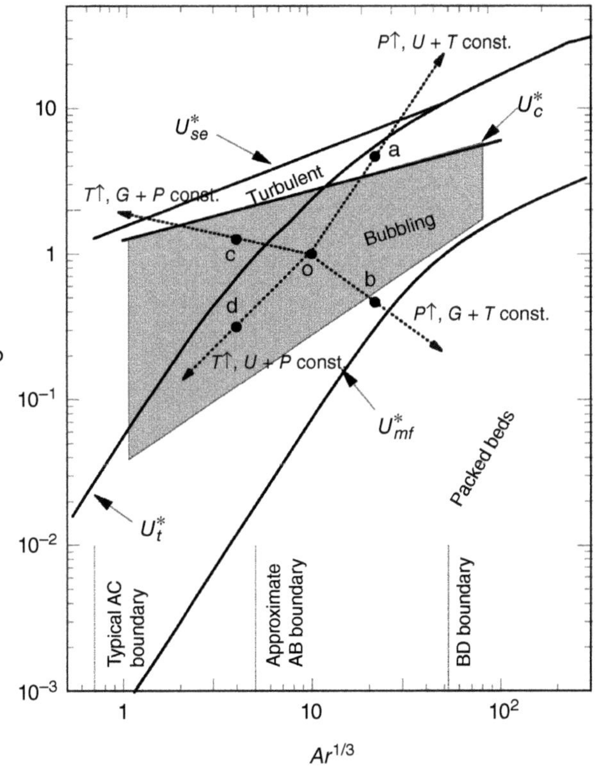

Figure 4.12 Flow regime map for batch gas–solid fluidization together with loci of operating points for increasing system temperature and pressure for constant superficial gas velocity and constant gas mass flow rate. Heavy lines indicate transition velocities. Shaded region is the typical operating range for bubbling fluidized beds. o: $P = 0.1$ MPa, $T = 300$ K; a: $P = 1$ MPa, $T = 300$ K, $U =$ constant; b: $P = 1$ MPa, $T = 300$ K, $G =$ constant; c: $P = 0.1$ MPa, $T = 1200$ K, $G =$ constant; d: $P = 0.1$ MPa, $T = 1200$ K, $U =$ constant. Source: Bi and Grace 1996 [31]. Reproduced with permission of John Wiley and Sons.

at a constant gas mass flow rate, the increase in pressure moves the operating condition away from the turbulent regime, leading to a decrease of bubble size and pressure fluctuations. An increase in temperature at a constant volumetric flow rate is predicted to cause a shift within the bubbling regime. At a constant mass flow rate, the operating condition moves towards the turbulent regime with increasing temperature. The diagram suggests that transitions from bubbling to turbulent and fast fluidization require lower superficial gas velocities, but higher gas mass flow rates, in pressurized systems, compared with systems operated at atmospheric pressure. This is in agreement with experimental data [17, 33, 34]. With increasing temperature, the transitions to turbulent and fast fluidization are predicted to be delayed if the superficial gas velocity is maintained constant or promoted when based on constant gas mass flow rate, again in agreement with experimental observations [17, 35–37].

Solved Problems

4.1 A batch fluidized bed of internal diameter 0.3 m is equipped with a multi-orifice distributor and two-stage cyclones to capture and return entrained particles. The column is 4 m tall. (1) If 40 kg of FCC particles of 60 μm mean size, 1600 kg/m^3 particle density, and 850 kg/m^3 bulk density are fluidized in this column at a superficial gas velocity of 0.3 m/s with air at ambient conditions, what is the operating flow regime? (2) It is desired to operate this column at high gas velocities. If the gas velocity is increased to 0.6 m/s, what is the flow regime? (3) What is the highest gas velocity at which this column can be operated under dense fluidization conditions?

Solution

(1) At ambient condition: air density $\rho_g = 1.25\ \text{kg/m}^3$, air viscosity $\mu_g = 1.8 \times 10^{-5}$ Pa s

$$d_p = 6 \times 10^{-5}\ \text{m},\ \rho_p = 1600\ \text{kg/m}^3$$

$$Ar = \rho_g(\rho_p - \rho_g)d_p^3 g/\mu_g^2 = 6.68$$

From Eq. (4.7), $Re_c = 0.565Ar^{0.461} = 1.356$; $U_c = 0.41$ m/s, which is higher than 0.3 m/s. The column therefore operates in either the bubbling or slugging flow regime.

For fine particles, Eq. (2.19), $U_{mf} = 0.000\,61 d_p^2(\rho_p - \rho_g)g/\mu_g = 0.0012$ m/s

From Eq. (4.3), $U_{ms} = U_{mf} + 0.07(gD)^{0.5} = 0.12$ m/s, which is less than 0.3 m/s. The maximum bubble diameter-to-column diameter ratio needs to be checked.

From Eq. (2.7), $v_t' = 10.71$ m/s calculated using a particle diameter of $2.7d_p$ (162 μm). Then from Eq. (4.6) $d_{b,max} = 0.104$ m. $d_{b,max}/D = 0.35 < 0.67$, so no slugs form in this column. Therefore, the column operates in the bubbling flow regime.

(2) $U = 0.6$ m/s is now greater than $U_c = 0.41$ m/s, so the column operates in the turbulent fluidization flow regime.

(3) The highest gas velocity for a dense fluidized bed is the significant entrainment velocity, U_{se}.

From Eq. (4.9), $Re_{se} = 1.53Ar^{0.5} = 3.95$. Thus $U_{se} = 0.95$ m/s.

4.2 A circulating fluidized bed of 0.2 m diameter and 10 m tall is operated at ambient temperature and pressure. FCC particles (d_p = 0.06 mm and ρ_p = 1600 kg/m^3) are used as the bed material. At a superficial gas velocity of 4 m/s, what is the lowest solids circulation flux in kg/m^{2}s required for the riser to operate in the fast fluidization flow regime, with a dense bottom region and a dilute upper region?

Solution

The solids circulation rate needs to be greater than the saturation gas carrying capacity, which can be estimated by Eq. (4.10):

$$G_s^* = 3.45 \times 10^{-3} \rho_g U Ar^{-0.194} \left(\frac{U}{\sqrt{g d_p}} \right)^{1.845} = 147 \text{ kg/m}^2\text{s}$$

Notations

Ar	Archimedes number $(= \rho_g(\rho_p - \rho_g) d_p^3 g / \mu_g^2)$
d_p	Sauter mean particle diameter (m)
d_p^*	dimensionless particle diameter (–)
D	column diameter (m)
d_b	bubble diameter (m)
$d_{b,max}$	maximum stable bubble diameter (m)
F_{45}	fraction of solids by mass smaller than 45 μm in diameter (–)
g	gravitational acceleration (= 9.81 m/s^2)
G_s	solids circulation flux (kg/m^{2}s)
G_s^*	saturated gas carrying capacity (kg/m^2 s)
$G_{s,F}$	solids flux at flooding (kg/m^2 s)
$G_{s,d}$	solids flux at onset of dense suspension upflow (kg/m^2 s)
H	expanded bed height (m)
H_{entry}	minimum bed height for slugging (m)
N_{or}	number of orifices in distributor plate (–)
P	pressure (Pa)
Re	Reynolds number $(= \rho_g U d_p / \mu_g)$
T	temperature (K)
U	superficial gas velocity (m/s)
U^*	dimensionless gas velocity $(= Re/Ar^{1/3})$
u_{b0}	rise velocity of an isolated bubble without any wall effect (m/s)
u_b	rising velocity of an isolated bubble with wall effect (m/s)
U_c	onset superficial gas velocity for turbulent fluidization (m/s)
U_{mb}	minimum bubbling velocity (m/s)
U_{mf}	minimum fluidization velocity (m/s)
U_{ms}	minimum gas superficial velocity for slug flow (m/s)
U_{se}	termination velocity of dense turbulent fluidization (m/s)
v_t	terminal settling velocity of single particles (m/s)
v_t'	terminal settling velocity of particles of diameter 2.7d_p (m/s)

Greek Letters

δ_c	void phase volume fraction at $U = U_c$ (–)
ε_c	voidage at $U = U_c$ (–)
ε_{mf}	voidage at $U = U_{mf}$ (–)
ε_s	solids holdup or fraction (–)
μ_g	gas viscosity (kg/m s)
θ	angle of internal friction of particles
ρ_D	density of the dense phase (kg/m^3)
ρ_g	gas density (kg/m^3)
ρ_p	particle density (kg/m^3)

References

1 Grace, J.R. and Bi, X.T. (1997). Introduction to circulating fluidized beds. In: *Circulating Fluidized Beds* (eds. J.R. Grace, A.A. Avidan and T.M. Knowlton). New York: Blackie Academic & Professional Press.

2 Bi, X.T. and Grace, J.R. (1995). Effects of measurement method on velocities used to demarcate the onset of turbulent fluidization. *Chem. Eng. J.* 57: 261–171.

3 Bi, X.T. (1994). Flow regime transitions in gas-solid fluidization and transport. PhD dissertation. University of British Columbia, Vancouver.

4 Abrahamsen, A.R. and Geldart, D. (1980). Behaviour of gas-fluidized beds of fine powders, Part I. Homogeneous expansion. *Powder Technol.* 26: 35–46.

5 Wallis, G.B. (1969). *One-dimensional Two-Phase Flow*. New York: McGraw-Hill.

6 Stewart, P.S.B. and Davidson, J.F. (1967). Slug flow in fluidized beds. *Powder Technol.* 1: 61–80.

7 Geldart, D. (1977). *Gas Fluidization, Short Course*. Bradford: University of Bradford.

8 Sun, G.L. and Grace, J.R. (1992). Effect of particle size distribution in different fluidization regimes. *AIChE J.* 38: 716–722.

9 Kai, T., Imamura, T., and Takahashi, T. (1995). Hydrodynamic influences on mass transfer between bubble and emulsion phases in a fine particle fluidized bed. *Powder Technol.* 83: 105–110.

10 Bi, X.T., Ellis, N., Abba, I.A., and Grace, J.R. (2000). A state-of-the-art review of gas–solid turbulent fluidization. *Chem. Eng. Sci.* 55 (21): 4789–4825.

11 Bi, X.T., Grace, J.R., and Lim, K.S. (1995). Transition from bubbling to turbulent fluidization. *Ind. Chem. Res. Dev.* 34: 4003–4008.

12 Chen, A.H. and Bi, X.T. (2003). Pressure fluctuations and transition from bubbling to turbulent fluidization. *Powder Technol.* 133: 237–246.

13 Davidson, J.F. (1961). Symposium on fluidization – discussion. *Trans. Inst. Chem. Eng.* 39: 230–232.

14 Thornton, C., Yang, F., and Seville, P.K. (2015). A DEM investigation of transitional behaviour in gas-fluidized beds. *Powder Technol.* 270: 128–134.

15 Wang, J.W., Tan, L.H., van der Hoef, M.A. et al. (2011). From bubbling to turbulent fluidization: advanced onset of regime transition in micro-fluidized beds. *Chem. Eng. Sci.* 66: 2001–2007.
16 Deza, M. and Battaglia, F. (2013). A CFD study of pressure fluctuations to determine fluidization regimes in gas–solid beds. *J. Fluids Eng.* 135 (10): 101301.
17 Cai, P., Chen, S.P., Jin, Y. et al. (1989). Effect of operating temperature and pressure on the transition from bubbling to turbulent fluidization. *AIChE. Symp. Ser.* 85 (270): 37–43.
18 Bi, X.T., Grace, J.R., and Zhu, J.X. (1995). Transition velocities affecting regime transitions in gas–solids suspensions and fluidized beds. *Chem. Eng. Res. Des.* 73A: 163–173.
19 Li, Y. and Kwauk, M. (1980). The dynamics of fast fluidization. In: *Fluidization* (eds. J.R. Grace and J.M. Matsen), 537–544. New York: Plenum Press.
20 Bi, X.T. and Fan, L.S. (1991). Regime transitions in gas solid circulating fluidized beds. *Paper #101e, AIChE Annual Meeting*, Los Angeles, CA (17–22 November).
21 Bi, X.T. and Zhu, J.X. (1993). Static instability analysis of circulating fluidized beds and the concept of high-density risers. *AIChE J.* 39: 1272–1280.
22 Bi, X.T., Grace, J.R., and Zhu, J.X. (1993). On types of choking in pneumatic systems. *Int. J. Multiphase Flow* 19: 1077–1092.
23 Grace, J.R., Issangya, A.S., Bai, D.R. et al. (1999). Situating the high-density circulating fluidized beds. *AIChE J.* 45: 2108–2116.
24 Kim, S.W., Kirbas, G., Bi, X.T. et al. (2004). Flow behaviour and regime transition in a high-density circulating fluidized bed riser. *Chem. Eng. Sci.* 59: 3955–3963.
25 Zenz, F.A. (1949). Two-phase fluidized-solid flow. *Ind. Eng. Chem.* 41: 2801–2806.
26 Yerushalmi, J. and Cankurt, N.T. (1979). Further studies of the regimes of fluidization. *Powder Technol.* 24: 187–205.
27 Squires, A.M., Kwauk, M., and Avidan, A.A. (1985). Fluid beds: at last, challenging two entrenched practices. *Science* 230: 1329–1337.
28 Grace, J.R. (1986). Contacting modes and behaviour classification of gas–solid and other two-phase suspensions. *Can. J. Chem. Eng.* 64: 353–363.
29 Bi, X.T. and Grace, J.R. (1995). Flow regime maps for gas–solids fluidization and upward transport. *Int. J. Multiphase Flow* 21: 1229–1236.
30 Papa, G. and Zenz, F.A. (1995). Optimize performance of fluidized-bed reactors. *Chem. Eng. Prog.* 91 (4): 32–36.
31 Bi, X.T. and Grace, J.R. (1996). Effects of pressure and temperature on flow regimes in gas – solid fluidization systems. *Can. J. Chem. Eng.* 74 (6): 1025–1027.
32 Bi, X.T. (2011). A generalized flow regime diagram for fluid-particle vertical transport. In: *Proceedings of the 10th Circulating Fluidized Bed Conference* (ed. T. Knowlton), 129–136. Sunriver, OR: Engineering Conferences International.

33 Tsukada, M., Nakanishi, D., and Horio, M. (1993). The effect of pressure on the phase transition from bubbling to turbulent fluidization. *Int. J. Multiphase Flow* 19: 27–34.
34 Kai, T., Murakami, M., Yamasaki, K.I., and Takahashi, T. (1991). Relationship between apparent bed viscosity and fluidization quality in a fluidized bed with fine particles. *J. Chem. Eng. Japan* 24: 494–500.
35 Kai, T. and Furusaki, S. (1985). Behaviour of fluidized beds of small particles at elevated temperatures. *J. Chem. Eng. Japan* 18: 113–118.
36 Knowlton, T.M. (1992). Pressure and temperature effects in fluid-particle systems. In: *Fluidization VII* (eds. O.E. Potter and D.J. Nicklin), 27–46. New York: Engineering Foundation.
37 Rapagna, S., Foscolo, P.U., and Gibilaro, L.G. (1994). The influence of temperature on the quality of gas fluidization. *Int. J. Multiphase Flow* 20: 305–313.

Problems

4.1 A stainless steel fluidized bed column has an internal diameter of 0.5 m. It is equipped with a multi-orifice distributor for uniform gas distribution and two-stage cyclones for capturing and returning entrained particles. The column is 4 m tall, filled with 100 kg sand particles of 150 μm in mean diameter and 2600 kg/m^3 in density. Assume that $\varepsilon_{mf} = 0.42$.

(a) The particles in this column are fluidized at a superficial gas velocity of 0.5 m/s with air at ambient conditions. In what flow regime is this column operated?

(b) At the same gas mass flow rate as in (a), the system is now heated from ambient temperature to 500 °C. Which flow regime is now applicable?

(c) At the same gas mass flow rate as in (b), the column now at 500 °C is pressurized to 1000 kPa. What flow regime is now operational?

(d) It is desired to operate this column in the turbulent fluidization regime. What is the minimum gas velocity required?

4.2 A circulating fluidized bed riser of 0.4 m diameter and 10 m tall is operated under ambient temperature and pressure. FCC particles ($d_p = 0.06$ mm and $\rho_p = 1600$ kg/m^3) are used as the bed material.

(a) At a superficial gas velocity of 7 m/s and solids circulation flux of 100 kg/m^{2}s, which is the flow regime: dilute flow or fast fluidization?

(b) At the same gas mass flow rate as in (a), the system is now pressurized to 500 kPa and 800 °C. What is now the flow regime in the riser?

(c) If you would like to operate this riser at a high solids flux of 1000 kg/m^{2}s, what bottlenecks could appear?

5

Experimental Investigation of Fluidized Bed Systems

Naoko Ellis

University of British Columbia, Department of Chemical and Biological Engineering, 2360 East Mall, Vancouver, Canada V6T 1Z3

5.1 Introduction

> Experiment and observation are the sole and ultimate judge of the truth of an idea.
>
> Richard P. Feynman [1]

This chapter deals with the practical aspects of designing, fabricating, operating, and understanding experimental fluidized beds. Hydrodynamic studies of fluidized beds are often conducted in small- to pilot-scale fluidization systems. The design and construction of such units depend on the overall objectives of the study. For example, fundamental characteristics of particles and their "fluidizability" and/or reactivity are often demonstrated in small-scale units. As multiphase systems are inherently complex, careful planning of small- to pilot-scale experimental units can often provide valuable insights into their operations.

5.1.1 Design Goals

Designing a fluidized bed system starts with "why." The goals of designing and building a fluidized bed system may be multifaceted. However, there is usually a key component that is of priority in a project. Here are some examples that can inform the thinking around design goals:

Scale-up: When the goal is to scale up a process, various scales of units may be required to conduct experiments. For a catalytic process, the initial step may involve testing reaction kinetics, catalyst deactivation, and particle behaviour in a small fixed bed reactor. Following catalyst testing, one may need to understand the hydrodynamics of the catalyst bed. The next step may be to understand the hydrodynamics of the small-scale catalyst bed under atmospheric conditions, followed by a "hot" unit in which operating temperature and pressure are varied, with or without chemical reactions occurring. To advance further along the scale-up path, the next step may be to build a hot unit, which may be a slender version of a demonstration reactor,

Essentials of Fluidization Technology, First Edition. Edited by John R. Grace, Xiaotao Bi, and Naoko Ellis.

running at the proposed industrial operating temperature and pressure. The goals of each stage should be clearly defined and complementary to the whole scale-up scheme. See Chapter 17 for more on scale-up.

Visualization: The flowability of particles or powders at ambient conditions is easily observable in a transparent fluidized bed. "Seeing is believing," and it may be justifiable to construct a fluidization column simply to visualize the motion of the particles. Having an acrylic column in which one can observe particle movement is useful, given that reactive fluidized beds are almost always constructed of metal, so one cannot see inside. Many research laboratories install transparent (e.g. acrylic) columns in order to be able to observe particle movement while measuring pressure drop and temperature. For example, one may wish to observe the mixing or segregation of particles for a wide particle size distribution or multiple particle systems. Careful attention is needed to assess the potential influence of electrostatic charges associated with acrylic columns. Particle–wall interactions can cause electrostatic charges to accumulate, requiring special measures (see Chapter 13). Furthermore, plastic columns can deteriorate over time, and the inner surface can be scratched by the particles, changing the transparency and affecting particle–wall interactions and hence the hydrodynamics.

Isolating part of a system: In some cases, one may wish to test a specific component or device in a fluidization system. For example, liquid injection in a fluidized bed can be visualized and tested in a column that features an isolated location. Another example would be a circulating fluidized bed (CFB) with one part having a transparent "half column" to visualize particle movement in that section. Clever experiments can be set up to acquire the required information.

5.1.2 Purpose of Experiments

Experiments conducted in small- and pilot-scale fluidization columns can serve various purposes such as:

- Hydrodynamic studies to determine:
  - Flow patterns by visualization or dynamic pressure measurement
  - Particle concentration (or voidage) distributions
  - Particle velocity characteristics
  - Solids mixing and circulation flux
  - Flowability of particles or powders
  - Gas mixing and concentration profiles
- Measurements to test the validity of models and correlations
- Finding yield and conversion of a laboratory- or pilot-scale reactor
- Assessing changes in experimental conditions (particle attrition, catalytic activity, etc.) over time

5.2 Configuration and Design

Columns of circular cross section are the most popular configuration for three-dimensional (3D) columns for ease of fabrication, symmetry of results, and

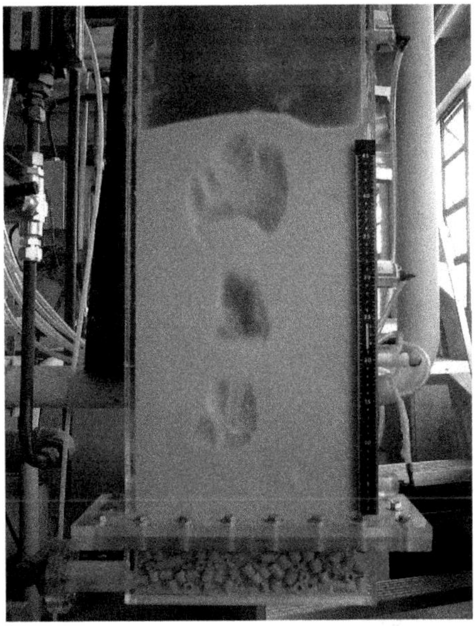

Figure 5.1 2D column with glass beads depicting gas bubbles.

withstanding elevated pressures. However, square or rectangular cross-sectional columns may be used for specific purposes, such as visual access and specific mixing studies where corner effects may be significant. So-called "two-dimensional" (2D) units are rectangular in cross section, with a relatively small gap width in the y-direction, as shown in Figure 5.1, allowing visual observation of bubbles or clusters and qualitative assessment of gross solid movement. However, considerable wall effects may need to be taken into consideration when interpreting results. For example, Sánchez-Delgado et al. [2] reported on the effect of 2D bed thickness (gap width) on measured U_{mf}, compared with values measured in 3D columns and concluded that wall effects can be neglected when the bed thickness to particle size ratio >100, provided that electrostatic effects are negligible. In another study, Li et al. [3], using computational fluid dynamics (CFD), investigated the particle–wall interactions in a 2D simulated bed using a particle-scale model, noting significant frictional interaction between particles and wall, and concluded that transition from 2D to 3D flow behaviour occurs when the gap thickness is between $20d_p$ and $40d_p$. Thus, while the horizontal dimension in the y-direction must be much less than the column width, it must also be at least 20 times the Sauter mean particle diameter to minimize interference with particle circulation.

The basic components of a fluidized bed experimental set-up are shown in Figure 5.2. The column is normally supported at the bottom on a structure. Depending on its size, it may be on a bench top with a lab frame, or on a metal frame (e.g. Unistrut) mounted to a floor. It is critical that the column be perfectly vertical, as any tilt is likely to cause preferential flow patterns. Hot or vibrating units may be suspended from the top in order to allow for thermal expansion and/or mechanical vibrations.

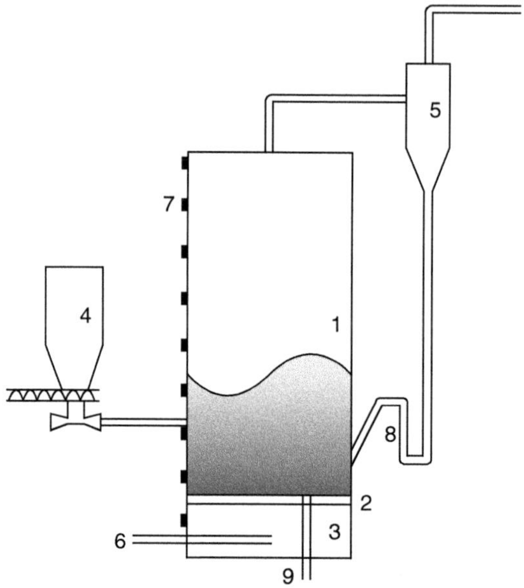

Figure 5.2 Basic components of a gas–solid fluidized bed system. (1) fluidized bed column, (2) distributor plate (grid), (3) plenum chamber, (4) solids feeding system, (5) gas solids separation system, cyclone, (6) gas feed, (7) instrument ports, (8) non-mechanical valve, loop seal, (9) solids extraction pipe.

Depending on the size of the column and the gases, the whole system may be housed in an enclosure for health and safety reasons. Pressurized vessels must obey safety codes. Hydrocarbons and other flammable gases require special measures such as explosion-proof casings. Gas leaving a cyclone or gas–solid separation system may need to be cooled and/or filtered prior to release. An afterburner, filter, or sorption media may be needed downstream of the column to combust or intercept harmful gases and fine particles carried over from a fluidized bed reactor.

5.2.1 Column Material

The main component of an experimental fluidized bed system is the fluidization column. Common materials used to fabricate these columns are metals (e.g. mild steel or stainless steel) and acrylic (e.g. Plexiglas), depending on the purpose of the experiments. For nonreactive media, the use of acrylic columns allows visual observations. When requiring numerous ports for probe insertion, such as for pressure or velocity measurements, Plexiglas is a desirable choice for its ease of gluing ports. However, depending on the roughness and hardness of powders, the inner surface of the column may be obscured by wear and scratches over time. The longevity of the column for Plexiglas is shorter than for metal columns. For high-temperature experiments, and/or oxidative or corrosive environments, materials resistant to the operating environmental conditions must be selected. For example, depending on the composition, stainless steel can be used for temperatures up to 927 °C (for SS304) or >1000 °C (for SS309/310). Inconel may be suitable for high temperatures and corrosive environments. For reactive environments, consideration must be given to potential catalytic effects of the column surface, depending on the material. With 3D printing, it is becoming possible to fabricate more complex configurations of fluidization columns and components [4].

5.2.2 Distributor Plate

The main purposes of a distributor or orifice plate at the bottom of fluidization columns are to distribute the fluid evenly, assure good solids and fluid contact at the bottom of the bed, support the weight of bed material during shutdown, minimize solids from falling into the plenum chamber (a.k.a. weeping or back-sifting), and prevent attrition of particles and wear of the column walls. Various distributor designs are available such as perforated plate, sparger, pierced sheet grid (top view in Figure 5.3a), tuyere, bubble cap, and shroud designs (side view in Figure 5.3b).

Some characteristics of the more common types of distributors are:

- Perforated plates (upward gas flow) are simple, easy to design and build, and inexpensive, but they may be subject to back-sifting and may buckle.
- Bubble caps (lateral flow) have less problem with back-sifting and good turndown, but they may give stagnant regions, are expensive, and are difficult to clean.
- Spargers (downward or obliquely downward flow) minimize back-sifting, offer good turndown, and withstand thermal expansion, but tend to cause defluidized zones.
- Pierced sheet grids (laterally directed flow) promote solids mixing, prevent stagnant build-up, are difficult to construct and require reinforcement under the sheet for support.

The most common geometry for experimental fluidized beds for ease of construction is the perforated plate. These distributor plates are normally designed

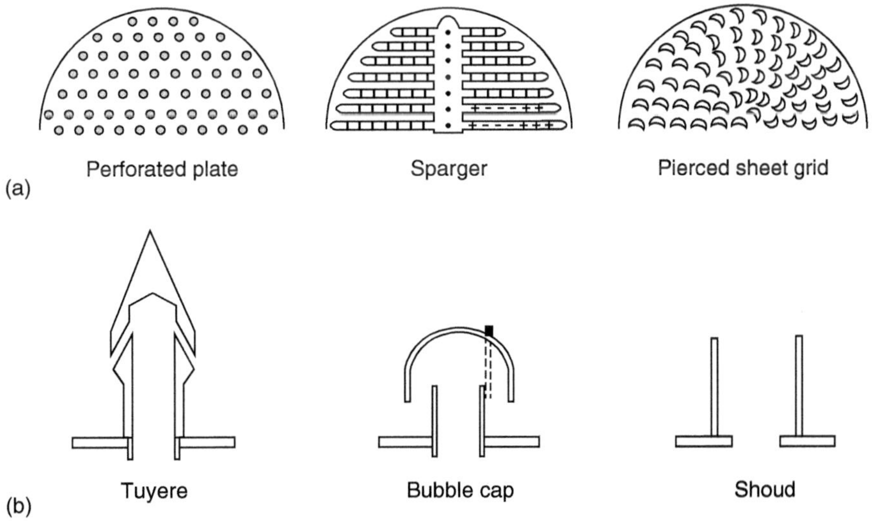

Figure 5.3 Some examples of distributor plates: (a) top view and (b) side view.

to operate with pressure drops ranging from 10% to 30% of the bed pressure drop in order to ensure good distribution of fluid, i.e.

$$0.3\Delta P_{bed} \geq \Delta P_{dist} \geq 0.1\Delta P_{bed} \tag{5.1}$$

The gas velocity through an orifice on a distributor plate is related to the pressure drop by

$$u_{or} = C_d\sqrt{2\Delta P_{dist}/\rho_{g,bed}} \tag{5.2}$$

where C_d, the discharge coefficient, depends on the grid plate thickness and hole pitch, and is often taken as 0.6–0.8 for Reynolds numbers of practical importance [5]. Here $\rho_{g,bed}$ is the fluidizing gas density at the pressure and temperature in the bed. The distributor pressure drop is based on the absence of solids. In practice, pressure drop with solids tends to be slightly higher due to solids plugging holes or weeping. The gas supply system, e.g. blower, must ensure that enough pressure is supplied to overcome the distributor and bed pressure drops during operation.

The onset of fluidization occurs when enough upflow of fluid is passed through a bed of particles where the weight of particles is suspended. The pressure drop across a fluidized bed then becomes the apparent weight of the particles per unit area of the bed:

$$\Delta P = H(1-\varepsilon)(\rho_p - \rho_g)g \tag{5.3}$$

The open area ratio, hole diameter, and pitch (triangular or square pitch) are commonly reported parameters that define a distributor. The volumetric flow rate of gas through the orifice is

$$Q = N_{or}\frac{\pi d_{or}^2}{4}u_{or} \tag{5.4}$$

where N_{or} is the number of orifices. Jet penetration lengths are important, especially in designing a reactor with internals, e.g. heat exchanger tubes, where impingement of jets is likely to cause excessive wear. Common correlations to predict jet penetration length for vertical and horizontal jets [6, 7] are listed in Chapter 19. Inclined, downward, or horizontal jets must not impinge on the side walls or on the distributor plate to prevent wear. Jet shields (a.k.a. shrouds), shown in Figure 5.4, can be used to guide the fluid jets and to minimize weeping and particle attrition. Shrouds can also be designed to protect reactor walls from erosion by jets. Short pipes can be centred around grid holes to guide jets.

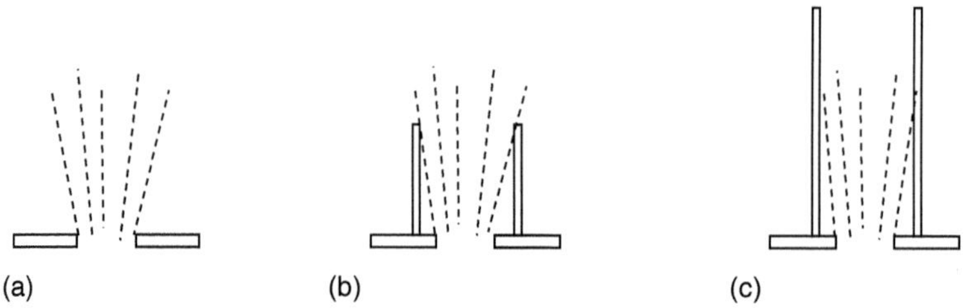

Figure 5.4 Side view of the effect of shroud on jets: (a) diverging jet, (b) shroud being too short, and (c) long enough shroud to contain jet.

However, if shrouds are too short relative to jets, it can cause more erosion and attrition than where there are no shrouds [5]. Longer shrouds are desirable to prevent adverse effects of diverging free jets and to minimize header erosion. Furthermore, gas expands to completely fill the tube, thereby reducing entry of particles into eddies near the orifice holes.

5.2.3 Plenum Chamber

The main function of a plenum chamber, a.k.a. windbox, as in Chapter 19, is to assure uniform distribution of fluid. This becomes particularly important for low distributor pressure drops that occur during operation at low superficial gas velocities. Introduction of fluid through a tube (shown in Figure 5.2) can be oriented in many directions, such as upward facing, downward facing, or oblique (see [5]). It is critical to avoid preferential paths of fluid in the windbox. For liquid fluidized beds, the windbox is often filled with ceramic Raschig rings or oversize particles to promote better radial uniformity as liquid enters the column. For gases, the windbox also serves as a dampening chamber with its relatively larger volume suppressing pressure fluctuations arising due to bubble formation. In modelling a fluidized bed system using CFD, the air feeding system and the plenum chamber may need to be included in the model [8].

Entry of particles into the plenum chamber by sifting from above is undesirable. It occurs mainly due to considerable pressure fluctuations at the distributor plate (above the windbox) when gas bubbles are formed. Higher distributor pressure drop can prevent weeping, but this cannot always be maintained, especially during shutdowns or when the gas flow rate is low. Weeping of particles results in loss of bed material, can be associated with plugging and wear of nozzles, and may lead to particle attrition. Furthermore, it can be especially detrimental if the particles are erosive and sticky or react exothermically, possibly leading to grid warping.

Mesh screens may be used to prevent fine particles from weeping through holes of the distributor plate. Allowance should be made for the additional pressure drop across the mesh, as well as for plugging of the mesh. To prevent solids from flowing or dropping into the plenum chamber, it is common practice to gradually reduce the fluid velocity to create a defluidized zone at the bottom of a fluidized bed column, before completely terminating the fluid flow. Abruptly evacuating the windbox may cause particles to be pushed below the plenum chamber due to pressure differences, especially for Geldart group A particles.

5.2.4 Feeding System (See Also Chapter 11)

For batch experiments, solids are normally added to a column from a port at the top. Continuous feeding of solids is commonly provided by a feeding system connected to the side or bottom of a column, as indicated in Figure 5.2, component 4. An auger can be used to move solids along a tube, before entering the bed. Alternatively, particles can be pneumatically conveyed into the bed.

Feeding systems can be challenging, especially if one is dealing with irregular and/or wide size distribution particles, such as many types of biomass. Furthermore, the pressure balance between the feeding system and the column needs to be considered for smooth and continuous feeding. Electrostatics can affect the flow of powders, hindering feeding. Increasing the relative humidity of gas or adding anti-electrostatic compounds, such as Larostat, may alleviate the problem. Nevertheless, feeding is often a challenging aspect of fluidization, especially for small units. Geldart group A and C particles may need to be fed with inert particles that prevent compression and allow smoother movement of the particles.

The solids feeding system indicated as component 4 in Figure 5.2 could consist of a hopper, auger, eductor, and feeding line. Not all of these components are necessary in every case. However, when feeding irregular particles, for example, biomass particles such as sawdust, one must ensure ways to alleviate plugging at various locations in the feeding system.

5.2.4.1 Hopper Design

For continuous operation, the total amount of solids required for the duration of planned experiments must be estimated, which then determines the size of solids storage units needed. A hopper can be used to store the solids to be fed (if not subject to spoilage) and must be carefully designed based on knowledge of the particle properties, such as angle of repose and angle of internal friction, for smooth operation. Disruption in operation due to unstable hopper flow is not uncommon in practice.

The desirable type of flow in hoppers is mass flow, where all particles are in motion during discharge, and the surface of solids remains nearly level until near the end of drainage. Mass flow results in a narrow residence time distribution of solids in the hopper, which is especially desirable for perishable material. Various correlations have been published on the predicted mass flow rate of hoppers, with some widely used correlations:

For coarse particles of $\gtrsim 400\,\mu m$,

$$M_p = C_d \rho_b \sqrt{g}(d_{or} - k\bar{d}_p)^{2.5} \tag{5.5}$$

where M_p is the mass flow rate in kg/s, C_d is the dimensionless empirical discharge coefficient (usually $0.55 < C_d < 0.65$), d_{or} is the orifice diameter at the bottom, and k is an empirical shape factor (with $1 < k < 2$ for many agricultural products tested, and $k = 2.9$ for sand) [9].

For fines of diameter $\lesssim 400\,\mu m$ [10]:

$$M_p = \rho_b \frac{\pi d_{or}^2}{4} v \tag{5.6}$$

where v is the average velocity of solids discharging, given by.

$$\frac{4v^2 \sin\theta}{d_{or}} + \frac{15\rho_g^{1/3} \mu_g^{2/3} v^{4/3}}{\rho_p \bar{d}_p^{\,5/3}} = g \tag{5.7}$$

Here θ is the angle that the cone makes with the vertical. Note that this angle must be smaller, i.e. steeper, than the angle of repose (see Section 5.3.1).

Draining solids from a fluidized bed can be achieved through a solids discharge tube at the bottom, through a distributor plate (depicted in Figure 5.2), or from the side of a column. Once the powders are fluidized with a superficial gas velocity close to the minimum fluidization velocity, material should flow freely upon opening the valve. For further details readers are directed to general hopper design information (e.g. [11, 12]).

5.2.5 Auxiliary Components

The previous sub-sections have covered the main components required in designing and constructing a fluidized bed unit. However, other components may be needed to meet specific requirements. For example, high-velocity fluidization requires continuous separation of particles from the gas. At the column exit where fluid leaves, a device is needed to capture and separate solids. It can range from a simple filter (e.g. vacuum cleaner bag) at the top to more elaborate systems including cyclones in series (Chapter 11), followed by an electrostatic separator. This depends on the operating conditions and flow regimes (see Chapters 4, 9, and 12) and estimating the amount of particles leaving the column. For laboratory cold tests at ambient conditions, a filter system and/or an internal or external cyclone is often used. For operations at higher superficial gas velocities where solids carry-over is expected to be high, a solids return leg and solids flow control device may be required (see Chapter 11).

Figure 5.5 depicts various auxiliary components for a fluidized bed system. The blower must be sized to overcome the pressure drop through the system. A dryer,

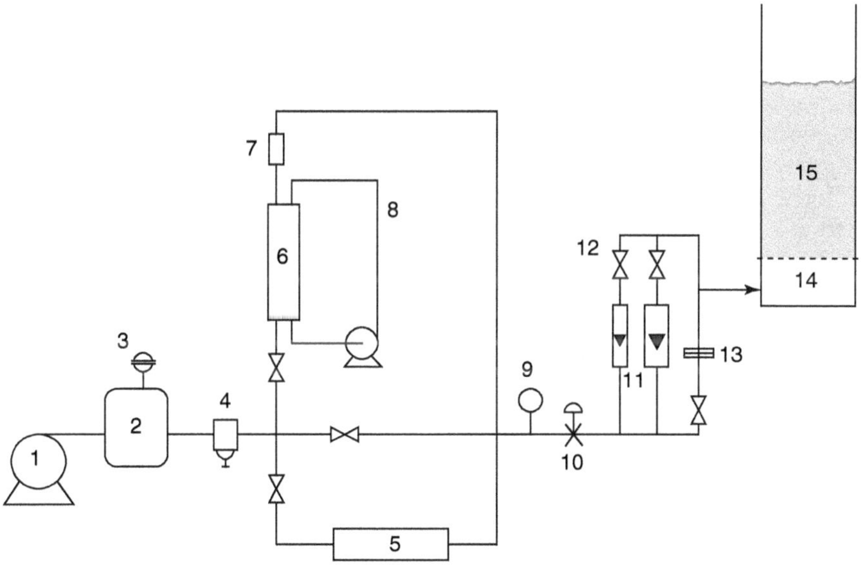

Figure 5.5 Schematic diagram of an air-supply system for a gas–solid fluidized bed. (1) blower, (2) surge tank (gas reservoir), (3) rupture disc, (4) filter (removes oil and water droplets), (5) drier, (6) humidification column, (7) mist eliminator, (8) water circulation, (9) humidity indicator, (10) pressure regulator (set at the calibration pressure of the rotameters), (11) rotameters, (12) control valves (rotameter valves should be downstream so that their calibrations can be used directly, without having to correct for pressure), (13) orifice meter, (14) windbox, (15) fluidized bed.

heater, or humidification component may be added to condition the fluidizing gas. Measurement of gas velocity can be provided by using a rotameter, orifice plate flow meter, or mass flow meter. Rotameters may be affected by downstream pressure fluctuations, causing them to not measure the flow accurately.

5.2.5.1 Secondary Injection of Fluid

In some cases, secondary fluid is introduced in a fluidized bed column to suspend particles (for gas–solid, gas–liquid–solid, liquid–solid, and gas–liquid systems.) This secondary flow is generally introduced at the bottom to a plenum chamber and through a distributor plate, both of which assist in dispersing fluid through the column. For reactive systems, the location of fluid or secondary fluid injection must be carefully considered to avoid initiating reactions in unintended locations within the reactor.

5.3 Fluidizability and Quality of Fluidization

Fluidization behaviour can be characterized qualitatively in terms of "flowability" and "fluidizability." Both parameters amalgamate various properties and conditions that influence how bulk solids behave. Some practical measurements are presented here to illustrate how these properties can be used to describe changes in particle properties.

5.3.1 Characterization of Flowability of Particles

Hausner ratio: Flowability of particles is influenced by the material physical properties and the material handling and storing (including duration of storage) of processing equipment [13]. As introduced in Chapter 2, flowability of particles can be expressed with the aid of the Hausner ratio, i.e. the ratio between tapped bulk density and poured bulk density. This can be measured using a graduated cylinder and a scale. Particles are poured into a graduated cylinder with the volume noted. Tapping the cylinder allows particles to become more interlocked. This is a quick measurement that can indicate how particles may behave when handled in hoppers and in laboratory systems.

Angle of repose: Another characterization parameter is the angle of repose where the forces of gravity and friction are in balance; however, the unconsolidated material maintains the maximum slope angle. An angle of repose $<30°$ indicates good flowability, with transition to more cohesive behaviour occurring around 30–45° [14]. Both static and dynamic angles of repose measurements are used, with results correlating consistently to the Hausner ratio [15]. A relatively simple set-up can be made to measure the angle, as shown in Figure 5.6. A practical application of the angle of repose is in estimating the amount of material in piles, e.g. sulphur piles in ports. Another application is in hopper design where the slope of the bottom section should be steeper than the drained angle of repose.

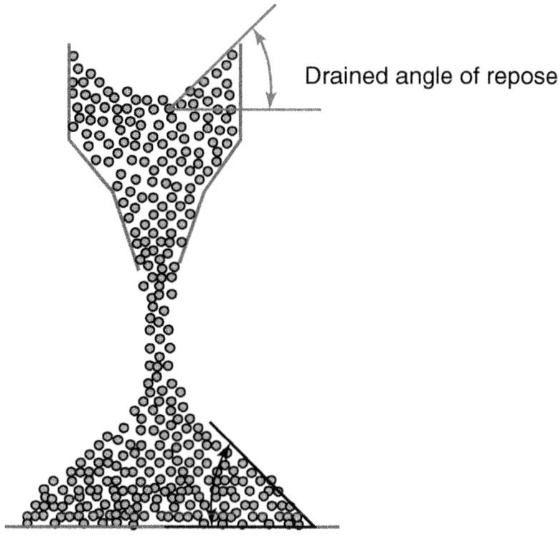

Figure 5.6 Depiction of angle of repose of particles.

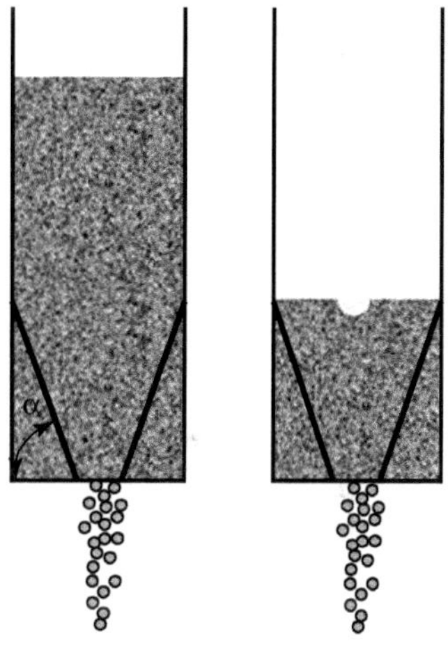

Figure 5.7 Depiction of angle of internal friction (α) indicated by particles draining from the bottom.

Angle of internal friction: The angle of internal friction is a measure of the force required to cause particles to move or slide over one another. It is influenced by particle surface friction, particle shape, hardness, and particle size distribution [16]. One way to measure this angle is to drain the particles from an orifice at the bottom of a bed and measure the angle to the horizontal of the particles once they stop draining, as depicted in Figure 5.7. Another method, called the piston test, is shown in Figure 5.8. Here particles are poured into a tube while moving a piston from the bottom. Above a critical depth of particles, the force exerted by the piston on the solids is not transmitted upwards, but to the wall. Thus, the piston is no longer able to move without exerting enough force to crack the tube. The angle of internal friction, α, is then given by

$$L = D \tan \alpha \tag{5.8}$$

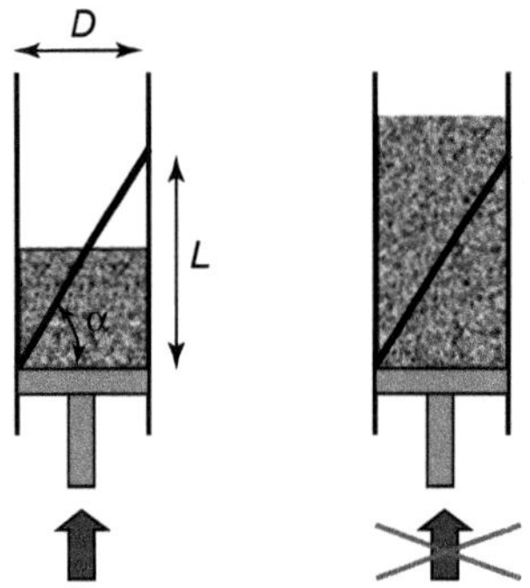

Figure 5.8 Piston set-up for measuring angle of internal friction.

Practically, the characteristic length, L, is the length of slugs in a relatively small tube (see Chapter 8). Furthermore, in designing distributors or auxiliary components involving tubes, one must measure or estimate the angle of internal friction to ensure that solids move smoothly.

5.3.2 Particle Properties and Fluidizability

The Geldart powder classification, summarized in Chapter 2, is a useful classification scheme for predicting the fluidizability of solid powders in gases.

In practice, the first information needed to set up a fluidized bed system comes from characterization of powder and fluid properties. Mean particle size can be obtained using a set of sieves and finding the particle size distribution (see Chapter 2). Particle density is not easy to measure accurately, especially for porous particles. By definition, particle density is the density that the fluid will encounter. In other words, the skeletal density of the material will equal the particle density for nonporous material. For porous materials, the porosity must be known (see Eqs. (2.4) and (2.5)). A pycnometer (gas or liquid) is used to measure the skeletal density. For a rough estimate, one can measure the bulk density of powders using a graduated cylinder and then use the interstitial voidage (typically 0.41–0.55) to estimate the particle density.

Irregular powders are difficult to fluidize as particles may interlock and become harder to mobilize. Interlocking may occur when irregular powders are transported and settle in containers. However, once they are fluidized, it may be easier to re-fluidize them. In practice, it is convenient to have a fluidization column available to test quickly the fluidizability of a broad range of particles.

5.3.3 Quality of Fluidization

Good quality fluidization is critical to attain and monitor, as the quality of fluidization affects heat and mass transfer, as well as reaction rates. Though it is a nebulous term and difficult to quantify, quality is commonly assessed by the uniformity of powder distribution, low pressure fluctuations, and absence of preferential fluid paths such as the channels that tend to form for Geldart group C solids. Various measuring methods can be used to assess and measure the quality of fluidization, with monitoring pressure fluctuations over time being the simplest. In cases where bed hydrodynamics are affected by reactions, particle agglomeration, attrition, or drying, continuous monitoring or detection of certain conditions in the bed may be challenging. For example, introduction of liquid droplets into a gas–solid system where bed wetness promotes agglomeration makes defining and monitoring the quality of fluidization challenging [17]. Furthermore, when fluidizing mixtures of multiple powders, the tendency to segregate may affect the overall quality of fluidization.

5.4 Instrumentation and Measurements

Instrumentation is a key component in fluidization experiments as it can provide insight into hydrodynamics and system monitoring. In general, small- and pilot-scale units have the luxury of being well instrumented, especially in units operating at ambient conditions, rather than at high pressure and temperature, or large-scale units. This section summarizes various invasive and non-invasive measurement systems. Each of these types of measurement requires careful calibration prior to conducting experiments. Furthermore, limitations of the instruments and data acquisition system must be considered to ensure that reliable data are collected. As various technologies advance, instrumentation continues to develop and improve rapidly.

A data acquisition system must be capable of capturing the dynamics of interest in a fluidized bed. In general, the frequency analysis of pressure fluctuations in fluidized beds shows predominant frequencies less than 10–15 Hz [18]. Thus, a typical sampling frequency of 100 Hz for pressure and temperature measurements should suffice for basic statistical time series analysis of signals according to the Nyquist sampling theorem, avoiding aliasing.

5.4.1 Pressure Measurements

Pressure measurement is the most common type of measurement for hydrodynamic studies in fluidized beds. Having measurement ports vertically aligned along the side of a column (as depicted in Figure 5.2, component 7), one can install pressure transducers to collect data for both time average system monitoring and dynamic pressure fluctuation measurements for further analysis. In general, for non-invasive measurements such as pressure measurements, one must prevent particles from contaminating the instrumentation. For example, pressure probes connected to a fluidized bed should either have a mesh or purge gas through a T connector to prevent particles from entering pressure transducers.

Absolute (gauge) pressure transducers capture pressure fluctuations in the bed. Gauge pressure is one of the simplest measurements to obtain, allowing the gross movement of particles and bubbles in the bed to be monitored. For example, if the fluidized bed is slugging, absolute pressure measurements show periodic fluctuations that indicate the frequency of slugs (Chapter 8). Figure 5.9 shows a time-average pressure profile along a fluidized bed. Pressure measurements from within the fluidized bed can be linearly correlated and extrapolated as shown.

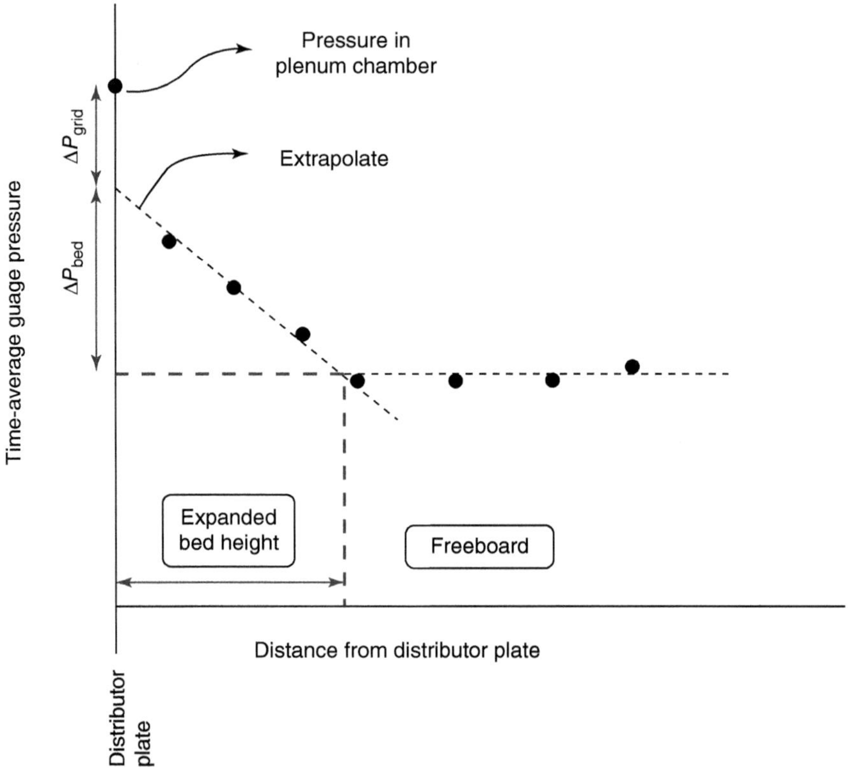

Figure 5.9 Typical axial distribution of time-average gauge pressure.

The difference between the pressure extrapolated at $z = 0$ and that in the plenum chamber allows the distributor (grid) pressure drop to be estimated. As noted in Eq. (5.1), the ratio in practice should be 10–30%. Having pressure ports in the plenum chamber, as well as dispersed along the column, allows one to experimentally measure the distributor plate pressure drop, expanded bed height, and fluidized bed pressure drop. Note that the blower feeding the fluidizing gas must overcome the upstream and downstream pressure drops through the system, in addition to the distributor and bed pressure drops.

Differential pressure transducers, on the other hand, have two measurement ports whose difference in signal can be considered to capture the movement of particles and bubbles between the ports. When acceleration, friction, and gas head are negligible, one can estimate the voidage, ε, from differential pressure measurements a distance L apart using

$$\frac{\Delta P}{L} = (1 - \varepsilon)(\rho_p - \rho_g)g \tag{5.9}$$

Various analysis methods have been employed to delineate hydrodynamic characteristics. Acrylic columns have the advantage of ease of port installation as well as transparency. Thus, small acrylic fluidized beds often have numerous ports. However, it is not uncommon to have only a few pressure ports in high-temperature or pressurized units. While most large-scale industrial units have only a few pressure measurements, most small experimental facilities should contain 10–20 transducers to facilitate a variety of measurements. The transducer volume (between transducer sensor and measurement points) must be minimized to prevent the reduction in transducer response amplitude [19]. Otherwise, dampening may affect the dynamic pressure data.

5.4.2 Thermal Sensors

Thermocouples can intrude into the fluidized bed column to measure time average and/or instantaneous temperatures. For high-temperature operations, the correct thermocouple type must be selected to cover the range of working temperatures. The diameter of a thermocouple affects its thermal response time. One must pay attention to the instrument response time and data acquisition frequency to collect data accurately. Other instruments such as optical pyrometers or infrared thermometers can also be used to measure temperatures over wide ranges.

Thermal conductivity probes operate by measuring the dielectric contact between two poles. When applied to fluidized beds, one can calibrate the signals to reflect the solids concentration between the poles [20]. The advantage of conductivity probes for voidage measurements compared with optical probes is their ability to be applied in high-temperature cases.

Thermal flux probes work by having two heat flux transducers: one directly measuring the total heat flux and the other shielded by a window and measuring only the radiant heat flux. The difference between these measurements then provides the net convective flux [21].

5.4.3 Optical Probes

Local voidage measurements can be acquired through probes that can differentiate particle concentrations at the tip of a probe. For example, optical probes (both emitting and reflecting types) have been applied extensively to reveal local voidage fluctuations in fluidized beds. Reflective optical probes use the light reflected from particles close to the probe tip, which is then correlated to pre-calibrated voidage values. As shown in Figure 5.10, consideration must be given to the optical fibre diameter relative to the particle diameter. In cases where optical probe fibres are orientated in a bundle detecting swarms of particles (Figure 5.10a), the signal reflects voidage at the tip of the probe. Careful calibration of these probes for voidage measurements is nontrivial, as the sample volume and light penetration change according to the particle concentration (or voidage) at the tip. For more information, see Ref. [22].

Detecting particle velocities using optical probes requires fibre diameters to be similar to the particle diameter (Figure 5.10b). Fibres can be orientated in order to capture the time lag between detecting fibres. For example, in the case where the cross section of the probe includes one emitting fibre (light source) and two receiving fibres, as in Figure 5.11, the signals captured by the two receiving fibres can provide information on the velocity at which a particle is moving

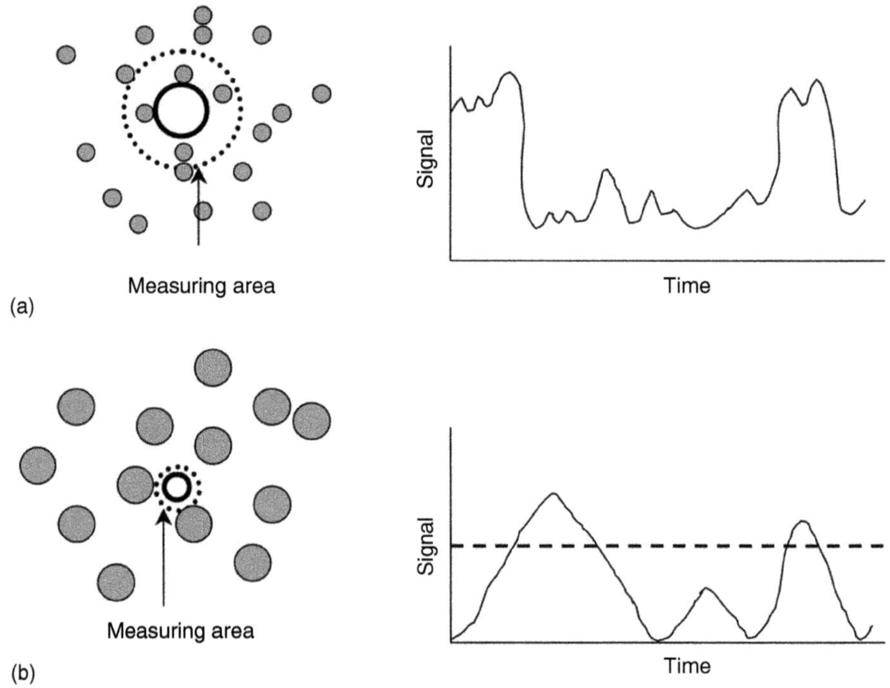

Figure 5.10 Top view of optical probe tips in relation to particle size and corresponding signal: (a) for particles much smaller than optical probe fibre, with signal reflecting swarm of particles and (b) for particles of similar size as optical probe fibre, with signal showing passing of individual particles.

in the region between the two fibres. Various configurations of fibres, including multiple fibres, have been reported [23]. Cross-correlation of the signals provides particle velocity information (see Section 5.6.2). Note that in the case of Figure 5.11, only those particles travelling parallel to the alignment of the fibres display higher cross-correlation coefficients. Calibration of these probes requires known and reliable movement of a mark, such as on a disc spinning at constant rotational speed, whose linear velocity can be used to correlate the signals.

Various light sources have been applied to optical probes, depending on their application. For example, halogen and light-emitting diodes (LEDs) can be used as visible light sources. A mercury lamp can be used to detect ozone gas or to excite fluorescent particles [23]. This flexibility gives opportunities for detecting targeted particles (e.g. tracers) or gases in multiphase systems, including fluidized beds. Simultaneous measurements of voidage and particle velocity using optical probes have been applied to dense phase fluidized beds [24].

5.4.4 Non-invasive Measurements

Non-invasive measurements in fluidized beds are challenging due to the opaque nature of particles blocking visual measurements. Laser sheet methods and laser Doppler velocimetry have been used to capture the dilute phase of fast fluidized

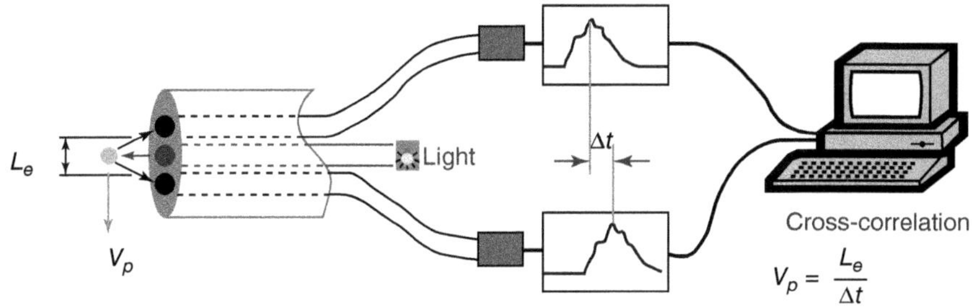

Figure 5.11 Tip of an optical fibre probe with one light-emitting and two light-receiving fibres and corresponding signals when a particle is falling down. L_e, effective length between two receiving fibres; v_p, particle velocity; and τ, time lag between signals.

beds, where a sheet of laser is illuminated through a clear column, while a visualization technique captures the solids concentration. This is possible for dilute regions of a fluidized unit (e.g. [25]). With further advances in high-speed image capturing devices, investigating the rotation of particles in high-velocity fluidized beds has become possible [26].

X-ray tomography, computed tomography (CT), and electric capacitance tomography (ECT) are also in the category of non-invasive measurements that are somewhat specialized instruments. ECT has been successfully applied as a nonintrusive, high-speed measurement technique, based on the difference between the relative electrical permittivities of gas and solid material in a riser [27]. Another application monitors the wetting and drying stages of a top-spray fluidized bed using ECT, coupled with pressure measurements [28]. Probe invasiveness has been examined using X-ray CT imaging in a 0.1 m diameter column, with recommendations for minimum influence from probe configuration and orientation [29]. A collaborative international study comparing various instruments for measuring voidage [30] and particle velocity [31] has raised interesting questions by comparing data from different measurement systems for identical operating conditions.

It is possible to follow tracer particles for considerable lengths of time to characterize the overall movement of particles. Positron emission particle tracking (PEPT) and radioactive particle tracking (RPT) have both been applied to determine the trajectories of tracer particles in fluidized beds. One must be careful in matching the tracer particles to the bed material for tracer particles to represent the gross movement of particles in the bed. Furthermore, differences in data analysis algorithms may result in differing results [31].

5.4.5 Visualization Measurements

With the advancement in high-resolution photography and high-speed videography, the ability to capture particle movement in fluidized beds has improved considerably. Boroscopes are devices that relay an image to the eyepiece. A high-speed camera has been attached to capture particle movement, thus delineating particle velocity, inside a fluidized bed. Particle image velocimetry captures the particle motion, which, when combined with digital image analysis, provides a powerful tool to delineate particle velocity information (e.g. [32]).

5.4.6 Acoustic Emission Measurements

A passive acoustic emission measurement system captures the acoustic signals generated from particle–particle and particle–wall collisions, bubble motion, and global fluidized bed movement [33]. An experimental study that analyzed passive acoustic emissions with pressure fluctuations from a 2D fluidized bed confirmed the ability of the acoustic system to capture the bubble motion [34]. More recently, it was used to identify flow regime transition through analyzing acoustic emission signals using Hurst and wavelet analyses (see Section 5.6.3) in resolving the micro-, meso-, and macroscale particle interactions [35].

5.4.7 Solids Circulation Flux

Solids circulation is critical to understand for higher-velocity fluidized beds such as those operating in turbulent or fast fluidization flow regime (see Chapters 9 and 10). The solids circulation flux, G_s, is a key parameter in the operation of fast or dual fluidized beds, affecting the extent of a reaction, as well as heat and mass transfer. Ideally, G_s is obtained in real-time measurement with minimal disturbance to the system. In principle, simultaneously measuring a velocity of descent of particles in a vertical standpipe of return leg and particle concentration can deduce G_s:

$$G_s = v \times \rho_b \tag{5.10}$$

A commonly employed method for a system operating at ambient conditions is to close a permeable valve in the return leg and measure the solids accumulation over time. However, at high-circulation rates, closing a valve upsets the system considerably. Another method is to divert particles moving down the return leg and measure the accumulation. This method requires shutting down the system and returning the collected particles to the system, interrupting continuous operation. Furthermore, measuring solids circulation rates in pilot or high-temperature systems is not trivial. Readers are directed to a literature survey [36] for further and more recent information. A review on mass flow measurement for the pneumatically conveying flow regime [37] highlights some measurement techniques applicable to dilute phase particle flow, such as in fast fluidization.

5.4.8 Gas or Solids Sampling

When the gas composition within and leaving a fluidized bed is of interest, one can install gas analyzers while ensuring that there is minimal disturbance of the gas flow. For invasive sampling, a ceramic porous cap can be installed at the tip of the sampling tube to minimize blockage and entry of particles into the gas analyzer. Isokinetic solids sampling ensures that the gas suction velocity is adjusted to match the local gas velocity, thus allowing representative solids samples to be extracted from a fluidized bed (e.g. [25]). Some cautionary notes when sampling gas from fluidized beds such as local pressure fluctuations due to bubble movement, two-phase flow averaging, hydrodynamic interference, radial distributions, and adsorption and trapping are documented elsewhere [38].

5.4.9 Mixing and Residence Time Distribution

When a system is disturbed with a stimulus, the analysis of response can provide the desired information about the system. Tracer studies allow analysis of gas, solids, or liquid mixing, or residence time distribution of a component in a multiphase system, such as those discussed in freely bubbling beds (Section 7.6), turbulent fluidized beds (Section 9.3), fast fluidization (Section 12.6), and three-phase fluidized beds (Section 22.5). The residence time distribution information, a widely applied parameter in reactor design, can be acquired through injection and detection of tracers. Injection of the input signal is provided through a random signal, a periodic signal, a step signal, or a pulse signal from which the information on mixing and dispersion of flow can be extracted [39]. Different signal analysis to interpret the measured residence time distribution signal is conducted according to the boundary conditions of tracer injection and detection.

Various tracer gases have been applied in fluidized beds, such as helium, argon, carbon dioxide, ozone, propane, sulphur dioxide, and carbon monoxide (e.g. [40–42]). Gas concentrations are often measured using thermoconductivity detectors, gas chromatography, or mass spectrometry. Care must be given to ensure that gas samples are free of solids prior to analysis, and pressure differences do not affect the concentration of tracer gas as it is compressible. A list of gas mixing studies is provided for CFB systems [41].

Solid tracers may include those bed material coated or impregnated with chemicals that may be detected, or those that may have similar densities and diameters to a bed material. Tracers such as radioactive material require specialized equipment for handling and detection (e.g. [43]). On the other hand, phosphorus or fluorescent tracer particles can be detected using photomultipliers, while ferromagnetic particles are detected through inductance measurement (e.g. [44, 45]). Solid carbon dioxide has also been used in conjunction with gas analysis and thermocouples to investigate the solids mixing in a CFB [46]. Extensive summaries of solid tracers used in fluidized beds are available (e.g. [45, 47]).

Simultaneous measurement of gas and solids mixing can attain further understanding of the mixing behaviour in fluidized beds [48].

5.5 Operation of Fluidized Beds

Once all the fluidization components have been commissioned, instruments calibrated, and data acquisition system tested, a leak test should be conducted to ensure that no gas nor particles leak from the system. This can be done by slightly pressurizing the system and ensuring that pressure holds and then applying leak test solutions (such as a soap solution) to every joint. Instruments requiring pre-warming, such as optical probe systems, must be turned on in advance to attain stable signals. When conducting experiments at high temperature, the heater or a burner may need to operate for several hours while the system is heating up to the desired temperature.

5.5.1 Start-Up and Shutdown

Starting up a fluidized bed experiment requires a gradual increase in gas velocity until a desirable condition is attained. It is important to ensure that there is no undesirable pressure build-up in the system while increasing the superficial gas velocity (see Chapter 18). If intrusive probes are used, they should normally be outside the column before start-up, if possible removed prior to the last shutdown. Since fluidized particles act like elements of a viscous liquid, moving components in a slightly fluidized state is easier than when the bed is defluidized.

When testing high-velocity fluidization, be aware that the gas–solid separation system (which may include cyclone(s), filters, bag filters, electrostatic precipitators, etc.) may contain leaks or deficiencies. The separation system is normally only tested once the fluidizing gas is at high flow, so small leaks can result in spectacular dust clouds. It is recommended that one prepare for a swift shutdown in such cases.

Once experimental data have been obtained, the fluidization system should be shut down according to a predetermined procedure. As noted above, intrusive probes should be removed or moved to the inner wall before the unit is shut down. If probes are left in the bed while starting up, there is a possibility that they will bend (depending on their dimensions, location, and rigidity). A dual fluidized bed system may require a special shutdown procedure that allows for accumulation of particles in certain parts of the system in preparation for subsequent smoother start-up.

5.5.2 Steady-State Operation

Designing, building, and commissioning fluidization experiments commonly require more time and effort than obtaining hydrodynamic or reactor data once the system is operational. In order to avoid the effects of hysteresis, it is recommended that runs be carried out in random order.

When operation of laboratory or pilot plant fluidized beds over extended periods is required, the system can be monitored through pressure drops or other means, such as comparing particle size distributions to ensure stable operation. One must keep in mind that over extended durations, particles may undergo attrition, with small fines escaping from the unit, or rounding of particle surfaces may occur. In a reactive system, gas analysis may be required to monitor the overall performance. Accumulation of unintended dust deposits on particles or reactor components may change the fluidization behaviour over time. One should also be aware that segregation of particles may occur over time. For reactive or high-temperature operation, agglomeration, due in most cases to sintering, may significantly change the hydrodynamics of a fluidized bed system, possibly even resulting ultimately in defluidization.

5.6 Data Analysis

Once quality data have been collected from an experimental fluidized bed, analysis and interpretation of data can lead to useful results. Time- or volume-averaged parameters alone may not adequately represent the complex hydrodynamics of multiphase flows for design and scale-up purposes. This section summarizes the various analysis techniques to gain further insight into the nature of the flow structure and to extract dynamically meaningful information from the data.

The standard deviation of the amplitude of a signal

$$\sigma = \sqrt{\frac{1}{N-1}\sum_{n-1}^{N}(x(n)-\overline{x})^2} \tag{5.11}$$

with the average

$$\overline{x} = \frac{1}{N}\sum_{n=1}^{N}x(n) \tag{5.12}$$

shows the deviation of data from the mean. The standard deviation of pressure measurements has been used to identify the onset of turbulent fluidization flow regime, as noted in Chapters 4 and 9. Higher statistical moments such as skewness and kurtosis numbers have also been applied to characterize hydrodynamics of fluidized beds from various signals [49].

5.6.1 Frequency Analysis

Pressure fluctuations in fluidized beds are often investigated owing to their relative ease of measurement and can infer stability, quality of fluidization, flow structure, and regime transitions [50]. Pressure signals reflect complex phenomena, including effects of absolute pressure, temperature, gas turbulence, and the formation, coalescence, and eruption of bubbles. The "natural frequency" of a fluidized bed characteristic of oscillations is generally considered to be a function of bed height [19], given [51] by.

$$f_n = \frac{1}{\pi}\sqrt{\frac{g}{H_{mf}}} \tag{5.13}$$

The Fourier transform (FT) transforms time domain functions into frequency domain representations. A function f can be expressed as an infinite sum of dilated cosine and sine functions, being approximated, in practice, by a finite sum as

$$f(x) = \frac{1}{2}a_0 + \sum_{j=1}^{J}(a_j\cos jx + b_j\sin jx) \tag{5.14}$$

The fast Fourier transform (FFT) reduces the time required to perform a Fourier analysis by expressing the coefficients as [52]

$$a_j + ib_j = \frac{2\left(\sum f(x)\exp(2\pi ijx/N)\right)}{N} \tag{5.15}$$

for $f(x)$ and N observations. FFT is commonly used to characterize the dominant frequency in a fluidized bed system reflective of the change in behaviour, such as in wetting and drying processes [53]. Pressure signals from a fluidized bed at $U = U_{mf}$ indicated distinct pressure waves, propagating upwards and downwards, originating from bubble formation and bubble eruptions at the bed surface [54]. Kage et al. identified three principal frequencies from the power spectra of pressure signals in a plenum chamber corresponding to the natural frequency of the fluidized bed, the bubble eruption frequency, and the bubble generation frequency. The latter two were confirmed from local voidages, determined by an optical probe, and visualization of bubble eruption by a video camera [55].

In practice, it is critical to ensure that the natural frequency of the structure differs appreciably from the major frequency range of the bed, for vibrations tend to become increasingly significant as units are scaled up [56].

5.6.2 Bivariate Time Series

The three main types of statistical functions in expressing the joint properties of two signals are joint probability density function, cross-correlation functions, and cross-spectral density functions, providing information regarding the joint properties in the amplitude, time, and frequency domains, respectively.

5.6.2.1 Joint Probability Density Function

The joint probability density function expresses the probability that both sample records will simultaneously assume values within some defined pair of ranges at any instant of time. It is used to establish a probabilistic description for an event that is associated with two sets of correlated random data [57]:

$$P(x, y) = \text{Prob}[x(t) \leq x \text{ and } y(t) \leq y] \tag{5.16}$$

$$p(x, y) = \lim_{\substack{\Delta x \to 0 \\ \Delta y \to 0}} \frac{1}{(\Delta x)(\Delta y)} \left[\lim_{T \to \infty} \frac{T_{x,y}}{T} \right] \tag{5.17}$$

5.6.2.2 Cross-Correlation Function

From bivariate time series, cross-correlation coefficients can provide empirical indications of possible relationships between variables. The cross-correlation function is defined as.

$$\varrho_{I_1 I_2} = \frac{C_{I_1 I_2}(\tau)}{\sqrt{C_{I_1}(0) C_{I_2}(0)}} \tag{5.18}$$

which satisfies, for all τ,

$$-1 \leq \varrho_{I_1 I_2} \leq 1$$

$C_{I_1 I_2}$ is the covariance functions at arbitrary fixed values of $t_1 = t$ and $t_2 = t + \tau$, defined by

$$C_{I_1 I_2} = E[(x(t) - \overline{x(t)})(y(t + \tau) - \overline{y(t + \tau)})] \tag{5.19}$$

$E[\ ]$ returns the product of the displacement of the product of the deviations of two functions from their mean values expected value over the interval between t and $t+\tau$. The value of τ at which $\varrho_{I_1I_2}$ is a maximum represents the lag between the two signals. Cross-correlation between two signals, for example, from optical probe signals, can deduce the rise velocity of bubbles or particle velocities [31]. Note that the cross-correlation function fails when the linearity assumption is not valid [57].

5.6.2.3 Cross-Spectral Density Function

The power spectral density function depicts the density of the spectral as a function of the frequency. Thus, the cross power spectral density examines the FT of both time series at a frequency that has a constant phase shift:

$$G_{xy}(f) = \frac{2}{T}P_x(f)P_y(f) \tag{5.20}$$

Coherence function was applied to interpret the time series for its coherent frequency:

$$\gamma_{xy}^2(f) = \frac{|G_{xy}(f)|^2}{G_{xx}(f)G_{yy}(f)} \leq 1 \tag{5.21}$$

Cai et al. [58] applied the coherence function to pressure data to further elucidate bubble behaviour between two sampling points, as shown in Figure 5.12. As U increased, region A–B shows a decrease in average coherence function, reflecting enhanced bubble variations [58]. The increase in gamma, the coherence function, in region B–C is interpreted to be the bubble size, of the same order of magnitude as the probe separation distance. This is overlaid with the standard deviation of pressure signal reaching its maximum, indicating the transition to the turbulent fluidization flow regime at $U = U_c$.

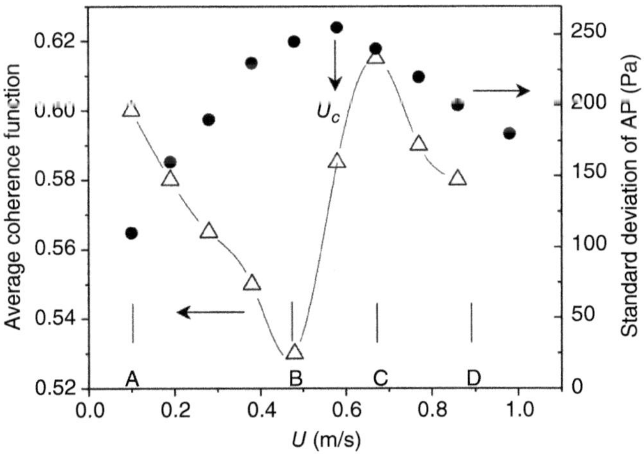

Figure 5.12 Coherence function and standard deviation vs. superficial gas velocity ($D = 0.139$ m, $d_p = 280$ μm, probe separation distance = 0.1 m). Source: Adapted from Cai et al. 1990 [58].

5.6.3 Other Signal Analyses

Various other analyses have been applied to data from fluidized bed experiments stemming from fluid dynamic modelling. There are many time, frequency, and time–frequency domain analyses used to characterize the flow behaviour. For example, in investigating fluidized beds, reflective of non-linear dynamics, chaos theory (e.g. [28]), and wavelet analysis [59] have been applied. Reconstructive attractors based on chaos theory have been monitored to observe the change in bed dynamics due to agglomeration [60]. Using wavelet analysis to decompose signals, Sotudeh-Gharebagh et al. [61] applied recurrence quantification analysis in the phase space to gain further insight into the multiscale structure deduced from pressure signals [61]. Readers interested in this area are encouraged to consult two comprehensive review papers [62, 63].

Solved Problem

5.1 Calculate the bed pressure drop at minimum fluidization for 25 kg of catalyst particles of density 2500 kg/m³ with a bulk density of 1350 kg/m³ in a 0.25 m diameter column. State your assumptions.

Solution

The volume occupied by particles at rest is expressed as

$$\text{Volume} = \text{AH} = \frac{M}{\rho_b}$$

from which a packed bed height is calculated:

$$H = \frac{M}{A\rho_b} = \frac{25\,\text{kg}}{(0.25\,\text{m})^2 \times \pi/4 \times 1350\,\text{kg/m}^3} = 0.37\text{m}$$

The packed bed voidage is then calculated from the particle and bulk densities:

$$\rho_b = (1-\varepsilon)\rho_p$$

Thus,

$$\varepsilon = 1 - \frac{\rho_b}{\rho_p} = 1 - \frac{1350\,\text{kg/m}^3}{2500\,\text{kg/m}^3} = 0.46$$

Assuming that voidage at minimum fluidization is equal to the packed bed voidage, the bed pressure drop is

$$\Delta P_{bed} = H(1-\varepsilon)(\rho_p - \rho_b)g = 0.37\,\text{m} \times (1 - 0.46)(2500 - 1350\,\text{kg/m}^3) \times 9.81\,\text{m s}^2 = 2254\,\text{Pa}.$$

Notations

A	cross-sectional area (m^2)
a, b	sample regression coefficients

C_d	discharge coefficient (–)
$C_{I_1 I_2}$	covariance function
D	column diameter (m)
d	diameter (m)
$E[..]$	expected value of [..]
f	frequency (s^{-1})
G	mass flux (kg/m^2 s)
$G_{xy}(f)$	cross-spectral density function
g	acceleration of gravity (m/s^2)
H	bed height (m)
k	empirical coefficient (–), Eq. (5.5)
L	length (m)
L_h	grid hole pitch (m)
M	mass (kg)
M_p	solids mass flowrate (kg/s)
N	number (–)
N_d	hole density (number of holes/area) (–)
n	number of data points (–)
P	pressure (Pa)
$P(x,y)$	probability distribution function
$p(x,y)$	joint probability density function
Q	volumetric flow rate (m^3/s)
R	radius (m)
r	radial coordinate (m)
T	total time (s)
t	grid thickness (m)
U	superficial gas velocity (m/s)
U_c	onset velocity of turbulent fluidization (m/s)
u	local gas velocity (m/s)
V	volume (m^3)
v	local particle velocity (m/s)
x,y,z	Cartesian coordinates, z vertical (m)
$x(t)$, $y(t)$	time-dependent variables

Greek Symbols

α	angle of internal friction
ΔP	pressure drop (Pa)
ε	voidage (–)
$\gamma_{xy}^2(f)$	coherence function
μ	dynamic viscosity (Pa s)
θ	angle
ρ	density (kg/m^3)
$\varrho_{I_1 I_2}$	cross-correlation function
σ	standard deviation
τ	time, lag (s)
υ	solids discharging velocity (m/s)

Subscript

b	bulk
bed	bed
dist	distributor
e	effective
g	gas
g,bed	gas at the bed condition
h	grid hole
mf	minimum fluidization
n	natural
or	orifice
p	particle
s	solids

Superscript

–	time or spatial average

Abbreviations

2D	two-dimensional
3D	three-dimensional
a.k.a.	also known as
CFB	circulating fluidized bed
CFD	computational fluid dynamics
CT	computed tomography
ECT	electric capacitance tomography
FCC	fluid cracking catalyst
FFT	fast Fourier transfer
FT	Fourier transfer
LED	light-emitting diode
SS	stainless steel
PEPT	positron emission particle tracking
RPT	radioactive particle tracking

References

1 Feynman, R.P. (2005). *Perfectly Reasonable Deviations from the Beaten Track: The Letters of Richard P. Feynman* (ed. M. Feynman). New York: Basic Books.

2 Sánchez-Delgado, S., Almendros-Ibáñez, J.A., García-Hernando, N., and Santana, D. (2011). On the minimum fluidization velocity in 2D fluidized beds. *Powder Technol.* 207: 145–153.

3 Li, T., Gopalakrishnan, P., Garg, R., and Shahnam, M. (2012). CFD-DEM study of effect of bed thickness for bubbling fluidized beds. *Particuology* 10: 532–541.

4 McDonough, J.R., Law, R., Reay, D.A., and Zivkovic, V. (2019). Fluidization in small-scale gas–solid 3D-printed fluidized beds. *Chem. Eng. Sci.* 200: 294–309.
5 Karri, S.B.R. and Werther, J. (2003). Gas distributor and plenum design in fluidized beds, Chapter 6. In: *Handbook of Fluidization and Fluid-Particle Systems* (ed. W.-C. Yang), 155–170. New York: Marcel Dekker.
6 Merry, J.M.D. (1975). Penetration of vertical jets into fluidized beds. *AIChE J.* 21: 507–510.
7 Merry, J.M.D. (1971). Penetration of a horizontal gas jet into a fluidised bed. *Trans. Inst. Chem. Eng.* 49: 189–195.
8 Johansson, K., van Wachem, B.G.M., and Almstedt, A.E. (2006). Experimental validation of CFD models for fluidized beds: influence of particle stress models, gas phase compressibility and air inflow models. *Chem. Eng. Sci.* 61: 1705–1717.
9 Beverloo, W.A., Leniger, H.A., and van de Velde, J. (1961). The flow of granular solids through orifices. *Chem. Eng. Sci.* 15: 260–269.
10 Carleton, A. (1972). The effect of fluid-drag forces on the discharge of free-flowing solids from hoppers. *Powder Technol.* 6: 91–96.
11 Schulze, D. (2008). *Powder and Bulk Solids: Behavior, Characterization, Storage and Flow* (English translation based on German edition). Berlin, Heidelberg: Springer.
12 Woodcock, C.R. and Mason, J.S. (1987). *Bulk Solids Handling: An Introduction to the Practice and Technology*. Dordrecht: Springer.
13 Prescott, J.K. and Hossfeld, R.J. (1994). Maintaining product uniformity and uninterrupted flow to direct-compression tableting presses. *Pharm. Technol.* 18 (6): 99–114.
14 Carr, R.L. (1965). Evaluating flow properties of solids. *Chem. Eng.* 72: 163–168.
15 Geldart, D., Abdullah, E.C., Hassanpour, A. et al. (2006). Characterization of powder flowability using measurement of angle of repose. *China Particuol.* 4: 104–107.
16 Gibilaro, L.G. (2001). *Fluidization Dynamics*. Woburn, MA: Butterworth-Heinemann.
17 McDougall, S.L., Saberian, M., Briens, C. et al. (2004). Characterization of fluidization quality in fluidized beds of wet particles. *Int. J. Chem. React. Eng.* 2: 1.
18 Bi, H.T. (2007). A critical review of the complex pressure fluctuation phenomenon in gas–solids fluidized beds. *Chem. Eng. Sci.* 62: 3473–3493.
19 van Omman, J.R., Schouten, J.C., Vander Stappen, M.L.M., and van den Bleek, C.M. (1999). Response characteristics of probe-transducer systems for pressure measurements in gas–solid fluidized beds: how to prevent pitfalls in dynamic pressure measurements. *Powder Technol.* 106: 199–218.
20 Louge, M. and Opie, M. (1990). Measurements of the effective dielectric permittivity of suspensions. *Powder Technol.* 62: 85–94.
21 Luan, W., Lim, C.J., Brereton, C.M.H. et al. (1999). Experimental and theoretical study of total and radiative heat transfer in circulating fluidized beds. *Chem. Eng. Sci.* 54: 3749–3764.

22 Xu, G., Liang, C., Chen, X. et al. (2013). Investigation on dynamic calibration for an optical-fiber solids concentration probe in gas–solid two-phase flows. *Sensors* 13 (7): 9201–9222.
23 Horio, M., Kobylecki, R.P., and Tsukada, M. (2003). Instrumentation and measurements, Chapter 25. In: *Handbook of Fluidization and Fluid-Particle Systems* (ed. W.-C. Yang), 643–704. New York: Marcel Dekker.
24 Zhu, H., Zhu, J., Li, G., and Li, F. (2008). Detailed measurements of flow structure inside a dense gas–solids fluidized bed. *Powder Technol.* 180: 339–349.
25 Lackermeier, U., Rudnick, C., Werther, J. et al. (2001). Visualization of flow structures inside a circulating fluidized bed by means of laser sheet and image processing. *Powder Technol.* 114: 71–83.
26 Wu, X., Wang, Q., Luo, Z. et al. (2008). Experimental study of particle rotation characteristics with high-speed digital imaging system. *Powder Technol.* 181: 21–30.
27 Wang, F., Marashdeh, Q., Fan, L.-S., and Warsito, W. (2010). Electrical capacitance volume tomography: design and applications. *Sensors* 10 (3): 1890–1917.
28 Xiang, J., Li, Q., Tan, Z., and Zhang, Y. (2017). Characterization of the flow in a gas–solid bubbling fluidized bed by pressure fluctuation. *Chem. Eng. Sci.* 174: 93–103.
29 Whitemarsh, E.A., Escudero, D.R., and Heindel, T.J. (2016). Probe effects on the local gas holdup conditions in a fluidized bed. *Powder Technol.* 294: 191–201.
30 Dubrawski, K., Tebianian, S., Bi, H.T. et al. (2013). Traveling column for comparison of invasive and non-invasive fluidization voidage measurement techniques. *Powder Technol.* 235: 203–220.
31 Tebianian, S., Dubrawski, K., Ellis, N. et al. (2015). Investigation of particle velocity in FCC gas-fluidized beds based on different measurement techniques. *Chem. Eng. Sci.* 127: 310–322.
32 Shaffer, F., Gopalan, B., Breault, R.W. et al. (2013). High speed imaging of particle flow fields in CFB risers. *Powder Technol.* 242: 86–99.
33 Villa Briongos, J., Aragón, J.M., and Palancar, M.C. (2006). Fluidised bed dynamics diagnosis from measurements of low-frequency out-bed passive acoustic emissions. *Powder Technol.* 162: 145–156.
34 Li, Y., Grace, J.R., Gopaluni, B. et al. (2011). Characterization of gas–solid fluidization – a comparative study of acoustic and pressure signals. *Powder Technol.* 214: 200–210.
35 Zhou, Y., Yang, L., Lu, Y. et al. (2018). Flow regime identification in gas–solid two-phase fluidization via acoustic emission technique. *Chem. Eng. J.* 334: 1484–1492.
36 Rahman, H.M. (2011). Novel measurement of solids circulation rate in pilot-scale dual fluidized bed gasifier at high temperature. PhD dissertation. University of British Columbia. https://open.library.ubc.ca/cIRcle/collections/ubctheses/24/items/1.0362393 (8 October 2019).
37 Zheng, Y. and Liu, Q. (2011). Review of techniques for the mass flow rate measurement of pneumatically conveyed solids. *Measurement* 44: 589–604.

38 Grace, J., Bi, H., and Zhang, Y. (2009). Pitfalls in gas sampling from fluidized beds. *Chem. Eng. Sci.* 64: 2522–2524.

39 Levenspiel, O. (1998). *Chemical Reaction Engineering*. Hoboken, NJ: Wiley.

40 Sterneus, J., Johnsson, F., and Leckner, B. (2000). Gas mixing in circulating fluidised-bed risers. *Chem. Eng. Sci.* 55: 129–148.

41 Mahmoudi, S., Seville, J.P.K., and Baeyens, J. (2010). The residence time distribution and mixing of the gas phase in the riser of a circulating fluidized bed. *Powder Technol.* 203: 322–330.

42 Zhong, W., Xiao, R., and Zhang, M. (2006). Experimental study of gas mixing in a spout-fluid bed. *AIChE J.* 52: 924–930.

43 Van de Velden, M., Baeyens, J., Seville, J.P.K., and Fan, X. (2008). The solids flow in the riser of a circulating fluidized bed (CFB) viewed by positron emission particle tracking (PEPT). *Powder Technol.* 183: 290–296.

44 Harris, A.T., Davidson, J.F., and Thorpe, R.B. (2003). Particle residence time distributions in circulating fluidised beds. *Chem. Eng. Sci.* 58: 2181–2202.

45 Guio-Perez, D.C., Proll, T., and Hofbauer, H. (2013). Measurement of ferromagnetic particle concentration for characterization of fluidized bed fluid-dynamics. *Powder Technol.* 239: 147–154.

46 Schlichthaerle, P. and Werther, J. (2001). Solids mixing in the bottom zone of a circulating fluidized bed. *Powder Technol.* 120: 21–33.

47 Kirbas, G (2004). Solids motion and mixing in high-density circulating fluidized beds. PhD thesis. The University of British Columbia. https://open.library.ubc.ca/cIRcle/collections/ubctheses/831/items/1.0058657 (8 October 2019).

48 Du, B., Fan, L.-S., Wei, F., and Warsito, W. (2002). Gas and solids mixing in a turbulent fluidized bed. *AIChE J.* 48: 1896–1909.

49 Salehi-Nik, N., Sotudeh-Gharebagh, R., Mostoufi, N. et al. (2009). Determination of hydrodynamic behavior of gas–solid fluidized beds using statistical analysis of acoustic emissions. *Int. J. Multiphase Flow* 35: 1011–1016.

50 Trnka, O., Veselý, V., Hartman, M., and Beran, Z. (2004). Identification of the state of a fluidized bed by pressure fluctuations. *AIChE J.* 46: 509–514.

51 Baskakov, A.P., Tuponogov, V.G., and Filippovsky, N.F. (1986). A study of pressure fluctuations in a bubbling fluidized bed. *Powder Technol.* 45: 113–117.

52 Chatfield, C. (1996). *The Analysis of Time Series*, 5e. London: Chapman and Hall.

53 Zhang, J.L., Mao, M.X., Ye, J.M. et al. (2016). Investigation of wetting and drying process in a gas–solid fluidized bed by electrical capacitance tomography and pressure measurement. *Powder Technol.* 301: 1148–1158.

54 van der Schaaf, J., Schoutin, J.C., and van den Bleek, C.M. (1998). Origin, propagation and attenuation of pressure waves in gas–solid fluidized beds. *Powder Technol.* 95: 220–233.

55 Kage, H., Agari, M., Ogura, H., and Matsuno, Y. (2000). Frequency analysis of pressure fluctuation in fluidized bed plenum and its confidence limit for detection of various modes of fluidization. *Adv. Powder Technol.* 11: 459–475.

56 Pell, M. (1990). *Gas Fluidization*. Amsterdam: Elsevier.

57 Bendat, J.S. and Piersol, A.G. (1986). *Analysis and Measurement Procedures.* New York: Wiley.
58 Cai, P., Jin, Y., Yu, Z.-Q., and Wang, Z.-W. (1990). Mechanism of flow regime transition from bubbling to turbulent fluidization. *AIChE J.* 36: 955–956.
59 Ellis, N., Briens, L.A., Grace, J.R. et al. (2003). Characterization of dynamic behaviour in gas–solid turbulent fluidized bed using chaos and wavelet analyses. *Chem. Eng. Sci.* 96: 105–116.
60 van Ommen, J.R., Coppens, M., van den Bleek, C.M., and Schouten, J.C. (2000). Early warning of agglomeration in fluidized beds by attractor comparison. *AIChE J.* 46: 2183–2197.
61 Tahmasebpour, M., Zarghami, R., Sotudeh-Gharebagh, R., and Mostoufi, N. (2013). Characterization of various structures in gas–solid fluidized beds by recurrence quantification analysis. *Particuology* 11: 647–656.
62 van Ommen, J.R., Sasic, S., van der Schaaf, J. et al. (2011). Time-series analysis of pressure fluctuations in gas–solid fluidized beds – a review. *Int. J. Multiphase Flow* 37: 403–428.
63 Sasic, S., Leckner, B., and Johnsson, F. (2007). Characterization of fluid dynamics of fluidized beds by analysis of pressure fluctuations. *Prog. Energy Combust. Sci.* 33: 453–496.

Problems

5.1 Sand is fluidized by air at ambient conditions in a fluidization column of 0.10 m diameter with a similar set-up as in Figure 5.1. The bed pressure drop is as shown in Figure P5.1. Given that the Sauter mean diameter of the sand is 260 μm, with particle and bulk densities of 2560 and 1250 kg/m^3, respectively, calculate the following:
(a) Voidage at minimum fluidization assuming it is the same as the voidage of a packed bed of the same material.
(b) Settled bed height.
(c) Minimum fluidization velocity and compare with the experimental value from Figure P5.1. Assume that the sand sphericity is 0.8.

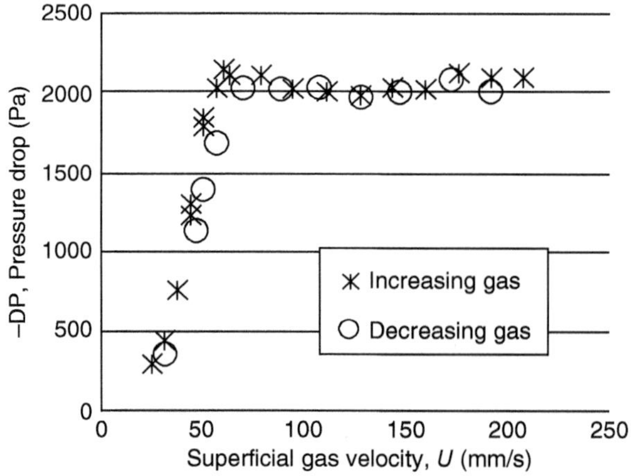

Figure P5.1 Time-average pressure drop plotted against U.

(d) Design a perforated distributor plate with its pressure drop 30% of the bed pressure drop when fluidized at $U = 0.15$ m/s. If the orifice diameter is 3 mm, calculate the number of holes required to meet the 30% pressure drop condition. Assume that the discharge coefficient of the plate is 0.70.

5.2 Design a perforated plate grid for introducing air and steam into a fluidized bed reactor for acrylonitrile production using a uranium- and antimony-based catalyst. What are the number of grid holes, pitch, and open area ratio of the distributor? How far from the grid should any internals (heat exchanger tubes, spargers, etc.) be placed for a safety factor of 10%, based on the simplified schematic of Figure P5.2? The discharge coefficient for the grid is shown in Figure P5.3.

Production rate: 56 700 metric tons/year (340 days)

Reactor temperature: 450 °C

Pressure above the bed: 2.38 atm abs

Pressure drop across the bed: 59 kPa

Reactor superficial velocity: 0.61 m/s above the bed (*note*: total product and reactant)

1.53 mol of steam and 11.97 mol of air are introduced per mole of C_3H_3N produced.

Reactor diameter: 6.8 m

Grid holes are 12.5 mm in diameter on an equilateral triangle pitch (which is expressed by $L_h = \frac{1}{\sqrt{N_d \sin 60°}}$) where N_d is the number of holes/unit area.

Grid pressure drop equal to 10% of the bed pressure drop

Thickness of distributor plate: 0.019 m

Mean particle diameter: 60 μm

Particle density: 1600 kg/m^3

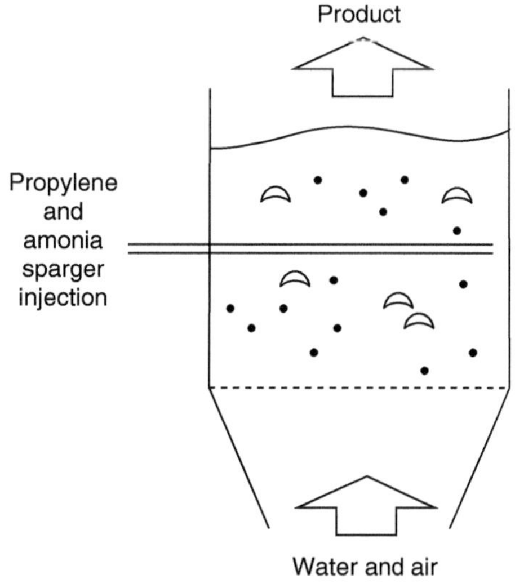

Figure P5.2 Simplified schematic of acrylonitrile fluidized bed reactor.

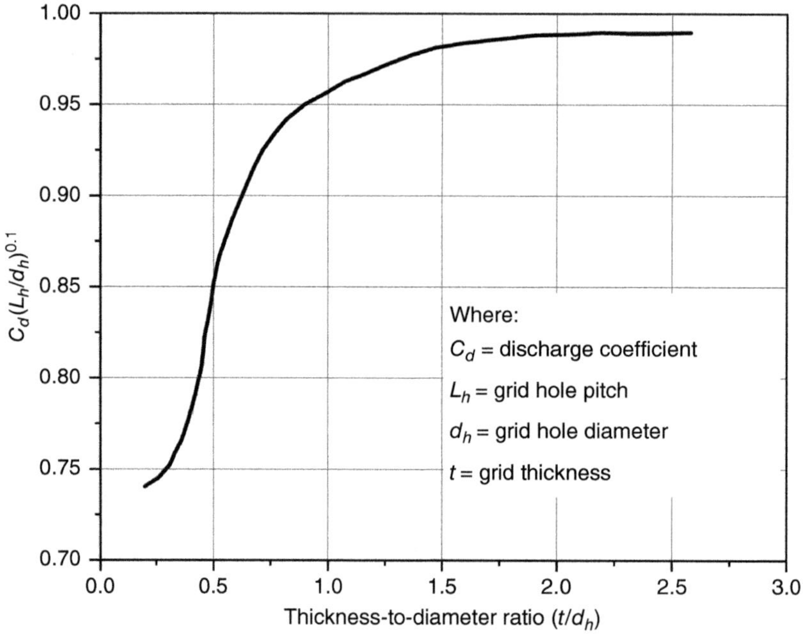

Figure P5.3 Grid hole discharge coefficient design chart. Source: Adapted from Karri 2003 [5].

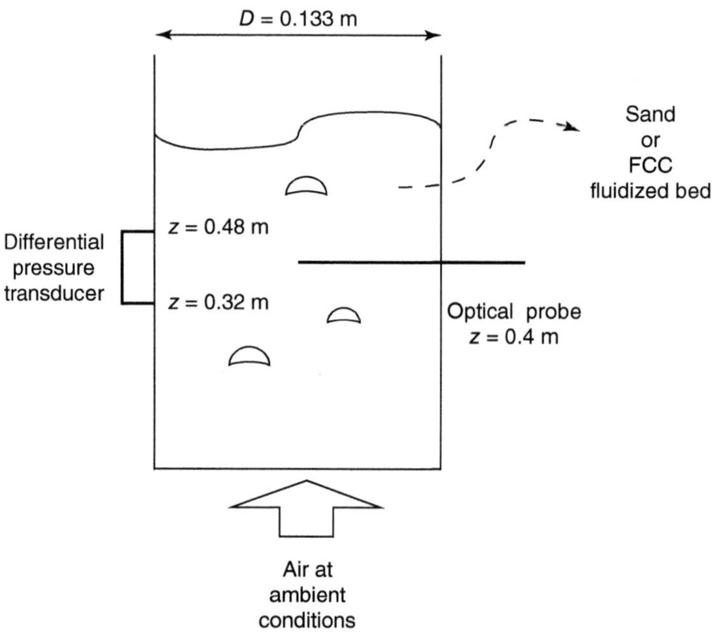

Figure P5.4 Schematic of experimental set-up.

5.3 A hopper is to be designed to feed biomass sawdust into a fluidized bed reactor. What is the sequence of steps required to design a hopper to feed at a certain feed rate? Number the steps in sequence. Furthermore, indicate a step that is not required by an X:
- Set the orifice diameter.
- Gather fundamental information related to gas and particle properties.
- Calculate the Hausner ratio by measuring the tapped and bulk densities.
- Measure the angle of internal friction.
- Calculate the necessary volumetric capacity of the hopper.
- Measure the angle of repose.
- Calculate the expected mass flow rate.

5.4 An optical probe is used to measure the radial profile of local voidage in a bubbling fluidized bed using two different solids, sand and fluid cracking catalyst (FCC), as shown in Figure P5.4. Given the time average voidage profile shown in Figure P5.5 and Table P5.1, calculate the cross-sectional local

Figure P5.5 Time-average radial voidage profile.

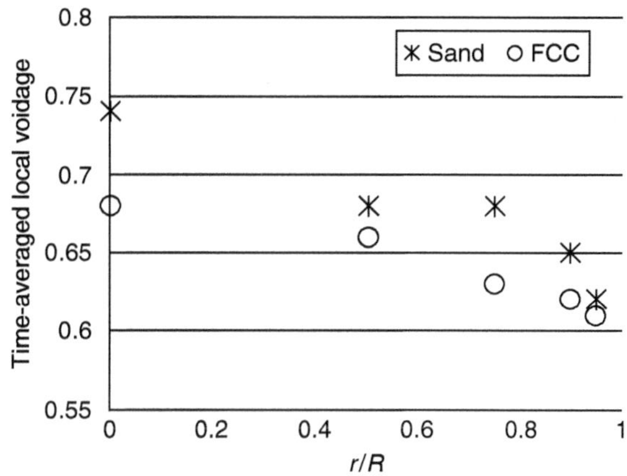

Table P5.1 Experimental data and conditions.

r/R	Sand	FCC
Time-average voidage		
0	0.74	0.68
0.5	0.68	0.66
0.75	0.68	0.63
0.9	0.65	0.62
0.95	0.62	0.61
Particle properties and time-average pressure drop		
ρ_p (kg/m^3)	2640	1560
ρ_b (kg/m^3)	1250	850
$-\Delta P$ (Pa)	1200	890

time average voidage and compare with those from a differential pressure measurement.

State the assumptions made in calculating the average voidages. Furthermore, obtain the cross-sectional average voidage using

$$\bar{\varepsilon} = \frac{1}{R^2} \int_0^R 2\varepsilon r \, dr$$

6

Computational Fluid Dynamics and Its Application to Fluidization

Tingwen Li[1] *and Yupeng Xu*[2,3]

[1] *SABIC Technology & Innovation, Engineering Science Department, 1600 Industrial Blvd, Sugar Land, TX 77478, USA*
[2] *National Energy Technology Laboratory, U.S. Department of Energy, 3610 Collins Ferry Road, Morgantown, WV 26507, USA*
[3] *Leidos Research Support Team, 3610 Collins Ferry Road, Morgantown, WV 26507-0880, USA*

Despite their long history of deployment and many applications, the design and scale-up of fluidized-bed reactors is still an empirical process, relying heavily on preliminary testing of pilot-scale model reactors, a time-consuming and expensive endeavour. Computational fluid dynamics (CFD) is an important auxiliary tool for understanding the physical phenomena involved in fluid flow and heat transfer using computer simulations, graphic displays, and numerical data analysis. CFD was created after World War II and was initially used for aerodynamic designs, gradually becoming useful in other areas. In the last few decades, considerable progress has been made in the hydrodynamic modelling of gas–solid flow and in its application to fluidization. Numerical simulations have been used to understand the complex dynamic behaviour of gas–solid suspensions and as a design tool to guide experimental setup and testing.

The gas–solid flows encountered in most applications exhibit dynamic structures of a wide range of spatial and temporal scales. For example, particle–particle or particle–wall collisions and energy dissipation occur at microscale and in microseconds or even faster. However, these microscale events are closely related to mesoscale structures such as bubbles and clusters, which further affect the macroscale flow behaviours at the device level. Because of the inherently unsteady and highly coupled multiscale characteristics of these flow structures, it is not computationally feasible to resolve all details using a single model.

Different modelling approaches have been developed to study the complex physics associated with multiphase flows. Some of these are portrayed in Figure 6.1. These models can simulate typical gas–solid multiphase flows at different scales with different levels of details. The figure illustrates the general relationship between accuracy and computational cost (or scale) for different approaches and illustrates the trade-off between accuracy and speed for different models. Based on how the solid phase is handled, two different classes of models can be distinguished: Eulerian–Lagrangian (E–L) models and Eulerian–Eulerian (E–E) models. E–L models track the solid phase on the particle level by following the

Essentials of Fluidization Technology, First Edition. Edited by John R. Grace, Xiaotao Bi, and Naoko Ellis.

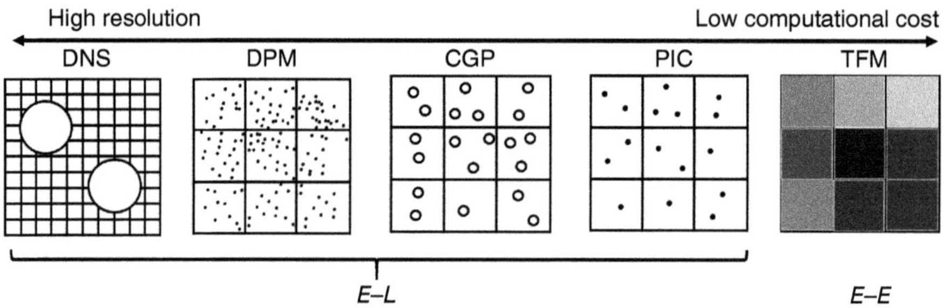

Figure 6.1 Different CFD models for gas–solid flows.

trajectory of each individual particle or swarm of particles. Examples are direct numerical simulation (DNS), discrete particle method (DPM), coarse-grained particle (CGP), and multiphase particle-in-cell (MP-PIC) methods. E–E models treat the dispersed solid phase and gas phase as interpenetrating continua (overlapping phases) with appropriate constitutive correlations for the flow rheology. The popular two-fluid model (TFM) is the best-known E–E model. Assisted by the continuous advancement in computer hardware and numerical algorithms, considerable improvements in both accuracy and speed have been achieved for both types of model. With such improvements continuing, CFD is bound to play a more and more significant role in addressing challenges in particulate multiphase flow.

This chapter focuses on the two most widely used models for gas–solid flows: the E–E TFM and the E–L CFD–DPM. Moreover, we limit our coverage to gas–solid systems.

6.1 Two-Fluid Model

The TFM, or more general multi-fluid model, treats both gas and solid phases in an Eulerian frame of reference as continuous and fully interpenetrating. The equations employed describe the conservation of mass, momentum, and energy. They can be derived from different averaging approaches, such as volume averaging, time averaging, and ensemble averaging.

In principle, TFM can be derived for different multiphase flow combinations, such as gas–liquid, gas–solid, etc., provided that proper closures are used to describe the rheology for each phase, as well as the interfacial interactions. The model accuracy is strongly dependent on such closure models. Despite that, the TFM is usually less computationally demanding than other popular modelling approaches. The TFM has been used widely for simulating large-scale gas–solid fluidized-bed systems.

6.1.1 Governing Equations

The governing equations are summarized in this section for gas–solid flows in typical fluidized-bed applications. In practice, a multi-fluid model can be used to model different solid phases with a unique combination of diameter, density, and other physical properties for each phase. However, development of the self-consistent constitutive closures for multi-fluid systems is nontrivial and therefore not covered in this chapter.

In the TFM, the phase volume fractions, written as continuous functions of space and time, are introduced to track the volume fraction occupied by gas and solids. The volume fractions must sum to one, so

$$\varepsilon_g + \varepsilon_s = 1 \tag{6.1}$$

where ε_g and ε_s are the volume fractions of gas and solid phases, respectively. The gas phase volume fraction is also widely known as the "void fraction" or "voidage".

The continuity equations for gas and solid phases are

$$\frac{\partial \varepsilon_g \rho_g}{\partial t} + \nabla \cdot (\varepsilon_g \rho_g \mathbf{u}_g) = -r_{gs} \tag{6.2}$$

and

$$\frac{\partial \varepsilon_s \rho_s}{\partial t} + \nabla \cdot (\varepsilon_s \rho_s \mathbf{u}_s) = r_{gs} \tag{6.3}$$

where ρ_g and ρ_s are the gas- and solid-phase densities, $\mathbf{u}_g$ and $\mathbf{u}_s$ are the gas- and solid-phase velocities, and r_{gs} is the mass transfer rate from gas phase to solid phase due to chemical reactions or physical processes, such as evaporation. All flow field variables here are averaged quantities according to the averaging approach, i.e. time average, space average, or ensemble average.

The momentum equations for gas and solid phases can be written as

$$\frac{\partial}{\partial t}\varepsilon_g \rho_g \mathbf{u}_g + \nabla \cdot (\varepsilon_g \rho_g \mathbf{u}_g \mathbf{u}_g) = \nabla \cdot \boldsymbol{\sigma}_g - \mathbf{f}_{gs} + \varepsilon_g \rho_g \mathbf{g} - \mathbf{m}_{gs} \tag{6.4}$$

and

$$\frac{\partial}{\partial t}\varepsilon_s \rho_s \mathbf{u}_s + \nabla \cdot (\varepsilon_s \rho_s \mathbf{u}_s \mathbf{u}_s) = \nabla \cdot \boldsymbol{\sigma}_s + \mathbf{f}_{gs} + \varepsilon_s \rho_s \mathbf{g} + \mathbf{m}_{gs} \tag{6.5}$$

where $\boldsymbol{\sigma}_g$ and $\boldsymbol{\sigma}_s$ are gas- and solid-phase stress tensors, $\mathbf{f}_{gs}$ is the interaction force between gas and solid phases, $\mathbf{g}$ is the gravity vector, and $\mathbf{m}_{gs}$ is the momentum transfer due to the mass transfer between phases.

The interaction force between the two phases can be further divided into a pressure gradient force and a drag force term:

$$\mathbf{f}_{gs} = \varepsilon_s \nabla P_g + \beta_{gs}(\mathbf{u}_g - \mathbf{u}_s) \tag{6.6}$$

The first term on the right side of Eq. (6.6) for pressure gradient force is straightforward, while the second term is closed by the drag correlation. Detailed discussion on the drag model is provided in Section 6.3.1 considering the significance of the drag force in fluidized beds.

The energy balance for the gas phase is written in term of gas temperature T_g as

$$\varepsilon_g \rho_g C_{pg} \left(\frac{\partial T_g}{\partial t} + \mathbf{u}_g \cdot \nabla T_g \right) = -\nabla \cdot \mathbf{q}_g - Q_{gs} - \Delta Q_{rg} + Q_r \tag{6.7}$$

where C_{pg} is the gas-phase heat capacity, $\mathbf{q}_g$ is the gas-phase conductive heat flux, and Q_{gs} describes the interphase heat transfer between gas and solid phases, and ΔQ_{rg} is the heat of reaction in the gas phase, which also accounts for the energy transfer associated with the interphase mass transfer. The last term accounts for heat transfer due to radiation. In this equation, the energy source terms due to viscous dissipation and interphase momentum transfer are neglected as they are of limited importance for most gas-fluidized bed applications.

The energy balance equation for the solid phase is written in term of solid temperature T_s as

$$\varepsilon_s \rho_s C_{ps} \left(\frac{\partial T_s}{\partial t} + \mathbf{u}_s \cdot \nabla T_s \right) = -\nabla \cdot \mathbf{q}_s + Q_{gs} - \Delta Q_{rs} - Q_r \tag{6.8}$$

where C_{ps} is the solid heat capacity, $\mathbf{q}_s$ is the solid-phase conductive heat flux, ΔQ_{rs} is the heat of reaction in the solid phase, and Q_r is the radiation heat transfer from the solid phase to the gas flow. For heterogeneous chemical reactions taking place at the surface of the catalyst particles or fuel particles in fluidized-bed reactors, it is important to distribute the heat of reaction properly to gas and solid phases. Usually, it is best that the heat of reaction be assigned to the solid phase to aid solver stability.

The gas and solid phases may contain an arbitrary number of chemical species. The transport equations for these species are, respectively,

$$\frac{\partial \varepsilon_g \rho_g X_{gi}}{\partial t} + \nabla \cdot (\varepsilon_g \rho_g X_{gi} \mathbf{u}_g) = \nabla \cdot (D_{gi} \nabla X_{gi}) + R_{gi} \tag{6.9}$$

and

$$\frac{\partial \varepsilon_s \rho_s X_{si}}{\partial t} + \nabla \cdot (\varepsilon_s \rho_s X_{si} \mathbf{u}_s) = \nabla \cdot (D_{si} \nabla X_{si}) + R_{si} \tag{6.10}$$

where X_{gi} and X_{si} are the mass fractions of the gas- and solid-phase species, D_{gi} and D_{si} are diffusivities of the gas and solid species, whereas R_{gi} and R_{si} are the rates of formation of gas and solid species.

6.1.2 Kinetic Granular Theory

To solve the governing equations for TFM, proper constitutive relations are needed for the gas and solid stresses. It is straightforward to close such terms for the gas phase as a standard Newtonian fluid, but it is challenging to develop proper constitutive closures for the solid phase. A variety of models have been developed, ranging from empirical correlations to kinetic theory-based models. Here, the widely used closure relations based on the granular kinetic theory are summarized.

In granular kinetic theory, it is assumed that the random motion of particles caused by collisions is analogous to the thermal motion of molecules in a gas. The granular temperature, which is proportional to the mean square of the random particle velocity, is defined to model the turbulent fluctuating energy of the solid phase. Constitutive relations for the solid-phase stress can, therefore, be derived based on the kinetic theory concepts, as detailed in Ref. [1].

A transport equation of granular temperature derived from kinetic theory is

$$\frac{3}{2}\left[\frac{\partial}{\partial t}(\varepsilon_s\rho_s\Theta)+\nabla\cdot(\varepsilon_s\rho_s\mathbf{u}_s\Theta)\right]=\boldsymbol{\sigma}_{s,vis}:\mathbf{u}_s+\nabla\cdot(\kappa_\Theta\nabla\Theta)-\gamma_\Theta+\phi_{gs} \tag{6.11}$$

where the four terms on the right side of Eq. (6.11) account for the generation of granular energy, diffusion of granular temperature, dissipation of granular energy due to particle–particle inelastic collisions, and energy exchange between the gas and solid phases.

The stress for the solid phase in the collisional regime is given by

$$\boldsymbol{\sigma}_{s,vis}=-P_{s,vis}\mathbf{I}+\boldsymbol{\tau}_{s,vis} \tag{6.12}$$

where $P_{s,vis}$ is the solid pressure and $\boldsymbol{\tau}_{s,vis}$ is the solid shear stress tensor.

The solid pressure $P_{s,vis}$, which represents the solid phase normal forces caused by particle collisions, consists of two terms: a kinetic term and a term resulting from particle collisions:

$$P_{s,vis}=\varepsilon_s\rho_s\Theta(1+2\varepsilon_s g_0(1+e)) \tag{6.13}$$

Here e is the restitution coefficient for particle collisions and g_0 is a radial distribution function, which takes into account the increase of collisions frequency when the volume fraction of particles approaches the packing limit. For example, the expression given by Gidaspow [1] is

$$g_0=\frac{3}{5}\left[1-\left(\frac{\varepsilon_s}{\varepsilon_{s,max}}\right)^{\frac{1}{3}}\right]^{-1} \tag{6.14}$$

where $\varepsilon_{s,\,max}$ is the maximum volume fraction of particles at close packing.

The solid shear stress tensor is given by

$$\boldsymbol{\tau}_{s,vis}=\mu_s[\nabla\mathbf{u}_s+(\nabla\mathbf{u}_s)^T]+\left(\lambda_s-\frac{2}{3}\mu_s\right)\mathbf{I}\cdot\nabla\cdot\mathbf{u}_s \tag{6.15}$$

with λ_s and μ_s being solid bulk viscosity and solid shear viscosity, respectively.

Among various expressions for solid shear viscosity and bulk viscosity, the model of Gidaspow [1] takes

$$\mu_s=\frac{4}{5}\varepsilon_s^2\rho_s d g_0(1+e)\sqrt{\frac{\Theta}{\pi}}+\frac{\frac{5\sqrt{\pi}}{48}\rho_s d\sqrt{\Theta}}{(1+e)g_0}\left[1+\frac{4}{5}g_0\varepsilon_s(1+e)\right]^2 \tag{6.16}$$

$$\lambda_s=\frac{4}{3}\varepsilon_s^2\rho_s d g_0(1+e)\sqrt{\frac{\Theta}{\pi}} \tag{6.17}$$

According to [1], the diffusion coefficient for granular temperature is

$$\kappa_\Theta=\frac{150\rho_s d\sqrt{\Theta\pi}}{384(1+e)g_0}\left[1+\frac{6}{5}\varepsilon_s g_0(1+e)\right]^2+2\rho_s\varepsilon_s^2 d(1+e)\sqrt{\frac{\Theta}{\pi}} \tag{6.18}$$

and the dissipation of granular energy, γ_Θ, is given by

$$\gamma_\Theta=\frac{12(1-e^2)g_0}{d\sqrt{\pi}}\rho_s\varepsilon_s^2\Theta^{\frac{3}{2}} \tag{6.19}$$

The transfer of kinetic energy of random fluctuations in particle velocity from the solid phase to the gas phase is modelled as

$$\phi_{gs} = -3\beta_{gs}\Theta \tag{6.20}$$

where β_{gs} is an interphase drag coefficient, discussed in Section 6.3.1.

Instead of solving the partial differential Eq. (6.11) for the granular temperature, an algebraic expression can be derived by assuming that the granular energy is dissipated locally, neglecting convection and diffusion and retaining only the generation and dissipation terms. Although it is more desirable to solve the full transport equation for best accuracy, the simplified algebraic expression for granular temperature has often been used in numerical simulations to reduce the computational cost while maintaining reasonable accuracy.

6.1.3 Frictional Model

In dense fluidized-bed applications, the solid volume fraction can be very high, usually >0.5 (50%). In this regime, particles may endure long, sliding, and rubbing contact, and surface friction between individual particles becomes significant. This reflects the transition between two limiting flow regimes, rapid and quasi-static regimes, as illustrated in Figure 6.2. The constitutive model developed from kinetic theory is not fully valid here, so a different model should be used. Savage [2] first presented the total solid stress as the sum of a rate-independent frictional part and a rate-dependent viscous part obtained from granular kinetic models:

$$\boldsymbol{\sigma}_s = \boldsymbol{\sigma}_{s,vis} + \boldsymbol{\sigma}_{s,fric} \tag{6.21}$$

where $\boldsymbol{\sigma}_s$ is the total solid stress appearing in Eq. (6.5), $\boldsymbol{\sigma}_{s,vis}$ is the viscous stress modelled with granular kinetic theory as shown in Eq. (6.14), and $\boldsymbol{\sigma}_{s,fric}$ stands for the frictional stress, for which the constitutive models are largely based on soil mechanics or model developed from discrete particle simulations. This ad hoc patching approach unifies the available models for the rapid flow regime and the quasi-static regime.

In the quasi-static regime, a granular material under a normal stress will shear only when the shear stress reaches a critical value, i.e. the yield condition. Many powders obey the yield condition as long as they undergo continuous deformation. This state is usually called "the critical state." The simplest frictional

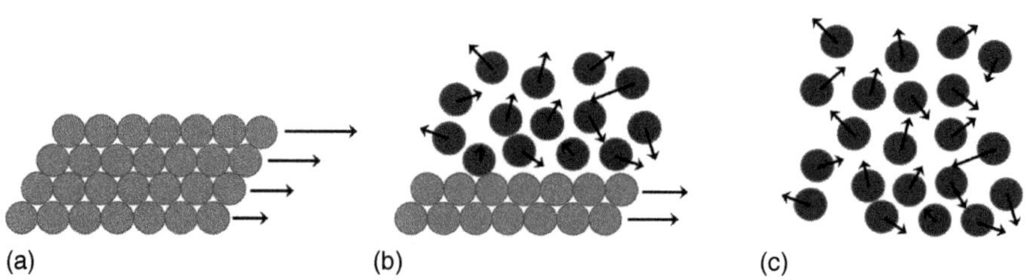

Figure 6.2 Regime map for granular flow rheology. (a) Quasi-static regime, (b) intermediate flow regime, and (c) inertial regime.

model, proposed by Schaeffer [3] and derived from the von Mises yield conditions, can be employed to calculate the frictional component of solid stress as

$$\boldsymbol{\sigma}_{s,fric} = -P_{s,fric}\left(\mathbf{I} - \frac{\sin\phi}{\sqrt{I_{2D}}}\mathbf{D}_s\right) \tag{6.22}$$

where $P_{s,fric}$ is the normal frictional stress or frictional pressure, usually calculated with empirical correlations; ϕ is the internal angle of friction; $\mathbf{D}_s$ is the rate of deformation tensor; and I_{2D} is the second invariant of $\mathbf{D}_s$.

Accordingly, the frictional viscosity for Schaeffer's model is of the form

$$\mu_{fric} = \frac{P_{s,fric}\sin\phi}{\sqrt{I_{2D}}} \tag{6.23}$$

The frictional pressure can be calculated based on the correlation suggested by Johnson et al. [4]:

$$P_{s,fric} = \begin{cases} 0 & \varepsilon_s < \varepsilon_{s,t} \\ Fr\dfrac{(\varepsilon_s - \varepsilon_{s,t})^n}{(\varepsilon_{s,max} - \varepsilon_s)^p} & \varepsilon_s \geq \varepsilon_{s,t} \end{cases} \tag{6.24}$$

With empirical material constants $Fr = 0.05$, $n = 2$, and $p = 5$. $\varepsilon_{s,t}$ is the solid volume fraction when frictional stress becomes important. All these parameters are usually dependent on the solid material properties. The frictional model is critical for certain fluidized-bed applications with a dense emulsion phase persisting, such as spouted beds and bubbling fluidized beds. To obtain the best model predictability, it is important to adopt a frictional model available in the literature, which is appropriate for the target application.

6.1.4 Limitations of TFM

The widely used kinetic theory-based TFM was originally developed for flow of smooth, frictionless spherical particles dominated by binary collisions. Its extension to the frictional flow regime is through simple patching as discussed in Section 6.1.3. Extension of the kinetic theory of granular flow for frictional spheres and incorporation of gas-phase contributions have been developed by different authors. However, this usually leads to sophisticated models that limit implementation and validation. In addition, the improvement in most fluidized-bed applications is only marginal. Furthermore, it is nontrivial to account for the particle size distribution and particle shape in the theory instead of adopting empirical closures for the leading-order effects. Despite these limitations, the model has been successfully applied for many applications with reasonable speed and accuracy. Because of that, TFM will continue to be used in the foreseeable future as a major modelling approach for fluidized-bed applications.

6.2 Discrete Particle Method

The other popular modelling approach is CFD-coupled DPM (or discrete element method), which is one type of E–L approach. The Newtonian equations of motion for each individual solid particle are then solved, with the acceleration, velocities, and trajectory of each particle tracked. In this manner, it is then straightforward to account for the particle size and density distributions. Particle–particle and particle–wall interactions can be described either by the soft-sphere model [5] or by the hard-sphere model [6]. In the soft-sphere model, the contact force between colliding particles is calculated based on the deformation of colliding particles using a combination of spring and dashpot. To achieve that, a small numerical time step is used to resolve the collisional process. For the hard-sphere model, the particles are assumed to interact through instantaneous binary collisions, and the post-collisional velocities are calculated based on momentum conservation. However, it is difficult to model multi-body contacts in the dense flow using the hard-sphere model. Because of the wide span of solid concentrations typically encountered in fluidized-bed systems, the soft-sphere model is usually adopted.

6.2.1 Governing Equations

In the DPM, position, linear, and angular velocities of each particle are tracked as $\mathbf{x}_i$, $\mathbf{c}_i$, and $\boldsymbol{\omega}_i$ with $\mathbf{c}_i = d\mathbf{x}_i/dt$. The translational and rotational motions of a particle with mass m_i and moment of inertia I_i are described by Newton's equations for rigid body motion:

$$m_i \frac{d\mathbf{c}_i}{dt} = m_i\mathbf{g} + \mathbf{F}_i^c + \mathbf{F}_i^d \tag{6.25}$$

$$I_i \frac{d\boldsymbol{\omega}_i}{dt} = \mathbf{T}_i \tag{6.26}$$

The terms on the right side of Eq. (6.25) are gravity, forces exerted by other particles including the contact force and non-contact forces such as van der Waals force, and the force due to interphase momentum exchange. In Eq. (6.26), $\mathbf{T}_i$ is the torque acting around the centre of mass of the particle.

The energy equation in the form of particle temperature is written as

$$m_i C_{p,i} \frac{dT_i}{dt} = q_{gs,i} - \Delta q_{rs,i} - q_{r,i} \tag{6.27}$$

where $C_{p,i}$ is the heat capacity of solid particle and T_i is the particle temperature assuming a uniform temperature distribution. On the right side are heat fluxes resulting from interfacial heat transfer, heat of reaction, and radiation, respectively.

Several models are available for describing the contact force between colliding particles. Compared with other more complex models, the linear model, despite its simplicity, has performed reliably in simulating different granular flows. The widely used linear-spring dashpot model of Cundall and Strack [7] is introduced here. A schematic of particle i in collision with particle j is shown in Figure 6.3.

In this model, the total contact force on particle i of radius R_i is given by a sum of normal and tangential forces with neighbouring particles:

$$\mathbf{F}_i^c = \sum_{j \in B} (\mathbf{F}_{n,ij} + \mathbf{F}_{t,ij}) \tag{6.28}$$

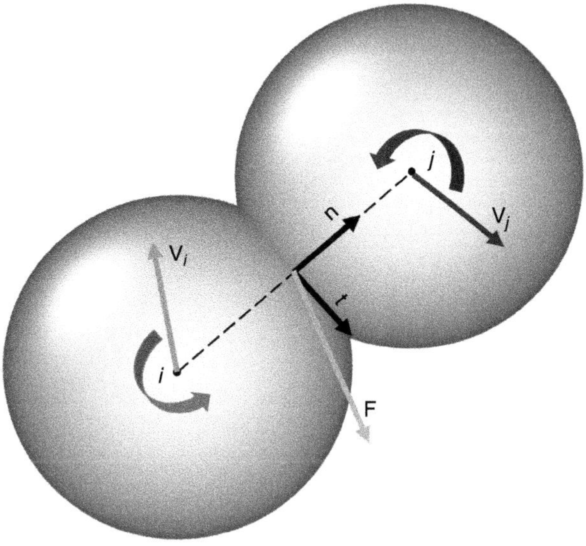

Figure 6.3 Schematic of soft-sphere model of particle collision.

where B is the set of particles in contact with particle i. The normal force, $\mathbf{F}_{n,ij}$, between particles i and j is calculated by

$$\mathbf{F}_{n,ij} = -k_n \delta_n \mathbf{n} - \eta_n \mathbf{c}_{n,ij} \tag{6.29}$$

where k_n is the normal spring stiffness and η_n is the normal damping coefficient. These quantities are related to material properties such as Young's modulus, stiffness, etc. The normal force depends linearly on the overlap between two particles, $\delta = R_i + R_j - |\mathbf{x}_i - \mathbf{x}_j|$, and the relative normal velocity, $\mathbf{v}_{n,ij} = (\mathbf{v}_{ij} \cdot \mathbf{n})\mathbf{n}$, where $\mathbf{n}$ is the unit vector pointing from the centre of particle i to the centre of particle j and $\mathbf{v}_{ij}$ is the relative velocity at the particle contact, given by

$$\mathbf{v}_{ij} = (\mathbf{c}_i - \mathbf{c}_j) + (R_i \boldsymbol{\omega}_i + R_j \boldsymbol{\omega}_j) \times \mathbf{n} \tag{6.30}$$

The tangential component of the contact force is given by

$$\mathbf{F}_{t,ij} = \begin{cases} -k_t \delta_t \mathbf{t} - \eta_t \mathbf{v}_{t,ij} & \text{if } |\mathbf{F}_{t,ij}| \leq \mu_f |\mathbf{F}_{n,ij}| \\ -\mu_f |\mathbf{F}_{n,ij}| \mathbf{t} & \text{if } |\mathbf{F}_{t,ij}| > \mu_f |\mathbf{F}_{n,ij}| \end{cases} \tag{6.31}$$

where k_t is the tangential spring stiffness, δ_t is the tangential displacement, η_t is the tangential damping coefficient, $\mathbf{v}_{t,ij}$ is the tangential relative velocity with $\mathbf{v}_{t,ij} = \mathbf{v}_{ij} - \mathbf{v}_{n,ij}$, μ_f is the friction coefficient, and $\mathbf{t}$ is the tangential unit vector defined by $\mathbf{t} = \mathbf{v}_{t,ij}/|\mathbf{v}_{t,ij}|$.

The tangential force leads to a torque $\mathbf{T}_{ij}$ acting on the particles:

$$\mathbf{T}_{ij} = R_i \mathbf{n} \times \mathbf{F}_{t,ij} \tag{6.32}$$

The hydrodynamic force due to the fluid phase can be written as

$$\mathbf{F}_i^d = -\nabla P_g(\mathbf{x}_i) V_i + \frac{\beta_i V_i}{\varepsilon_s} (\mathbf{v}_g(\mathbf{x}_i) - \mathbf{c}_i) \tag{6.33}$$

where the pressure gradient at the particle location is used to calculate the pressure gradient force acting on the particle volume of V_i. The drag force induced by the relative velocity between the particles and the local gas flow is calculated by the second term.

The governing equations for the gas phase are the same as those in TFM as shown in Eqs. (6.2) and (6.4), except that the solid volume fraction is directly calculated as

$$\varepsilon_s = \frac{1}{V_c}\sum_{i=1}^{N} W_i V_i \tag{6.34}$$

where V_i and V_c are the volumes of the particle and computational cell, respectively; and W_i is the weight function. There are different approaches to calculate the distribution of solid volume fraction from the dispersed particle locations to determine the continuous flow field. For simple centroid-based methods, W_i is 1 if the particle centre is within the computational cell; otherwise, it is zero. Complicated interpolation algorithms can be used to calculate a smoother field for solid concentration by distributing the mass of a particle over many computational cells.

The interphase momentum exchange in the gas phase is calculated as

$$\mathbf{f}_{gs} = \frac{1}{V_c}\sum_{i=1}^{N} W_i \mathbf{F}_i^d \tag{6.35}$$

If Eq. (6.33) is substituted into Eq. (6.35), the interphase momentum exchange differs slightly from that in Eq. (6.6) for TFM. The difference is mainly caused by the mapping of continuous flow field to a dispersed location and then mapping back to the continuous field. Different methods are available for this purpose.

Similarly, the energy equation for the gas phase as shown in Eq. (6.8) is solved with the source terms calculated as

$$Q_{gs} = \frac{1}{V_c}\sum_{i=1}^{N} W_i q_{gs,i} \tag{6.36}$$

$$\Delta Q_{gr} = \frac{1}{V_c}\sum_{i=1}^{N} W_i \Delta q_{gr,i} \tag{6.37}$$

$$Q_r = \frac{1}{V_c}\sum_{i=1}^{N} W_i q_{r,i} \tag{6.38}$$

In CFD–DPM simulations, it is generally recommended that the CFD grid size be at least three to four times the particle diameter so that a smooth flow field of solid concentration can be obtained. There are methods to overcome this limitation to facilitate smaller numerical grids, but they usually demand more computing resources. The interested reader is referred to the literature for different methods calculating the continuous solid distribution. Consistent mapping methods are needed for the coupling of mass, momentum, and heat transfer.

For DPM, the time step must be small enough to resolve the collisional process and is usually from one-twentieth to one-fiftieth of the collisional time. For a linear-spring system, the collision time can be estimated as $t_{col} \propto \sqrt{m/k_n}$. As a result, to speed up the simulation, a small spring constant is usually applied to facilitate a large time step. However, the spring constant must still be large enough to prevent exaggerated overlap during collisions. For typical fluidized-bed applications, a maximum overlap of <1% of particle diameter is recommended.

6.2.2 Limitations of CFD–DPM

The major challenge associated with CFD–DPM is the high computational cost to track each individual particle within the system. This can easily be excessively costly for industrial-scale fluidized-bed reactors. In addition, the simulation has to be unsteady, with very small time steps to resolve the collisional process. Numerous ways have been explored to speed up DPM simulations, including hybrid particle contact model coupling hard- and soft-sphere approaches, coarse-grained method to simulate many particles together using numerical parcels, and statistical collision detection, instead of deterministic collision detection.

Another important E–L method is the MP-PIC method, which tracks a group of particles as a point, with and particle–particle interaction only considered in a statistical way [8]. Without resolving the interparticle contact, this method provides superior performance in speed, but with the compromise of losing accuracy by treating particles as points without accurate interaction.

6.3 Gas–Solid Interaction

In both TFM and CFD–DPM, closure models are needed for the gas–solid interactions for momentum and energy equations. For the gas–solid momentum transfer, many mechanisms have been investigated, including the effects of drag, Saffman lift, virtual mass, Basset history, and Magnus forces. For most gas–solid flows in fluidized beds of practical interest, the density ratio between solid and gas greatly exceeds unity so that the drag becomes the dominant mechanism characterizing the momentum exchange and driving the solid flow. It has been shown [9] that different drag models can lead to quantitatively different results, such as bed expansion and gas holdup in bubbling fluidized beds and bed inventory in circulating fluidized beds.

6.3.1 Gas–Solid Drag

The drag force on a single spherical particle in gas flow is well understood. However, the drag force experienced by the particle in an ensemble of particles is very different because of the influence of neighbouring particles. Numerous drag models have been developed to describe the momentum exchange between gas and solid particles covering the full range of solid concentrations encountered in the real-world applications.

In early studies, the average drag force between gas and solid particles was obtained from pressure drop measurements in packed beds or measurements of terminal settling velocity of solid sedimentation in quiescent fluid. The drag force acts in the direction opposite to the relative motion of particle and fluid and is simply written as

$$\mathbf{f}_{gs,d} = \beta_{gs}(\mathbf{u}_g - \mathbf{u}_s) \tag{6.39}$$

where β_{gs} is the drag coefficient. Different drag models are available in the literature. For example, the popular Gidaspow drag model [1], which is a combination of Ergun equation from fixed-bed experiment and Wen and Yu model from terminal velocity data of liquid–solid system, gives

$$\beta_{gs} = \begin{cases} 150\dfrac{\varepsilon_s(1-\varepsilon_g)\mu_g}{\varepsilon_g d^2} + 1.75\dfrac{\rho_g\varepsilon_s|\mathbf{u}_g - \mathbf{u}_s|}{d} & \text{if } \varepsilon_g \le 0.8 \\ \dfrac{3}{4}C_D\dfrac{\rho_g\varepsilon_g\varepsilon_s|\mathbf{u}_g - \mathbf{u}_s|}{d}\varepsilon_g^{-2.65} & \text{if } \varepsilon_g > 0.8 \end{cases} \tag{6.40}$$

where C_D is the single particle drag coefficient:

$$C_D = \begin{cases} \frac{24}{Re_p}(1 + 0.15(Re_p)^{0.687}) & \text{if } Re_p \le 1000 \\ 0.44 & \text{if } Re_p > 1000 \end{cases} \tag{6.41}$$

with the particle Reynolds number defined as

$$Re_p = \frac{\rho_g\varepsilon_g|\mathbf{u}_g - \mathbf{u}_s|d}{\mu_g} \tag{6.42}$$

In recent years, DNS has been widely used to study the interaction between gas flow and solid particles. In the DNS, the flow around individual particle is fully resolved by solving Navier–Stokes equations, while particle motion is tracked by solving Newton's equations of motion. The drag force can be directly calculated with this method. Compared with the empirical models developed from experimental data, the DNS provides idealized well-controlled flow conditions with a wide range of parameter space.

More and more drag models are being developed from DNS using different methods. In most DNS studies, simulations of homogeneous systems in a periodic domain are conducted. Particle positions are randomly seeded and remain stationary to save computational cost. The drag force experienced by the bulk covering a wide range of solid concentrations and flow conditions, i.e. Reynolds numbers, are simulated. In the end, the mean drag force experienced by the particles is usually developed as a function of the Reynolds number and solid volume fraction. For example, the drag model for monodisperse gas–solid system developed by Tenneti et al. [10] using particle-resolved DNS for solid volume fraction up to 0.5 and mean flow Reynolds number up to 300 is

$$\begin{aligned} \beta_{gs} &= \frac{18\mu_g\varepsilon_s(1-\varepsilon_s)^2}{d_p^2}\left(\frac{f_{iso}(Re_p)}{(1-\varepsilon_s)^3} + f_1(\varepsilon_s) + f_2(\varepsilon_s, Re_p)\right) \\ f_{iso}(Re_p) &= 1 + 0.15Re_p^{0.687} \\ f_1(\varepsilon_s) &= \frac{5.81\varepsilon_s}{(1-\varepsilon_s)^3} + 0.48\frac{\varepsilon_s^{1/3}}{(1-\varepsilon_s)^4} \\ f_2(\varepsilon_s, Re_p) &= \varepsilon_s^3 Re_p\left(0.95 + \frac{0.61\varepsilon_s^3}{(1-\varepsilon_s)^2}\right) \end{aligned} \tag{6.43}$$

Drag forces experienced by particles of various sizes subjected to interstitial flow have also been studied with DNS, and several drag models for polydisperse particle systems can be found in the literature [11, 12]. DNS has also been used to develop drag models for nonspherical particles. Currently, most drag models were developed for the average drag force experienced by particles in a homogeneous suspension. The slip velocity is based on the mean gas and solid velocities in the control volume in an E–E frame. When applied to CFD–DPM simulations, the local gas and particle velocities are used, in addition to the local interpolated solid volume fraction, to calculate the drag force for each individual particle. In this regard, a dedicated drag model should be developed for the E–L framework, which needs to account for the effect of neighbouring particles that are beyond the solid volume fraction.

The drag models discussed so far all rely on homogeneous flow assumptions, i.e. they assume that the solid concentration is more or less constant in the control volume. In fluidized-bed reactors, the formation of heterogeneous structures, such as clusters and bubbles, plays a major role in affecting all transport phenomena inside the system. The grid resolution used in the CFD simulation needs to be sufficient to resolve those structures to account accurately for their effects. However, clusters and bubbles can be very small, requiring an extremely fine grid size of a few particle diameters, making the simulation of industrial-scale problems intractable. It has been shown in many grid sensitivity studies that the grid resolution required for grid-independent results is strongly dependent on the particle properties, e.g. size and density. Finer grid size is required for small particles such as Geldart group A particles. When such grid resolution is impossible, an alternative approach must be used to account for the effect of heterogeneous flow structures.

To overcome the grid refinement required by the homogeneous drag model, heterogeneous drag models are being developed to account for the existence of small-scale structures such as clusters and bubbles. Different sub-grid models based on the energy minimization multiscale (EMMS) model [13] or coarse grid filtering of fine grid simulations [14] have been developed, with successes reported for both approaches. However, the performances of heterogeneous drag models are inconsistent and tend to be case dependent, requiring considerable development and extensive validation.

Drag models for gas–solid flow are of great interest to the fluidization community because of the important role of gas–solid interactions. Many efforts are being made to enhance the predictability of numerical models to account for additional physics, such as the effect of particle shape. Readers are encouraged to check the latest literature for state-of-the-art drag model development and for best practices in their applications.

6.3.2 Gas–Solid Heat Transfer

High heat transfer rate between the fluidizing medium and the particles is a key feature of fluidized-bed reactors. Proper closure is needed to describe the heat transfer between particles and interstitial gas flow. The average heat transfer between gas and solid phases without interphase mass transfer is typically written as

$$Q_{gs} = \frac{6k_g\varepsilon_s Nu}{d^2}(T_s - T_g) \tag{6.44}$$

where T_s and T_g are the average temperatures of gas and solids in the small control volume, respectively; k_g is the thermal conductivity of the gas phase; d is the particle diameter; and Nu is the Nusselt number, typically closed by an empirical correlation.

Gunn [15] developed a model based on Nusselt number for average gas–solid heat transfer by fitting experimental data from many sources. His correlation covers a wide range of solid volume fractions from 0 to 0.65:

$$Nu = (7 - 10\varepsilon_g + 5\varepsilon_g^2)(1 + 0.7Re_p^{0.2}Pr_g^{1/3}) + (1.33 - 2.4\varepsilon_g + 1.2\varepsilon_g^2)Re_p^{0.7}Pr_g^{1/3} \tag{6.45}$$

where $Pr_g = C_{p,g}\mu_g/k_g$ is the Prandtl number.

Sun et al. [16] proposed a new correlation for the average gas–solid heat transfer using DNS to cover a wide range of Reynolds number (1–100) and solid volume fraction (0–0.5):

$$Nu = (-0.46 + 1.77\varepsilon_g + 0.69\varepsilon_g^2)/\varepsilon_g^3 + (1.37 - 2.4\varepsilon_g + 1.2\varepsilon_g^2)Re_p^{0.7}Pr_g^{1/3} \tag{6.46}$$

The bulk temperature was used to develop this model [16]. However, the average temperature was used in the TFM for calculation. As a result, a correction was introduced to rectify the inconsistency, and the heat flux is calculated as

$$Q_{gs} = \frac{6\pi k_g \varepsilon_s Nu}{4d^2\theta}(T_s - T_g) \tag{6.47}$$

$$\theta = 1 - 1.6\varepsilon_s(1 - \varepsilon_s) - 3\varepsilon_s(1 - \varepsilon_s)^4 \exp(-Re_p^{0.4}\varepsilon_s) \tag{6.48}$$

It is straightforward to calculate the heat flux for each individual particle in the CFD–DPM simulation as

$$q_{gs,i} = \pi k_g dNu(T_g - T_i) \tag{6.49}$$

where T_g is the gas temperature interpolated at the particle location and T_i is the particle temperature. Usually the correlations provided above are used.

6.4 Boundary Conditions

In CFD simulations, it is critical to define the proper computational domain and then supply corresponding boundary conditions (BCs). Usually, the computational domain for a fluidized-bed reactor is simplified as much as possible to reduce the computational cost. For this purpose, a large volume of modelling work assuming two-dimensional flow can be found in the literature, reducing three-dimensional columns into two-dimensional planes. Two-dimensional flow simulation is not recommended for quantitative prediction, but it provides useful qualitative information with short turnaround time. Two-dimensional (strictly speaking pseudo-two-dimensional) fluidized beds are widely used in experimental work for visual observation. Despite the overall two-dimensional flow behaviour, three-dimensional simulation is needed to capture the strong wall effect [17]. Whether the computational domain is two- or three-dimensional, a typical fluidized-bed simulation requires boundary conditions for inflow, outflow, and column wall.

For batch reactors, only gas is introduced into the system mainly through the bottom distributor at certain conditions, e.g. pressure and temperature. Usually a uniform velocity is assigned, which is a good assumption for most lab-scale fluidized beds with considerable pressure drop through the distributor to maintain uniform flow distribution. For industrial large-scale reactors, the distributor design is complicated and of critical importance to the flow distribution. As a result, it would be beneficial to include the full distributor geometry in the inflow boundary by resolving the key distributor nozzles/orifices. For circulating fluidized beds, a continuous solid inflow can be imposed in the same fashion if the mass flow rate is known.

For gas–solid flows encountered in fluidized-bed applications, the wall effect is mainly dominant by particle–wall interactions, which transfer much higher momentum than gas–wall interactions. Different solids wall boundary conditions, including free-slip, no-slip, and partial-slip walls, have been used in TFM simulations of fluidized beds. A partial-slip wall boundary is preferential to describe particle–wall interactions. The widely used Johnson and Jackson [18] partial-slip wall boundary condition relies on a specularity coefficient to quantify the tangential momentum transfer. This is a semiempirical parameter affected by particle–wall properties and operating conditions [19]. In recent years, several advanced wall boundary conditions have been developed based on the kinetic granular theory to better characterize particle–wall interactions.

A pressure outlet boundary condition is usually applied at the outlet to allow gas and particles to leave the system freely. For a specific application, it may be necessary to impose a porous plate near the exit to prevent solids from leaving the system and to mimic the baghouse or an internal cyclone without explicitly modelling those components.

For CFD–DPM simulations, it is straightforward to consider the particle–wall interactions by providing the proper physical parameters including particle–wall restitution coefficient and friction coefficient. It is also possible to include the effect of wall roughness, which can be explicitly captured or statistically accounted for. The inflow and outflow boundary conditions are similar to those in the TFM simulations.

6.5 Example and Discussion

Two example problems are presented in this section to illustrate the general features of TFM and CFD–DPM. These simulations were carried out using MFIX, an open-source CFD software developed at US Department of Energy's National Energy Technology Laboratory. MFIX provides state-of-the-art modelling capabilities for particulate multiphase flows, including TFM, CFD–DPM, MP-PIC, and hybrid models.

6.5.1 TFM Simulation of a Bubbling Fluidized Bed with Tube Bundle

The first example is TFM simulation of a bubbling fluidized bed with an immersed tube bundle as shown in Figure 6.4. The experimental setup simulated is a rectangular column with dimensions of 0.48 m width, 0.34 m depth, and 0.6 m height. The submerged tube bank contains 25 horizontal tubes of diameter

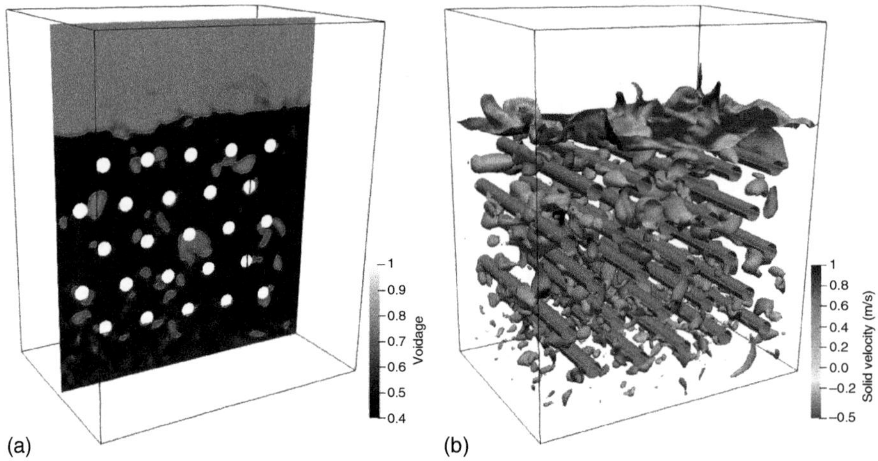

Figure 6.4 TFM simulation of a bubbling fluidized bed with immersed tube bundle. (a) Slice view of voidage distribution and contour line for voidage of 0.8. (b) Iso-surface for voidage of 0.8 with distribution of vertical solid velocity.

25 mm arranged in a triangular pitch. The bed material is silica sand with a mean diameter of 240 μm and a density of 2582 kg/m^3. More details on the system setup and experimental measurements can be found in Kim et al. [20].

Numerical results of a three-dimensional TFM simulation are presented here for a superficial gas velocity of 0.11 m/s. In Figure 6.4, a slice view of the voidage distribution is shown to illustrate the bubbling behaviour in the bed as well as the bed surface. Specifically, the contour plot of voidage of 0.8 is highlighted to depict the bubble interface. The complete contour plot of voidage at 0.8 is shown in Figure 6.4 to present the three-dimensional bubbles. The bubble surface is further coloured by the solid vertical velocity to show the solid motion. The tubes tend to split large bubbles as they pass. The flow characteristics near the tube, including bubble frequency and bubble phase fraction, can be analyzed to study the heat transfer characteristics between bed and immersed tube. The detailed analysis of bubble properties, as well as the effect of horizontal tube, was reported by Li et al. [21].

6.5.2 CFD–DPM Simulation of a Small-Scale Circulating Fluidized Bed

The second example shows a CFD–DPM simulation of high-density polyethylene particles in a small-scale full-loop circulating fluidized bed. The experimental facility was designed for validation purpose with different types of experimental instrumentation. The small-scale Circulating Fluidized Bed (CFB) system consists of a 1.2 m tall riser with 5 cm inner diameter and a 2.5 cm diameter standpipe connected by a cyclone and crossover pipe. In total, about 1 million High-Density Polyethylene (HDPE) particles with a mean diameter of 871 μm and a density of 863 kg/m^3 were tested.

Figure 6.5 shows a snapshot of the transient result predicted by the CFD–DPM simulation. In the middle, a slice view of voidage depicts the solid distribution in

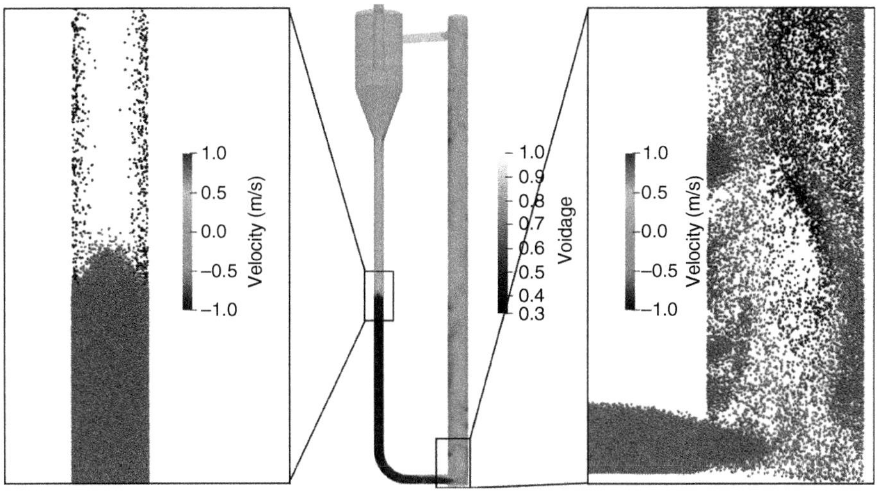

Figure 6.5 CFD–DPM simulation of a small-scale circulating fluidized-bed system (overall voidage distribution and zoom-in views of particles in standpipe and riser).

different parts of the CFB system including standpipe, L-valve, riser, and cyclone, which exhibits distinct distributions of solid concentration, hence different fluidization regimes. Close views in the middle of standpipe and near the bottom of the riser are presented where the vertical velocity component for each particle is shown. For better illustration, only a thin layer of particles is shown in the figure. The dense frictional flow in the lower standpipe and L-valve can be easily identified. In the riser section, downward solids flow in the form of clusters along the wall as observed experimentally.

Different drag models were tested. The simulation results showed that proper selection of drag is a crucial factor in the quantitative validation study. With the selected drag, numerical results compared well with experimental data, including pressure drop at separate locations, overall solid circulation rate, and the bed height in the standpipe, as reported in Xu et al. [22]. In addition, the simulation method provided very detailed information on each individual particle, such as position, velocity, etc. Thus the average particle residence time in distinct regions can be obtained directly. In this work, CFD–DPM demonstrated its value as a high-fidelity simulation tool that can predict key performance parameters for challenging CFB flow conditions.

6.5.3 Discussion

Two examples are presented in this section to briefly demonstrate TFM and CFD–DPM simulations. As shown in these examples, CFD modelling provides a great amount of information on the detailed flow behaviours in fluidized-bed reactors. Benefiting from that, CFD simulation has been used to study different aspects of fluidized-bed systems including gas–solid interaction, heat and mass transfer, gas mixing, solids segregation and mixing, and transition of different fluidization regimes. It has also been used as a complementary tool to experiment and theoretical analysis to help interpret experimental finding, guide experimental design, verify/improve measurement technique, and testify new theory. In addition, CFD modelling is playing an increasingly important role in design and operation of fluidized-bed reactors and scale-up of commercial processes. Extensive research work can be found in the open literature.

6.6 Conclusion and Perspective

CFD modelling is a valuable tool for helping to understand the complex flows encountered in various applications. Different numerical methods have been developed to model multiphase flows at different scales with different levels of detail. The most widely used numerical methods for simulating fluidized-bed reactors, i.e. TFM and CFD–DPM, are briefly described in this chapter. Both approaches have advantages and disadvantages. For example, the TFM provides good speed for gas–solid flow modelling, with reasonable accuracy, and it is still the workhorse for large-scale industrial applications. However, it is difficult to account accurately for the particle size distribution and other particle-scale physics. The CFD–DPM provides superior model accuracy and flexibility in many respects, but it is expensive in terms of computational cost. So it is limited to small-scale systems.

In spite of the continuous development in theory and numerous successes in applications, CFD modelling of gas–solid flow still faces great challenges because of the inherent multiscale flow characteristics. No single model is capable of capturing all the physics and features of large-scale systems. It is believed that a multiscale modelling approach should be developed for the multiscale flow problems encountered in fluidized-bed reactors. By coupling the models at different scales with effective communication among models, multiscale models should be able to address the challenges in multiphase flow modelling without too much compromise between accuracy and speed. In addition, considerable improvements in both accuracy and speed have been achieved for each type of model leading to increased overlaps among models. This provides a great opportunity for cross-verification and validation to further enhance the fidelity of multiscale modelling. To achieve this goal, a large number of reliable and high-quality experimental data are needed for model development, validation, and uncertainty quantification.

Solved Problem

6.1 Please describe the general steps of setting up a CFD simulation of fluidized-bed reactor.

Solution

The typical steps of conducting a CFD simulation for a fluidized-bed reactor includes:

1. **Problem definition**: To define scope of the simulation, quantity of interest, etc.

2. **Computational domain**: To build/import the geometry to be included in the computational domain for simulating
3. **BC and IC**: To establish proper boundary condition (BC) and initial condition (IC) to best represent the real system
4. **Mesh generation**: To generate high-quality mesh with proper resolution to be used in the simulation
5. **Model selection**: To select the proper modelling approach and sub-models for the simulation
6. **Execution of simulation**: To run the simulation until a converged solution is obtained
7. **Post-processing and analysis**: To process the numerical results and conduct necessary analysis to extract information of interest.

Notations

Symbols

β	drag coefficient
$\mathbf{c}$	velocity
C_p	heat capacity
d	particle diameter
D	diffusivity
ε	volume fraction
e	restitution coefficient
$\mathbf{f}$	force
$\mathbf{F}$	force
$\mathbf{g}$	gravitational acceleration
g_0	radial distribution function
I	moment of inertia
k	thermal conductivity
m	mass
Nu	Nusselt number
μ	viscosity
$\boldsymbol{\omega}$	angular velocity
P	pressure
Pr	Prandtl number
$\mathbf{q}$	heat flux
Q	heat
ρ	density
r	mass transfer rate
R	radius
Re	Reynolds number
$\boldsymbol{\sigma}$	stress tensor
Θ	granular temperature
t	time
T	temperature
$\mathbf{T}$	torque

$\mathbf{u}$	velocity
V	volume
$\mathbf{x}$	position
X	mass fraction

Subscripts

c	computational cell
$fric$	frictional
g	gas
i	individual particle
max	maximum
p	particle
s	solid phase
vis	viscous

References

1 Gidaspow, D. (1994). *Multiphase Flow and Fluidization: Continuum and Kinetic Theory Descriptions*. New York: Academic Press.

2 Savage, S.B. (1983). Granular flows down rough inclines-review and extension. *Stud. Appl. Mech.* 7: 261–282.

3 Schaeffer, D.G. (1987). Instability in the evolution equations describing incompressible granular flow. *J. Differ. Equ.* 66 (1): 19–50.

4 Johnson, P.C., Nott, P., and Jackson, R. (1990). Frictional–collisional equations of motion for participate flows and their application to chutes. *J. Fluid Mech.* 210: 501–535.

5 Tsuji, Y., Kawaguchi, T., and Tanaka, T. (1993). Discrete particle simulation of two-dimensional fluidized bed. *Powder Technol.* 77 (1): 79–87.

6 Hoomans, B.P.B., Kuipers, J.A.M., Briels, W.J., and van Swaaij, W.P.M. (1996). Discrete particle simulation of bubble and slug formation in a two-dimensional gas-fluidized bed: a hard-sphere approach. *Chem. Eng. Sci.* 51 (1): 99–118.

7 Cundall, P.A. and Strack, O.D.L. (1979). A discrete numerical model for granular assemblies. *Géotechnique* 29 (1): 47–65.

8 Andrews, M.J. and O'rourke, P.J. (1996). The multiphase particle-in-cell (MP-PIC) method for dense particulate flows. *Int. J. Multiphase Flow* 22 (2): 379–402.

9 Gao, X., Li, T., Sarkar, A. et al. (2018). Development and validation of an enhanced filtered drag model for simulating gas-solid fluidization of Geldart A particles in all flow regimes. *Chem. Eng. Sci.* 184: 33–51.

10 Tenneti, S., Garg, R., and Subramaniam, S. (2011). Drag law for monodisperse gas–solid systems using particle-resolved direct numerical simulation of flow past fixed assemblies of spheres. *Int. J. Multiphase Flow* 37 (9): 1072–1092.

11 Beetstra, R., van der Hoef, M.A., and Kuipers, J.A.M. (2007). Drag force of intermediate Reynolds number flow past mono- and bidisperse arrays of spheres. *AIChE J* 53 (2): 489–501.

12 Rong, L.W., Dong, K.J., and Yu, A.B. (2014). Lattice-Boltzmann simulation of fluid flow through packed beds of spheres: effect of particle size distribution. *Chem. Eng. Sci.* 116: 508–523.

13 Wang, W. and Li, J. (2007). Simulation of gas–solid two-phase flow by a multi-scale CFD approach – of the EMMS model to the sub-grid level. *Chem. Eng. Sci.* 62 (1–2): 208–231.

14 Igci, Y., Andrews, A.T. IV, Sundaresan, S. et al. (2008). Filtered two-fluid models for fluidized gas-particle suspensions. *AIChE J* 54 (6): 1431–1448.

15 Gunn, D.J. (1978). Transfer of heat or mass to particles in fixed and fluidised beds. *Int. J. Heat Mass Transf.* 21 (4): 467–476.

16 Sun, B., Tenneti, S., and Subramaniam, S. (2015). Modeling average gas–solid heat transfer using particle-resolved direct numerical simulation. *Int. J. Heat Mass Transf.* 86: 898–913.

17 Li, T., Grace, J., and Bi, X. (2010). Study of wall boundary condition in numerical simulations of bubbling fluidized beds. *Powder Technol.* 203: 447–457.

18 Johnson, P.C. and Jackson, R. (1987). Frictional-collisional constitutive relations for granular materials, with application to plane shearing. *J. Fluid Mech.* 176: 67–93.

19 Li, T., Grace, J.R., and Bi, H. (2010). Numerical investigation of gas mixing in gas-fluidized beds. *AIChE J* 56 (9): 2280–2296.

20 Kim, S.W., Ahn, J.Y., Kim, S.D., and Lee, D.H. (2003). Heat transfer and bubble characteristics in a fluidized bed with immersed horizontal tube bundle. *Int. J. Heat Mass Transf.* 46 (3): 399–409.

21 Li, T., Dietiker, J.F., Zhang, Y., and Shahnam, M. (2011). Cartesian grid simulations of bubbling fluidized beds with a horizontal tube bundle. *Chem. Eng. Sci.* 66 (23): 6220–6231.

22 Xu, Y., Musser, J., Li, T. et al. (2018). Numerical simulation and experimental study of the gas-solid flow behavior inside a full-loop circulating fluidized bed: evaluation of different drag models. *Ind. Eng. Chem. Res.* 57 (2): 740–750.

7

Hydrodynamics of Bubbling Fluidization

John R. Grace

University of British Columbia, Department of Chemical and Biological Engineering, 2360 East Mall, Vancouver, Canada V6T 1Z3

7.1 Introduction

As discussed in Chapter 4, the bubbling flow regime begins at a superficial velocity of U_{mb} (called the minimum bubbling velocity). For Geldart groups B and D particles, this is equal to the minimum fluidization velocity, U_{mf}, but for group A powders, by definition, $U_{mb} > U_{mf}$. The bubbling flow regime then extends to either the slug flow regime for columns of relatively small cross-sectional area (Chapter 8), or to the turbulent fluidization flow regime (Chapter 9). Even if the vessel containing a fluidized bed is opaque, an observer can usually recognize that a fluidized bed is operating in the bubbling bed flow regime from dynamic pressure signals that are relatively large in amplitude, are irregular (not periodic) in nature, and have frequencies of order 1–5 Hz.

The bubbles in gas-fluidized beds are also sometimes called "voids." They have been studied extensively in two-dimensional (thin) columns (see Chapter 5), as well as by using non-invasive techniques like X-ray photography [1], magnetic resonance imaging (MRI) [2], electrical capacitance tomography [3], and X-ray tomography [4]. Images of isolated bubbles in fluidized beds of different particulate materials, as viewed by X-rays [1, 5], are shown in Figure 7.1. All of these bubbles are of spherical-cap shape, but it is seen that the wake angles differ significantly for the different particulate media in which the bubbles were rising. As bubbles rise through fluidized beds, they grow, mostly as a result of coalescence with other bubbles but also, to a small extent, due to the reduction in hydrostatic pressure with increasing height. Bubble shapes change as bubbles interact with each other or with internal tubes or baffles. For example, the shapes of a pair of vertically aligned bubbles in a late stage of coalescing are shown in Figure 7.2. Bubbles may also undergo splitting, which involves disturbances at the top of the bubble growing to appear like a descending knife as they also move around towards the equator [1], slicing off smaller bubbles, as portrayed in Figure 7.3.

Within the bubbling flow regime (see Chapter 4), rising bubbles are the dominant feature, both visually and in terms of determination of many of the

Essentials of Fluidization Technology, First Edition. Edited by John R. Grace, Xiaotao Bi, and Naoko Ellis.

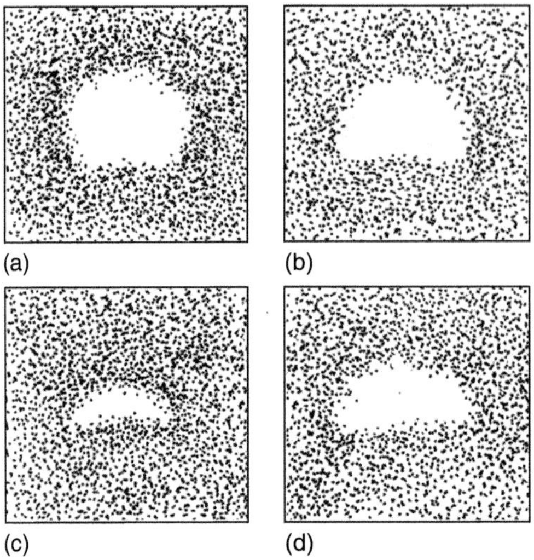

Figure 7.1 Shapes of isolated bubbles, viewed by X-rays [1], rising in fully three-dimensional fluidized bed of (a) crushed coal, (b) magnesite, (c) synclyst catalyst, and (d) acrylic granules. Source: From Rowe and Partridge 1965 [1].

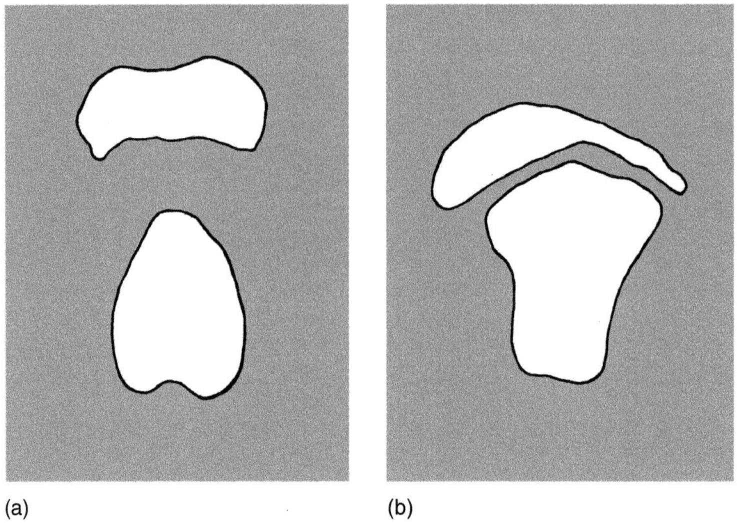

Figure 7.2 Tracing of images of a pair of vertically aligned bubbles in late stages of coalescing in a two-dimensional (thin) fluidized bed of 300 μm glass beads. (a) Front bubble widens, while back one elongates; (b) both bubbles are substantially distorted just before the coalescence. Source: From Clift and Grace 1970 [6].

key properties of a given fluidized bed. The affected properties considered in this chapter include bed expansion, gas mixing, solids segregation and mixing, and gas–solid contacting. An understanding of the bubble properties and their influence on other features also helps to explain other aspects of the bubbling flow regime, such as heat and mass transfer (Chapter 14) and jet penetration (Chapter 19). This chapter summarizes knowledge of the properties of bubbling beds gained from experimental and theoretical (including Computational Fluid

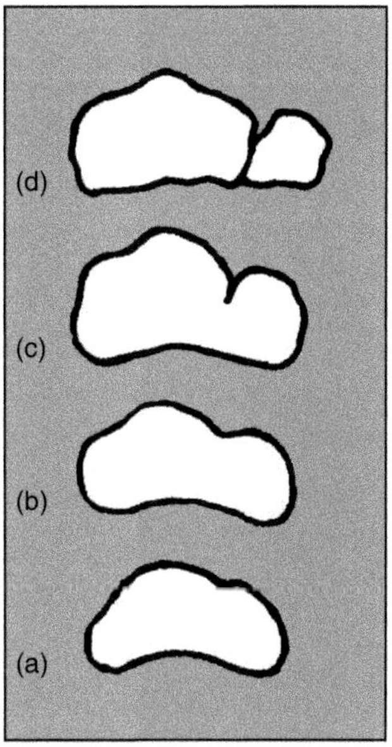

Figure 7.3 Tracings from photographs of a bubble splitting into two smaller bubbles as a result of growth of instability on the roof in a three-dimensional fluidized bed viewed by X-rays. The frames are at times (a) 0.0 seconds; (b) 0.04 seconds; (c) 0.08 seconds; (d) 0.12 seconds. Source: From Rowe and Partridge 1965 [1].

Dynamics (CFD) studies). We consider only cases where the volume-equivalent diameters of the bubbles are much smaller than the diameter or width of the column. Cases where bubbles become "slugs" with volume-equivalent diameters comparable with, or larger than, the vessel cross-sectional dimension are considered in Chapter 8.

7.2 Why Bubbles Form

Bubbles form in gas-fluidized beds due to the instability of the state of uniform fluidization. This situation is analogous to the development of turbulence in single-phase (i.e. gas or liquid) flow in pipes, in which smooth laminar flow satisfies the governing equations and boundary conditions for steady fully developed flow at any Reynolds number, but, in practice, turbulent flow is observed beyond a critical Reynolds number because smooth laminar flow becomes unstable to perturbations. In the case of pipe flow, the instabilities grow in amplitude, leading to turbulent structures such as eddies. In fluidized beds growth of instabilities in an expanded homogeneously fluidized bed grows in amplitude, leading to production of bubbles [7]. This growth of disturbances is normally very rapid in gas-fluidized beds, so for the bubbling fluidization flow regime, an observer finds bubbles immediately above the distributor. On the other hand, for liquid-fluidized beds (Chapter 3), the disturbances grow much more slowly (over a considerable height above the distributor) in response to inevitable perturbations introduced by pumps or blowers, vibrations, and distributors, so an observer finds much more homogeneous flow patterns in practice.

7.3 Analogy Between Bubbles in Fluidized Beds and Bubbles in Liquids

Bubbles in liquids are nearly spherical if small enough (e.g. volume-equivalent diameter, d_e, $\lesssim$0.8 mm for air bubbles in water) because of the dominant roles of surface tension and, to a lesser extent, viscous forces. As they become larger, bubbles in liquids dilate in a periodic manner and become distorted (approximately ellipsoidal in time-mean shape) as inertial effects lessen the relative importance of the surface tension and viscous forces. When they are large enough (e.g. $d_e \gtrsim$ 17 mm for bubbles in water), bubbles in liquids adopt spherical-cap shapes [8].

There is then a striking similarity between bubbles observed in gas-fluidized beds and spherical-cap bubbles in liquids (e.g. in gas–liquid contactors called bubble columns) [9]. This occurs because the dense phase in fluidized beds, made up of particles and their interstitial gas, acts like a viscous liquid of negligible surface tension and density $\approx\rho_p\,(1-\varepsilon_{mf})$. Although the dense phase of a bubbling fluidized bed is undoubtedly somewhat non-Newtonian, to a first approximation it can treated like a Newtonian liquid of effective viscosity 0.4–1.3 Pa s, with this value increasing with increasing particle size and as the particles become more angular (i.e. decreasing sphericity in Chapter 2) [10].

The analogy between bubbles in fluidized beds and large bubbles in relatively viscous (apparent viscosity of order 1 Pa s) Newtonian liquids then applies. Their analogous features include the following:

- Bubbles are of spherical-cap shape in both cases.
- Time-average rising velocity of single bubbles with negligible wall effects is given in both cases by

$$u_b = 0.67\sqrt{ga_b} \tag{7.1}$$

  where g is the acceleration of gravity and a_b is the radius of curvature of the bubble at its nose (top).
- Dense phase or liquid travels tangentially around the bubble upper surface, away from the nose, as could be observed by an observer travelling with the bubble.
- Wake structure and circulation patterns of dense phase in the wake are similar.
- Displacement of particles or liquid due to similar "drift" trajectories [11] occurs in both types of system.
- Coalescence behaviour, with bubbles drawn in behind and overtaking bubbles rising ahead of them, is strikingly similar. In both cases, the overtaking bubble elongates significantly as it catches up to the leading one, while the leading bubble slightly broadens, but its shape does not change very much.
- Bubble breakup mechanisms are similar, with splitting originating due to downward growth of disturbances at the upper surface of the bubble, while the top of each of these disturbances is carried around towards the equator.

Note that the analogy does not apply to the gas motion, with the gas circulating almost entirely (except for diffusional mass transfer) within the bubble in a liquid, whereas the gas percolates through the surface of the particulate phase, which forms the top of bubbles in fluidized beds. In fact, this percolation (often referred to as "throughflow") is critical to stabilizing the roof of bubbles in fluidized beds [12]. This same gas throughflow also contributes in an important manner to interphase (bubble-to-dense-phase) mass transfer in fluidized beds, whereas the throughflow mechanism is not available for bubbles in liquids. In addition, surface tension, an important property in gas–liquid systems, is not present in gas-fluidized beds. Moreover, interparticle capillary forces when gas–solid systems encounter small volumes of liquid [13] are not encountered in gas–liquid systems.

7.4 Hydrodynamic Properties of Individual Bubbles

7.4.1 Rising Velocity of Single Bubbles

In Eq. (7.1) the rising velocity of a single bubble is written in terms of the radius of curvature at the bubble nose. However, it is usually more convenient to express the bubble size in terms of its volume or the diameter of a sphere with the same volume as the bubble:

$$d_e = \left(\frac{6V_b}{\pi}\right)^{1/3} \tag{7.2}$$

For single bubbles with negligible wall effects, like that in Figure 7.1, the included wake angle, θ_w, (see Figure 7.1) is typically about 50°, and the bubble rising velocity can then be related to the bubble volume or volume-equivalent diameter by

$$u_b = 0.792 g^{0.5} V_b^{1/6} = 0.711\sqrt{g d_e} \tag{7.3}$$

Experimental rise velocities are, on average, in reasonable agreement with Eqs. (7.1) and (7.3) (e.g. [14]) but show considerable scatter, as well as periodic fluctuations related to shape dilations. However, different constants have been fitted by different authors. For example, Rowe and Everett [15] fitted constants from 0.7 to 1.0 instead of 0.711 above in a study of transition from two-dimensional to three-dimensional bubbles, and Huang et al. [16] found this constant to lie between 0.2 and 0.6 for bubbles in supercritical water.

7.4.2 Bubble Wakes

The wake region behind a rising bubble plays a very important role with respect to the mixing, coalescence, and entrainment of particles in bubbling fluidized beds. As a first approximation, the wakes of small bubbles in fluidized beds can be assumed to complete the sphere of which the spherical-cap bubble occupies the front portion. Figure 7.4 shows the wake of a two-dimensional bubble after a bubble in a two-dimensional bed has passed through an interface between darker particles (below) and lighter ones above. Based on bubbles in fluidized beds observed by X-rays [1, 10], the wake fraction, defined as

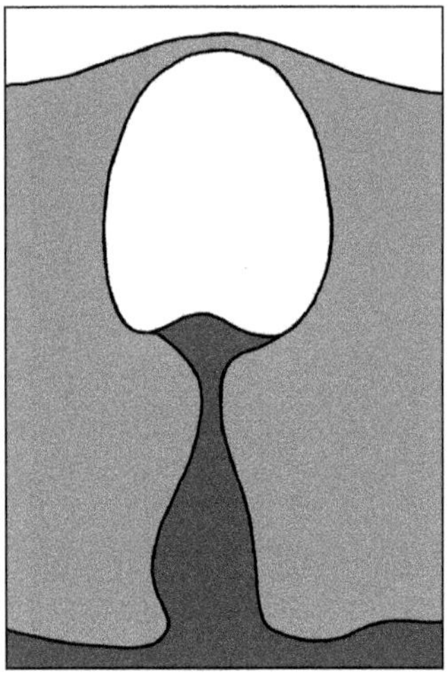

Figure 7.4 Bubble rising in a thin "two-dimensional" fluidized bed showing wake at the bottom and particles drawn upwards from below [1]. The interface between the black and grey particles was originally horizontal before the bubble was injected and passed through the interface. Source: From Rowe and Partridge 1965 [1].

the wake volume divided by the bubble volume, is plotted against the mean particle diameter in Figure 7.5 for four different types of particle. It is clear that smooth spherical particles (glass beads) result in larger wake fractions, smaller wake angles, and lower effective dense-phase viscosities than for angular sand and magnesite particles, with rounded catalyst particles showing intermediate values. It is also seen that particles of larger diameter result in larger apparent

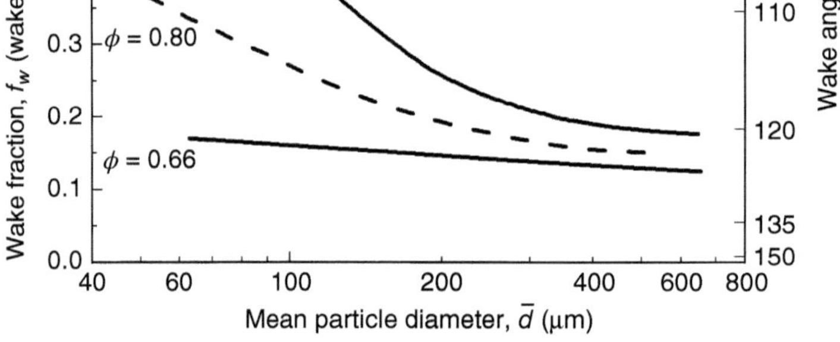

Figure 7.5 Wake volume fraction and wake angle as functions of particle size and shape, based on X-ray photographs [1]. These data are useful in estimating dense phase viscosities based on analogy with spherical-cap bubbles in liquids [10]. Source: From Rowe 1970 [10] and Rowe and Partridge 1965 [1].

viscosities, smaller wake fractions, and larger wake angles. As bubbles become larger, for example, in columns of greater diameter, wakes become "open," [17] with periodic shedding of "packets" of particles that affect the particle mixing behaviour and transition to the turbulent fluidization flow regime (see Chapter 4). Periodic shedding also results in secondary motion of the bubbles, leading to periodic low-amplitude shape dilations and fluctuations of bubble rise velocity at frequencies of order 1 Hz.

Since the wakes travel with the rising bubbles, they provide an important mechanism for vertical displacement (and mixing) of particles in bubbling beds [18]. However, periodic wake shedding can occur, causing some wake particles to be released along the way as bubbles rise, rather than being transported all the way to the bed surface. As bubbles reach the upper bed surface, some wake particles are ejected upwards into the freeboard, providing an important component of the mechanism of particle entrainment (see Chapter 10).

In addition to the movement of particles induced by bubble wakes, particles are vertically displaced by "drift" [11], where particles undergo looping motion, upward and then downward, as bubbles pass, ending up at different vertical levels, depending on their initial location relative to the vertical axis of the bubble. Wake transport and drift are the two most important mechanisms of solids mixing in bubbling fluidized beds, discussed in Section 7.6.7.

7.4.3 Bubble Breakup and Maximum Stable Size

Bubble splitting occurs by a process analogous to Rayleigh–Taylor instability in gas–liquid two-phase systems [19]. In this process, the upper surface of a bubble (overlying a void space) is unstable when perturbed, so a disturbance begins to grow downwards, as the position on the roof travels tangentially around towards the equator (see Figure 7.3). The growth process soon becomes nonlinear, making predictions difficult. Bubble splitting, together with coalescence, plays a major role in determining the bubble size distribution in bubbling beds.

In practice, bubbles in fluidized beds are believed to be limited by a maximum stable bubble size. Probably because the effective dense-phase viscosity is, as discussed above, greater for large particles than for small ones, causing instabilities to grow more slowly, bubbles grow to larger sizes in beds of larger particles. An empirical procedure, introduced by Geldart [20], to predict the maximum stable bubble diameter for given particles, is described in Eq. (4.6).

7.4.4 Interphase Mass Transfer and Cloud Formation

Exchange of elements of gas between bubbles and the surrounding dense phase occurs by two major mechanisms:

(a) The gas throughflow, discussed in Section 7.3, results in gas flow into the bottom of the bubble and out the top, at a superficial velocity of order U_{mf}.
(b) Molecular diffusion of the transferring gaseous component occurs.

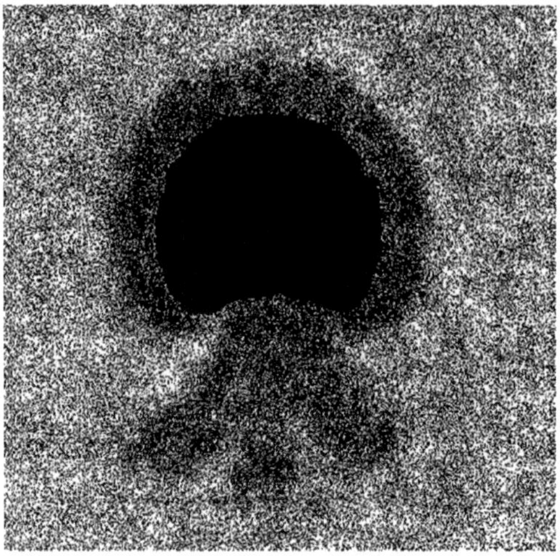

Figure 7.6 Nitric oxide bubble rising in a two-dimensional fluidized bed showing the surrounding cloud for $\alpha = 2.4$. Source: Adapted from Rowe et al. 1964 [23].

These two components are accounted for in interphase mass transfer equations. With some allowance for augmentation of the throughflow component due to bubble shape change accompanying bubble interaction and coalescence, the recommended interphase mass transfer coefficient equation [21] is

$$k_{bd} = \frac{U_{mf}}{3} + \sqrt{\frac{4D_{AB}\varepsilon_{mf}\overline{u}_b}{\pi\overline{d}_b}} \tag{7.4}$$

where D_{AB} is the binary molecular diffusivity of component A, the diffusing species in gas B.

When throughflow gas emerges from the top surface of bubbles in fluidized beds, it encounters particles sliding tangentially around the upper bubble surface from the nose towards the equator and beyond. These particles exert a drag force on the interstitial gas, tending to carry the gas around the top of the bubble. If the ratio $\alpha = u_b\varepsilon_{mf}/U_{mf} > 1$, the gas recirculates in a region called the "cloud" that surrounds the bubble. Clouds were first predicted by Davidson [22] and then observed experimentally by Rowe et al. [23] by injecting NO_2 (brown) gas bubbles into a bed fluidized by air. The tracing in Figure 7.6 from a photograph published by Rowe et al. [23] of an NO_2 bubble rising in a two-dimensional bed clearly shows the cloud region travelling with a rising bubble.

For large α, the cloud becomes thin enough that it is of the same order or less than the particle diameter. For $\alpha < 1$, no cloud forms, but gas tends to short circuit into the base of and out the top of gas bubbles. Clouds play a major role in some fluidized bed reactor models, e.g. the well-known model of Kunii and Levenspiel [24] (see Chapter 15). The ratio of the cloud volume to the volume of the bubble with which the cloud is associated depends on α, as well as the modelling assumptions.

7.5 Bubble Interactions and Coalescence

Bubbles coalescence in fluidized beds results in growth of bubbles with increasing height. Coalescence is associated with an increase in the vertical component of bubble velocity and hence a reduction of bed expansion and shorter residence time of gas elements in the bed, while also influencing the extent of particle mixing and solids entrainment. To a good approximation, the trajectories of interacting bubbles can be approximated by the vector sum of two components: (i) the velocity of each bubble if it were rising in isolation from other bubbles, and (ii) the velocity at the position of the bubble nose caused by all the other bubbles present if the bubble in question were not present [6], with each bubble represented by a doublet and particle velocities determined by potential flow equations. Consistent with observations of coalescing bubbles in two- and three-dimensional beds, this simple coalescence model results in bubbles being drawn in behind and then overtaking bubbles whose noses are further ahead. The coalescence model is also consistent with observations that the shapes and velocities of chasing bubbles are affected much more during interactions and coalescence than the shapes and velocities of leading bubbles being overtaken from behind. Figure 7.2 demonstrates the outlines of a pair of bubbles in the late stages of coalescence, showing substantial elongation of the trailing bubble while the leading bubble underwent some broadening.

When a trailing bubble overtakes one ahead just as the leading one reaches the bed surface, particles travelling at high speed in the wake of the overtaking (back) bubble can punch through at high velocity into the freeboard, causing particularly energetic eruptions of particles at the bed surface, which then lead to unusually high particle rise in the freeboard region, increasing both the transport disengagement height (TDH) and particle entrainment (see Chapter 10).

The porous nature of the dense phase in fluidized beds means that gas can, under some circumstances, flow from one bubble to a neighbouring bubble without coalescence occurring as predicted by Clift et al. [25] and observed with NO_2 (brown gas) bubbles by Geldart [20].

7.6 Freely Bubbling Beds

7.6.1 Flow of Gas by Translation of Bubbles

In freely bubbling fluidized beds, bubbles are generated at the gas distributor, coalesce with each other as they rise, split in some cases, carry particles upwards, and erupt at the upper beds surface, ejecting particles into the freeboard region. Figure 7.7 shows bubbles in a thin "two-dimensional" fluidized beds at three different superficial gas velocities based on photographs by Zenz and Othmer [26]. From these images one can see that the average bubble diameter increases with height. However, for Geldart group A particles, the average bubble diameter is likely to approach an asymptotic value as coalescence and splitting reach a

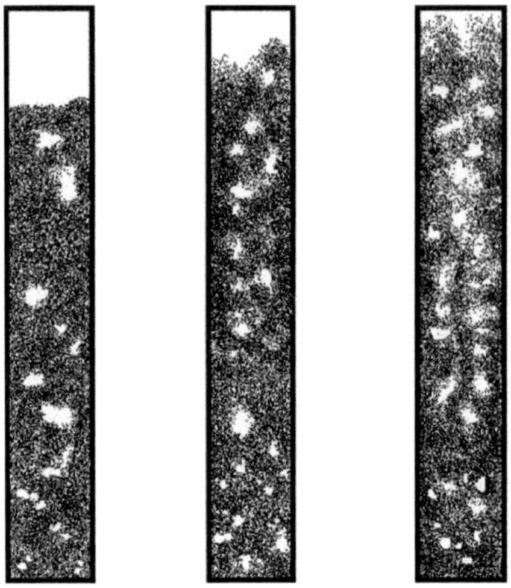

(a) (b) (c)

Figure 7.7 View of bubbles rising in a two-dimensional (thin) fluidized bed of Fluid Catalytic Cracking (FCC) catalyst at three superficial gas velocities: (a) 0.3 m/s, (b) 0.6 m/s, and (c) 0.9 m/s. Source: Adapted from Zenz and Othmer 1960 [26].

dynamic equilibrium, whereas net bubble growth continues for group B and D particles.

Toomey and Johnstone [27] proposed a simple idea that, while not precisely accurate [28], is commonly applied in the form of two reasonable approximations, which together are widely referred to as "the two-phase theory of fluidization":

1. A bubbling fluidized bed is composed of two phases, a bubble phase with a voidage close to 1.0 and a dense phase whose voidage remains ε_{mf}.
2. The flow translated as bubble volumes rising in the bed, G_b, is equal to the total volumetric flow through the bed minus the flow through the bed at minimum fluidization, i.e.

$$G_b = (U - U_{mf})A \tag{7.5a}$$

This equation is commonly adopted as an approximation for Geldart group A particles. However, if, instead, one introduces a correction factor, Y, so that

$$G_b = Y(U - U_{mf})A \tag{7.5b}$$

then it is typically found that $Y \approx 0.8$–1.0 for Geldart group A particles, 0.6–0.8 for group B particles, and 0.25–0.6 for group D solids.

Note also that there is evidence (e.g. [29]) that the voidage immediately surrounding bubbles in gas-fluidized beds exceeds ε_{mf} resulting in an increase in overall bed permeability and hence accounting for some of the gas flow not carried by bubbles.

7.6.2 Mean Bubble Diameter as a Function of Height and Gas Velocity

Several empirical correlations and semi-empirical models are available for the prediction of mean bubble diameter as a function of height. Of these, two are widely regarded as superior to others. These two are of roughly equal accuracy, so it is recommended that both be utilized and the predictions averaged:

(a) Model of Mori and Wen [30]

$$d_b(z) = d_{bm} - (d_{bm} - d_{b0})e^{-0.3z/D} \tag{7.6}$$

Here d_{bm} would be the bubble diameter approached if unrestrained coalescence occurred in a tall column of the same diameter:

$$d_{bm} = 2.59\left[\frac{(U - U_{mf})A}{g^{0.5}}\right]^{0.4} \tag{7.7}$$

whereas d_{b0} is the diameter of bubbles formed at the distributor, predicted for multi-orifice plates or other distributors containing N_{or} discrete gas entry points by

$$d_{b0} = 1.38\left[\frac{(U - U_{mf})A}{g^{0.5}N_{or}}\right]^{0.4} \tag{7.8}$$

In practice, Eq. (7.6) typically predicts that the bubble diameter reaches a maximum at a vessel diameter, D, of about 1.2 m. Since this does not appear to happen, D and the corresponding cross-sectional bed area, $A = \pi D^2/4$, should be based on $D = 1.2$ m for all $D \geq 1.2$ m. For columns of non-circular cross section, D in Eq. (7.6) is replaced by the hydraulic diameter, $D_H = 4A/$(perimeter of fixed surface exposed to particles). The Mori and Wen [30] model was later extended by Horio and Nonaka [31] to allow for both vertical jets at the gas distributor connected to forming bubbles and for bubble splitting, as well as bubble coalescence.

(b) Correlation of Darton et al. [32], based on a model of lateral coalescence of vertical chains of bubbles:

$$d_b(z) = 0.54(U - U_{mf})^{0.4}\left[z + 4\sqrt{\frac{A}{N_{or}}}\right]^{0.8} g^{-0.2} \tag{7.9}$$

While these two approaches are mechanistic and appear to offer the most rational choices for predicting bubble sizes, a comparison of alternate correlations for predicting bubble size [33] found that the empirical correlation of Cai et al. [3]:

$$d_b(z) = 0.138z^{0.8}(U - U_{mf})^{0.42}\exp\{-2.5\times 10^{-5}(U - U_{mf})^2 - 10^{-3}(U - U_{mf})\} \tag{7.10}$$

provided the best overall predictions, especially for Geldart A and D particles. For group B particles, the correlation of Choi et al. [34] and that of Mori and Wen [30] featured above were most accurate. Empirical correlations (e.g. [35]) continue to be produced claiming improved accuracy of their predictions.

The above correlations are all for fluidized beds without any baffles or interior fixed surfaces. When such surfaces are present, they tend to cause bubbles splitting and to shield bubbles from coalescence. Both of these mechanisms result in smaller bubbles, with the shape, location, and spacing of the immersed fixed surfaces then playing an important role in determining the bubble size and spatial distribution in the bed. A rational approach is to assume that bubbles will grow in the usual manner (i.e. as predicted by Eqs. (7.6), (7.9), or (7.10)) below and above any tube bank or region containing fixed surfaces. Within such a region, bubbles will be limited in size, e.g. to the inter-tube spacing within banks of horizontal tubes. See Chapter 18 for further discussion of the influence of internal surfaces on bubbles, as well as consideration of the forces exerted on fixed surfaces due to the passage of bubbles.

7.6.3 Rising Velocity of Bubbles in Freely Bubbling Bed

The characteristic rising velocity of bubbles in a freely bubbling gas-fluidized bed is estimated by applying the relationship for a single bubble from Eq. (7.3) and adding on a term $(U - U_{mf})$ so that

$$u_b(z) = 0.711\sqrt{gd_b(z)} + (U - U_{mf}) \tag{7.11}$$

There has been debate about the origin and correctness of the $(U - U_{mf})$ term [36]. It can be regarded as an approximate correction for bubbles being accelerated as a result of bubble interactions and coalescence [37].

Alternatively, a number of empirical correlations have been proposed, with a comparative evaluation [33] favouring a simple equation of Werther [38]:

$$u_b(z) = k\sqrt{gd_b(z)} \tag{7.12}$$

with $k = 2.5$ and 1.8, for Geldart A and B particles, respectively, for the column diameter, D, $\geq$100 mm.

7.6.4 Bubble Volume Fraction (Holdup)

The bubble volume fraction varies with height according to

$$\varepsilon_b = \left(\frac{G_b}{A}\right)/u_b(z) \approx (U - U_{mf})/u_b(z) \tag{7.13}$$

As recommended by Fryer and Potter [39], an average value, $(\varepsilon_b)_{av}$, over the expanded bed depth is evaluated at a height of $z = 0.4H$.

7.6.5 Bed Expansion

Given the two-phase assumption above that the dense phase retains a constant void fraction equal to ε_{mf}, the bubble volume fraction (often referred to as holdup) is

$$(\varepsilon_b)_{av} = \frac{H - H_{mf}}{H} \tag{7.14}$$

so that the expanded bed depth is given by

$$H = H_{mf}/[1 - (\varepsilon_b)_{av}] \tag{7.15}$$

Combining the above relationships leads to

$$H = H_{mf} \times \frac{u_b(0.4H)}{u_b(0.4H) - (U - U_{mf})} \tag{7.16}$$

Equation (7.15) must be solved iteratively: guess an expanded bed height, H. Then evaluate a better estimate of H based on Eqs. (7.9), (7.11), and (7.16). Repeat until H converges.

7.6.6 Radial Nonuniformity of Bubbles and Its Effect on Mixing

As bubbles grow by coalescence in freely bubbling beds, the spatial distribution of bubbles becomes nonuniform, even if the distributor delivers bubbles uniformly at the bottom of the bed. This is related to the constraint imposed by the column wall on lateral coalescence causing a deficit of bubbles adjacent to the wall at the bottom, with a surplus of bubbles a short distance further into the bed [40–42]. As the distribution evolves with height, the surplus migrates inwards, reaching the middle at a depth, z, of typically $\sim 2D$. If the bed is deep enough for slug flow to be reached (Chapter 8), then the slugs occur symmetrically in the core of the bed.

The development of the spatial distribution described above normally results in a weak up-the-outside, down-at-the-centre flow pattern of time-mean solids flow at the bottom of the bed, giving way to a stronger up-the-middle, down-the-outside "gulf-streaming" circulation pattern higher in the bed. This pattern, shown schematically in Figure 7.8, is self-reinforcing, because both wake transport and drift cause particles to move predominantly and strongly upwards in the regions where the bubbles are more concentrated. Solids mixing and gas mixing, both axial and radial, are therefore affected by the flow patterns. Since the mixing is primarily related to overall "cells," axial and radial dispersion mixing models, which instead rely on short-range random diffusion-like steps, tend to do poorly in representing both solids and gas mixing, which are instead strongly influenced by the overall circulation patterns.

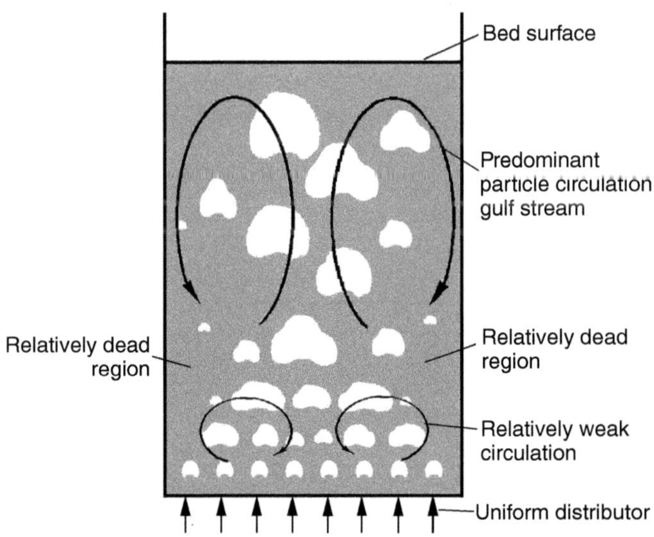

Figure 7.8 Schematic showing development of characteristic nonuniform spatial time-average bubble patterns and resultant particle circulation patterns in bubbling fluidized beds.

In deep bubbling beds ($H/D \gtrsim 4$), upsets in flow can lead to nonuniform and asymmetric flow patterns, with bypassing of gas through one sector of the cross section, sometimes shifting to another sector [43]. This tendency appears to be related to the interparticle forces that play an important role for Geldart group A particles, as increasing the proportion of fines has been found [43] to diminish the extent of jet streaming and bypassing. When channelling and large bubbles occur, they tend to cause poor reactor performance [44].

7.6.7 Turnover Time, Solids Mixing, and Particle Segregation

Solids mixing in the vertical direction is dominated by displacement of particles due to transport by bubble wakes and drift, as discussed in Section 7.4.2. If there are vertical jets at the distributor, some particles will also be entrained into the jets, providing some additional vertical mixing. A characteristic particle mixing time, often called the "turnover time," provides an estimate of the average time taken by a single representative particle, subject to transport by bubble wakes and drift, to return to its original height after being displaced and making one circuit. The solids mass flux in the vertical direction in a bubbling bed due to wake transport and drift is given by

$$J_s = \rho_p(1 - \varepsilon_{mf})(f_w + f_d) \times Y(U - U_{mf}) \tag{7.17}$$

where f_w is the wake fraction that can be estimated from Figure 7.5; f_d accounts for the drift fraction and can be taken [11] as $1.6f_w$. The turnover time can then be approximated as

$$t_{turnover} = \frac{H_{mf}}{2.6f_w Y(U - U_{mf})} \tag{7.18}$$

The turnover time, estimated in this manner, should be compared with the particle mean residence time

$$\tau = \frac{\text{Mass of particles in bed}}{\text{Mass flow rate into bed}} \tag{7.19}$$

If, as is usually the case, $\tau \gtrsim 3t_{turnover}$, it can be assumed that particles are perfectly mixed in the bed, so the fluidized bed solids residence time distribution can be represented by

$$E(t) = \frac{1}{\tau}e^{-t/\tau} \tag{7.20}$$

If the above requirement is not met, the residence time distribution can instead be fitted to a simple model, such as N identical well-stirred tanks in series:

$$E(t) = \frac{t^{N-1}}{(N-1)!\tau_i^N}e^{-t/\tau_i} \tag{7.21}$$

where $\tau_i = \tau/N$ is the mean residence time in each of the N identical compartments in series. Typically N can be taken as 2–4.

Particles are displaced horizontally by lateral bubble motion as bubbles coalesce and when particles are ejected (and scattered) by bubbles bursting at the bed surface. This horizontal motion is, however, much less than the vertical displacement by bubble wakes and drift, with the result that the extent of lateral mixing is generally about an order of magnitude less than for vertical mixing. As a result, the composition of particles and the bed temperature can usually be assumed to be uniform in the vertical direction, whereas appreciable gradients exist horizontally. Hence, if it is desirable to achieve particle residence times that approach plug flow (e.g. for dryers or gas–solid reactions), long narrow shallow fluidized beds can be adopted, with particles fed at one end and product particles removed at the other end. Alternatively, several beds can be located side by side in series, with particles flowing from one to the next. Baffles or other fixed internal surfaces that block particle motion in a given direction will decrease mixing in that direction. Gas distributors that promote internal circulation or horizontal gradients in gas velocity will affect the extent of lateral mixing.

It is common in chemical engineering to represent mixing by means of "effective dispersion coefficients," which assume an analogy between the system in question and diffusion-like mixing. Like diffusion coefficients (also called diffusivities), the resulting effective dispersion coefficients, D_{ax}, have units of m^2/s and are often expressed in dimensionless form as Peclet numbers, $Pe = HU/D_{ax}$, or as reciprocals of Peclet numbers. However, diffusion-like mixing processes involve small random mixing steps, whereas axial mixing of solids in bubbling fluidized beds, as we have seen, is dominated by large-scale displacements due to bubble wakes and drift. Given the marked difference in the nature of the mixing, axial dispersion models should be avoided. On the other hand, lateral and radial dispersion models may be suitable in view of the smaller and more random horizontal displacements of particles in bubbling fluidized beds.

The two preceding paragraphs consider mixing related to the distribution of residence time spent by different particles with the fluidized bed or beds. Also important in many cases is the extent to which particle segregation occurs, causing larger or denser (jetsam) particles to congregate in the lower regions of fluidized beds, with smaller lighter (flotsam) particles then occupying positions towards the top. Extensive work on segregation for binary solids mixtures was reported by Rowe et al. [45, 46]. An important finding is that particle segregation is much more likely to occur when particles differ in density than when they differ to the same degree in size.

7.6.8 Gas Mixing

Gas mixing in a bubbling bed is strongly affected by and related to solids mixing [47, 48]. For example, downward stick-slip motion of particles along the inner wall of the column tends to drag gas upstream, resulting in backmixing. Particles that adsorb the gas tracer on their surfaces and porous particles with large pore volumes have greater capacity for transporting gas and therefore cause more gas mixing than nonporous, non-sorbent particles. A gas that adsorbs on the particles likewise will show more dispersion than a non-adsorbing gas.

For group A and B particles, gas carried by and passing through rising bubbles tends to travel vertically faster than gas elements percolating through the dense phase, resulting in a spread of residence times in the bed and hence in greater vertical mixing. Interphase gas exchange [49] then plays a significant role in determining the overall mixing. Residence time distributions obtained from gas tracer studies reflect these mixing processes, in addition to the displacement of gas molecules due to solids circulation, as discussed in the previous paragraph, and turbulence in the gas motion in jets, bubbles, and the freeboard.

For superficial gas velocities barely in excess of U_{mb}, the residence time distribution will not differ greatly from plug flow. At high gas flow rates, there will be extensive gas mixing in bubbling beds due to the mechanisms identified above, but perfect mixing is not a very good approximation [48], although it is sometimes adopted for simplicity. The mixing can instead be modelled by adapting two-phase reactor models, like those presented in Chapter 15, to include time-derivative terms, while excluding generation (reaction) terms. Axial dispersion models are often applied for practical comparisons of large-scale industrial systems [48, 50], despite their shortcomings in terms of misrepresentation of the correct mechanisms causing mixing, as discussed above.

Horizontal gas mixing is again linked closely to lateral solids mixing. As expected, it increases with increasing superficial gas velocity and is sensitive to the design of the gas distributor [50].

7.7 Other Factors Influencing Bubbles in Gas-Fluidized Beds

Bubbles in fluidized beds can also be influenced by a number of other factors and properties not mentioned so far in this chapter. Some of these are identified in Table 7.1, with references that can serve as starting points for investigating the nature and extent of the influence of each of these factors.

Table 7.1 Other influences on bubbles in gas-fluidized beds.

Factor	References	Comment or details
Acoustic (sound) field	[51]	Promotes fluidization/bubbling of group C powder
Electrical field	[52]	Allows control of bubble size and uniformity
Electrostatics	[53], Chapter 13	Bubbles elongate and rise more quickly
Extreme gas properties	[54]	Supercritical water widens range of properties
Fines content	[55]	More fines result in smaller bubbles
Magnetic field	[56]	Stabilization delaying the onset of bubbling
Pressure increase	[57]	Greater bed expansion, smaller bubbles
Pulsations	[58]	Affect bubble size, mixing, heat transfer
Temperature increase	[59]	Complex relationship with bubbling
Vertical internal surfaces	[60], Chapter 18	Smaller bubbles rising more quickly
Vibrations	[61]	Affect bubble size, bed expansion, homogeneity
Wetness	[62]	Bubbles wider, less stable

Solved Problem

7.1 Particles having a minimum fluidization velocity of 0.020 m/s and a voidage at minimum fluidization velocity of 0.48 are being fluidized in a cylindrical column of diameter 1.3 m by air at a superficial velocity of 0.50 m/s. The expanded bed depth is 2.6 m. Assuming that the bed is operating in the bubbling flow regime, estimate the bed depth at minimum fluidization, the fraction of the bed volume occupied by bubbles, the mean bubble diameter at height 0.4H, and the corresponding bubble velocity. Compare the Mori and Wen predictions with those of the Darton correlation. Assume bubbles form at the distributor with an initial diameter of 10 mm. How do you expect ε_B, f_B, d_B, and u_B to vary with increasing height, z.

Solution

Mori and Wen:

From Eq. (7.7), d_{bm} = 1.369 m; d_b = 0.362 m at $z = H/2$; from Eq. (7.11), u_b = 1.821 m/s at $z = H/2$; $\varepsilon_B \approx (\varepsilon_b)_{av} = (0.50 - 0.02)/1.821 = 0.264$; $H_{mf} = H[1 - (\varepsilon_b)_{av}] = 1.914$ m from Eq. (7.14).

Darton et al.:

From Eq. (7.9), d_b = 0.368 m at $z = H/2$; from Eq. (7.11), u_b = 1.831 m/s at $z = H/2$; from Eq. (7.13), $\varepsilon_B \approx (\varepsilon_b)_{av} = (0.50 - 0.02)/1.831 = 0.262$; $H_{mf} = 1.919$ m from Eq. (7.14).
Note that the Mori and Wen and Darton et al. correlations give very similar predictions for this case. With increasing height, we expect that the bubble volume fraction will slightly decrease as bubbles grow in size and therefore rise more quickly. Bubble frequency will decrease sharply, whereas average bubble diameter and rising velocity will increase, due to bubble coalescence.

Notations

A	cross-sectional area of column (m^2)
a_b	radius of curvature of upper surface of bubble at the nose (m)
D	diameter of column (m)
D_{AB}	molecular diffusivity of component A in gas B (m^2/s)
D_{ax}	axial dispersion coefficient (m^2/s)
D_H	hydraulic diameter of column (m)
d_b	bubble diameter (m)
d_{bm}	bubble diameter reached in a tall column (m)
d_{bo}	diameter of bubbles formed at the distributor (m)
d_e	volume-equivalent bubble diameter (m)
$E(t)$	residence time distribution function (s^{-1})
f_b	bubble frequency (s^{-1})
f_d	drift constant (–)
f_w	wake fraction = volume of wake/volume of associated bubble (–)
G_b	volumetric bubble flow due to displacement of bubbles (m^3/s)

g	acceleration of gravity (m/s^2)
H	expanded bed depth (m)
H_{mf}	bed depth at minimum fluidization (m)
J_s	solids mass flux due to passage of bubbles (kg/m^2 s)
k	constant in Eq. (7.12) (–)
k_{bd}	bubble-to-dense-phase mass transfer coefficient (m/s)
N	number of tanks in series (–)
N_{or}	number of orifices (–)
Pe	Peclet number $= HU/D_{ax}$ (–)
t	time (s)
t_{turn}	solids turnover time (s)
U	superficial gas velocity (m/s)
U_{mb}	superficial gas velocity at minimum bubbling (m/s)
U_{mf}	superficial gas velocity at minimum fluidization (m/s)
u_b	bubble velocity (m/s)
V_b	bubble volume (m^3)
Y	fitted constant equal to true bubble flow/that predicted by two-phase theory (–)
z	axial coordinate (m)

Greek Letters

α	dimensionless bubble velocity $= u_b\varepsilon_{mf}/U_{mf}$ (–)
ε_b	fraction of bed volume occupied by bubbles (–)
ε_{mf}	bed voidage at minimum fluidization
ρ_p	particle density (kg/m^3)
τ	particle mean residence time (s)
τ_i	mean residence time of particles in the ith tank in series (s)

References

1 Rowe, P.N. and Partridge, B.A. (1965). An X-ray study of bubbles in fluidized beds. *Trans. Inst. Chem. Eng.* 43: 157–175.

2 Müller, C.R., Holland, D.J., Davidson, J.F. et al. (2007). Rapid two-dimensional imaging of bubbles and slugs in a three-dimensional gas-solid, two-phase flow system using ultrafast magnetic resonance. *Phys. Rev. E* 75: 020302.

3 Cai, P., Schiavetti, M., Michele, G.D., and Grazzini, G.C. (1994). Quantitative estimation of bubble size in PFBC. *Powder Technol.* 80: 99–109.

4 Mudde, R.F. (2010). Double X-ray tomography of a bubbling fluidized bed. *Ind. Eng. Chem. Res.* 49: 5061–5065.

5 Rowe, P.N. (1971). Experimental properties of bubbles, Chapter 4. In: *Fluidization* (eds. J.F. Davidson and D. Harrison), 121–191. London: Academic.

6 Clift, R. and Grace, J.R. (1970). Bubble interaction in fluidized beds. *Chem. Eng. Progr. Symp. Ser.* 66 (105): 14–27.

7 Jackson, R. (1985). Hydrodynamic stability of fluid-particle systems, Chapter 2. In: *Fluidization*, 2e (eds. J.F. Davidson, R. Clift and D. Harrison), 47–72. London: Academic Press.
8 Clift, R., Grace, J.R., and Weber, M.E. (1978). *Bubbles, Drops and Particles*. New York: Academic Press.
9 Abbasi, M., Grace, J.R., Sotude-Gharebagh, R. et al. (2015). Numerical comparison of gas-liquid bubble columns and gas-solid fluidized beds. *Can. J. Chem. Eng.* 93: 1838–1848.
10 Grace, J.R. (1970). The viscosity of fluidized beds. *Can. J. Chem. Eng.* 48: 30–33.
11 Baeyans, J. and Geldart, D. (1973). Particle mixing in a gas fluidized bed. In: *La Fluidization et ses Applications*, Cepaduese (ed. H. Angelino), 182–195. Toulouse: Société de Chimie Industrielle.
12 Rowe, P.N. (1964). A note on the motion of a bubble rising through a fluidized bed. *Chem. Eng. Sci.* 19: 75–77.
13 Makkawi, Y.T. and Wright, P.C. (2004). Tomographic analysis of dry and semi-wet bed fluidization: the effect of small liquid loading and particle size on the bubbling behavior. *Chem. Eng. Sci.* 59: 201–213.
14 Boyce, C.M., Penn, A., Lehnert, M. et al. (2019). Magnetic resonance imaging of single bubble injected into incipiently fluidized beds. *Chem. Eng. Sci.* 200: 147–166.
15 Rowe, P.N. and Everett, D. (1972). Fluidized bed bubbles viewed by X-rays. Part II: the transition from two to three dimensions of undisturbed bubbles. *Trans. Inst. Chem. Eng.* 50: 49–54.
16 Huang, J., Lu, Y., and Wang, H. (2019). Fluidization of particles in supercritical water: a comprehensive study on bubble hydrodynamics. *Ind. Eng. Chem. Res.* 58: 2036–2051.
17 Bi, X.T., Grace, J.R., and Lim, K.S. (1995). Transition from bubbling to turbulent fluidization. *Ind. Chem. Res. Dev.* 34: 4003–4008.
18 Rowe, P.N., Partridge, B.A., Cheney, A.G. et al. (1965). The mechanisms of solids mixing in fluidized beds. *Trans. Inst. Chem. Eng.* 43: 271–286.
19 Clift, R., Grace, J.R., and Weber, M.E. (1974). Stability of bubbles in fluidized beds. *Ind. Eng. Chem. Fundam.* 13: 45–51.
20 Geldart, D. (1977). *Gas Fluidization*, short courses. Bradford, UK: University of Bradford and Center for Professional Advancement, New Jersey.
21 Sit, S.P. and Grace, J.R. (1981). Effect of bubble interaction on interphase mass transfer in gas fluidized beds. *Chem. Eng. Sci.* 36: 327–335.
22 Davidson, J.F. (1961). Symposium on fluidization – discussion. *Trans. Inst. Chem. Eng.* 39: 230–232.
23 Rowe, P.N., Partridge, B.A., and Lyall, E. (1964). Cloud formation around bubbles in fluidized beds. *Chem. Eng. Sci.* 19: 973–985.
24 Kunii, D. and Levenspiel, O. (1991). *Fluidization Engineering*, 2e. Boston: Butterworth-Heinemann.
25 Clift, R., Grace, J.R., Cheung, L., and Do, T.H. (1972). Gas and solids motion around deformed and interacting bubbles in gas-fluidized beds. *J. Fluid Mech.* 51: 187–205.

26 Zenz, F.A. and Othmer, D.F. (1960). *Fluidization and Fluid-Particle Systems.* New York: Reinhold.
27 Toomey, R.O. and Johnstone, H.F. (1952). Gaseous fluidization of particles. *Chem. Eng. Prog.* 48: 220–226.
28 Grace, J.R. and Clift, R. (1974). On the two-phase theory of fluidization. *Chem. Eng. Sci.* 29: 327–334.
29 Almendros-Ibanez, J.A., Pallares, D., Johnsson, F., and Santana, D. (2010). Voidage distribution around bubbles in a fluidized bed: influence on through-flow. *Powder Technol.* 197: 73–82.
30 Mori, S. and Wen, C.Y. (1975). Estimation of bubble diameter in gaseous fluidized beds. *AIChE J.* 21: 109–115.
31 Horio, M. and Nonaka, A. (1987). A generalized bubble diameter correlation for gas-solid fluidized beds. *AIChE J.* 33: 1865–1872.
32 Darton, R.C., Lanauze, R.D., Davidson, J.F., and Harrison, D. (1977). Bubble growth due to coalescence in fluidized beds. *Trans. Inst. Chem. Eng.* 55: 274–280.
33 Karimpour, S. and Pugsley, T. (2011). A critical evaluation of literature correlations for predicting bubble size and velocity in gas-solid fluidized beds. *Powder Technol.* 205: 1–14.
34 Choi, J.H., Son, J.F., and Kim, S.D. (1998). Generalized model for bubble size and frequency in gas-fluidized beds. *Ind. Eng. Chem. Res.* 37: 2559–2564.
35 Agu, C.E., Pfeifer, C., Eikeland, M. et al. (2018). Models for predicting average bubble diameter and volumetric bubble flux in deep fluidized beds. *Ind. Eng. Chem. Res.* 57: 2658–2469.
36 Turner, J.C.R. (1966). On bubble flow in liquids and fluidized beds. *Chem. Eng. Sci.* 21: 971–974.
37 Grace, J.R. and Harrison, D. (1969). The behavior of freely bubbling fluidized beds. *Chem. Eng. Sci.* 24: 497–508.
38 Werther, J. (1978). Effect of gas distributor on the hydrodynamics of gas-fluidized beds. *German Chem. Eng.* 1: 166–174.
39 Fryer, C. and Potter, O. (1972). Bubble size variation in two-phase models of fluidized bed reactors. *Powder Technol.* 6: 317–322.
40 Grace, J.R. and Harrison, D. (1968). The distribution of bubbles within a gas-fluidized bed. *Inst. Chem. Eng. Symp. Ser.* 30: 105–113.
41 Werther, J. (1974a). The influence of the bed diameter on the hydrodynamics of fluidized beds. *AIChE Symp. Ser.* 70 (141): 52–62.
42 Werther, J. (1974b). Bubbles in gas fluidized beds. *Trans. Inst. Chem. Eng.* 52: 149–169.
43 Issangya, A.S., Knowlton, T.M., and Reddy Karri, S.B. (2007). Detection of gas bypassing due to jet streaming in deep fluidized beds of group A particles. In: *Fluidization XII* (eds. X. Bi, F. Berruti and T. Pugsley), 775–782. Brooklyn: Engineering Conferences International.
44 Zhang, C., Li, P.L., Lei, C. et al. (2018). Experimental study of non-uniform bubble growth in deep fluidized beds. *Chem. Eng. Sci.* 176: 515–521.
45 Rowe, P.N. and Nienow, A.W. (1976). Particle mixing and segregation in gas-fluidized beds: a review. *Powder Technol.* 15: 141–147.

46 Rowe, P.N., Nienow, A.W., and Agbim, A.J. (1972). The mechanisms by which particles segregate in gas fluidized beds. *Trans. Inst. Chem. Eng.* 50: 310–323.

47 May, W.G. (1959). Fluidized bed reactor studies. *Chem. Eng. Prog.* 555 (12): 49–56.

48 Van Deemter, J.J. (1980). Mixing patterns in large-scale fluidized beds. In: *Fluidization* (eds. J.R. Grace and J.M. Matsen), 69–89. New York: Plenum.

49 Sit, S.P. and Grace, J.R. (1981). Effect of bubble interaction on interphase mass transfer in gas fluidized beds. *Chem. Eng. Sci.* 36: 327–335.

50 Van Deemter, J.J. (1985). Mixing, Chapter 9. In: *Fluidization*, 2nd ed. (eds. J.F. Davidson, R. Clift and D. Harrison), 331–355. London: Academic.

51 Herara, C.A. and Levy, E.K. (2001). Bubbling characteristics of sound-assisted fluidized beds. *Powder Technol.* 119: 229–240.

52 van Willigen, F.K., van Ommen, J.R., and van Turnhout, J. (2003). Bubble size reduction in a fluidized bed by electric fields. *Int. J. Chem. Reactor Eng.* 1: A21.

53 Jalalinejad, F., Bi, X.T., and Grace, J.R. (2012). Effect of electrostatic charges on a single bubble in gas–solid fluidized beds. *Int. J. Multiphase Flow* 44: 15–28.

54 Hernandez-Jimenez, F., Garcia-Gutierez, L.M., and Acosta-Ibora, A. (2019). Numerical study of the effects of temperature and pressure on the fluidization of particles with air and (supercritical) CO_2. *J. Supercrit. Fluids* 147: 271–283.

55 Brouwer, G.C., Wagner, E.C., van Ommen, J.R., and Mudde, R.F. (2012). Effects of pressure and fines content on bubble diameter in a fluidized bed studied using fast X-ray tomography. *Chem. Eng. J.* 207–298: 711–717.

56 Valverde Millán, J.M. (2013). *Fluidization of Fine Particles*. Dordrecht: Springer.

57 Li, H., Yan, W.P., Wu, W., and Wang, C. (2013). Characteristics of fluidization behavior in a pressurised bubbling fluidized bed. *Can. J. Chem. Eng.* 215: 479–490.

58 Nishimura, A., Deguchi, S., Matsuda, H. et al. (2000). Bubble characteristics in a pulsated fluidized bed under intermittent fluidization. *Kagaku Kogaku Ronbunshu* 26: 88–93.

59 Sanaei, S., Mostoufi, N., Radmanesh, R. et al. (2010). Hydrodynamic characteristics of gas-solid fluidization at high temperature. *Can. J. Chem. Eng.* 88: 1–11.

60 Maurer, S., Wagner, E.C., van Ommen, J.R. et al. (2015). Influence of vertical internals on a bubbling fluidized bed characterized by X-ray tomography. *Int. J. Multiphase Flow* 75: 237–249.

61 Mawatari, Y., Akune, T., Tatemoto, Y., and Noda, K. (2002). Bubbling and bed expansion behavior under vibration in a gas-solid fluidized bed. *Chem. Eng. Technol.* 25: 1095–1100.

62 Boyce, C.M., Penn, A., Lehnert, M. et al. (2019). Effect of liquid bridging on bubbles injected into a fluidized bed: a magnetic resonance imaging study. *Powder Technol.* 343: 813–820.

Problems

7.1 Compare the bubble diameter at mid-height predicted by the Cai et al., correlation (Eq. (7.10)) with the values predicted by the Mori and Wen and Darton et al. equations in Solved Problem 7.1.

7.2 Estimate the turnover time for the conditions of Solved Problem 7.1. How large would the mass feed rate of particles need to be for the assumption of perfect solids mixing not to be an appropriate assumption? Assume a particle density of 2800 kg/m^3 and the particles to be of rounded shapes.

8

Slug Flow

John R. Grace

University of British Columbia, Department of Chemical and Biological Engineering, 2360 East Mall, Vancouver, Canada V6T 1Z3

8.1 Introduction

As discussed in Chapter 4, the slug flow (or slugging) regime arises in gas-fluidized beds when bubbles grow sufficiently large that their shapes and rising velocities are mainly determined by the diameter of the column in which they are found. Conditions that tend to promote slug flow include the following:

(a) Although slug flow can occur for Geldart group A particles in very small columns (e.g. <100 mm diameter), slug flow is most important for Geldart group B or D solids, particles that are big enough to allow bubbles to grow to be large before reaching a maximum stable size (see Chapter 7).
(b) The superficial gas velocity, U, exceeds the minimum slugging velocity (see Chapter 4).
(c) H/D = bed height to column diameter > 2, again facilitating bubble growth.
(d) There are no immersed horizontal heat transfer surfaces or other fixed internal surfaces that tend to break up bubbles (see Chapter 18).

While these conditions are seldom all satisfied in commercial fluidized bed reactors and contactors, it is quite common for slug flow to occur in laboratory- and pilot-scale columns. This can then affect experimental data obtained in small-scale equipment, contributing to scale-up difficulties (see Chapter 17), as the flow regime then tends to change as equipment is scaled up from laboratory scale (typically $D \leq 0.15$ m) to pilot scale (typically $0.2\ \text{m} \lesssim D \lesssim 0.6\ \text{m}$) and then to commercial scale ($D \gtrsim 0.9$ m). For an excellent review of early work on slug flow in fluidized beds, see Hovmand and Davidson [1].

8.2 Types of Slug Flow

Two types of gas slugs are encountered in fluidized bed studies and applications:

(a) **Round-nosed slugs** (axisymmetric and wall slugs): These are elongated bullet-shaped bubbles, usually axially symmetric, as shown in Figure 8.1a,

Essentials of Fluidization Technology, First Edition. Edited by John R. Grace, Xiaotao Bi, and Naoko Ellis.

with the volume-equivalent bubble diameter, $d_e \gtrsim 0.5D$. These slugs rise at a velocity greater than the superficial gas velocity, U. If there is significant turbulence, some round-nosed slugs attach themselves to the column wall, forming asymmetric "wall slugs" [2] as shown in Figure 8.1b.

(b) **Square-nosed slugs**: A combination of smooth column walls, smooth Geldart group D or B solid particles, high superficial velocities, and small columns (diameters typically <150 mm) can lead to a mode of slugging, shown schematically in Figure 8.1c, where "square-nosed slugs" rise relatively slowly (at a velocity $< U$). Particles peel off and fall from the underside of dense moving bed plugs separating cylindrical gas slugs. These particles rain downwards over the entire column cross section onto the top of the next plug of solids. Gas rises inside the gas slugs.

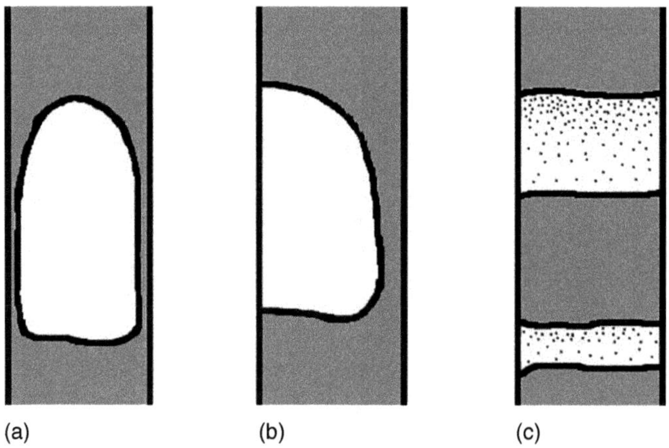

Figure 8.1 Profiles of types of slug flow in vertical columns. (a) Round-nosed axisymmetric slugs. (b) Round-nosed asymmetric wall slugs clinging to the column wall. (c) Square-nosed slug.

Table 8.1 Factors that tend to favour each of the three modes of slugging.

Variable	Round-nosed slugs: Axisymmetric	Round-nosed slugs: Wall slugs	Square-nosed slugs
Particle diameter	Small and medium particles	Large particles	Large particles
Bed height	Broad range	Tall	Tall
Gas velocity	$U > U_{ms}$, broad range	High	High
Column diameter	Wide range		Small
Wall roughness	Somewhat rough	Rough	Smooth
Particle shape	Rounded	Irregular	
Particle roughness	Smooth	Rough	
Fines content	Large content		Few fines

Blanks indicate either that the variable in question is unimportant or that the influence is not clear at this stage.

Conditions that have been reported to favour each of these slugging modes are identified in Table 8.1. Although square-nosed slugs have received some attention in experimental studies of small-scale fluidized bed hydrodynamics (e.g. [3–6]), this mode of slugging is normally undesirable, therefore avoided and not of significant practical interest. Hence, the rest of this chapter is devoted almost entirely to slugging of the round-nosed variety.

8.3 Analogy Between Slugs in Fluidized Beds and Slugs in Liquids

Section 7.3 examined the analogy between gas bubbles in liquids and bubbles in gas–solid fluidized beds. The analogy between gas–liquid and gas–solid flows can be extended to round-nosed slug flow[1], the flow regime of primary interest in this chapter. In both cases, slug flow can be considered to occur for gas rising through vertical columns when the average volume-equivalent diameter of bubbles exceeds half the column diameter. Two-phase (gas–liquid) flows are encountered in boiling heat transfer in vertical tubes and in production of gas from drilled deep vertical wells, providing examples of slug flow in analogous gas–liquid systems.

In this chapter we consider the transition from bubbling fluidization to (round-nosed) slugging, before presenting the properties of fully developed slug flow in tall columns. A number of the most important hydrodynamic properties of fully developed slug flow, such as bed expansion, slug frequency, and slug length, can be predicted quite accurately based on the simple hydrodynamics (without the need for complex calculations or computational fluid dynamics). These properties provide upper or lower limits for the performance of fluidized beds for some practical applications.

8.4 Experimental Identification of the Slug Flow Regime

The slug flow regime can be readily identified in physical experiments. When it is possible to view the bed through transparent walls, an observer sees periodic oscillations of the bed surface, surging upwards relatively slowly, ejection of packets of particles each time a slug reaches and breaks through the bed surface, and then rapid collapse of the upper surface, leading to a new cycle of slow rise, ejection, and collapse. If visualization is not possible, then a differential pressure transducer connected across the bed, or pressures measured in the upper section of the bed, will reveal a distinctive pattern, with pressure drop increasing relatively slowly as the top surface of the bed rises, then dropping back sharply as voids burst at the bed surface. The frequency of the oscillations in both cases is typically of order 0.5–1 Hz, depending on the column diameter, as determined by ultrafast magnetic resonance imaging (MRI) as detailed below. These oscillations

---

1 Note that the wall boundary condition can differ – no slip for slugs in liquids vs. partial slip in fluidized beds. In addition, fluidized beds are more likely to exhibit non-Newtonian rheology.

are larger in amplitude, more regular (nearly periodic), and lower in frequency than those corresponding to the bubbling, turbulent, and fast fluidization flow regimes. Hence slug flow is easily distinguishable from other flow regimes, even if the walls of the column are opaque. The distinctive features of slug flow also permit modern advanced experimental techniques such as spectral analysis [7] and MRI [8].

8.5 Transition to Slug Flow

The rise of bubbles in fluidized beds is appreciably restrained by wall effects whenever the volume-equivalent bubble diameter, $d_e \gtrsim 0.2D$. As bubbles grow beyond this size, the retarding influence of the column wall becomes greater, and slug flow can be assumed to take over once $d_e \gtrsim 0.5D$. This implies that U_{ms}, the superficial velocity for minimum slugging [9], given by Eq. (4.3) also be exceeded. In a normal bubbling bed, some distance in the vertical direction is required for bubbles to grow sufficiently for slug flow to be reached. Based on their coalescence model for bubble growth, considered in Chapter 7, Darton et al. [10] suggested that the minimum bed depth required for slugging for a multi-orifice gas distributor to be established is given by

$$H_{ms} = 3.5D\left(1 - \frac{1}{\sqrt{N_{or}}}\right) \tag{8.1}$$

The region between the distributor plate and this level can be considered to be occupied by a bubbling bed, with bubbles increasing in volume and decreasing in frequency as they coalesce in the developing (or entry) region, subject to the hydrodynamic relationships detailed in Chapter 7.

8.6 Properties of Single Slugs

In this chapter, this book, and most multiphase flow literature, the term “slug” refers to the successive discrete bullet-shaped gaseous entities that rise, one after the other, to the surface of fluidized beds in which slug flow occurs. Readers should, however, be cautioned that in some of the literature originating from North America, the term “slug” is instead used to identify the succession of dense-phase regions or packets separating the gaseous entities. Here we use the word “plug” to refer to these dense entities.

Single (gas) slugs adopt the shape shown in Figure 8.1a. Relative to the rising slug, the dense phase flows over and downwards around the nose and upper surface and then streams downwards through the thin gap separating the inner wall of the column from the outer surface of the slug. If the slug volume is increased, the slug then simply becomes longer, with more gas added at the bottom. The minimum length of a slug is about $D/2$.

Since there is no surface tension in gas–solid systems, the rising velocity of a single slug is independent of its length, instead being only dependent on the column diameter:

$$u_{s0} = 0.35\sqrt{gD} \tag{8.2}$$

Equation (8.2) applies only to axially symmetric slugs [11]. For wall slugs, the velocity is larger, with the slug acting as if it were in a column of twice the diameter, so the 0.35 coefficient is replaced by $0.35 \times \sqrt{2} = 0.49$. Square-nosed slugs rise significantly more slowly [12]:

$$u_{s0} = 0.18\sqrt{gD} \tag{8.3}$$

Clouds form for slugs when $\alpha = u_s\varepsilon_{mf}/U_{mf} > 1$, similar to what applies to bubbles [13], but with the slug velocity replacing the bubble velocity, which appeared in α in Chapter 7. The cloud around the slug is distorted by the containing vessel (compare Figures 8.2 and 7.6).

Slugs can split by a mechanism similar to that described for bubbles in Chapter 7. Again, this involves perturbations of the upper surface of the bubble growing in amplitude as the top slips around the upper surface towards the nearly vertical sides of the slug, slicing off, knifelike, a portion of the slug. The resulting smaller bubble is, however, likely to slip behind, then accelerate, catch up to, and re-coalesce with, the back of the larger slug from which it originated.

Figure 8.2 Cloud formation for a single slug in a column of circular cross section. Adapted from photograph of a single NO_2 slug injected into a fluidized bed by Rowe et al. [13]. Source: Rowe et al. [13]. Reproduced with permission of Elsevier.

8.7 Hydrodynamics of Continuous Slug Flow

8.7.1 Slug Rising Velocity

In continuous fully developed slugging, the rise velocity of individual slugs, not in the process of overtaking another slug, is given by

$$u_s = 0.35\sqrt{gD} + \kappa(U - U_{mf}) \tag{8.4}$$

where the coefficient, κ, is usually taken as 1.0 based on continuity considerations. For gas slugs in liquids, κ may instead be taken as 1.2 or 2.0 to reflect velocity profiles in the liquid ahead of slugs corresponding to fully developed turbulent or laminar flow in a channel with no slip at the inner surface of the column [14]. Experiments at temperatures from room temperature to 400 °C [15] have shown dependence of slug properties on temperature.

8.7.2 Slug Spacing and Length

Axisymmetric round-nosed slugs rising vertically in line tend to have trailing slugs overtake leading slugs until the slugs are separated by a distance of about $2D$ (referring to the distance between the nose of one slug and the trailing edge of the slug immediately ahead of it). The assumed sequence of slugs is portrayed in Figure 8.3. For the $2D$ spacing, Hovmand and Davidson [16] derived a quadratic equation in $\sqrt{(\lambda_s/D)}$:

$$\frac{\lambda_s}{D} - 0.495\,(1+\beta)\sqrt{\frac{\lambda_s}{D}} + 0.061 - 1.94\beta = 0 \tag{8.5}$$

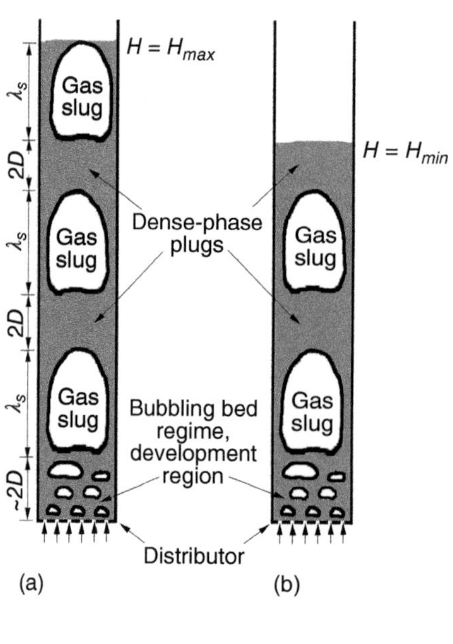

Figure 8.3 The assumed sequence of slugs is portrayed in Figure 8.3, with the leading slug on the point of bursting at the bed surface in (a) and the array immediately after the particles around the bursting slug have collapsed in (b).

where

$$\beta = \frac{(U - U_{mf})}{u_{s0}} > 0.2 \tag{8.6}$$

from which the fully slug length, λ_s, for fully developed slug flow can be estimated.

8.7.3 Time Between Successive Slugs and Slug Frequency

The time between successive slugs can now be calculated by dividing the distance between successive slug noses ($\lambda_s + 2D$) by the slug rising velocity, u_s. Hence, the slug frequency at a given level is simply the reciprocal, i.e.

$$f_s = \frac{u_s}{\lambda_s + 2D} \tag{8.7}$$

where u_s and λ_s are obtained from Eqs. (8.4) and (8.5).

8.7.4 Bed Expansion

As noted above, the upper surface of the continuously slugging fluidized bed rises and falls in a periodic manner. The relevant dimension for design of the overall equipment height is the maximum dense bed height achieved as each slug arrives and erupts at the top. This is given [17] by

$$H_{max} = H_{mf}(1 + \beta) + \beta(2D - H_e) \tag{8.8}$$

Since the entry length, H_e, is approximately $2D$, then

$$H_{max} \approx H_{mf}(1 + \beta) \tag{8.9}$$

an equation suggested by Matsen et al. [18].

8.7.5 Uniformity and Symmetry of Flow

Chapter 7 indicated that the development of bubbles in bubbling gas-fluidized beds normally causes bubbles to establish up-the-centre, down-at-the-wall flow patterns in relatively deep fluidized beds. This means that slugs tend towards axial symmetry when they form. However, local disturbances, like those that can lead to preferential flow through one side of deep beds [19], can also lead to wall slugs (see Section 8.2), jet streaming, and considerable asymmetry of the flow. Increasing the fines content of the fluidized particles tends to reduce the extent of jet streaming, wall slugs, and asymmetric flow.

8.8 Mixing of Solids and Gas in Slugging Beds

Slug flow implies fluidized beds of high H/D ratio and small diameter or width. Hence mixing in the radial or horizontal direction is seldom an issue, while axial mixing is of substantial importance, with respect to both in-bed segregation of particles of different properties and residence time distributions of particles and gas passing through slugging bed reactor. Axial mixing of solids is governed by similar mechanisms as for bubbling beds in Chapter 7, i.e. vertical transport of particles in slug wakes and due to drift, supplemented to a small extent by small random displacements [20]. However, because the wakes are stunted and paths for drift are restricted in the thin gap between the edge of round-nosed axisymmetric slugs and the column wall, vertical transport of particles is slower than for bubbling beds.

Axial mixing of gas can be represented by well-mixed tanks in series or by axial dispersion in the bottom flow-development (bubbling) region ($z \leq 2D$), followed by a number of well-mixed tanks in series, one tank for each slug length + plug length (i.e. for each $\lambda_s + 2D$).

8.9 Slugging Beds as Chemical Reactors

As indicated above, laboratory and pilot plant fluidized bed reactors often operate in the slug flow regime due to their small diameters. For efforts to provide reactor models of catalytic gas-phase solid-catalyzed reactors, see Hovmand et al. [21]. Yates and Grégoire [22] found that it was essential to model separately the bottom bubbling bed region in which slug flow is not yet applicable ($z \lesssim 2D$), as well as for the fully developed slugging region ($z \gtrsim 2D$).

Solved Problem

8.1 20 kg of particles of mean diameter 228 μm and density 2800 kg/m^3, with U_{mf} = 0.061 m/s and ε_{mf} = 0.40, is fluidized in a tall cylindrical column of inside diameter 0.10 m. (a) Show that slug flow is occurring for U = 0.152 m/s. (b) Estimate the slug length, slug spacing, slug frequency, and top surface fluctuation amplitude. (c) What is the maximum expanded bed height?

Solution

(a) H_{mf} = 1.5 m; u_{s0} = 0.345 m/s; U_{ms} = 0.13 m/s. Hence $U > U_{ms}$; $H_{mf}/D >> 2$; maximum stable bubble diameter ≫ 0.1 m for these Geldart group B particles, and, from Chapter 4, $U < U_c$ needed for turbulent fluidization. Therefore the system meets all requirements for slug flow.

(b) Inter-slug spacing = $2D$ = 0.20 m; β = 0.264; quadratic Eq. (8.5) gives $x^2 - 0.495 \times 1.264x + (0.061 - 1.94 \times 0.264) = 0$ with solution $x = 1.053$ so that $\lambda_s = Dx^2 = 0.111$ m. Slug frequency = {0.345 + (0.152 − 0.061)}/{0.111 + 0.20} = 1.41 s^{-1}. By continuity, top surface rises at velocity $(U - U_{mf})$. So amplitude of top surface fluctuation = (0.152 − 0.061)/1.41 m = 0.065 m.

(c) Maximum expanded bed height = 1.5 m × (1 + 0.264) = 1.9 m.

Notations

D	column diameter (m)
d_e	volume-equivalent bubble diameter (m)
f_s	slug frequency (s^{-1})
g	acceleration of gravity ($m\ s^{-2}$)
H	expanded bed depth (m)
H_{max}	bed depth reached as slugs burst at bed surface (m)
H_{mf}	bed depth at minimum fluidization velocity (m)
N_{or}	number of orifices in multi-orifice gas distributor plate (–)
U_{mf}	superficial gas velocity at minimum fluidization ($m\ s^{-1}$)
U_{ms}	superficial gas velocity at minimum spouting ($m\ s^{-1}$)
u_s	rising velocity of slugs during continuous slugging ($m\ s^{-1}$)
u_{s0}	rising velocity of a single slug in a bed operated at $U = U_{mf}$ ($m\ s^{-1}$)
z	height coordinate (m)

Greek Letters

α	dimensionless group that determines cloud/slug volume ratio (–)
β	dimensionless superficial gas velocity = $(U - U_{mf})/u_{s0}$ (–)
ε_{mf}	bed voidage at minimum fluidization (–)
κ	dimensionless weighting factor in slug velocity equation (–)
λ_s	slug length for fully developed slugging (m)

References

1 Hovmand, S. and Davidson, J.F. (1971). Pilot plant and laboratory scale fluidized reactors at high gas velocities: the relevance of slug flow, Chapter 5. In: *Fluidization* (eds. J.F. Davidson and D. Harrison), 193–259. London: Academic Press.

2 Kehoe, P.W.K. and Davidson, J.F. (1970). Continuously slugging fluidized beds. Proceedings of Chemeca Conference, Butterworths, Melbourne, pp. 97–116.

3 Thiel, W.J. and Potter, O.E. (1977). Slugging in fluidized beds. *AIChE J.* 16: 242–247.

4 Noordergraaf, I.W., van Dijk, A., and van den Bleek, C.M. (1987). Fluidization and slugging in large-particle systems. *Powder Technol.* 52: 59–68.

5 Van Putten, I.C., van Sint Annaland, M., and Weickert, G. (2007). Fluidization behavior in a circulating slugging fluidized bed reactor. *Chem. Eng. Sci.* 62: 2522–2534.

6 Tebianian, S., Dubrawski, K., Ellis, N. et al. (2016). Comparison of particle velocity measurement techniques in a fluidized bed operating in the square-nosed slugging flow regime. *Powder Technol.* 296: 45–52.

7 Van der Schaaf, J., Schouten, J.C., Johnsson, F., and van der Bleek, C.M. (2002). Non-obtrusive determination of bubble and slug length scales in fluidized beds by decomposition of the power spectral density of pressure time series. *Int. J. Multiphase Flow* 28: 865–880.

8 Boyce, C.M., Davidson, J.F., Holland, D.J. et al. (2014). The origin of pressure oscillations in slugging fluidized beds: comparison of experimental results from magnetic resonance imaging with a discrete element model. *Chem. Eng. Sci.* 116: 611–622.
9 Stewart, P.S.B. and Davidson, J.F. (1967). Slug flow in fluidized beds. *Powder Technol.* 1: 61–80.
10 Darton, R.C., Lanauze, R.D., Davidson, J.F., and Harrison, D. (1977). Bubble growth due to coalescence in fluidized beds. *Trans. Inst. Chem. Eng.* 55: 274–280.
11 Müller, C.R., Davidson, J.F., Dennis, J.S. et al. (2007). Rise velocities of bubbles and slugs in gas-fluidized beds: ultra-fast magnetic resonance imaging. *Chem. Eng. Sci.* 62: 82–93.
12 Rudolph, V. and Judd, M.R. (1986). Circulation and slugging in a fluid bed gasifier fitted with a draft tube. In: *Circulating Fluidized Bed Technology* (ed. P. Basu), 437–442. Toronto: Pergamon.
13 Rowe, P.N., Partridge, B.A., and Lyall, E. (1964). Cloud formation around bubbles in fluidized beds. *Chem. Eng. Sci.* 19: 973–985.
14 Grace, J.R. and Clift, R. (1979). Dependence of slug rise velocity on tube Reynolds number in vertical gas-liquid flow. *Chem. Eng. Sci.* 34: 1348–1350.
15 Choi, J.H., Kim, T.W., Moon, Y.S. et al. (2003). Effect of temperature on slug properties in a gas fluidized bed. *Powder Technol.* 131: 15–22.
16 Hovmand, S. and Davidson, J.F. (1968). Chemical conversion in a slugging fluidized bed. *Trans. Inst. Chem. Eng.* 46: 190–203.
17 Grace, J.R., Krochmalnek, L.S., Clift, R., and Farkas, E.J. (1971). Expansion of liquids and fluidized beds in slug flow. *Chem. Eng. Sci.* 26: 617–628.
18 Matsen, J.M., Hovmand, S., and Davidson, J.F. (1969). Expansion of fluidized beds in slug flow. *Chem. Eng. Sci.* 24: 1743–1754.
19 Issangya, A.S., Knowlton, T.M., and Reddy Karri, S.B. (2007). Detection of gas bypassing due to jet streaming in deep fluidized beds of group A particles. In: *Fluidization XII* (eds. X. Bi, F. Berruti and T. Pugsley), 775–782. Brooklyn: Engineering. Conferences International.
20 Chen, H.Z. and Guo, Z.K. (2012). Characteristics of mixing/segregation in a bubbling/slugging fluidized bed with binary mixtures. *Adv. Mater. Res.* 396–398: 322.
21 Hovmand, S., Freedman, W., and Davidson, J.F. (1971). Chemical reaction in a pilot-scale fluidized bed. *Trans. Inst. Chem. Eng.* 49: 149–162.
22 Yates, J.G. and Grégoire, J.Y. (1980). An experimental test of slugging-bed reactor models. In: *Fluidization* (eds. J.R. Grace and J.M. Matsen), 581–588. New York: Plenum Press.

9

Turbulent Fluidization

Xiaotao Bi

University of British Columbia, Department of Chemical and Biological Engineering, 2360 East Mall, Vancouver, Canada V6T 1Z3

9.1 Introduction

As shown in Chapter 4, the turbulent fluidization regime is located between the bubbling and fast fluidization flow regimes, representing a transition of two-phase flow from a state with a distinctive dispersed bubble phase to that with a dispersed dense phase as the dilute phase gradually transforms into a continuous phase, characterized by decreasing amplitude of pressure fluctuations, resulting from the disappearance of large bubbles/voids. The first photograph of a turbulent fluidized bed, distinctly different from bubbling fluidization, was published by Matheson et al. in 1949 [1]. A turbulent fluidization regime was first included in the flow regime diagram of Zenz [2]. The first quantitative study on the flow structure of a turbulent fluidized bed was performed by Lanneau [3] who measured local voidage, voidage fluctuations, and pierced void length in a 76 mm internal diameter fluidized bed with fine catalyst particles at high gas velocities. Kehoe and Davidson [4] extended their work on slugging to higher-velocity operation and identified the transition from bubbling to turbulent fluidization, based on visualization of a two-dimensional bed and measured void rise velocities and capacitance signal traces in a three-dimensional column. Later the turbulent fluidization regime was reported by Massimilla [5], Thiel and Potter [6], and Crescitelli et al. [7]. In these early studies, transition to turbulent fluidization was generally determined based on visual observations and local voidage or pressure signal traces.

The first criterion for transition from bubbling or slugging to turbulent fluidization was proposed by Yerushalmi et al. [8]. As described in Chapter 4, the gas velocity, U_c, at which the standard deviation of pressure fluctuations reached a maximum marked the onset of the transition to turbulent fluidization, while U_k, where the standard deviation of the pressure fluctuations levels off, was denoted as the end of this transition. The turbulent fluidization flow regime has been controversial and not always accepted as a separate flow regime (e.g. [3, 9–13]), with the controversy originating largely from having two transition velocities, U_c and U_k. Confusion arose because the early authors [8] proposed that U_k marked the

Essentials of Fluidization Technology, First Edition. Edited by John R. Grace, Xiaotao Bi, and Naoko Ellis.

onset of turbulent fluidization, whereas subsequent authors have found either that U_k does not exist, partly because it is highly influenced by the design of a solids capture and return system, or that it marks the end (not the beginning) of the turbulent flow regime, i.e. the transition from turbulent fluidization to fast fluidization [13]. Turbulent fluidization is now widely accepted as extending from U_c to the onset of fast fluidization, U_{se}, defined in Chapter 4, beyond which the solids flow rate starts to significantly influence the gas–solid flow structure, and this is the turbulent fluidization regime considered here.

At least three possible flow patterns have been identified in turbulent fluidized beds [13]. For a non-slugging system with Geldart group A powders in a relatively large column, with the maximum stable bubble size, $d_{b,max}$, substantially smaller than the column diameter, D, i.e. $d_{b,max}/D < 0.7$, the transition from bubbling to turbulent fluidization is relatively sharp, associated with significant breakdown of large bubbles, as observed by Kehoe and Davidson [4], Yerushalmi et al. [8], Crescitelli et al. [7], Yang and Chitester [14], and Tsukada et al. [15]. In a slugging system of deep bed with group B particles, a gradual transition involving intermittent slug-like structures interspersed with periods of fast fluidization-like behaviour was observed. The latter becomes predominant with increasing superficial gas velocity [7, 9, 16, 17]. In a relatively shallow bed of group B or D particles ($H_{mf}/D \ll 2$), the penetration of gas jets through beds of large particles where fully developed slug flow is never achieved due to the limited bed depth, and transition to turbulent fluidization simply occurs when intermittent jets develop into a continuous phase with increasing superficial gas velocity, in a manner similar to what occurs in spouted or jetting fluidized beds [18–20].

Because of superior gas–solid contact in the turbulent fluidization flow regime, turbulent fluidized beds have been widely adopted in commercial fluidized bed reactors, either by increasing the superficial gas velocity beyond U_c in existing bubbling bed reactors or by building dedicated turbulent fluidized bed reactors since the 1970s. Some industrial applications of turbulent fluidized beds in chemical and metallurgical processes are listed in Table 9.1.

Table 9.1 Some commercial turbulent fluidized bed reactors.

Process	Geldart particle classification	Typical superficial gas velocity (m/s)
FCC regenerators	Group A	0.5–1.5
Mobil methanol to gasoline process (MTG) reactors	Group A	~0.5
Acrylonitrile	Group A	~0.5
Maleic anhydride	Group A	~0.5
Phthalic anhydride	Group A	~0.5
Ethylene dichloride	Group A	~0.5
Roasting of zinc sulfide	Group B	~1.5

In this chapter, the overall hydrodynamics, local flow structure, and mixing of gas and solid phases in turbulent fluidized beds are described. Heat and mass transfer and reactor performance are covered in Chapters 14 and 15, respectively.

9.2 Flow Structure

9.2.1 Axial and Radial Voidage Distribution

As the superficial gas velocity, U, increases from bubbling to the turbulent flow regime, the bed surface separating the upper freeboard region and the lower dense bed region becomes more and more diffuse, eventually disappearing as more and more particles are entrained. The axial voidage distribution in the turbulent regime is characterized by a smooth decrease of solids holdup with height in the freeboard region, with the voidage remaining ~0.65–0.75 in the dense bed region [21]. As the gas velocity increases, the average solids concentration in the bed decreases, while solids holdup in the freeboard region increases, as shown by Figure 9.1.

In the dense bed region, downward flow of particles along the column wall persists, with upward flow of particles at higher voidage in the interior of the column, and gas voids rising preferentially in the interior of the bed [23–26]. The nonuniform radial distribution of local voidage, ε, can be correlated to the cross-sectional average voidage, $\overline{\varepsilon}$, [27], by

$$\frac{1-\varepsilon}{1-\overline{\varepsilon}} = 0.908 + 0.276\left(\frac{r}{R}\right)^4 \tag{9.1}$$

where r is the radial coordinate and R is the column radius.

In the transition region and the freeboard region above the turbulent bed surface, a core–annulus flow structure develops when the superficial gas velocity increases, with particles descending in the near wall region, as shown in Figure 9.2

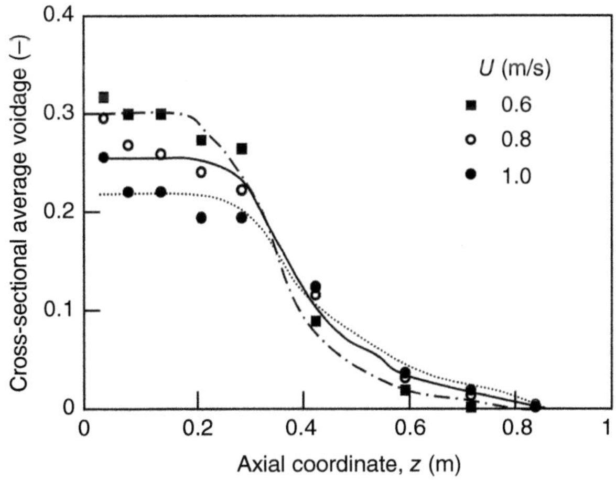

Figure 9.1 Axial voidage profiles in turbulent fluidized beds. Source: From Werther and Wein 1994 [22].

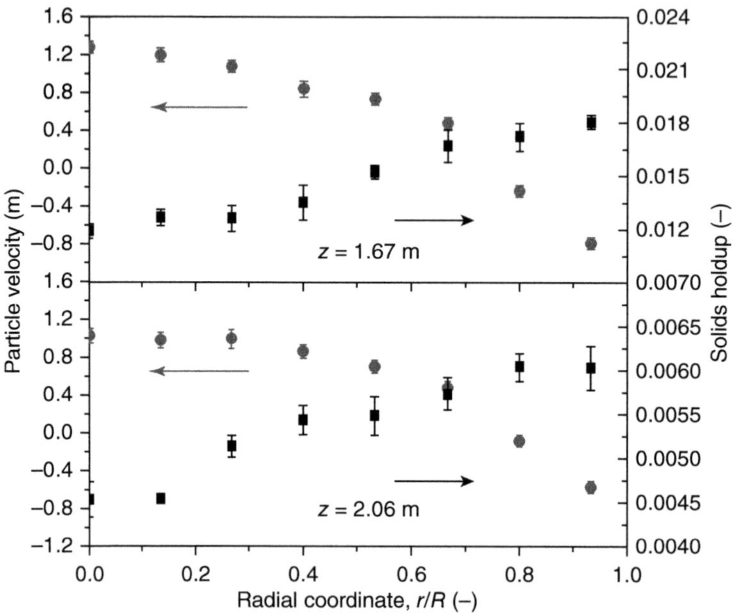

Figure 9.2 Radial profiles of solids holdup and particle velocity in turbulent fluidized beds (sand particles, $U = 0.79$ m/s, $H_0 = 0.5$ m). Source: Li et al. 2017 [28]. Reproduced with permission of Elsevier.

[28]. This core–annulus flow structure is similar to what is observed in a fast fluidized bed (see Chapter 12), with the thickness of the falling dense layer increasing with increasing superficial gas velocity [28].

9.2.2 Local Void Size and Rise Velocity

In turbulent fluidized beds, deformed voids exist, distinguishable from the dense phase, as in the bubbling flow regime. Local void size and rise velocities have been measured using optical and capacitance probes [3, 29–33]. Figure 9.3 shows typical void sizes and rise velocities for type I turbulent fluidized beds with group A particles, with void splitting dominant in the turbulent regime [29]. The void size and its rise velocity are seen to reach maximum values at U_c, followed by decreases with further increases in gas velocity [29, 31, 33]. The bubble coalescence and splitting model of Horio and Nonaka [34] appears to capture the change in void size for group A powders in turbulent fluidized beds [33, 35]. The void rise velocity in a turbulent fluidized bed is more sensitive to the void size, but less to the superficial gas velocity, than in bubbling fluidized beds, likely because of the suppression of "gulf streaming" and more intimate void–void interactions in turbulent fluidized beds [33]. For a type II turbulent fluidization system with group B particles, where voids break up gradually, while their separation distance becomes less than the bubble size with increasing gas velocity beyond U_c, both void size and its rise velocity can be predicted with reasonable accuracy from correlations developed for bubbling beds [22].

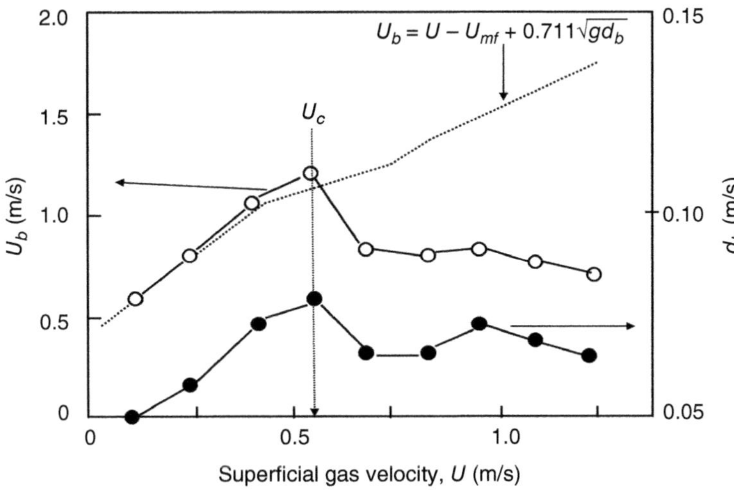

Figure 9.3 Bubble/void diameter and its rise velocity as a function of superficial gas velocity for group A particles. Source: From Yamazaki et al. 1991 [29].

9.2.3 Void Phase Volume Fraction, Void Phase, and Dense Phase Solids Holdup

For two-phase flow with a dilute void phase and a dense emulsion phase, the local voidage, ε, void phase volume fraction, δ_v, void phase voidage, ε_v, and dense phase voidage, ε_d, are related by

$$1 - \varepsilon = \delta_v(1 - \varepsilon_v) + (1 - \delta_v)(1 - \varepsilon_d) \tag{9.2}$$

In bubbling fluidized beds, the bubble phase volume fraction increases with increasing superficial gas velocity, while the dense phase solids holdup remains almost constant at the minimum fluidization value, i.e. $\varepsilon_d = \varepsilon_{mf}$ [29, 36, 37]. However, beyond U_c, the coalescence and splitting of voids causes the dense phase to expand, reducing its solids holdup [22, 25, 29, 38–40], based on analyses of the probability density distribution of capacitance and optical probe signals. Solids holdup in the dilute void phase also tends to increase with increasing superficial gas velocity in the turbulent regime [22, 25, 29, 38, 39, 41]. Both increased gas flow in the dense phase and higher solids holdup in the dilute void phase contribute to the improved gas–solid contacting, relative to the bubbling flow regime, as discussed in Chapter 15.

The local void fraction is also distributed nonuniformly in the radial direction in the turbulent flow regime, with higher concentrations near the axis. The heterogeneity increases with increasing superficial gas velocity, as shown in Figure 9.4 [42]. One expects that the heterogeneity should reach a maximum in the fast fluidization regime as the core–annulus flow structure develops and then declines as the flow transforms to a dilute flow regime at much higher gas velocities.

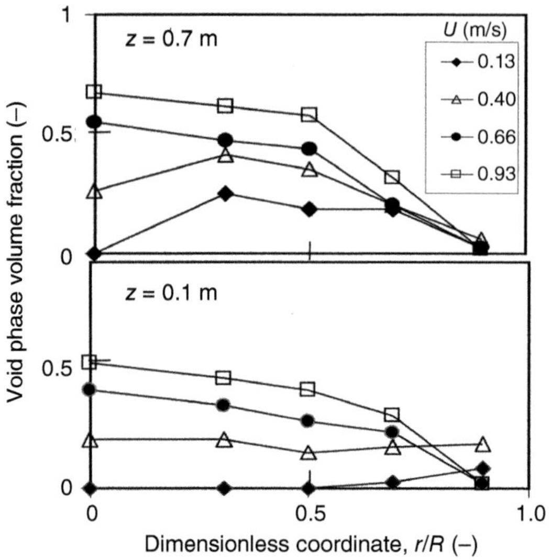

Figure 9.4 Radial profiles of local void fraction in bubbling and turbulent fluidized beds for FCC particles. Source: From Nakajima et al. 1991 [42].

9.3 Gas and Solids Mixing

9.3.1 Gas Mixing

Gas flow in the turbulent flow regime becomes intermittent as it develops from dispersed to a continuous phase, so it is more appropriate to use the pseudo-homogeneous axial dispersion model to characterize the gas mixing and dispersion in the turbulent regime than in the bubbling regime.

Axial gas dispersion in turbulent fluidized beds is commonly studied by a dynamic tracer method, with a pulse or a stepwise change of gas tracer injected into the bottom of the bed followed by detection of tracer concentrations at an upper level of the bed. By fitting the measured gas residence time distribution (RTD) curve with a one-dimensional pseudo-homogeneous axial dispersion model, one obtains an axial gas dispersion coefficient, $D_{z,g}$.

$D_{z,g}$ in turbulent fluidized beds typically ranges from 0.1 to 1 m^2/s and, similar to the bubbling regime, increases with increasing column diameter [13]. With increasing superficial gas velocity, $D_{z,g}$ increases in the bubbling regime, as shown in Figure 9.5 [43, 44], reaches a maximum at $U \approx U_c$, and then decreases in the turbulent regime as the superficial gas velocity increases further [43–45].

Several correlations are available for estimating $D_{z,g}$ in turbulent fluidized beds. Li and Wu [46] correlated experimental $D_{z,g}$ data from a 90 mm diameter column to the cross-sectional average voidage,

$$D_{z,g} = 0.1835\overline{\varepsilon}^{-4.4453}\ \mathrm{m^2/s} \tag{9.3}$$

Foka et al. [43] correlated their data from a 100 mm diameter column to particle and column properties,

$$Pe_g = \frac{UH}{D_{z,g}} = 0.071Ar^{0.32}\left(\frac{d_p}{D}\right)^{0.4} \tag{9.4}$$

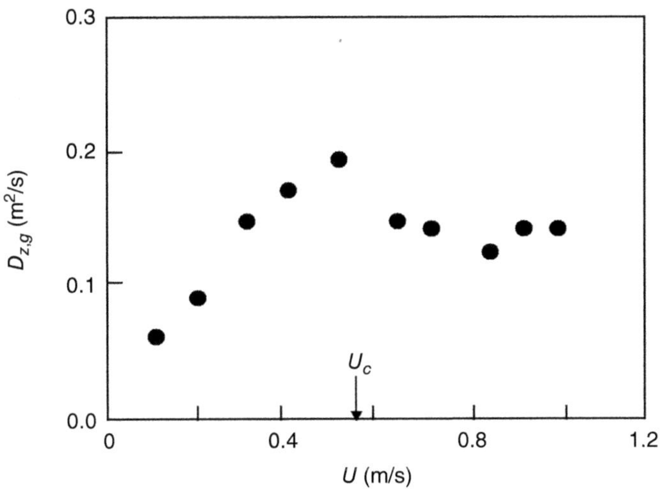

Figure 9.5 Axial gas dispersion coefficient as a function of superficial gas velocity in 100 mm diameter bubbling and turbulent fluidized bed. Source: Adapted from Foka et al. 1996 [43].

where d_p is the mean particle diameter, H is the bed height, and Ar is the Archimedes number given by Eq. (2.18).

Bi et al. [13] correlated literature data within the range of $D \leq 0.6$ m, $55 < d_p < 360$ μm, by

$$Pe_g = \frac{UH}{D_{z,g}} = 3.47 Ar^{0.149} Re^{0.0234} Sc^{-0.231} \left(\frac{H}{D}\right)^{0.285} \tag{9.5}$$

where Pe is the Péclet number, Re is the Reynolds number $(=\rho_g \cdot U \cdot d_p / \mu_g)$, and Sc is the Schmidt number $[=\mu_g/(\rho_g \cdot D_m)]$, with D_m being the molecular diffusivity of the gas component being mixed.

In view of the persistence of the emulsion phase–void phase two-phase flow structure in the turbulent fluidization regime, axial gas dispersion in each phase is needed in a two-phase model of turbulent fluidized bed reactors. Gas dispersion coefficients in each phase can be extracted by fitting the measured RTD curve to a one-dimensional two-phase dispersed flow model with interphase mass exchange [47]. As shown in Figure 9.6, the axial dispersion coefficient of the emulsion phase is higher than that for the dilute void phase, because of the downflow and strong backmixing of particles in the emulsion phase, which drags gas downwards, spreading the RTD.

Gas movement opposite to the flow direction (i.e. downwards) is characterized as gas backmixing. It can be measured by injecting a tracer gas steadily at a downstream location and then detecting tracer concentrations at locations upstream of the injection level. Tracer gas would not be detected if there were no gas backmixing. By fitting detected steady tracer upstream concentrations to a one-dimensional homogeneous model with backmixing, the backmixing

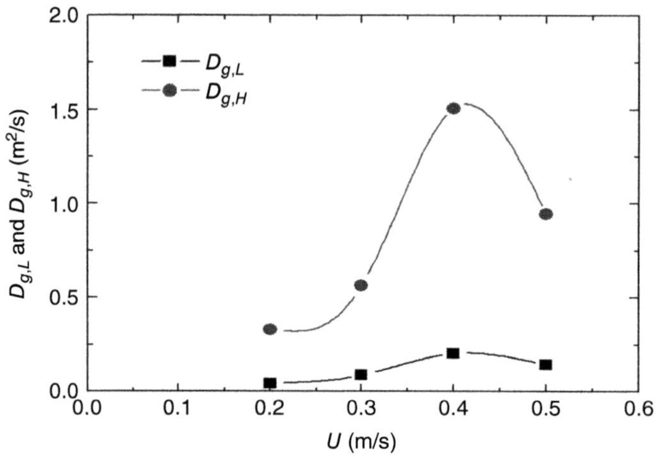

Figure 9.6 Axial gas dispersion coefficient in dense emulsion phase and dilute void phase in bubbling and turbulent fluidization regimes. Source: Adapted from Abba 2001 [47].

coefficient, $D_{b,g}$, is obtained:

$$\ln\left[\frac{C_{zp} - C_{zi}}{C_{zi}}\right] = \frac{U}{\bar{\varepsilon}D_{b,g}}(Z_p - Z_i) \tag{9.6}$$

where C_{zi} is the tracer concentration at the injection height, C_{zp} is the tracer concentration detected at the vertical coordinate Z_p, and Z_i is the injection height.

As in the bubbling flow regime, gas backmixing in the turbulent regime is strongly correlated with particle backmixing because gas is dragged downwards by descending particles. It has been reported [48, 49] that, similar to the axial gas dispersion coefficient $D_{z,g}$, $D_{b,g}$ increases with increasing gas velocity in the bubbling regime and reaches a maximum at U_c followed by a decrease with further increase of gas velocity in the turbulent flow regime. Gas backmixing is also stronger near the wall than in the core region, because of the higher tendency for particles to flow downwards in the annular region [50].

Axial gas dispersion ($D_{z,g}$) includes both gas backmixing ($D_{b,g}$) and gas spreading resulting from radial gas velocity distribution and radial gas dispersion ($D_{r,g}$), related by

$$\frac{D_{z,g}}{UD} = \frac{D_{b,g}}{UD} + b\frac{UD}{D_{r,g}} \tag{9.7}$$

where b is a constant characterizing the radial gas velocity distribution [51, 52], ranging from 5×10^{-4} to 5×10^{-3} in turbulent fluidized beds [48]. There have been very few studies on radial gas dispersion in turbulent fluidized beds. For group A particles, it was found [44, 53] (see Figure 9.7) that the radial dispersion coefficient increased with increasing superficial gas velocity, reaching a maximum at U_c, and then decreasing with further increase in superficial gas velocity. $D_{r,g}$ is about five times smaller in magnitude than $D_{z,g}$. For group B particles, radial dispersion $D_{r,g}$ is about 1 order of magnitude smaller than axial dispersion $D_{z,g}$ [54]. For both group A and B particles in small diameter units, e.g. $D < 0.2$ m, the

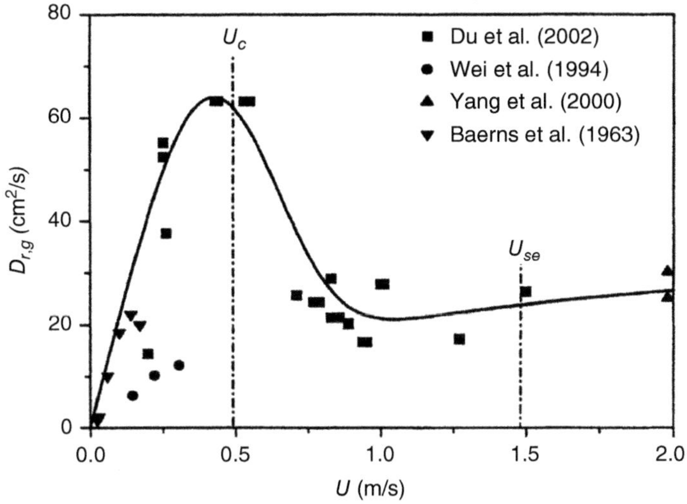

Figure 9.7 Radial gas dispersion coefficient as a function of superficial gas velocity in bubbling and turbulent fluidized beds at ambient conditions. Source: Adapted from Du et al. 2002 [44].

second term on the right side of Eq. (9.7) is generally much smaller than the first term. Therefore, $D_{z,g}$ is approximately equal to $D_{b,g}$ in columns commonly used in laboratory studies.

9.3.2 Solids Mixing

Solids mixing in fluidized beds has a direct impact on gas-particle contacting, gas–solid reactor performance, and catalyst deactivation. Particle mixing is commonly studied by solid tracer techniques, using such tracers as salt-coated, ferromagnetic, fluorescent, and phosphorescent particles. While many solids mixing studies have been conducted in the bubbling and fast fluidization regimes, applying various experimental techniques, relatively few have been done in the turbulent regime. Axial solids dispersion was found [44] to continue to increase with increasing superficial gas velocity in the turbulent fluidization regime beyond U_c, then started to decrease in the fast fluidization regime, as shown in Figure 9.8.

Figure 9.8 Axial solids dispersion coefficient as a function of superficial gas velocity in bubbling, turbulent, and fast fluidization regimes. Source: Adapted from Du et al. 2002 [44].

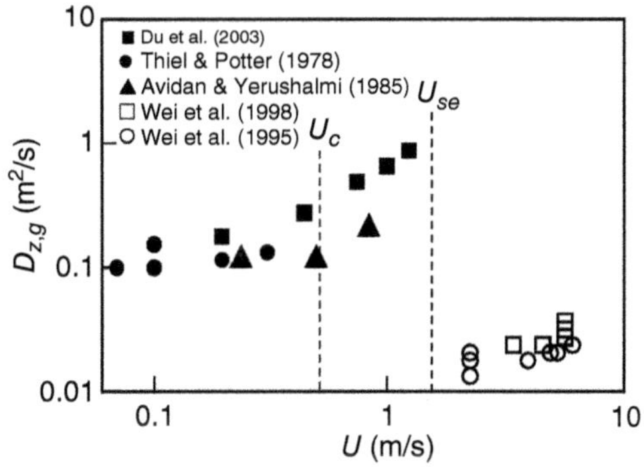

The particle dispersion coefficient is larger for smaller particles, likely associated with the large wake-to-void volumetric ratio for small particles.

For group B particles, a correlation was developed by Lee and Kim [54] to estimate the effective axial solids dispersion coefficient, $D_{z,s}$, in turbulent fluidized beds:

$$\frac{U \cdot D}{D_{z,s}} = 4.22 \times 10^{-3} Ar \tag{9.8}$$

For group A particles in the bubbling and turbulent fluidization flow regimes, Lee et al. [55] proposed

$$\frac{D_{z,s}}{[g(U - U_{mf})]^{1/3} D^{3/4}} = 0.364 Re_t^{-0.368} \tag{9.9}$$

where Re_t is the particle Reynolds number based on particle terminal velocity $(= \rho_g \cdot v_t \cdot d_p / \mu_g)$.

9.4 Effect of Column Diameter

In turbulent fluidized beds of group A particles where a type I transition takes place, with void splitting dominant in the turbulent fluidization regime, the column diameter is expected to have less effect on the flow behaviour when the column diameter is at least two to three times larger than the maximum stable bubble size (e.g. >300 mm). Based on a comparison of literature data obtained from larger [27, 30] and smaller [23, 38, 56, 57] columns (see Figure 9.9), the radial profile of voidage seems to be steeper for smaller columns, and flatter in the central core region for larger columns, due to the reduced wall effect in large

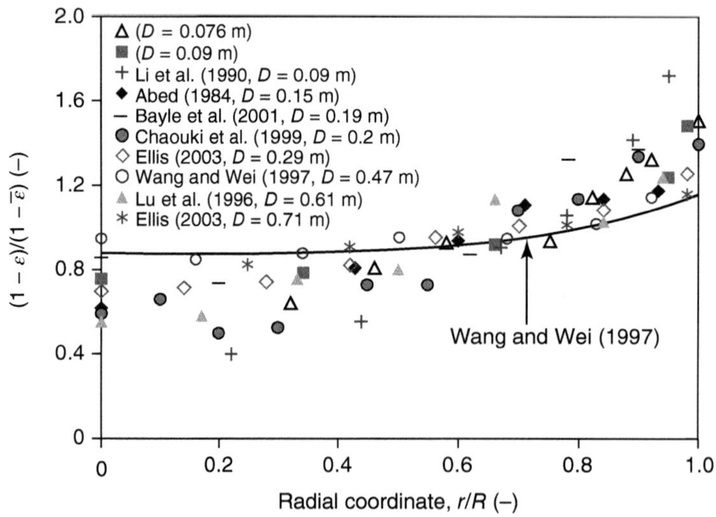

Figure 9.9 Normalized radial voidage profiles in turbulent fluidized beds of different sizes. Source: Adapted from Ellis 2003 [58].

columns. In columns of diameter 0.05, 0.1, and 0.3 m, the bubble/void holdup was lower in the two small-diameter units [59] due to the occurrence of slugging, which increased the gas rising velocity. In relatively large turbulent fluidized beds of 0.3, 0.6, 0.9, and 1.5 m diameter, pressure fluctuations showed similar trends, with only slight differences [60], and the bed expansion was greater in columns of larger diameter than in small columns [60, 61]. For small-diameter columns, U_c decreased with increasing column diameter [59, 62]. For large columns, instead of column diameter, static bed height appears to have had a significant impact on pressure fluctuations, with larger fluctuations in a taller column [60], and the U_c transition velocity, determined from pressure fluctuations, increased with increasing expanded bed height or height-to-diameter ratio [61].

Since void size and rising velocity increase with increasing column diameter because of less wall restriction, solids and gas backmixing are both expected to intensify with increasing column diameter. This is supported by limited solids mixing data [44, 54, 63–65], which indicate that particle mixing is faster in large units than in small ones. Correspondingly, the axial solids dispersion coefficient is greater in columns of larger diameter.

9.5 Effect of Fines Content

As for the bubbles in bubbling fluidized beds, voids in turbulent fluidized bed can be altered significantly by varying the fine particle content of the bed. Using three particle size distributions (narrow, normal, and wide) of the same mean diameter and nearly the same particle density and Brunauer–Emmett–Teller (BET) specific surface area analysis, it was found [66] that voids were smaller and U_c was achieved earlier for a wide distribution powder than for particles of narrow size distribution. Similarly, increasing the fines content (mass fraction of particles <45 μm) increased the bed expansion, decreased pressure fluctuations, and lowered U_c [40, 60].

Gas and solids mixing are also strongly influenced by the fines content [44]. Both axial and radial gas dispersion coefficients increase with the addition of fines, reach a maximum at ~15% fines content, and then decrease gradually, eventually levelling off with further increases in fines content (see Figure 9.10). A similar trend is followed by solids dispersion, with both axial and radial solids dispersion coefficient reaching a maximum at a fines content of ~15%. The initial increase in gas and solids dispersion is related to the breakdown of large bubbles into small energetic voids, while the gradual decrease at higher fines content may be caused by the agglomeration of fines into clusters, reducing the mobility of particles and increasing the effective diameter of the bed particles.

Solved Problem

9.1 A cylindrical fluidization column of internal diameter 0.3 m is equipped with a porous plate distributor and two-stage cyclones. The 4 m tall column

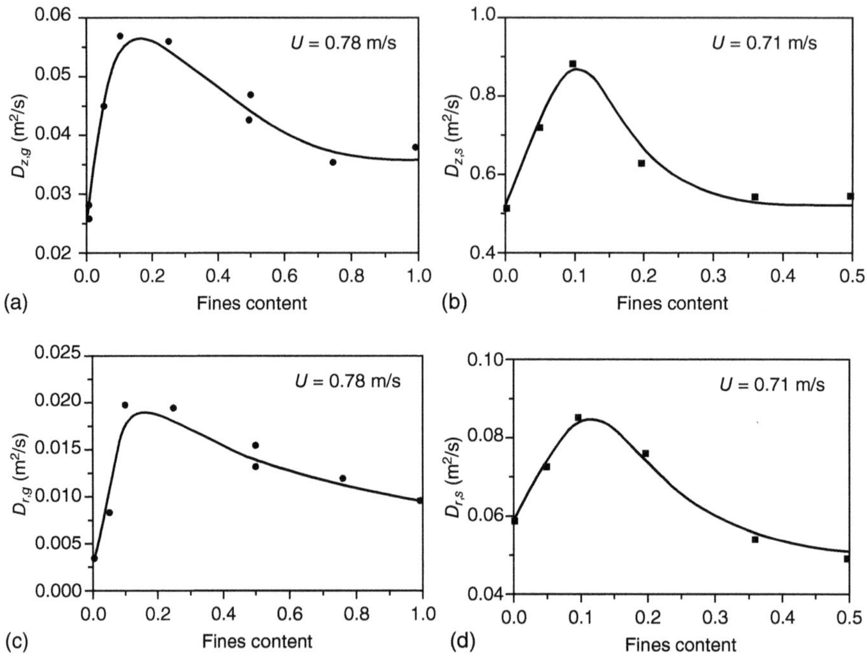

Figure 9.10 Effect of fines content on axial and radial dispersion coefficient of gas and particles in turbulent fluidized beds at ambient conditions. (a) Axial gas dispersion, (b) axial solids dispersion, (c) radial gas dispersion, and (d) radial solids dispersion. Source: Adapted from Du et al. 2002 [44].

is filled with 50 kg of fluid catalytic cracking (FCC) particles of 60 μm mean diameter and 1600 kg/m³ particle density. The loose-packed voidage is 0.42. If particles are fluidized in this column at a superficial velocity of 0.5 m/s by air at ambient conditions, estimate the expanded bed height based on ideal two-phase theory. Assume that the bubbles reach maximum stable size, as estimated from Eq. (4.6). Assume also that the solids holdup inside the voids is negligible.

Solution

$$\text{Static bed height} = \frac{m/[\rho_p(1-\varepsilon_0)]}{\left(\frac{\pi}{4}\right)D^2} = 0.762\text{ m} \approx H_{mf}$$

From Eq. (4.6), the maximum stable bubble size $d_{b,\max} = 2.0v_t'^2/g = 0.104$ m
$U_{mf} = 0.0012$ m/s from Chapter 2.
At $U = 0.5$ m/s, the height for the bubbles to reach the maximum size (0.104 m) is estimated to be ~0.10 m above the distributor. This is much smaller than the static bed height (0.762 m). Therefore, the void size along the entire bed height is assumed to be constant at 0.104 m.
From Eq. (7.11), the average void rise velocity is $u_b = 0.71\sqrt{gd_b} + U - U_{mf} = 1.20$ m/s.

Based on the two-phase theory (Eq. (7.13)), $\delta_{v,av} = \frac{U-U_{mf}}{u_b} = \frac{0.50-0.0012}{1.20} = 0.416$.
From Eq. (7.15), $H = \frac{H_{mf}}{1-\delta_{v,av}} = \frac{0.762\ \text{m}}{1-0.416} = 1.3\ \text{m}$.

Notations

Ar	Archimedes number ($=\rho_g(\rho_p - \rho_g)d_p{}^3 g/\mu_g{}^2$)
d_p	Sauter mean particle diameter (m)
C_z	tracer concentration (mol/m^3)
C_{zp}	tracer concentration at probe location (mol/m^3)
D	column diameter (m)
$D_{b,g}$	gas backmixing coefficient (m^2/s)
$D_{g.h}$	gas backmixing coefficient in the dense phase (m^2/s)
$D_{g,l}$	gas backmixing coefficient in the lean phase (m^2/s)
$D_{r,g}$	gas radial dispersion coefficient (m^2/s)
$D_{z,g}$	gas axial dispersion coefficient (m^2/s)
$D_{z,s}$	particle axial dispersion coefficient (m^2/s)
D_m	molecular diffusivity of gas mixture (m^2/s)
$d_{b,max}$	maximum stable bubble diameter (m)
d_b	bubble diameter (m)
g	gravitational constant (=9.81 m/s^2)
G_s	solids circulation flux (kg/m^2 s)
H	expanded bed height (m)
H_{mf}	dense bed height at $U = U_{mf}$ (m)
m	mass of bed materials (kg)
Pe	Péclet number ($=UH/D_z$) (–)
Pe_g	Péclet number of gas (–)
r	radial coordinate (m)
R	column radius (m)
Re	Reynolds number ($=\rho_g U d_p/\mu_g$) (–)
Re_c	Reynolds number at U_c ($=\rho_g U_c d_p/\mu_g$) (–)
Re_t	particle Reynolds number based on particle terminal velocity ($=\rho_g \cdot v_t \cdot d_p/\mu_g$) (–)
Sc	Schmidt number ($=\mu_g/(\rho_g \cdot D_m)$) (–)
T	temperature (K)
U	superficial gas velocity (m/s)
U_c	onset velocity of turbulent fluidization (m/s)
U_k	velocity at which pressure fluctuations level off (m/s)
U_{mf}	minimum fluidization velocity (m/s)
U_{se}	termination velocity of dense turbulent fluidization (m/s)
U_b	bubble rise velocity (m/s)
v_t	terminal settling velocity of particles (m/s)
z	axial position (m)
Z_p	axial position of sampling probe (m)
Z_i	axial position of tracer injection (m)

Greek Letters

ε	local voidage (–)
$\overline{\varepsilon}$	cross-sectional average voidage (–)
ε_d	dense phase voidage (–)
ε_{mf}	voidage at U_{mf} (–)
ε_v	void phase voidage (–)
δ_v	void phase volume fraction (–)
μ_g	gas viscosity (kg/m s)
ρ_g	gas density (kg/m^3)
ρ_p	particle density (kg/m^3)

References

1 Matheson, G.L., Herbst, W.A., and Holt, P.H. (1949). Characteristics of fluid-solid systems. *Ind. Eng. Chem.* 41: 1099–1104.

2 Zenz, F.A. (1949). Two-phase fluidized-solid flow. *Ind. Eng. Chem.* 41: 2801–2806.

3 Lanneau, K.P. (1960). Gas-solids contacting in fluidized beds. *Trans. Inst. Chem. Eng.* 38: 125–137.

4 Kehoe, P.W.K. and Davidson, J.F. (1970). Continuously slugging fluidized beds. In: *Chemeca '70, Institution of Chemical Engineers Symposium Series, 33*, 97–116. Australia: Butterworths.

5 Massimilla, L. (1973). Behavior of catalytic beds of fine particles at high gas velocities. *AIChE. Symp. Ser.* 69 (128): 11–12.

6 Thiel, W.J. and Potter, O.E. (1977). Slugging in fluidized beds. *Ind. Eng. Chem. Fundam.* 16: 242–247.

7 Crescitelli, S., Donsi, G., Russo, G., and Clift, R. (1978). High velocity behaviour of fluidized beds: slugs and turbulent flow. In: *Selected Papers Presented at the Sixth International CHISA '78 Congress* (ed. S. Schlosser), 1–11. Prague: Czech Society of Chemical Engineering.

8 Yerushalmi, J., Cankurt, N.T., Geldart, D., and Liss, B. (1978). Flow regimes in vertical gas-solid contact systems. *AIChE. Symp. Ser.* 174 (176): 1–12.

9 Rowe, P.N. and MacGillivray, H.J. (1980). The structure of a 15 cm diameter gas fluidized bed operated at up to 1 m/s and seen by X-rays. In: *Fluidization* (eds. J.R. Grace and J.M. Matsen), 545–553. New York: Plenum Press.

10 Rhodes, M.J. and Geldart, D. (1986). Transition to turbulence? In: *Fluidization V* (eds. K. Østergaard and A. Sørensen), 281–288. New York: Engineering Foundation.

11 Rhodes, M.J. (1996). What is turbulent fluidization? *Powder Technol.* 88: 3–14.

12 Rhodes, M.J. (1997). Turbulent fluidization: myth or reality? In: *Circulating Fluidized Bed Technology V* (eds. M. Kwauk and J. Li), 254–259. Beijing: Science Press.

13 Bi, X.T., Abba, I.A., Ellis, N., and Grace, J.R. (2000). A state-of-the-art review of gas–solids turbulent fluidization. *Chem. Eng. Sci.* 55: 4789–4825.

14 Yang, W.C. and Chitester, D.C. (1988). Transition between bubbling and turbulent fluidization at elevated pressure. *AIChe. Symp. Ser.* 84 (262): 10–21.
15 Tsukada, M., Nakanishi, D., and Horio, M. (1993). The effect of pressure on the phase transition from bubbling to turbulent fluidization. *Int. J. Multiphase Flow* 19: 27–34.
16 Brereton, C.M.H. and Grace, J.R. (1992). The transition to turbulent fluidization. *Chem. Eng. Res. Des.* 70: 246–251.
17 M'chirgui, A., Tadrist, L., and Radev, S. (1999). Analysis of the hydrodynamic instabilities in a gas fluidized bed. In: *Circulating Fluidized Bed Technology VI* (ed. J. Werther), 137–142. Frankfurt: Dechema.
18 Canada, G.S., McLaughlin, M.H., and Staub, F.W. (1978). Flow regimes and void fraction in gas fluidization of large particles in beds without tube banks. *AIChe. Symp. Ser.* 74 (176): 14–26.
19 Chyang, C.-S. and Huang, W.-C. (1988). Characteristics of large particle fluidization. *J. Chin. Inst. Chem. Eng.* 19: 81–89.
20 Dunham, G.E., Mann, M.D., and Grewal, N.S. (1993). Dependence of transition to turbulent fluidization on static bed depth in a fluidized bed. In: *Preprints of the Fourth International Conference Circulating Fluidized Beds* (ed. A.A. Avidan). Somerset, NJ: Engineering Conferences International.
21 Bai, D., Shibuya, E., Nakagawa, N., and Kato, K. (1996). Characterization of gas fluidization regimes using pressure fluctuations. *Powder Technol.* 87: 105–111.
22 Werther, J. and Wein, J. (1994). Expansion behavior of gas fluidized bed in the turbulent regime. *AIChe. Symp. Ser.* 90 (301): 31–44.
23 Abed, R. (1984). The characterization of turbulent fluid bed hydrodynamics. In: *Fluidization IV* (eds. D. Kunii and R. Toei), 137–144. New York: Engineering Foundation.
24 Zhu, H.Y., Zhu, J., Li, G.Z., and Li, F.Y. (2008). Detailed measurements of flow structure inside a dense gas–solids fluidized bed. *Powder Technol.* 180: 339–349.
25 Andreux, R. and Chaouki, J. (2008). Behaviors of the bubble, cloud, and emulsion phases in a fluidized bed. *AIChE J* 54: 406–414.
26 Holland, D.J., Mueller, C.R., Dennis, J.S. et al. (2010). Magnetic resonance studies of fluidization regimes. *Ind. Eng. Chem. Res.* 49: 5891–5899.
27 Wang, Z. and Wei, F. (1997). Similarity of the particles concentration distribution between bubbling and turbulent fluidized beds (in Chinese). In: *First Annual Conference of Chinese Society of Particle Technology*, 396–400. Beijing: Chinese Society of Particuology.
28 Li, Y.Z., Li, T., Zhang, H.T. et al. (2017). LDV measurements of particle velocity distribution and annular film thickness in a turbulent fluidized bed. *Powder Technol.* 305: 578–590.
29 Yamazaki, R., Asai, M., Nakajima, M., and Jimbo, G. (1991). Characteristics of transition regime in a turbulent fluidized bed. In: *Proceedings of the fourth China-Japan Fluidization Conference*, 720–725. Beijing: Science Press.
30 Lu, C.X., Xi, Y., Shi, M. et al. (1996). Radial voidage distribution in a large scale gas–solids turbulent fluidized bed. *J. Petrochem.* 25 (5): 315–317.

31 Farag, H.I., Ege, P.E., Grislingås, A., and de Lasa, H.I. (1997). Flow patterns in a pilot plant-scale turbulent fluidized bed reactor: concurrent application of tracers and fiber optic sensors. *Can. J. Chem. Eng.* 75: 851–860.
32 Taxil, I., Guigon, P., Archimbault, F., and Gauthier, T.A. (1998). Gas flow characterization in turbulent fluidization for FCC catalyst. In: *Fluidization IX* (eds. L.S. Fan and T.M. Knowlton), 69–76. New York: Engineering Foundation.
33 Zhang, X.P. and Bi, X.T. (2010). Study on void behavior in a turbulent fluidized bed with catalyst powders. *Ind. Eng. Chem. Res.* 49: 6862–6869.
34 Horio, M. and Nonaka, A. (1987). A generalized bubble diameter correlation for gas–solid fluidized beds. *AIChE J* 33: 1865–1872.
35 Lu, C., Xu, Y., Shi, M., and Shen, F. (1997). Two-area model for bubble distribution in a turbulent fluidized bed of fine particles. *Chin. J. Chem. Eng.* 5 (4): 287–295.
36 Lee, G.S. and Kim, S.D. (1989). Gas mixing in slugging and turbulent fluidized beds. *Chem. Eng. Commun.* 86: 91–111.
37 Wang, C., Wang, Y., Jin, Y., and Yu, Z. (1997). Effect of particle mean size on the performance of fluidized beds with FCC particles (in Chinese). In: *Symposium of the First Annual Conference of Chinese Society of Particle Technology* (eds. Y. Jin, Z. Yu and H. Li), 333–337. Beijing: Chinese Society of Particuology.
38 Chaouki, J., Gonzalez, A., Guy, C., and Klvana, D. (1999). Two-phase model for a catalytic turbulent fluidized-bed reactor: application to ethylene synthesis. *Chem. Eng. Sci.* 54: 2039–2045.
39 Bi, X.T. and Su, P. (2001). Local phase holdups in gas–solids fluidization and transport. *AIChE J* 47: 2025–2031.
40 Du, B., Warsito, W., and Fan, L.S. (2003). Bed nonhomogeneity in turbulent gas-solid fluidization. *AIChE J* 49: 1109–1126.
41 Cui, H.P. and Chaouki, J. (2004). Effects of temperature on local two-phase flow structure in bubbling and turbulent fluidized beds of FCC particles. *Chem. Eng. Sci.* 59: 3413–3422.
42 Nakajima, M., Harada, M., Asai, M. et al. (1991). Bubble fraction and voidage in an emulsion phase in the transition to a turbulent fluidized bed. In: *Circulating Fluidized Bed Technology III* (eds. P. Basu, M. Hasatani and M. Horio), 79–85. Oxford: Pergamon Press.
43 Foka, M., Chaouki, J., Guy, C., and Klvana, D. (1996). Gas phase hydrodynamics of a gas–solid turbulent fluidized bed reactor. *Chem. Eng. Sci.* 55: 713–723.
44 Du, B., Fan, L.S., Wei, F., and Warsito, W. (2002). Gas and solids mixing in a turbulent fluidized bed. *AIChE J* 48: 1896–1909.
45 Zhang, Y.M., Lu, C.X., Grace, J.R. et al. (2008). Gas back-mixing in a two-dimensional baffled turbulent fluidized bed. *Ind. Eng. Chem. Res.* 47: 8484–8491.
46 Li, Y. and Wu, P. (1991). A study on axial gas mixing in a fast fluidized bed. In: *Circulating Fluidized Bed III* (eds. P. Basu, M. Horio and M. Hasatani), 581–586. Oxford: Pergamon Press.

47 Abba, I.A. (2001). A generalized fluidized bed reactor model across the flow regimes. PhD dissertation. University of British Columbia, Vancouver, Canada.
48 Cankurt, N.T. and Yerushalmi, J. (1978). Gas backmixing in high velocity fluidized beds. In: *Fluidization* (eds. J.F. Davidson and D.L. Keairns), 387–393. Cambridge: Cambridge University Press.
49 Venderbosch, R.H. (1998). The role of clusters in gas–solids reactors. PhD dissertation. University of Twente, Netherlands.
50 Li, J. and Weinstein, H. (1989). An experimental comparison of gas backmixing in fluidized beds across the regime spectrum. *Chem. Eng. Sci.* 44: 1697–1705.
51 Schügerl, K. (1967). Experimental comparison of mixing processes in two- and three-phase fluidized beds. In: *Proceedings of the International Symposium on Fluidization* (ed. A.A.H. Drinkenburg), 782–796. Amsterdam: Netherlands University Press.
52 van Deemter, J.J. (1980). Mixing patterns in large-scale fluidized beds. In: *Fluidization* (eds. J.R. Grace and J.M. Matsen), 69–76. New York: Plenum Press.
53 Dang, T.Y.N., Gallucci, F., and van Sint Annaland, M. (2014). Gas mixing study in freely bubbling and turbulent gas–solid fluidized beds with a novel infrared technique coupled with digital image analysis. *Chem. Eng. Sci.* 116: 38–48.
54 Lee, G.S. and Kim, S.D. (1990). Axial mixing of solids in turbulent fluidized beds. *Chem. Eng. J.* 44: 1–9.
55 Lee, G.S., Kim, S.D., and Baird, M.H.I. (1991). Axial mixing of fine particles in fluidized bed. *Chem. Eng. J.* 47: 47–50.
56 Li, J., Reh, L., Tung, Y., and Kwauk, M. (1990). Comparison of fluidization behaviour in different flow regimes (in Chinese). In: *Proceedings of the Fifth National Conference on Fluidization* (eds. Y. Jin and S. Liu), 110–113. Chinese Society of Particuology.
57 Issangya, A.S. (1998). Flow dynamics in high density circulating fluidized beds. PhD dissertation. University of British Columbia, Vancouver, Canada.
58 Ellis, N. (2003). Hydrodynamics of gas-solid turbulent fluidized beds. PhD dissertation. University of British Columbia, Vancouver, Canada.
59 Du, B., Warsito, W., and Fan, L.-S. (2005). ECT Studies of gas-solid fluidized beds of different diameters. *Ind. Eng. Chem. Res.* 44: 5020–5030.
60 Issangya, A., Reddy Karri, S.B., Knowlton, T., and Cocco, R. (2013). Effects of bed diameter, baffles, fines content and operating conditions on pressure fluctuations in fluidized beds of FCC catalyst particles. In: *Fourteenth International Conference on Fluidization – From Fundamentals to Products*, ECI Symposium Series (eds. J.A.M. Kuipers, R.F. Mudde, J.R. van Ommen and N.G. Deen), 477–484. New York: Engineering Conferences International (ECI).
61 Ellis, N., Bi, X.T., Lim, C.J., and Grace, J.R. (2004). Hydrodynamics of turbulent fluidized beds of different diameters. *Powder Technol.* 141: 124–136.
62 Cai, P., Chen, S.P., Jin, Y. et al. (1989). Effect of operating temperature and pressure on the transition from bubbling to turbulent fluidization. *AIChe. Symp. Ser.* 85 (270): 37–43.
63 May, W.G. (1959). Fluidized-bed reactor studies. *Chem. Eng. Prog.* 55: 49–56.

64 Yerushalmi, J. and Avidan, A.A. (1985). High velocity fluidization. In: *Fluidization* (eds. J.F. Davidson, R. Clift and D. Harrison), 225–291. London: Academic Press.
65 Wei, F., Lin, S., and Yang, G. (1993). Gas and solids mixing in a commercial FCC regenerator. *Chem. Eng. Technol.* 16: 109–113.
66 Sun, G.L. and Grace, J.R. (1992). Effect of particle size distribution in different fluidization regimes. *AIChE J* 38: 716–722.

Problems

9.1 In Solved Problem 9.1, bubble growth near the distributor region is neglected in the calculation. If the bubble growth in the near distributor region is considered, redo the calculation for the expanded bed height and compare the result with that from Solved Problem 9.1.

9.2 In the same column as in Solved Problem 9.1 with an internal diameter of 0.3 m, now 80 kg sand particles of 150 μm in mean diameter and 2600 kg/m^3 in density are tested. Loosely packed voidage, $\varepsilon_0 = 0.42$. If particles are fluidized in this column at a superficial gas velocity of 0.8 m/s with air at ambient conditions, estimate the overall bed expansion based on ideal two-phase theory. The solids holdup inside the voids is assumed to be negligible in all calculations.

9.3 A fluidized bed has an internal diameter of 0.3 m, equipped with a porous plate distributor and two-stage cyclones. The column is 4 m tall, filled with 50 kg FCC particles of 60 μm in mean diameter and 1600 kg/m^3 in density. Loosely packed voidage, $\varepsilon_0 = 0.42$. If particles are fluidized in this column at a superficial gas velocity of 0.5 m/s with air at ambient conditions,

(a) Estimate axial gas and solids dispersion coefficients using proper correlations, and compare their order of magnitude.
(b) Estimate the radial gas dispersion coefficient, and compare it with the axial dispersion coefficient.

9.4 A fluidized bed of internal diameter 0.3 m is equipped with a porous plate distributor and two-stage cyclones. The column is 4 m tall, filled with 80 kg sand particles of 150 μm in mean diameter and 2600 kg/m^3 in density. Loosely packed voidage, $\varepsilon_0 = 0.42$. If particles are fluidized in this column at a superficial gas velocity of 1.0 m/s with air at ambient conditions,

(a) Estimate axial gas and solids dispersion coefficients using proper correlations, and compare their order of magnitude.
(b) Estimate the radial gas dispersion coefficient, and compare it with the axial dispersion coefficient.

10

Entrainment from Bubbling and Turbulent Beds

Farzam Fotovat

Sharif University of Technology, Chemical and Petroleum Engineering Department, P.O. Box 11365-9466, Azadi Avenue, 14588, Tehran, Iran

10.1 Introduction

In vigorously fluidized bed systems, solid particles are ejected and entrained from the dense bed. They enter the space between the bed surface and the gas outlet known as the freeboard. Depending on the size of the freeboard and velocity of particles, these particles may also leave the freeboard and escape from the unit. For most processes this is a disadvantage due to loss of valuable materials, such as catalysts, as well as potential release of fine particulates into the atmosphere. However, loss of fines by solids entrainment is exploited in some processes such as fluidized bed classifiers, incineration of sewage sludge, and ore roasting, where the entrained fines are removed and collected as fly ash or product [1].

When a single particle falls under gravity in a static gas in the absence of boundary constraints, it approaches its terminal settling velocity where the gravitational, buoyancy, and drag forces are balanced. If this particle encounters a gas of large extent moving upwards at a velocity equal to its terminal settling velocity, it will remain stationary. When the gas travels upwards in an enclosed vessel like a column with an average velocity equal to the particle's terminal velocity, due to the velocity profile of the gas (no slip at the wall, maximum velocity at the centre), the particle may move upwards or downwards depending on its radial position. If a number of particles of different sizes and densities are injected into the moving gas stream, particles may fall or rise depending on their sizes, densities, and radial positions. Moreover, due to interparticle forces, such as the electrostatic or van der Waals forces, the motion of the particles is hard to predict.

The bubbling characteristics of gas–solid fluidized beds enhance the complexity of the entrainment process. For instance, the size and velocity of bubbles bursting at the surface of the bed cause packets of particles to be ejected into the gas stream from the bed, cause gas turbulence in the freeboard, and distort the gas velocity profile immediately above the bed surface. These features cause the entrainment to be complex, difficult to characterize, and, in the fluidization literature, difficult to predict.

Essentials of Fluidization Technology, First Edition. Edited by John R. Grace, Xiaotao Bi, and Naoko Ellis.

10.2 Definitions

10.2.1 Transport Disengagement Height (TDH)

As particles travel upwards in the freeboard, some particles return to the dense bed, while others continue on upward trajectories, thereby resulting in decreasing entrainment of particles with increasing height. The lowest part of the freeboard, called the splash zone, accommodates the majority of coarse particles, whereas fine particles are prevalent in the upper part of the freeboard. As shown in Figure 10.1, the volumetric concentration of particles (often referred to as "solids holdup") monotonically decreases with increasing height, but becomes nearly constant at a certain height, called the "transport disengagement height" (TDH). The TDH is the effective freeboard height (H_f), measured from the surface of the dense bed below, beyond which the entrainment rate or particle concentration does not change appreciably. In practice, the TDH is the freeboard height commonly provided by fluidized bed designers above the estimated expanded bed height, since a shorter freeboard height than the TDH increases the carry-over from the bed [3], while any greater height is ineffective in reducing

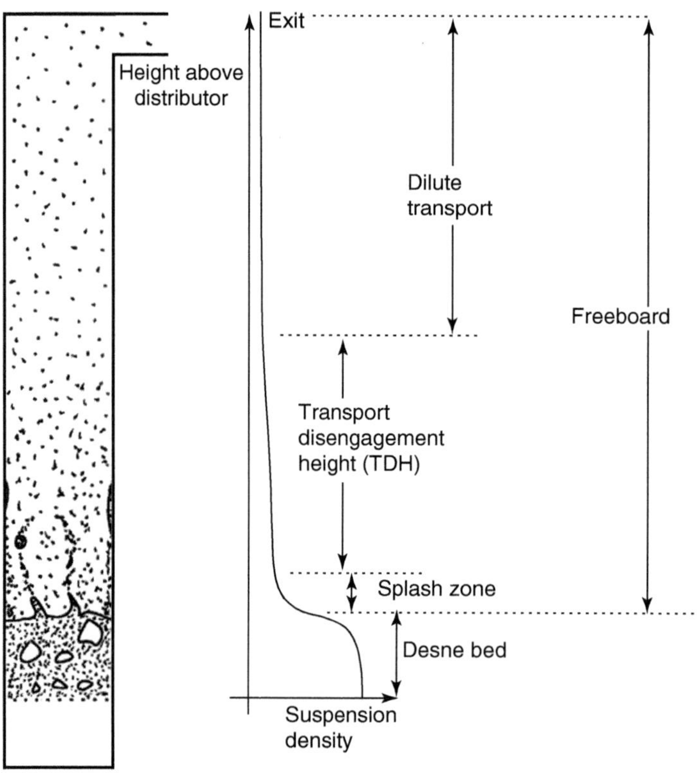

Figure 10.1 Solids concentration profiles in different zones of a gas–solid fluidized bed freeboard. Source: Rhodes 2013 [2]. Reproduced with permission of John Wiley and Sons.

the carry-over. The TDH is then an important parameter to ensure efficient and sustainable performance of one or more cyclones separating entrained particles from the gas downstream of fluidization units.

Experimental evidence shows that the TDH is affected by several factors, including particle properties (size, density, shape), gas properties (density, viscosity), superficial gas velocity, bubble characteristics at top of dense bed (diameter, velocity), column dimension (diameter, height), internals (e.g. tubes, baffles), solid mass, gas distributor, and operating conditions (i.e. temperature, pressure) [4, 5]. At low superficial gas velocities, the TDH is determined by ejection of coarse particles from the dense bed into the freeboard, while at high gas velocities entrainment of fines has a significant influence on the TDH [5]. A graphical correlation by Zenz and Othmer [6], based on industrial expertise for fluid catalytic cracking (FCC) or similar solids, is commonly used to provide a conservative estimate of the TDH. This correlation, plotted in SI units, is reproduced in Figure 10.2.

Several other approaches are also available to predict the TDH, but there are reportedly up to 5 orders of magnitude discrepancies among the values predicted by the TDH predictive correlations [5]. Moreover, some of these correlations predict unphysical phenomena, such as negative TDH values. The disagreement among the correlations stems primarily from the lack of a clear understanding of the microscale phenomena occurring in the freeboard zone, such as the interparticle (e.g. cohesion or clustering effects) and interspecies interactions (e.g. collisional momentum transfer effects) [4].

As shown in Figure 10.1, the region above the TDH is called the dilute transport zone, in which all particles are assumed to be carried upwards. For a freeboard of constant cross-sectional area, this region is characterized by time-average solids concentration and flux that do not vary with height.

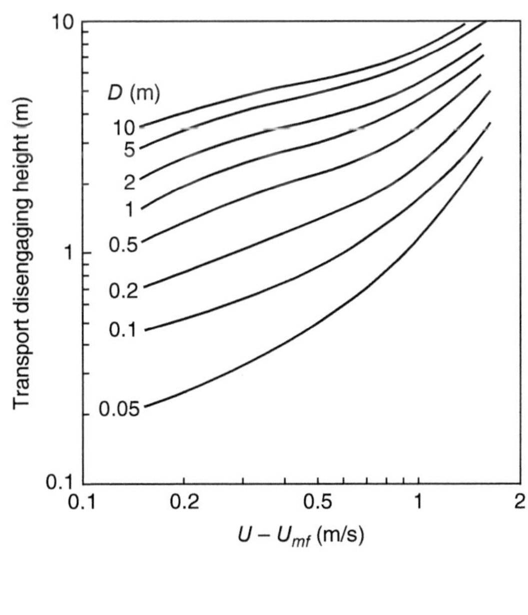

Figure 10.2 Transport disengagement height (TDH) graphical correlation. Source: Adapted from Zenz and Othmer 1960 [6].

10.2.2 Elutriation

Due to the dominance of the drag force relative to the gravitational force for very fine particles in a polydisperse system (where there is a range of particle sizes and/or particle densities), the fine particles are preferentially entrained from the bed, while coarse particles are likely to fall back onto the bed surface. In an empirical approach, the fine particles may be defined as particles whose terminal settling velocities are less than the superficial gas velocity (i.e. $v_{ti} < U$), whereas coarse particles have terminal settling velocities greater than the gas velocity ($v_{ti} > U$). The preferential entrainment of finer and lighter particles from the bed in a polydisperse system is called "elutriation." Thus, in general, the terms "entrainment" and "elutriation" should not be used interchangeably [7]. The classifying effect of the elutriation process results in selective removal of particles of smaller size or lower density from polydisperse fluidized beds. Unless attrition in the bed is sufficient to make up for the losses of elutriated particles, the size and density distribution of the particles remaining in the bed therefore shift to larger or denser particles as a result of the elutriation.

10.3 Ejection of Particles into the Freeboard

Once the bubbles travelling upwards in a bubbling fluidized bed reach the bed surface, they burst, ejecting clumps of particles into the freeboard. These particles may rise in the freeboard due to their initial momentum and either reach the gas exit assisted by drag exerted by the upward-flowing gas or disengage from the gas and fall back to the bed, almost always landing at some other horizontal position on the bed surface, thereby contributing to lateral solids mixing [3]. Bubbles are therefore responsible for initiating entrainment by ejecting particles into the freeboard. However, the ejected particles may originate at the nose or wake of bubbles, depending on the particle size, the fluidizing velocity, and bed geometry [8]. Bubbles bursting at the bed surface can eject particles from their nose into the freeboard [9]. In this case particles located between the bubble nose and the surface of the bed are ejected into the freeboard as a result of bubble eruption, with pressure higher than the surface pressure [10]. When bubbles coalesce as they reach the bed surface, the ejected particles come predominantly from the wake of overtaking bubbles [11]. In wake ejection, particles travel in the wake of bubble at the same velocity as the overtaking bubble and are ejected into the freeboard due to inertia when the bubble erupts at the bed surface, with the gas leaving the bed surface and then entraining these ejected particles in the freeboard. Particularly energetic bursts of particles occur when two bubbles coalesce just as the leading one reaches the bed surface, almost doubling the initial vertical component of particle velocity in the freeboard.

Ejection of particles from nose or wake of bubbles can affect the characteristic features of entrainment. Particles ejected from the bubble nose tend to be fines with smaller mean size than the bed material particles and are well dispersed into the freeboard after ejection. On the other hand, particles originating from the bubble wake are in the form of packets of bulk solids that may have been picked up from deep within the bed with nearly the same size distribution as bed material particles. Moreover, in the case of a bubble wake model, the solids flux at the bed surface is about 1 order of magnitude larger than predicted by the bubble nose model [1].

According to the literature [12, 13], elutriation of some nanoparticles can occur without the emergence of bubbles, i.e. a fluidized bed of such particles transits directly from the non-bubbling phase to elutriation, particularly when the fluidizing gas is highly viscous.

10.4 Entrainment Beyond the Transport Disengagement Height

The optimal design and the efficiency of cyclones or other gas–solid separation devices placed downstream of fluidized beds depend on the flux of particles entrained from the system. Failure to accurately estimate the entrainment rate could lead to underdesign or overdesign of the cyclone recovery system suffering from gas bypassing to the cyclone(s) or plugging of diplegs. Moreover, the extent of conversion of solids in gas–solid fluidized bed reactors depends on the residence time of particles in the bed, and this is a strong function of the entrainment rate, i.e. the higher entrainment rate of particles from fluidized beds, the shorter their residence time. In dual fluidized beds used for combustion, gasification, and chemical looping purposes, the mass and heat integration of the process greatly depend on the bed material cycle rate, which is directly related to the entrainment rate of particles. Hence, being able to predict the entrainment is critical for the technical and economic success of fluidized bed applications.

Since most industrial fluidized bed units have freeboards taller than the TDH, determining the entrainment rate above the TDH is of practical importance. To determine the elutriation rate of particles from vessels with freeboard heights greater than the TDH, it is generally assumed that the flux of size i of solid particle (G_i) is proportional to the mass fraction (x_i) of particles of size i in the bed (M_i) divided by the total mass of particles (M) in the bed:

$$G_{si} = K_i^* x_i \tag{10.1}$$

K_i^* (kg/m^2 s) is called the elutriation rate constant. It can be considered to be the largest flux of particles of size i that can be entrained with gas out of a vessel with a freeboard zone taller than the TDH, known as the saturation carrying capacity of the gas for that particular size of solid particle [14]. K_i^* is independent of bed height and inventory. Thus, it can also be used to estimate elutriation of particles from columns under batch or unsteady-state conditions as follows:

$$G_{si} = -\frac{1}{A}\frac{dM_i}{dt} = K_i^* x_i \tag{10.2}$$

where A and t are the cross-sectional area of the column and the duration of entrainment, respectively. For coarse particles not elutriated from the column, $K_i^* = 0$. It is generally assumed that the elutriation fluxes of fine species are cumulative. Accordingly, if the bed contains n sizes of elutriable particles ($U > v_{ti}$), then the total entrainment rate is given by

$$G_s = \sum_{i=1}^{n} K_i^* x_i \tag{10.3}$$

If all the bed solids are elutriable, then $\sum_{i=1}^{n} x_i = 1$. Otherwise, when coarse particles are in the system, $\sum_{i=1}^{n} x_i < 1$. The method used to experimentally determine K_i^* depends on the variability of bed composition over time. Under steady-state conditions, e.g. when the entrained particles are recirculated to the bed or the system is constantly fed so that the bed composition remains unchanged despite entrainment of fines, K_i^* can be simply calculated through Eq. (10.2) after measuring the carry-over rate and mass fraction of particles with size i in the bed.

For batch experiments, the composition and total mass of the bed change with time. By replacing $x_i = M_i/M$ in Eq. (10.2), it can be written as

$$-\frac{1}{A}\frac{dM_i}{dt} = K_i^* \left(\frac{M_i}{M}\right) \tag{10.4}$$

If the total mass of bed changes slightly (<20%) due to elutriation during a given period of time, then integration of Eq. (10.4) with M as a constant gives

$$\frac{M_i}{M_{i0}} = \exp\left(-\frac{K_i^* At}{M}\right) \tag{10.5}$$

which can be rearranged to

$$K_i^* = \frac{M}{At} \ln\left(\frac{M_{i0}}{M_i}\right) \tag{10.6}$$

Equation (10.6) suggests that K_i^* can be simply calculated by knowing the initial mass of particles of size i in the bed (M_{i0}) and the corresponding value (M_i), as well as the total mass of bed material (M) at time t.

If entrainment of particles leads to a significant loss of bed solids so that the bed inventory changes significantly (>20%) during the experimental run, integration of Eq. (10.4) for each elutriable species and dividing $\frac{M_1}{M_{10}} = \exp\left(-\frac{K_1^* At}{M}\right)$ by the corresponding equation of other elutriable species gives

$$\frac{M_1}{M_{10}} = \left(\frac{M_2}{M_{20}}\right)^{\frac{K_1^*}{K_2^*}} = \left(\frac{M_3}{M_{30}}\right)^{\frac{K_1^*}{K_3^*}} = \ldots = \left(\frac{M_i}{M_{i0}}\right)^{\frac{K_1^*}{K_i^*}} \tag{10.7}$$

By measuring the initial and final compositions of a bed subject to Eq. (10.7), one can determine the ratios of K_i^* values. By accounting for the mass of non-elutriating components in the bed (M_{ne}), integration of Eq. (10.4) also gives [14]

$$K_1^* At = M_{ne} \ln\left(\frac{M_{10}}{M_1}\right) + \sum_{i=1}^{n} M_{i0} \frac{K_1^*}{K_i^*} \left[1 - \left(\frac{M_1}{M_{10}}\right)^{\frac{K_i^*}{K_1^*}}\right] \tag{10.8}$$

K_1^* can be obtained from Eq. (10.8) that when replaced in Eq. (10.7) gives other K_i^* values.

In view of the complexity of the entrainment process, empirical approaches, rather than theoretical models, have generally been adopted to predict the rate of entrainment from fluidized beds. A number of empirical correlations have been suggested in the literature for K_i^* based on experimental studies on entrainment of particles from fluidized beds. However, some of these correlations predict unphysical phenomena, such as preferential carry-over of larger particles [7]. This is because the experimental data underlying the empirical correlations are only relevant in narrow ranges of conditions utilized in obtaining data and may not be appropriate for extrapolation. Thus, the most appropriate correlation for a specific system is the one based on conditions that most closely match those of the system of interest. However, Chew et al. [7] noted that whether a particular correlation was developed for a specific Geldart group or for monodisperse/polydisperse systems seems to have little or no influence on the accuracy of the predictions.

Table 10.1 provides empirical entrainment correlations for the widest ranges of tested conditions such as particle size distribution, fluidization velocity, and column diameter. According to Chew et al. [7], these correlations do not predict any unphysical entrainment trend for monodisperse, binary, and continuous particle size distribution systems for Geldart group A and B classifications. Among these correlations, the one reported by Choi et al. [17] is the most comprehensive empirical entrainment correlation, as it is based on >1000 experimental data points covering wide ranges of operating pressure (101–3200 kPa) and temperature (12–600 °C). Unlike other entrainment correlations, this correlation is based on the hydrodynamic forces exerted on the entrained particles, rather than individual variables.

There is a huge discrepancy, up to 20 orders of magnitude, in K_i^* values predicted by the empirical correlations [2]. No recommendation can be given on the most appropriate entrainment correlation due to the complexity and lack of fundamental understanding of the entrainment phenomena. The major contributors to the discrepancy in the underlying experimental data of the empirical correlations, which in turn affect the accuracy of the correlations, are as follows:

- Entrainment of species from a polydisperse system that may differ from the circumstances when particles of one species are entrained alone.
- Uncertainty regarding the TDH in different studies due to the discrepancies in the TDH prediction. This uncertainty results in the collection of experimental data at heights that may be below the actual TDH.
- Dissimilarity in the operating conditions, such as operating pressure and temperature, under which the experimental data were collected.
- Dissimilarity in the geometry and specifications of the columns used for the entrainment experiments, e.g. dissimilar geometry of the column exit, rectangular vs. circular cross section, different column diameters, wall roughness, etc.
- Overlooking the impact of interparticle interactions on the entrainment process.

In view of the above sources of discrepancy, it is not surprising that the empirical correlations obtained from fitting the experimental data of a specific system often fail to successfully predict the entrainment rate for systems of different

Table 10.1 Selected correlations for elutriation rate constant.

Investigators	Correlation	Geldart group	ρ_p (kg/m^3)	d_p (μm) (wt% of fines)	U (m/s)	D_c (m)	H_c (H_s) (m)
Zenz and Weil [9]	$K_i^* = 1.26\times10^7\rho_g U\left(\frac{U^2}{gd_{pi}\rho_p^2}\right)^{1.88}$ for $\frac{U^2}{gd_{pi}\rho_p^2} < 3\times10^{-4}$ $K_i^* = 4.31\times10^4\rho_g U\left(\frac{U^2}{gd_{pi}\rho_p^2}\right)^{1.18}$ for $\frac{U^2}{gd_{pi}\rho_p^2} > 3\times10^{-4}$	A		10–150	0.3–0.72	2-D column; 0.051 × 0.61	4.3
Merrick [15]	$K_i^* = \rho_g U\left[A + 130\exp\left(-10.4\left(\frac{v_{ti}}{U}\right)^{0.5}\left(\frac{U_{mfi}}{U-U_{mfi}}\right)^{0.25}\right)\right]$ $A = 10^{-4}, \ldots, 10^{-3}$	A, B	2400	125–2000	0.61–2.44	Square cross section; 0.91 × 0.91, 0.91 × 0.45	3.96 (0.61–1.22)
Sciążko et al. [16]	$K_i^* = 1.6\rho_g\frac{U^2}{U_{ti}}\left(1-\frac{v_{ti}}{U}\right)$	C, A	1250	80–2000	0.51–1.23	0.3	

Choi et al. [17]	$K_i^* = \frac{\mu_g Ar^{0.5} exp\left(6.92 - 2.11F_g^{0.303} - \frac{13.1}{F_d^{0.902}}\right)}{d_{pi}}$ $F_g = gd_{pi}(\rho_p - \rho_g)$ $F_d = \frac{C_d \rho_g U^2}{2}$ $C_d = \begin{cases} \frac{24}{Re_p} \text{ for } Re_p \leq 5.8 \\ \frac{10}{Re_p^{0.5}} \text{ for } 5.8 < Re_p \leq 540 \\ 0.43 \text{ for } Re_p > 540 \end{cases}$	A, B	2400–6158	30–710	0.152–2.8		1.97–9.1
Choi et al. [18]	$K_i^* = 0.314 C_d \rho_g U P^{0.946} \frac{F_d^{3.92}}{F_g^{2.04}}$ $F_g = gd_{pi}(\rho_p - \rho_g)$ $F_d = \frac{C_d \rho_g U^2}{2}$ $P = \sum x(U - v_{ti})$ $C_d = \begin{cases} \frac{24}{Re_p} \text{ for } Re_p \leq 5.8 \\ \frac{10}{Re_p^{0.5}} \text{ for } 5.8 < Re_p \leq 540 \\ 0.43 \text{ for } Re_p > 540 \end{cases}$	A, B	2465–2562	12–650	0.3–0.7	0.1	1.97 (0.15)

particles, operating conditions and experimental set-ups. The *ad hoc* inclusion or exclusion of parameters or dimensionless groups is not effective to improve empirical data fitting and generate a comprehensive entrainment correlation.

Accounting for particle-scale phenomena affecting the motion of particles in the freeboard zone is expected to reduce the discrepancy in the predictive entrainment correlations. For example, understanding and allowing for the interparticle effects, e.g. cohesive agglomeration of smaller particles, dynamic clusters due to hydrodynamic instabilities, and inter-species collisional momentum transfer, can lead to the development of universally applicable predictive equations with greatly improved accuracy [7]. An instance of such attempts is the continuum theory developed by Kellogg et al. [19] in which the effects of cohesive forces on rapid solid particle flows have been incorporated into a continuum framework. This theory has been recently applied to an unbounded gas–solid riser and successfully predicted the variation of the entrainment rate with solid volume fraction and the critical velocities of agglomeration and breakage, which, respectively, correspond to the velocity below which all collisions result in agglomeration and the velocity above which collisions with agglomerates result in breakage of the agglomerate [20].

Below the TDH, the particle flux decays approximately exponentially with height:

$$G_{si} = K_i^* x_i + G_{so} \exp\{-\alpha_i z'\} \tag{10.9}$$

where z' is the vertical coordinate measured from the bed surface. The decay constant, α_i, has been found [21] to vary from 3.5 to 6.4 m^{-1} and can be taken as 4.0 m^{-1} in the absence of specific information [3]. At high gas velocities $\alpha_i U$ is approximately constant rather than α_i, indicating that α_i decreases with increasing gas velocity. $\alpha_i U$ reportedly increases with increasing particle size. G_{so}, the entrainment flux at the bed surface, can be approximated [21] by

$$G_{so} = 9.6A(U - U_{mf})^{2.5}\, g^{0.5} \rho_g^{3.5} \mu_g^{-2.5} \tag{10.10}$$

where U_{mf} is the minimum fluidization velocity of the bed material and ρ_g and μ_g denote the density and viscosity of the fluidizing gas, respectively.

10.5 Entrainment from Turbulent Fluidized Beds

Increasing the superficial gas velocity and shifting from bubbling to the turbulent flow regime (see Chapter 5) sharply intensifies particle entrainment. Compared with the bubbling fluidization regime, particle entrainment due to wake ejection is much more pronounced in the turbulent regime because of the formation of more energetic voids rising faster in the bed. The elutriation rate constant above the TDH for turbulent fluidized beds is often predicted by means of an equation developed by Tasarin and Geldart [22]:

$$K_i^* = 14.5\rho_g U^{2.5} \exp\left\{-5.4\frac{v_{ti}}{U}\right\} \quad \text{for } Re_p > 3000 \tag{10.11}$$

For turbulent fluidized beds, the TDH can be predicted as for bubbling beds, i.e. based on Figure 10.2, or from other correlations in the literature [5].

A further increase in gas velocity results in fast fluidization, where the carry-over of solids is very large. Hence, fresh solids have to be recirculated or introduced continuously at a sufficient rate to make up for the loss of the bed solids and to achieve steady-state operation [14].

10.6 Parameters Affecting Entrainment of Solid Particles from Fluidized Beds

10.6.1 Properties of Particles

The entrainment rate constant increases with decreasing particle size for particles with a terminal velocity v_{ti} below the gas velocity in the freeboard. However, there is a critical particle size below which the elutriation rate constant K_i^* decreases with decreasing size. This has also been observed for fluidized beds containing binary particle mixtures involving fine and coarse particles [23]. The concept of critical particle size has been ascribed to the tendency of particles smaller than the critical size to adhere together or to adhere to bigger particles due to van der Waals attractive interactions [24]. Such particles are generally categorized as Geldart C particles for which the adhesion forces are comparable to the gravitational forces.

According to Ma and Kato [23] and Baeyens et al. [25], the critical particle diameter (d_{crit}) is only a function of particle density (ρ_p):

$$d_{crit} = \frac{101\,000}{\rho_p^{0.731} g} \text{ Ma and Kato [24], } (\rho_p \text{ in kg/m}^3, g \text{ in m/s}^2, d_{crit} \text{ in µm}) \tag{10.12}$$

$$d_{crit} = \frac{10\,325}{\rho_p^{0.725}} \text{ Baeyens et al. [26], } (\rho_p \text{ in kg/m}^3, d_{crit} \text{ in µm}) \tag{10.13}$$

These equations were based on experimental work carried out on a variety of fine materials with a broad range of particle size and density, $10 < d_p < 78$ µm, $1320 < \rho_p < 9442$ kg/m^3.

Although it is often assumed that coarse particles with a terminal settling velocity v_{ti} greater than the superficial gas velocity in the freeboard cannot be entrained from a freeboard that is larger than TDH, there is experimental evidence that coarse particles may be entrained, too, in the presence of elutriating fines with a considerable flux. The amount of elutriated coarse particles is significantly influenced by the flux of fines as a result of the momentum transferred from fines to coarse particles [26].

In a fluidized bed of fine–coarse particle mixture, the Geldart classification of the constituent species influences the entrainment of each component. For example, the weight fraction of fine particles and the mean diameter of the particle mixture affect the elutriation rate constant of group C powders mixed with group A particles [27–29]. However, for fine particles belonging to other Geldart groups, elutriation has been reported to be related to the initial loading of fines, but nearly independent of the size distribution of bed particles [18, 30]. In general, the impact of fines on the entrainment of coarse particles diminishes as the gas velocity increases [31].

Angular-shaped particles reportedly have lower entrainment rates compared with spherical particles of similar size and density, likely due to the bubble-splitting property of nonspherical particles resulting in a reduced size and activity of bubbles in fluidized beds involving such particles [32]. The impact of particle sphericity on its elutriation rate has also been shown by 3-D numerical simulations of char entrainment from a bubbling fluidized bed reactor [33]. Moreover, the surface properties of fines such as roughness can affect the entrainment process. This is explained due to the key role of van der Waals forces, which heavily depend on surface properties of particles, in the adherence and agglomeration of fines [34].

10.6.2 Geometry and Shape of Freeboard

Experimental findings suggest that entrainment is independent of column diameter (D_c) when $D_c > 0.1$ m. For columns with $D_c < 0.1$ m, the entrainment increases with increasing column diameter. In the case of an expanded freeboard, where the freeboard diameter > dense bed diameter, the gas velocity is reduced by the square of the ratio of bed diameter (D_{bed}) to freeboard diameter (D_{FB}) leading to a significant reduction in entrainment flux, so [35]

$$G_{si} \propto \left(\frac{D_{bed}}{D_{FB}}\right)^4 \tag{10.14}$$

This proportionality applies when the freeboard is sufficiently long that a uniform gas velocity profile applies across the whole cross-sectional area of the freeboard [1]. There is no consensus in the literature on the impact of the geometry and shape of the top gas exit on the particle entrainment rate.

10.6.3 Dense Bed Height

The impact of the static dense bed height on entrainment rate is not clear. Baron et al. [36] observed that the flux of entrained particles below and above the TDH is increased by increasing the static bed height in a column 0.61 m in diameter; however, Choi et al. [37] found no influence of dense bed height on entrainment rate of particles from two large-scale fluidized bed combustors with different rectangular cross-sectional areas. According to Zheng et al. [38], increasing the ratio of static bed height to column diameter significantly enhanced the entrainment of ultrafine silicon powder ($d_p = 2.7$ μm) from a stainless steel column with an inner diameter of 30 mm. This enhancement was attributed to a longer contact between gas and particles resulting in an increase in the amount of ultrafine particles carried by bubbles.

10.6.4 Internals

Inserting internals into the dense bed of a fluidized bed system can lead to a reduction in the entrainment rate due to the reduced bubble size [39]. The impact of internals and baffles inserted into the freeboard of a fluidized bed on entrainment depends on the design, orientation, and dimensions of the internals [40]. While they may promote gas–solid separation and therefore reduce the entrainment flux, their presence may lead to increased superficial gas velocity, due to a smaller cross section left for gas flow, or breakup of particle clusters [41], which in turn enhances entrainment flux. Thus, it is not surprising that different investigators have reported negative [42], positive [39], and even no influence [43] of the bed internals inserted into the freeboard of fluidized beds on the entrainment of particles.

10.6.5 Pressure and Temperature

Increasing the operating pressure leads to an increase in both the TDH and entrainment rate [44, 45]. These increases are attributed to the greater density of the gas at higher gas pressure, which reduces the particle terminal settling velocity. However, the rate of increase in the entrainment rate depends on the range of pressure [44].

Since most fluidized beds operate at elevated temperatures, it is important to consider the influence of temperature on entrainment. However, most investigations of entrainment have been performed under ambient conditions. Choi et al. [37] observed that, in the case of relatively coarse coal particles, increasing the bed temperature up to about 900 °C led to a reduction in solids entrainment rate. This was likely due to the lower gas density at high temperatures resulting in a reduced drag force acting on particles, associated with the bubble breakup and formation of smaller bubbles at high temperatures in a bed of coarse particles [46].

Later studies of Choi et al. [17, 47] on the impact of temperature on entrainment of sand (ρ_p = 2509 kg/m^3), emery (ρ_p = 3891 kg/m^3), and cast iron (ρ_p = 6158 kg/m^3) particles showed the occurrence of a minimum entrainment for temperatures of about 450–700 K when the temperature varied from ~300 to 900 K. As the temperature increases, the gas density decreases, while gas viscosity increases. This opposite impact of temperature on gas properties, which influences the drag force exerted on particles, likely results in a decreasing-minimum-increasing trend for the entrainment rate vs. temperature. This is in agreement with findings of studies in which the properties of the fluidizing gas were varied independently. For instance, Findlay and Knowlton and coworkers [48] observed a significant increase in the entrainment of char and limestone particles from high-temperature fluidized beds in which only the gas viscosity was changed while the gas density was kept constant by adjusting the system pressure. In addition to the influence of temperature on raising the gas viscosity and therefore augmenting the drag force exerted on entraining particles in the freeboard, increasing the viscosity of fluidizing gas by increasing the temperature also raises the viscosity of the emulsion phase of a fluidized bed composed of solid particles and interstitial gas. Consequently, larger bubbles with more momentum form in the bed, augmenting bed turbulence and entrainment of fines at a given superficial gas velocity [46].

To manipulate the fluidizing gas properties, Ellis et al. [49] varied the concentration of fluidizing air from 0 to 96 vol.%. by adding helium to the air used to fluidize FCC particles in a circulating fluidized bed. Increasing the helium concentration decreased both the fluidizing gas viscosity and density but surprisingly increased the circulation rate of FCC particles. This is in agreement with observations by Hoekstra and Sookai [50], who noticed similar entrainment rates of Geldart A particles with air and helium as fluidizing gas up to superficial velocities of 1 m/s under ambient conditions.

In the case of ultrafine (Geldart group C) particles, a relatively small increase in the fluidizing gas temperature reportedly resulted in a significant increase in particle entrainment rate [51].

The above findings suggest that the impact of gas temperature on entrainment cannot be solely explained based on the changes in the physical properties of the gas with changes in temperature. Operating temperature also influences other factors such as the distribution of gas between the bubble and emulsion phases, interparticle interactions, attrition and agglomeration of particles, and growth or shrinkage of particles as a result of chemical reactions, all of which could affect entrainment.

10.6.6 Electrostatic Charges

There is extensive evidence that the entrainment of electrically charged fines is reduced relative to uncharged particles due to particle–particle and particle–wall attractive interactions hindering free and independent particle motion [52]. It has also been demonstrated that the electrical properties of the fluidizing particles and the column wall material, which affect electrostatic interactions, can influence the entrainment tendency of particles from gas–solid fluidized beds [53, 54].

Figure 10.3 depicts the normalized entrainment flux of fines vs. electrostatic-to-gravity-force (F_e/F_g) ratio for a variety of binary mixtures composed of fine and coarse particles. As this figure suggests, regardless of the ratio between the superficial gas velocity and terminal settling velocity of fine particles (U/v_t), the

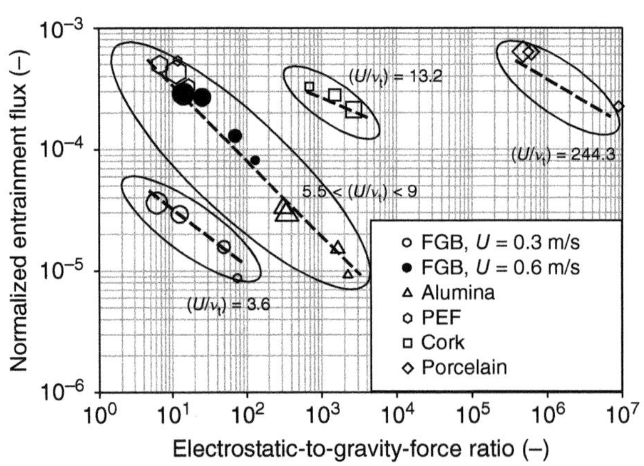

Figure 10.3 Normalized entrainment flux vs. electrostatic-to-gravity-force ratio of entrained fine particles from binary mixtures of coarse (90–95 wt%) and fine particles. (Coarse particles in all cases were glass beads; fines are identified in the figure; FGBs, fine glass beads; PEF, polyethylene furanoate.). Source: Adapted from Fotovat et al. 2016 [55].

entrainment flux decreases with increasing electrostatic force. Accounting for the impact of the electrostatic forces in an entrainment correlation markedly improves its accuracy [55].

10.7 Possible Means of Reducing Entrainment

According to the literature, the following measures can be taken to reduce the entrainment of particles from gas–solid fluidized beds:

- Decreasing the superficial velocity [1]
- Expanding the freeboard cross section [1]
- Using a conical geometry for the fluidization column [56]
- Increasing freeboard length if <TDH [14]
- Incorporating internals or baffles into the dense bed or freeboard [57]
- Utilizing swirling gas flow [58]
- Applying a centrifugal field to particles in a rotating fluidized bed [59]
- Applying mechanical and acoustic vibration in fluidized beds of fine particles [60, 61]
- Applying an electromagnetic field to the bed [60, 62]
- Enhancing the electrostatic charges carried by particles [63]
- Pulsating gas flow in fluidized bed of fine particles [60]
- Reducing attrition [34]

Solved Problem

10.1 Calculate the K_i^* values for a batch of coarse and fine sand particles when fines have a binary particle size distribution. The particles are fluidized with air at $U = 1.5$ m/s for 30 minutes in a pilot-scale column 0.8 m in diameter. Specifications of the fluidized solids are as follows:

Material	Species	ρ_p (kg/m^3)	d_p (μm)	v_t (m/s)	M_{i0} (kg)	M_i (kg)
Fine particles	1	2650	10	0.01	200	40
	2	2650	50	0.20	300	180
Coarse particles	3	2650	200	3.19	500	500

Solution

Assuming $H_f >$ TDH, the elutriation rate constants of fine species are calculated by Eq. (10.8) since 28% of the bed content has been entrained from the system. K_1^*/K_2^* ratio in Eq. (10.8) is obtained from Eq. (10.7) as follows:

$$\frac{M_1}{M_{10}} = \left(\frac{M_2}{M_{20}}\right)^{\frac{K_1^*}{K_2^*}} \rightarrow \frac{40}{200} = \left(\frac{180}{300}\right)^{\frac{K_1^*}{K_2^*}} \rightarrow \frac{K_1^*}{K_2^*} = 3.15$$

Equation (10.8) in its expanded format is written as

$$K_1^* = \frac{1}{At}\left(M_{ne}\ln\left(\frac{M_{10}}{M_1}\right) + M_{10}\left[1-\left(\frac{M_1}{M_{10}}\right)\right] + M_{20}\frac{K_1^*}{K_2^*}\left[1-\left(\frac{M_1}{M_{10}}\right)^{\frac{K_2^*}{K_1^*}}\right]\right)$$

Thus K_1^* is

$$K_1^* = \frac{1}{\left(\frac{\pi}{4}0.8^2\right)(30\times 60)}\left(500\ \ln\left(\frac{200}{40}\right) + 200\left[1-\left(\frac{40}{200}\right)\right]\right.$$
$$\left. + 300(3.15)\left[1-\left(\frac{40}{200}\right)^{\frac{1}{3.15}}\right]\right)$$
$$= 1.48\ \text{kg/m}^2\ \text{s}$$

and K_2^* is given by

$$K_2^* = \frac{K_1^*}{3.15} \rightarrow K_2^* = 0.47\ \text{kg/m}^2\ \text{s}$$

Note that the coarse sand particles (species 3) have not been entrained from the system.

Notations

A	cross-sectional area of the column (m²)
Ar	particle Archimedes number (dimensionless)
C_d	drag coefficient (dimensionless)
D_{bed}	inner diameter of bed (m)
D_c	inner diameter of fluidization column (m)
d_{crit}	critical particle diameter (μm)
D_{FB}	inner diameter of freeboard (m)
d_p	particle diameter (m)
F_d	drag force exerted on particles (kg/m s²)
F_e	electrostatic force exerted on particles (kg/m s²)
F_g	gravitational force exerted on particles (kg/m s²)
g	gravitational constant (m/s²)
G_i	entrainment flux of particles having d_i as average diameter (kg/m² s)
G_{si}	entrainment flux below TDH (kg/m² s)
G_{so}	entrainment flux at bed surface (kg/m² s)
H_c	height of fluidization column (m)
H_f	effective freeboard height (m)
H_s	height of settled (dense) fluidized bed (m)
K_i^*	elutriation rate constant of particles having d_i as average diameter (kg/m² s)
M	total mass of particles
M_i	mass of particles having d_i as average diameter (kg)
M_{i0}	initial mass of particles having d_i as average diameter (kg)
M_{ne}	mass of non-elutriating components in the bed (kg)
Re_p	particle Reynolds number (dimensionless)

t	duration of entrainment (s)
U	superficial gas velocity (m/s)
U_{mf}	minimum fluidization velocity (m/s)
v_{ti}	terminal settling velocity of particle having d_i as average diameter (m/s)
x_i	mass fraction of particles having d_i as average diameter (dimensionless)
z'	vertical coordinate measured from the bed surface (m)

Greek letters

ρ_p	particle density (kg/m^3)
ρ_g	gas density (kg/m^3)
μ_g	gas viscosity (kg/m s)
α_i	decay constant in Eq. (10.9) for entrainment in TDH, (m^{-1})

Subscripts

c	fluidized bed column
f	freeboard
g	gas
i	particles having d_i as average diameter
ne	non-elutriating components in the bed
p	particle

Abbreviations

TDH	transport disengagement height
FGB	fine glass beads
PEF	polyethylene furanoate
FCC	fluid catalytic cracking

References

1 Werther, J. and Hartge, E.U. (2003). Elutriation and entrainment. In: *Handbook of Fluidization and Fluid-Particle Systems* (ed. W.C. Yang), 113–128. CRC Press.
2 Rhodes, M. (2013). Fluidization. In: *Introduction to Particle Technology*, 2e, 169–210. Wiley.
3 Grace, J.R., Leckner, B., Zhu, J., and Cheng, Y. (2006). Fluidized beds. In: *Multiphase Flow Handbook* (ed. C.T. Crowe), 5-1–5-93. Florida: Taylor & Francis.
4 Cahyadi, A., Neumayer, A.H., Hrenya, C.M. et al. (2015). Comparative study of transport disengaging height (TDH) correlations in gas–solid fluidization. *Powder Technol.* 275: 220–238.
5 Tannous, K., Donida, M.W., and Bata, L.A. (2008). Entrainment of heterogeneous particles from gas-fluidized bed. *Part. Sci. Technol.* 26 (3): 222–234.
6 Zenz, F.A. and Othmer, D.F. (1960). *Fluidization and Fluid-Particle Systems.* Reinhold.

7 Chew, J.W., Cahyadi, A., Hrenya, C.M. et al. (2015). Review of entrainment correlations in gas–solid fluidization. *Chem. Eng. J.* 260: 152–171.

8 Pemberton, S.T. and Davidson, J.F. (1986). Elutriation from fluidized beds – I. Particle ejection from the dense phase into the freeboard. *Chem. Eng. Sci.* 41 (2): 243–251.

9 Zenz, F.A. and Weil, N.A. (1958). A theoretical-empirical approach to the mechanism of particle entrainment from fluidized beds. *AIChE J.* 4 (4): 472–479.

10 Fan, L.S. and Zhu, C. (2005). Dense-phase fluidized beds. In: *Principles of Gas-Solid Flows*, 371–420. Cambridge University Press.

11 George, S.E. and Grace, J.R. (1978). Entrainment of particles from aggregative fluidized beds. *AIChE Symp. Ser.* 74 (176): 67–74.

12 Zhu, C., Yu, Q., Dave, R.N., and Pfeffer, R. (2005). Gas fluidization characteristics of nanoparticle agglomerates. *AIChE J.* 51 (2): 426–439.

13 Valverde, J.M. and Castellanos, A. (2008). Bubbling suppression in fluidized beds of fine and ultrafine powders. *Part. Sci. Technol.* 26 (3): 197–213.

14 Kunii, D. and Levenspiel, O. (2013). Entrainment and elutriation from fluidized beds. In: *Fluidization Engineering*, 165–192. Elsevier.

15 Merrick, D. (1974). Particle size reduction and elutriation in a fluidized bed process. *AIChE. Symp. Ser.* 70: 366–378.

16 Sciążko, M., Bandrowski, J., and Raczek, J. (1991). On the entrainment of solid particles from a fluidized bed. *Powder Technol.* 66 (1): 33–39.

17 Choi, J.H., Chang, I.Y., Shun, D.W. et al. (1999). Correlation on the particle entrainment rate in gas fluidized beds. *Ind. Eng. Chem. Res.* 38 (6): 2491–2496.

18 Choi, J.H., Suh, J.M., Chang, I.Y. et al. (2001). The effect of fine particles on elutriation of coarse particles in a gas fluidized bed. *Powder Technol.* 121 (2–3): 190–194.

19 Kellogg, K.M., Liu, P., Lamarche, C.Q., and Hrenya, C.M. (2017). Continuum theory for rapid cohesive-particle flows: general balance equations and discrete-element-method-based closure of cohesion-specific quantities. *J. Fluid Mech.* 832: 345–382.

20 Kellogg, K.M., Liu, P., and Hrenya, C.M. (2019). Continuum prediction of entrainment rates and agglomeration of gas-fluidized, lightly-cohesive particles. *Chem. Eng. Sci.* 199: 249–257.

21 Wen, C.Y. and Chen, L.H. (1982). Fluidized bed freeboard phenomena: entrainment and elutriation. *AIChE J.* 28 (1): 117–128.

22 Tasirin, S.M. and Geldart, D. (1998). Entrainment of FCC from fluidized beds – a new correlation for the elutriation rate constants $K_{i\infty}$. *Powder Technol.* 95 (3): 240–247.

23 Ma, X. and Kato, K. (1998). Effect of interparticle adhesion forces on elutriation of fine powders from a fluidized bed of a binary particle mixture. *Powder Technol.* 95 (2): 93–101.

24 Tasirin, S.M. and Geldart, D. (1999). The elutriation of fine and cohesive particles from gas fluidized beds. *Chem. Eng. Commun.* 173 (1): 175–195.

25 Baeyens, J., Geldart, D., and Wu, S.Y. (1992). Elutriation of fines from gas fluidized beds of Geldart A-type powders – effect of adding superfines. *Powder Technol.* 71: 71–80.

26 Geldart, D. and Pope, D.J. (1983). Interaction of fine and coarse particles in the freeboard of a fluidized bed. *Powder Technol.* 34 (1): 95–97.

27 Li, J., Yamashita, A., and Kato, K. (2000). Elutriation of very fine particles from fluidized bed – effect of mean diameter of bed particles. *J. Chem. Eng. Jpn.* 33 (5): 730–739.

28 Mahmoud, E.A., Nakazato, T., Kohama, K., and Kato, K. (2003). Total solid entrainment rate in a circulating powder-particle fluidized bed. *J. Chem. Eng. Japan* 36 (11): 1405–1409.

29 Nakazato, T. and Kato, K. (2008). Entrainment rates and elutriation rate constants of bed particles and group C fine powder in a fluidized bed and in a circulating fluidized bed. *J. Chem. Eng. Japan* 41 (7): 678–685.

30 Monazam, E.R., Breault, R.W., Weber, J., and Layfield, K. (2017). Elutriation of fines from binary particle mixtures in bubbling fluidized bed cold model. *Powder Technol.* 305: 340–346.

31 Choi, J.H., Kim, S.D., and Grace, J.R. (2008). Entrainment rate of coarse particles at different temperatures in gas fluidized beds. *Can. J. Chem. Eng.* 85 (2): 151–157.

32 Sookai, S. (2007). The effect of particle properties on fluidized bed hydrodynamics and entrainment. MSc thesis. University of Kwazulu-Natal, Durban, South Africa.

33 Papadikis, K., Gu, S., and Bridgwater, A.V. (2010). 3D simulation of the effects of sphericity on char entrainment in fluidised beds. *Fuel Process. Technol.* 91 (7): 749–758.

34 Maurer, S., Durán, S.R., Künstle, M., and Biollaz, S.M.A. (2016). Influence of interparticle forces on attrition and elutriation in bubbling fluidized beds. *Powder Technol.* 291: 473–486.

35 Smolders, K. and Baeyens, J. (1997). Elutriation of fines from gas fluidized beds: mechanisms of elutriation and effect of freeboard geometry. *Powder Technol.* 92 (1): 35–46.

36 Baron, T., Briens, C.L., Galtier, P., and Bergougnou, M.A. (1990). Effect of bed height on particle entrainment from gas-fluidized beds. *Powder Technol.* 63 (2): 149–156.

37 Choi, J.H., Son, J.E., and Kim, S.D. (1989). Solid entrainment in fluidized bed combustors. *J. Chem. Eng. Japan* 22 (6): 597–606.

38 Zheng, X., Yin, S., Ding, Y., and Wang, L. (2019). Experimental study on high concentration entrainment of ultrafine powder. *Powder Technol.* 344: 133–139.

39 Tweddle, T.A., Capes, C.E., and Osberg, G.L. (1970). Effect of screen packing on entrainment from fluidized beds. *Ind. Eng. Chem. Process. Des. Dev.* 9 (1): 85–88.

40 Harrison, D., Aspinall, P.N., and Elder, J. (1974). Suppression of particle elutriation from a fluidized-bed. *Trans. Inst. Chem. Eng.* 52 (2): 213–216.

41 Cocco, R., Hays, R., Reddy Karri, S.B.K., and Knowlton, T.M. (2010). The effect of cohesive forces on catalyst entrainment of fluidized bed regenerators. In: *Advances in Fluid Catalytic Cracking: Testing, Characterization, and Environmental Regulations* (ed. M.L. Occelli), 155–172. CRC Press.

42 Baron, T., Briens, C.L., and Bergougnou, M.A. (1987). Reduction of particle entrainment from gas-fluidized beds with a screen of floating balls. AIChE Annual Meeting, November 1987.

43 George, S.E. and Grace, J.R. (1981). Entrainment of particles from a pilot scale fluidized bed. *Can. J. Chem. Eng.* 59 (3): 279–284.

44 Chan, I.H. and Knowlton, T.M. (1984). The effect of pressure on entrainment from bubbling gas-fluidized beds. In: *Fluidization* (eds. D. Kunii and R. Toei), 283–290. New York: Engineering Foundation Conferences.

45 Chan, I.H. and Knowlton, T.M. (1984). The effect of system pressure on the transport disengaging height (TDH) above bubbling gas-fluidized beds. *AIChE. Symp. Ser.* 80 (241).

46 Shabanian, J. and Chaouki, J. (2017). Effects of temperature, pressure, and interparticle forces on the hydrodynamics of a gas–solid fluidized bed. *Chem. Eng. J.* 313: 580–590.

47 Choi, J.H., Choi, K.B., Kim, P. et al. (1997). The effect of temperature on particle entrainment rate in a gas fluidized bed. *Powder Technol.* 92 (2): 127–133.

48 Foscolo, P.U., Gibilaro, L.G., Findlay, J.G., and Knowlton, T.M. (1985). Final Report for US Department of Energy. Project DE-AC21-83MC20314.

49 Ellis, N., Xu, M., Lim, C.J. et al. (2011). Effect of change in fluidizing gas on riser hydrodynamics and evaluation of scaling laws. *Ind. Eng. Chem. Res.* 50 (8): 4697–4706.

50 Hoekstra, E. and Sookai, S. (2014). The effect of gas density on fluidized-bed entrainment. *S. Afr. J. Chem. Eng.* 19 (3): 90–98.

51 Yang, F.M., Wang, L., Yin, S.W., and Li, Y.H. (2012). Experimental study on the relationship between carrier gas temperature and entrainment characteristics of ultrafine silicon powder in fluidized bed. *Adv. Mater. Res.* 482–484: 2587–2591.

52 Fotovat, F., Bi, X.T., and Grace, J.R. (2017). Electrostatics in gas–solid fluidized beds: a review. *Chem. Eng. Sci.* 173: 303–334.

53 Fotovat, F., Grace, J.R., and Bi, X.T. (2016). Particle entrainment from gas-solid fluidized beds: conductive vs dielectric fines. *AIChE J.* 63 (4): 1194–1202.

54 Fotovat, F., Gill, K., Grace, J.R., and Bi, X.T. (2017). Impact of column material on electrostatics and entrainment of particles from gas-solid fluidized beds. *Chem. Eng. Sci.* 167: 120–134.

55 Fotovat, F., Alsmari, T.A., Grace, J.R., and Bi, X.T. (2016). The relationship between fluidized bed electrostatics and entrainment. *Powder Technol.* 316: 157–165.

56 Shin, M.K., Kim, E.M., Koo, B.S. et al. (2007). Entrainment characteristics of fine particles in cylindrical and conical inert-medium fluidized beds. *Ind. Eng. Chem. Res.* 46 (4): 1408–1414.

57 Callen, A., Moghtaderi, B., and Galvin, K.P. (2007). Use of parallel inclined plates to control elutriation from a gas fluidized bed. *Chem. Eng. Sci.* 62 (1–2): 356–370.
58 Zhu, S. and Lee, S.W. (2005). Co-combustion performance of poultry wastes and natural gas in the advanced Swirling Fluidized Bed Combustor (SFBC). *Waste Manag.* 25 (5): 511–518.
59 Saunders, J.H. (1986). Particle entrainment from rotating fluidized beds. *Powder Technol.* 47 (3): 211–217.
60 Zhang, W. (2009). A review of techniques for the process intensification of fluidized bed reactors. *Chin. J. Chem. Eng.* 17 (4): 688–702.
61 Zhu, C., Liu, G., Yu, Q. et al. (2004). Sound assisted fluidization of nanoparticle agglomerates. *Powder Technol.* 141: 119–123.
62 Wang, J.S. and Colver, G.M. (2003). Elutriation control and charge measurement of fines in a gas fluidized bed with ac and dc electric fields. *Powder Technol.* 135–136: 169–180.
63 Yang, Y., Zi, C., Huang, Z. et al. (2017). CFD–DEM investigation of particle elutriation with electrostatic effects in gas-solid fluidized beds. *Powder Technol.* 308: 422–433.

Problems

10.1 A batch fluidized column bed, 1 m in inner diameter, containing a mixture of sand and olivine particles is operated for four hours. Using the initial and final mass of each species shown in the table below, calculate the ratio of the entrainment flux rate above the TDH to that 9 m above the bed surface. Can you conclude based on the calculated ratio that the TDH of this fluidized bed is larger than 9 m?

$\rho_g = 1.22\ \text{kg/m}^3$	$\mu_g = 1.80 \times 10^{-5}\ \text{kg/m s}$
$U = 0.5\ \text{m/s}$	$U_{mf} = 0.2\ \text{m/s}$

Material	M_0 (kg)	M (kg)
Olivine	4000	3800

10.2 FCC particles with a size distribution given below and a particle density of 1700 kg/m^3 are fluidized at 1200 K in a circulating fluidized bed at $U = 1.5$ m/s. Entrained FCC particles are fully separated from the fluidizing air by an efficient cyclone and returned to the bed. Assuming that $H_f >$ TDH, calculate the size distribution of the particles in the cyclone return leg based on:
(a) Zenz and Weil [9] correlation
(b) Choi et al. [17] correlation

$\rho_g = 0.29\ \text{kg/m}^3$	$\mu_g = 4.63 \times 10^{-5}\ \text{kg/m s}$

Size range number (i)	Size range (μm)	Mass fraction
1	40–60	0.20
2	60–80	0.60
3	80–100	0.15
4	100–120	0.05

10.3 Monodispersed fine particles ($\rho_p = 2000\ \text{kg/m}^3$, $d_p = 40\ \mu\text{m}$) are elutriated by air from a batch fluidized bed. Using the Choi et al. [17] correlation, estimate how long it will take for 40 wt% of the bed content to be depleted by entrainment if the bed is operated at:

(a) Room temperature ($\rho_g = 1.18\ \text{kg/m}^3$, $\mu_g = 1.85 \times 10^{-5}\ \text{kg/m s}$)
(b) 700 K ($\rho_g = 0.50\ \text{kg/m}^3$, $\mu_g = 3.33 \times 10^{-5}\ \text{kg/m s}$)

11

Standpipes and Return Systems, Separation Devices, and Feeders

Ted M. Knowlton and Surya B. Reddy Karri

Particulate Solid Research, Inc., 4201 West 36th Street, Building A, Chicago, IL 60632, USA

This chapter discusses solids return systems in fluidized bed processes, feed systems to controllably feed and discharge solids from fluidized beds, and different ways in which particulate solids are separated from exit gases.

11.1 Standpipes and Solids Return Systems

In many chemical processes, solid particles flow around a circulation loop, such as that shown in Figure 11.1 for a fluidized catalytic cracking (FCC) process in which the solids are transported from high pressure to low pressure in a riser reactor where oil is cracked catalytically into a product gas. During its passage up the riser, the catalyst is coated with carbon, reducing its reactivity. Its reactivity is restored in a regenerator, which burns off the carbon on the surface of the catalyst. The catalyst is transferred from the pressure at the top of the riser to a higher pressure in the regenerator via a standpipe. After its reactivity is restored, the catalyst is transferred back to a higher pressure at the bottom of the riser via another standpipe.

Standpipes in the FCC loop are pipes through which solids flow by gravity. The primary purpose of a standpipe is to transfer solids from low pressure to a higher pressure. Standpipes can be vertical, angled, or a combination of angled and vertical pipes called hybrid standpipes.

Solids can be transferred by gravity from a low pressure to a higher pressure if gas flows upwards *relative* to the downward-flowing solids. This relative gas–solid flow generates the sealing pressure drop required for the system. The direction of the gas flow in the standpipe relative to the standpipe wall can be upwards or downwards and still have the relative gas–solid velocity, v_r, directed upwards. This can be seen with the aid of Figure 11.2, and the definition of relative velocity is

$$v_r = |v_s - v_g| \tag{11.1}$$

where v_s is the solid velocity:

$$v_s = G_{s/}\rho_{susp} = (W_s/A)/(\rho_p(1-\varepsilon)) \tag{11.2}$$

Essentials of Fluidization Technology, First Edition. Edited by John R. Grace, Xiaotao Bi, and Naoko Ellis.

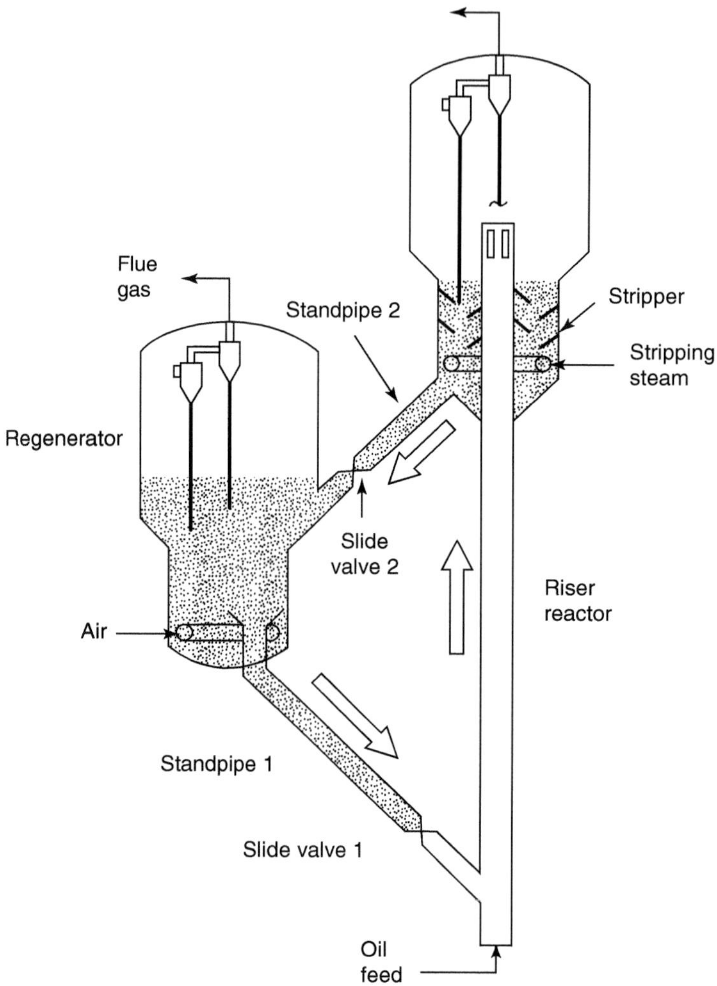

Figure 11.1 Schematic drawing of FCC process solids circulation loop.

and v_g is the interstitial gas velocity:

$$v_g = U/\varepsilon = W_g/(A \cdot \varepsilon \cdot \rho_g) \tag{11.3}$$

The positive direction for determining v_r in this chapter is downwards, i.e. the direction in which the solids are flowing. In Figure 11.2 solids are being transferred downwards in a standpipe from pressure P_1 to a higher pressure P_2. The solid velocity is denoted by the length of the bold arrows, gas velocity by the length of the dashed arrows, and the relative velocity by the length of the gray arrows.

For Case I solids are travelling downwards, and gas is flowing upwards relative to the standpipe wall. The relative velocity is directed upwards and is equal to the sum of the solid velocity and the gas velocity:

$$v_r = |v_s - (-v_g)| = |v_s + v_g| \tag{11.4}$$

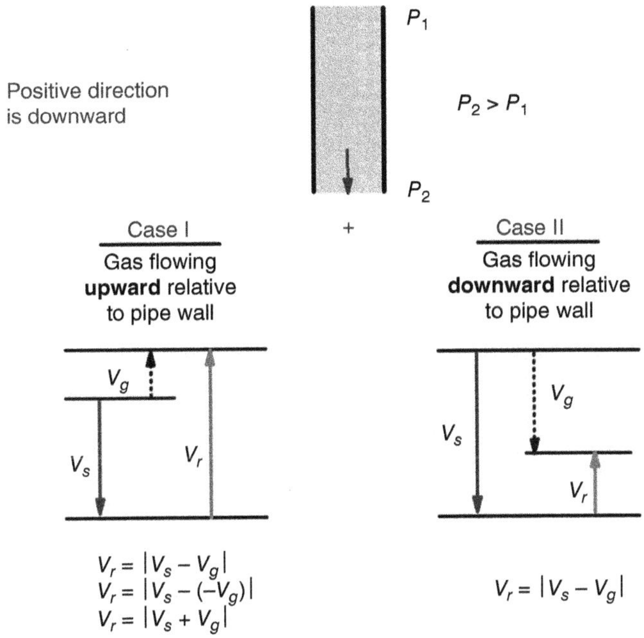

Figure 11.2 Standpipe relative gas velocity flow cases.

In Case II, solids are moving downwards relative to the standpipe wall. Gas is also flowing downwards relative to the standpipe wall, but at a velocity less than that of the solids. For this case, v_r is directed upwards and is equal to the difference between v_s and v_g:

$$v_r = |v_s = v_g| \tag{11.5}$$

In both cases, if one were riding down the standpipe with the solids, the gas would appear to be moving upwards. Gas flowing upwards relative to the solids generates a frictional pressure drop. The relationship between the pressure drop per unit length ($\Delta P/(Lg)$) and v_r for a material is determined by the fluidization curve for that material. This fluidization curve is generated in a fluidization column and applies for solids flowing in a standpipe.

Most standpipe transfer systems use either group A or group B solids (see Chapter 2). Group C solids have difficulties flowing in standpipes because the cohesive nature of the solids causes the particles to bridge and stop flowing. Group D solids can be used in standpipes, but processes using group D solids in standpipes are uncommon. The fluidization curve for group B solids differs from that for group A solids. For group A and group B solids, as the relative gas velocity through the bed increases from zero, the $\Delta P/(Lg)$ through the bed increases linearly with v_r (packed bed region). At some v_r, the ΔP generated by the gas flowing through the solids is equal to the weight of the solids per unit area, and the solids become fluidized. The relative velocity at this point is called the interstitial minimum fluidization velocity, v_{mf}, equal to U_{mf}/ε_{mf}. $\Delta P/(Lg)$ at v_{mf} is designated as $\Delta P/(Lg)_{mf}$. Increases in v_r above v_{mf} do not lead to further increases in $\Delta P/(Lg)$. Instead, as v_r increases beyond v_{mf}, $\Delta P/(Lg)$ remains almost constant and then begins to decrease slowly as the bubble volume in the bed increases.

For group A materials, as v_r increases above v_{mf}, the solids expand without bubble generation over a small velocity range and $\Delta P/(Lg)$ decreases. The velocity where bubbles begin to form in group A materials is called the minimum bubbling velocity, v_{mb}.

Standpipes generally operate in one of two basic flow regimes – packed bed flow and fluidized bed flow.

11.1.1 Packed Bed Flow

In packed bed flow, $v_r < v_{mf}$, and the voidage in the standpipe is more or less constant. As v_r increases, $\Delta P/(Lg)$ increases linearly.

11.1.2 Fluidized Bed Flow

In fluidized bed flow, $v_r \geq v_{mf}$. The voidage in the standpipe can change along the length of the standpipe, and $\Delta P/(Lg)$ does not change significantly with increasing v_r. There are two kinds of fluidized bed flow:

(a) Bubbling fluidized bed flow
(b) Non-bubbling fluidized bed flow

Fluidized group B solids always operate in the bubbling fluidized bed mode because bubbles are formed at all $v_r > v_{mf}$. For group A solids, there is a velocity window corresponding to v_r between v_{mf} and v_{mb}, where the solids are fluidized, but no bubbles are formed in the standpipe. Above v_{mb} group A solids operate in the bubbling fluidized bed mode. Bubbles that are large relative to the standpipe diameter are undesirable in a standpipe. When a standpipe is operating in the bubbling fluidized bed mode where v_s is less than the bubble rise velocity, u_b, bubbles rise and grow by coalescence. Bubbles rising against downflowing solids hinder and limit the solids flow rate [1, 2] and also reduce the $\Delta P/(Lg)$ or "density" of the solids in the standpipe. When v_s in the standpipe $> u_b$, the solids carry the bubbles down the standpipe.

For optimum fluidized standpipe operation for group A, B, and D solids, v_r should be maintained just slightly above v_{mf} or v_{mb}.

11.1.3 Types of Standpipes

There are two standpipe configurations – the overflow standpipe (Figure 11.3(a)) and the underflow standpipe (Figure 11.3(b)). In an overflow standpipe the solids "overflow" from the top of the fluidized bed into the standpipe, and there is *no bed of solids above the standpipe.* In the underflow standpipe, solids are introduced into the standpipe from the underside of the bed or hopper, and *a bed of solids is present above the standpipe.* With these definitions, a cyclone dipleg is a fluidized overflow standpipe because there is no bed of solids above the dipleg. With two

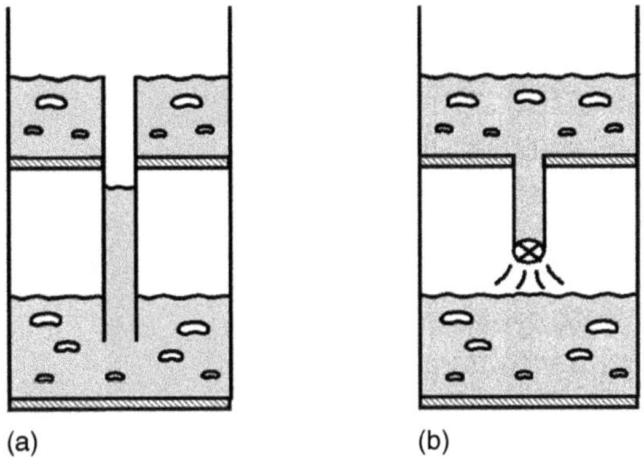

Figure 11.3 (a) Overflow and (b) underflow standpipe configurations.

standpipe configurations and two standpipe flow regimes, there are four different types of standpipes:

1. Underflow packed bed standpipe
2. Underflow fluidized bed standpipe
3. Overflow fluidized bed standpipe
4. Overflow packed bed standpipe

These standpipes are used in industry, except for the overflow packed bed standpipe. It is possible for this standpipe to operate, but it would require an extensive and complicated control scheme, so it is not used.

11.1.3.1 Overflow Fluidized Standpipe

In the overflow fluidized standpipe shown in Figure 11.4, solids are transferred from an upper fluidized bed to a lower fluidized bed against a pressure differential $P_2 - P_1$. This pressure drop consists of the sum of the pressure drop in the lower fluidized bed from the standpipe exit to the top of the bed (ΔP_{lb}), the pressure drop across the distributor (ΔP_d), and the pressure drop across the upper fluidized bed (ΔP_{ub}):

$$P_2 - P_1 = \Delta P_{lb} + \Delta P_d + \Delta P_{ub} \tag{11.6}$$

This pressure drop must be balanced by the pressure drop generated in the overflow standpipe. If the solids in the standpipe are fluidized, the solids height in the standpipe, H_{sp}, adjusts so that the pressure build-up in the standpipe, ΔP_{sp}, equals the product of $(\Delta P/Lg)$ and H_{sp}, i.e.

$$\Delta P_{sp} = P_2 - P_1 = \Delta P_{lb} + \Delta P_d + \Delta P_{ub} = \Delta P/(Lg)\,(H_{sp}) \tag{11.7}$$

If the gas flow through the two beds is increased, ΔP_d increases, while the pressure drops across the two fluidized beds remain essentially constant, and $P_2 - P_1$ increases to $P_2' - P_1$. The pressure drop across the overflow standpipe also increases to $P_2' - P_1$ because the height of fluidized solids in the standpipe increases from H_{sp} to H_{sp}':

$$\Delta P_{sp} = P_2' - P_1 = \Delta P/(Lg)\,(H_{sp}') \tag{11.8}$$

This is shown in Figure 11.4 and also in the pressure diagram. If the increase in the pressure drop across the distributor is such that H_{sp} increases to a value

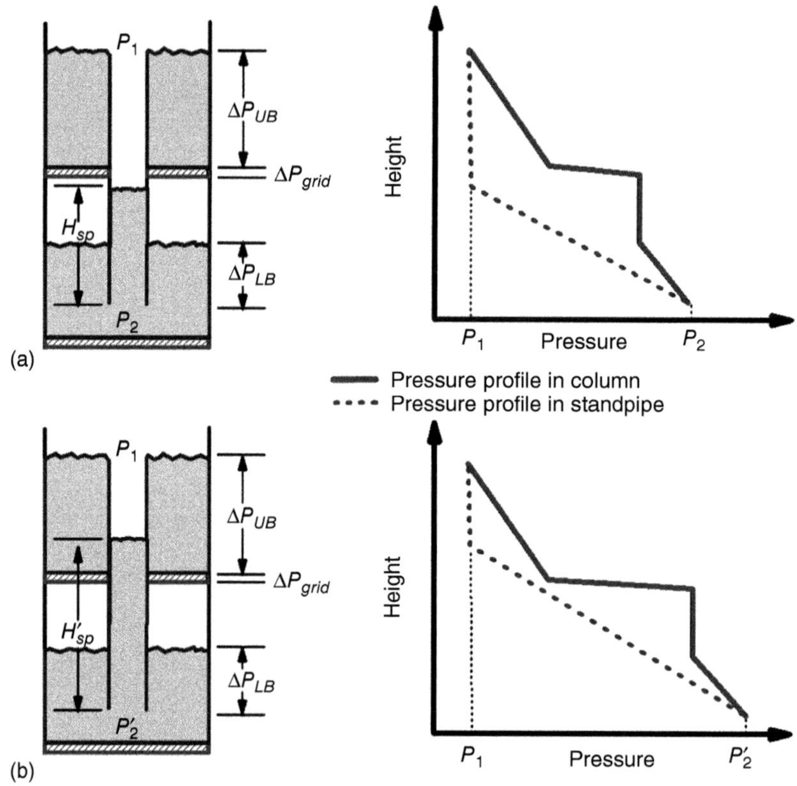

Figure 11.4 Operation of an overflow fluidized standpipe. (a) low pressure drop case, (b) high pressure drop case.

greater than the standpipe height available, the standpipe has reached its limit of operation.

11.1.3.2 Underflow Packed Bed Standpipe

In Figure 11.5 solids are being transferred through an underflow packed bed standpipe against a differential pressure $P_2 - P_1$. $P_2 - P_1$ equals the pressure drop across the gas distributor, ΔP_d. There is also a pressure drop across the solids flow control valve, ΔP_v. The standpipe pressure drop, ΔP_{sp}, is then

$$\Delta P_{sp} = \Delta P/(Lg)\, H_{sp} = \Delta P_d + \Delta P_v = P_2 - P_1 + \Delta P_v \tag{11.9}$$

This is shown as Case I in Figure 11.5.

If the gas flow through the column is increased, ΔP_d will increase. If ΔP_v remains constant, then ΔP_{sp} must also increase to balance the pressure drop (Case II in Figure 11.5). The solids level in the standpipe cannot rise to increase the pressure drop across the standpipe. However, the $\Delta P/(Lg)$ in the underflow packed bed standpipe increases because of an increase in v_r in the standpipe. This can be visualized with the aid of Figure 11.5(a) and the gas–solid flow arrows above Figure 11.5(b).

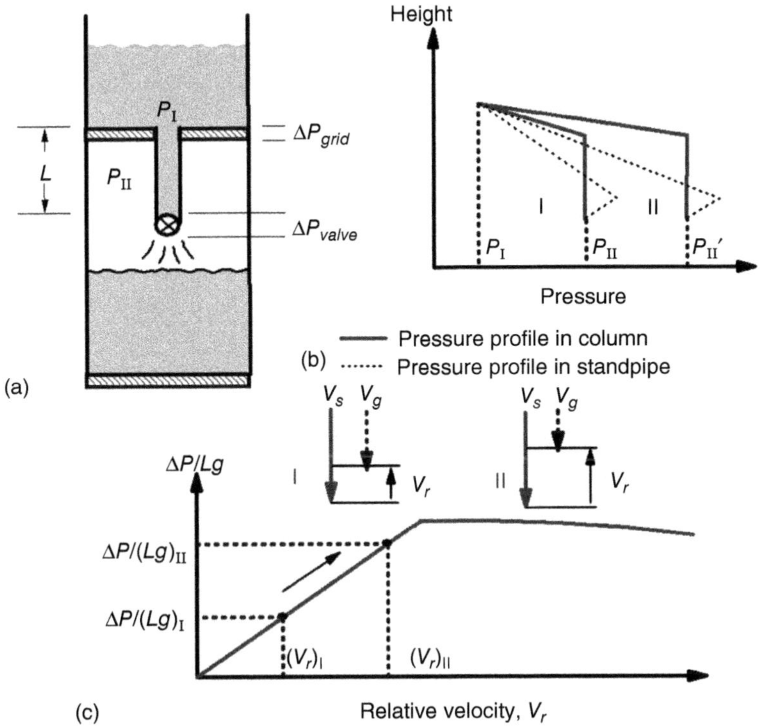

Figure 11.5 Operation of an underflow packed bed standpipe. (a) underflow standpipe, (b) pressure-height diagram, (c) fluidization diagram.

Assume that the underflow packed bed standpipe for the conditions of Case I is operating with solids and gas flowing down the standpipe, as shown in the arrow diagram above Figure 11.5(c). The relative velocity is directed upwards. The arrow diagram for Case I corresponds to Point I on the $\Delta P/(Lg)$ vs. v_r curve. At Point I, $(v_r)_I$ generates the required $\Delta P/(Lg)_I$ across the standpipe. When the distributor pressure drop increases because of the increased gas flow, v_r in the standpipe increases to $(v_r)_{II}$ to generate the higher $(\Delta P/(Lg)_{II})$, required to balance the higher pressure drop across the distributor. v_r increases because the amount of gas flowing down the standpipe decreases, as shown in the arrow diagram for Point II. In an overflow fluidized bed standpipe, the standpipe level adjusts to balance changes in pressure across it. In an underflow packed bed standpipe, the amount of gas flowing down the standpipe adjusts to produce the required pressure drop across it.

If the pressure drop across the distributor is increased so that the product of $\Delta P/(Lg)_{mf}$ (the maximum $\Delta P/(Lg)$ possible in the standpipe) and the standpipe length, H_{sp}, is less than the sum of ΔP_v and ΔP_d, then this standpipe is beyond its limit of operation, and it will not seal. To seal a pressure drop higher than $\Delta P/(Lg)_{mf} \cdot H_{sp}$, a longer standpipe is required.

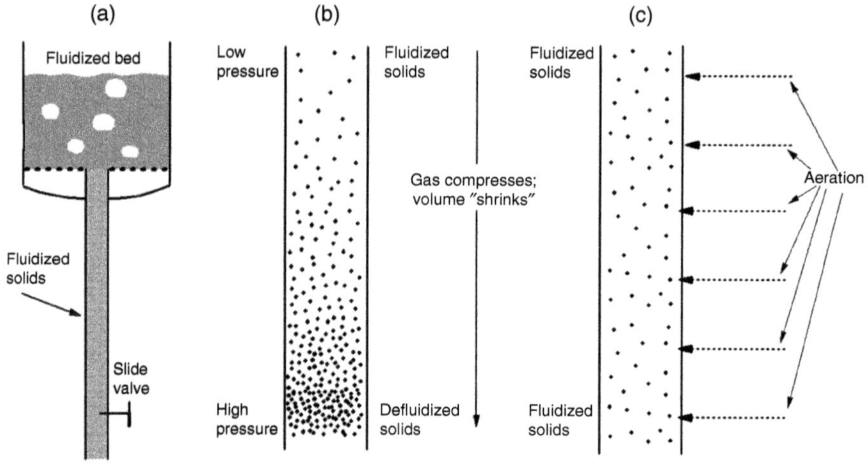

Figure 11.6 Adding aeration prevents defluidization in underflow fluidized standpipes. (a) standpipe plus bed, (b) gas compression in standpipe, (c) where to add aeration.

11.1.3.3 Underflow Fluidized Standpipes

In many catalytic processes, fluidized underflow standpipes (Figure 11.6) are used. This is one of the most widely used standpipes in industry. Unlike the packed bed underflow standpipe, changes in v_r do not cause changes in the pressure drop in the standpipe. Because it is full of solids, changes in solids height to balance the pressure drop are not possible. Therefore, this type of standpipe is designed so that it is long enough to generate more pressure than required. The excess pressure is then "burned up" across the solids control valve to balance the required pressure drop. As the solids flow from a low to a higher pressure, the gas in the standpipe is compressed, causing the solids to move closer together. If aeration is not added to the standpipe to replace the volume lost because of compression, the solids can defluidize near the bottom of the standpipe (Figure 11.6).

To maintain solids in a fluidized underflow standpipe in a fluidized mode, aeration gas is added to the standpipe. Adding the correct amount of gas to a fluidized underflow standpipe prevents defluidization at the bottom of the standpipe (Figure 11.6(c)). If the material flowing in the standpipe is a group A material, aeration should be added uniformly along the standpipe. If aeration is added only at the bottom of the standpipe with group A solids, a large bubble will form in the standpipe at the aeration point (Figure 11.7(a)). If the bubble is large enough, it can restrict the flow of solids down the standpipe. The large bubble forms because it is difficult for the aeration gas to permeate through the fine solids in the standpipe. It requires a significant area for the gas to dissipate through the fine particles at the same rate that aeration is added. However, if the aeration gas is added at several locations, the bubble size is significantly reduced, and standpipe operation improves (Figure 11.7(b)).

For group B and D solids, it is unnecessary to add aeration uniformly along the standpipe. Adding aeration at the bottom of the standpipe is generally sufficient. Gas can permeate through larger particles much more easily than through

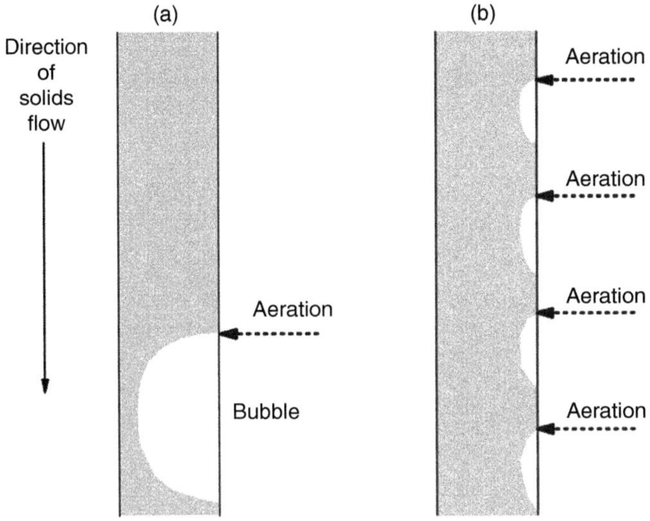

Figure 11.7 Why aeration is added uniformly in group A underflow fluidized standpipes. (a) aeration added at one point, (b) aeration added at multiple points.

group A particles, which have significantly larger surface area per unit volume and produce more drag for the same gas flow.

The aeration flow required to maintain solids in a fluidized state in the standpipe is [3]

$$Q = 1000\left[\frac{P_b}{P_t}\left(\frac{1}{\rho_{mf}} - \frac{1}{\rho_{sk}}\right) - \left(\frac{1}{\rho_t} - \frac{1}{\rho_{sk}}\right)\right] \tag{11.10}$$

In a fluidized underflow standpipe operating with group A solids, the amount of aeration theoretically required is added in equal increments via aeration taps located approximately 2 m apart. Care should be taken not to over-aerate the standpipe. If this occurs, large bubbles are generated, which hinder solids travelling downwards through the standpipe.

In underflow fluidized standpipes, it has been found that the optimum amount of aeration added to the standpipe should be about 70% of that predicted. One theory is that the gas only fluidizes the solids at the wall of the standpipe. Because the aeration only fluidizes a portion of the standpipe, not all of the theoretical aeration is required. When solids flow from a fluidized bed into the top of an underflow fluidized standpipe, the solids are accelerated from a low velocity in the bed to as much as 2 m/s in the standpipe. This increase in velocity can carry bubbles down the standpipe and degrade standpipe operation. To prevent this, a cone is added to the top of the standpipe to minimize the solid velocity at the standpipe entrance and minimize bubble "carry-under." Experience has shown that the diameter of the cone should be 4–6.5 times the area of the standpipe [4].

In many fluidized beds, a sparger gas distributor is used. A sparger consists of several pipes with holes in the pipe. Solids flow between the sparger pipes into the standpipe below. Another design can be used to prevent bubbles from entering the standpipe with sparger grids. With this design the standpipe entrance is located below the sparger grid. As the solids flow through the defluidized region between the sparger and the standpipe entrance, bubbles dissipate and do not enter the standpipe. The standpipe entrance should be located 1.5–2 times the standpipe diameter below the jet penetration from downward-pointing gas nozzles in the sparger.

11.2 Standpipes in Recirculating Solids Systems

Many fluidized bed catalytic processes recirculate catalyst through a reaction/regeneration cycle. A circulating fluidized bed combustor (CFBC) recirculates fuel and ash to completely burn the fuel. All recirculation systems use standpipes.

There are two basic solids recirculation systems: (i) automatic and (ii) controlled. In the automatic system solids are recirculated around a loop at their "natural" recirculation rate without being controlled. In the controlled recirculation system, a valve is used to control the solids flow rate. In the automatic system, overflow fluidized standpipes are almost always employed. In controlled recirculation systems, the standpipe is an underflow standpipe that can be fluidized or non-fluidized.

11.2.1 Automatic Solids Recirculation Systems

One of the simplest types of automatic solids recirculation systems is the cyclone/dipleg system. This system is shown in Figure 11.8 for a fluidized bed containing both primary and secondary internal cyclones. The dipleg in this system is a fluidized overflow standpipe. External cyclone return systems are also automatic recycle systems. All solids entering the primary and secondary cyclones are returned to the bed automatically via primary and secondary diplegs.

The pressure balance around a primary cyclone is

$$\Delta P_{bed} + \Delta P_{cy1} = \Delta P_{dip1} = \rho_{dip1} H_{dip1} g \tag{11.11}$$

The pressure drop across the cyclone in Eq. (11.11) is the pressure drop between the cyclone inlet and the bottom of the cyclone. The pressure drop across the gas outlet tube (GOT) of the primary cyclone (ΔP_{GOT1}) is not required for this pressure balance.

The required solids height in the primary cyclone dipleg is

$$H_{dip1} = (\rho_{bed} I_{1bed} g + \Delta P_{cy1})/(\rho_{dip1} g) \tag{11.12}$$

where I_{1bed} is the immersion depth of the primary cyclone dipleg in the bed.

The pressure balance around the system for returning solids to the bed via a secondary cyclone dipleg is

$$\Delta P_{bed} + \Delta P_{cy1} + \Delta P_{GOT1} + \Delta P_{cy2} = \Delta P_{dip2} = \rho_{dip2} H_{dip2} g \tag{11.13}$$

The secondary cyclone dipleg is the most critical dipleg because it has to seal both cyclone pressure drops and the pressure drop across the primary cyclone

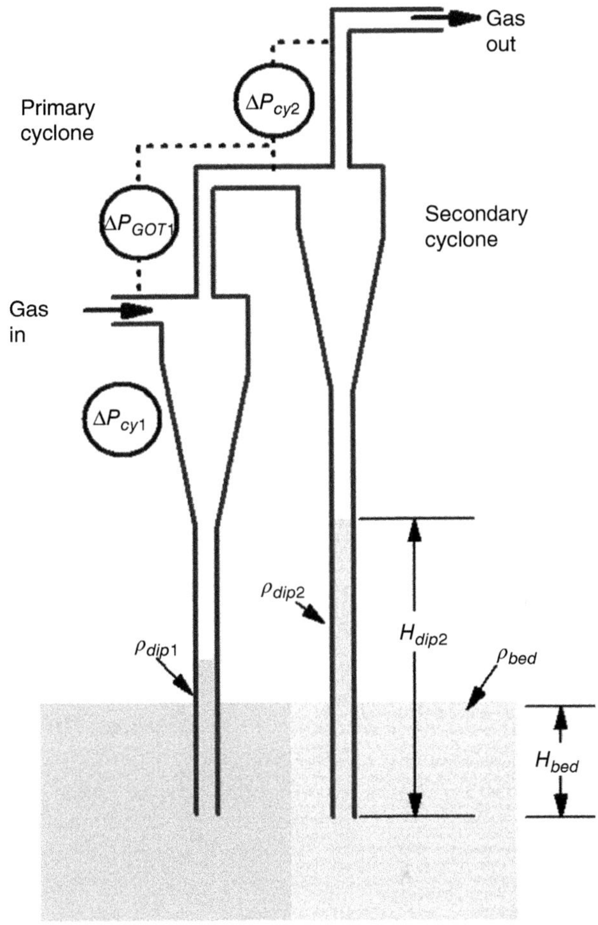

Figure 11.8 Fluidized bed two-stage cyclone recirculation system.

GOT and the bed immersion pressure drop. The seal height can also be high because the solids density in the secondary dipleg will be lower than the solids density in the bed (because the particle size in the secondary dipleg is small). For solids to be returned to the bed in the secondary dipleg, the solids height in the secondary cyclone dipleg length, H_{dip2}, is

$$H_{dip2} = (\rho_{bed} I_{2bed} g + \Delta P_{cy1} + \Delta P_{GOT1} + \Delta P_{cy2})/(\rho_{dip2} g) \tag{11.14}$$

where I_{2bed} is the immersion depth of the secondary dipleg in the bed.

The design dipleg length for both primary and secondary cyclones should be greater than the calculated height as a contingency for changes in operating conditions. The height between the solids level in a dipleg and the bottom of the cyclone cone should be at least 1 m.

11.2.2 Controlled Solids Recirculation Systems

In a controlled solids recirculation system, a valve is used to control the solids flow rate around the unit. In controlled systems, the standpipe is almost always an underflow fluidized or non-fluidized standpipe. A typical controlled recirculation system is that used by FCC units. A similar recirculation system is used by other processes.

Primary standpipes in FCC units are underflow fluidized bed standpipes. Nearly all FCC units incorporate two standpipes in their flow loops.

A typical FCC riser/recirculation system developed by the UOP company is shown in Figure 11.1. A pressure drop balance around this unit gives

$$\Delta P_{regen} + \Delta P_{sp1} - \Delta P_{sv1} - \Delta P_{riser} + \Delta P_{stripper} + \Delta P_{sp2} - \Delta P_{sv2} = 0 \quad (11.15)$$

In most FCC units, the solids flow rate around the system is controlled by a slide valve. In Standpipe 1, Slide Valve 1 controls the temperature at the outlet of the riser by varying the flow of catalyst to the riser. The slide valve in Standpipe 2 controls the fluidized catalyst level in the stripper.

Underflow fluidized standpipes are operated in a vertical configuration, a completely angled configuration or a hybrid configuration, in which both vertical and angled sections are present. Angling a standpipe is a convenient way to transfer solids between two points that are separated both horizontally and vertically. However, it has been found [3, 5] that long angled underflow fluidized standpipes do not perform as well as vertical standpipes.

Viewing angled standpipe flow in a transparent standpipe [3] showed that gas and solids separated in the standpipe, with gas bubbles flowing along the upper portion of the standpipe and solids flowing along the bottom of the standpipe as shown in Figure 11.9. As a result, the pressure build-up in a hybrid standpipe was lower than in a vertical standpipe. The rising bubbles in the angled section of the hybrid standpipe become relatively large at a low solids flow rate. At a certain solids mass flux, the bubbles become large enough to bridge across the vertical section at the top of the angled standpipe, hindering solids flow. When this occurs, the maximum solids flow rate in the hybrid angled standpipe has been achieved.

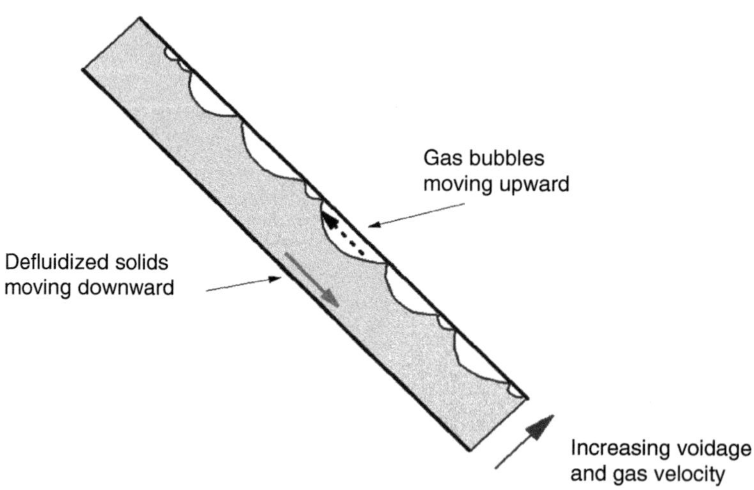

Figure 11.9 Gas–solid flow in an angled standpipe.

Even though vertical standpipes can transfer solids more efficiently than hybrid angled standpipes, angled standpipes containing no vertical sections are commonly operated satisfactorily in units with group A solids. These standpipes are short and are designed so that the mass flux through them is high. Long angled standpipes have a limited solids circulation rate relative to vertical standpipes [5]. Thus, when operating a hybrid angled standpipe or a true angled standpipe, it is essential to keep the solids mass flux through the standpipe at a high value and the standpipe as short as possible, so large gas slugs are minimized.

11.2.3 Function of a Standpipe

The primary purpose of a standpipe is to build pressure in a solids circulation loop. However, the standpipe is also the part of the solids circulation loop that automatically adjusts to balance the pressure drops around a loop.

There are three components in a pressure balance loop:

1. A fixed pressure drop component (riser, fluidized bed, or cyclone)
2. A variable pressure drop component (slide valve)
3. An automatically variable pressure drop component where the pressure drop across it automatically varies to ensure that the pressure balance is satisfied (standpipe)

Consider the schematic drawing of an FCC unit in Figure 11.10. The pressures in various parts of the circulation loop are shown in this figure. A pressure balance

Figure 11.10 Pressures around an FCC unit.

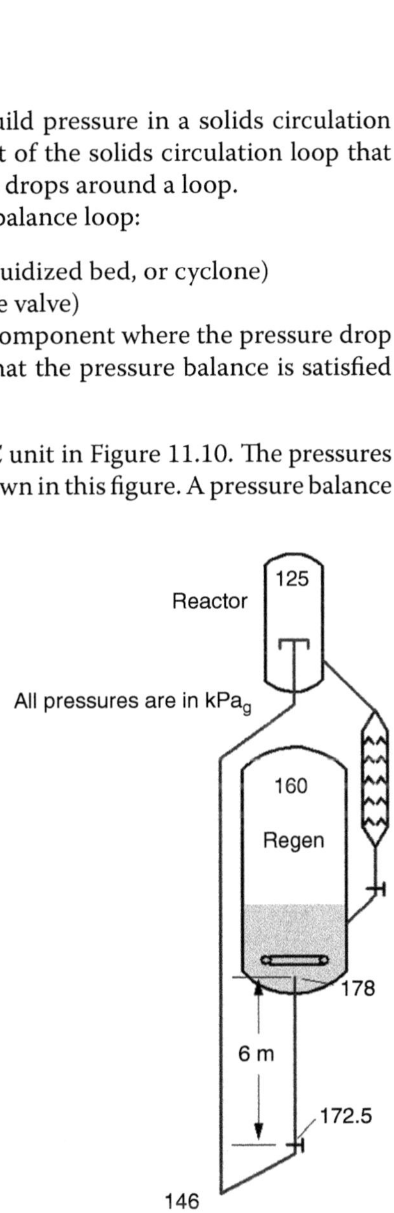

around the unit gives

$$\Delta P_{sv} + \Delta P_{riser} - \Delta P_{react/regen} - \Delta P_{bed} = \Delta P_{sp} \tag{11.16}$$

or

$$26.5\,\text{kPa} + 21\,\text{kPa} - 35\,\text{kPa} - 18\,\text{kPa} = -5.5\,\text{kPa}$$

The pressure build-up across the standpipe is negative (i.e. the pressure at the top is greater than the pressure at the bottom). When this occurs, the standpipe operates as a packed bed.

11.3 Standpipes Used with Nonmechanical Solids Flow Devices

A nonmechanical solids flow device uses only aeration gas in conjunction with its geometrical shape to cause solids to flow through it. Nonmechanical devices can operate in two modes:

(1) A valve mode to control the flow rate of solids
(2) An "automatic" solids flow-through mode

The type of standpipe used for each mode is different.

11.3.1 Nonmechanical Solids Control Mode Operation

In the solids control mode of operation, solids flow through the nonmechanical device is controlled by the amount of aeration gas flow added to it. Nonmechanical devices are shown schematically in Figure 11.11. The most common nonmechanical is the L-valve [1, 6]. These devices differ in their shape and in the direction in which they discharge solids. Because the principle of operation of nonmechanical valves is the same, nonmechanical valve operation for solids control is presented here through discussion of the L-valve.

When aeration gas is added to a nonmechanical valve, gas flows downwards through the particles and around the constricting bend. This relative gas–solid

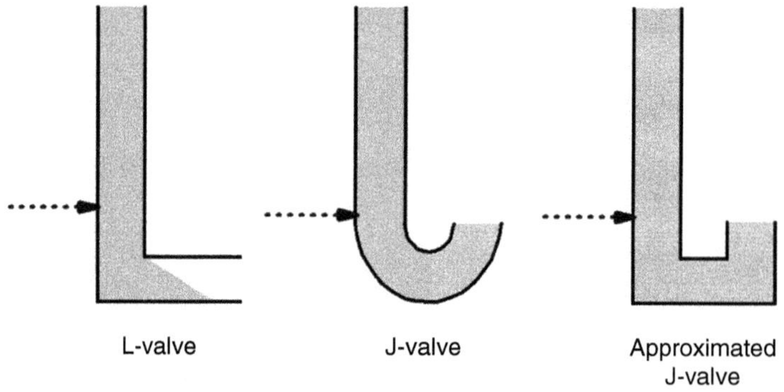

Figure 11.11 Most common types of nonmechanical valves used for solids control.

flow produces a frictional drag force on the particles in the direction of flow. When the drag force exceeds the force required to overcome the resistance to solids flow around the bend, solids flow through the valve.

An L-valve operating in the solids control mode ***must*** be located at the bottom of an *underflow packed bed standpipe.* The amount of aeration gas required to operate the L-valve depends on the operation of the underflow packed bed standpipe above it.

The gas flow that causes the solids to flow around the L-valve is not the amount of aeration gas added to the valve, Q_A. If gas is travelling down the standpipe with the solids, the amount of gas flowing around the L-valve bend, Q_T, is the sum of the standpipe gas flow, Q_{sp}, and the aeration gas flow, Q_A (Figure 11.12(a)). If gas is flowing up the standpipe, the gas flowing around the bend, Q_T, is the difference between the aeration flow and the gas flow up the standpipe (Figure 11.12(b)).

Nonmechanical valves work best with materials having average particle sizes between 120 and 5000 μm. These materials are in groups B and D. Group A materials (with average particle sizes from approximately 30 to 120 μm) do not work in L-valves. Group A materials retain air in their interstices and remain fluidized for a substantial period of time. Because they remain fluidized, they flow through the constricting bend like water, and the L-valve cannot control the solids. Group C materials are very cohesive and do not flow well in L-valves.

A nonmechanical device operating in the valve control mode is always located at the bottom of an underflow packed bed standpipe.

The operation of a nonmechanical valve depends on the pressure balance and geometry of the system. Consider the circulating fluidized bed (CFB)/L-valve system shown in Figure 11.13. The high-pressure point in this recycle loop is at the L-valve aeration point. The low-pressure point is at the bottom of the cyclone. The pressure drop balance around the recycle loop is

$$\Delta P_{L-valve} + \Delta P_{CFB} + \Delta P_{cy} = \Delta P_{sp} + \Delta P_{surge\ hopper} \tag{11.17}$$

The surge hopper pressure drop is generally negligible, so the above equation becomes

$$\Delta P_{sp} \cong \Delta P_{L-valve} + \Delta P_{CFB} + \Delta P_{cy} \tag{11.18}$$

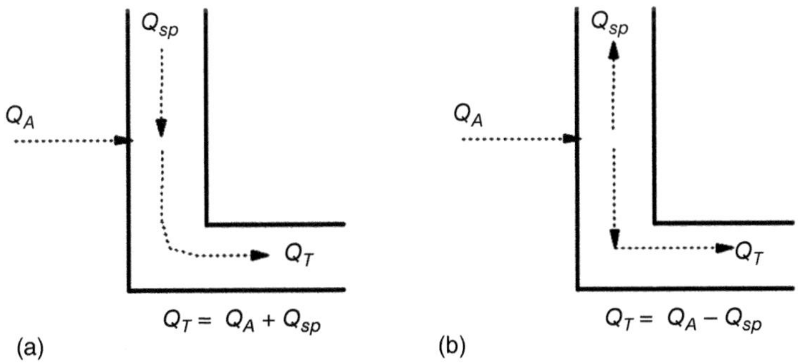

(a) (b)

Figure 11.12 Gas flows around an L-valve bend. (a) Gas flow down standpipe and (b) gas flow up standpipe.

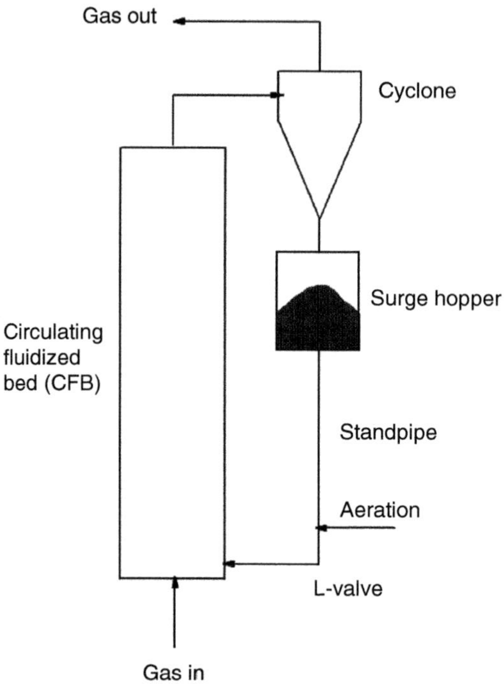

Figure 11.13 L-valve in a CFB loop configuration.

The packed bed standpipe is the dependent part of the pressure drop loop, so its pressure drop adjusts to balance the pressure drops on the independent side of the loop.

The independent pressure drop can be increased by increasing the pressure drops across the CFB, cyclone, or the L-valve. For a constant gas velocity in the riser, as the solids flow rate increases, the independent pressure drop increases. The packed bed standpipe pressure drop then increases to balance this increase. It does this by automatically increasing v_r in the packed bed standpipe. Further increases in the solids flow rate can occur until the $\Delta P/(Lg)$ in the packed bed standpipe reaches the maximum value of $\Delta P/(Lg)_{mf}$ for the solids. A short standpipe reaches its maximum $\Delta P/Lg$ at a lower solids flow rate than a longer standpipe. Thus, the maximum solids flow rate through a nonmechanical valve depends on the length of standpipe above it.

When v_r in the standpipe above the L-valve reaches the value needed to fluidize the solids, a transition from packed bed to fluidized bed flow occurs. Any further increase in v_r results in fluidizing the standpipe and the formation of bubbles. These bubbles hinder the flow of solids through the standpipe and cause a decrease in solids flow rate.

It is desirable to add aeration to an L-valve as low in the standpipe as possible. This results in maximizing standpipe length and minimizing L-valve pressure drop. Both of these factors increase the maximum solids flow rate through the L-valve. If aeration is added too low in the standpipe, gas bypassing results and solids flow control is ineffective. L-valve aeration is most effective if it is added at a length-to-diameter (L/D) ratio of 1.5 above the centre line of the horizontal section of the L-valve.

11.3.2 Automatic Solids Flow Devices

Nonmechanical devices can also be used to automatically pass solids through them. In the automatic mode, they serve as simple flow-through devices without controlling the solids. Automatic solids flow devices can be used with groups A, B, and D.

If the solids flow rate to an automatic nonmechanical device is changed, it automatically adjusts to accommodate the changed solids flow rate. There are several types of these devices. The most common are the straight cyclone dipleg, loop seal, and L-valve. The standpipe above each automatic nonmechanical device should be an *overflow fluidized bed standpipe.* This is a different standpipe than the underflow packed bed standpipe that must be used for the control mode of nonmechanical devices.

In CFB systems, the most frequent application of automatic nonmechanical devices is to recycle collected cyclone solids back to the CFB. The simplest nonmechanical device used to recycle cyclone solids is a straight vertical cyclone dipleg. In order to perform properly, this dipleg must be immersed in a fluidized dense-phase bed or fitted with a valve at its discharge end. Since cyclones in CFBs are generally external to the bed and located some distance horizontally from the CFB riser, a vertical cyclone dipleg cannot be used. Angled diplegs tend to slug and perform poorly for essentially the same reason as angled standpipes. Therefore, loop seals, seal pots, and automatic L-valves (Figure 11.14) are used to return the solids to the bed.

11.3.2.1 Cyclone Diplegs

The solids flow rates in a primary cyclone dipleg are significantly greater than in a secondary cyclone dipleg. If there are 10 000 kg/min of solids entering a primary cyclone operating at an efficiency of 99.9%, the solids flow rate down the primary dipleg is 9990 kg/min, whereas the solids flow rate into the secondary cyclone is only 10 kg/min (Figure 11.15). If the secondary cyclone is operating at an efficiency of 90%, the solids flow rate in the secondary dipleg is 9 kg/min, and the solids loss rate from the cyclone system is 1 kg/min. The solids flow rate in

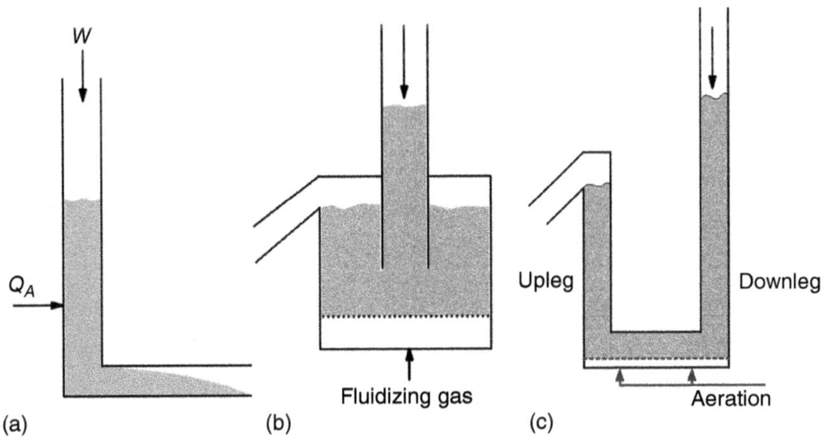

Figure 11.14 Automatic (a) L-valve, (b) seal pot, and (c) loop seal.

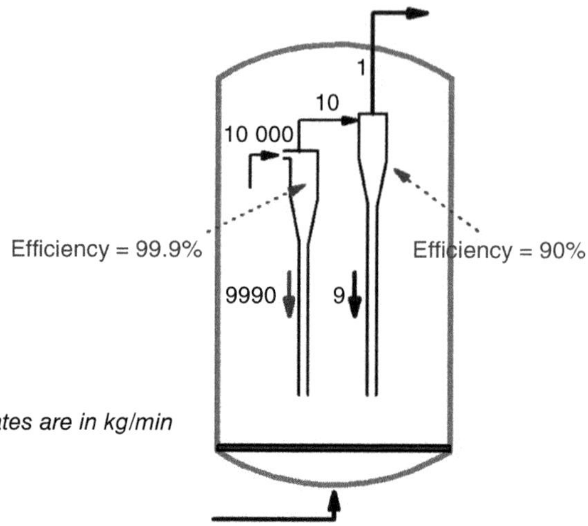

Figure 11.15 Relative solids flow rates in primary and secondary diplegs.

the secondary dipleg is then approximately 1000 times less than in the primary dipleg. In many processes, the primary cyclone efficiency can be 99.99%, and secondary cyclone efficiencies can be 98%. For these efficiencies, the solids flow rate down the secondary dipleg can be 10 000 times less than in the primary dipleg.

Primary cyclone diplegs give few operational problems because the solids flux down them is high and the particle size is large. Typical design fluxes for primary cyclone diplegs range from 350 to 750 $kg/s/m^2$. These fluxes are so high that the gas cannot flow upwards in the primary dipleg and, instead, will be carried down the dipleg with the solids [7]. Secondary diplegs tend to have flow problems because the solids in secondary diplegs are fine (often approaching cohesive size) and because the solids mass flux is low. It is not practical to reduce the size of the secondary cyclone diplegs to give the same solids fluxes as in primary diplegs. The secondary dipleg diameter used for most commercial secondary diplegs is about 200 mm, large enough to allow large refractory chunks to pass through the dipleg.

In many processes, gas flow through the fluidized bed vessel is established before solids are added to build the bed. For this case, the amount of gas flowing up the secondary dipleg upon start-up is so great that it prevents solids from building up in the secondary dipleg. Therefore, a trickle valve or a counterweighted flapper valve (Figure 11.16) is added to the end of the dipleg to prevent most of the gas from flowing up the dipleg at start-up. These valves do not prevent all of the gas from flowing up the secondary cyclone dipleg, but they reduce the gas flow to such an extent that a solids seal can be developed. This problem does not occur with a highly loaded cyclone dipleg because the solids flux down the dipleg is so great that gas will not flow upwards against it. However, even with a trickle valve or counterweighted flapper valve at the bottom of a secondary dipleg, some gas will still flow up the secondary dipleg (Figure 11.17) [7].

In many fluidized bed processes, a recurring problem is the loss of fine particles from the bed. Often this is a problem with secondary dipleg operation. In

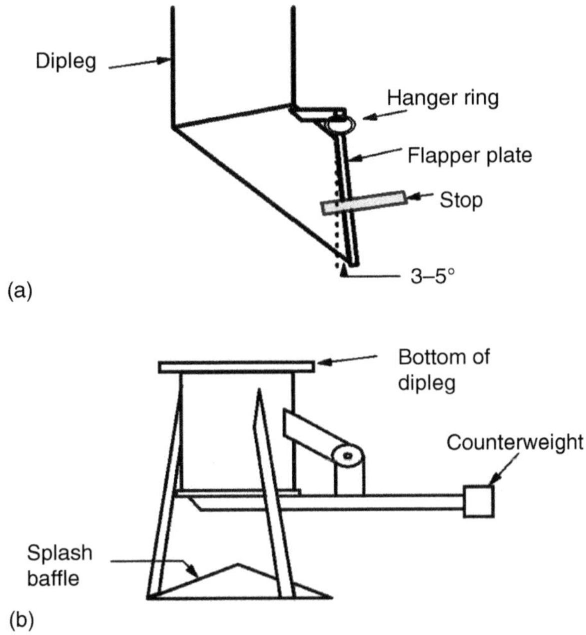

Figure 11.16 (a) Trickle valve and (b) counterweighted flapper valve.

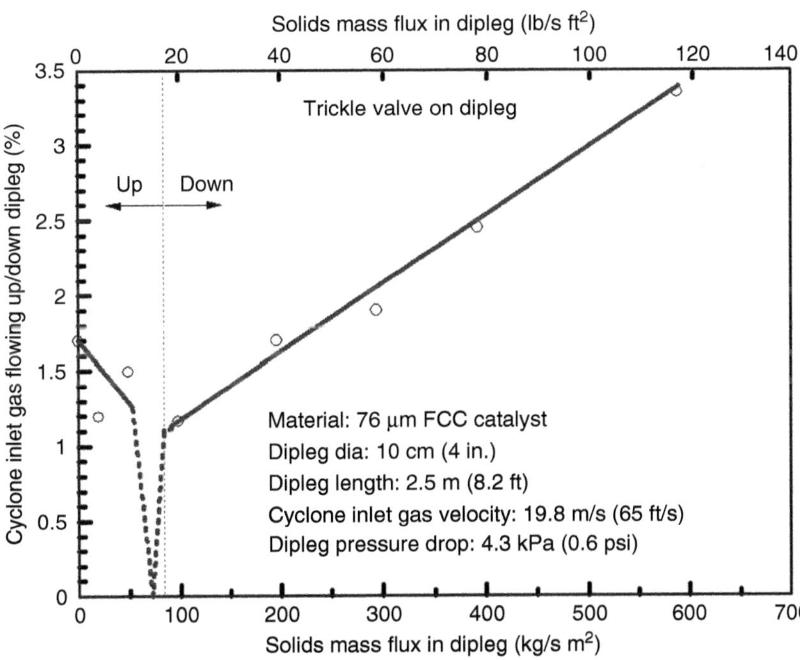

Figure 11.17 Gas flow direction in a dipleg as a function of solids mass flux in the dipleg.

this dipleg, the solid flux is so low, and the particle size so small, that solids flow through this dipleg can be blocked, or so much gas can flow up the secondary cyclone dipleg that a significant amount of material can be lost. Frequently the problem is with the trickle or flapper valve plate not sealing well (due to poor design or warping), causing too much gas to flow up the dipleg and negatively affecting cyclone efficiency. Immersing the trickle or flapper valve in the bed a short distance can mitigate this problem.

Normally, a trickle or a flapper valve on the outlet of a secondary cyclone dipleg operates in an intermittent discharge mode. Fluidized solids build up in the dipleg to a height, which will provide enough "head" to generate enough force to open the flapper plate. Often there will be a continuous, small trickling flow of the solids through the trickle valve or flapper valve constituting approximately 10% of the solids flow rate [8]. This constant trickling is superimposed on the intermittent discharge mode.

An automatic L-valve can also be used at the bottom of the *overflow fluidized bed dipleg*. How the automatic L-valve works is described in [9]. Group A solids will work in an automatic L-valve if it is discharging into a dense bed. An automatic L-valve works well with group B solids when discharging into either a dense phase or a dilute phase.

Loop seals at the bottom of overflow fluidized cyclone diplegs are very reliable. They can be used with both group A and group B solids. For a loop seal to operate correctly, the upleg of the loop seal must be fluidized (Figure 11.14). For best operation the downleg of the loop seal should also be fluidized.

11.4 Solids Separation Devices

There are three primary reasons for separating gases from an exit stream (i) to minimize emissions, (ii) to protect processing equipment (turbines, heat transfer surfaces) from gas-particle streams, and (iii) to stop unwanted gas–solid reactions from occurring. Cyclones are by far the most common devices used to remove solids from fluidized bed exit gas streams, so they are described here in much more detail than other devices.

11.4.1 Cyclones

A cyclone is a device that separates particulate solids from a gas stream by centrifugal force.

Cyclone diameters range in size from 2 cm for laboratory units to 10 m in CFB combustion units. Cyclone solids loadings range from about 0.0002 kg/m^3 (0.000 13 kg_s/kg_g) to nearly 50 kg/m^3 (42 kg_s/kg_g), a factor of 250 000! The number of cyclones deployed in a single fluidized bed varies from one to 22 sets of primary and secondary cyclones (44 cyclones total).

The advantages of cyclones are that they have no moving parts, are inexpensive to construct and maintain, and have relatively low pressure drops. The main limitation of most cyclones is that they have reduced collection efficiencies for particles smaller than about 20 μm. Special cyclones called multiclones, consisting of many small cyclones in parallel, can collect particles down to 3 μm with efficiencies approaching 90%.

11.4.1.1 Cyclone Types

There are two primary types of cyclones: reverse-flow and uniflow cyclones. Reverse-flow cyclones (Figure 11.18(b)) are by far the most common. They are called reverse flow because the gas–solid mixture enters the cyclone tangentially at the top and spirals around the barrel and then the gas reverses flow and exits through the GOT, while the solids exit at the bottom of the cone.

In the uniflow cyclone (Figure 11.18(a)), the gas–solid mixture enters the top of the cyclone, and the gas exits at the bottom in the centre of the cyclone. The captured solids leave at the bottom, along the wall. A small amount of gas (typically 1–3% of the inlet gas flow) is withdrawn with the solids to improve the collection efficiency.

The most common reverse-flow cyclone geometries are the volute inlet cyclone, tangential inlet cyclone, and axial inlet cyclone (Figure 11.19). The axial cyclone uses swirl vanes to transform the gas–solid mixture into a centrifugal motion.

Cyclones are generally classified as low-loaded or high-loaded cyclones. The difference between high and low loading is arbitrary. High-loaded cyclones are generally primary cyclones, whereas low-loaded cyclones are secondary, tertiary, or even fourth-stage cyclones. Particulate Solid Research Inc. (PSRI) classifies cyclones with loadings >1 kg_s/kg_g, as high-loaded cyclones.

When solids are introduced to a cyclone, the cyclone pressure drop initially decreases, often resulting in a reduction of >50%. As the solids loading increases, the pressure drop goes through a minimum and then increases [10]. When solids are present at the wall of the cyclone, the increased drag on the gas due to the

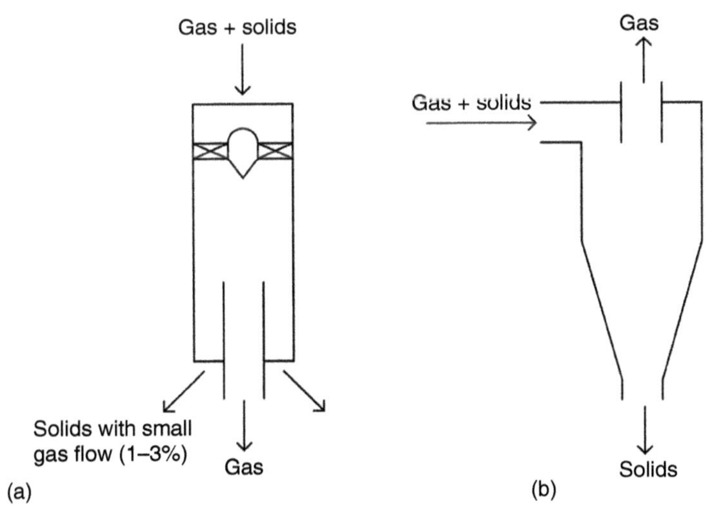

Figure 11.18 (a) Uniflow and (b) reverse-flow cyclones.

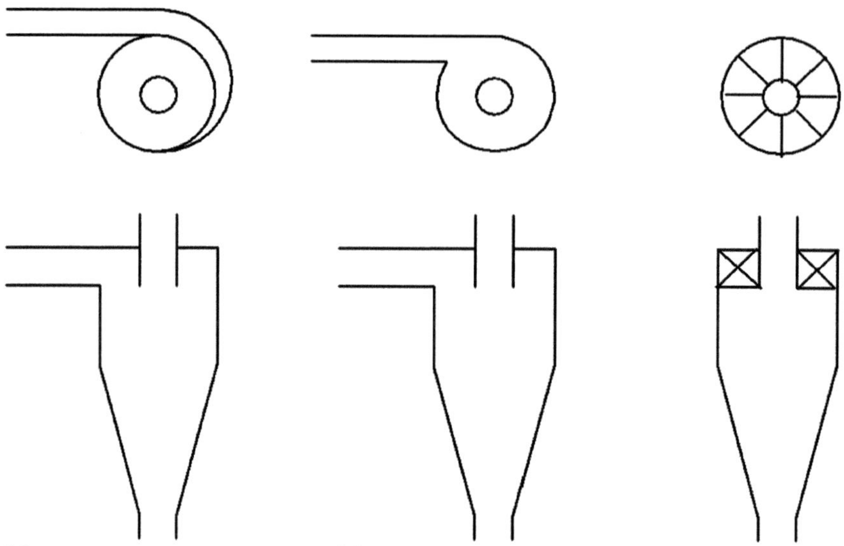

Figure 11.19 (a) Volute, (b) tangential, and (c) axial inlet reverse-flow cyclones.

solids at the wall causes the cyclone tangential velocity to decrease, resulting in the reduction of the cyclone pressure drop [11]. At high solids loadings, the pressure drop due to solids acceleration becomes large, and cyclone pressure drop then increases with solids loading (Figure 11.20).

The most common cyclones are the tangential and volute inlet cyclones (Figure 11.19). A volute inlet has been found to be more efficient for high-loaded

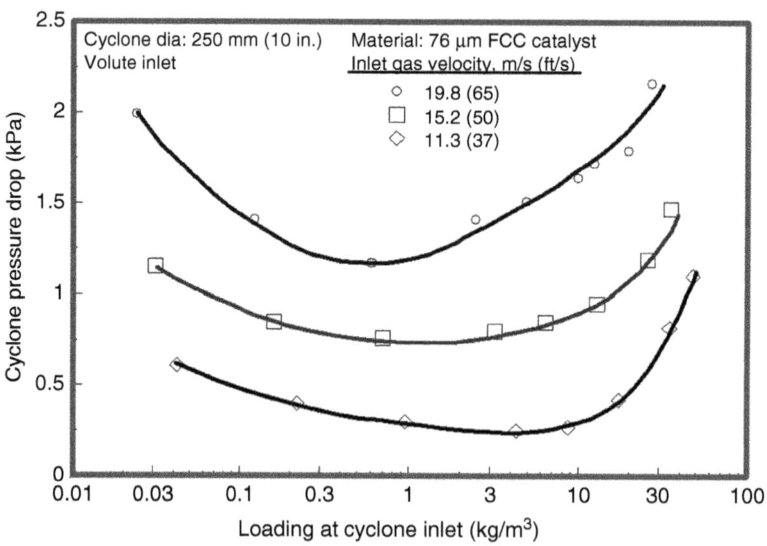

Figure 11.20 Effect of loading on cyclone pressure drop.

cyclones than a tangential inlet. A tangential inlet cyclone can produce an interference eddy near its inlet. This eddy causes fluctuations in the inlet solids stream that can cause the inlet solids to strike the GOT and the rotating solids stream. This results in lower efficiency and possible erosion of the GOT. If a tangential cyclone is used for a high-loaded cyclone, the inlet should be located so that the distance between the cyclone inlet and the wall of the GOT is great enough that the fluctuating solids do not impact the GOT.

A high-loaded volute inlet cyclone does not have this problem. With a volute inlet the solids enter the cyclone at an angle, and the entering solids do not experience interference from the rotating solids stream. This is shown in Figure 11.21. Therefore, a volute inlet is used in almost all high-loaded cyclones. For low-loaded cyclones, interference with the inlet solids stream does not occur, so tangential inlets are satisfactory.

11.4.1.2 Flow Patterns in Cyclones

In a reverse-flow cyclone, the cyclone inlet transforms the linear inlet gas–solid flow into a rotating outer vortex. The gas flow then reverses and forms an inner vortex below the GOT.

The inner vortex rotates at a much higher angular velocity than the outer vortex. The radius of the inner vortex has been measured to be about 85% of the diameter of the GOT. With axial inlet cyclones, the inner core vortex is aligned with the axis of the GOT. With tangential or volute cyclone inlets, the vortex is not aligned with the axis. The asymmetric entry of the tangential or volute inlet causes the axis of the vortex to be eccentric to the cyclone axis. This means that the bottom of the vortex is displaced some distance from the axis and can

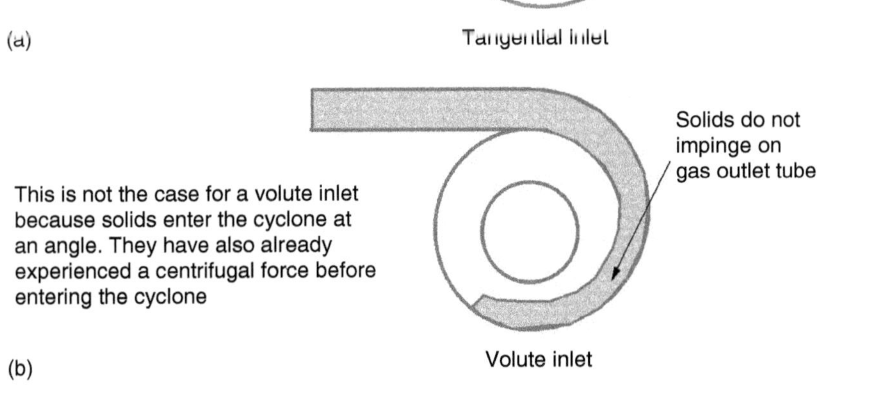

Figure 11.21 (a) Tangential and (b) volute inlet operation at high solids loading.

re-entrain solids if the vortex approaches the cone wall too closely. It is important that the cyclone be long enough to prevent this.

With primary cyclones, the solids loading is generally so high that primary cyclones operate as inertial separators [12]. Gas can carry only a maximum loading of solids (critical loading). At a solids loading in excess of this critical loading, the solids are inertially separated from the gas at the inlet to the cyclone. The solids remaining in the gas are then separated in the inner vortex.

Because solids at high loading are separated largely by inertial separation, the efficiency of a high-loaded, large-diameter cyclone can approach that of a smaller cyclone. High loading also affects the vortex below the GOT. The large mass of solids reduces the spinning rate of the vortex [13], making it shorter.

11.4.1.3 Cyclones in Series

Solids collection efficiency can be enhanced by placing cyclones in series. Cyclones in series are necessary for most processes to minimize particulate emissions and to limit the loss of expensive solids. Two cyclones in series are most common, but three cyclones in series are also often used. Cyclones in series can be very efficient. For example, in FCC processes, two stages of cyclones can give efficiencies greater than 99.999%.

Typically, primary cyclones have an inlet gas velocity less than secondary cyclones. The lower inlet velocity of primary cyclones results in lower particle attrition rates and reduced wall erosion rates. After most of the solids are collected in the primary cyclone, a higher velocity is used in secondary cyclones to increase the centrifugal force on the solids, thereby increasing collection efficiency. Erosion rates are low because of the vastly reduced flux of solids into the secondary cyclone.

11.4.1.4 Cyclones in Parallel

Several small cyclones are placed in parallel when it is not possible to fit a single large cyclone into the available freeboard height or when extremely high centrifugal forces are required. When using parallel cyclones, the gas and solids flow into each cyclone is almost always unequal. This can lead to cyclone inefficiencies and increased wear on the cyclones that handle most of the solids flow. Increasing the pressure drop across the cyclones improves solids distribution, but does not ensure equal solids flow to each parallel cyclone.

When the number of parallel cyclones is <3–4, each cyclone has an individual inlet and outlet duct. When the number of cyclones exceeds this number, the cyclones have common inlet and outlet plenums and a common collection bin. Parallel arrangements of this type result in the operating problem of equalizing the gas and solids flow rates into each cyclone, preventing backflow from the common chamber into one or more cyclones and re-entrainment of solids from the common collection hopper into one or more cyclone outlets.

Placing many cyclones in parallel in a common inlet plenum results in distribution problems because it is easier for the gas and solids to flow through the closest cyclone than one located a significant distance away. Multiple inlets reduce this problem but result in increased complexity and cost.

Because of problems in operation, many parallel cyclones have collection efficiencies less than a large, single cyclone processing the same amount of gas and solids [14–18]. Koffman [16] found that the efficiency of a single cyclone for a system was 96%, while 7 cyclones in parallel and 14 cyclones in parallel gave efficiencies of 94.1% and 92.2%, respectively. Solids flow rates in parallel cyclones can vary by a factor between two [15] and four [18].

11.4.1.5 Internal vs. External Cyclones

Cyclones can either be placed in the freeboard above a fluidized bed or outside the fluidized bed vessel. The optimum placement depends on what is best for a process. Internal cyclones have the advantages that they require no inlet piping (their inlets are open to the freeboard), no high-pressure shell, and have straight cyclone diplegs. Internal cyclones are smaller in diameter than external cyclones because their size is limited by the headspace in the freeboard, which commonly results in using several smaller cyclones in parallel instead of one large cyclone.

The advantages of external cyclones are that they can be much larger than internal cyclones and they are more accessible than internal cyclones. The disadvantages of external cyclones are that they require a pressure shell, and their diplegs generally require an angled or a horizontal pipe to return the solids to the bed, which can result in poor dipleg operation.

11.4.1.6 Cyclone Inlet Design

It is generally desirable for the width of the cyclone inlet (L_w) to be as narrow as possible so that the entering solids will be close to the cyclone wall where they are collected. However, narrow inlet widths require a greater height of the inlet (H) in order to give an inlet area required for the desired inlet gas velocities. Low-loaded cyclones have height-to-width ratios (H/L_w) of between 2.5 and 3. For high-loaded cyclones, the inlet aspect ratio can be increased to 7, but this requires increased cyclone lengths.

A good inlet design is one in which the inner wall of the inlet is angled towards the cyclone outer wall. This forces solids to flow towards the inner wall of the cyclone. The bottom wall of this inlet is angled downwards so that the decrease in area along the inlet flow path is not too rapid and the solids acceleration is controlled.

11.4.1.7 Effect of Solids Loading

Cyclone loadings are reported as kg of solids per m^3 of gas (kg/m^3) or kg of solids per kg of gas (kg_s/kg_g). Loading based on mass is probably not the best way to report solids loadings because the volume of solids processed by a cyclone at the same mass loading can vary greatly, depending on the density of the solids. Polymers have a bulk density of approximately 400 kg/m^3, whereas ores have a bulk density of approximately 2400 kg/m^3 – higher by a factor of 6. A cyclone operating with a polymer would have to process six times the volume of solids of a cyclone handling ore at the same mass loading. If the cyclone operating with the polymer were designed to operate at high loadings on a mass basis, it could plug. In addition, the diplegs below the cyclone operating with the polymer may experience problems because of the high volumetric loading. Despite this situation, almost all cyclone loadings are reported on a mass basis, so they are reported in this manner in this chapter.

Cyclone efficiencies increase with increasing solids loading. At higher loadings, very fine particles are trapped in the interstices of the larger particles, increasing the collection efficiency of the small particles.

11.4.1.8 Gas Outlet Tube

In most cyclones, a GOT extends into the cyclone to a level equal to or below the bottom of the inlet. This prevents solids from bypassing directly into the outlet of the cyclone. The GOT increases the efficiency of a cyclone relative to a cyclone that does not have a GOT.

Decreasing the diameter of the GOT increases the efficiency of a cyclone. This is because it increases the length and angular velocity of the vortex below the GOT.

11.4.1.9 Inlet Gas Velocity

The inlet gas velocity for cyclones can range from a low of approximately 12 m/s to as high as 40 m/s. When cyclones are used to separate a fragile, expensive material from a gas stream, high inlet gas velocities can result in excessive attrition of the material. In these cases, the inlet velocity of the cyclone is reduced to minimize attrition.

In secondary cyclones, the inlet gas velocity is higher than on the primary cyclone. In a secondary cyclone, the loading is low, and an increased gas velocity is required for high collection efficiencies. Because the loading in the secondary cyclone is low, erosion and attrition are small, and high gas velocities can be used.

11.4.1.10 Cyclone Dimensions and Design

A small-diameter cyclone should be able to collect small particles because it can develop a high centrifugal force. However, small cyclones mean that many have to be in parallel to accommodate large gas flows. It is difficult to get even distribution of solids in parallel cyclones causing inefficiencies that negate the advantage of the small cyclones.

An equation [19] that has been used extensively in cyclone design is

$$D_{p,th} = \sqrt{\frac{9\mu L_w}{\pi N_s U(\rho_p - \rho_g)}} \tag{11.19}$$

in which $D_{p,th}$ is a particle collected with an efficiency of 50%. According to this equation, the higher the inlet gas velocity, the greater the cyclone collection efficiency. However, in an actual cyclone, efficiency increases up to a limiting value of gas inlet velocity and then begins to decrease. This is because the inner vortex lengthens as the inlet gas velocity increases. At some vortex length, the vortex approaches the wall of the cyclone cone, re-entraining solids and causing the efficiency to decrease.

The length of the inner vortex, L_v, is called the vortex length of the cyclone. If the cyclone is too short, the bottom of the vortex can touch the cyclone wall and re-entrain solids. Therefore, it is important to estimate the length of the vortex.

Bryant et al. [20] developed an equation for estimating vortex length that applies at low solids loadings:

$$L_v/D_B = 2/(A_o/A_i)^{1/2} \tag{11.20}$$

For most cyclones, the minimum practical value of A_o/A_i is 0.4.

As described above, the vortex of a cyclone precesses around the centre axis of the cyclone. A long vortex can lead to re-entrainment of the solids and cone erosion. Sometimes an inverted cone or similar device is added to a secondary cyclone near the cone bottom. The vortex then attaches to the cone and is stabilized, minimizing efficiency loss due to re-entrainment of solids and erosion in the cone.

Rough cyclone walls reduce cyclone efficiency and decrease the pressure drop in a cyclone [21]. If solids encounter a large protuberance such as a weld bead, this can act like a "ski jump," causing solids to be deflected into the vortex. In very small cyclones, solids can bounce off the cyclone wall across from the cyclone inlet and be deflected into the vortex. To counteract this effect, the barrel diameter can be increased to prevent solids from bouncing into the vortex.

11.4.2 Other Separation Devices

Although the cyclone is the most commonly used device to separate solids from the exit gas from fluidized beds, other devices are sometimes deployed. These other devices are discussed in Subsections 11.4.2.1–11.4.2.5 below. A more detailed description of these devices is given elsewhere [22].

11.4.2.1 Particulate Scrubbers

Scrubbers are a class of devices in which a spray of liquid (usually water) is used to remove small particles from a gas stream. Such devices have been in use for over 100 years.

The dominant collection mechanism of scrubbers is inertial impaction. Sprayed liquid droplets contact the particles and remove them from the gas stream. Particulate scrubbers are suitable for high-temperature operation and are small. They are constant pressure drop devices and are less expensive than electrostatic precipitators and filters. However, a slurry stream is generated from scrubbers, and they can foul easily. They also require a significant energy consumption to remove fine particles.

11.4.2.2 Fabric Filters

Fabric filters ("baghouses") remove dry particles from a gas stream by passing particles and gas through a filter media such as woven cloth, felt, or porous membranes. The filter media can be flexible or rigid. Fabric filters are used because of their high collection efficiencies for particles over broad ranges of particle size. They have an infinite turndown ratio and nearly complete collection of particles larger than 2 μm and are insensitive to process upsets. One problem is that very fine particles can penetrate and blind the filter. The pressure drop across a filter is primarily due to gas flow through the particle cake on the surface of the filter and, therefore, varies with time.

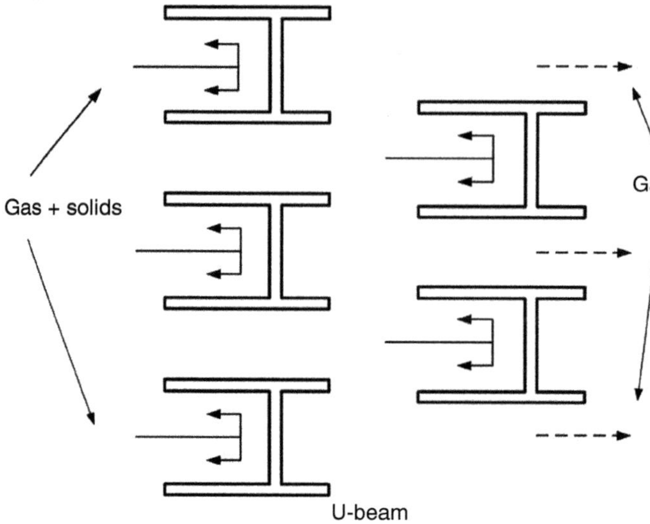

Figure 11.22 Schematic drawing of top view of U-beam separator.

11.4.2.3 Granular Bed Filters

Granular bed filters collect particles by passing them through a bed of larger particles (granules). Granular bed media include sand, gravel, and ceramics. Granular bed filters are often moving bed filters. They are capable of operating at high temperature (>800 °C) and removing abrasive, sticky, and combustible particles.

11.4.2.4 Electrostatic Precipitators

When particles suspended in a gas are exposed to gas ions in an electrostatic field, they can become charged and migrate to collection plates under the action of an electrostatic field. Electrostatic precipitators can give high collection efficiencies for fine particles, with efficiencies up to 91% achievable for very fine materials such as fly ash. Precipitators have high initial cost and can have high maintenance costs as well. They are often installed downstream of secondary cyclones to achieve a clean exit gas stream.

11.4.2.5 U-Beams

Another approach to separate particles from a gas stream is to use impingement separators. These devices cause solids to impinge on a surface, which separates the solids from the gas. An example of an impingement device is a U-beam separator (Figure 11.22). The collection efficiency of a U-beam separator is generally between 95% and 97%, not as high as a cyclone (which is about 99.9%). U-beams are used because of their low pressure drops, which can be as low as 1 kPa.

11.5 Solids Flow Control Devices/Feeders

There are several solids flow control devices that also serve as feeders. Schematic drawings of these devices are shown in Figure 11.23. One of the most common

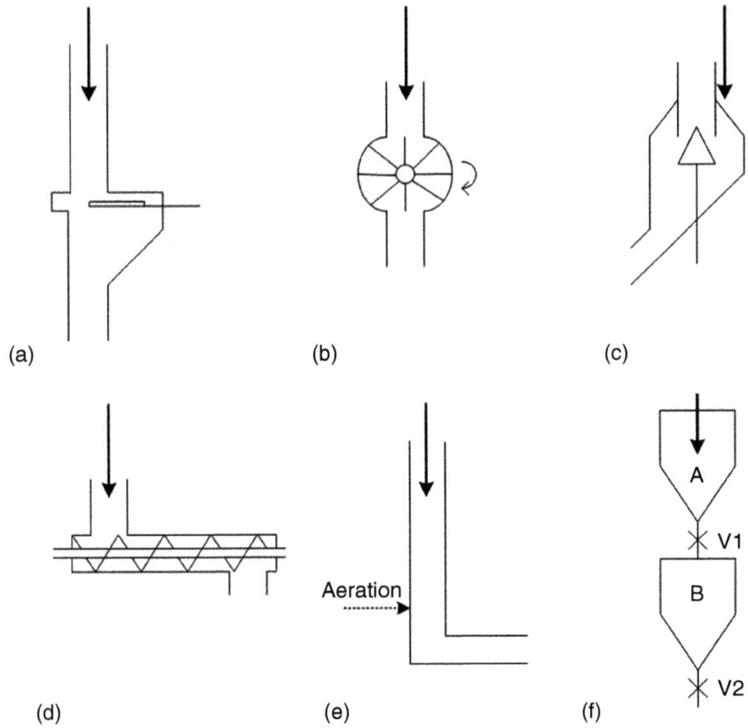

Figure 11.23 Control valve/feeders for particulate solids. (a) Slide valve, (b) rotary valve, (c) cone valve, (d) screw feeder, (e) L-valve, and (f) lock hopper.

devices to control solids flow rates is the slide valve shown in Figure 11.23(a). This device is installed in fluidized bed processes to control solids flows as high as 40 tonnes/min. They control solids in pipes as large as 1.5 m in diameter. The valve features a horizontal plate that is slid across the cross section of the pipe, varying the area available for flow, and, therefore, the solids flow rate.

Rotary valves (Figure 11.23(b)) are also used to control solids flow, but not with such high solid flows as with slide valves. The valve consists of a series of rotating pockets that partially fill with solids at the top of the valve and then discharge solids at the bottom of the valve. The returning pockets can "pump" gas back into the solids feeding the pockets at the top. This returning gas can cause flow difficulties in some applications, so often a gas bypass is added to the valve to prevent the returning gas from disturbing the entering solids.

Another solids flow control device is the cone valve (Figure 11.23(c)). This device uses a vertical, movable cone to obstruct the flow area of the solids. Often the surface of the cone is covered with refractory or ceramic to minimize erosion. This type of valve can be used in high solids flow situations and sometimes is used instead of a slide valve.

Another very common solids flow control device is the screw feeder (Figure 11.23(d)). This device moves solids using an "auger" composed of several flights rotating in a horizontal pipe, although screw conveyors can be used vertically, as well as on an angle. This device is very reliable but operates at lower throughputs than other devices such as cone and slide valves.

Unlike other solids flow control devices shown in Figure 11.23, the L-valve is nonmechanical. The operation of the L-valve is described in Section 11.3.1. The L-valve can be used in high-temperature situations where other valves would have to be made of expensive alloys. It is also a very high throughput device.

The lockhopper (Figure 11.23(f)) is a device used to feed solids into, or extract them from, high-pressure systems. To feed solids into a high-pressure system, solids from hopper A would be added to hopper B. Valves V1 and V2 would be closed, and hopper B pressurized to match the pressure of the system. Valve V2 would then be opened to allow the solids to flow into the pressurized system.

To discharge solids from high pressure, hopper B would be pressurized to the pressure of the system. Valve V1 would then be opened to allow the solids to flow into hopper B. Valve V1 would then be closed, and hopper B depressurized (usually to ambient pressure). Valve V1 would then be opened to allow the solids to flow out of hopper B into ambient pressure.

Solved Problems

11.1 Required primary dipleg length
Calculate the required length of a primary cyclone dipleg for the following operating conditions:
Bed density in the fluidized bed = 1200 kg/m^3
Fluidized bed density in the primary dipleg = 1100 kg/m^3
Primary cyclone pressure drop = 2.5 kPa = 2500 Pa
Dipleg immersion = 0.95 m

The required height of a primary cyclone dipleg is given by Eq. (11.12):

$$H_{dip1} = (\rho_{bed}\, I_{1bed}\, g + \Delta P_{cy1})/(\rho_{dip1}\, g) \tag{11.12}$$

Therefore, the required primary dipleg height for this example is

$$\mathrm{H}_{\mathrm{dip1}} = (1200 \times 0.95 \times 9.81 + 2500)/(1100 \times 9.81) = 1.27\ \mathrm{m}$$

As discussed, the dipleg length below both primary and secondary cyclones should be at least 1 m more than the calculated length to take into account various contingencies such as upsets, changes in bed height, and changes in particle size distribution. Therefore, the primary dipleg length should be at least 1.27 m + 1 m = 2.27 m long.

11.2 Pressure Balance.
In the figure on the right, catalyst is flowing out of a regenerator into a fluidized bed catalyst cooler, through a standpipe below the catalyst cooler, and through a slide valve, which controls the catalyst flow rate around the catalyst cooler loop. After passing through the slide valve, the catalyst is fed into a riser, which conveys the catalyst back into the fluidized bed of the regenerator. Pressures in kPa$_g$ at various points around the catalyst flow

loop are shown on the figure. Set up a pressure balance around the catalyst cooler flow loop and calculate the pressure drop across the standpipe. Is the flow in the standpipe fluidized or non-fluidized?

A pressure balance around the catalyst cooler flow loop is

$$\Delta P_{sv} + \Delta P_{riser} + \Delta P_{regen} = \Delta P_{cc} + \Delta P_{sp}$$

so that

$$\begin{aligned}\Delta P_{sp} &= \Delta P_{sv} + \Delta P_{riser} + \Delta P_{regen} - \Delta P_{cc}\\ &= (230.5 - 210.5) + 210.5 - 200.1) + 200.1 - 188.9)\\ &\quad - 244.3 - 188.9) = 20 + 10.4 + 11.2 - 55.4 = -13.8\ \text{kPa}\end{aligned}$$

The pressure drop across the standpipe is negative, which means that it is not building pressure, but dissipating pressure. When a standpipe dissipates pressure, it cannot operate as a fluidized standpipe. Therefore, the standpipe is operating in the non-fluidized bed mode, and in this mode the gas is flowing downward at higher speed than the solids.

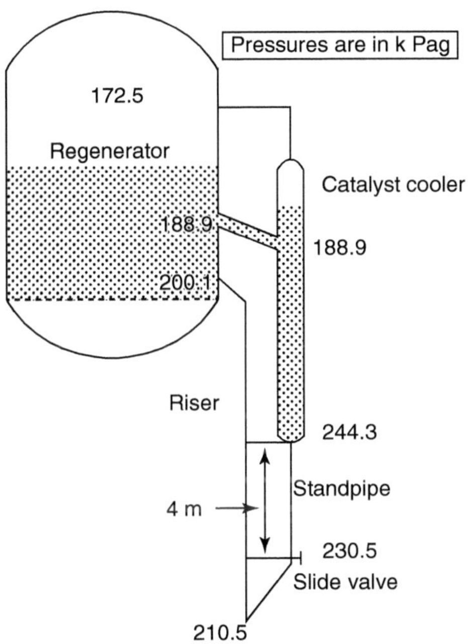

Notations

A	area (m^2)
A_i	inlet area (m^2)
A_o	outlet area (m^2)
D_B	cyclone barrel diameter (m)
$D_{p,th}$	particle diameter for 50% collection efficiency (m)

g	gravitational constant (9.81 m/s^2)
G_s	solids mass flux (kg/m^{2}s)
H	cyclone inlet height (m)
H_{dip1}	primary cyclone dipleg length (m)
H_{dip2}	secondary cyclone dipleg length (m)
H_{sp}	standpipe height (m)
H'_{sp}	changed standpipe height (m)
I_{1bed}	immersion of primary dipleg in the bed (m)
I_{2bed}	immersion of secondary dipleg in the bed (m)
L_v	vortex length (m)
L_w	cyclone inlet width (m)
N_s	number of spirals in a cyclone (–)
P	pressure (Pa)
P'	changed pressure (Pa)
P_b	pressure at bottom of standpipe (Pa)
P_t	pressure at top of standpipe (Pa)
ΔP	pressure drop (Pa)
ΔP_{CFB}	CFB pressure drop (Pa)
ΔP_{cy}	cyclone pressure drop (Pa)
ΔP_{cy1}	primary cyclone pressure drop (Pa)
ΔP_{cy2}	secondary cyclone pressure drop (Pa)
ΔP_d	distributor pressure drop (Pa)
ΔP_{dip1}	pressure drop across primary cyclone dipleg (Pa)
ΔP_{dip2}	pressure drop across secondary cyclone dipleg (Pa)
ΔP_{GOT1}	gas outlet tube pressure drop (Pa)
ΔP_{lb}	lower fluidized bed pressure drop (Pa)
$\Delta P_{L\text{-}valve}$	L-valve pressure drop (Pa)
$\Delta P_{react/regen}$	reactor/regenerator pressure drop (Pa)
ΔP_{regen}	regenerator pressure drop (Pa)
ΔP_{riser}	riser pressure drop (Pa)
ΔP_{sp}	standpipe pressure drop (Pa)
ΔP_{sp1}	first standpipe pressure drop (Pa)
ΔP_{sp2}	second standpipe pressure drop (Pa)
$\Delta P_{stripper}$	stripper pressure drop (Pa)
$\Delta P_{surge\ hopper}$	pressure drop across surge hopper (Pa)
ΔP_{sv1}	first slide valve pressure drop (Pa [psi])
ΔP_{sv2}	second slide valve pressure drop (Pa)
ΔP_{ub}	upper fluidized bed pressure drop (Pa)
ΔP_v	valve pressure drop (Pa)
$\Delta P/Lg$	pressure drop per unit length (Pa/m)
$\Delta P/(Lg)_{mf}$	pressure drop per unit length at minimum fluidization (Pa/m)
$\Delta P/(Lg)_{\mathrm{I}}$	pressure drop per unit length at Point I (Pa/m)
$\Delta P/(Lg)_{\mathrm{II}}$	pressure drop per unit length at Point II (Pa/m)
Q_A	aeration added to L-valve (m^3/s)
Q_{sp}	volumetric flow in standpipe (m^3/s)
Q_T	total L-valve aeration (m^3/s)

u_b	bubble rise velocity (m/s)
U	superficial gas velocity (m/s) (g subscript removed as per common notation.)
U_{mf}	superficial gas velocity at minimum fluidization (m/s)
v_g	interstitial gas velocity, U_g/ε (m/s)
v_{mb}	minimum bubbling velocity (m/s)
v_{mf}	interstitial minimum fluidization velocity, U_{mf}/ε_{mf} (m/s)
v_r	relative gas/solid velocity (m/s)
$(v_r)_{\text{I}}$	relative velocity at Point I (m/s)
$(v_r)_{\text{II}}$	relative velocity at Point II (m/s)
v_s	solid velocity (m/s)
W_g	gas flow rate (kg/s)
W_s	solids flow rate (kg/s)

Greek Letters

ε	voidage (−)
ρ_{dip}	primary cyclone dipleg density (kg/m^3)
ρ_{dip2}	secondary cyclone dipleg density (kg/m^3)
ρ_g	gas density (kg/m^3)
ρ_{mf}	fluidized bed density at minimum fluidization (kg/m^3)
ρ_p	particle density (kg/m^3)
ρ_{sk}	skeletal density (kg/m^3)
ρ_{susp}	suspension density (kg/m^3)
ρ_t	fluidized bed density at top of standpipe (kg/m^3)
μ	gas viscosity (kg/m s)

Abbreviations

CFBC	circulating fluidized bed combustion
FCC	fluid catalytic cracking
GOT	gas outlet tube
PSRI	Particulate Solids Research Inc

References

1 Knowlton, T.M. and Hirsan, I. (1978). L-valves characterized for solids flow. *Hydrocarb. Process.* 57: 149–161.

2 Eleftheriades, C.M. and Judd, M.R. (1978). The design of downcomers joining gas-fluidized beds in multistage systems. *Powder Technol.* 21: 217–225.

3 Karri, S.B.R. and Knowlton, T.M. (1993). Comparison of group A solids flow in hybrid angled and vertical standpipes. In: *Circulating Fluidized Bed Technology IV* (ed. A.A. Avidan), 253–259.

4 King, D. (1991). Fluidized catalytic crackers, an engineering review. In: *Fluidization VII* (eds. O.E. Potter and D.J. Nicklin), 15–26.

5 Yaslik, A.D. (1993). Circulation difficulties in long angled standpipes. In: *Circulating Fluidized Bed Technology IV* (ed. A.A. Avidan), 484–492.

6 Knowlton, T.M. (1080) Coal conversion apparatus. US Patent 4,202,673, filed 17 August 1978 and issued 13 May 1980.

7 Karri, S.B.R. and Knowlton, T.M. (2001). Streaming flow in cyclone diplegs. *Proceedings of 10th International Fluidization Conference*, Beijing, China (20–25 May 2001).

8 Geldart, D. and Kerdoncuff, A. (1992). The behaviour of secondary and tertiary cyclone diplegs. Presented at AIChE Annual Meeting, Miami Beach (November 1992).

9 Chan, I., Findlay, J., Knowlton, T.M. (1988). Operation of a nonmechanical L-valve in the automatic mode. Presented at Fine Particle Society Meeting, Santa Clara, CA (19 June 1988).

10 Knowlton, T. and Karri, S.B.R. (2008). Differences in cyclone operation at low and high solids loading. *Proceedings of the International Fluidization South Africa Conference*. Southern African Institute of Mining and Metallurgy, Johannesburg (November 2008), pp. 119–160.

11 Yuu, S., Tomosada, J., Yuji, T., and Yoshida, K. (1978). The reduction of pressure drop due to dust loading in a conventional cyclone. *Chem. Eng. Sci.* 33: 1573–1580.

12 Muschelknautz, E. and Krambrock, W. (1996). Pressure drop and separation efficiency in cyclones. In: *VDI Heat Atlas*, 6 and 1 edition (English), chap. Lj, 1–8.

13 Hoffman, A.C., de Jonge, R., Arends, H., and Hanrats, C. (1995). Evidence of the "natural vortex length" and its effect on the separation efficiency of gas cyclones. *Filt. Sep.* 32: 799.

14 Whiton, L. Jr., (1941). *Trans. Am. Soc. Mech. Eng.*: 213–218.

15 Smellie, J. (1942). *Iron and Coal Trades Rev.* 144: 169.

16 Koffman, L. (1953). *Gas and Oil Power* 48: 89–94.

17 Grace, J. (2008). Maldistribution of flow through parallel cyclones in circulating fluidized beds. In: *Circulating Fluidized Bed Technology IX* (eds. J. Werther, N. Nowak, K.-E. Wirth and E.-U. Hartge), 969–977. Hamburg, Germany: TuTech Innovation GmbH.

18 Knowlton, T., Findlay, J., Hackman, L., and McKnight, C. (2016). Solids maldistribution in parallel cyclones. In: *Presented at Fluidization XV* (eds. J. Chaouki, X. Bi, F. Berruti and R. Cocco). New York: Engineering Conferences International.

19 Lapple, C.E. (1940). Calculation of particle trajectories. *Ind. Eng. Chem.* 32: 605–617.

20 Bryant, H.S., Silverman, R.W., and Zenz, F.A. (1983). How dust in gas affects cyclone pressure drop. *Hydrocarb. Process.* 62: 87–90.

21 Iinoya, K. (1953). Memoirs of Faculty of Engineering, Nagoya University, Faculty of Engineering. Nagoya University, 5, (2).

22 Knowlton, T.M. and Dhodapkar, S. (2018). Gas-solid operations and equipment. In: *Perry's Chemical Engineering Handbook*, 9e, Section 17 – 17-21-70.

Problems

12.1 Required secondary dipleg length
A secondary cyclone dipleg is immersed into a fluidized bed a distance of 2 m. The pressure drop across the primary cyclone from the cyclone inlet to the top of the primary dipleg is 2.5 kPa, and the pressure drop across the GOT of the primary cyclone is 0.5 kPa. The pressure drop across the secondary cyclone from the inlet to the top of the secondary cyclone dipleg is 2 kPa. The pressure drop across the GOT of the secondary cyclone is 0.4 kPa. The density in the fluidized bed is 800 kg/m^3, and the suspension density in the secondary dipleg is 300 kg/m^3. What is the minimum dipleg height required for the secondary cyclone dipleg?

12.2 Solids are flowing down a vertical, 100-mm-diameter moving-packed bed (non-fluidized) standpipe at a rate of 1131 kg/h. The bulk density of the solids flowing in the standpipe is 800 kg/m^3, and $V_{mf}(=U_{mf}/\varepsilon_{mf})$ for this material is 0.08 m/s. A pressure drop measurement over a 1.33-m-length of the standpipe gives a reading of 7.85 kPa. What is the interstitial gas velocity in the standpipe? Which direction is the gas flowing – upward or downward relative to the wall of the standpipe? Assume that the density at minimum fluidization conditions is equal to the bulk density in the standpipe. The pressure at the bottom of the standpipe is greater than the pressure at the top of the standpipe (the standpipe is building pressure).

12

Circulating Fluidized Beds

Chengxiu Wang[1] *and Jesse Zhu*[2]

[1] *China University of Petroleum, State Key Laboratory of Heavy Oil Processing, No. 8 Fuxue Road, Changping, Beijing 102249, P.R. China*
[2] *Western University, Department of Chemical and Biochemical Engineering, 1151 Richmond Street, London, Canada N6A 5B9*

12.1 Introduction

12.1.1 What Is a Circulating Fluidized Bed?

A circulating fluidized bed (CFB) forms when the superficial gas velocity in a vertical column is increased beyond the terminal velocity of particle aggregates in a fluidized bed, causing so much particle entrainment to occur that continuous particle recirculation into the bottom of the column becomes essential to maintain a gas–solid suspension inside the column. The flow regime in the column is then fast fluidization, which, as covered in Chapter 4, corresponds to superficial gas velocities exceeding those corresponding to the bubbling and turbulent fluidization flow regimes.

Yerushalmi et al. [1] first used the circulating bed and fast fluidization flow terminology, although the concept had been applied in the petroleum industry for fluid catalytic cracking (FCC) in the 1960s under the name of a transport riser reactor [2] and in calcination of aluminum trihydrate at industrial scale beginning in 1971 [3].

In a CFB riser, solids must be continuously fed into the bed and carried upwards by high-velocity fluid to maintain the required solids holdup. Solids are captured and sent back to the bottom of the riser through a recirculation system, or fresh particles may be continuously added into the riser. A typical configuration for a CFB reactor is shown schematically in Figure 12.1. It consists of a riser, in which fast fluidization is achieved and where most reactions happen, a gas–solid separator (most commonly a cyclone) to capture the entrained solids, a solids recirculation system (e.g. a downcomer), and a solids feed system (e.g. an L-valve or loop seal) to return the particles to the bottom of the riser. In some processes, particles separated from the top of the riser are not returned directly, but go through another reactor before being returned (e.g. FCC process where catalyst particles are regenerated first).

Essentials of Fluidization Technology, First Edition. Edited by John R. Grace, Xiaotao Bi, and Naoko Ellis.

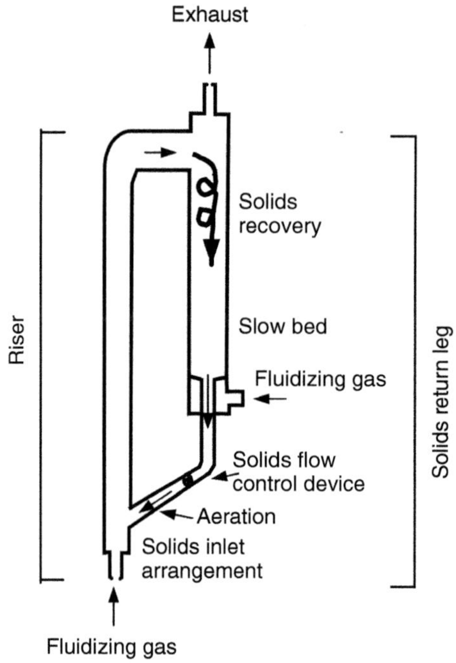

Figure 12.1 Typical configuration for a circulating fluidized bed (CFB).

Different terminologies have been used over the years. The term "fast fluidized bed" was commonly used in the late 1970s and 1980s. Later, "fast" was considered too vague a term, but particle recirculation was considered essential to maintain such a fluidized bed, so discussion at the First International Circulating Fluidized Bed Technology conference in 1985 (CFB-1) suggested "circulating fluidized bed" when referring to the reactor, while "fast fluidization" denotes the operating flow regime. At the same time, the terms "riser" and "riser reactor," from the transport riser in the FCC process, have continued to be used.

12.1.2 Key Characteristics of Circulating Fluidized Beds

In a CFB, while many particles are suspended by upflowing gas inside a riser, particles are continuously entrained from the riser by the high-velocity gas, replaced by recirculated and/or fresh particles added lower in the riser by solids return and/or feeding devices. This implies two complementary characteristics for CFB systems [4, 5]:

(1) A configuration where particles, entrained at a considerable flux from a tall main reactor or "riser," are separated efficiently from the carrying fluid, usually external to the reactor, and returned to the bottom of the riser, forming a recirculation loop.
(2) Operation at high superficial gas velocity (typically 2–12 m/s) and high particle flux (typically 10–1200 kg/m^2 s) so that there is no distinct interface in the riser between a dense bed and a dilute upper region.

Compared with conventional low-velocity bubbling and turbulent fluidized bed (TFB) reactors, CFB reactors offer several key advantages:

Favourable gas–solid contacting efficiency due to high slip between gas and solids
More uniform distribution of solids due to reduced gas bypassing
Reduced axial gas and solids backmixing
Higher production capacity due to higher gas throughput
Independent control of gas and solids retention time
Improved turndown ratio
Excellent intraparticle and interparticle heat and mass transfer rates
Less particle segregation
Possible separate gaseous reactant entrances to control solids flow
Ease of solids circulation for continuous catalyst regeneration or solids reheating

These advantages are achieved at the expense of some reduction in heat transfer coefficients between heat transfer surfaces and the bed (or suspension) and somewhat greater temperature gradients than in dense beds.

The above advantages make CFB reactors extremely useful for:

(1) Short contact time reactions, e.g. reactions where the intermediate is the desired product given the short contact time, narrow residence time distribution (RTD), and quick separation of gas product and solids
(2) Reactions with rapidly deactivating catalysts, given the reduced solids backmixing and ease of solids transportation
(3) Reactions involving large heat introduction or release where the solids serve as efficient heat carriers
(4) Gas–solid reactions where high gas-particle contact efficiency is essential
(5) Processes with varying feed and product requirements, given the flexibility of operation

In recent decades, major efforts have been devoted to the study of the fast fluidization regime and the development of CFB reactors. This can be seen from the increasing share of CFB-related papers in the international fluidized bed conferences and a book devoted solely to CFBs [6]. With the expanded interest in these high-velocity systems have also come new developments of CFB reactors such as cocurrent downflow (Chapter 20), internal CFB [7], high-density CFB [8], circulating turbulent fluidized bed (CTFB) [9], and liquid–solid and three-phase CFBs [10, 11].

12.2 Basic Parameters

The two key operating parameters are the superficial gas velocity, U_g, and solids circulation flux, G_s, defined as

$$Ug = \frac{\text{Volumetric flowrate of gas}}{\text{Bed cross-sectional area}} = \frac{Q_g}{A} \tag{12.1}$$

$$Gs = \frac{\text{Mass flowrate of solids}}{\text{Bed cross-sectional area}} = \frac{m_s}{A} \tag{12.2}$$

These two operating parameters and the gas and particle properties (gas density and viscosity; particle size, size distribution, shape, and density) determine the gas and solids flow conditions inside the riser. Other parameters that also affect the flow patterns include the operating temperature and pressure, total solids inventory inside the CFB system, and the geometry and dimensions. Note that both U_g and G_s fluctuate and that the holdup and velocities discussed below are time-averaged values.

The flow conditions inside the riser are generally characterized by voidage or local gas holdup, ε, defined as the fraction occupied by the gas phase at the location of concern, and two average voidages, $\overline{\varepsilon}$, the cross-sectional average voidage at a given vertical location, and, $\overline{\varepsilon}_{bed}$, the overall voidage over the whole riser, here assumed to be cylindrical:

$$\varepsilon = \frac{1}{\pi R^2}\int_0^R 2\pi r\varepsilon\, dr = \frac{2}{R^2}\int_0^R \varepsilon r\, dr \tag{12.3}$$

$$\overline{\varepsilon}_{bed} = \frac{2}{HR^2}\int_0^H\int_0^R \varepsilon r\ dr\ dz \tag{12.4}$$

where H and R are the total height and radius of the riser column, respectively.

The solids holdup, ε_s, is defined as the fraction of the volume occupied by the solid phase in the vicinity of the location of concern. Note that

$$\varepsilon + \varepsilon_s = 1 \tag{12.5}$$

$$\overline{\varepsilon} + \overline{\varepsilon}_s = 1$$

$$\overline{\varepsilon}_{riser} + \overline{\varepsilon}_{s,riser} = 1 \quad \text{or} \quad \overline{\varepsilon}_{bed} + \overline{\varepsilon}_{s,bed} = 1 \tag{12.6}$$

The suspension density, ρ_{susp}, is the density of the gas–solid mixture or suspensions inside the riser given by

$$\rho_{susp} = \rho_s(1-\varepsilon) + \rho_g\varepsilon \tag{12.7}$$

Depending on which voidage is used, one can define a local suspension density, cross-sectional average suspension density, and overall suspension density in the entire riser. When the friction between the gas–solid suspension and the wall as well as the particle acceleration/deceleration are neglected, the cross-sectional average suspension density can be determined from the pressure drop, ΔP, across a vertical section of length z:

$$\rho_{susp} = \frac{\Delta P}{zg} \tag{12.8}$$

Note that Eq. (12.8) cannot be applied to the bottom particle acceleration region and the top exit region if a more restrictive exit is used where a significant part of the pressure drop is due to the acceleration or deceleration of particles.

The cross-sectional average interstitial gas velocity, u_g, and particle velocity, v_p, inside a riser can be obtained from the equations

$$\overline{u}_g = \frac{U_g}{\overline{\varepsilon}} \tag{12.9}$$

$$\overline{v}_p = \frac{G_s}{(1-\overline{\varepsilon})\rho_p} \tag{12.10}$$

The local particle velocity, v_p, varies with radial position. It is more difficult to measure local particle velocities than their local concentrations. Particle velocities can be detected by several techniques. Fibre-optic probes are usually used to minimize disturbance of the gas–solid flow and interference by temperature, humidity, electrostatics, and electromagnetic fields [12].

When two optical probes are separated by a small distance ΔL, the cross-correlation function of the two signals f_1 and f_2 registered by probes 1 and 2 over time τ is given by

$$\phi_{12}(\tau) = \lim_{T\to\infty} \frac{1}{T} \int_0^T f_1(t)f_2(t+\tau)dt \tag{12.11}$$

The mean particle velocity can then be estimated as

$$v_p = \Delta L/\tau_{\max} \tag{12.12}$$

where τ_{max} is the delay time at which $\phi_{12}(\tau)$ reaches a maximum.

The difference between the actual gas and particle velocities is defined as the slip velocity u_{slip}:

$$u_g - v_p = u_{slip} \tag{12.13}$$

12.3 Axial Profiles of Solids Holdup/Voidage

Axial and radial distributions (profiles) of the various parameters are often used to characterize the hydrodynamics and reactor performance inside risers, including solids holdups, gas and particle velocities, etc. Along a CFB, the axial flow structure in the riser, including solids holdup (or bed voidage), particle velocity, and gas–solid contact, is generally nonuniform [13].

Figure 12.2 displays the axial distribution of the cross-sectional average solids holdup in a CFB riser for superficial gas velocities of 5, 7, and 9 m/s and solids flux, G_s, up to 1000 kg/m^2 s. As shown in Figure 12.2, significantly different axial profiles can be seen for various operating conditions. In general, the axial profile is initially linear at low solids flux and is replaced by a relatively nonuniform exponential shape with a dense region growing at the riser base, as the solids circulation rate increases. At even higher solids fluxes, >500 kg/m^2 s, an S-shaped profile begins to appear. At the very high G_s of 1000 kg/m^2 s, the axial solids holdup profiles become more uniform, with over 30% of the riser volume occupied by particles. For constant U_g, the solids holdup increases with increasing G_s (Figures 12.2), whereas for constant G_s, the solids holdup decreases with increasing U_g (Figure 12.2). When the gas velocity is low, the transition to the denser S-shape occurs at lower solids flux; for example, the axial profile is clearly S-shape for $G_s = 400$ kg/m^2 s, at $U_g = 5$ m/s, but nearly linear at 7 m/s.

For the nearly linear profile conditions, particles introduced at the bottom of the riser are first quickly accelerated upwards by the fluidization gas (Figure 12.2, $G_s < 200$ kg/m^2 s). At higher solids flux, $G_s = 200$–400 kg/m^2 s, particle acceleration continues to an axial location where the particle velocity becomes constant, i.e. the particle acceleration becomes negligible, leading to a steep profile with

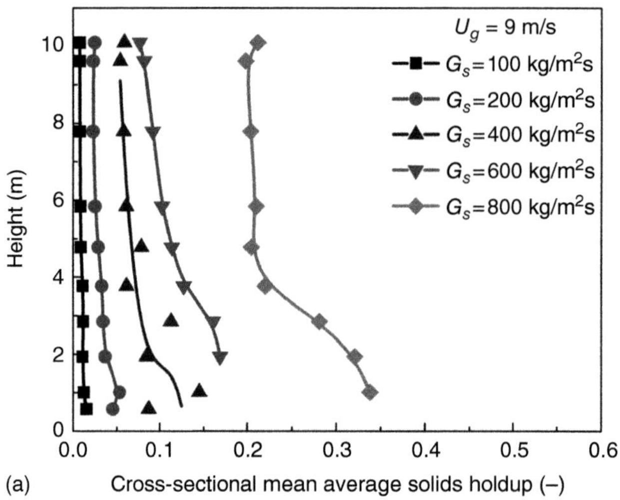

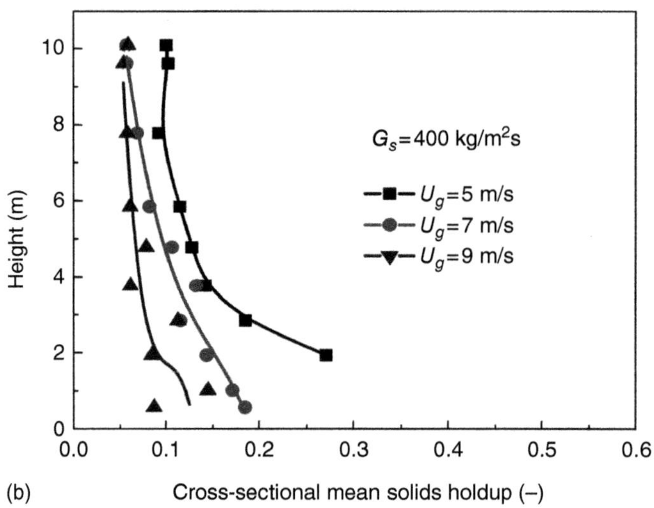

Figure 12.2 Typical solids holdup profiles along circulating fluidized bed: (a) at constant superficial gas velocity; (b) at constant solids mass flux. Source: Adapted from Wang et al. 2014 [14].

denser bottom. The dense phase at the bottom then becomes more "prominent" with increasing solids flux to $G_s = 500–600\ \text{kg/m}^2\ \text{s}$. With a dilute phase at the top of the riser, the axial profile is typical of S-shape as reported by many researchers [15–17], although the transition to S-shape may occur at lower G_s due to the lower gas velocities employed in most earlier studies.

To summarize, with a transition section in between, the riser may be divided into three regions, namely, a dense region at the bottom, a dilute region at the top, and a transition region between these two, resulting in an overall S-shaped axial profile. For lower G_s, the dense bottom is either non-existent (linear profile) or insignificant (exponential profile). When the riser exit is more restrictive, e.g.

for a sharp bend or T-shape, the entrained particles may slow down, causing an increase in suspension density. This leads to a C-shaped axial solids profile when superimposed upon an exponential profile, or a C + S profile at higher G_s [18, 19].

Due to equipment limitations and lack of awareness [20], most CFB studies until the mid-1990s were limited to low solids circulation fluxes ($G_s < 200\,\text{kg/m}^2\,\text{s}$) and thus very low solids concentration in the developed upper region of risers ($\varepsilon > 97\%$) corresponding to the operating conditions of circulating fluidized bed combustors (CFBCs). However, risers in the other major CFB application, FCC, operate at much higher solids circulation fluxes, as well as higher gas velocities, significantly differing from CFBC (Table 12.1). Bi and Zhu [8, 20] proposed a high-density operating mode for circulating fluidized beds, distinguishing high-density risers or high-density circulating fluidized beds (HDCFBs) corresponding to the operating conditions in FCC risers from low-density circulating fluidized beds (LDCFBs) corresponding to conditions in CFBC, which had been the main subject of studies until then.

In general, low-density low-flux CFBs are used in gas–solid reaction processes with Geldart group B particles because gas–solid processes such as coal combustion, alumina calcination, and iron ore reduction usually have low reaction rates, requiring neither high solids circulation nor high gas velocity. In addition,

Table 12.1 Operating conditions of CFB combustors and FCC risers.

Operating conditions	FCC riser reactor	CFB combustor
Particle density (kg/m^3)	1100–1700	1800–2600
Particle mean size (mm)	0.04–0.08	0.10–0.30
Superficial gas velocity (m/s)	6–28 (increase with height)	5–9
Net solids flux (kg/m^2 s)	400–1200	10–100
Apparent suspension density in the developed region (solids holdup) (kg/m^3)	50–160 (3–15%)	10–40 (mostly <1%)
Average solids residence time (single pass) (s)	2–4	20–40
Cross-sectional geometry	Circular	Mostly rectangular or square
Riser diameter (m)	0.7–0.15	4–8 equiv
Height-to-diameter ratio	>20	<5–10
Solids return column/regenerator to riser diameter ratio	≫1	<1
Solids inventory	High	Low
Solids exit structure	Smooth	Abrupt
Solids feeding device	Mechanical valves (e.g. slide valve)	Nonmechanical valve (e.g. L-valve)

Source: Adapted from Zhu and Bi 1995 [20].

since the solids are the product and the extent of gas backmixing is unlikely to be critical, the unit can operate at relatively low gas velocity and low solids circulation flux. The high-density and high-flux CFB, on the other hand, represents gas-phase catalytic reaction processes, where the gas phase is the product and the catalyst may deactivate quickly, so axial dispersion is undesirable. This requires a relatively high gas velocity in the riser to promote plug flow. High reaction rates also require short contact time between the gas and solid catalyst (e.g. several seconds), so a high gas velocity is needed in the riser. For a given catalyst-to-feed ratio, the solids circulation flux needs to be high as well. High gas velocity and solids circulation flux also increase the capacity of a given reactor. Major differences in typical operating conditions of FCC and combustion are summarized in Table 12.1.

Comprehensive studies on HDCFB began in the mid-1990s [20–23]. More recent experiments by Wang et al. [14, 24–27] have achieved extremely high solids fluxes (G_s up to 1000 kg/m^2 s) by carefully adjusting the solids flow control. Wang et al. [14] showed that up to 30% particle holdup can be realized in the fully developed region of the riser using FCC particles (Figure 12.2). Correspondingly, much greater catalytic conversions of ozone were achieved under high-density operation [25].

12.4 Radial Profiles of Solids Distribution

The nonuniformity of CFB radial solids flow structures has long been recognized. The radial solids holdup profiles in Figure 12.3 cover a wide range of operating conditions [14]. Under all operating conditions, radial distributions are nonuniform with low solids holdup (dilute region) in the centre and much higher holdup close to the wall. The radial profile is relatively flat in the core of the riser. The solids holdup increases towards the wall, with a maximum value at the wall. Operating conditions dramatically affect the radial solids

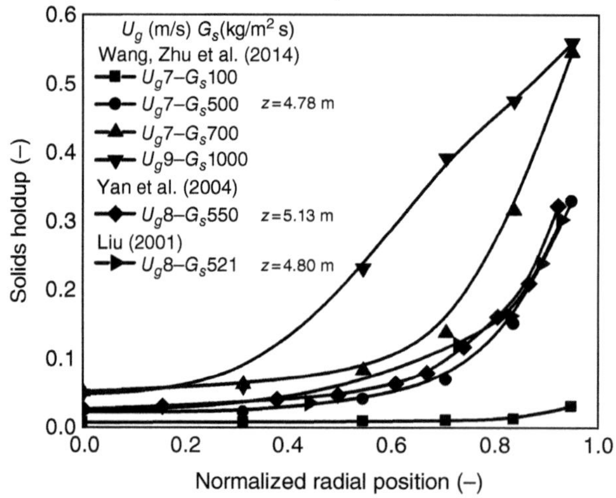

Figure 12.3 Typical radial profile of solids holdup. Source: Adapted from Wang et al. 2014 [14].

holdup distribution. By feeding more solids at a fixed superficial velocity, the riser became progressively denser towards the centre. The solids holdup in HDCFB risers differs markedly from those in low-density systems. For lower G_s (<500 kg/m^2 s) with U_g = 7–9 m/s, the radial profile is of typical "core–annulus" structure, with a dilute uniform core region surrounded by a dense annulus zone [28]. The radial variation can often be considered to have three parts: a central region up to $r/R \approx 0.5$–0.6 with a low nearly constant holdup and a slight increase towards the wall, an intermediate region between $r/R \approx 0.6$–0.7 and 0.8–0.9 where solids holdup increases appreciably, and a wall region ($r/R > 0.9$) where the solids holdup is high, but no more than 0.35. Similar results have been reported by many researchers under comparable operating conditions [23]. However, as G_s exceeds 700 kg/m^2 s, the dilute region shrinks ($r/R = 0$–0.2, occupying <20% of the cross-sectional area), and solids holdup increasing rapidly, approaching $(1 - \varepsilon_{mf})$, under most dense conditions.

The radial flow structure is often described by a core–annulus model. The original core–annulus model [28] postulated a dilute central region surrounded by a dense annulus zone: in the core region, both gas and particles rise quickly, while in the annulus, gas still moves upwards, while most particles descend slowly due to streamers. A comprehensive core–annulus model was proposed to correlate the flow structures inside a riser for various operating conditions [29]. However, such a model fails at very high G_s conditions when the solids holdup shows a continuous increasing trend towards the wall, as shown in Figure 12.4.

The nonuniformity of radial solids flow structure can also be seen from the distribution of local particle velocities, as plotted in Figure 12.4. Generally, the particle velocity reaches a maximum value at the riser axis and decreases with increasing radial position. The shapes of the radial profiles of particle velocity change as the solids circulation flux increases. At low solids circulation flux (G_s = 100 kg/m^2 s), the radial profile of the particle velocity is typically "core–annulus," with a wide flat region near the centre. At around $0.7 < r/R < 0.85$, the particle velocity decreases sharply towards the wall. For $r/R > 0.85$, the time mean particle velocity may be downwards, with negative values nearly constant around 0.5–1 m/s. The flatter velocity region in the central area is the core region, with the rest considered the annulus region. With further increases of solids circulation, the core–annulus structure again becomes unclear. These characteristics correspond well with the radial solids holdup profiles shown in Figure 12.4.

When G_s reaches 400 kg/m^2 s, the radial profiles of particle velocity deviate greatly from the "core–annulus" structure to a more parabolic shape with a maximum at the centre line and substantial decrease towards the wall [15, 24]. When G_s is increased further to 1000 kg/m^2 s, the parabolic shape transforms to a steeper and approximately linear radial profile. The particle velocities are higher in the centre of the riser for extremely high solids fluxes compared with the low fluxes at U_g = 9 m/s. The reason could be that more solids occupy the wall region, restricting the gas flow under high flux conditions. In order to "maintain" the superficial gas velocity at 9 m/s, the gas velocity has to be

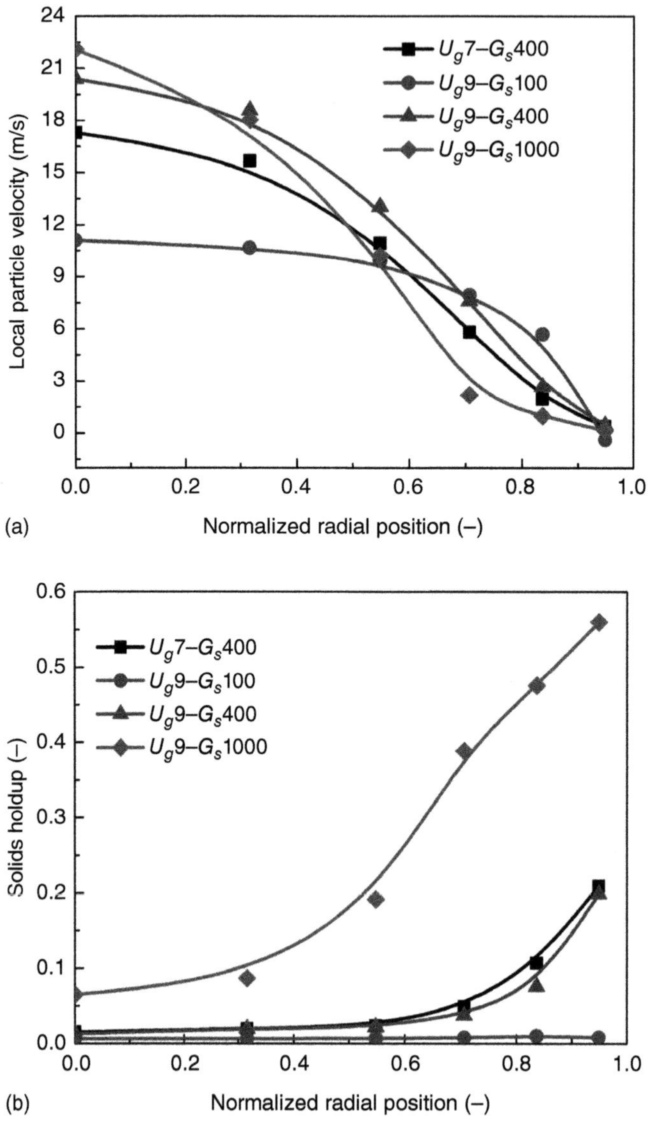

Figure 12.4 Radial profiles of (a) particle velocity and (b) corresponding solids holdup. Source: Adapted from Wang et al. 2014 [24].

correspondingly higher in the riser centre [30, 31], leading to a higher particle velocity in the centre. In addition, the particle velocity at all radial positions of the riser is positive for high solids circulation fluxes, completely different from low flux conditions where net negative particle velocities are often seen in a wide region near the wall. Similar results have also been reported by other researchers [15, 32]. Finally, the particle velocities in the wall region are all small in magnitude at high G_s, no doubt affected by the gas velocity no-slip condition at the wall.

12.5 The Circulating Turbulent Fluidized Bed

Although solids holdup in CFB risers can be increased by increasing the solids circulation and/or decreasing superficial gas velocity, the solids holdup in HDCFB is still significantly lower than in TFB (see Chapter 9). This is a major drawback of CFB, meaning that the reactor intensity tends to be lower, for example, than the TFB. The TFB, on the other hand, has higher bed density and more uniform radial distribution of gas and particles, so it has been widely applied in industry. To carefully examine those two fluidized bed reactors, typical advantages and drawbacks of TFBs and CFBs are summarized in Table 12.2.

Despite the many advantages of the TFB reactor, such as more vigorous gas–solid contacting, better bed-to-surface heat transfer, and higher solids concentrations (typically 25–35% by volume), the TFB cannot readily replace CFB reactors for existing CFB applications such as the FCC process because of inherent shortcomings such as extensive solids backmixing and, to a lesser extent, gas backmixing. For the FCC process, because catalyst particles deactivate quickly, approaching solids plug flow is essential so that deactivated catalyst can be removed in a timely manner for regeneration. Serious backmixing of solids phases in TFBs would lead to broad RTD of the solids, resulting in inferior reaction selectivity in TFBs for gas-phase catalytic reactions when catalysts deactivate rapidly.

In an effort to combine the benefits of both CFB and TFB, Zhu and Zhu [9, 33] proposed a CTFB that operates in the gas velocity range for TFB (U_g = 0.5–1.5 m/s for FCC particles) but allows up to 300 kg/m^2 s solids circulations, comparable to CFB operation. As shown in Figure 12.5, this first CTFB operation was realized in an existing CFB set-up by reducing the diameter of the lower section and injecting secondary air about halfway up from the bottom, quickly entraining particles through the top exit of the riser. Under such conditions, the high solids inventory in the downcomer combined with the lower solids holdup in the top portion of the riser provides enough pressure to enable significant solids circulation. Experimental results have clearly demonstrated that the CTFB successfully combines the advantages of TFBs (e.g. high solids holdup, uniform flow structure) and CFBs (e.g. significant solids circulation and

Table 12.2 Typical advantages and disadvantages of turbulent and circulating fluidized beds.

	Turbulent fluidized beds	Circulating fluidized beds
Advantages	Higher solids holdup More gas–solid interaction Uniform temperature distribution Excellent heat transfer	Reduced solids backmixing Easier to input and discharge large amounts of particles
Disadvantages	Serious solids backmixing Some gas mixing Limited operating flexibility	Low solids holdup Nonuniform flow structure Lower turndown ratio

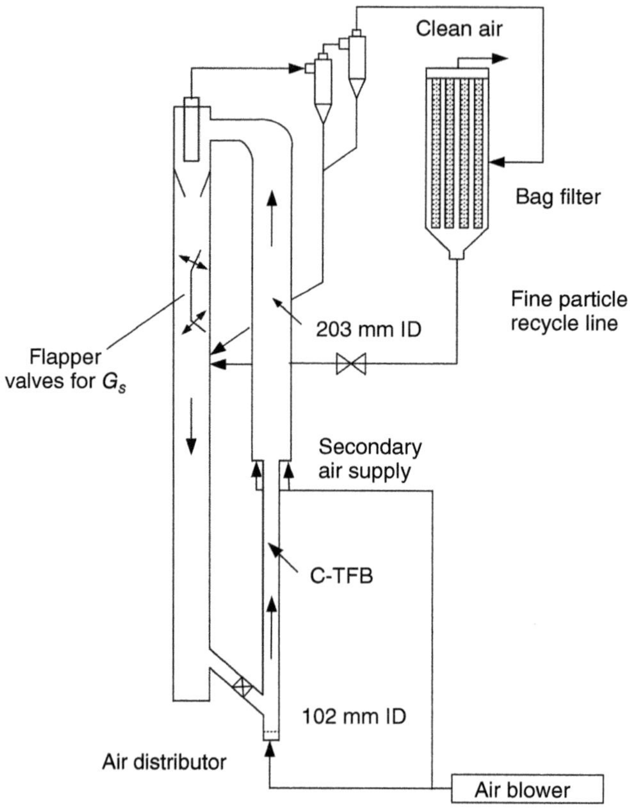

Figure 12.5 Schematic of circulating turbulent fluidized bed. Source: Adapted from Zhu and Zhu 2008 [33].

reduced solids backmixing), while overcoming some key disadvantages of CFB (e.g. insufficient solids holdup) [33].

Figure 12.6 compares the solids holdup distribution in the TFB, CTFB, and HDCFB as a function of dimensionless height, h/H. An extremely uniform axial solids holdup distribution in the CTFB is seen in Figure 12.6, compared with substantial variations in solids concentrations with both TFB and CFB. Figures 12.6 provides typical radial profiles of time mean solids concentration for CTFB, TFB, and HDCFB at the same level. The radial solids concentration distributions in the CTFB and TFB are seen to be nearly identical, but both differ significantly from that in the HDCFB. This shows that the CTFB is essentially still a TFB, with all its features, but with substantial net solids circulation, giving the CTFB additional benefits.

12.6 Micro-flow Structure

As outlined above, the gas–solid macro flow in CFB can be characterized by a highly heterogeneous structure with a dense phase at the bottom and a dilute

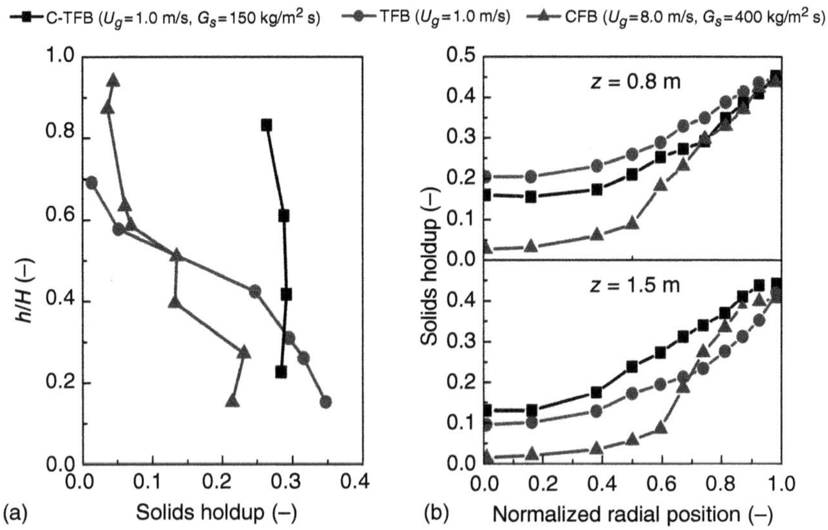

Figure 12.6 Comparison of solids concentration distribution in three fluidization systems. (a) axial distribution of solids concentration; (b) radial distribution of solids concentration. Source: Adapted from Zhu and Zhu 2008 [33].

phase at the top, as well as a core–annulus structure in the radial direction. Such complex flow with nonuniform hydrodynamics is mainly due to interactions between the gas and solids, in particular the formation of particle aggregates, commonly called clusters. As in bubbling beds (Chapter 7) where gas and solid particles form dilute gas bubbles and dense particle-rich regions, CFBs also feature a gas-rich dilute phase and a particle-rich cluster phase. The formation of clusters is a key contributor to nonuniform hydrodynamics [34, 35].

Clusters are defined as dense assemblies of particles having significantly more particles per unit volume than the surrounding dilute regions [36]. These clusters were used by early researchers to analyze the flow structure [17, 37]. The existence of aggregates has been confirmed in many published studies based on a variety of experimental techniques such as optical fibre probes, laser sheet techniques, capacitance probes, and video cameras [34, 35, 38, 39].

Li et al. [34] observed that clusters have irregular shapes with highly variable sizes as shown in Figure 12.7. Clusters were observed to transform from strands at the centre of the column into near spheres adjacent to the wall. Horio and Kuroki [35] captured typical images of clusters of parabolic or horseshoe shape heading downwards enclosing a gas wake on the upper side (Figure 12.7). Particles were shed from clusters to the dilute phase continuously from the periphery and absorbed by other clusters. Clusters in the annulus tended to descend, while those in the core usually travelled upwards. Typical downward velocities in the annulus region are in the range of 0.5–2 m/s. Clusters at the wall of a riser tend to form, descend, break up, travel laterally from the annulus to the core, and then be re-entrained in the upward-flowing suspension in the core. Images of clusters by Xu and Zhu [39] in the annulus and core regions in Figure 12.8 suggest that low

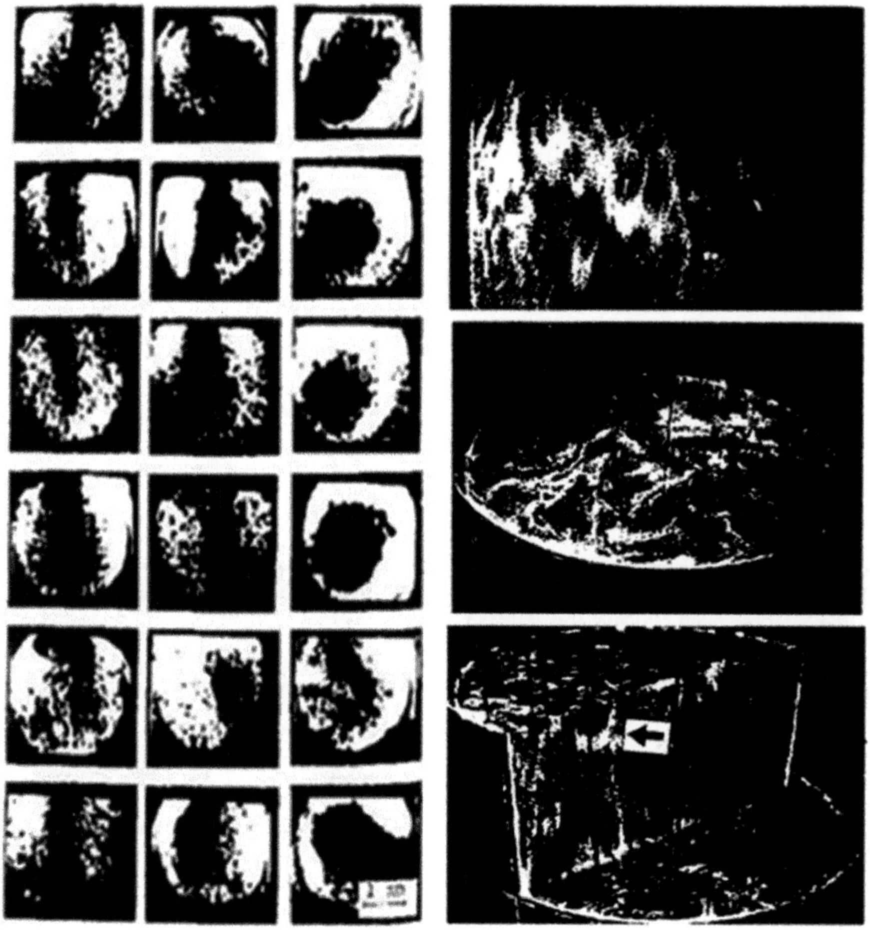

(a) (b)

Figure 12.7 Photographs of particle clusters in risers: (a) images of clusters in vertical cross-section; (b) three dimensional visualization of clusters. Source: Adapted from Li et al. 1991 [34] and Horio et al. 1994 [35].

particle density and small particle size enhance cluster formation. Descriptions of clusters from the published literature are summarized in Table 12.3.

As shown in Table 12.3, many terms have been used to describe the aggregates of particles in the riser such as "agglomerates," "clusters," "streamers," "strands," "ribbons," "swarms," and "dense packets." Yerushalmi et al. [1] used "streamers," "strands," "ribbons," and "dense packets" interchangeably, while Horio and Clift [46] suggested that there are differences between agglomerates and clusters, with agglomerates consisting of particles held together by interparticle forces (e.g. van der Waals, liquid bridges, electrostatic), while clusters form and travel solely as a result of hydrodynamic factors. In an effort to classify the different types of particle aggregation and based on direct observation of particle aggregates in a

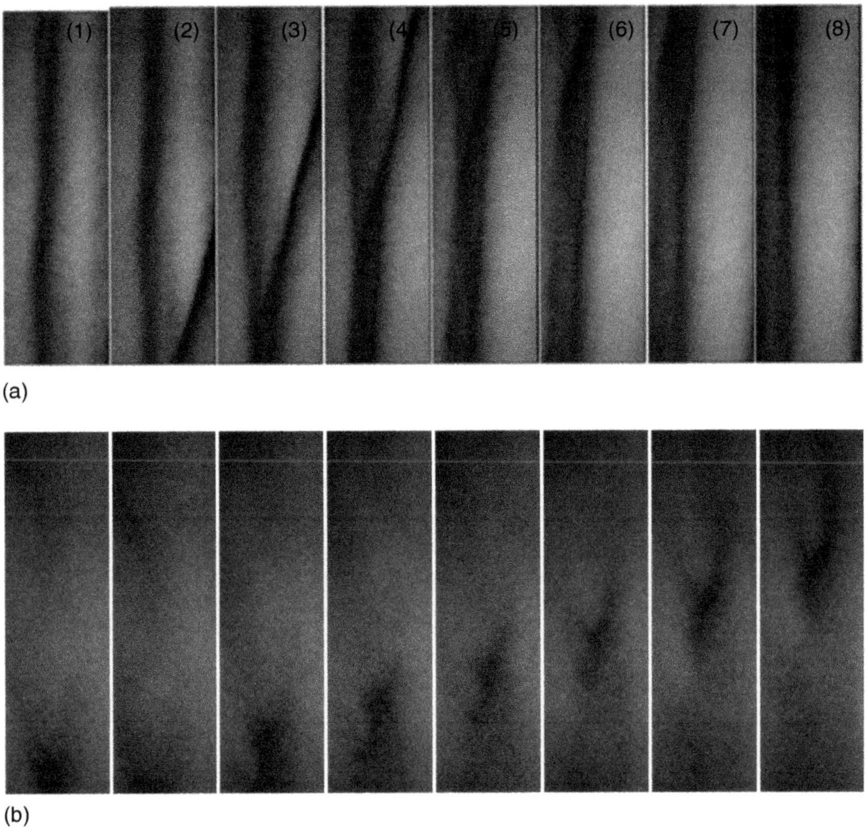

Figure 12.8 Sequences of clusters in CFB riser: (a) cluster on wall side; (b) cluster in core region. Source: Adapted from Xu and Zhu 2012 [39].

two-dimensional CFB, using a high-speed video camera, Bi et al. [47] proposed four categories of clusters as summarized in Table 12.4:

(1) **Clusters**: Groups of several to dozens of particles congregated to reduce the effective drag force exerted on them. These assemblies may be in the form of strings, triangles, and other shapes.
(2) **Particle streamers/strands**: Dense strips of larger size than clusters in the core region and along the walls of CFBs.
(3) **Particle sheets**: Two-dimensional layers of densely populated particles falling near the column wall.
(4) **Particle swarms**: Relatively dense particle aggregates at the wall, especially when a riser is operated under relatively dilute conditions. They are generally approximately spherical in shape.

A possible particle aggregation process postulated by Bi et al. [47] is shown in Figure 12.9.

Table 12.3 Experimental studies on characteristics of particle aggregates in gas–solid flow.

Author	Particles	Particle properties		Operating conditions		Riser dimensions		Measurement technique
		d_p (μm)	ρ_p (kg/m^3)	G_s (kg/m^2 s)	U_g (m/s)	D (m)	H (m)	
Xu and Zhu [39]	Glass beads, FCC	67–288	1877–2453	50–200	3–8	0.02 × 0.114	7.6	High-speed video
Shi et al. [40]	Glass beads	250, 333, 420	2379, 2478, 2514	4.35, 10.5, 45	5.2, 5.47	0.2 × 0.2	4	Particle Imaging Velocimetry (PIV)
Manyele et al.[41]	FCC	67	1500	100, 300–500	5.5–10	0.076	n/a	Optic fibre probe
Lackermeier et al. [42]	Sand	140	n/a	6	3	0.4	15.6	High-speed video and laser sheet
Takeuchi et al [43]	FCC	57	930	45.6	2.03–2.42	0.1	5.5	High-speed video
Soong et al.[44]	Sand	251	n/a	45.0	5.0	0.15	11	Dual capacitance probe
Rhodes et al.[45]	FCC	74.9	2456	2–80	3–5	0.305	6.6	High-speed video analysis at the wall
Li et al. [34]	FCC	54	930	7.3–94.6	1.3–3.5	0.09	10	Video camera
Horio and Clift [46]	FCC	61	1780	0.016–16.5	0.15–1.3	0.2	1.6	Laser sheet

Table 12.4 Summary of solids aggregation forms in CFB risers.

Form of aggregation	Region	Shape	Characters	Scale	Origin
Particle clusters	Core and annulus	Sphere	$D_e < 1$ cm $\varepsilon \approx \varepsilon_{mf}$	Microscale	1. Interparticle force 2. Particle–wake interaction
Particle stream/strands	Core	Stripe	$L_e > 1$ cm $\varepsilon \approx 0.7$–0.95 $U_e > 0$	Mesoscale	1. Resemble clusters 2. Breakup of particle sheets 3. Nonuniform introduction of particles
Particle swarms	Annulus	Swarm	$L \approx 1$–1.5 cm $U_e \approx -0.3$ to 0.4 m/s	Mesoscale	Wall–particle interaction
Particle sheets	Annulus	Layer	$\delta_e \approx 0$–3 cm $U_e \approx -1$ m/s	Macroscale	1. Resemble particle swarms 2. Particles diffusing to the wall region and swarms

D_e, equivalent size; L_e, equivalent length; U_e, average velocity; ε, average voidage; δ_e, average thickness of particle sheets.
Source: From Bi et al. 1993 [47].

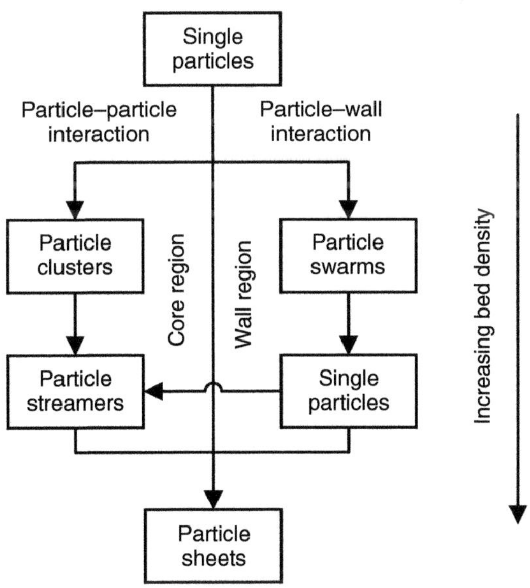

Figure 12.9 Illustration of proposed particle aggregation processes. Source: Adapted from Bi et al. 1993 [47].

As cluster formation is mainly due to the gas–solid interactions, slip velocity is another important parameter in CFB hydrodynamic studies. Slip velocity is defined as the velocity of the gas minus the velocity of the particles, i.e. $U_{slip} = u_g - v_p$. Very few studies have been carried out to measure local slip velocity, so a "one-dimensional" slip velocity [48] or apparent slip velocity [49], calculated from the superficial gas velocity, solids circulation rate, and cross-sectional average voidage, is commonly used:

$$\overline{U}_{slip} = \overline{u}_g - \overline{v}p = \frac{U_g}{\overline{\varepsilon}} - \frac{G_s}{\rho_p(1-\overline{\varepsilon})} \tag{12.14}$$

This apparent slip velocity has been found to be much higher than the terminal velocity of single particles [50], attributed to the existence of the particle clusters or streamers [34, 36, 50]. Yang et al. [49, 51] appear to be the only authors to report a systematic study of local slip velocity by measuring local gas and particle velocities simultaneously in a 140 mm diameter with 11 m high laboratory-scale CFB using fibre-optic probe Laser Doppler Velocimetry (LDV) technique. Based on a "uniform voidage slip velocity" that considered the effect of cluster formation, but not that of the radial nonuniformity, Yang et al. [49] separated the relative contributions of particle aggregation and radial particle segregation, concluding that the latter is three to five times larger than the former.

12.7 Gas and Solids Mixing

The downward motion of particles in the wall region described above in CFB risers causes backmixing of solids and may also lead to downflow (backmixing) of gas in the annulus. This results in nonuniform RTDs, which are undesirable for most chemical reactions. On the other hand, radial dispersion of gas and solids is usually beneficial to overall conversion, because it helps to reduce the radial gradients of gas and particle properties and therefore the effect of radial segregation. Thus vertical and horizontal mixing are of considerable importance in predicting conversions and selectivities in CFB reactors, especially for catalytic reactions, although one is more favourable and the other is undesirable. To properly design a gas–solid two-phase CFB reactor, the RTD for both gas and solids must be known.

Gas mixing in CFB has received considerable attention [52–60]. The axial gas flow deviates significantly from plug flow [54–56]. One-dimensional dispersed plug flow models have been employed by several researchers to correlate experimental data [61–63], with reported axial gas dispersion coefficients mostly in the range of 0.1–0.6 m^2/s [31, 54, 55, 61]. Gas backmixing increases with increasing solids circulation flux [54, 56, 61, 64] while decreasing with increasing superficial gas velocity [56, 61]. For radial dispersion, except for the work of Yang et al. [52] who reported extremely low dispersion coefficients, other reported radial dispersion coefficients [31, 53, 54, 58, 59, 63] were found to be of order 0.002–0.02 m^2/s, 1–2 orders of magnitudes less than axial dispersion coefficients. The influences of the various operating conditions on radial gas dispersion appear to be complex. In the core region of the dilute upper part of the riser, Werther et al. [63]

reported that radial gas dispersion was independent of superficial gas velocity and also independent of solids circulation rate for solids circulation fluxes from 0 to 70 kg/m^2 s. Zheng et al. [58], on the other hand, suggested that there is a threshold solids circulation flux below which radial gas dispersion increases with superficial gas velocity. Bi [65] examined the results from some HDCFB risers, with the effect of superficial gas velocity on axial gas dispersion plotted in Figure 12.10.

Axial and radial solids mixing have also been studied [31, 54, 64–71]. Solids backmixing is mainly due to the downflow of particles in the annular region [54] so that operating conditions influence solids backmixing. Unfortunately, there are few data available. From the results of both Bai et al. [64] and Wei et al. [72], axial solids dispersion increases somewhat at high solids circulation rates, but a correlation of Rhodes et al. [68] suggests the opposite. Increasing gas velocity was seen to increase the axial solids dispersion [72], a trend different from that of axial gas dispersion. Again, further study is needed to clarify this point. Bai et al. [64] found that finer particles have a more uniform RTD than coarser particles, since fines are more uniformly distributed in the riser than coarser particles.

Bai et al. [64] and Martin et al. [31] concluded that axial solids mixing is greater than axial gas mixing due to the significant internal solids circulation given the core–annulus structure. Bai et al. [64] showed that the RTDs for gas are relatively narrow with high peak values, while solids have wider distributions, with lower peaks and longer tails, indicating more axial dispersion and backmixing of solids. This difference, however, diminishes as the superficial gas velocity increases, due to more uniform suspension of solids [64]. At high gas velocity, the gas–solid flow patterns approach those of dilute-phase transport where the slip velocity tends to be constant and close to the particle terminal velocity and the gas and particles have similar radial flow structures.

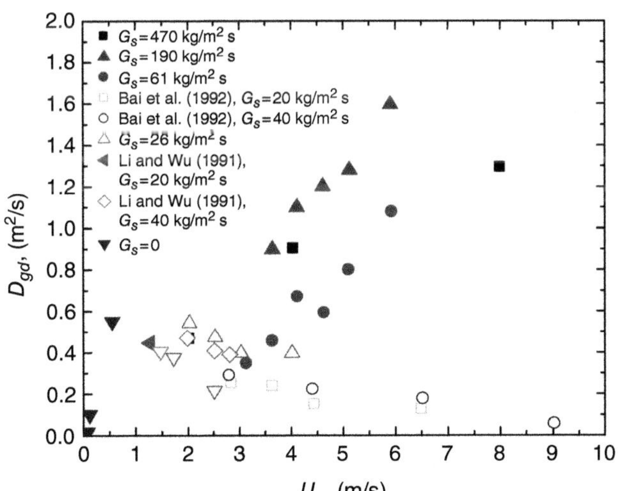

Figure 12.10 Axial gas dispersion at different superficial gas velocities. Source: Adapted from Bi 2004 [65].

Little information is available on the radial exchange of particles between the core and annulus, and no data have been reported on the radial solids dispersion coefficient. Mori et al. [70] used fluorescent particles to trace particle movement and concluded that there is considerable lateral particle exchange between the upflow core region and the downflow wall region. This was confirmed by Wei and Zhu [71] using a phosphorus particle tracer technique. Bolton and Davidson [73] and Rhodes et al. [74] argued that there must be significant movement of particles from the core to the annulus given the radial solids concentration profiles shown in Figure 12.4. The large lateral fluxes of solids measured by Qi and Farag [75] in the near wall region also indicate extensive radial solids mixing. These findings are consistent with the observations by Martin et al. [31] in an FCC riser reactor where a flat radial distribution profile of coke yield suggested considerable radial solids mixing.

12.8 Reactor Performance of Circulating Fluidized Beds

Various experimental methods have been explored to observe the hydrodynamic behaviour, which is important for gas–solid contacting efficiency in CFB systems. Contact efficiency is related to gas bypassing due to the nonuniform structure, especially in the radial direction, in both LDCFB and HDCFB reactors. In a CFB reactor, not all the gas fed into the reactor comes into close contact with the solids: some of the reactant gas experiences intimate contact with dispersed particles, while some reaches the outlet without having made substantial contact with any solids. Of the gas that comes into close contact with the solids, a fraction is converted, and the rest is unchanged. The contact efficiency then can be determined based on the conversion of a special species under carefully chosen conditions. Gas-phase conversion of CFB reactors reflects a combination of hydrodynamic and reaction kinetics.

Chemical reaction results provide direct information on CFB reactors. Catalytic decomposition of ozone, a first-order reaction taking place at room temperature, has been used to characterize gas–solid contacting in fluidized bed reactors [76]. Li et al. [74] reported axial and radial distributions of ozone concentration in a CFB riser reactor. The ozone concentration decreased rapidly near the riser bottom, then more slowly with increasing elevation. This trend corresponds well to the axial profiles of solids holdup, showing strong correlation between the reaction rate and solids holdup (Figure 12.11). However, these experiments were limited to low solids circulation flux ($<100\ \mathrm{kg/m^2\ s}$).

Wang et al. [25] investigated the reactor performance of the same CFB riser, but with solids circulation fluxes as high as $800\ \mathrm{kg/m^2\ s}$ (Figure 12.12). The ozone concentration decreased monotonically with increasing distance from the bottom inlet. Beyond the entrance region of the column, where the ozone concentration decreased quickly, the concentration changed more gradually with increasing height.

Typical radial distribution profiles of ozone concentration at different axial positions in the riser are plotted in Figure 12.12. The higher solids holdup

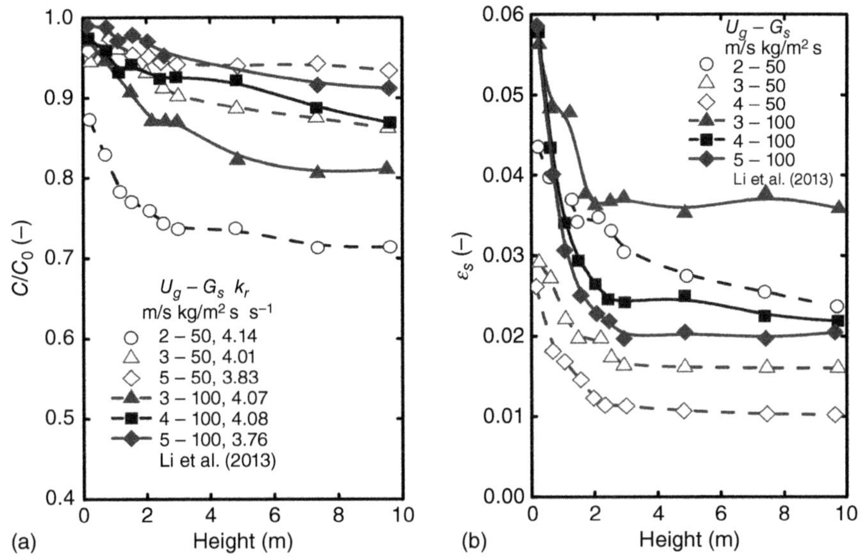

Figure 12.11 Concentrations along CFB reactor: (a) ozone concentration; (b) corresponding solids concentration. Source: Adapted from Nowak et al. 1999 [77].

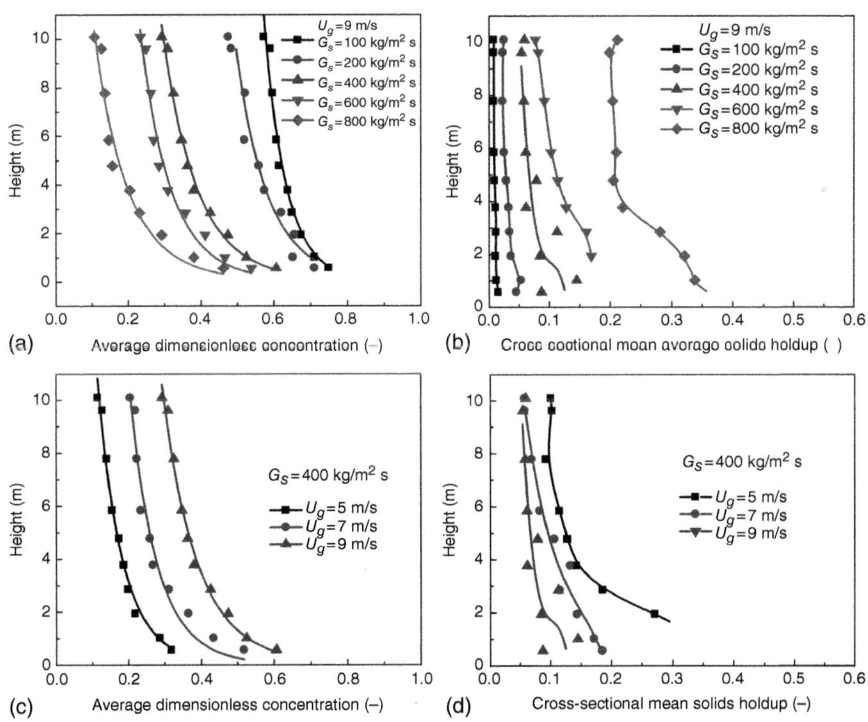

Figure 12.12 Ozone concentration in axial direction for very high density conditions: (a) ozone concentration at constant superficial gas velocity; (b) ozone concentration at constant solids circulation rate; (c) solids concentration at constant superficial gas velocity; (d) solids concentration at constant solids circulation rate. Source: Adapted from Wang et al. 2014 [25].

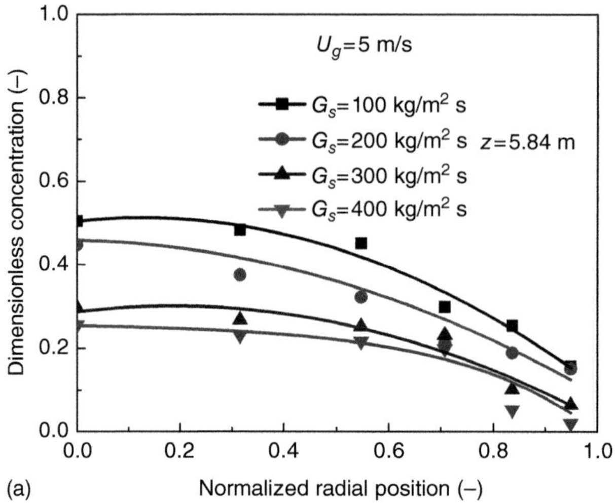

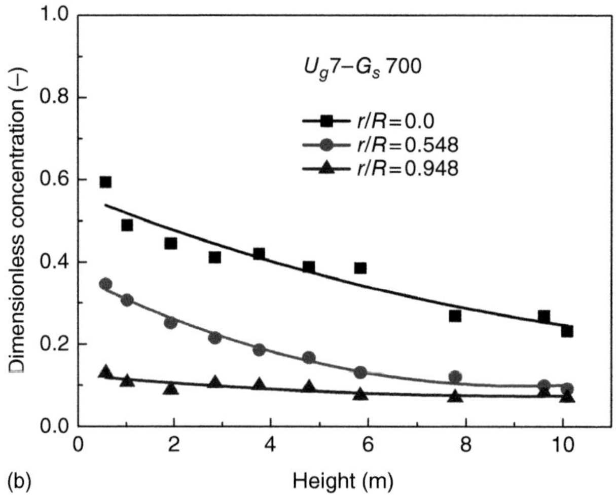

Figure 12.13 Distribution of the ozone concentration inside a CFB riser, (a) radial profiles of ozone concentration for various G_s and (b) axial profiles ozone concentration at different radius. Source: Adapted from Wang et al. 2014 [25].

near the wall results in higher reaction rate [78]. However, the radial nonuniformity could be mitigated at higher G_s [79]. Ozone concentration decreased with increasing solids circulation flux and the associated increase of solids holdup.

As shown in Figure 12.13, the ozone concentration in the central region dramatically decreased with increasing height, whereas that in the wall region remained almost constant, due to the severe lateral gas/solids segregation. Jiang et al. [80] reported that baffles can improve conversions by disturbing the core–annular flow structure.

12.9 Effect of Reactor Diameter on CFB Hydrodynamics

Typical diameters of the laboratory-scale CFB reactors are in the range of 50–150 mm, while full-scale FCC reactors used in the petrochemical industry have diameters up to 1.0–2.0 m [2] and CFBC are even bigger, with bed cross-sectional areas reaching 200 m^2 [77]. As equipment size increases, characteristic changes take place in the gas–solid flow that can decisively affect reactor performance. Such changes result either directly from the geometry or indirectly from design changes as the unit is enlarged. In particular, experience has shown that reactor diameter and its cross section shape affect hydrodynamics and reactor performance.

There have been a number of efforts to understand the effects of reactor diameter on CFB hydrodynamics. For example, recent studies [81, 82] have directly compared the effects of diameter on the distribution of solids holdup and particle velocity at $U_g \approx 5$ m/s and $G_s = 100$ and 600 kg/m^2 s (see Figure 12.14). The results show that a larger-diameter riser has a higher solids holdup at every axial and radial position for both low and high solids flux. Radial profiles of the solids holdup for large-diameter risers increase more sharply towards the riser wall than those of small-diameter risers. The same trend can also be found for the distribution of the local particle velocity.

The geometry of CFB reactors has been found to have considerable influence on the hydrodynamics [83]. Figure 12.15 shows the flow development observed from the axial solids holdup distribution. It is found that, at almost the same solids circulation flux and gas velocity, a riser of rectangular cross section has a higher solids holdup than a riser of circular cross section at every vertical level. Corners

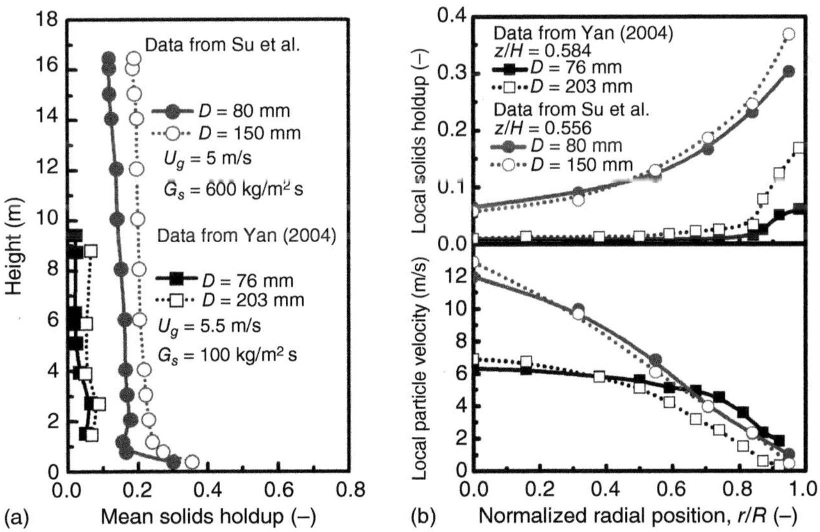

Figure 12.14 Effects of reactor diameter on solids concentration and particle velocity distributions: (a) axial distribution of solids concentration; (b) radial distribution of solids concentration and particle velocity. Source: Adapted from Yan 2004 [81] and Su et al. 2019 [82].

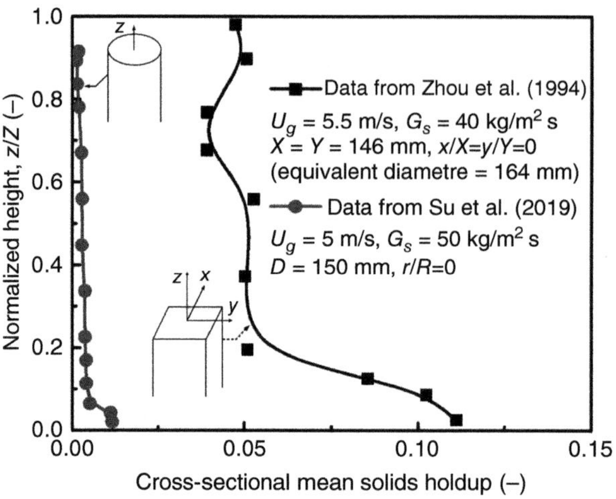

Figure 12.15 Effects of reactor geometry on solids holdup distribution. Source: Adapted from Zhou et al. [83] and Werther [84].

can shelter particles rebounding from the top of the riser, while descending particles are more difficult to strip from the corners than those travelling downwards along a smooth surface. Hence particles descend further in the corners of a rectangular column. Therefore, exit effects tend to be more prominent in rectangular columns. For the same reason, solids holdup tends to be higher in risers of rectangular cross section for a given superficial gas velocity and solids circulation flux.

From the above discussion, it is clear that reactor size affects performance. As mentioned in Section 12.7, ozone concentration decreases with increasing solids circulation flux and associated increase of solids holdup. Hence ozone concentration in a riser of larger diameter tends to be lower than in a smaller reactor for a given gas velocity and solids circulation flux. This is indirectly proven by a modelling study carried out by Werther [84], who found out that the reaction efficiency of a larger fluidized bed reactor was much lower than for a small unit. For example, in a bench-scale reactor, a conversion of 95% was achieved with an expanded bed height of 0.65 m, whereas a bed height of 2 m was just sufficient to achieve 85% conversion in a larger unit.

Notations

C	ozone concentration (ppm)
C_0	inlet ozone concentration (ppm)
D	riser or downer diameter (m)
G_s	solids circulation flux (kg/(m^2 s))
h	axial coordinate in riser or downer (m)
H	height of reactor (m)
k_r	reaction rate constant based on particle volume, first-order reaction (s^{-1})
k_r'	Damköhler number, $k_r\overline{\varepsilon}_s(1-\overline{\varepsilon}_s)H/U_g$ (–)
P	pressure (Pa)
r	radial coordinate (m)
R	column radius (m)

U_g	superficial gas velocity (m/s)
U_{slip}	slip velocity ($=u_g - v_p$ for riser and $=v_p - u_g$ for downer) (m/s)
u_g	local gas velocity (m/s)
v_p	local particle velocity (m/s)
z	axial coordinate, i.e. distance above the gas distributor (m)

Greek Letters

ε_s	solids holdup (–)
$\overline{\varepsilon}_s$	average solids holdup (–)

Subscripts

g	gas
p	particle
s	solids

Abbreviations

CFB	circulating fluidized bed
CTFB	circulating turbulent fluidized bed
FCC	fluid catalytic cracking
HDCFB	high-density circulating fluidized bed
LDCFB	low-density circulating fluidized bed
LDV	laser Doppler velocimeter
RTD	residence time distribution
TFB	turbulent fluidized bed

References

1 Yerushalmi, J., Turner, D.H., and Squires, A.M. (1976). The fast fluidized bed. *Ind. Eng. Chem. Process Des. Dev.* 15: 47–53.
2 Zhu, J. and Cheng, Y. (2005). Fluidized bed reactors and applications, Chapter 5. In: *Multiphase Flow Handbook* (ed. C. Crowe), 55–93. New York: CRC Press.
3 Reh, L. (1971). Fluidized bed processing. *Chem. Eng. Prog.* 67: 58–63.
4 Reh, L. (1980). The circulating fluidized bed reactor – its main features and applications. *Chem. Eng. Process.* 20: 117–127.
5 Grace, J.R., Bi, H., and Golriz, M. (2003). Circulating fluidized beds. In: *Handbook of Fluidization and Fluid-Particle Systems* (ed. W.-C. Yang), 485–545. New York: CRC Press.
6 Grace, J.R. and Bi, H. (1997). Introduction to circulating fluidized beds. In: *Circulating Fluidized Beds* (eds. J.R. Grace, A.A. Avidan and T.M. Knowlton), 1–20. Springer.

7 Milne, B.J., Berruti, F., Behie, L.A., and De Bruijn, T.J.W. (1992). The internally circulating fluidized bed (ICFB): a novel solution to gas bypassing in spouted beds. *Can. J. Chem. Eng.* 70: 910–915.
8 Bi, H. and Zhu, J. (1993). Static instability analysis of circulating fluidized beds and concept of high-density risers. *AIChE J.* 39: 1272–1280.
9 Zhu, H. and Zhu, J. (2008). Comparative study of flow structures in a circulating-turbulent fluidized bed. *Chem. Eng. Sci.* 63 (11): 2920–2927.
10 Zhu, J., Zheng, Y., Karamanev, D.G., and Bassi, A.S. (2000). (Gas–)liquid–solid circulating fluidized beds and their potential applications to bioreactor engineering. *Can. J. Chem. Eng.* 78: 82–94.
11 Atta, A., Razzak, S.A., Nigam, K., and Zhu, J. (2009). (Gas)–liquid–solid circulating fluidized bed reactors: characteristics and applications. *Ind. Eng. Chem. Res.* 48: 7876–7892.
12 Zhang, H., Johnston, P.M., Zhu, J., and Bergougnou, M.A. (1998). A novel calibration procedure for a fiber optic solids concentration probe. *Powder Technol.* 100: 260–272.
13 Bai, D., Jin, Y., Yu, Z.-Q., and Zhu, J. (1992). The axial distribution of the cross-sectionally averaged voidage in fast fluidized beds. *Powder Technol.* 71: 51–58.
14 Wang, C., Zhu, J., Barghi, S., and Li, C. (2014). Axial and radial development of solids holdup in a high flux/density gas–solids circulating fluidized bed. *Chem. Eng. Sci.* 108: 233–243.
15 Pärssinen, J. and Zhu, J. (2001). Axial and radial solids distribution in a long and high-flux CFB riser. *AIChE J.* 47: 2197–2205.
16 Schnitzlein, M.G. and Weinstein, H. (1988). Flow characterization in high-velocity fluidized beds using pressure fluctuations. *Chem. Eng. Sci.* 43: 2605–2614.
17 Li, Y. and Kwauk, M. (1980). The dynamics of fast fluidization. In: *Fluidization* (eds. J.R. Grace and J.M. Matsen), 537–544. New York: Plenum Press.
18 Jin, Y., Yu, Z., Qi, C., and Bai, D. (1988). The Influence of exit structure on the axial distribution of voidage in fast fluidized bed. In: *Fluidization: Science and Technology* (eds. M. Kwauk and D. Kunii), 165–173. Beijing: Science Press.
19 Brereton, C.M.H. and Grace, J.R. (1994). End effects in circulating fluidized bed hydrodynamics. In: *Circulating Fluidized Bed Technology IV* (ed. A.A. Avidan), 137–144. New York: AIChE.
20 Zhu, J. and Bi, H.T. (1995). Distinctions between low density and high density circulating fluidized beds. *Can. J. Chem. Eng.* 73: 644–649.
21 Issangya, A., Bai, D., Bi, H. et al. (1999). Suspension densities in a high-density circulating fluidized bed riser. *Chem. Eng. Sci.* 54: 5451–5460.
22 Karri, S.B.R. and Knowlton, T.M. (1999). A comparison of annulus solids flow direction and radial solids mass flux profiles at low and high mass fluxes in a riser. In: *Circulating Fluidized Bed Technology VI* (ed. J. Werther), 71–76. Frankfurt: DECHEMA.
23 Yan, A., Zhu, H., and Zhu, J. (2004). Comparison of FCC particle flow in a twin-riser system. In: *Fluidization XI* (eds. U. Arena, R. Chirone, M. Miccio and P. Salatino), 387–394. New York: Engineering Foundation.

24 Wang, C., Zhu, J., Li, C., and Barghi, S. (2014). Detailed measurements of particle velocity and solids flux in a high density circulating fluidized bed riser. *Chem. Eng. Sci.* 114: 9–20.

25 Wang, C., Wang, G., Li, C. et al. (2014). Catalytic ozone decomposition in a high density circulating fluidized bed riser. *Ind. Eng. Chem. Res.* 53: 6613–6623.

26 Wang, C., Barghi, S., and Zhu, J. (2014). Hydrodynamics and reactor performance evaluation of a high flux gas-solids circulating fluidized bed downer: experimental study. *AIChE J.* 60: 3412–3423.

27 Wang, C., Zhu, J., and Barghi, S. (2015). Performance evaluation of high density riser and downer: experimental study using ozone decomposition. *Chem. Eng. J.* 262: 478–489.

28 Harris, B.J. and Davidson, J.F. (1994). A core/annulus deposition model for circulating fluidized beds. In: *Circulating Fluidized Bed Technology IV* (ed. A.A. Avidan), 32–39. New York: AIChE.

29 Bai, D., Zhu, J., Jin, Y., and Yu, Z. (1995). Internal recirculation flow structure in vertical upward flowing gas–solids suspensions, part I. A core/annular model. *Powder Technol.* 85: 171–178.

30 Liu, J., Grace, J.R., Bi, H. et al. (1999). Gas dispersion in fast fluidization and dense suspension upflow. *Chem. Eng. Sci.* 54: 5441–5449.

31 Martin, M.P., Tulier, P., Bernard, J.R., and Wild, G. (1992). Gas and solid behaviour in cracking circulating fluidized beds. *Powder Technol.* 70: 249–258.

32 Grace, J.R., Issangya, A.S., Bai, D. et al. (1999). Situating the high-density circulating fluidized bed. *AIChE J.* 45: 2108–2116.

33 Zhu, H. and Zhu, J. (2008). Gas–solids flow structures in a novel circulating-turbulent fluidized bed. *AIChE J.* 54: 1213–1223.

34 Li, H., Xia, Y., Tung, Y., and Kwauk, M. (1991). Micro-visualization of clusters in a fast fluidized bed. *Powder Technol.* 66: 231–235.

35 Horio, M. and Kuroki, H. (1994). Three-dimensional flow visualization of dilutely dispersed solids in bubbling and circulating fluidized bed. *Chem. Eng. Sci.* 49: 2413–2421.

36 Grace, J.R. and Tuot, J. (1979). A theory for cluster formation in vertically conveyed suspension of intermediate density. *Trans. Inst. Chem. Eng.* 57: 49–54.

37 Yerushalmi, J., Kolodney, M., Graff, R.A. et al. (1975). Agglomeration of ash in fluidized beds gasifying coal: the Godel phenomenon. *Science* 187: 646–648.

38 Syamlal, M. and O'Brien, T. (1989). Computer simulation of bubbles in a fluidized bed. *AIChE Symp. Ser.* 85: 22–31.

39 Xu, J. and Zhu, J. (2012). A new method for the determination of cluster velocity and size in a circulating fluidized bed. *Ind. Eng. Chem. Res.* 51: 2143–2151.

40 Shi, H., Wang, Q., Xu, L. et al. (2008). Visualization of clusters in a circulating fluidized bed by means of particle-imaging velocimetry (PIV) technique. In: *Circulating Fluidized Bed Technology IX* (ed. J. Werther), 1013–1019. Hamburg, Germany: TuTeck Innovation.

41 Manyele, S.V., Pärssinen, J.H., and Zhu, J. (2002). Characterizing particle aggregates in a high-density and high-flux CFB riser. *Chem. Eng. J.* 88: 151–161.

42 Lackermeier, U., Rudnick, C., Werther, J. et al. (2001). Visualization of flow structures inside a circulating fluidized bed by means of laser sheet and image processing. *Powder Technol.* 114: 71–83.

43 Takeuchi, H., Pyatenko, A.T., and Hatano, H. (1998). Gross behavior of parabolic strands in a riser. In: *Fluidization IX*, 177–182. New York: Engineering Foundation.

44 Soong, C.H., Tuzla, K., and Chen, J. (1994). Identification of particle clusters in circulating fluidized bed. In: *Circulating Fluidized Bed Technology IV* (ed. A.A. Avidan), 615–620. New York: AIChE.

45 Rhodes, M., Mineo, H., and Hirama, T. (1992). Particle motion at the wall of circulating fluidized bed. *Powder Technol.* 70: 207–214.

46 Horio, M. and Clift, R. (1992). A note on terminology: 'clusters' and 'agglomerates'. *Powder Technol.* 70: 196–196.

47 Bi, H.T., Zhu, J., Jin, Y., and Yu, Z.Q. (1993). Forms of particle aggregations in CFB (in Chinese). In: *Proceedings of 6th Chinese Conference on Fluidization, Wuhan, China*, 162–167.

48 Sinclair, J.L. and Jackson, R. (1989). Gas-particle flow in a vertical pipe with particle–particle interaction. *AIChE J.* 35: 1473–1486.

49 Yang, Y., Jin, Y., Yu, Z. et al. (1993). Local slip behaviour in the circulating fluidized bed. *AIChE Symp. Ser.* 89 (296): 81–90.

50 Yerushalmi, J. and Cankurt, N. (1979). Further studies of the regimes of fluidization. *Powder Technol.* 24: 187–205.

51 Yang, Y., Jin, Y., Yu, Z., and Wang, Z. (1992). Investigation on slip velocity distributions in the riser of dilute circulating fluidized bed. *Powder Technol.* 73: 67–73.

52 Yang, G., Huang, Z., and Zhao, L. (1984). Radial gas dispersion in a fast fluidized bed. In: *Fluidization IV* (eds. D. Kunii and R. Toei), 145–152. New York: Engineering Foundation.

53 Adams, C.K. (1988). Gas mixing in fast fluidised beds. In: *Circulating Fluidized Bed Technology II* (eds. P. Basu and J.F. Large), 299–306. Toronto: Pergamon.

54 Bader, R., Findlay, J., and Knowlton, T.M. (1988). Gas/solids flow patterns in a 30.5-cm-diameter circulating fluidized bed. In: *Circulating Fluidized Bed Technology II* (eds. P. Basu and J.F. Large), 123–137. Toronto: Pergamon.

55 Brereton, C.M.H., Grace, J.R., and Yu, J. (1988). Axial gas mixing in a circulating fluidized bed. In: *Circulating Fluidized Bed Technology II* (eds. P. Basu and J.F. Large), 307–314. Toronto: Pergamon Press.

56 Li, Y. and Wu, P. (1991). A study on axial gas mixing in a fast fluidized bed. In: *Circulating Fluidized Bed Technology III* (eds. P. Basu, M. Horio and M. Hasatani), 581–586. Toronto: Pergamon Press.

57 White, C.C., Dry, R.J., and Potter, O.E. (1992). Modelling gas-mixing in a 9 cm diameter circulating fluidized bed. In: *Fluidization VII* (eds. O.E. Potter and D.J. Nicklin), 265–273. New York: Engineering Foundation.

58 Zheng, Q., Wei, X., and Lou, F. (1992). Experimental study on radial gas dispersion and its enhancement in circulating fluidized beds. In: *Fluidization*

VII (eds. O.E. Potter and D.J. Nicklin), 285–293. New York: Engineering Foundation.
59 Amos, G., Rhodes, M.J., and Mineo, H. (1993). Gas mixing in gas–solids risers. *Chem. Eng. Sci.* 48: 943–949.
60 Arena, U., Marzocchella, L.A., Bruzzi, V., and Massimilla, L. (1994). Mixing between a gas-solids suspension flowing in a riser and a lateral gas stream. In: *Circulating Fluidized Bed Technology IV* (ed. A.A. Avidan), 195–202. New York: AIChE.
61 Bai, D., Jin, Y., and Yu, Z. (1992). Circulating fluidized bed technology-(IV) mixing and residence time distributions. *Chem. React. Eng. Technol.* 8: 101–112.
62 Martin, M.P., Derouin, C., Tulier, P. et al. (1991). Catalytic cracking in riser reactors: core-annulus and elbow effects. *Chem. Eng. Sci.* 47: 2319–2324.
63 Werther, J., Hartge, E.-U., and Krause, M. (1992). Radial gas mixing in the upper dilute core of a circulating fluidized bed. *Powder Technol.* 70: 293–301.
64 Bai, D., Yi, J., Shi, G. et al. (1992). Backmixing and residence time distribution of gas in circulating fluidized bed (in Chinese). *J. Chem. Eng. Chin. Univ.* 6: 258–263.
65 Bi, X.T. (2004). Gas and solid mixing in high-density CFB risers. *Int. J. Chem. Eng.* 2: A12.
66 Ambler, P.A., Milne, B.S., Berruti, F., and Scott, D.S. (1990). Residence time distribution of solids in a circulating fluidized bed: experimental and modelling studies. *Chem. Eng. Sci.* 45: 2179–2186.
67 Milne, B.J. and Berruti, F. (1991). Modelling the mixing of solids in circulating fluidized beds. In: *Circulating Fluidized Bed Technology III* (eds. P. Basu, M. Horio and M. Hasatani), 575–580. Toronto: Pergamon Press.
68 Rhodes, M.J., Zhou, S., Hirama, T., and Cheng, H. (1992). Effects of operating conditions and scale on solids mixing in risers of circulating fluidized beds. In: *Fluidization VII* (eds. O.E. Potter and D.J. Nicklin), 249–256. New York: Engineering Foundation.
69 Weinell, C.E. and Dam-Johansen, K. (1992). Single particle behaviour in circulating fluidized beds. In: *Fluidization VII* (eds. O.E. Potter and D.J. Nicklin), 295–304. New York: Engineering Foundation.
70 Mori, S., Kashima, N., Wakazono, Y., and Tsuchiya, M. (1993). Particle behaviour and flow pattern in a circulating fluidized bed. In: *Preprint for Circulating Fluidized Bed Technology IV* (ed. A. Avidan), 308–314. New York: AIChE.
71 Wei, F. and Zhu, J.-X. (1996). Effect of flow direction on axial solids dispersion in gas–solids cocurrent upflow and downflow systems. *Chem. Eng. J.* 64 (3): 345–352.
72 Wei, F., Jin, Y., Yu, Z. et al. (1995). Lateral and axial mixing of the dispersed particles in CFB. *J. Chem. Eng. Japan* 28: 506–510.
73 Bolton, L.W. and Davidson, J.F. (1988). Recirculation of particles in fast fluidized risers. In: *Circulating Fluidized Bed Technology II* (eds. P. Basu and J.F. Large), 139–146. Toronto: Pergamon Press.
74 Li, D., Ray, A.K., Ray, M.B., and Zhu, J. (2013). Catalytic reaction in a circulating fluidized bed riser: ozone decomposition. *Powder Technol.* 242: 65–73.

75 Qi, C.-M. and Farag, I.H. (1993). Lateral particle motion and its effect on particle circulation distribution in the riser of CFB. *AIChE Symp. Ser.* 89 (296): 73–80.
76 Syamlal, M. and O'Brien, T. (2003). Fluid dynamic simulation of O_3 decomposition in a bubbling fluidized bed. *AIChE J.* 49: 2793–2801.
77 Nowak, W., Bis, Z., Laskawiec, J. et al. (1999). *Circulating Fluidized Bed Technology VI* (ed. J. Werther), 1003–1008. Frankfurt: DECHEMA.
78 Jiang, P.J., Inokuchi, K., Jean, R.-H. et al. (1991). Ozone decomposition in a catalytic fluidized bed reactor. In: *Circulating Fluidized Bed Technology III* (eds. P. Basu, M. Horio and M. Hasatani), 557–562. Toronto: Pergamon Press.
79 Schoenfelder, H., Kruse, M., and Werther, J. (1996). Two-dimensional model for circulating fluidized-bed reactors. *AIChE J.* 42: 1875–1888.
80 Jiang, P., Bi, H., Jean, R.H., and Fan, L.S. (1991). Baffle effects on performance of catalytic circulating fluidized bed reactor. *AIChE J.* 37: 1392–1400.
81 Yan, A. J. (2004). Hydrodynamics and scale up of circulating fluidized bed. PhD thesis. London, Canada: University of Western Ontario.
82 Su, X., Wang, C., and Zhu, J. (2019). Scale up effects of the high density gas-solids circulating fluidized bed riser. In preparation.
83 Zhou, J., Grace, J.R., Qin, S. et al. (1994). Voidage profiles in a circulating fluidized bed of square cross-section. *Chem. Eng. Sci.* 49: 3217–3226.
84 Werther, J. (1992). Scale-up modeling for fluidized bed reactors. *Chem. Eng. Sci.* 47: 2457–2462.
85 Horio, M., Morishita, K., Tachibana, O., and Murata, N. (1988). Solid distribution and movement in circulating fluidized beds. In: *Circulating fluidized bed technology*, 147–154. Toronto: Pergamon Press.

Problems

12.1 In a circulating fluidized bed system, similar to that shown in Figure 12.1 with a 0.40 m Inner Diameter (ID), 10 m high reactor column, the circulation flux of solids is kept constant at $G_s = 100\ \mathrm{kg/m^2\ s}$, while the gas velocity is increased further. Determine the distribution of solids in the vessel, i.e. find the solids holdup distribution vs. bed height (ε_s vs. z). Assume that the solids holdup in the lower part of the bed, ε_{sd}, is 0.13, 0.12, and 0.10 for $U = 7$, 8, and 10 m/s, respectively.
Data: Column: $D = 0.4$ m, $H = 10$ m. Particles: catalyst, $\rho_p = 1400\ \mathrm{kg/m^3}$, $\overline{d}_p = 60\ \mu\mathrm{m}$. Gas: air at ambient conditions; the decay constant, a, is related to the gas velocity by: $a \cdot U = 3\ \mathrm{s^{-1}}$.

12.2 For a fast chemical reaction, we intend to use fine catalyst ($\overline{d}_p = 60\ \mu\mathrm{m}$) in a fast fluidized circulating bed system, similar to that shown in Figure 12.1. At the planned operating conditions, the superficial gas velocity will be 10 m/s in a 0.40 m ID and 10 m high reactor riser. For the desired chemical conversion, it is essential to keep 300 kg of solids in the column at all times. From this information determine the necessary solids circulation rate. Data are the same as for Problem 12.1.

13

Operating Challenges

Poupak Mehrani and Andrew Sowinski

University of Ottawa, Chemical and Biological Engineering Department, 161 Louis Pasteur, Ottawa, Ontario, Canada K1N 6N5

13.1 Electrostatics

When two solid surfaces are brought into contact and then separated, charge transfer tends to occur due to the phenomenon known as triboelectric charging or contact electrification, making one surface positively charged and the other negatively charged. Depending on the nature of the surfaces in contact, induction charging can also occur when a charged object creates an oppositely charged image of itself on a conductor surface.

As a result of contact electrification, an attractive or repulsive force acts between two point charges, described by Coulomb's law:

$$\mathbf{F}_E = \frac{1}{4\pi\varepsilon_f}\frac{q_1 q_2}{r^2} \tag{13.1}$$

where ε_f is the permittivity of the medium or fluid, q_1 and q_2 are the two charges, and r is the distance between the points. If the distance is sufficiently large, this law can be applied to non-point charges, such as for spheres with uniform surface charges. In this case, the total charge of the sphere can be assumed to reside at its centre point for the purpose of the calculation. Coulomb's law as written above considers only two charges. The force acting on a charge due to multiple surrounding charges can be calculated by summing the forces due to each of the other charges.

In the case of induction charging, an attractive force can act between the charged object and the conductive surface that can be estimated by an equation proposed by Masuda et al. [1]:

$$\mathbf{F}_I = \frac{1}{4\pi\varepsilon_w}\frac{q_1^2}{(2Z_{PW})^2} \tag{13.2}$$

where q_1 is the charge of the object, Z_{pw} is the distance between the charged object and the conductive surface, and ε_w is the permittivity of the conductive

Essentials of Fluidization Technology, First Edition. Edited by John R. Grace, Xiaotao Bi, and Naoko Ellis.

surface material. Tribocharging also results in the generation of an electric field, **E**, that affects the charge distribution:

$$\nabla \cdot \mathbf{E} A = \frac{\rho_C}{\varepsilon_o} \tag{13.3}$$

where ρ_C is the charge density and ε_o is the permittivity of free space.

What is still not clear in triboelectric charging, especially when insulators are involved, are the mechanisms underlying charge transfer, which may be due to electron transfer, ion transfer, and/or material transfer [2].

Due to the nature of gas–solid fluidized beds, providing continuous collisions among the particles and between particles and column walls, triboelectric charging is inevitable. In addition, for metallic columns, induction charging can occur. Therefore, particle motion inside a fluidized bed is affected not only by gravitational, buoyancy, drag, and van der Waals forces but also by electrostatic and image forces. Major nuisances can arise from electrostatic charge generation in gas–solid fluidized beds, including particle agglomeration causing deviation from the desired particle properties, adhesion of particles to the column wall and other surfaces necessitating regular shutdowns for clean-up, and electrostatic discharges endangering operators and equipment. In any gas–solid fluidized bed, the attractive electrostatic force between charged fluidizing particles of opposite polarity, as well as the image forces between the charged particles and metallic column walls, are dominant forces causing particle agglomeration and accumulation on the column wall [3]. An example of an industrial fluidized bed process that has faced significant operational challenges due to electrostatics is the gas-phase polymerization of ethylene to produce polyethylene [4]. In this process, triboelectric charging of catalyst and polyethylene particles causes them to adhere to the reactor wall, a phenomenon known as "sheeting" [4] that results in major economic losses due to forced shutdowns. Electrostatic forces have also been acknowledged as a probable source of particle cluster formation in circulating fluidized bed (CFB) risers.

The first recognition of electrostatic charges in gas–solid fluidized beds dates back to 1949 [5]. Since then, significant research has been carried out in an effort to understand electrostatic charging mechanisms in gas–solid fluidized beds and to find methods for its elimination or mitigation. Overall, the fluidization system characteristics, including particle size and material, gas velocity, column wall material, system pressure, and temperature, as well as the amount of moisture inside the fluidized bed, can influence the extent of bed electrification [2, 6, 7]. In general, fluidized particle size has an inverse relationship, whereas fluidizing gas velocity and system pressure are directly related to the particle charge-to-mass ratio. Particles smaller than 100 μm typically generate significant charges, especially when the particles are semi-conductors or insulators. Hence, for applications in which there is significant entrainment, it is important to consider and assess electrostatic effects.

As for the effect of gas velocity, it is important to realize that the transition from one flow regime to another changes the degree of charging of the fluidizing particles due to changes in the mixing behaviour inside the bed. For example, one study [8] reported that the turbulent flow regime generates more solids charging than the bubbling regime. The surface chemistry of the fluidized particles and column walls is also important in determining the extent of tribocharging. For instance, fluidization of glass beads (i.e. semi-conductors) in a Plexiglas column results in significant charging of the particles and, in turn, their adherence to the fluidization column wall, whereas fluidization of the same particles in a same sized stainless steel column generates negligible charge and no wall accumulation. On the other hand, fluidization of polyethylene particles in the same stainless steel column results in significant tribocharging and particle–wall coating, particularly for particles smaller than 600 μm [2]. Overall, three important aspects must be considered in relation to tribocharging in fluidized beds: First, although tribocharging could exist in any gas–solid fluidized bed, its extent and influence on the bed operation varies significantly from one system to another. Second, processes upstream of a fluidized bed must be evaluated for their potential to charge particles prior to their entrance into a fluidization unit. Third, when studying electrostatic charging using bench-scale experimentation, the materials must be the same as those utilized in the full-scale fluidization system of interest.

13.1.1 Measurement and Prediction of Electrostatic Charge

At times it is desired to predict the extent of the electrostatic effects on a fluidized bed based on the fluidization system characteristics, where knowledge of the direction of charge transfer between surfaces, as well as its magnitude, would be important. The direction of charge transfer can be qualitatively predicted based on the triboelectric series, which arranges materials according to their charging tendency, with those listed higher expected to have positive polarity relative to any material situated below [9, 10]. However, the triboelectric series should be employed with caution, as conflicting series are available in the literature, since various factors, such as surface finish, purity, and moisture content, can influence the results. It is also important to note that the direction of charge transfer between surfaces in contact cannot be easily predicted, because, as described by Mehrani et al. [2], the sources of charge carriers are typically unknown, making prediction of the direction and the extent of charge transfer uncertain.

Bench-scale testing is therefore recommended in relation to materials involved in any specific fluidization system to determine the potential for bed electrification. This testing could include single and multiple contact charging, as well as small-scale fluidization tests. Single particle contact charging is typically conducted by contacting a single particle with a plate [11–17] or repeated contacts in a container [16]. Multiple particle contact charging tests can be carried out when a number of particles are shaken in a container [16]. In all cases, the material of the plate or container should be similar to that of the fluidization column in which the testing is conducted. Results from such testing provide information on charge polarity, as well as approximate degree of charging of particles. In all types of testing, visual observations of the container after a shaking period or the fluidization column wall after a period of fluidization also indicate the potential for particle–wall fouling.

Two methods are used to quantify the electrostatic charge on solids:

(1) A Faraday cage consists of two concentric metallic (e.g. copper) containers that are electrically isolated from each other. The outer container is grounded to form an electric shield from external fields, while the inner container is connected to an electrometer that measures charge on the solids inside it. The electrometer measures the charge and polarity of the object inside the Faraday cage inner container. This measurement technique can be used for any particle contact charge testing and has been implemented in fluidized beds in various ways, including simply placing a sample of particles from the bed inside the Faraday cage [18, 19]. However, two challenges arise from such an application: (a) altering particle charges due to handling and (b) the samples that may not be representative of the whole bed. Later applications of the Faraday cage technique include those employed by Mehrani et al. [19] who designed the fluidization column as a Faraday cage, enabling the measurement of the charge of particles entrained from the fluidization column. Sowinski et al. [20] placed a Faraday cage below the distributor plate to measure the charge on the bed particles at the end of a fluidization period and a second Faraday cage at the exit of the column to measure the charge of the entrained particles during fluidization. This technique provides further information about the charge and extent of the particles fouled on the fluidization column wall.

(2) The other technique employed for charge measurement, especially in fluidized bed testing, is the electrostatic probe. A variety of set-ups have been employed in fluidized beds including induction probes and collision probes [2, 6]. Such probes are inserted into the bed at various locations and are aimed at providing an online measurement of electrostatic charge via measuring current or the electric potential. Although electrostatic probes are preferred for large-scale operations, there are issues, and thus research is still progressing to make this technique more reliable for monitoring charges inside fluidized beds [2, 6].

To investigate triboelectric charging in gas–solid fluidized beds, it is also desirable to use models to predict this effect as a cost-effective alternative to pilot-scale experiments. Although computational fluid dynamic (CFD) modelling has been widely employed for fluidized bed hydrodynamic predictions, the inclusion of models to predict tribocharging has only received attention in recent years. Both Eulerian–Lagrangian and Eulerian–Eulerian models (see Chapter 6) have been considered [6] with the latter initiated by Rokkam et al. [21] who incorporated a fixed charge for fluidizing particles based on experimental data. However, the most recent models by Kolehmainen et al. [22] and Ray et al. [23] account for charge generation and transport due to particle–particle and particle–wall interactions.

13.1.2 Mitigation Techniques

In cases where tribocharging has the potential to disturb the fluidized bed operation, charge mitigation or elimination may be needed. This can be achieved by eliminating and/or reducing charge generation, or by dissipating charges [2, 6]. The main techniques attempted include:

- Increasing the relative humidity of the system to augment charge dissipation
- Adding fine particles to moderate contacts among the fluidized particles and collisions between the particles and the column wall
- Adding antistatic agents that adhere to the particles or column wall surface, in turn changing the surface chemistry of the particles and/or the fluidization column wall or diminishing the level of electrostatic forces within the fluidized bed
- Coating the fluidization column inner wall to alter the wall surface chemistry, resulting in less charge generation or reduction of image forces in the case of metallic columns

Note that the choice between these options is dependent on the specifics of each system. At times none may be workable or effective. For instance, raising the system relative humidity may be impossible in cases where this could hinder the process (e.g. due to catalyst poisoning in polyolefin production), or when hydrophobic surfaces are involved. If the addition of antistatic or even prostatic agents is considered, the specifics of the system must be considered since one agent may work in one system, but not in another.

13.1.3 Summary

Electrostatics can occur in any fluidized bed depending on the system characteristics, altering bed behaviour and resulting in operational challenges. Over the past decade, several review articles were published [2, 6, 7] providing comprehensive assessments of electrostatics in gas–solid fluidized beds. They focus on fundamental understanding of triboelectric charging of solids in general, as well as in fluidized beds, methods to measure particle charges, and experiments to elucidate particle charging processes in fluidized beds. Research is still underway to better understand the phenomena, to develop viable measurement techniques enabling quantification, and to find suitable prevention and control methods.

13.2 Agglomeration

Agglomerates are inherent characteristics of gas–solid fluidization systems that greatly influence the bed performance in many applications. In many gas–solid fluidized beds, particles travel in groups either due to various forces (e.g. van der Waals and electrostatics) holding them together, or because of issues such as particle sintering. As explained by Cahyadi et al. [24], agglomerates form when particles are physically bonded together, whereas clusters are formed when hydrodynamic and weak interparticle forces are involved. High-temperature operations are prone to sintering depending on the particle softening temperature. For instance, in fluidized bed combustors, softening of the ash or other bed materials initiates formation of agglomerates as particles begin to adhere to each other. Another example of a system greatly affected by particle agglomeration is in polyolefin processes (making polyethylene and polypropylene.) In these reactors, the generation of electrostatic charge initiates the tendency of particles to agglomerate, which can then progress to "sheets" of foulant on the inner walls of the reactors aided by the highly exothermic nature of the reactions involved. Similarly, agglomerates and fouling of walls and baffles occur in fluid cokers, initiated by particle stickiness and interparticle forces.

For high-temperature operations, particle agglomeration can cause bed defluidization and thus costly process shutdowns. Bartels et al. [25] published a comprehensive review of particle agglomeration in high-temperature fluidized beds. In processes such as solids fuel combustion and gasification, the potential for agglomeration can be evaluated offline, by characterizing the fuel ash rheological properties, or detected online, for instance, by monitoring the bed temperature and pressure. Evaluation of the standard deviation and variance of pressure drop can be a suitable tool for monitoring deviations in bed particle size, facilitating detection of agglomeration. However, caution should be exercised since this method is also sensitive to variations in fluidization gas velocity. Local temperature measurements can also provide an indication of the extent of solids mixing in the bed and thus can provide an appropriate agglomeration detection technique. The increase in temperature difference between various locations in the bed is an indication of the decline in solids mixing due to agglomeration. Therefore, in designing new high-temperature systems, information on material sintering tendency (e.g. materials softening and melting temperature) should be gathered early in the process development stages. In addition, as suggested by Bartels et al. [25], the presence of gaseous alkali compounds can also be a potential source for agglomeration and thus should be detected and evaluated if possible.

For processes in operation, a number of techniques have been proposed to reduce or nullify agglomeration effects, including lowering the bed temperature below the melting or softening temperature of bed materials, increasing the gas flow, and controlling the particle size in the bed by replacing spent particles with fresh bed material. In fluidized beds where the generation of electrostatic charges is expected, the potential of particle agglomeration should also be considered. For example, electrostatics causes the formation of particle agglomerates in polyethylene reactors, altering the bed particle size distribution, in turn causing blockage of the distributor plate and product discharge pipe, and causing process shutdowns [4]. When electrostatic charging is likely as explained in Section 13.1., testing should be carried out to determine the approximate extent of particle charging, followed by determination of possible mitigation techniques.

13.3 Attrition

Particle attrition occurs when particles break into smaller particles. Many industries encounter operational challenges due to particle attrition, such as catalyst attrition in fluid catalytic cracking (FCC) in the petroleum industry, and limestone (sorbent) attrition in CFBs [26]. The main mechanisms for attrition are [27]

- **Stresses due to impacts**: Typically occur from compression, tensile, and shear stresses exerted onto the particle due to their collision with the column wall, other particles, or fluidized bed internals. These stresses can be further divided into static and kinetic, where static stresses occur at no or low velocities, while the kinetic occurs at high particle velocities.
- **Abrasion**: Abrasion occurs when rough particles rub against surfaces, detaching the roughness elements (commonly called asperities) to form particles that are much smaller than the original particles.
- **Stresses due to reaction**: Chemical reactions can cause change in the lattice of structure of particles further weakening the particle structure and reduce the particle size [28].
- **Thermal stresses due to changes in internal pressure**: The uneven contraction or expansion of a particle due to changes in temperature can cause particle breakage, for example, when cold particles are added to a hot bed.

The major contributors to attrition in most fluidized bed processes are impact attrition and abrasion, and fragmentation. Abrasion occurs when many fine particles wear off a larger particle, whereas fragmentation occurs when a large particle completely breaks into a limited number of smaller particles. Abrasion leads to a bimodal particle size distribution where fragmentation leads to a shift towards smaller particles in the particle size distribution, as shown in Figure 13.1. The design of a fluidized bed contributes to attrition in different areas such as bubble formation, jets from the distributor plate, and splashing of ejected particles at the bed surface [29]. Regardless of the location, attrition alters the number of particles and their size distribution in the system, with an impact on bed hydrodynamics and particle entrainment/elutriation [30].

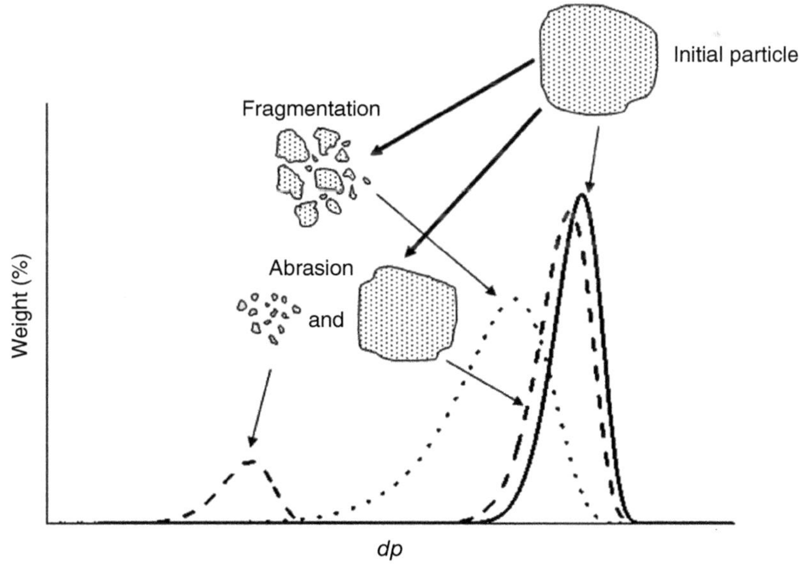

Figure 13.1 Change of particle size distribution from different modes of attrition where d_p is the particle diameter.

The rate of attrition of particles in a fluidized bed is influenced by the type of particle, as well as the process conditions, such as fluidization time and gas temperature, but particularly the excess gas velocity. Parameters that characterize the resistance to abrasion of a particle include [27]:

- **Hardness**: The resistance to plastic deformation on the surface of the particle and can be described by three types: scratch, static indentation, rebound hardness.
- **Fracture toughness**: The measure of a particle to resist fracturing under a given stress.

To have high resistance to attrition, particles should have high hardness, a large resistance to fracture, and low attrition rates due to wear. However, high hardness means that wear in many cases, the choice of the fluidizing particles is limited, and it is often difficult to measure particle hardness. Many researchers have therefore studied attrition rates in small-scale units to estimate its impact in larger fluidization systems. One way of doing this is with a three-jet system introduced by Gwyn [31] later adopted as an ASTM standard measurement method [32]. The jets entrain particles, which then collide with each other causing attrition over a prescribed time interval. Product fines are then carried over, collected in a filter, and weighed. A simple attrition index (AI) is then used to characterize the particles. AI can have many definitions, but one that usually employed is the change of a certain particle weight fraction from before to after testing. While this method was first proposed for testing catalyst particles at room temperature, it can also be used for other materials and at elevated temperatures [33].

Another attrition testing method is a small Davison Jet cup [34], a cup with a tangential gas inlet attached to a larger disengagement section. The goal of the jet cup method is to mimic the particles flow and the particle–wall impacts in cyclones, fluidized beds, and risers. This method is primarily used to rank particles in terms of the AI. During the testing of attrition, particles that are small leave the system and are trapped by a filter. The loss of these particles is directly related to the attrition loss. Although the Davidson cup method has been used in many cases, researchers at Particulate Solid Research Inc. (PSRI) [35] found through experimentation and CFD modelling that the standard cylindrical apparatus did not achieve proper particle mobility. PSRI has thus proposed a conical cup, which gave good agreement with a fluidized cyclone test [34]. Instead of using the jet cup method, other researchers have also resorted to bench- or pilot-scale fluidized beds and cyclones [36].

13.3.1 Modelling Attrition

Despite attrition being present in many industries, and the availability of many different attrition models, no general model exists, most likely due to the difficulty of quantifying the impact of each of the attrition mechanisms outlined above. A number of models attempting to quantify attrition are summarized here:

Lowrison [37] correlated the degree of particle size reduction with the energy consumed that is proportional to the excess gas velocity, U_e, and the mass of the bed particles, M. The total mass rate of attrition, R_t, is then considered to be proportional to the effective breaking energy:

$$R_t = aE_b \tag{13.4}$$

where a is an attrition coefficient (mass of fines generated per pass of effective breaking energy) and E_b is the effective breaking energy. E_b can be related to the kinetic energy as follows:

$$R_t = a\eta E_k = a\eta MgU_e \tag{13.5}$$

where η is the efficiency of transferring kinetic energy to breaking energy and g is the gravitational acceleration. For a distribution of particles, the rate of attrition, R_{ti}, of particles of radius r_i, is proportional to the mass of the component, the mass fraction of components smaller than r_i, and the excess gas velocity:

$$R_{ti} = k_1 U_e M x_i w_i \tag{13.6}$$

where k_1 is a constant, x_i is the mass fraction of component i, and w_i is the total mass fraction of components of particles with radii smaller than r_i. Due to the w_i factor, this formula implies that larger particles have larger attrition rates. The total attrition rate is then the summation of individual rates:

$$R_t = \sum R_{ti} \tag{13.7}$$

Ulerich et al. [38] concluded that for a single material with a particle distribution, R_t is independent of particle size with a fixed U_e. Ray et al. [28] also assumed that the rate of effective breaking energy of component i, E_{bi}, is proportional to the total surface area of component i, SA_i, and defined the energy-sharing coefficient for component i, b_i, as

$$b_i = \frac{E_{bi}}{SA_i} \tag{13.8}$$

For a multicomponent mixture with wide size distributions, Ray et al. [28] created a generalized attrition model. They defined the individual component rate of attrition as

$$R_{ti} = a_i \frac{\frac{b_i x_i}{e_i r_i \rho_i}}{\sum_{j=1}^{N} \frac{b_j j}{e_j r_j \rho_j}} E_b \tag{13.9}$$

where e_i is the shape factor, ρ_i is the density of component i, and N is the number of components.

Lin and Wey [39] investigated the attrition rate of silica at high temperatures (maximum of 800 °C) and studied the effects of temperature, particle size, gas velocity, and static bed height on the rate of attrition. Their final experimental correlation was

$$R_t = k_0 \left[\frac{(U_e)d_p}{Ar\mu_g}\right]^{\frac{1}{2}} \frac{W^{\frac{3}{2}}}{A} \tag{13.10}$$

where Ar is the Archimedes number, k_0 is the proportionality coefficient, W is weight of the bed material, μ_g is the gas viscosity, A is the cross-sectional area of the bed, and d_p is the average particle diameter.

Chen et al. [26] studied limestone particle attrition at different temperatures in a small CFB and proposed the following correlation:

$$\frac{d_0}{d_t} = k_0 e^{\frac{-E_a}{RT}} U_e^2 t + 1 \tag{13.11}$$

where d_0 is the initial Sauter mean particle diameter, d_t is the Sauter mean diameter after a cumulative attrition time t, k_0 is a pre-exponential factor, E_a is the activation energy, R is the universal gas constant of the Arrhenius equation, and T is the temperature of the fluidized bed. At constant temperatures and gas velocity, Eq. (13.11) becomes linear and simplifies to

$$\frac{d_0}{d_t} = k_A t + 1 \tag{13.12}$$

where k_A is the attrition rate constant, which needs to be determined experimentally. Note that the activation energy is usually negative, meaning that the attrition tends to decrease with increasing temperature due to the dependence of material properties on temperature.

Additional empirical correlations exist in the literature; however, they all include similar parameters as aforementioned models that affect the rate of particle attrition, namely, the temperature, the gas velocity, the weight of the bed, and the fluidization time. For more information, see Ray et al. [28], Werther and Reppenhagen [36], Knight et al. [33], and Bayham et al. [27]. From the models presented above, it can be concluded that in systems prone to particle attrition, the best approach to minimize the attrition would be to operate at high temperature and low excess gas velocities. However, these conditions are not optimal for many fluidized bed applications, and thus research on attrition mitigation is ongoing.

13.4 Wear

Wear (often referred to as erosion or wastage) can be described as the continual loss of material from the surface of a solid through continual contact by gas, solid particles, or liquid, possibly promoted also by chemical reactions through scaling, erosion, and corrosion [40]. In any fluidized bed, just as particles that are subjected to deformation and breakage within the bed, the fluidization internal components and surfaces are prone to wear. Examples of fluidization systems facing problem of wear include combustion and gasification units with in-bed heat exchanger tubes and CFB boilers. Basu [41] provided a detailed look at wear in CFBs and other fluidized bed boilers and showed that erosion due to particles impacts can occur in the combustor, water–wall interface, distributor plates, vortex find in the cyclone, secondary air nozzle, thermocouples, and temperature elements. As with attrition, wear tends to be especially detrimental where local gas and entrained particle velocities are high, such as in regions near distributor jets and cyclone inlets, with attrition favoured when the fixed surfaces are harder than the particles and wear favoured when the particles are harder.

For any particular fluidized bed, wear typically occurs due to continual particle impacts and thus is affected by gas properties, structural and morphological properties of particles and the surfaces, and the angular direction of particle impacts on the surfaces [40]. The largest effect typically occurs through particle impacts. Thus, the angle of the impact and the particle velocity have the most influence on the erosion of materials. The force normal to the surface causes the surface to deform, while the force parallel to the surface causes small parts of the surface to break off and cause wear [41]. The angle of impact has the largest effect where generally the highest damage occurs at ~45°, while the minimum erosion occurs parallel to the surface such as with vertical tubes in a CFB boiler [42]. The operating temperature and pressure typically do not have a large impact on erosion, unless the temperatures are high enough to make the solid surfaces soft [41]. For example, inclined evaporator tubes in a steam-water circulation system lost 4.8 mm of material in a 50 mm diameter tube after ~254 days of service [43]. An example of tube wear can appear in Figure 13.2 that shows cross-sectional areas of a horizontal tubes [45] at two positions, 125 mm apart, from a CFB coal combustor after 4000 h of service. Note the marked difference between the two positions with one showing severe wastage, while the other shows negligible loss. Note also the much greater loss of material at the bottom of the damaged tube than at the top. Anderson et al. [44] reported similar enhanced wear at the bottom of tubes immersed in the pressurized bubbling fluidized bed combustion power generation plant at Grimethorpe (UK) (see Figure 13.2).

Physical and mechanical properties of particles also have a major influence on the degree of wear, as the particle hardness and abrasiveness, usually defined through the abrasion index, can lead to more erosion. Similarly, particles size and sphericity have a major influence on wear, with large particles and more angular particles leading to more erosion [46]. Zhu et al. [47] proposed a model for the wear loss, $\dot{E}$:

$$\dot{E} = C f_v \rho_p d_p^{1.2} U_v^{2.1} (1.04 - \Phi_P) \tag{13.13}$$

where C (μm s$^{3.1}$/kg h m$^{0.3}$) is a parameter related to the material and physical properties, such as hardness and Young's modulus, f_v the void (bubble or slug) frequency, U_v the void velocity (m/s), and Φ_P a particle shape factor.

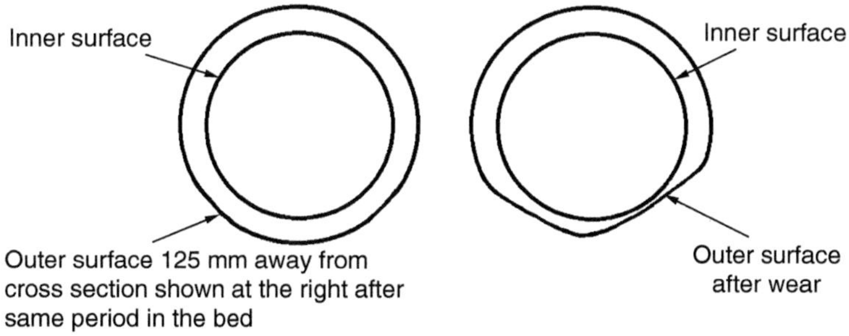

Figure 13.2 Cross section of horizontal heat transfer tube after being immersed in a pressurized power generation coal-burning bubbling fluidized bed combustor facility in Grimethorpe (UK). Source: Adapted from Anderson et al. 1987 [44].

The greatest wear in fluidization systems tends to occur within bed internals, such as immersed heat exchanger tubes, especially if these are horizontal. There are two main types of erosion mechanisms that have been defined for the internals of a fluidized bed, including two-body and three-body wear [41]. In two-body wear particles impact the wall of the tubes under their own momentum, whereas in the three-body wear, the particles moving along the tube wall are struck by nearby particles not in direct contact with the tube wall. Numerous methods are proposed and employed to reduce erosion in fluidized beds. These include the use of fins, ball studs, or rods, which are welded directly onto in-bed tubes to minimize direct contact of particles [41]. Coating the tubes with a hard attrition-resistant coating can also extend the life of the tubes. The ceramic Cr_3C_2 applied by the hypersonic velocity oxygen fuel (HVOF) process has a high erosion and corrosion resistance.

Corrosion and scaling also have a large role in the wear of fluidized beds. Usually, in a fluidized bed boiler or combustor, corrosion is triggered to the high chlorine, sulphur, and vanadium content in fuels, the lack of combustion control, high gas and/or metal temperatures, and a reducing atmosphere [41]. Different methods have been used to minimize corrosion on the fire sides of a fluidized bed boiler, particularly using low oxygen combustion, providing a uniform distribution of combustion air and fuel in the combustion chamber, avoiding local hot spots, injecting additives, controlling the flue temperature at the furnace outlet, and utilizing corrosion-resistant alloys.

Solved Problems

13.1 As shown in Figure 13P.1, in a fluidized bed where tribocharging effects are present a single charged polyethylene particle is adhering to a metallic grounded column wall. For the illustrated case and with the data provided in Table 13P.1, (a) perform a force balance on the charged particle, and (b) using the data provided, compare the frictional force with the sum of the other forces in the z-direction, and plot the results as a function of increasing particle diameter from 10 to 1000 μm. Assume that the static friction coefficient, μ_s, is equal to 0.5.

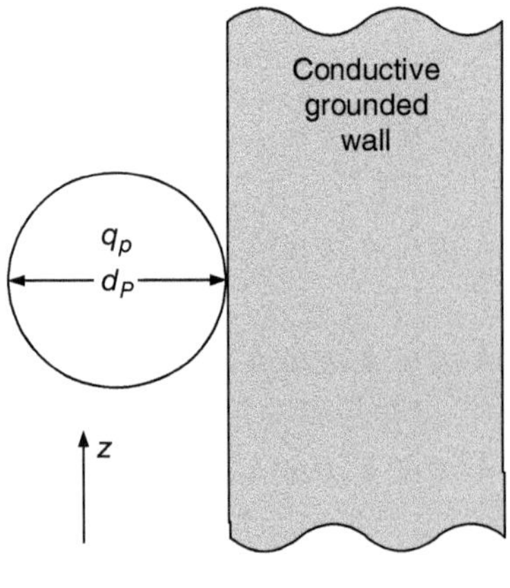

Figure 13P.1 Single charged polyethylene particle adhering to a fluidization column wall.

Table 13P.1 Particle and fluid properties.

Particle properties				Fluid properties			
Density	ρ_P	**918**	**kg/m³**	**Density**	ρ_f	**1.8**	**kg/m³**
Diameter	d_P	500	μm	Viscosity	μ	1.86E-05	Pa s
Charge	q_P	0.005	nC	Permittivity	ε_f	8.8×10^{-12}	F/m
Critical gap between particle and the wall	z	1	μm	Superficial gas velocity	u	0.2	m/s
Sphericity	ϕ	1					

Solution

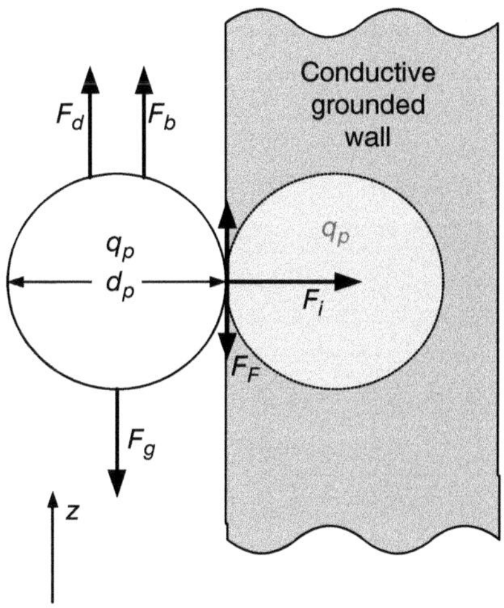

(a) **Force of gravity:**

$$\mathbf{F}_g = m_P g = \frac{1}{6}\rho_p \pi d_P^3 g \tag{13.14}$$

Buoyancy force, F_b:

$$\mathbf{F}_b = m_f g = \frac{1}{6}\rho_f \pi d_P^3 g \tag{13.15}$$

Drag force, F_d:

$$\mathbf{F}_d = \frac{1}{2}\rho_f u^2 A_p C_D \tag{13.16}$$

where A_P is the cross-sectional area of the particle and C_D is the drag coefficient. There are many different models for C_D. A generalized one [48] is

$$C_D = \frac{24}{Re_P}\left[1 + (8.1716e^{-4.0655\phi})Re_P^{0.0964+0.5565\phi}\right] + \frac{3.69(e^{-5.0748\phi})Re_P}{Re_P + 5.378e^{6.2122\phi}} \tag{13.17}$$

where $Re_P = \frac{\rho_f d_P u}{\mu_f}$

Electrostatic force, F_E: The electrostatic force, caused by induction charging, is important due to the frictional force being proportional to the normal force. It is assumed that only the normal force acting against the wall contributes to the electrostatic force.

$$\mathbf{F}_E = \frac{q_p^2}{4\pi\varepsilon_f(d_P + z)^2} \tag{13.18}$$

Frictional force, F_F: Assuming that only static friction is occurring at the wall, the frictional force is

$$\mathbf{F}_F = \mu_s \mathbf{F}_E \tag{13.19}$$

Using the above equations, the forces acting on the particle are as follows:

Vertical forces		
$\mathbf{F}_g$	5.89×10^{-7}	N
$\mathbf{F}_b$	1.16×10^{-9}	N
$\mathbf{F}_d$	2.84×10^{-8}	N
$\mathbf{F}_F$	4.49×10^{-7}	N
Horizontal forces		
$\mathbf{F}_E$	8.98×10^{-7}	N

(b) The same procedure can be used as for part (a) and completed in Excel. Let the summation of the vertical force, except for the frictional force, be equal to

$$\mathbf{F}_V = \mathbf{F}_g - \mathbf{F}_d - \mathbf{F}_b \tag{13.20}$$

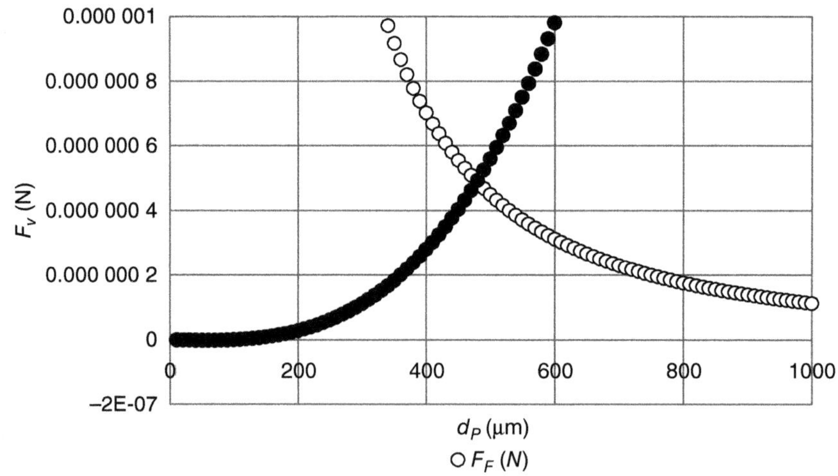

As can be seen from the graph above, at a certain particle size, for the conditions above, the gravitational force overcomes all other forces, and thus the particle no longer adheres to the column wall.

13.2 Attrition measurements for limestone particles were conducted in a lab-scale fluidization system. The tests were carried out at different fluidizing gas velocities and temperatures. The particle size distribution of samples collected from the fluidized bed was measured after one, three, and six hours of fluidization. The Sauter mean diameter of each sample, calculated for each operating condition, is reported in Table 13P.2. Determine the activation energy, E_a, and pre-exponential factor, k_0, for the Chen et al. [26] attrition model.

Table 13P.2 Attrition testing conditions and Sauter mean diameter results.

T (K)	298	400	600	298	400	600	298	400	600
U_e (m/s)	1	1	1	3	3	3	5	5	5
t (h)	d_t (μm)	d_t (μm)	d_t (μm)	d_t (μm)	d_t (μm)	d_t (μm)	d_t (μm)	d_t (μm)	d_t (μm)
0	250.8	250.8	250.8	250.8	250.8	250.8	250.8	250.8	250.8
1	244.0	246.3	247.8	248.3	249.1	249.7	250.5	250.5	250.7
3	231.5	238.0	242.3	243.4	246.0	247.6	249.9	250.2	250.4
6	215.1	226.4	234.2	236.5	241.4	244.6	249.1	249.7	250.1

Solution

The Chen et al. [26] particle size attrition model is as follows:

$$\frac{d_0}{d_t} = k_0 e^{\frac{-E_a}{RT}} U_e^2 t + 1 \tag{13.21}$$

At a constant temperature and excess gas velocity, the model has a linear fit.

$$\frac{d_0}{d_t} = k_A t + 1 \tag{13.22}$$

where k_A is expressed as

$$k_A = k_0 e^{\frac{-E_a}{RT}} U_e^2 \tag{13.23}$$

Therefore the steps for this solution is as follows:

1. Convert the particles Sauter mean diameter from the experimental data into d_0/d_t.
2. For a given temperature and excess gas velocity determine k_A.

3. Linearize Eq. (13.23), and solve for E_a and k_0.
   Step 1
   The initial Sauter diameter, d_0, occurs at $t = 0$; therefore the data can be easily transformed into d_0/d_t as follows:

T (K)	298	400	600	298	400	600	298	400	600
U_e (m/s)	1	1	1	3	3	3	5	5	5
t (h)	d_0/d_t	d_0/d_t	d_0/d_t	d_0/d_t	d_0/d_t	d_0/d_t	d_0/d_t	d_0/d_t	d_0/d_t
0	1.0000	1.0000	1.0000	1.0000	1.0000	1.0000	1.0000	1.0000	1.0000
1	1.0279	1.0183	1.0121	1.0101	1.0068	1.0044	1.0012	1.0012	1.0004
3	1.0834	1.0538	1.0351	1.0304	1.0195	1.0129	1.0036	1.0024	1.0016
6	1.1660	1.1078	1.0709	1.0605	1.0389	1.0253	1.0068	1.0044	1.0028

Step 2
The data above can be plotted as follows:

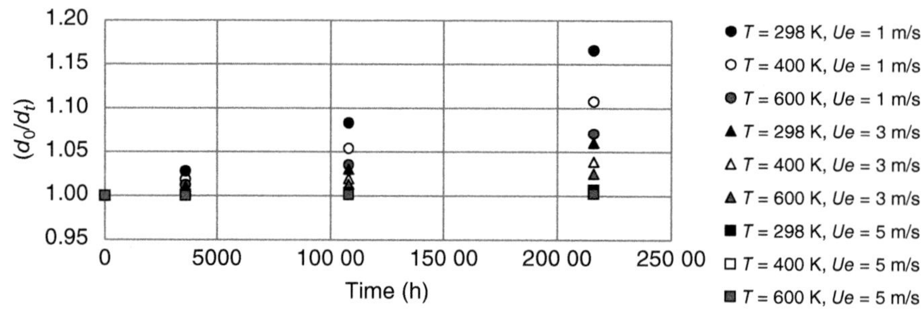

The graph is forming a linear trend with the y-intercept equal to 1. The slopes of each line are given in the table below.

Condition	Slope (k_A)	Intercept
$T = 298$ K, $U_e = 1$ m/s	7.68E-06	1.0002
$T = 400$ K, $U_e = 1$ m/s	4.98E-06	1.0001
$T = 600$ K, $U_e = 1$ m/s	3.27E-06	1.0001
$T = 298$ K, $U_e = 3$ m/s	2.80E-06	1.0000
$T = 400$ K, $U_e = 3$ m/s	1.80E-06	1.0002
$T = 600$ K, $U_e = 3$ m/s	1.17E-06	1.0001
$T = 298$ K, $U_e = 5$ m/s	3.16E-07	1.0001
$T = 400$ K, $U_e = 5$ m/s	1.96E-07	1.0002
$T = 600$ K, $U_e = 5$ m/s	1.32E-07	1.0000

It is clear that the intercept is close to 1, which means that the data are in good agreement with the model.
Step 3
Equation (13.23) can be linearized as follows:

$$k_A = k_0 e^{\frac{-E_a}{RT}} U_e^2 \Rightarrow \ln\left(\frac{k_A}{U_e^2}\right) = \underbrace{\left(\frac{-E_a}{R}\right)}_{m}\left(\frac{1}{T}\right) + \underbrace{\ln k_0}_{b} \tag{13.24}$$

Plotting $\ln\left(\frac{k_A}{U_e^2}\right)$ vs. $\left(\frac{1}{T}\right)$ yields a nearly straight line, where $E_a = -mR$ and $k_0 = e^b$.
The data and graph are:

$1/T$	$\ln(k_A/U_e^2)$
K^{-1}	–
0.003 356	−11.777
0.0025	−12.210
0.001 667	−12.630
0.003 356	−14.983
0.0025	−15.427
0.001 667	−15.854
0.003 356	−18.186
0.0025	−18.664
0.001 667	−19.058

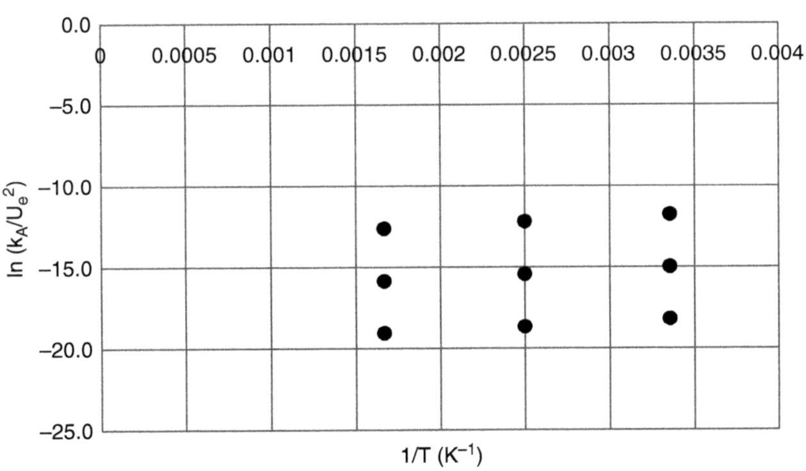

The slope and the intercept from the data are as follows:
$m = 512.36$ and $b = -16.71$
Therefore $E_a = -4.26$ kJ/mol and $k_0 = 5.6 \times 10^{-8}\ s^{-2}$.

Notations

A	cross-sectional area (m^2)
a	attrition coefficient (mass of fines generated per pass of effective breaking energy)
Ar	Archimedes number (–)
b_i	energy-sharing coefficient (–)
C	erosion material parameter ($\mu m\ s^{3.1}/kg\ h\ m^{0.3}$)
d_0	initial Sauter mean particle diameter (m)
d_p	particle diameter (m)
d_t	Sauter mean diameter after a cumulative attrition time (m)
$\vec{E}$	electric field (V/m)
E_a	activation energy (J/mol)
E_b	breaking energy (J)
E_k	kinetic energy (J)
e	shape factor
$\mathbf{F}_E$	electrostatic force (N)
$\mathbf{F}_I$	image force (N)
f_v	void (bubble or slug) frequency (–)
g	gravitational constant (9.81 m/s^2)
k_0	proportionality coefficient
k_0	pre-exponential factor (m^2/s)
k_1	constant
k_A	attrition rate constant (s^{-1})
M	mass of bed of particles (kg)
N	number of components (–)
q_1	charge (C)
q_2	charge (C)
r	distance between two points, or radius (m)
R	universal gas constant (8.314 J/mol K)
R_t	total mass rate of attrition (kg/s)
SA	surface area (m^2)
t	fluidization time (s)
T	temperature (K)
U_e	excess gas velocity (m/s)
U_v	void velocity (m/s)
w	mass fraction of components smaller than radius r (–)
W	weight of the bed material (N)
x	mass fraction (–)
Z_{pw}	distance between charged object and conductive surface (m)

Greek Letters

ε_f	permittivity of medium or fluid (F/m)
ε_w	permittivity of conductive surface material (F/m)
ε_o	permittivity of free space (F/m)
η	efficiency of transferring kinetic energy to breaking energy (–)
μ_g	gas viscosity (kg/m s)
Φ_p	particle shape factor (–)
ρ	density (kg/m^3)
ρ_C	charge density (C/m^3)

References

1 Masuda, H., Gotoh, K., Higashitani, K., and Matsusaka, S. (2006). Adhesive force of a single particle. In: *Powder Technology Handbook*, 3e (eds. H. Masuda, K. Higashitani and H. Yoshida), 157–163. Boca Raton: CRC Press LLC.

2 Mehrani, P., Murtomaa, M., and Lacks, D.J. (2017). An overview of advances in understanding electrostatic charge buildup in gas-solid fluidized beds. *J. Electrostat.* 87: 64–78.

3 Song, D. and Mehrani, P. (2017). Mechanism of particle build-up on gas-solid fluidization column wall due to electrostatic charge generation. *Powder Technol.* 316: 166–170.

4 Hendrickson, G. Electrostatics and gas phase fluidized bed polymerization reactor wall sheeting. *Chem. Eng. Sci.* 61: 1041–1064.

5 Lewis, W.K., Gilliland, E.R., and Bauer, W.C. (1949). Characteristics of fluidized particles. *Ind. Eng. Chem.* 41 (6): 1104–1117.

6 Fotovat, F., Bi, X.T., and Grace, J.R. (2017). Electrostatics in gas-solid fluidized beds: a review. *Chem. Eng. Sci.* 173: 303–334.

7 Matsusaka, S., Maruyama, H., Matsuyama, T., and Ghadiri, M. (2010). Triboelectric charging of powder: a review. *Chem. Eng. Sci.* 65: 5781–5807.

8 Song, D. and Mehrani, P. (2017). Comparison of electrostatic charge generation in gas-solid fluidized beds in turbulent versus pre-turbulent flow regime. *Powder Technol.* 319: 426–433.

9 Cross, J. (1987). *Electrostatics: Principles, Problems and Applications.* Bristol: Adam Hilger.

10 Tanue, K. and Masuda, H. (2006). Electrostatic separation. In: *Powder Technology Handbook*, 3e (eds. H. Masuda, K. Higashitani and H. Yoshida), 653–656. Boca Raton: CRC Press LLC.

11 Matsuyama, T. and Yamamoto, H. (2006). Impact charging of particulate materials. *Chem. Eng. Sci.* 61: 2230–2238.

12 Itakura, T., Masuda, H., Ohtsuka, C., and Matsusaka, S. (1996). The contact potential difference of powder and the tribo-charge. *J. Electrostat.* 38: 213–226.

13 Matsusaka, S. and Masuda, H. (2003). Electrostatics of particles. *Adv. Powder Technol.* 14: 143–166.
14 Ema, A., Yasuda, D., Tanoue, K., and Masuda, H. (2003). Tribo-charge and rebound characteristics of particles impact on inclined or rotating metal target. *Powder Technol.* 135–136: 2–13.
15 Ireland, P.M. (2010). Triboelectrification of particulate flows on surfaces: part I—experiments. *Powder Technol.* 198: 189–198.
16 Chowdhury, F., Sowinski, A., Ray, M. et al. (2018). Charge generation and saturation on polymer particles due to single and repeated particle-metal contacts. *J. Electrostat.* 91: 9–15.
17 Ali, F., Ali, M., Ali, R., and Inculet, I. (1998). Minority charge separation in falling particles with bipolar charge. *J. Electrostat.* 45 (2): 139–155.
18 Zhao, H., Castle, G., Inculet, I., and Bailey, A. (2003). Bipolar charging of poly-disperse polymer powders in fluidized beds. *IEEE Trans. Ind. Appl.* 39 (3): 612–618.
19 Mehrani, P., Bi, H., and Grace, J.R. (2005). Electrostatic charge generation in gas-solid fluidized beds. *J. Electrostat.* 63 (2): 165–173.
20 Sowinski, A., Miller, L., and Mehrani, P. (2010). Investigation of electrostatic charge distribution in gas-solid fluidized beds. *Chem. Eng. Sci.* 65 (9): 2771–2781.
21 Rokkam, R.G., Fox, R.O., and Muhle, M.E. (2010). Computational fluid dynamics and electrostatic modeling of polymerization fluidized-bed reactors. *Powder Technol.* 203: 109–124.
22 Kolehmainen, J., Ozel, A., and Sundaresan, S. (2018). Eulerian modelling of gas–solid flows with 735 triboelectric charging. *J. Fluid Mech.* 848: 340–369.
23 Ray, M., Chowdhury, F., Sowinski, A. et al. (2018). An Euler-Euler model for mono-dispersed gas-particle flows incorporating electrostatic charging due to particle-wall and particle-particle collisions. *Chem. Eng. Sci.* 197: 327–344.
24 Cahyadi, A., Anantharaman, A., Yang, S. et al. (2017). Review of cluster characteristics in circulating fluidized bed (CFB) risers. *Chem. Eng. Sci.* 158: 70–95.
25 Bartels, M., Lin, W., Nijenhuis, J. et al. (2008). Agglomeration in fluidized beds at high temperatures: mechanisms, detection and prevention. *Prog. Energy Combust. Sci.* 34: 633–666.
26 Chen, Z., Grace, J.R., and Lim, C.J. (2008). Limestone particle attrition and size distribution in a small circulating fluidized bed. *Fuel* 87: 1360–1371.
27 Bayham, S.C., Breault, R., and Monazam, E. (2016). Particulate solid attrition in CFB systems—an assessment for emerging technologies. *Powder Technol.* 302: 42–62.
28 Ray, Y.-C., Jiang, T.-S., and Wen, C.Y. (1987). Particle attrition phenomena in a fluidized bed. *Powder Technol.* 49: 193–206.
29 Vaux, W.G. and Fellers, A.W. (1981). Measurement attrition tendency in fluidization. *AlChE Symp. Ser.* 205: 107–115.
30 Wen, C.Y. and Chen, L.H. (1982). Fluidized bed freeboard phenomena: entrainment and elutriation. *AlChE J.* 28: 117–128.
31 Gwyn, J.E. (1969). On the particle size distribution function and the attrition of cracking catalysts. *AlChE J.* 15: 35–39.

32 ASTM Standard D5757 (2006). *Determination of Attrition and Abrasion of Powdered Catalysts by Air Jets*, 1–3. West Conshohocken, PA: ASTM International.
33 Knight, A., Ellis, N., Grace, J.R., and Lim, C.J. (2014). CO_2 sorbent attrition testing for fluidized bed systems. *Powder Technol.* 266: 412–423.
34 Cocco, R., Karri S.B.R., Arrington, Y. et al. (2010). Particle attrition measurements using a jet cup. *Conference Proceedings of the 13th International Conference on Fluidization—New Paradigm in Fluidization Engineering*, Gyeong-ju, South Korea, pp. 1–8.
35 Cocco, R., Arrington, Y., Hays, R. et al. (2010). Jet cup attrition testing. *Powder Technol.* 200: 224–233.
36 Werther, J. and Reppenhagen, J. (2003). Attrition. In: *Handbook of Fluidization and Fluid-Particle Systems* (ed. W.-C. Yang), 198–234. New York, USA: Marcel Dekker Inc.
37 Lowrison, G.C. (1974). *Crushing and Grinding*, 1–286. Cleveland, OH: CRC Press.
38 Ulerich, N.H., Vaux W.G., Newby, R.A., Keairns, D.L. (1980). Experimental/Engineering Support for EPA's FBC Program: Final Report. Vol. 1. Sulfur Oxide Control, EPA-600/7-80-015a.
39 Lin, C.-L. and Wey, M.-Y. (2005). Influence of hydrodynamic parameters on particle attrition during fluidization at high temperature. *Korean J. Chem. Eng.* 22 (1): 154–160.
40 Di Natale, F. and Nigro, R. (2016). An experimental procedure to estimate tube erosion rate in bubbling fluidised beds. *Powder Technol.* 287: 96–107.
41 Basu, P. (2006). *Material Issues in Combustion and Gasification in Fluidized Beds*, 299–336. Boca Raton: Taylor & Francis.
42 BEI (1992). *Modern Power Station Practice*, 3e, 431–438. London: Pergamon Press.
43 Buchanan, E.R. (1995). Evaluation of several coating after extended FBS service. In: *Proceedings of the 13th International Conference on Fluidized Bed Combustion*, vol. I (ed. K. Heinshel), 191–199. New York: ASME.
44 Anderson, J.S., Carls, E.L., Mainhardt, P.J. et al. (1987). Wastage of in bed heat transfer surfaces in the pressurized fluidized bed combustor at Grimethorpe. *J. Eng. Gas Turb. Power* 109: 298–304.
45 Kobro, H. (1984). A discussion of the operation and performance of a 2.6 MW fast fluidized bed combustor and a 16 MW bubbling bed combustor. *Third International Fluidised Conference*, Institute Energy, London, pp. 110–120.
46 Wang, B.Q., Geng, G.Q., Levy, A.V., and Mack, W. (1992). Erosivity of particles in circulating fluidized bed combustors. *Wear* 152: 201–222.
47 Zhu, J., Lim, C.J., Grace, J.R., and Lund, J.A. (1991). Tube wear in gas fluidized beds—II. Low velocity impact erosion and semi-empirical model for bubbling and slugging fluidized beds. *Chem. Eng. Sci.* 46 (4): 1151–1156.
48 Haider, A. and Levenspiel, O. (1989). Drag coefficient and terminal velocity of spherical and non-spherical particles. *Powder Technol.* 58: 63–70.

Problem

13.1 Estimate the erosion rate of a stainless steel Grade 316 horizontal heat transfer tube when subjected to impacts of silica sand particles ($d_p = 1$ mm, $\Phi_p = 0.91$) with a particle impact velocity of 5 m/s. Assume that the erosion material parameter, C, is 0.0097 μm s$^{3.1}$/kg h m$^{0.3}$ [44].

14

Heat and Mass Transfer

Dening Eric Jia

Building and Thermal Sciences Group, National Renewable Energy Laboratory, 15013 Denver West Parkway, Golden, CO 80401, USA

14.1 Heat Transfer in Fluidized Beds

Fluidized beds are used in a variety of processes due to their excellent heat and mass transfer properties. Heat transfer in fluidized beds can refer to interphase (fluid-to-particle or bubble-to-dense phase) heat transfer, or more commonly to bed-to-surface heat transfer. Bed-to-surface heat transfer, in some cases referred to as bed-to-wall heat transfer, occurs between the dense bed of particles and interstitial gas or liquid and any surface it may be in contact with, such as the column wall, heating or cooling coils, and bare and finned tube banks. Bed-to-surface heat transfer coefficients in fluidized beds are usually an order of magnitude higher than those in fixed bed reactors, which are in turn an order of magnitude higher than those of an empty vessel with air passing through it. The relative magnitudes of heat transfer coefficients are portrayed in Figure 14.1.

Another key advantage of fluidized beds is temperature uniformity. In practice, heat transfer is generally rapid enough that the temperature difference between bubbles and dense phase is no more than a few degrees under most operating conditions. Therefore, in most practical cases, a gas-fluidized beds may be treated as isothermal. It is not normally necessary to consider interphase heat transfer. As thermal equilibrium can be usually assumed to be achieved within about 25 mm of the bottom of the fluidized bed [2]. However, for fluidized beds with jets entering through a perforated plate or nozzles, a greater distance may be required.

Although the temperature is usually very uniform in the dense phase of gas fluidized beds, substantial temperature gradients can exist in the interior of fluidized beds when solid circulation is impeded (e.g. due to severe channelling, or when large particles are used in combination with tightly packed tubes or other heat transfer surfaces) or when heat is generated internally (e.g. chemical reaction within fluidized beds, or electrothermal fluidized beds). Large temperature gradients can also occur in the freeboard region. A typical temperature profile is shown in Figure 14.2, in which a uniform temperature is achieved in the dense phase of a fluidized bed combustor [3], while a significant increase in temperature

Essentials of Fluidization Technology, First Edition. Edited by John R. Grace, Xiaotao Bi, and Naoko Ellis.

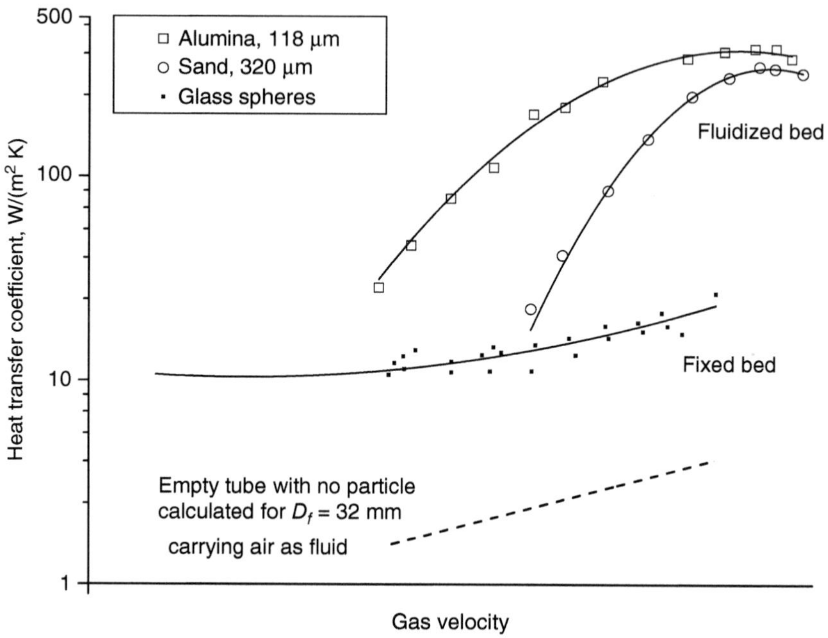

Figure 14.1 Relative orders of magnitude of heat transfer coefficients for empty tube, fixed bed, and fluidized bed reactors, each with flow of air at room temperature and atmospheric pressure. Source: Adapted from Leva 1959 [1].

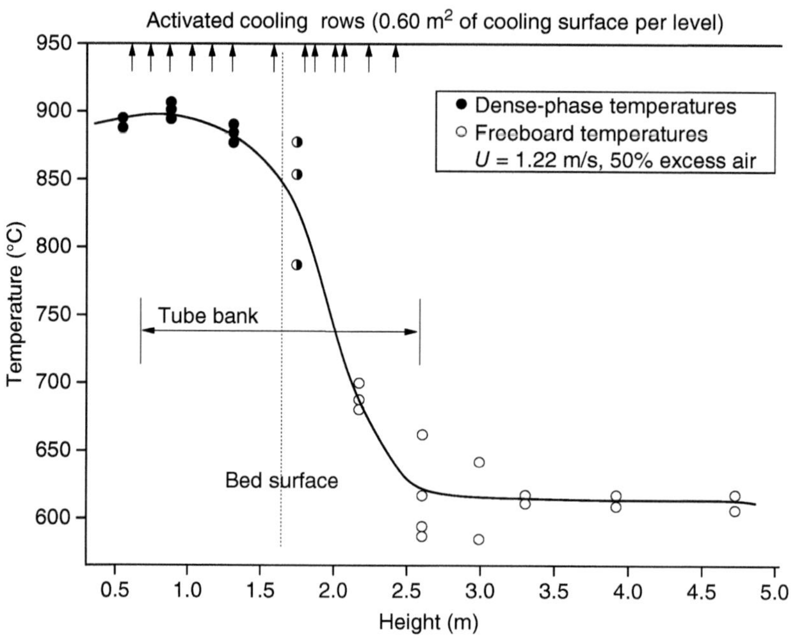

Figure 14.2 Temperature profile in a fluidized bed combustor with heavily cooled freeboard region. Source: Adapted from Byam et al. 1981 [3].

is observed in the splash zone and freeboard region of a bed containing cooling surfaces at a number of levels (shown by arrows).

14.1.1 Interphase Heat Transfer

As stated in previous section, for most practical applications, a gas fluidized bed may be treated as isothermal, except when the bed material has very poor heat transfer properties such as biomass [4, 5] or when extremely exothermic reactions are taking place in the fluidized bed. The isothermal nature of gas fluidized bed can be demonstrated though a simple heat transfer analysis. From a differential heat balance for a horizontal slice of height dz through the dense bed,

$$c_{pg}\rho_g UA \cdot dT_g = haA(T_p - T_g) \cdot dz \tag{14.1}$$

where T_p and T_g denote the particle and gas temperature, respectively, c_{pg} is the gas specific heat, ρ_g the gas density, a the surface area per volume of the particles, h the heat transfer coefficient between the particle and gas, U the superficial gas velocity, and A the cross-sectional area. By integration of Eq. (14.1), the temperature of the gas phase T_g is obtained as

$$T_g = T_p + (T_{g0} - T_p)\exp\left(\frac{-haz}{\rho_g c_{pg} U}\right) \tag{14.2}$$

In a typical fluidized bed with particles of density 3000 kg/m^3 and diameter of 200 μm fluidized by air at standard condition (15 °C, 101.3 kPa), the Reynolds number based on a single particle is of order 10^{-2}. The corresponding Nusselt number is therefore low and less than 2. It is noteworthy that even if the Nusselt number should be as low as 0.01, the required bed height z in which $(T_g - T_p)$ decreases by 90% is of the order of 10 mm or less. Gas in the dense phase therefore reaches the bed (particle) temperature very quickly due to high surface area per unit volume and low interstitial gas velocities, despite the possibly poor heat transfer coefficients for individual particles.

Heat transfer between bubble phase and dense phase can be characterized by an interphase (bubble-to-dense phase) heat transfer coefficient, h_{bd}. The derivation of h_{bd} is achieved by an analogy to mass transfer, elaborated later in this chapter. In practice, heat transfer coefficients tend to be higher in freely bubbling beds than for single bubbles. For an isolated bubble, the interphase heat transfer coefficient is given as

$$h_{bd} = \frac{U_{mf}\rho_g c_{pg}}{4} + 0.95\sqrt{k_g \rho_g c_{pg} \varepsilon_{mf}}\left(\frac{g}{d_b}\right)^{0.25} \tag{14.3}$$

where the first item on the right is due to convection (throughflow) and the second arises from conduction. In a freely bubbled bed, the first term is augmented so that the interphase heat transfer coefficient is given by

$$h_{bd} = \frac{U_{mf}\rho_g c_{pg}}{3} + 0.95\sqrt{k_g \rho_g c_{pg} \varepsilon_{mf}}\left(\frac{g}{d_b}\right)^{0.25} \tag{14.4}$$

The enhancement in interphase heat transfer from a single bubble to a freely bubbling bed is more significant for large bubbles. Nevertheless, heat transfer between bubbles and dense phase is generally rapid enough that the temperature difference between the bubble and dense phase is no more than a few degrees under most operating conditions. In view of the above considerations, temperature uniformity exists in most gas–solid fluidized beds, and the area of interest mostly lies in heat transfer between fluidized beds and fixed surface.

14.1.2 Bed-to-Surface Heat Transfer

14.1.2.1 General Considerations

Heat transfer between fluidized beds and solid surfaces such as heating tubes, cooling coils, and bed walls is of great interest. The design and smooth operation of fluidized beds often depend heavily on the bed-to-surface heat transfer as it is an integral part of heat balance considerations. The bed-to-surface heat transfer is usually very favourable, of the order of 3×10^2 W/(m^2 K).

The bed-to-surface heat transfer coefficient can be considered to consist of three components:

$$h = h_{pc} + h_{gc} + h_{rad} \tag{14.5}$$

The particle convective component h_{pc} arises from conductive heat transfer through stagnant gas and particles during the times that particles are of the heating surface. The term particle convection refers to the fact that time-averaged values rely on frequent particle renewal on the surface, i.e. convection by particles, while in conductive heat transfer takes place via conduction through the thin film of gas separating the particle and surface. The gas convection term h_{gc} includes heat transfer to and from interstitial gas and gas bubbles. The radiative component h_{rad} is only relevant at high temperatures, typically >500 °C. It is a common simplification to assume that the three components can be predicted separately and are additive. However, one must proceed with caution if two or more of the components are of comparable magnitude.

14.1.2.2 Particle Convective Component

There have been a few detailed mechanistic approaches to characterizing bed-to-surface heat transfer in fluidized beds. Mechanistic models can often be classified into the following categories:

(1) Single particle models where heat is transferred through unsteady-state conduction to a single particle [6–8], two particles [9], a chain of four particles [10], or a string of particles [11]. Examples are illustrated in Figure 14.3a.
(2) Alternating layers of gas and solids theorized by Gabor [11] can be used to approximate gas and solid phases. Good agreement with experimental results was found when the thickness of the layers representing the solid phase was taken to be (2/3)d, while thicknesses of 0.1d for the gas phase between particles and 0.05d between the wall and first layer of particles were chosen, as depicted in Figure 14.3b.
(3) Packet theory proposed by Mickley and Fairbanks [12] in which heater surfaces are contacted alternatively by gas bubbles and an emulsion of closely packed particles called packets. A schematic is shown in Figure 14.3c. Packet theory has been widely accepted and utilized in heat transfer modelling.

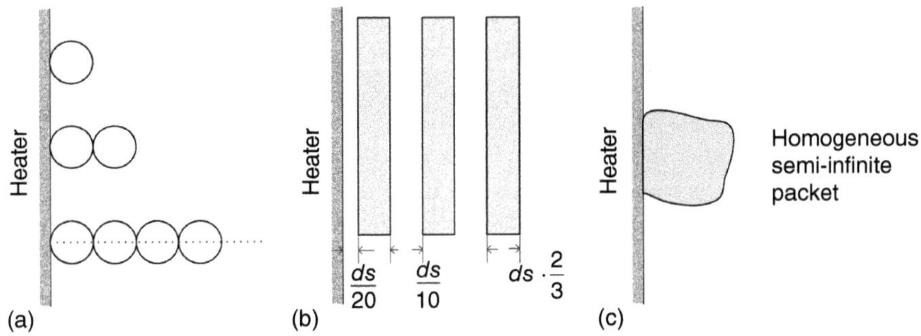

Figure 14.3 Schematics of unsteady-state gas convective heat transfer: (a) single and multiple spheres; (b) alternating layers of gas and solids; (c) Packet theory.

Probes capable of differentiating between bubble and dense phases according to their dielectric properties were utilized by Ozkaynak and Chen [13] and Chandran and Chen [14] to measure the transient particle concentrations on heater surfaces. It was discovered that the instantaneous solids concentration alternated between almost pure gas (bubbles) and high concentration particle cloud similar to that of packed beds, as proposed by Mickley and Fairbanks [12]. The favourable time-average heat transfer coefficients between gas-fluidized bed and fixed surfaces are often associated with frequent renewal of particles. For a constant surface temperature, the instantaneous heat transfer coefficient due to unsteady-state conduction of thick packets of particles with average voidage ε_{mf}, where $t = 0$ at the initial contact between the packet and surface

$$h_i = \sqrt{\frac{k_{mf}\rho_{mf}c_{p,mf}}{\pi t}} \tag{14.6}$$

Here k_{mf}, ρ_{mf}, and $c_{p,mf}$ are the effective thermal conductivity, density, and heat capacity of the dense phase, respectively. The instantaneous heat transfer coefficient h_i then decreases with time, but equilibrium is not established if the renewal of particles is sufficiently fast, driven by the repeated passage of bubbles. While the surface is covered by a bubble, the instantaneous heat transfer rate drops sharply, as verified by Mickley [15], Baeyens and Goossens [16], Kubie [17], Bernis et al. [18], and Selzer and Thomson [19].

The overall magnitude of the heat transfer coefficient is heavily influenced by the rate of local heat transfer and the residence time of particles. The residence time of a particle (t_r) is proportional to the time fraction during which the surface is in contact with the dense phase, f_p, and the time fraction in contact with the bubble phase, f_b. Roughly, one can assume

$$t_r \approx \frac{1}{f_b} \tag{14.7}$$

in which t_r is usually between 0.3 and 0.5 s. Decker and Glicksman [20] proposed a thermal time (τ_s) constant given by

$$\tau_s = \frac{\rho_p c_{pp} d_p^2}{36 k_g} \tag{14.8}$$

Figure 14.4 depicts three scenarios on a heating surface, where the relative magnitudes of t_r and τ_s vary. More details are provided in Table 14.1.

In the original packet theory, Mickey and Fairbanks proposed that the local heat transfer coefficient be estimated by integrating Eq. (14.6), with an age distribution function to account for different times spent by different packets at the surface. For random renewal by bubbles and an average bubble frequency, $\overline{f_B}$, the result

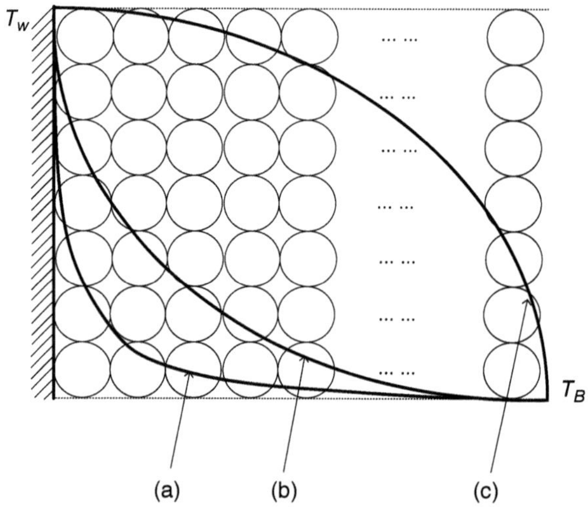

Figure 14.4 Different temperature profiles between heating surface and surrounding particles: (a) $\tau_s \gg 1/f_b$; (b) $\tau_s \approx 1/f_b$; (c) $\tau_s \ll 1/f_b$

Table 14.1 Regimes of heat transfer for different particle sizes.

Particle size	Characteristic time	Dominant mechanism	Notes	Important variables
$d_p < 200\ \mu m$ (small)	$\tau_s \ll \frac{1}{f_b}$	Particle convection	Heat penetrates many particle layers; packet theory works reasonably well	Renewal frequency, k_g, ρ_s, c_{ps}
$200\ \mu m < d_p < 1$ mm (intermediate)	$\tau_s \approx \frac{1}{f_b}$	Particle convection	Heat penetrates only a few particle layers; Botterill-type model required [9, 21]	Renewal frequency, d_p, k_g, ρ_s, c_{ps}
$d_p > 2$ mm (large)	$\tau_s \gg \frac{1}{f_b}$	Gas convection	Little penetration of heat during particle residence time; steady-state conduction can be assumed	ρ_g, c_{pg}, and other related variables

Radiation may also be important at high temperatures, especially for large particles.

was shown to be

$$h_p = \sqrt{(1-\varepsilon_b)\rho_p(1-\varepsilon_{mf})c_{pp}k_{eff}\bar{f}_B} \tag{14.9}$$

In bubbling fluidized beds, renewal of packets is driven by bubbles. The residence time of packets on a heating/cooling surface at length L can be approximated by the velocity of the bubble wake, which of the same as the bubble rise velocity. Hence

$$\tau = \frac{L}{U_b} \tag{14.10}$$

For longer surfaces where packets are swept away (renewed) before reaching the end of the heater, L is replaced by l, which may be taken as about half the equivalent bubble diameter, d_b [22]. The average heat transfer coefficient between the surface and the dense phase is then

$$h_p = \int_0^{\tau} \frac{h_i}{\tau} dt = 2\sqrt{\frac{k_{mf}(1-\varepsilon_b)\rho_p(1-\varepsilon_{mf})c_{pp}U_b}{\min\{l, L\}}} \tag{14.11}$$

For fluidized beds with small particles, the contribution of the bubble phase is usually negligible. However, for systems with large particles, this assumption is no longer valid. Instead, the particle convective component only covers the fraction of the bed volume occupied by the dense phase:

$$h_{pc} = h_p(1-\varepsilon_b) \tag{14.12}$$

Baskakov [23] modified the original packet theory by adding a time-dependent resistance to adjust for discrepancies between the original packet theory and experimental results. Modifications to the packet theory proposed by other authors include replacing constant surface temperature with constant heat flux and adding thin layers of gas in series with packets.

To evaluate the particle convective heat transfer in a given system, the effective thermal conductivity, k_{eff}, must be estimated. It is generally assumed to consist of two components, k_0 and k_t. The former represents heat transfer in a fixed bed reactor with motionless fluid, while the latter is associated with heat transfer induced by fluid flow:

$$k_{eff} = k_0 + k_t \tag{14.13}$$

Deissler and Boegli [24] reported the following correlation for estimating k_0:

$$\frac{k_0}{k_g} = \left(\frac{k_p}{k_g}\right)^{0.28-0.757\log_{10}\varepsilon-0.057\log_{10}\left(\frac{k_p}{k_g}\right)} \tag{14.14}$$

Values of k_0 from Eq. (14.14) agree well with experimental results from Dietz [25]. Ranz [26] identified the contribution of k_t, derived mostly from the mixing of fluids not parallel to the direction of fluid flow. For a system with normally packed spheres and small d_p/D ratio, Ranz [26] showed that k_t could be expressed as

$$k_t = 0.1\rho_g c_{pg} d_p U_{mf} \tag{14.15}$$

Consequently, the effective thermal conductivity of the dense phase or packet has the form of

$$k_{mf} = k_0 + 0.1\rho_g c_{pg} d_p U_{mf} \tag{14.16}$$

In cases where unsteady-state heat transfer from surface to particles outpaces the renewal of particles, i.e. $\tau_s \ll 1/f_b$, allowing heat to penetrate through multiple layers of particles, the effective thermal conductivity has been shown [27] to be

$$k_{eff} = \frac{0.9065}{\frac{0.13}{k_g} + \frac{0.667}{k_p}} \tag{14.17}$$

Since k_g in gas fluidized beds is usually much smaller than k_s, the thermal properties of the gas play an important role. The influence of the volumetric heat capacity of the solids $\rho_s c_{ps}$ is also critical. Good agreement with experimental results can generally be obtained if one makes allowance for a thin layer of stagnant gas on the heater surface. The overall particle convective heat transfer coefficient is then

$$h_{pc} = \frac{1 - \varepsilon_b}{\frac{1}{h_p} + \frac{1}{h_f}} \tag{14.18}$$

in which h_f represents the gas layer (film) resistance. Experimental results have identified particle diameter, gas thermal conductivity, and λ = film thickness divided by particle diameter as the main contributors to the thin gas film heat transfer:

$$h_f = k_g \lambda / d_p \tag{14.19}$$

Baskakov et al. [28] proposed $\lambda = 8$ based on experimental results, whereas Zabrodsky et al. [29] suggested $\lambda = 7.2$. Botterill et al. [9] fitted a value of 10 in a two-phase heat transfer model. Xavier [30] found that $\lambda = 4$ worked well for horizontal surfaces. For design purposes, $\lambda = 6$ is commonly utilized.

Another simplification applies to large particles where particle renewal at a surface is much faster than unsteady-state heat transfer from surface to particles, i.e. $\tau_s \gg 1/f_b$. This usually applies to fluidized beds of large particles, i.e. $d_p > 1$ mm. Little change occurs to temperature of these large particles while they are in contact with the surface, so that steady-state heat transfer can be assumed. Decker and Glicksman [31] derived a Nusselt number of 11.2 based on an average conduction path length. Furthermore, the particle convective heat transfer coefficient was derived assuming that heat transfer during coverage by bubbles is negligible, so that

$$\overline{h}_{pc} = 11.2(1 - \varepsilon_b)\frac{k_g}{\overline{d}_p} \tag{14.20}$$

14.1.2.3 Gas Convection Component

The contribution to heat transfer from gas percolating through the dense phase and inside gas bubbles becomes relevant in fluidized beds with $d_p > 1$ mm and dominant for $d_p \geq 3$ mm. The influence of bubble passage on heat transfer is clearly demonstrated in Figure 14.5, where the bed voidage and the heat transfer coefficient in a gas fluidized bed of dolomite particles of diameter 4.0 mm, at superficial gas velocities well above the minimum fluidization velocity, were measured simultaneously. A decline in the capacitance signal indicated the presence of voids or bubbles, while a rise reflected the arrival of particles. The capacitance signal correlated well with the corresponding heat transfer signal. When the particle phase contacted the measurement probe, the instantaneous heat transfer coefficient rose sharply. The peaks of heat transfer coefficient occurred approximately every second, so the bubble frequency was indicated by the local voidage fluctuations.

Baskakov et al. [33] correlated the gas convective heat transfer coefficient by

$$\frac{h_{gc}\overline{d}_p}{k_g} = 0.009Ar^{0.5}Pr^{0.33} \tag{14.21}$$

in which $Ar = \frac{\rho_g(\rho_p-\rho_g)g\overline{d}_p^3}{\mu^2}$ and $Pr = \frac{c_{pg}\mu_g}{k_g}$. From Eq. (14.21) $h_{gc} \propto d_p^{1/2}$. An alternative was proposed by Denloye and Botterill [34] by correlating the dimensional group, $h_{gc}d_p^{1/2}/k_g$, leading to

$$\frac{h_{gc}\overline{d}_p^{-1/2}}{k_g} = 0.664\left[\frac{(U_b + 3U_{mf})\overline{d}_p\rho_g}{\mu_g}\right]^{0.5} \cdot Pr^{0.33} \cdot \left(\frac{\overline{d}_p}{L}\right)^{0.5} \tag{14.22}$$

In a system where both particle convection and gas convection are important, the overall heat transfer coefficient between a heat transfer surface and a fluidized bed, excluding the influence of radiative heat transfer, can be approximated by

$$\overline{h}_c = (1-\varepsilon_b)h_{pc} + \varepsilon_b h_{gc} \tag{14.23}$$

14.1.2.4 Maximum Heat Transfer Coefficient

Zabrodsky [35] compiled bed-to-surface heat transfer coefficients in fluidized beds with particle size ranging from 82.5 to 1160 μm. Results are shown in Figure 14.6, in which heat transfer coefficients are plotted against the superficial gas velocity. As the superficial gas velocity increases, the overall bed-to-surface heat transfer coefficient increases, passes through a maximum, and then decreases. At low gas flow rates, the major impact of increasing the gas velocity is to increase the frequency of bubbles, increasing the renewal frequency of packets at the heat transfer surface, therefore improving heat transfer. Eventually the increase in packet renewal frequency is outweighed by the fraction of time during which the heat transfer surface is occupied by bubbles, leading to the decline in h. As the particle size increases, the impact of solid renewal or particle convection diminishes, so the maximum in the heat transfer coefficient is less pronounced. Nevertheless, a maximum bed-to-surface heat transfer coefficient exists for almost all systems. A correlation for maximum heat transfer coefficient proposed by Zabrodsky [35] is

$$\frac{\overline{h}_{max}\overline{d}_p}{k_g} = 0.88Ar^{0.213}\ (10^2 < Ar < 1.4\times10^5) \tag{14.24}$$

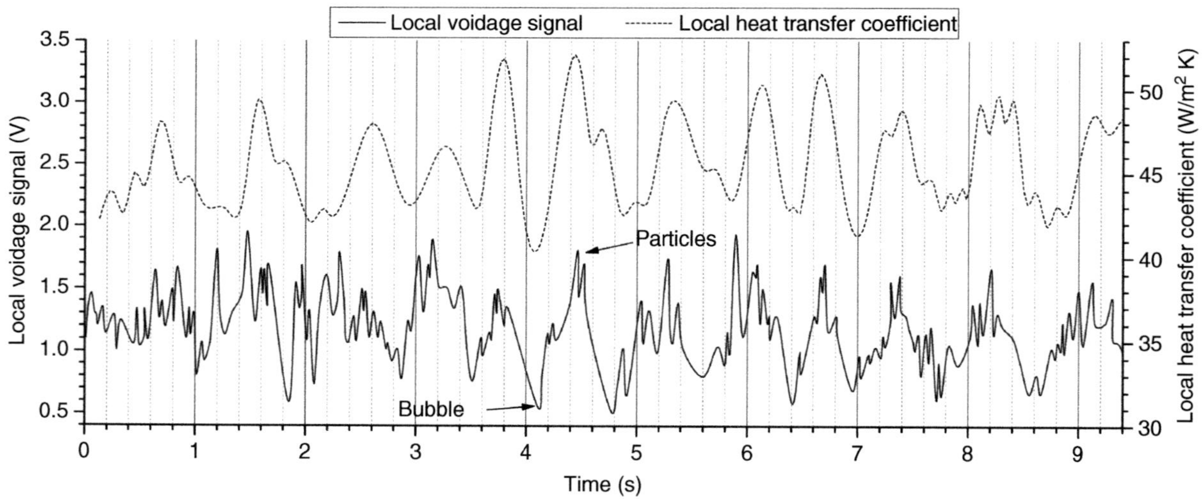

Figure 14.5 Simultaneous measurement of local bed voidage and local heat transfer coefficient in a fluidized bed of dolomite particles with $d_p = 4.0$ mm, $U/U_{mf} = 1.28$, $U_{mf} = 1.83$ m/s. Source: Adapted from Catipovic 1979 [32].

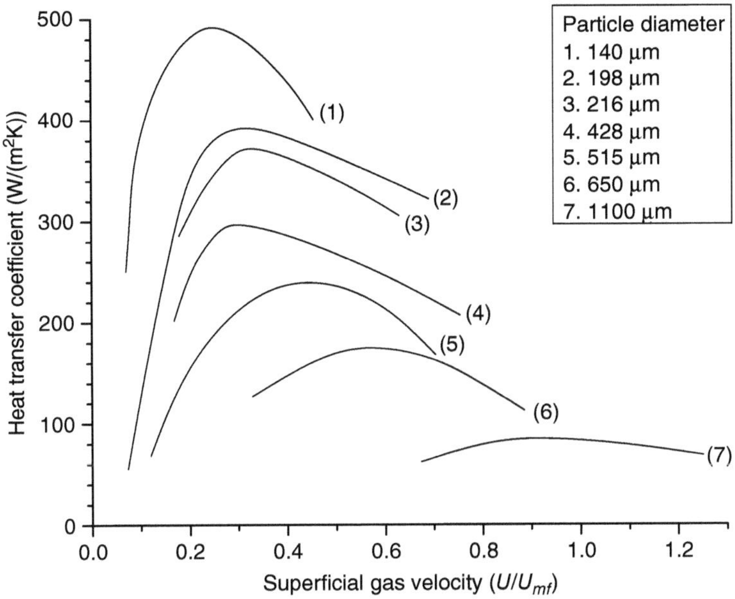

Figure 14.6 Effect of gas flow rate and particle size on heat transfer. Source: Adapted from Zabrodsky 1966 [35] and experimental data from Varygin and Martyushin 1959 [36].

Equation (14.24) can be applied to both horizontal and vertical tubes, as well as to heat transfer or from the inside surface of fluidization columns. But it is only suitable for particles of diameter ≤1 mm. For larger particles, gas convection is no longer negligible. Instead, as the particle diameter increases, the dominant heat transfer mechanism slowly shifts from particle convection to gas convection, and the maximum heat transfer coefficient decreases. As illustrated in Figure 14.7, a minimum value of h_{max} appears to be reached at $d_p \approx 3$ mm.

For extremely fine particles with $d_p \leq 30$ μm, heat transfer is often hindered by poor fluidization quality, resulting from the cohesive nature of Geldart group C particles. Similar trends have been observed for low density particles such as biomass, where interparticle forces such as van der Waals forces reach magnitudes comparable with the gravitational forces on particles [4, 37].

For particles of 30 μm $< d_p <$ 1000 μm, for which particle convection dominates bed-to-surface heat transfer, the packet theory works reasonably well in predicting particle convective heat transfer. For larger particles, gas convection, and hence gas-phase properties, must be taken into consideration.

Once the particle size exceeds ~1000 μm, gas convection must be taken into consideration. It becomes dominant for $d_p >$ 3000 μm, with the properties of the gas phase then playing a major role in determining the bed-to-surface heat transfer. For large particles the correlation from Maskaev and Baskakov [38] is recommended:

$$\frac{\overline{h}_{max}\overline{d}_p}{k_g} = 0.21Ar^{0.32} \; (1.4 \times 10^5 < Ar < 3 \times 10^8) \tag{14.25}$$

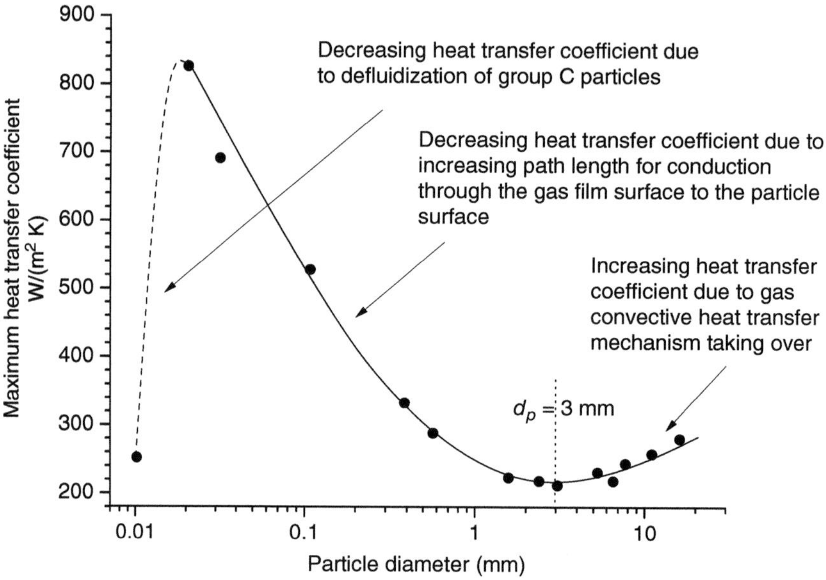

Figure 14.7 Dependence of maximum heat transfer coefficient on particle size. Source: Adapted from Baskakov et al. 1973 [28].

Borodulya et al. [39] measured bed-to-surface heat transfer coefficients of a vertical cylinder immersed in fluidized beds with particle diameter ranging from 126 to 3100 μm and pressure from 0.6 to 8.1 MPa. Results are displayed in Figure 14.8. The maximum heat transfer coefficient increased with increasing operating pressure. The influence of pressure on h_{max} is more pronounced for larger particles, as gas convection and gas density depend heavily on pressure. For smaller particles the controlling factor of bed-to-surface heat transfer is particle convection, which is much less influenced by pressure. As a result, increasing operating pressure has little effect on heat transfer for smaller particles, as is clearly demonstrated in Figure 14.8. The slight increase in h with increasing pressure for smaller particles in this figure may be attributable to improved fluidization quality and altered renewal frequency at elevated pressures.

To increase the heat transfer performance in fluidized beds of small particles, it is good practice to enhance particle convection. This can be achieved by using gases of high thermal conductivity, k_g, (e.g. hydrogen) or increasing the operating temperature so that the effective thermal conductivity of the packets and resulting heat flux are increased. For larger particles, operating at elevated pressure or improving fluidization quality is likely to augment bed-to-surface heat transfer.

14.1.3 Heat Transfer Correlations for Fluidized Beds

The general features of bed-to-surface heat transfer in fluidized beds have been investigated, and mechanisms are outlined above. But when it comes to predicting heat transfer rates for design and operation purposes, empirical

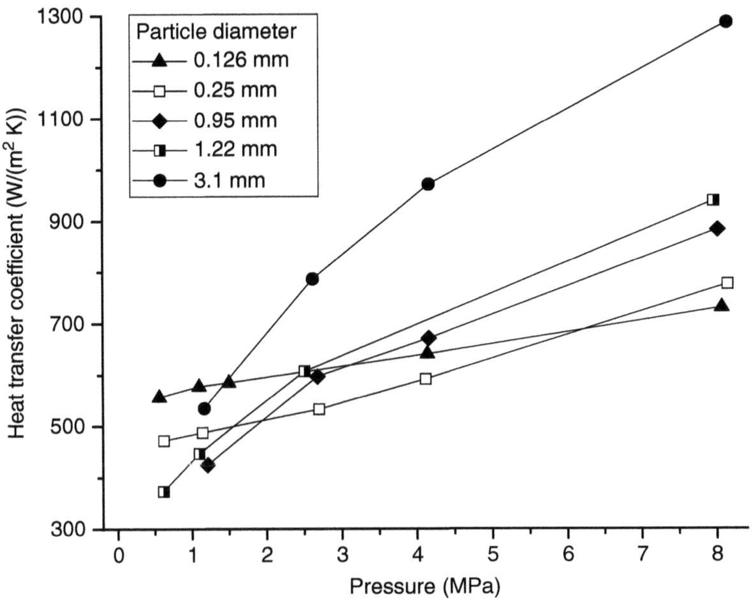

Figure 14.8 Effect of pressure on heat transfer. Source: Adapted from Borodulya et al. 1980 [39].

and semi-empirical correlations are still widely used. Saxena [40] reviewed more than 200 publications on fluidization, mostly on heat transfer, in which 11 correlations were included for heat transfer to and from smooth horizontal tubes in fluidized beds containing small particles. The predicted heat transfer coefficients from different correlations disagree widely. It is of great importance not to extrapolate these correlations or apply them in situations outside the range in which they were derived. Separate correlations and related studies can be found for heat transfer to external walls, horizontal tubes, vertical tubes, and tube banks. Some correlations seek to depict the influence of superficial gas velocity, whereas others are for the maximum heat transfer coefficient. The most successful correlations and close alternatives are identified below. These tend to be semi-empirical mechanistically based correlations. It should be noted that the accuracy of these correlations should not be assumed to be better than ±30% within their designated range of applications. Larger discrepancies must be expected should the correlations be extrapolated beyond their ranges.

14.1.3.1 Correlations for Bed-to-Surface Heat Transfer

The most successful correlation for bed-to-surface heat transfer is that of Molerus and Wirth [41], who analyzed experimental data from Wunder [42] and determined that for different flow regimes in a bubbling fluidized bed, the controlling factor for heat transfer can vary. Therefore, the prediction of heat transfer coefficient could not be accurately performed until the underlying mechanism was identified. Parameters related to fluidized bed heat transfer were divided into five dimensionless groups:

$$\begin{aligned}
\pi_1 &= d_p^3 g(\rho_p - \rho_g)/\mu^2 \equiv Ar \\
\pi_2 &= \rho_p/\rho_g \\
\pi_3 &= c_{pg}\mu/k_g \equiv Pr_g \\
\pi_4 &= c_{pp}\mu/k_g \\
\pi_5 &= k_p/k_g
\end{aligned} \tag{14.26}$$

For example, for group A particles in the laminar flow regime ($Re \leq 0.5$), fluidized beds usually have an Archimedes numbers (Ar) <100 and particle diameter of 15 μm < d_p < 100 μm at ambient conditions. Under such circumstances the influences of c_{pg}, ρ_g, and k_p are pronounced compared with k_g and the dimensionless heat transfer coefficient, Nu ∝ (π_1, π_4). By combining dimensional analysis with experimental results, the maximum heat transfer coefficient has the form

$$\frac{1}{h_{max}} = \frac{l_L/0.09}{k_g} + \frac{l_L/0.09}{2c_{pp}\mu} \quad (Ar < 10^2) \tag{14.27}$$

where l_L is the laminar flow length scale

$$l_L = \left(\frac{\mu}{\sqrt{g(\rho_p - \rho_g)}}\right)^{2/3} \tag{14.28}$$

By evaluating all factors controlling fluidized bed heat transfer and correlating experimental data, the following semi-empirical correlation was proposed [41] for bubbling fluidized beds of $Ar \leq 10^8$ with negligible radiative heat transfer, i.e. bed temperature less than about 500 °C:

$$\frac{hl_L}{k_g} = \frac{0.125(1-\varepsilon_{mf})\left[1 + 33.3\left(\sqrt[3]{\frac{U-U_{mf}}{U_{mf}}} \cdot \sqrt[3]{\frac{\rho_p c_{pp}}{k_g g}} \cdot (U - U_{mf})\right)^{-1}\right]^{-1}}{1 + \frac{k_g}{2c_{pp}\mu}\left[1 + 0.28(1-\varepsilon_{mf})^2\left(\frac{\rho_g}{\rho_p-\rho_g}\right)^{0.5}\left(\sqrt[3]{\frac{\rho_p c_{pp}}{k_g g}}\right)(U - U_{mf})^2 \frac{U_{mf}}{U-U_{mf}}\right]} + 0.165 Pr^{1/3}\left(\frac{\rho_g}{\rho_p - \rho_g}\right)^{1/3}\left[1 + 0.05\left(\frac{U - U_{mf}}{U_{mf}}\right)^{-1}\right]^{-1} \quad (Ar \leq 10^8) \tag{14.29}$$

Good agreement has generally been obtained with the above correlation, which includes both the particle convective and gas convective components and covers a wide range of physical properties and operating conditions. Selected comparisons between correlation predictions and experimental data are plotted in Figure 14.9.

14.1.4 Heat Transfer Between Fluidized Bed and Immersed Surfaces

By adding tubes to a fluidized bed, the heat transfer area dramatically increases, which is beneficial to heat addition or removal. Most commonly, immersed tubes

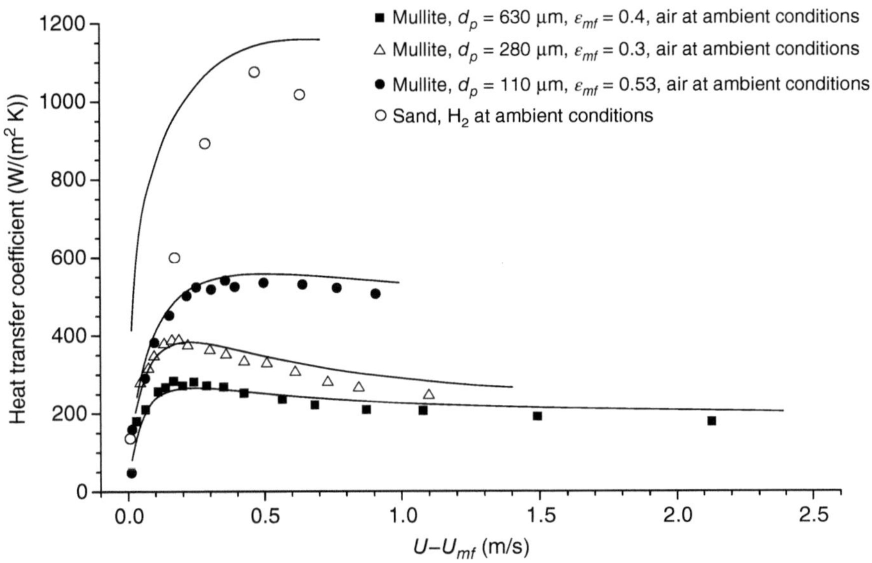

Figure 14.9 Measurements (symbols) of heat transfer coefficients and predictions (curves) according to Eq. (14.29). Source: Adapted from Molerus and Wirth 1997 [43].

are oriented either horizontally or vertically. Few studies have been carried out on inclined tubes, as orientations other than vertical and horizontal, for a given surface area, generally result in less favourable heat transfer. Figure 14.10 shows experimental values of local instantaneous heat transfer coefficients around the periphery of a horizontal tube immersed in fluidized bed, with three different bubble-tube interactions illustrated.

It is evident that the instantaneous heat transfer coefficient is reduced significantly when the tube surface is enveloped by bubbles. Installing heat transfer tubes horizontally often results in higher local voidage beneath the tubes, while stagnant solids build up on the top of the tubes. A more detailed investigation of local heat transfer coefficient in relation to the angle on an immersed horizontal tube is displayed as a polar plot in Figure 14.11. The average heat transfer coefficient is again seen to increase and then decrease as the superficial gas velocity increases. The profile of the local heat transfer coefficient also changes, with the maximum shifting from the bottom to the top of the tube with increasing superficial gas velocity.

Orienting heat transfer tubes vertically in a fluidized bed tends to reduce the impact of scale-up on heat transfer performance. However, vertical and horizontal tubes generally yield similar overall average heat transfer coefficients. Hence the orientation of heat transfer tubes is heavily influenced by thermal expansion, possible erosion of tube surfaces exposed in the bed or the freeboard, and the cost associated with the configuration. Horizontal tubes are more likely to be chosen for $H/D < 1$, and vertical tubes for $H/D > 1$, where H is the expanded bed depth.

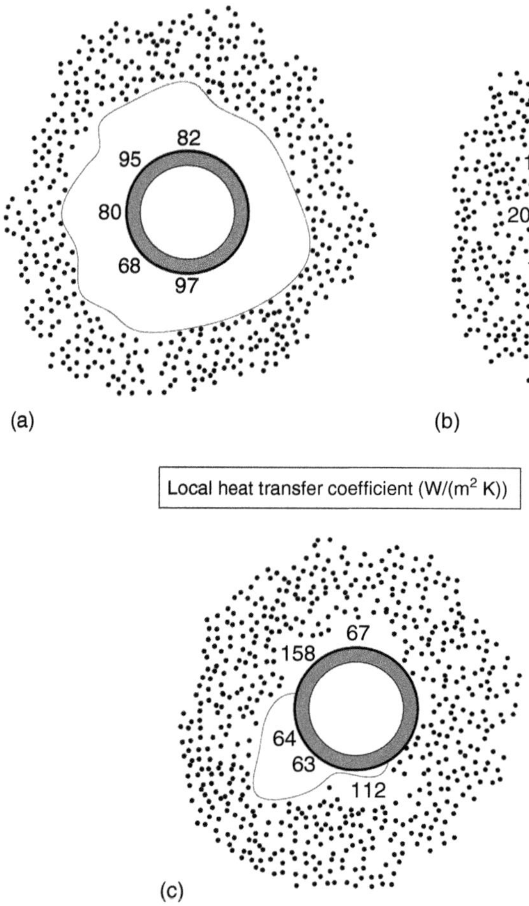

Figure 14.10 Experimental instantaneous heat transfer coefficient near a horizontally installed tube in a fluidized bed. (a) Tube completely inside a bubble, $d_p = 4$ mm, $U/U_{mf} = 1.35$, (b) tube surrounded by moving dense phase, $d_p = 4$ mm, $U/U_{mf} = 1.35$, (c) tube partially contacted by a bubble, $d_p = 2.8$ mm, $U/U_{mf} = 1.15$. Source: Adapted from Catipovic 1979 [32].

14.1.4.1 Correlations for Vertical Tubes

For vertical tubes the average heat transfer coefficient decreases with increasing tube length, tube diameter, and particle size, which is consistent with the influence of these parameters on the residence time of packets at heat transfer surfaces, especially for particles of $d_p < 600$ μm. For larger particles the influence of particle size and particle residence time diminishes, so heat transfer is less dependent on tube length. It is noteworthy that in a tube bundle the spacing between adjacent tubes also affects heat transfer. The heat transfer coefficient tends to increase as distance between tubes increases. The closest distance between tubes should be at least $20 - 30\ d_p$ to prevent particles from jamming in the gaps (Figure 14.12).

Gunn and Hilal [45] reported that fluidization quality was affected by the number and relative position of tubes in a horizontal bundle immersed in a fluidized

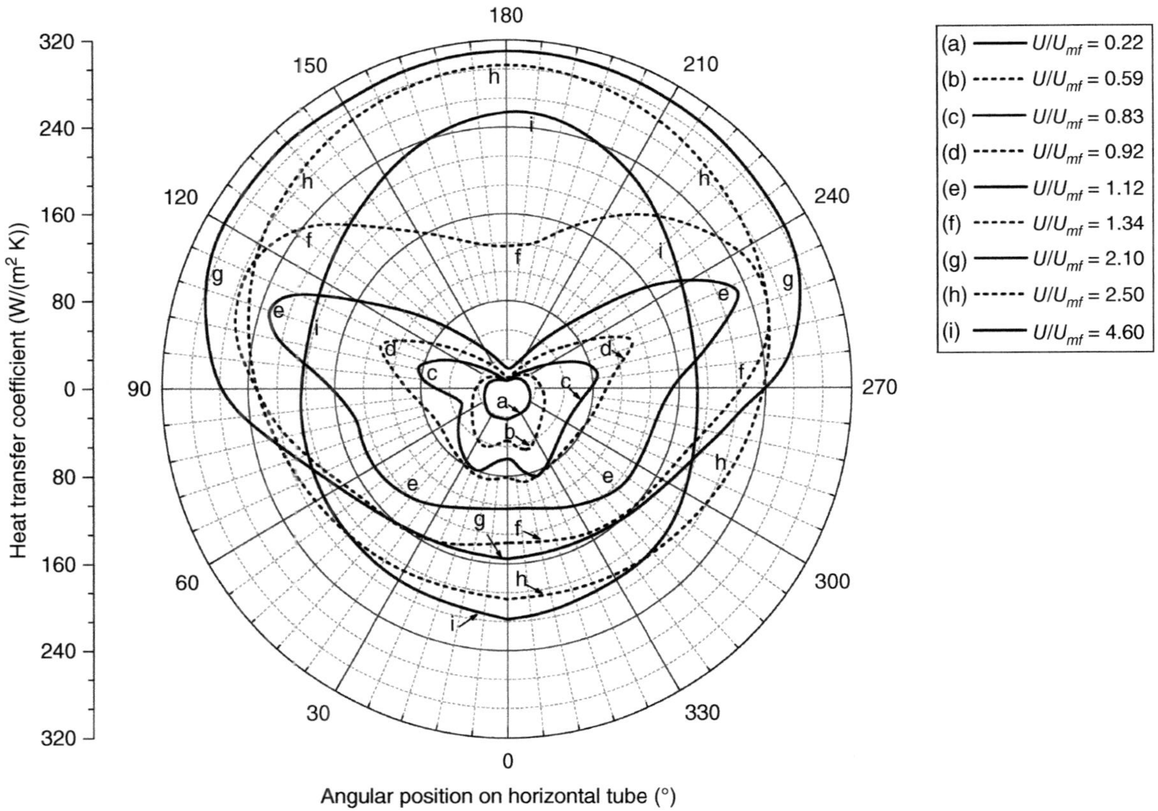

Figure 14.11 Polar plot of local heat transfer coefficient vs. angle around a horizontally installed heat transfer tube immersed in fluidized bed, $d_p = 600\ \mu m$. Source: Adapted from Noack 1970 [44].

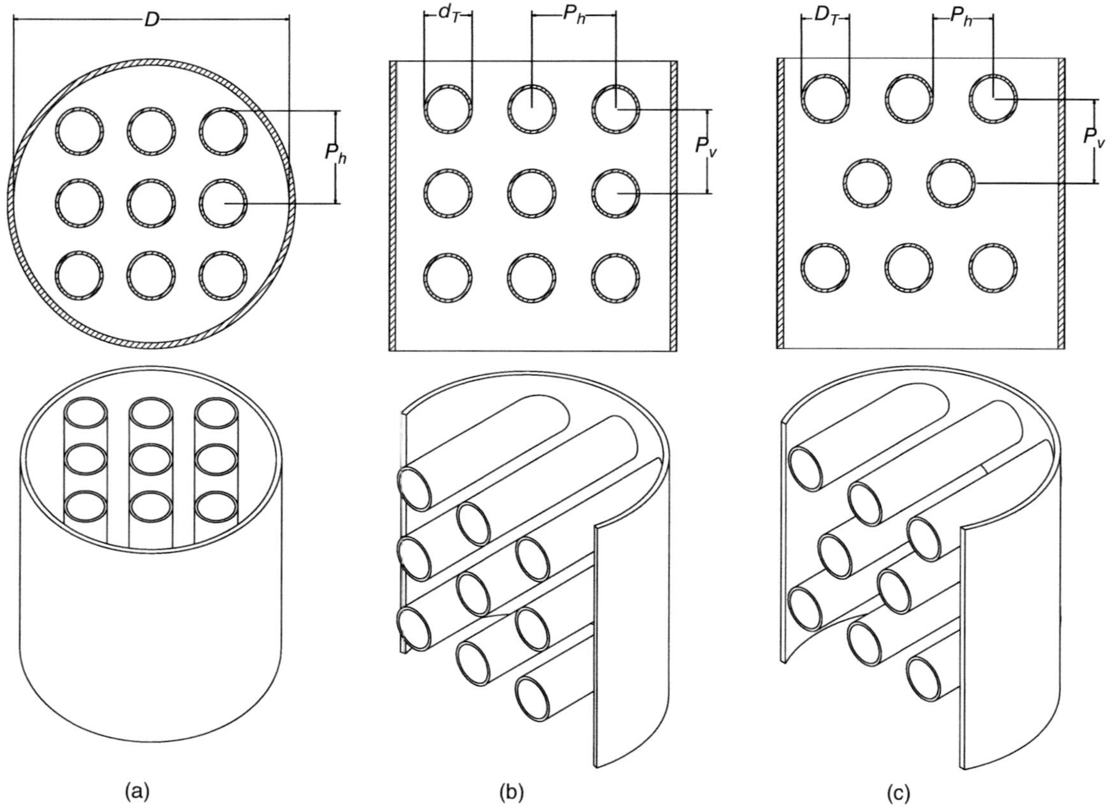

Figure 14.12 Common tube bundle configurations in fluidized beds and parameters, (a) vertical tube bundles, (b) horizontal tube bundles in rectangular pitch, and (c) horizontal tube bundles in triangular pitch.

bed, but the overall heat transfer coefficient was not greatly affected. They calculated the maximum heat transfer coefficient by

$$Nu_{max} = 0.005\,15Ar^{0.184} \cdot \left(\frac{\rho_p c_{pp}}{\rho_g c_{pg}}\right)^{0.6} \cdot \left(\frac{d_T}{d_{ref}}\right) \quad (10^2 < Ar < 10^5) \qquad (14.30)$$

where $d_{ref} = 0.09$ m is a reference diameter. Another well-known empirical (and early) correlation for heat transfer to vertical unfinned tubes is that of Vreedenberg [46], but the Molerus and Wirth [41, 43] and Martin [47, 48] mechanistic approaches are today widely regarded as more reliable, both for horizontal and vertical surfaces.

For off-axis single vertical tubes, a correction factor shown in Figure 14.13 should be applied.

As shown in Figure 14.12, horizontal heat transfer tubes in fluidized beds can be stacked with rectangular or triangular array. For the former case, heat transfer coefficients among tubes are primarily affected by the horizontal pitch, P_h, while, for the latter, both the horizontal pitch, P_h, and the vertical pitch, P_v, influence the heat transfer.

14.1.4.2 Martin's Correlations for Heat Transfer to Immersed Surfaces

Another popular predictive approach for heat transfer between a gas fluidized bed and immersed tubes, proposed by Martin [47, 48], applies to both horizontal and vertical tubes. In this model heat transfer is believed to consist of three

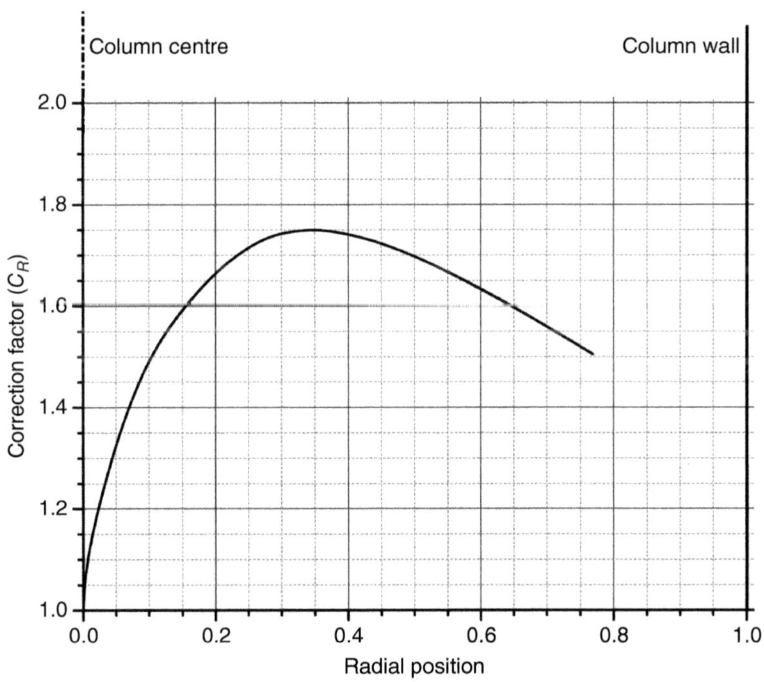

Figure 14.13 Dimensionless correction factor for heat transfer coefficients between bed and vertical tubes. Source: Adapted from Wender and Cooper 1958 [49].

components, corresponding to particle convection, gas convection, and radiation. The gas convection part can be obtained from Baskakov et al. [28], as shown in Eq. (14.21). Radiative heat transfer can be obtained from standard approaches, as addressed in Section 14.1.5.

A novel approach was proposed to predict the particle convective heat transfer. Single particle transfer at a heat transfer surface was said to be the rate-controlling mechanism, even for dense fluidized beds, with the understanding that the thickness of thermal boundary layers was of the same order of magnitude as the particle diameter. An analogy with the kinetic theory of gases was made to suggest a relationship for average random velocity of particles. Heat transfer in the interstitial gas was estimated from a relationship proposed by Schlunder [50]:

$$Nu = hd_p k_g = 4\left[(1 + Kn) \cdot \ln\left(1 + \frac{1}{Kn}\right) - 1\right] \tag{14.31}$$

where Nu is the Nusselt number for particle-to-surface heat transfer, h is the corresponding heat transfer coefficient, and Kn is the Knudsen number given by

$$Kn = \frac{4}{d_p}\left(\frac{2}{\gamma} - 1\right)\frac{k_g\sqrt{2\pi RT/M}}{p(2c_{pg} - R/M)} \tag{14.32}$$

in which M is the molecular weight and R is the universal gas constant. The accumulation coefficient, γ, for gas can be calculated from

$$\gamma = \left[1 + 10\frac{0.6B - 1 - 1000/T}{B}\right]^{-1} \tag{14.33}$$

in which

$$B = \frac{1000/298.15 - 1}{0.6 - \log_{10}\left[\frac{1}{\gamma_{25}} - 1\right]} \tag{14.34}$$

Values of γ for common gases at 298.15 K (25 °C) are listed in Table 14.2. $B = 2.80$ for air.

Table 14.2 Accumulation coefficients for common gases at room temperature.

Gas	γ_{25}	Gas	γ_{25}
H_2	0.200	Ar	0.876
He	0.235	CO_2	(0.90)
CH_4	(0.70)	Kr	0.933
NH_3	(0.90)	CCl_2F_2	(1.0)
Ne	0.573	Xe	0.953
Air	0.900		

Values in brackets ( ) are estimated values.

For most cases the thermal resistance within the particles can be ignored. Martin's particle convection component can then be calculated from

$$\frac{h_{pc}d_p}{k_g} = (1-\varepsilon)Z[1-\exp(-Nu/\kappa Z)] \tag{14.35}$$

where Z is a characteristic parameter for particle convection

$$Z = \frac{\rho_p c_{pp}}{6k_g}\sqrt{\frac{gd_p^3(\varepsilon-\varepsilon_{mf})}{5(1-\varepsilon_{mf})(1-\varepsilon)}} \tag{14.36}$$

and κ is a constant, which should range from 2 to 4. A value of $\kappa = 2.6$ was fitted based on the data of Wunder [42]. For the prediction of ε, Martin [47] suggested

$$\frac{\varepsilon-\varepsilon_{mf}}{1-\varepsilon_{mf}} = \frac{U-U_{mf}}{\overline{U}_b+(U-U_{mf})} \tag{14.37}$$

with the average bubble rise velocity, $\overline{U}_b$, given by

$$\overline{U}_b = 0.71\sqrt{gd_b} \tag{14.38}$$

Comparisons between Martin's model and some experimental data from the literature are shown below. Good agreement was found for h vs. U results at different temperatures (see Figure 14.14).

Despite Martin's model being able to offer reasonable estimation of heat transfer coefficients, it should be noted that this model tends to under-predict heat

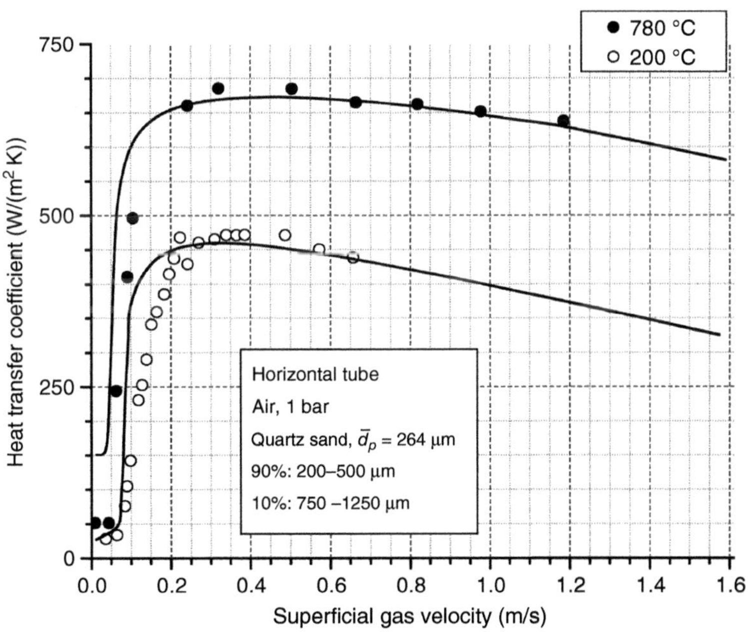

Figure 14.14 Comparison between predictions of Martin's model and heat transfer coefficient data from Janssen et al. 1974 [51]. Source: Adapted from Martin 1984 [47].

transfer for higher voidage systems. Parameters such as tube spacing, orientation, radial position, and tube diameter that can play significant roles are not reflected in this model.

14.1.4.3 Finned Tubes and Non-cylindrical Tubes

Extended heat transfer surfaces introduced by fins to bare tubes increases heat transfer rates [52–56], but the increases tend to level off as the fin height is increased. Genetti et al. [56] found that copper fins or spines gave higher heat transfer rates than steel ones. Additionally, in a typical case, the maximum heat transfer coefficient on a bare tube basis was increased to 2340 W/(m^2 K) compared with a base value of 256 W/(m^2 K) on a total area basis. The fins became much less effective if particles were unable to circulate freely between fins. Decrease in heat transfer performance was noticed for d_p greater than 1/10 of fin spacing. Given the absence of correlations for reliable predictions of heat transfer rates under different conditions, the best procedure, if pilot plant data are unavailable, is to use bare tube correlations and to apply correction factors to allow for the presence of fins, using the studies cited above to suggest the degree of augmentation likely for a given configuration and tube material. The performance of two types of finned tubes is compared with bare tubes in Table 14.3 and Figure 14.15.

It has also been demonstrated that higher average heat transfer coefficients can be obtained by flattening the sides of horizontal tubes so that a reduced

Table 14.3 Heat transfer performances between two different types of finned tubes and bare tube.

	Number of fins per m	L_{fin} (mm)	P_{fin} (mm)	d_i (mm)	d_o (mm)	Thermal performance (W/(m K))
Finned – Type I	98	2.7	7.5	25	51	45
Finned – Type II	280	1.5	2.1	25	51	132
Bare tube	0	N/A	N/A	25	51	19

Source: Data from Staub and Canada 1978 [53].

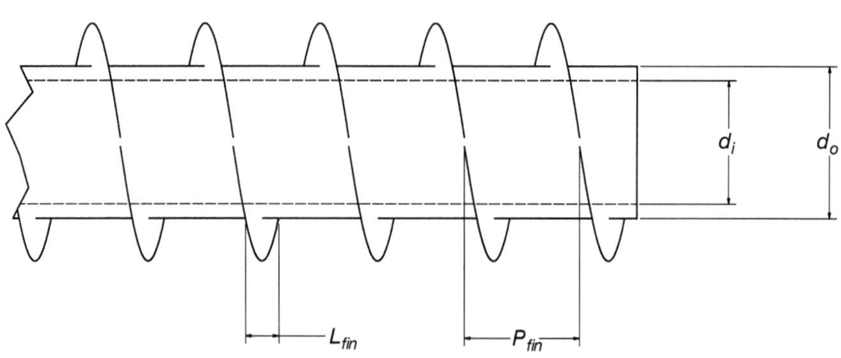

Figure 14.15 Key dimensions of a finned tube.

portion of the surface area faces upwards and downwards [57]. This improvement is attributable to the relatively low heat transfer at the top and bottom of a horizontal cylindrical tube, as portrayed in Figure 14.11.

14.1.4.4 Tubes in Freeboard Region

Horizontal tubes may also be placed in the freeboard region for waste heat recovery, quenching of reaction products, or because the bed depth varies. Heat transfer coefficients can be nearly as favourable in the splash zone as in the dense phase of the bed. However, the heat transfer coefficient is likely to decline sharply as the distance between a tube and the bed surface increases. George and Grace [58] correlated their own results and those of other workers by the equation:

$$h_{fbd} = \frac{h_\infty + (h_{dense} - h_\infty)}{\left[1 + 34\left(\frac{3.5z}{TDH}\right)\right]^n} \tag{14.39}$$

in which z is the vertical distance above the expanded bed surface, h_{dense} is the corresponding heat transfer coefficient in the bed, and h_∞ is the limiting value, which can be the heat transfer coefficient in the gas phase alone, or gas–solid suspension, depending on whether or not appreciable entrainment is present beyond the transport disengagement height (TDH). The index n was found to have a value between three and five for particles in 100–900 μm size range. A comparison between the correlation and available data is presented in Figure 14.16.

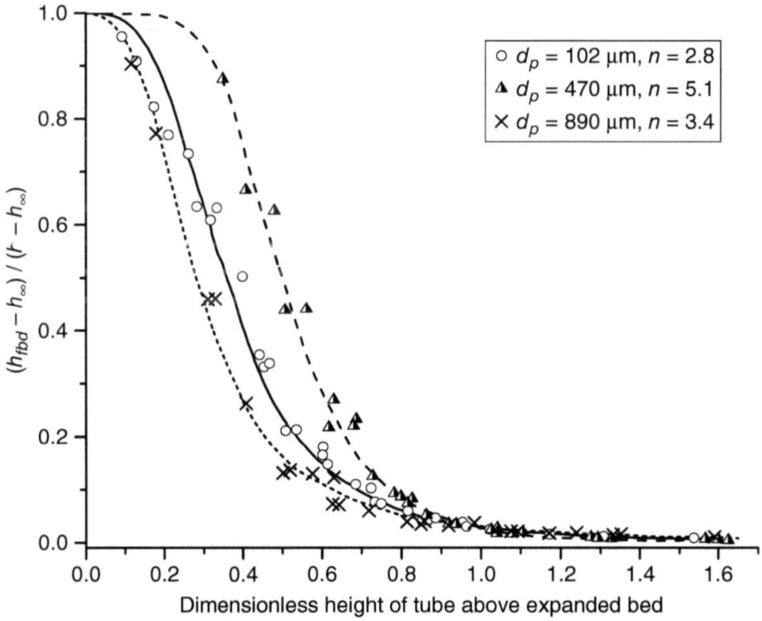

Figure 14.16 Comparison of predicted and experimental heat transfer coefficients of horizontal tubes in freeboard region, tube outer diameter 62 mm. Source: Adapted from George 1980 [59].

14.1.4.5 Methods of Augmenting Bed-to-Surface Heat Transfer

For most fluidized beds, heat transfer is rapid enough that no additional means are necessary to augment the heat transfer rate. In addition to using finned tubes and flattened tubes, the following approaches can also be implemented to increase heat transfer rates:

1. Vibrated fluidized beds (vibrofluidized beds) or vibrating surfaces.
2. Pulsation in the flow rate of fluidizing gas (pulsed fluidized beds).
3. Stirring paddles or other impellers.
4. Sonic or ultrasonic vibrations.
5. Alternating electric fields normal to a heat transfer surface.
6. Baffles or jets designed to enhance particle movement near a heat transfer surface.

Greater enhancements are easier to achieve in the vicinity of minimum fluidization velocities rather than higher velocities. Moreover, small particles ($d_p < 500$ μm) benefit more from such measures than large particles, for which the heat transfer is dominated by gas, rather than solids, movement.

14.1.5 Radiative Heat Transfer

As previously mentioned, the overall heat transfer coefficient in fluidized beds consists of particle convection, gas convection, and radiation. The contribution of radiation, h_{rad}, is only relevant for systems at elevated temperatures ($T_B > 500\,°\text{C}$). For larger particles radiative heat transfer is straightforward as their temperature does not very appreciably during exposure to a hot or cold surface. In this case radiation can be predicted from the net flux between two isothermal planes. For smaller particles, particle temperature changes during its contact with heat transfer surfaces. The renewal frequency of particles is also important. Radiation heat transfer is also complicated by the fact that particles adjacent to the heat transfer surface often possess different temperatures than the bulk of the bed. By treating both the wall and particle surface as grey bodies

$$h_{rad} = \frac{\sigma\varepsilon_r(T_W^4 - T_B^4)}{T_W - T_B} \tag{14.40}$$

in which σ is the Stefan–Boltzmann constant, $\sigma = 5.67 \times 10^8$ W/(m^2 K^4). ε_r is the effective emissivity, which is related to both the emissivity of the heat transfer surface ε_W, and the emissivity of the particle layer next to the surface ε_B

$$\varepsilon_r = \frac{1}{\frac{1}{\varepsilon_W} + \frac{1}{\varepsilon_B} - 1} \tag{14.41}$$

Since temperature of the boundary layer is often unavailable, the emissivity ε_B is difficult to obtain. Baskakov et al. [28] proposed a modified bed emissivity ε_m, to replace ε_B, which is a function of temperature of the bulk. The modified emissivity usually ranges from 0.6 to 0.7 and can be obtained from Figure 14.17. Alternatively, Botterill [60] suggested

$$\varepsilon_B \approx 0.5(1 + \varepsilon_S) \tag{14.42}$$

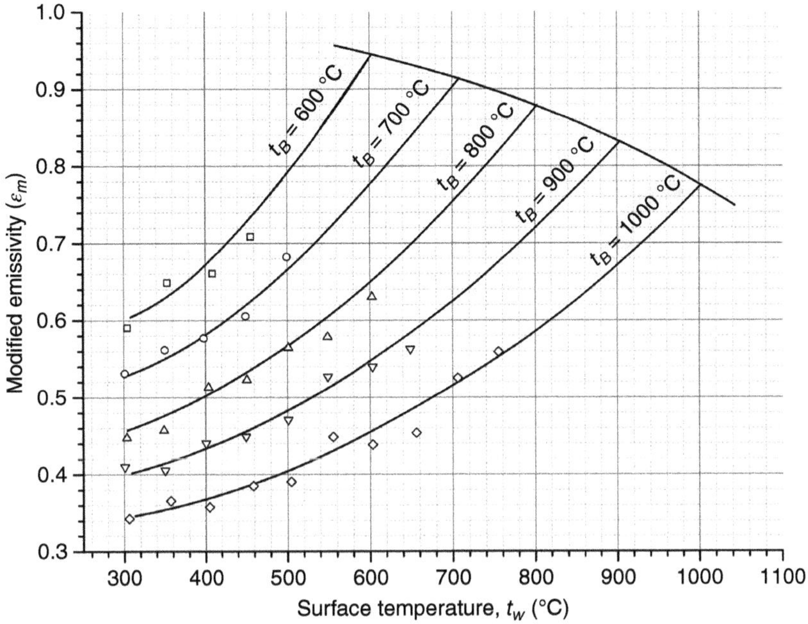

Figure 14.17 Modified emissivity vs. surface temperature. Source: Adapted from Baskakov et al. 1973 [28].

in which ε_S is the particle emissivity and ε_B is greater than ε_S due to multiple reflections. Ozkaynak et al. [61] summarized the contribution of radiative heat transfer as percentages of overall heat transfer, as listed in Table 14.4.

Note that the relative importance of radiative heat transfer is greater for larger particles and at higher bed temperatures, with roughly 15–30% of the transfer being due to radiation at a bed temperature of 800 °C [28]. Much lower ratios were found with catalyst particles of 180 μm [62].

Table 14.4 Radiation percentages of overall heat transfer.

t_W (°C)	t_B (°C)	d_p (μm)	q_{rad}/q_{total}	Methodology
15–251	779–980	≤1600	0.33–0.42	Cylindrical calorimeter
30	160–1050	340–1600	Negligible	Spherical calorimeter
100–200	430–1430	570–1750	0.18–0.5	Radiometer probe
40–100	315–377	100–150	Insignificant	Transparent wall
N/A (water-cooled)	500–1400	200–300	0.05–0.60	Radiometer probe
150–800	850	350–1250	0.06–0.32	Emissivity difference
30–150	500–1500	180	0.01–0.06	Emissivity difference
20–30	800–900	280	0.05–0.10	Transparent heater wall
30–200	500–800	1030	0.15–0.35	Radiometer probe

Source: Adapted from Ozkaynak et al. 1983 [61].

14.1.6 Heat Transfer in Fast and Circulating Fluidized Beds

When hold-up of bubbles or slugs reaches ~40–45% of the bed volume, fluidized beds enter the turbulent flow regime, where transient darting voids or intermittent slug-like structures are interspersed with periods resembling fast fluidization. In fast fluidization, which operates at even higher gas velocities than turbulent fluidization, a bed surface disappears with high particle carryover, e.g. 10–1000 kg/(m^2 s), at superficial gas velocities of typically 4–20 m/s for catalyst particles ($\rho_p \approx 1800\,\text{kg/m}^3$). In order to continuously operate fast fluidized beds, entrained particles are constantly captured and returned (see Chapter 12). A typical design of circulating fluidized bed (CFB) showing heat transfer surfaces is shown schematically in Figure 14.18. The cross-sectional averaged solids concentration usually decreases with increasing height, and the solids fractions at the bottom and the top of a fast fluidized bed are approximately 15% and 3%, respectively. The solids concentration varies significantly in the radial/lateral direction,

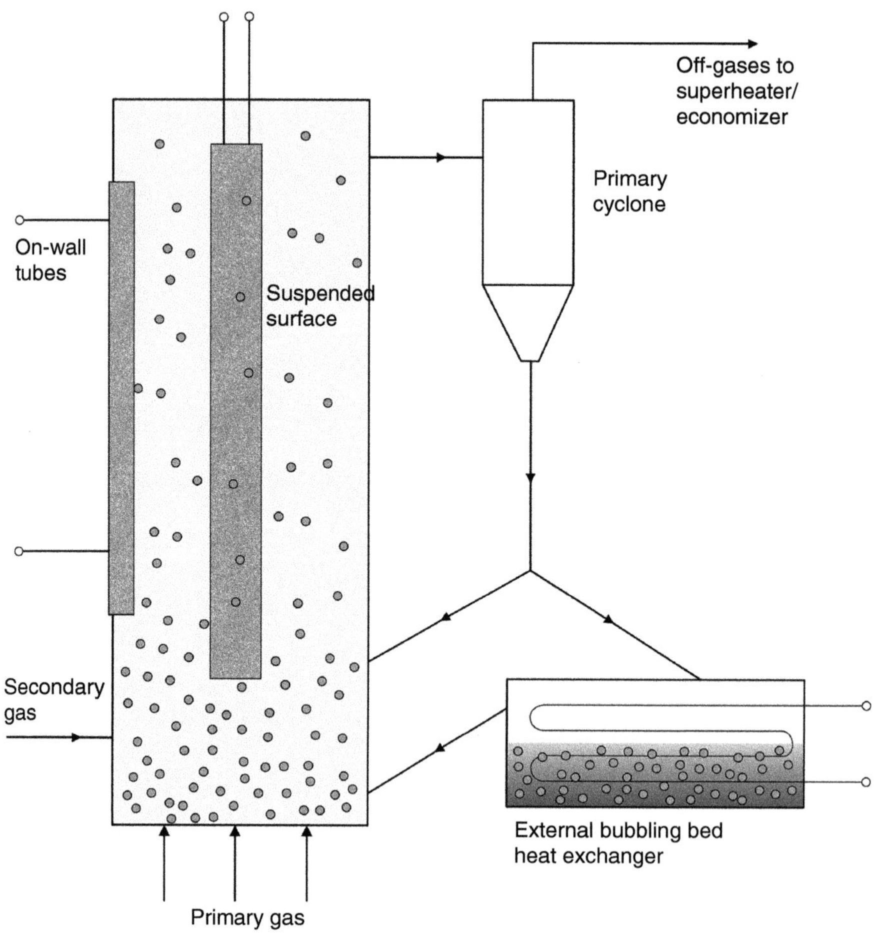

Figure 14.18 Schematic diagram of circulating fluidized bed with heat transfer surface.

as well as the vertical direction. Beaud and Louge [63] measured local solids concentrations in a CFB coal combustor and found that the volumetric solids concentration next to the wall was 30%, while in the centre it was only 1–3%. The complicated axial and radial solids distributions, as well as particle aggregation in fast fluidized beds, have made the prediction of heat transfer performance in this flow regime rather difficult. There has been a lack of suitable correlations for fast fluidized bed heat transfer.

The commonly implemented heat transfer tubes in bubbling fluidized beds are not often employed in fast fluidized beds due to the higher gas and particle velocities that could promote wear. Possible locations and formats of heat transfer surfaces in fast fluidized beds include the following:

(1) Surfaces forming part of the wall of the riser, e.g. for membrane wall.
(2) Surfaces inside the riser, e.g. wing walls.
(3) Surfaces of cyclones.
(4) An external, low-velocity (e.g. bubbling bed) fluidized bed heat exchanger providing additional surface area and higher heat transfer coefficient.
(5) Downstream surfaces, e.g. superheater or economizer tubes.

Heat transfer surfaces of types (i) and (ii) in a fast fluidized bed are shown in Figure 14.18. While such surfaces can take many shape and form, cooling surfaces employed in boiler applications or for other exothermic reactions are generally membrane water–wall surfaces, similar to those used in conventional boilers, consisting of steel tubes of 40–80 mm in diameter joined together by welding them to longitudinal fins or webs. Such a water–wall design is portrayed in Figure 14.19. Note that tubes are oriented vertically in membrane walls.

A number of investigators have measured heat transfer coefficients in fast fluidized beds [64–69]. Parameters that have been found to affect bed-to-surface heat transfer coefficient include superficial gas velocity, solids circulation rate,

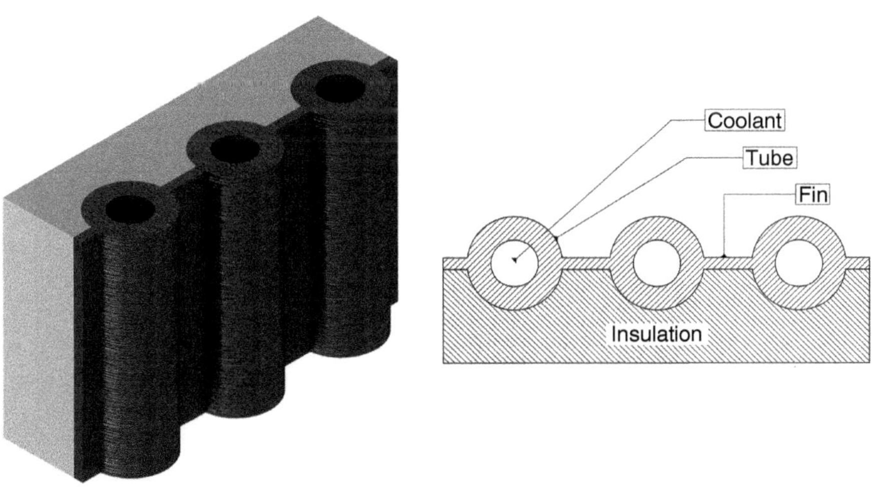

Figure 14.19 3D and sectional view of part of membrane water–wall assembly.

average particle diameter, axial location of the heat transfer surface along the height of the reactor, radial location of the heat transfer surface across the bed, and geometric dimension of the heat transfer surface. A few basic observations from previous studies include the following:

(1) Bed-to-surface heat transfer coefficient in fast fluidized beds, h_t , tends to be higher than gas convective heat transfer coefficient, but lower than the particle convective heat transfer coefficient in bubbling fluidized beds.
(2) h_t decreases with increasing particle diameter.
(3) h_t increases with increasing solids circulation rate.
(4) h_t decreases with increasing elevation along the bed.
(5) h_t is heavily influenced by the solids concentration. At lower elevations, h_t decreases with increasing superficial gas velocity; at higher elevations, h_t increases with increasing superficial gas velocity.
(6) h_t is often higher for smaller and shorter heat transfer surfaces; the effect of heater dimension is more prominent for particles smaller than 200 μm.
(7) h_t increases significantly with increasing temperature, especially for temperatures greater than 500 °C, mostly due to radiative heat transfer.

A more detailed discussion of fast fluidized bed heat transfer can be found at Grace [70].

14.2 Mass Transfer in Fluidized Beds

Mass transfer in fluidized beds usually refers to gas-to-particle mass transfer, or interphase mass transfer, where a component migrates from particles to fluids, or vice versa. Interphase mass transfer consists of two major components: (i) particle and fluid mass transfer in the dense phase and (ii) bubble- to dense-phase mass transfer. Bed-to-surface mass transfer is rarely considered, with an exception being in membrane fluidized bed reactors [71–77], in which case the best approach is to use the analogy between heat transfer and mass transfer to estimate the Sherwood number from equations like (14.29) and (14.39) by replacing the Nusselt number by a Sherwood number and Prandtl number by a Schmidt number.

14.2.1 Particle and Fluid Mass Transfer in the Dense Phase

It is noteworthy that there are two mass transfer coefficients of interest regarding interphase mass transfer, the single particle (or local) mass transfer coefficient $k_{d,p}$, and the overall mass transfer coefficient, $k_{d,\,bed}$.

For a single particle containing removable species A surrounded by gas free of A, the mass transfer process could be represented by

$$N_A = k_d A_p (C_{Af} - C_{Ap}) \tag{14.43}$$

in which C_{Ap} and C_{Af} represent concentration of A in the particle and in the fluid, respectively. Additional parameters for this bubbling bed approach are listed in Figure 14.20.

In a well-mixed bubbling fluidized bed, gas flow within the emulsion phase is rather small. It is reasonable to assume all gas phase enters the fluidized bed as bubbles. At steady-state, the overall mass transfer rate equals to the change of C_A with height in the bubble phase:

$$-\frac{dN_A}{dz} = a'' k_{d,bed}(C_{Af} - C_{Ab}) \tag{14.44}$$

in which a'' is the overall particle surface area in unit bed volume $(A \cdot dz)$.

In a fluidized bed with fine particles, gas phase travels through the dense phase in the form of bubbles. The mass transfer between gas and solid phase include mass transfer between bubbles and the dense phase and the mass transfer between the interstitial gas and the particles. Due to the added mass transfer resistance, the overall mass transfer coefficient is often smaller than that of a single particle. In fluidized bed with large particles, the interphase mass transfer resistance between bubbles and the dense phase is negligible. Consequently, $k_{d,\,bed} \approx k_{d,\,p}$.

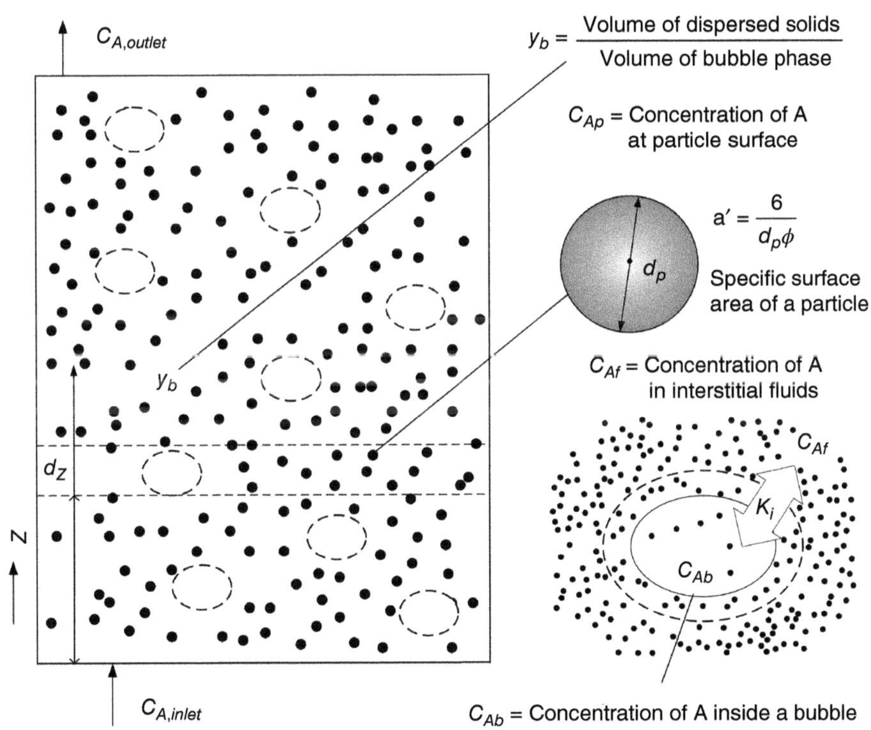

Figure 14.20 Schematic of the bubbling bed approach to mass transfer calculation in a fluidized bed. Source: Adapted from Kunii and Levenspiel 1991 [78].

14.2.2 Bubble to Dense-Phase Interphase Mass Transfer

In gas–solid fluidized beds operated at low gas velocities, the system could be treated as a packed bed, where the interphase mass transfer coefficient can be predicted as a packed bed with voidage of ε_{mf} and relative gas velocity of U_{mf}/ε_{mf}. The introduction of bubbles enhances interphase mass transfer, resulting from bubble interactions and bubble coalescence. The increase in mass transfer is especially significant immediately before the merging of two bubbles [79]. The influence of bubble interaction is more prominent for large particles than for smaller particles, which is consistent with the fact that for large particles the dominant mass transfer mechanism is gas convection. The rate-controlling step in bubbling bed can be the exchange of gas between voids and dense phase.

As previously mentioned, between the two major mass transfer mechanisms, convective mass transfer due to throughflow is dominant for Geldart B and D particles, where diffusion across particle boundary is dominant for Geldart A particles. Sit and Grace [79] studied the influence of bubbles on interphase mass transfer and proposed the most widely used expression for interphase mass transfer coefficient across the boundary. The interphase mass transfer coefficient in a free bubbling fluidized bed K_i is given as

$$K_i = \frac{U_{mf}}{3} + \sqrt{\frac{4D_A\varepsilon_{mf}\overline{U}_b}{\pi\overline{d}_b}} \tag{14.45}$$

where $\overline{U}_b$ and $\overline{d}_b$ are average bubble rise velocity and bubble diameter, respectively.

The interphase mass transfer rate can be estimated by

$$N_A = K_i a'(C_{Ad} - C_{Ab}) \tag{14.46}$$

Here a' is the specific surface area between bubbles and dense phase specified in Figure 14.20, C_{Ad} is the concentration in the dense phase, and C_{Ab} is the concentration in the bubbles.

Solved Problem

14.1 Bed-to-surface heat transfer coefficient in a fluidized bed with biomass particles

Biomass has become increasingly popular as a feedstock for fluidized beds. Figure 14P.1 depicts the experimental set-up for measuring the bed-to-surface heat transfer coefficient of Douglas-fir sawdust. A heat transfer probe is placed in the centre of the fluidized bed. Both ends of the probe are made of Teflon caps to reduce heat loss, whereas the middle is made of Ni–Cr heating wire wrapped by copper tube to evenly dissipate the heat. The radius of the probe is 7.9 mm, and the length is 50.8 mm. The centre of the probe is 20 mm above the distributor plate. While the probe is able to measure the heat transfer coefficient, it could also be predicted from correlations listed in this section. Fluidization medium is air, at

Figure 14P.1 Heat transfer measurement in a rectangular fluidized bed.

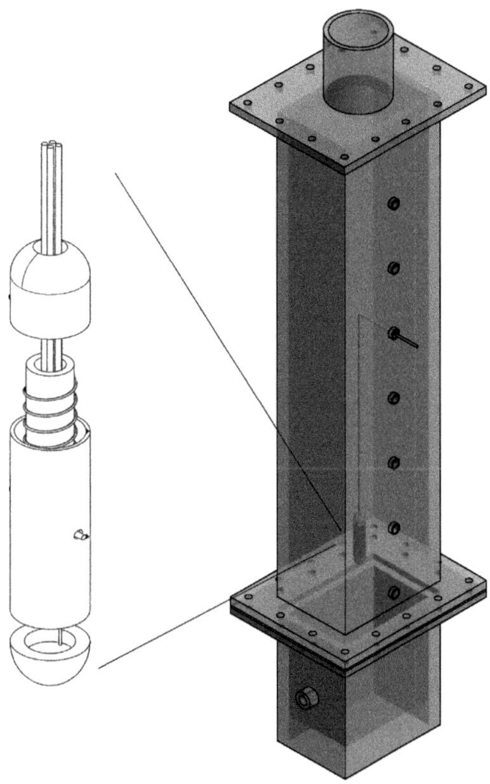

1 atm and 25 °C, and the superficial gas velocity is 0.271 m/s. Minimum fluidization velocity has been measured to be 0.238 m/s. The bed has a cross-sectional area of 0.015 m^2 and a hydraulic diameter of 0.12 m. Bulk density of the particle is 164 kg/m^3. The fluidized bed has an orifice plate distributor; the total number of orifices is 190. Calculate both the particle convective and gas convective heat transfer component and the overall bed-to-surface heat transfer coefficient of the system.

Solution

In a fluidized bed with multi-orifice distributor, the bubble size at bed height z can be estimated as follows:

$$d_{Bm} = 1.64[A(U - U_{mf})]^{0.4} = 0.078\ \text{m}$$

$$d_{B0} = \frac{1.38}{g^{0.2}}\left[\frac{A(U - U_{mf})}{N_{or}}\right]^{0.4} = 0.005\ \text{m}$$

$$d_B = d_{Bm} - (d_{Bm} - d_{B0})\exp\left(-\frac{0.3z}{D_t}\right) = 0.009\ \text{m}$$

$$U_B = 0.71\sqrt{gd_B} + (U - U_{mf}) = 0.241\ \text{m/s}$$

The overall length of the probe $L = 0.051$ m, while the diameter of the gas bubble at $z = 0.02$ m or 0.0045 m. Therefore, the average heat transfer coefficient between the dense phase and heating surface is

$$h_p = \int_0^{\tau} \frac{h_i}{\tau} dt = 2\sqrt{\frac{k_{mf}(1-\varepsilon_b)\rho_p(1-\varepsilon_{mf})c_{pp}U_b}{\min\{l, L\}}} = 115.42\ \mathrm{W/(m^2\,K)}$$

in which k_{eff} is calculated according to

$$k_{eff} = k_{es} + k_{ef} = 0.068\ \mathrm{W/(m\,K)}$$

Here

$$k_{es} = \left(\frac{k_p}{k_g}\right)^{0.28-0.757\log_{10}\varepsilon-0.057\log_{10}(k_p/k_g)} \cdot k_g = 0.068\ \mathrm{W/(m\,K)}$$

whereas

$$k_{ef} = 0.1\rho_g c_{pg} d_p U_{mf} = 8.2 \times 10^{-7}\mathrm{W/(m\,K)}$$

$$\varepsilon_b = \frac{U - U_{mf}}{U_b} = \frac{0.271 - 0.238}{0.241} = 0.14$$

Since the probe surface is alternatively covered by dense phase and bubbles, the particle convective heat transfer component is

$$h_{pc} = h_p(1-\varepsilon_b) = 115.42 \times (1 - 0.14) = 99.4\ \mathrm{W/(m^2\,K)}$$

For the gas convective heat transfer component, the overall formula is given in Eq. (14.18). The integral mean gas convective heat transfer coefficient over the entire length is

$$h_s = \sqrt{\frac{4k_{eff}\rho_g c_{pg} U}{\pi L_S}} + \frac{k_{eff}}{D_s} = 598.2\ \mathrm{W/(m^2\,K)}$$

And the contribution of a thin gas layer is calculated according to

$$h_w = 6k_g/d_p = 108.49\ \mathrm{W/(m^2\,K)}$$

Consequently, the gas convective heat transfer component can be obtained:

$$h_{gc} = \frac{1}{1/h_s + 1/h_w}\varepsilon_B = 12.71\ \mathrm{W/(m^2\,K)}$$

Finally, the overall heat transfer coefficient is

$$h = h_{gc} + h_{pc} = 112.15\ \mathrm{W/(m^2\,K)}$$

It can be seen that due to the poor thermal properties of biomass, the overall heat transfer coefficient from fluidized bed-to-heat transfer surfaces is rather poor. Attentions should be paid when dealing with such uncommon feedstocks.

Notations

Ar	Archimedes number (–)
A	cross-sectional area of fluidized bed (m^2)
a	particle surface area per unit volume (m^2/m^3)
a'	specific area of a particle (m^2/m^3)
a''	surface area of solids per unit bed height (m^2/m)
B	parameter in Martin's heat transfer coefficient (–)
C_{Ab}	concentration of A in the bubbles ($kmol/m^3$)
C_{Af}	concentration of A in bulk of interstitial fluid ($kmol/m^3$)
C_{Ap}	concentration of A at gas-particle interphase ($kmol/m^3$)
C_{pg}	heat capacity of gas phase (J/(kg K))
$C_{p,mf}$	effective heat capacity of dense phase (J/(kg K))
C_{pp}	heat capacity of the particles (J/(kg K))
C_R	correction factor (–)
C'	parameter in Eq. (14.33) (–)
C	parameter in Eq. (14.31) (–)
D_A	molecular diffusion coefficient of species A (m^2/s)
D	column diameter (m)
d_b	bubble diameter (m)
d_i	tube inner diameter (m)
d_o	tube outer diameter (m)
$\overline{d}_p$	average particle diameter (m)
d_p	particle diameter (m)
d_r	reference diameter (0.09 m)
d_T	tube outer diameter (m)
$\overline{f}_B$	average bubble fraction (–)
f_b	time fraction of a surface occupied by bubble (–)
f_p	time fraction of a surface occupied by dense phase (–)
g	gravitational constant (9.81 m/s^2)
h_{bd}	bed-to-particle (interphase) heat transfer coefficient ($W/(m^2\,K)$)
$\overline{h}_c$	average convective heat transfer coefficient ($W/(m^2\,K)$)
h_{dense}	dense-phase heat transfer coefficient ($W/(m^2\,K)$)
h_{fbd}	heat transfer coefficient in freeboard ($W/(m^2\,K)$)
h_f	heat transfer coefficient in thin gas film ($W/(m^2\,K)$)
h_i	instantaneous heat transfer coefficient ($W/(m^2\,K)$)
h_{gc}	gas convective heat transfer coefficient ($W/(m^2\,K)$)
$\overline{h}_{max}$	average maximum heat transfer coefficient ($W/(m^2\,K)$)
$\overline{h}_{pc}$	average particle convective heat transfer coefficient ($W/(m^2\,K)$)
h_{pc}	particle convective heat transfer coefficient ($W/(m^2\,K)$)
h_p	heat transfer coefficient between dense phase and heating surface ($W/(m^2\,K)$)
h_{rad}	radiative heat transfer coefficient ($W/(m^2\,K)$)
h_t	heat transfer coefficient in high-velocity fluidized beds ($W/(m^2\,K)$)
h_∞	limiting heat transfer coefficient ($W/(m^2\,K)$)
h	heat transfer coefficient ($W/(m^2\,K)$)

K_i	interchange coefficient between bubble-cloud and dense phase (s^{-1})
Kn	Knudsen number (–)
k_g	thermal conductivity of gas (W/(m K))
k_{mf}	effective thermal conductivity of dense phase (W/(m K))
k_d	mass transfer coefficient of well-dispersed single spheres (m/s)
k_t	effective thermal conductivity of packed beds with motionless fluids (W/(m K))
k_0	effective thermal conductivity of due to turbulent eddy conduction (W/(m K))
L_{fin}	length of the fin (mm)
L	length of heat transfer surface (m)
l_T	laminar length scale (m)
l	characteristic length of heater (m)
M	molecular weight (kg/mol)
N_A	molar flux of A (kmol/m^2 s)
Nu_{max}	maximum Nusselt number (–)
Nu	Nusselt number (–)
n'	exponent in Eq. (14.33) (–)
n	exponent in Eqs. (14.31) and (14.43) (–)
P_{fin}	pitch of the fin (mm)
P_h	horizontal pitch (m)
Pr	Prandtl number (–)
P_v	vertical pitch (m)
q_{rad}	heat flux due to radiation (kJ/(m^2 s))
q_{total}	total heat flux (kJ/(m^2 s))
Re_p	particle Reynolds number (–)
R	universal gas constant (8.314 J/(mol K))
TDH	transport disengagement height (m)
T_B	bed temperature (K)
T_g	gas-phase temperature (K)
T_{g0}	gas temperature at entrance (K)
T_p	temperature of particles (K)
T_w	wall temperature (K)
T	temperature (K)
t_B	bed temperature (°C)
t_r	residence time of packet (s)
t_w	wall temperature (°C)
t	time (s)
U_b	bubble rise velocity (m/s)
U_{mf}	minimum fluidization velocity (m/s)
U	superficial gas velocity (m/s)
y	mole fraction of inert component (–)
Z	characteristic groups for particle convection (–)
z	vertical coordinate (m)

Greek Letters

κ	constant in Martin's heat transfer model ($\kappa = 2$–4, –)
γ_{25}	accumulation constant at room temperature of 25 °C (–)
γ	accumulation constant (–)
ε_B	emissivity of the bed (–)
ε_b	fraction of bed volume occupied by bubbles or slugs (–)
ε_{mf}	voidage in a bed at minimum fluidization velocity (–)
ε_m	modified bed emissivity (–)
ε_r	effective diffusivity (–)
ε_s	emissivity of the particle (–)
ε_w	emissivity of the wall (–)
λ	thickness modifier of the thin gas film relative to particle diameter (-)
μ	gas-phase viscosity ($N\ s/m^2$)
$\pi_1 - \pi_5$	groups of parameters that influence bed-to-surface heat transfer (–)
ρ_g	density of the gas phase (kg/m^3)
ρ_{mf}	effective density of the dense phase (kg/m^3)
ρ_p	particle density (kg/m^3)
σ	Stefan–Boltzmann constant ($5.67 \times 10^8\ W/(m^2\ K^4)$)
τ_s	thermal time constant (s)
τ	residence time of packet on heater (s)
ϕ	sphericity (–)

References

1 Leva, M. (1959). *Fluidization*. New York: McGraw-Hill.

2 Zenz, F.A. and Othmer, D.F. (1960). *Fluidization and Fluid-Particle Systems*. New York: Reinhold.

3 Byam, J., Pillai, K., and Roberts, A. (1981). Heat transfer to cooling coils in the splash zone of a pressurized fluidized bed combustor. *AlChE Symp. Ser.*: 351–358.

4 Jia, D., Bi, X., Lim, C.J. et al. (2017). Heat transfer in a pulsed fluidized bed of biomass particles. *Ind. Eng. Chem. Res.* 56: 3740–3756.

5 Moreno, R.M., Antolín, G., and Reyes, A.E. (2016). Heat transfer during forest biomass particles drying in an agitated fluidised bed. *Biosystems Eng.* 151: 65–71.

6 Botterill, J., Redish, K., Ross, D., and Williams, J. (1962). The mechanism of heat transfer to fluidized beds. *Inst. Chem. Eng. Symp.*: 183–189.

7 Ziegler, E.N., Koppel, L.B., and Brazelton, W.T. (1964). Effects of solid thermal properties on heat transfer to gas fluidized beds. *Ind. Eng. Chem. Fundam.* 3: 324–328.

8 Couderc, J.P., Angelino, H., Enjalbert, M., and Guiglion, C. (1967). Echanges thérmiques en fluidisation gazeuse II—Etude mathématique à l'aide d'un modèle simple. *Chem. Eng. Sci.* 22: 99–107.

9 Botterill, J., Butt, M., Cain, G. et al. (1967). The effect of gas and solids thermal properties on the rate of heat transfer to gas-fluidized beds. *Proceedings of the International Symposium on Fluidization*, Netherlands University Press, Amsterdam, pp. 442–457.
10 Brusenback, R.A. (1963). A microscopic analysis of fluidized-bed heat transfer. Ph.D. thesis. Northwestern University, Ann Arbor, p. 156.
11 Gabor, J. (1970). Wall-to-bed heat transfer in fluidized and packed beds, Chem. *Eng. Prog. Symp. Ser.*: 76–86.
12 Mickley, H.S. and Fairbanks, D.F. (1955). Mechanism of heat transfer to fluidized beds. *AlChE J.* 1: 374–384.
13 Ozkaynak, T.F. and Chen, J.C. (1980). Emulsion phase residence time and its use in heat transfer models in fluidized beds. *AlChE J.* 26: 544–550.
14 Chandran, R. and Chen, J.C. (1982). Bed-surface contact dynamics for horizontal tubes in fluidized beds. *AlChE J.* 28: 907–914.
15 Mickley, H. (1961). The relation between the transfer coefficient and thermal fluctuations in fluidized-bed heat transfer. *Chem. Eng. Prog. Symp. Ser.*: 51–60.
16 Baeyens, J. and Goossens, W.R.A. (1973). Some aspects of heat transfer between a vertical wall and a gas fluidized bed. *Powder Technol.* 8: 91–96.
17 Kubie, J. (1976). Bubble induced heat transfer in gas fluidized beds. *Int. J. Heat Mass Transfer* 19: 1441–1453.
18 Bernis, A., Vergnes, F., and Le Goff, P. (1977). Influence du passage d'une bulle sur le coefficient instantané de transfert de chaleur à une paroi immergée dans un lit fluidisé. *Powder Technol.* 18: 267–276.
19 Selzer, V. and Thomson, W. (1977). Fluidized bed heat transfer: the packet theory revisited. *AlChE Symp. Ser.* 73: 29–37.
20 Decker, N. and Glicksman, L.R. (1983). Heat transfer in large particle fluidized beds. *Int. J. Heat Mass Transfer* 26: 1307–1320.
21 Botterill, J.S.M., Redish, K.A., Ross, D.K., and Williams, J.R. (1962). The mechanism of heat transfer to fluidised beds. In: *Proceedings of the Symposium on the Interaction between Fluids and Particles* (ed. P.A. Rottenburg), 183. London: Institution of Chemical Engineers.
22 Darton, R. (1977). Bubble growth due to coalescence in fluidized bed. *Trans. Inst. Chem. Eng.* 55: 274–280.
23 Baskakov, A. (1964). The mechanism of heat transfer between a fluidized bed and a surface. *Int. Chem. Eng.* 4: 320–324.
24 Deissler, R. and Boegli, J. (1958). An investigation of effective thermal conductivities of powders in various gases. *ASME Trans.* 8: 1417–1425.
25 Dietz, P.W. (1979). Effective thermal conductivity of packed beds. *Ind. Eng. Chem. Fundam.* 18: 283–286.
26 Ranz, W. (1952). Friction and transfer coefficients for single particles and packed beds. *Chem. Eng. Prog.* 48: 247–253.
27 Gabor, J.D. (1970). Heat transfer to particle beds with gas flows less than or equal to that required for incipient fluidization. *Chem. Eng. Sci.* 25: 979–984.
28 Baskakov, A.P., Berg, B.V., Vitt, O.K. et al. (1973). Heat transfer to objects immersed in fluidized beds. *Powder Technol.* 8: 273–282.

29 Zabrodsky, S.S., Epanov, Y.G., Galershtein, D.M. et al. (1981). Heat transfer in a large-particle fluidized bed with immersed in-line and staggered bundles of horizontal smooth tubes. *Int. J. Heat Mass Transfer* 24: 571–579.
30 Xavier, A.M. (1977). Heat transfer between a fluidised bed and a surface. Ph.D. thesis. University of Cambridge (United Kingdom), Ann Arbor, p. 1.
31 Decker, N. and Glicksman, L. (1981). Conduction heat transfer at the surface of bodies immersed in gas fluidized beds of spherical particles. *AIChE Symp. Ser.*: 341–349.
32 Catipovic, N.M. (1979). Heat transfer to horizontal tubes in fluidized beds: experiment and theory. Ph.D. thesis. Chemical Engineering, Oregon State University, Corvallis, OR, p. 194.
33 Baskakov, A., Vitt, O., Kirakosyan, V. et al. (1974). Investigation of heat transfer coefficient pulsations and of the mechanism of heat transfer from a surface immersed into a fluidized bed. *Proceedings of International Symposium on Fluid Application*, Capadeus, Toulouse, p. 293.
34 Denloye, A.O.O. and Botterill, J.S.M. (1978). Bed to surface heat transfer in a fluidized bed of large particles. *Powder Technol.* 19: 197–203.
35 Zabrodsky, S.S. (1966). *Hydrodynamics and Heat Transfer in Fluidized Beds.* Cambridge, MA: Massachusetts Institute of Technology Press.
36 Varygin, N. and Martyushin, I. (1959). A calculation of heat transfer surface area in fluidized bed equipment. *Khimcheskaya Mashinostr (Moscow)* 5: 6–9.
37 Seville, J.P.K., Willett, C.D., and Knight, P.C. (2000). Interparticle forces in fluidisation: a review. *Powder Technol.* 113: 261–268.
38 Maskaev, V.K. and Baskakov, A.P. (1973). Characteristics of external heat transfer in a fluidization bed of coarse particles. *J. Eng. Phys.* 24: 411–414.
39 Borodulya, V.A., Ganzha, V.G., and Podberezsky, A.I. (1980). Heat transfer in a fluidized bed at high pressure. In: *Fluidization* (eds. J.R. Grace and J.M. Matsen), 201–207. Boston, MA: Springer US.
40 Saxena, S.C. (1989). Heat transfer between immersed surfaces and gas-fluidized beds. In: *Advances in Heat Transfer* (eds. J.P. Hartnett and T.F. Irvine), 97–190.
41 Molerus, O. and Wirth, K.-E. (1997). *Heat Transfer in Fluidized Beds*, 35–47. Dordrecht: Springer Netherlands.
42 Wunder, R. (1980). Wärmeübergang an vertikalen Wärmeaustauscherflächen in Gas-swirbelschichten. Ph.D. thesis. Technische Universität München.
43 Molerus, O. and Wirth, K.-E. (1997). *Heat Transfer in Fluidized Beds*, 1e. Dordrecht: Chapman & Hall.
44 Noack, R. (1970). Lokaler Wärmeübergang an horizontalen Rohren in Wirbelschichten (Local heat transfer at horizontal tubes in fluidized beds). *Chem. Ing. Tech.* 42: 371–376.
45 Gunn, D.J. and Hilal, N. (1996). Heat transfer from vertical inserts in gas-fluidized beds. *Int. J. Heat Mass Transfer* 39: 3357–3365.
46 Vreedenberg, H.A. (1959). Heat transfer between a fluidized bed and a vertical tube. *Chem. Eng. Sci.* 11: 274–285.
47 Martin, H. (1984). Heat transfer between gas fluidized beds of solid particles and the surfaces of immersed heat exchanger elements, Part II. *Chem. Eng. Process. Process Intensification* 18: 199–223.

48 Martin, H. (1984). Heat transfer between gas fluidized beds of solid particles and the surfaces of immersed heat exchanger elements, part I. *Chem. Eng. Process. Process Intensification* 18: 157–169.

49 Wender, L. and Cooper, G.T. (1958). Heat transfer between fluidized-solids beds and boundary surfaces—correlation of data. *AlChE J.* 4: 15–23.

50 Schlunder, E.U. (1971). Heat transfer in moving packings of spherical particles in contact for short periods. *Chem. Ing. Tech.* 43: 651.

51 Janssen, K., Hd, S., and Peters, W. (1974). Calculation of heat-transfer coefficients between fluidized-bed and submerging heat-exchange surfaces therein depending on flow conditions of nonhomogeneous fluidization. *Chem. Ing. Tech.* 46: 42–43.

52 Bartel, W.J. (1971). Heat transfer from a horizontal bundle of tubes in an air fluidized bed. Ph.D. thesis. Mechanical Engineering. Montana State University, Ann Arbor, p. 177.

53 Staub, F. and Canada, G. (1978). Effect of tube bank and gas density on flow behavior and heat transfer in fluidized beds. In: *Fluidization* (eds. J.F. Davidson and D.L. Keairns), 339–344. London: Cambridge University Press.

54 Zabrodsky, S.S., Tamarin, A.I., Dolidovich, A.F. et al. (1980). Heat transfer of single horizontal finned tubes and their bundles in a fluidized bed of large particles. In: *Fluidization* (eds. E.J.R. Grace and J.M. Matsen), 195–200. Boston, MA: Springer US.

55 Bartel, W. and Genetti, W. (1973). Heat transfer from a horizontal bundle of bare and finned tubes in an air fluidized bed. *AlChE Symp. Ser.*: 85–93.

56 Genetti, W., Schmall, R., and Grimmett, E. (1971). The effect of tube orientation on heat transfer with bare and finned tubes in a fluidized bed. *AlChE Symp. Ser.*: 90–96.

57 Andeen, B.R., Glicksman, L.R., and Bowman, R. (1978). Heat transfer from flattened horizontal tubes. In: *Fluidization* (eds. J.F. Davidson and D.L. Keairns), 345. Cambridge: Cambridge University Press.

58 George, S.E. and Grace, J.R. (1982). Heat transfer to horizontal tubes in the freedboard region of a gas fluidized bed. *AlChE J.* 28: 759–765.

59 George, S.E. (1980). Heat transfer to tubes in the freeboard region of a fluidized bed. Ph.D. thesis. McGill University (Canada), Ann Arbor.

60 Botterill, J.S.M. (1975). *Fluid-Bed Heat Transfer. Gas-Fluidized Bed Behaviour and its Influence on Bed Thermal Properties*. New York: Academic Press.

61 Ozkaynak, T., Chen, J., and Frankenfield, T. (1983). An experimental investigation of radiation heat transfer in a high temperature fluidized bed. In: *Fluidization* (eds. D. Kunii and R. Toei), 371–378. New York: Engineering Foundation.

62 Yoshida, K., Ueno, T., and Kunii, D. (1974). Mechanism of bed-wall heat transfer in a fluidized bed at high temperatures. *Chem. Eng. Sci.* 29: 77–82.

63 Beaud, F. and Louge, M. (1997). Similarity of radial profiles of solid volume fraction in a circulating fluidized bed. *Fuel Energy Abstr.* 38: 43.

64 Kiang, K., Liu, K., Nack, H., and Oxley, J. (1976). Heat transfer in fast fluidized beds. In: *Fluidization Technology*, 2e (eds. D.L. Keairns and J.F. Davidson), 471–483.

65 Dou, S., Herb, B., Tuzla, K., and Chen, J. (1992). Dynamic variation of solid concentration and heat transfer coefficient at wall of circulating fluidized bed. In: *Fluidization VII* (eds. O.E. Potter and D.J. Nicklin), 793–801. New York: Engineering Foundation.
66 Dou, S., Herb, B., Tuzla, K., and Chen, J.C. (1991). Heat transfer coefficients for tubes submerged in circulating fluidized bed. *Exp. Heat Trans.* 4: 343–353.
67 Ebert, T.A., Glicksman, L.R., and Lints, M. (1993). Determination of particle and gas convective heat transfer components in a circulating fluidized bed. *Chem. Eng. Sci.* 48: 2179–2188.
68 Wu, R.L., Grace, J.R., Lim, C.J., and Brereton, C.M.H. (1989). Suspension-to-Surface heat transfer in a circulating-fluidized-bed combustor. *AIChE J.* 35: 1685–1691.
69 Furchi, J., Goldstein, L. Jr., Lombardi, G., and Mohseni, M. (1988). Experimental local heat transfer in a circulating fluidized bed. In: *Circulating Fluidized Bed Technology* (eds. P. Basu and J.F. Large), 263–270. Compiègne, France: Elsevier.
70 Grace, J.R. (1990). Heat transfer in high velocity fluidized beds. *Proceedings of 9th International Heat Transfer Conference*, Jerusalem, Israel. pp. 329–339.
71 Chen, Y., Mahecha-Botero, A., Lim, C.J. et al. (2014). Hydrogen production in a sorption-enhanced fluidized-bed membrane reactor: operating parameter investigation. *Ind. Eng. Chem. Res.* 53: 6230–6242.
72 Rakib, M.A., Grace, J.R., Lim, C.J., and Elnashaie, S.S.E.H. (2010). Steam reforming of heptane in a fluidized bed membrane reactor. *J. Power Sources* 195: 5749–5760.
73 Grace, J., Elnashaie, S., and Lim, C.J. (2005). Hydrogen production in fluidized beds with in-situ membranes. *Int. J. Chem. Reactor Eng.* 3: 18.
74 Golriz, M.R. and Grace, J.R. (2002). Augmentation of heat transfer by deflectors on circulating fluidized bed membrane walls. *Int. J. Heat Mass Transfer* 45: 1149–1154.
75 Adris, A.M., Lim, C.J., and Grace, J.R. (1997). The fluidized-bed membrane reactor for steam methane reforming: Model verification and parametric study. *Chem. Eng. Sci.* 52: 1609–1622.
76 Zhou, J., Grace, J.R., Brereton, C.M.H., and Lim, C.J. (1996). Influence of membrane walls on particle dynamics in a circulating fluidized bed. *AIChE J.* 42: 3550–3553.
77 Deshmukh, S.A.R.K., Heinrich, S., Mörl, L. et al. (2007). Membrane assisted fluidized bed reactors: potentials and hurdles. *Chem. Eng. Sci.* 62: 416–436.
78 Kunii, D. and Levenspiel, O. (1991). *Fluidization Engineering*, 2e. Stoneham, MA: Butterworth-Heinemann.
79 Sit, S.P. and Grace, J.R. (1981). Effect of bubble interaction on interphase mass transfer in gas fluidized beds. *Chem. Eng. Sci.* 36: 327–335.

Problem

14.1 A bubbling fluidized bed heat exchanger is shown in Figure 14P.2 below. The fluidized bed contains hot sand particles, with an average temperature

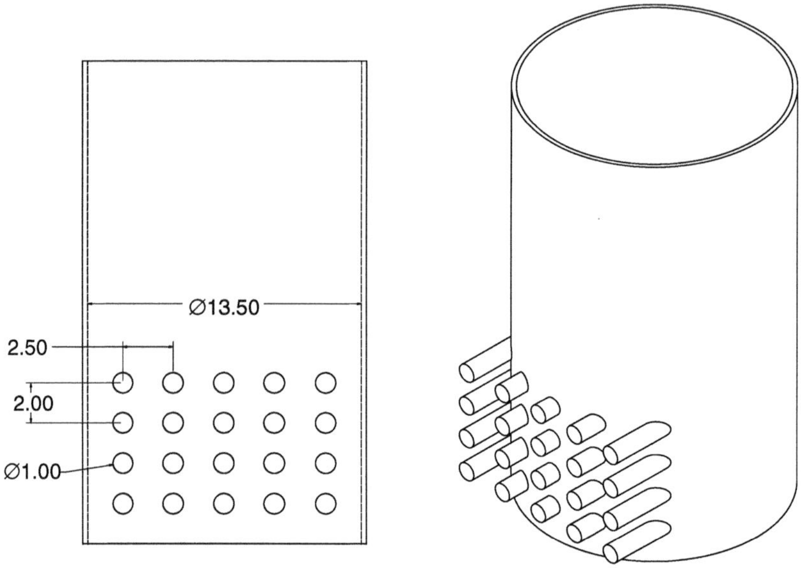

Figure 14P.2 Dimensions of the fluidized bed particle heat exchanger.

Table 14P.1 Particle sieving analysis result.

Size (μm)	% Retained	% Cumulative
850	0	100
600	0.7	100
425	60.7	99.3
300	34.5	38.6
212	3.8	4.1
150	0.3	0.3
106	0	0
75	0	0
53	0	0

of approximately 600 °C. In order to cool down the sand particles, a tube bank is installed in the dense phase of the fluidized bed. The tubes are running cooling water, with an average temperature of 50 °C. The density of the sand is 2650 kg/m^3, with sieving analysis results listed in Table 14P.1.

The dimensions of the tube, horizontal and vertical distances are shown in Figure 14P.2. Please calculate the following,

(1) Calculate the Sauter mean diameter of the sand sample
(2) Estimate the minimum fluidization velocity of this sand sample in air, under standard conditions.

(3) Design a perforated plate distributor that could evenly and effectively fluidize the system under above conditions
(4) Calculate the bed-to-surface heat transfer coefficient
(5) If the temperature of sand particles increases to 900 °C, calculate the influence of radiative heat transfer.

15

Catalytic Fluidized Bed Reactors

Andrés Mahecha-Botero [1,2]

[1] *NORAM Engineering and Constructors Ltd., 200 Granville Street, Suite 1800, Vancouver, Canada V6C 1S4*
[2] *University of British Columbia, Department of Chemical and Biological Engineering, Vancouver, Canada V6T 1Z3*

15.1 Introduction

Decisions associated with the design of chemical reactors determine the environmental performance, productivity, and economics of many industrial plants. Issues experienced by the reactor have a large impact on downstream unit operations. Improvements in reactor performance can result in major dividends to industry. Fluidized bed reactors play a major role in a variety of industries, and, as discussed in Chapter 1, their utilization has a number of advantages that make them attractive for large-scale operation. Although many process industries still utilize packed bed reactors due to their simplicity and easier scale-up, there is increasing interest in fluidized beds for many industrial applications.

The design of fluidized bed reactors requites multidisciplinary elements from chemical process engineering (modelling, mass and energy balances, piping and instrumentation diagrams, etc.), chemistry (catalysis, chemical kinetics, stoichiometry, equilibrium), physics (hydrodynamics, mass and heat transfer, flow regimes), simulation (differential equations, numerical techniques, programming, advanced software), and collaboration (with mechanical engineering, materials science, process control, etc.).

Fluidized bed reactors play an important role in the evolution of chemical industries toward improved sustainability since they are utilized in key processes associated with challenges faced by mankind. This chapter focuses on fluidized bed reactors and provides tools, techniques and strategies for design, technology development, and project execution.

Essentials of Fluidization Technology, First Edition. Edited by John R. Grace, Xiaotao Bi, and Naoko Ellis.

15.2 Reactor Design Considerations

15.2.1 Suitability of Fluidized Beds for Catalytic Processes

Fluidized beds are not the best choice for all processes. The selection of catalytic reactor systems is a multidisciplinary effort that requires balancing process performance, energy efficiency, and plant emissions with capital and operating costs. In the preliminary stages of design (e.g. pre-feasibility and feasibility studies), a number of alternatives and operating conditions are commonly explored. Figure 15.1 summarizes some requirements of fluidized bed catalytic reactors. A design must pass multiple stages of review to be fully implemented. Key review stages include a check of emissions and efficiency, process and mechanical design reviews, study of operating conditions, review of materials of construction, hazards and operability studies (HAZOP), and analysis of the project economics. These reviews are based on available technical information, previous experience, and reactor modelling.

Table 15.1 summarizes some requirements of fluidized bed catalytic reactors.

15.2.2 Reactor Types by Flow Regime and Phase

The great majority of applications for fluidized bed catalytic reactors are for gas–solid systems. Typical operating parameters of catalytic and non-catalytic gas–solid fluidized bed reactors are provided in Table 15.2.

15.3 Reactor Modelling

The ultimate objective of reactor modelling is to understand what occurs inside reactors and to predict their performance without the need for additional

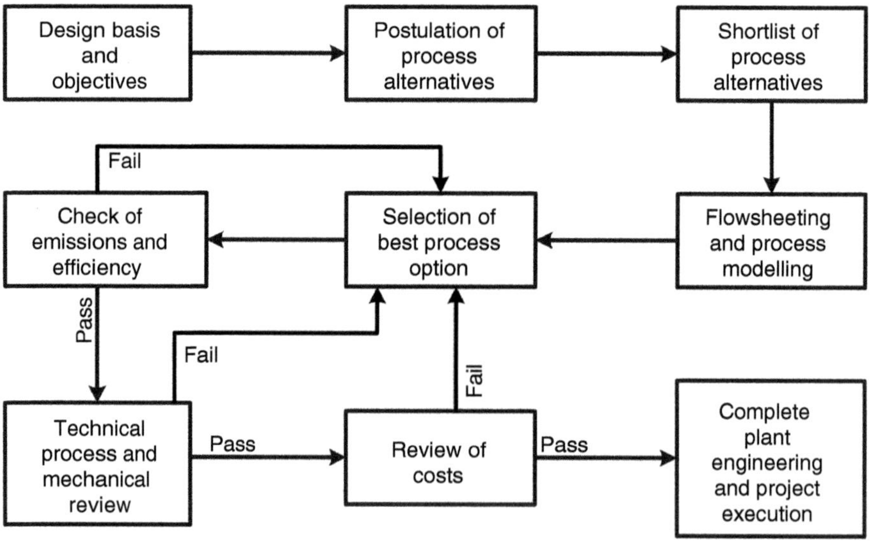

Figure 15.1 Design workflow for selection of process alternatives and reactor configurations.

Table 15.1 Some requirements of fluidized bed catalytic reactors.

Process consideration	Requirement for fluidized bed configuration	Preferred process features
Catalyst mechanical strength	Catalyst must be mechanically robust	Catalyst attrition rates and losses should be low
Catalyst fluidization properties	Mean particle diameters should not be less than ~50 μm. Minimum gap width inside the reactor and internals should never be less than 20–30 times mean particle diameter	Good fluidization properties of catalyst particles Smooth rounded particle shapes Broad size distribution preferred (e.g. 0–200 μm) Minimum gap width inside the reactor and internals >50 times mean particle diameter
Catalyst stability	Catalyst is stable under fluidization conditions. Catalyst attrition rates are low	Low cost to replenish catalyst losses
Catalyst activity	Catalyst activity maintained or regenerated continuously	Processes not well suited to packed beds since catalyst requires continuous regeneration
Fouling considerations	A fluidized bed configuration can be used to manage fouling issues	Processes not well suited to packed beds due to fouling (inability to handle carbon and tar formation and regeneration)
Erosion	Equipment wear manageable under required process conditions	Low wear rates
Heat balance considerations	N/A	Highly endothermic or exothermic processes that benefit from fluidization to maintain a homogeneous temperature profile and effective input or removal of heat
Mass transfer limitations	N/A	Use of fine particles allows for increased catalyst effectiveness factors, higher reaction rates, and lower catalyst volume
Process intensification	N/A	Fluidized beds are preferred when they allow process intensification and combination of unit operations, such as reaction, separation of species, heating, and cooling
Can the process be carried out in a packed bed?	N/A	If it is difficult to carry out the process in a packed bed, the fluidized bed reactor becomes more attractive. Examples include processes where the catalyst activity decreases rapidly, requiring continuous catalyst regeneration
Reaction rate	N/A	Fast reactions benefit from fluidized bed contacting

Table 15.2 Comparison of typical fluidized bed catalytic reactors and gas–solid reactors.

	Catalytic reactors (this chapter)	Gas–solid reactors (Chapter 16)
Solids	Catalytic	Non-catalytic. Solids are consumed or modified by the reactions
Mean, d_p	50–100 μm	150 μm–3 mm Usually >250 μm
Particle size distribution	Broad distribution preferred (e.g. 0–200 μm)	Not too narrow preferred
Column diameter	Up to ~7 m	Up to ~12 m
Pressure	Up to ~80 bar	Up to ~20 bar
Temperature	Up to ~600 °C	Up to ~1000 °C
Superficial gas velocity for bubbling and turbulent regime	~0.3 to ~0.8 m/s	Up to 2 m/s
Superficial gas velocity for fast fluidization or dense suspension upflow	3–12 m/s	Up to 7 m/s
Static bed depth	~1–10 m	~0.3–5 m
In-bed surfaces for heat transfer	Horizontal or vertical tubes	Horizontal tubes
Gas–solid separation	Cyclones	Cyclones, filters, dry electrostatic precipitators (ESPs), wet scrubbers
Bed voidage (bubbling and turbulent regime)	~0.7 to ~0.8	~0.65 to ~0.75
Bed voidage (fast fluidization regime)	~0.8 to ~0.95	~0.9 to ~0.98
Geldart particle group	*A* preferred	*B*, *D*, or *A*
Feeding solids	Some make-up required to replace catalyst losses	Requires continuous feeding and continuous or periodic removal

experimentation or physical testing. In a perfect world, a comprehensive reactor model would correctly map in space and time the local concentrations of solids and all chemical species inside the reactor, as well as the temperature and pressure, to predict the reactor profiles for C_i, T, and P as a function of the spatial coordinates and time. However, it is extremely difficult to create such a model due to the complexity of fluidized bed reactors. In practice reactor modelling requires simplifications and assumptions to produce meaningful results.

Reactor modelling requires the application of conservation principles of mass and energy and quantification of the interactions of transport phenomena and reaction kinetics to relate reactor performance to operating conditions and feed variables. Conservation principles may be applied to the entire reactor, or to one or more differential element within it, requiring understanding of the microscale (molecular level), mesoscale (transport phenomena, reaction kinetics), and macroscale (process engineering, plant design).

Although modelling fluidized bed reactors is complex and time consuming, it can provide significant technical and economic benefits in practical applications. This is of particular importance for large industrial systems, because even small improvements in performance and limited reductions in technical risks can have major economic implications.

15.3.1 Model Development

To develop a reactor model, it is necessary to understand the most important phenomena taking place inside the reactor and to translate this into mathematical expressions that can be coupled to represent complex systems. Figure 15.2 provides a schematic of the model development process. Some of the steps required when developing a new model include:

- **System characterization**: Identify the system and its limits. For a chemical reactor this requires understanding the chemistry, physics, and boundary conditions needed to develop mole and energy balances.
- **Identification of main process variables (state variables)**: These are variables that, if known throughout the reactor, define the state of the system completely. The intent of a reactor model is to predict these profiles and use them to calculate the system performance. The number of state variables defines the number of equations to be solved. The number of linearly independent equations should equal the number of state variables.
- **Identification of independent variables**: These are variables that can be changed, controlled, and manipulated in a reactor model. Their selection determines how difficult it is to solve numerically. Models can be three-dimensional (3-D) and dynamic. However, for practical reasons, most models are one-dimensional (1-D) and steady state, where the independent variable is the reactor height, reactor volume, or weight of catalyst. If the number of independent variables is zero, the resulting equations are algebraic. When only one independent variable is considered, the system can be described by ordinary differential equations (ODEs). If two or more independent variables are included, partial differential equations (PDEs) must be solved.
- **Model development**: This refers to the development of mole, energy, and pressure balances, as well as ancillary equations to estimate all reactor parameters.
- **Model simplification**: Simplifications and assumptions are introduced to tailor the model based on the development needs and available information. The simplifications are adopted based on experience.
- **Coding**: Different algorithms are developed to solve the model equations using analytical and numerical techniques. The code must be chosen to solve the type of equations resulting from the steps above.
- **Simulation**: Solve the model using analytical and computational tools.
- **Verification of modelling results**: Check for consistency, mathematical correctness, and numerical issues, and make sure that the intent of the model is captured without errors.

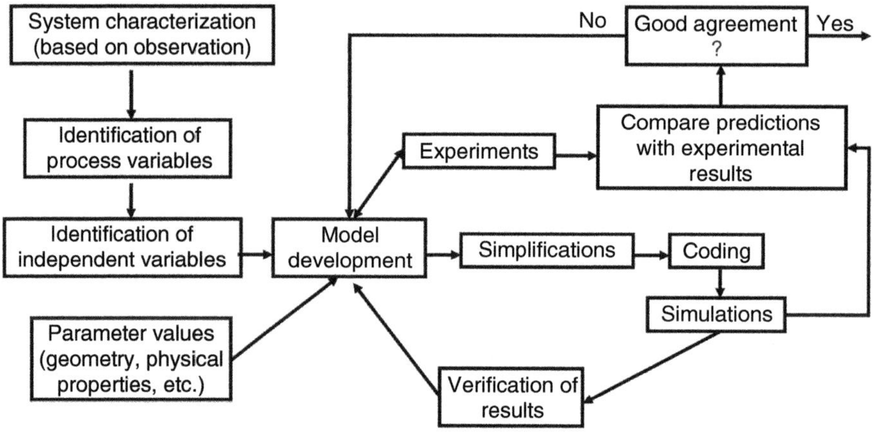

Figure 15.2 Steps for model development and validation.

- **Numerical experimentation**: Numerical experiments can be carried out to confirm certain parts of the model or to verify that the complete model works.
- **Comparison of model results with physical experimental findings**: Every model should be tested against experimental (e.g. pilot plant) data. The ultimate test for a model is to be able to predict experimental results without any adjustable parameters.

The same model equations can be used to model a new reactor or an existing operating unit. However, the required information to define the model and the conclusions extracted from the model depend on the objective of the modelling.

15.3.2 Model Structure and Reaction Considerations

Reactor models utilize simplifying assumptions to balance simplicity and accuracy. Aspects that determine the structure of the model include:

- Dynamic models can predict variations over time and can be useful to study start-up and shutdown scenarios or process upsets. Steady-state models assume that all reactor properties remain unchanged, usually representing normal operation.
- 3-D models consider variations in all three spatial coordinates. While this is desirable, it is often unnecessary. Many fluidized bed reactor models are 1-D, assuming that all properties are homogeneous over any reactor cross section.
- Non-isothermal models solve an energy balance along the reactor, while isothermal models assume constant temperature. Non-isobaric models solve a pressure balance equation, while isobaric models consider the reactor pressure to be constant.

15.3.2.1 Flow Regime Considerations

Selection of the flow regime for a reactor application can determine the model's performance and feasibility. Each fluidization flow regime has advantages and

Table 15.3 Comparison of major fluidization flow regimes for most reactor applications.

Subject	Bubbling regime, $U_{mb} < U < U_c$	Turbulent regime, $U_c < U < U_{se}$	Fast fluidization regime, $U > U_{se}$
Mass transfer rate	Slowest	Fastest	Medium
Required catalyst volume to achieve a certain conversion	Highest	Lowest	Low
Surface-to-bed heat transfer	Medium	Highest	Lowest
Axial dispersion of gas and solids	Highest	Medium	Lowest
Erosion and wear of surfaces	Lowest	Medium	Highest
Entrainment and catalyst losses	Lowest	Medium	Highest
Catalyst attrition	Lowest	Medium	Highest
Freeboard region	Required		Not required
Vertical and radial temperature gradients	Significant in freeboard region only		Significant in entire bed
Height-to-diameter ratio	Lowest	Medium	Highest
Used for laboratory-scale and technology development	Most common	Used	
Superficial gas velocity	~0.3 to ~0.8 m/s		3–12 m/s
Bed voidage	~0.6 to ~0.8		~0.8 to ~0.95
Residence time for particles	Long (could be months for highly stable catalysts)		Short (could be of the order of seconds for rapidly deactivating catalysts)
Gas throughput	Lowest	Medium	Highest
Radial homogeneity	Relatively high		Relatively low (core–annulus structure)

disadvantages compared with other flow regimes. Table 15.3 compares competing factors to be considered when selecting a flow regime for a reactor application. Note that not all aspects apply to all systems.

15.3.2.2 Reaction Equilibrium Considerations

It is important to consider the thermodynamic equilibrium for reversible processes early in the development stage. Reaction equilibrium constraints are independent of the reactor configuration.

Le Chatelier's principle indicates that whenever a system at equilibrium is disturbed, it adjusts itself in such a way that the effect of the change is reduced. Hence

- If a reactant is fed to the system, the forward reaction is favoured. If, on the other hand, a product is fed to the system, the reverse reaction is favoured.
- If heat is supplied, the system tries to consume heat. Thus endothermic reactions are favoured, while exothermic reactions are inhibited. If heat is removed, the system tries to produce heat, so exothermic reactions are favoured, and endothermic reactions inhibited.
- If pressure is increased, the system tries to reduce its pressure so that reactions that reduce the total number of moles are favoured. If pressure is decreased, the system tries to increase its pressure, so reactions that increase the total number of moles are favoured.
- For an endothermic reaction, the conversion and equilibrium constant, K_{eq}, increase as the reactor temperature increase. For an exothermic reaction, the conversion and K_{eq} decrease as the temperature increases.

Example: Consider two environmentally relevant reactions that determine the emissions of pollutants SOx and NOx. Two reversible exothermic gas-phase reactions take place in many industrial reactors as the main reaction or as part of complex reaction networks:

$SO_2 + \frac{1}{2}O_2 \leftrightarrow SO_3$	$(\Delta H^\circ_{298} = -98.9\,\text{kJ/mol})$
$NO + \frac{1}{2}O_2 \leftrightarrow NO_2$	$(\Delta H^\circ_{298} = -57.2\,\text{kJ/mol})$

The equilibrium constant (and composition) at a given temperature can be calculated by

$$K_p = \frac{P_{SO_3}}{P_{SO_2}(P_{O_2})^{0.5}}(\text{bar})^{-1}\text{; a similar expression applies to NO} \tag{15.1}$$

An equilibrium model can be used to calculate the maximum conversion achieved by a catalytic reactor. The results agree with Le Chatelier's principle. Some thoughts on the model results are as follows:

- As shown in Figure 15.3, increasing the concentration of reactant O_2 increases the maximum conversion achievable by the reactor.
- Decreasing the reactor temperature increases the maximum achievable conversion. However, reducing the reactor temperature also reduces the reaction rate, increasing the required catalyst volume and residence time. Moreover, commercial catalysts have a minimum operating temperature (e.g. ignition temperature), and therefore if the reactor is too cold, little reaction takes place.
- Increasing the reaction temperature reduces the maximum achievable conversion, but the resulting higher temperature may allow a higher conversion to be reached because the rate constant increases. However, the catalyst has a maximum operating temperature of ~640°C that must not be exceeded to prevent damage to equipment and the catalyst.

15.3.2.3 Reaction Kinetics Considerations

Reaction kinetics determine how fast a reaction takes place for given operating conditions. The rate of reaction generally increases as the concentration of

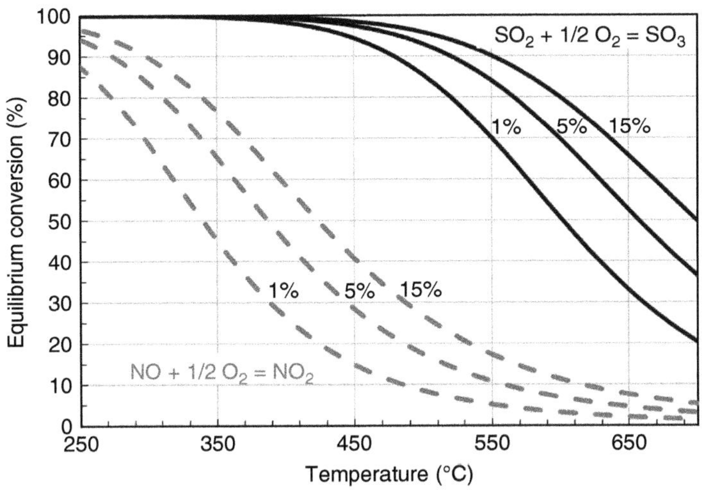

Figure 15.3 Equilibrium reactor simulation for NO and SO_2 oxidation. System pressure = 1.013 bar_a. Line tags correspond to feed concentration of O_2. Dashed lines: equilibrium conversion of NO with NO feed concentration of 1200 ppmv. Continuous lines: equilibrium conversion of SO_2, with SO_2 feed concentration of 700 ppmv.

the reactants increases, as the concentration of products decreases, as the temperature increases, and with the presence of catalyst. Catalysts accelerate both the forward and reverse chemical reactions, but are not consumed or changed chemically, do not affect the nature of the product, nor do they alter the thermodynamics. Catalysts often modify the reaction pathway, resulting in a smaller activation energy. In some cases, catalysts may speed up reactions by a factor of $\sim 10^6$ or more.

The following steps are taken for heterogeneous catalysis:

(1) External mass transfer to outer catalyst surface from reactor bulk
(2) Internal diffusion of reactants inwards into the catalyst pores
(3) Adsorption onto active sites along the catalyst internal pores
(4) Chemical reaction on the catalyst surface (reactants become products)
(5) Desorption of products from the catalyst surface
(6) Outward diffusion of products along the catalyst internal pores
(7) External mass transfer from catalyst outer surface to the bulk of the reactor

Quantifying reaction rates is essential for reactor design. Rate equations that take into account steps 3, 4, and 5, independent of all other steps, are considered "intrinsic reaction kinetics." The rate equations that consider other steps (1–7), and sometimes other aspects of the reactor operation, are called "apparent reaction kinetics." Any of the steps above may be much slower than the others, then becoming the "rate-limiting step."

The net conversion achieved in a catalytic reactor depends on:

- Reactor size and catalyst loading of the catalytic reactor
- Residence time
- Design and hydrodynamic performance of the reactor

- Catalyst activity
- Reaction network and side reactions
- Process conditions (such as temperature and pressure)
- Plant production rate

Apparent Reaction Kinetics When all factors listed above are considered, we obtain a resultant rate of reaction referred to as an "apparent" reaction rate. This rate takes into account all phenomena encountered in the reactor, including several aspects unrelated to reaction chemistry. Apparent reaction kinetics lump the effects of mass transfer, heat transfer, intrinsic reaction kinetics, catalyst activity, hydrodynamics, and other phenomena into an effective rate term. It is important to recognize that apparent reaction rates are only useful for a certain reactor configuration and specific reaction conditions. Engineers sometimes use apparent reaction rates based on correlation of plant data for a specific reactor to create simple models that can be computed quickly to provide approximate estimates of reactor performance. However, these simple models must be treated with care since they only can be applied to narrow ranges of operating conditions.

Intrinsic Reaction Kinetics Intrinsic reaction kinetics consider phenomena related to reaction chemistry, without the effects of other phenomena in a reactor. The intrinsic kinetics of a reaction system are not affected by the reactor type, hydrodynamics, mass transfer, and other phenomena. For this reason, intrinsic kinetics are a major building block required to develop a reactor model. A reactor model should use intrinsic kinetic rates for each of the chemical reactions in a reaction network and account for other phenomena separately.

It may be challenging to find out whether an expression for reaction rates found in the literature is an intrinsic or apparent rate. To address this issue, it is important to understand how the original kinetic expressions were developed experimentally and how the original laboratory data were analyzed. Table 15.4 shows some factors that should be considered to understand which kinds of kinetic expressions are available.

Special caution is recommended when:

- Kinetics are utilized outside the range of operating conditions for which they were derived.
- Kinetic data derived from one catalyst are assumed to apply to different catalysts.
- Kinetics are based on limited data. For instance, beware if experiments were done at two temperatures only or if no error bar or goodness of fit is presented.
- Kinetics are utilized for a mixture that includes a catalyst poison.

It is often difficult to find kinetics that obey all of these recommendations.

15.4 Fluidized Bed Catalytic Reactor Models

Engineering principles are needed to take advantage of the benefits of fluidized bed reactors. To predict their performance and to design them for different scales,

Table 15.4 How to identify apparent and intrinsic kinetics.

Characteristic	Apparent kinetics	Intrinsic kinetics
Scale of experimental reactor	Any scale. Sometimes large pilot scale or industrial scale	Laboratory scale
Rate equation	Typically only considers a few reactor parameters. Conversion may be a function of reaction time or a single parameter such as temperature. Rate may not account for the influence of all chemical species involved in the reaction (i.e. be zero order for some reactants). May ignore reverse reactions	Takes into account the local temperature and pressure, concentration of reactants, and concentration of products (maybe in terms of partial pressures). Rates typically use Arrhenius-type equations: $k = k_0 e^{\left(\frac{-E}{RT}\right)}$ or $k = k_0 T^a e^{\left(\frac{-E}{RT}\right)}$, where E is the activation energy and R is the gas constant. Considers forward and reverse reactions
Catalyst activity	May be lumped with rate equation	Catalyst activity is explicitly mentioned (use fresh catalyst or monitor catalyst activity carefully). It is desirable to have expressions for the catalyst activity as a function of time or key species concentrations, such as carbon
Units	May or may not have consistent units	Reaction rate has consistent units such as (mol/m^3 s), (mol kg/s)
Mass and heat transfer	Reaction rate affected by mass and heat transfer	Reaction rate independent of mass and heat transfer. Use small particles ($\lesssim$100 μm) to achieve high effectiveness factors
Reaction network	May lump all reactions into a single expression	Typically provide a dedicated reaction rate equation for each reaction step
Limitations	Dependent on reactor conditions such as geometry, particle size, gas velocity, fluidization regime, etc.	Can be used for any reactor type, as long as it is within range of concentrations and temperatures of original experimental work. May be used for any reactor type. May be used in models of any complexity or structure
Best practices	No specific requirements	Careful analysis of the data needed. Provide clear rationale for proposed reaction mechanisms. Indicate goodness of fit of the data to the proposed rate expressions. Extra care with experimental reaction rate derivations is needed when: • Catalyst particles are large • Reaction kinetics are very fast • Effective diffusivity in catalyst pores is small • Gas velocities are low (external mass transfer could be rate limiting)
Use for predictive fluidized bed reactor modelling and simulation?	No	Yes

Table 15.5 Comparison of simple and sophisticated reactor models.

Item	Simple models	Sophisticated models
Phenomena considered	Lump multiple phenomena into simple equations	Take into account multiple phenomena separately
Reaction kinetics	May use apparent or intrinsic kinetics	Consider intrinsic reaction kinetics and mass transfer effects
Solution methods, numerical techniques, and computational time	Easy to develop and solve. Relatively quick solutions	More difficult and time consuming to set up and to solve numerically
Applicability	May only be accurate for a specific process, a specific reactor, or a narrow range of process conditions	Useful for multiple applications

reactions, and operating conditions, reactor models must be created. This section explains several models and describes their most important assumptions, advantages, and shortcomings. Many fluidized bed reactor models have been proposed [1–12]. A comprehensive review and classification [13] of all major models available in the literature starting from the 1950s provided an overview of differences and capabilities of the most popular models. Table 15.5 compares simple and more complex models.

The following aspects are important to understand model capabilities and limitations:

- **Phases**: Most models divide the reactor into phases or "interpenetrating continua," where control volumes for conservation balances may include both gas and solid particles. The exchange of mass and energy between the phases plays a major role in determining the model performance. Models typically consider a single phase, two phases (high and low density), or three phases.
- **Flow regime**: Most models assume a specific flow regime (see Chapter 4). Although most industrial catalytic fluidized bed reactors operate in the turbulent fluidization flow regime, most models were originally developed for bubbling beds. Some recent models consider more than one flow regime to be present.
- **Process or reaction**: Earlier models were specific for processes with simple reaction chemistries (single reactions, elementary kinetic expressions). However, these models can be readily applied to a variety of reaction schemes and nonlinear kinetics.
- **Reactor regions**: Some models take into account additional regions such as the distributor zone and freeboard.
- **Mixing characteristics**: Models may assume perfect mixing, plug flow, or specific expressions within the phases. A high-density phase (H-phase) is likely to be modelled as perfectly mixed, whereas a low-density phase (L-phase) is commonly treated as being in plug flow. Well-mixed compartments in series are also sometimes used.

- **Chemical reaction**: Models can be developed to take into account catalyst activity, effectiveness factors, and complex kinetic expressions.
- **Mass transfer**: Models with more than one phase must consider mass transfer between the phases, simulated based on equimolar exchange, and/or balancing contributions (function of volumetric flow).
- **Independent variables**: Differential equations may be time dependent, distributed along the reactor height, or take two or three dimensions into account.
- **Energy and pressure balance**: Simpler models assume isothermal and isobaric behaviour, while more sophisticated models specify balances for energy and pressure.

15.4.1 Mass/Mole and Energy Balances

A mass/mole balance keeps track of the process changes in the system by conserving mass or moles. An energy balance applies the first law of thermodynamics, also known as the law of conservation of energy. Figure 15.4 provides a schematic of basic mole and energy balances, typically applied to control volumes to develop comprehensive balance equations.

15.4.2 Reaction Rate Expressions

Depending on the basis of calculation, reaction rates may be expressed on a volumetric basis or on a catalyst mass basis. The following general terms can be used to calculate reaction rate:

$$\left[\sum_{j=1}^{N_R} \upsilon_{ji} \cdot \Omega_j \cdot a_j \cdot r_j\right] = \left[\sum_{j=1}^{N_R} \upsilon_{ji} \cdot \Omega_j \cdot a_j \cdot \rho_{cat} \cdot (1-\varepsilon) \cdot r_j'\right], \quad i = 1, 2, \ldots, N_C \tag{15.2}$$

where υ_{ji} is the stoichiometric coefficient of species i in reaction j, Ω_j is the overall effectiveness factor, a_j is the catalyst activity (equal to 1.0 for fresh catalyst), ρ_{cat} is the catalyst density, ε is the void fraction, r_j is the volumetric rate of reaction

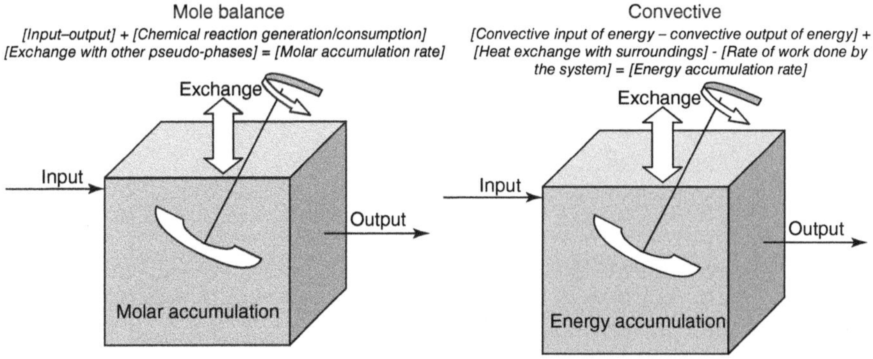

Figure 15.4 Basic formulation of mass/mole and energy balances.

j, and r'_j is the rate of reaction j per unit mass of catalyst. N_C is the number of chemical species, whereas N_R is the number of reactions.

The overall effectiveness factor accounts for external and internal mass transfer resistances:

$$\Omega = \frac{\text{Actual rate of reaction}}{\text{Rate ignoring mass transfer resistance}} = \frac{\text{Overall rate of reaction}}{\text{Rate if entire particle were at concentration } C_{i_{Bulk}}} \quad (15.3)$$

In fluidized bed catalytic reactors, overall effectiveness factors are typically close to 1 due to the small catalyst particle size. In contrast, fixed bed reactors with large catalyst pellets have much smaller effectiveness factors requiring that internal diffusion equations be solved.

15.4.2.1 Single-Phase Models

Single-phase models are useful for flowsheeting and can give good results when the interphase mass transfer is large or reaction rates are slow. However, these models usually fail to give good predictions due to poor representation of gas-fluidized bed mixing and phase separation.

Equilibrium Models Equilibrium models identify limits of possible performance of a process. They are also useful when reactor characteristics have not yet been defined. They can also provide process screening. Single-phase models may provide useful predictions for reactors at equilibrium (Gibbs reactors) and can be useful for flowsheeting, assuming an "approach to equilibrium" based on experience.

Single-Phase Lumped Model (Including CSTR Models) A simple lumped model can quickly and approximately represent a bubbling or turbulent bed reactor. This simple model assumes that concentrations, temperature, and pressure are all uniform within the reactor boundaries. The **molar balance** of a lumped reactor can be written as

$$[Q_f \cdot C_{if} - Q \cdot C_i] + \left[V \cdot \sum_{j=1}^{N_R} \upsilon_{ji} \cdot \Omega_j \cdot a_j \cdot r_j\right] = \left[V\frac{d}{dt}(C_i) + C_i\frac{d}{dt}(V)\right] \quad i = 1, 2, \dots, N_C \quad (15.4)$$

where Q is the volumetric flow rate, C_i the concentration of species i, and V the total reactor volume. To account for changes in the volume and/or number of moles, the mole balances can be rewritten in terms of molar flow rates. Sometimes it is convenient to express the concentration of species i as $C_i = \left(\frac{F_i}{Q}\right)$ where F_i is the molar flow rate of species i. One mole balance equation is required for each reacting species so that N_C equations should be solved. The units of all terms in the mole balance equations correspond to mol/s.

It is commonly assumed that the reactor volume does not change with time. The resulting model contains a set of ODEs:

$$[F_{if} - F_i] + \left[V \cdot \sum_{j=1}^{N_R} \upsilon_{ij} \cdot \Omega_j \cdot a_j \cdot r_j\right] = V \cdot \frac{d}{dt}\left(\frac{F_i}{Q}\right), \quad i = 1, 2, \dots, N_C \quad (15.5)$$

If the dynamics of the system are to be considered, the species concentrations (or molar flows) must be specified at $t = 0$. The initial conditions are expressed as follows:

$$\text{At } t = 0, C_i = (C_i)_{t=0} = C_{i_0}, \quad i = 1, 2, \dots, N_C \tag{15.6}$$

If the model assumes steady state, the time derivatives are equal to 0. The resulting steady-state mole balance for a continuous stirred tank reactor (CSTR) is given by an algebraic equation, which can be solved analytically in terms of reactor volume:

$$[F_{if} - F_i] + \left[V \cdot \sum_{j=1}^{N_R} \upsilon_{ij} \cdot \Omega_j \cdot a_j \cdot r_j \right] = 0, \quad i = 1, 2, \dots, N_C \tag{15.7}$$

For a single-reaction system (reactant "A"), the model can be defined by the following algebraic equation, where r_A is a net reaction rate:

$$[F_{Af} - F_A] + [V \cdot r_A] = 0;$$

Note that although such a model is simple, it still requires understanding of the stoichiometry and intrinsic kinetics of the reaction system.

The **energy balance** of a lumped (well-mixed) reactor can be written as an algebraic equation:

$$-\sum_{i=1}^{N_C} Q_f \cdot C_{if} \cdot Cp_i \cdot (T - T_f) + V \cdot \sum_{j=1}^{N_R} (-\Delta H_{RXj}) \cdot \Omega_j \cdot a_j \cdot r_j = 0 \tag{15.8}$$

where Cp_i is the heat capacity of species i, T is the reactor temperature, and $(-\Delta H_{RXj})$ is the heat of reaction for reaction j. Note that this equation assumes adiabatic operation (zero heat losses or heat input) and that additional terms for heat transfer from/to the surroundings can be added of the type $U_{SR} \cdot V \cdot a_{SR} \cdot (T_\infty - T_S) + \epsilon_M \cdot V \cdot a_{SR} \cdot \sigma \cdot (T_\infty^4 - T_S^4) + U_{Cool} \cdot V \cdot a_{Cool} \cdot (T_{Cool} - T)$, where U is a global heat transfer coefficient, a is the effective area of heat transfer surface in the appropriate units, and T is an appropriate temperature to estimate the heat transfer driving force.

Single-Phase One-Dimensional Reactor Model (Including Plug Flow Reactor PFR Models)

A single-phase model can be developed as an educational tool, although it is typically not suitable for making fluidized bed reactor predictions. The **model balance** of a general 1-D single-phase reactor can be written as

$$-\frac{1}{A} \cdot \frac{\partial}{\partial z}(v \cdot C_i) + \frac{\partial}{\partial z}\left(D_i \cdot \frac{\partial C_i}{\partial z} \right) + \sum_{j=1}^{N_R} \upsilon_{ji} \cdot \Omega_j \cdot a_j \cdot r_j = \frac{\partial}{\partial t}(C_i), \quad i = 1, 2, \dots, N_C \tag{15.9}$$

where D_i is the diffusion coefficient (or molecular diffusivity) of species i. This equation corresponds to the mole balance for distributed systems in terms of concentrations. The units of all terms in this equation correspond to [Transported entity/(Volume × Time)], or [mol/(m^3 s)].

This equation can instead be written in terms of volumetric flows, F_i, as

$$-\frac{1}{A}\cdot\frac{\partial}{\partial z}(F_i)+\frac{\partial}{\partial z}\left(D_i\cdot\frac{\partial}{\partial z}\left(\frac{F_i}{Q}\right)\right)$$
$$+\sum_{j=1}^{N_R} \upsilon_{ji}\cdot\Omega_j\cdot a_j\cdot r_j=\frac{\partial}{\partial t}\left(\frac{F_i}{Q}\right),\quad i=1,2,\ldots,N_C \tag{15.10}$$

The boundary conditions for this equation would correspond to a two-point boundary value problem along the reactor length. If the dynamics of the system are considered, the species concentrations (or molar flows) must be specified at $t = 0$. The initial conditions are expressed as

$$\text{At } t=0, C_i=(C_i)_{t=0}=C_{i_0},\quad i=1,2,\ldots,N_C \tag{15.11}$$

Often the dynamic component and the effect of axial dispersion are neglected to simplify the model solution. The resulting model equations are

$$-\frac{1}{A}\cdot\frac{d}{dz}(F_i)+\sum_{j=1}^{N_R} \upsilon_{ji}\cdot\Omega_j\cdot a_j\cdot r_j=0,\quad i=1,2,\ldots,N_C \tag{15.12}$$

The boundary conditions for the equation above are

$$\text{At } z=0, C_i=(C_i)_{z=0}=C_{if},\quad i=1,2,\ldots,N_C \tag{15.13}$$

For a single-reaction system (for reactant A), the model can be defined by the following equation, where r_A is a net reaction rate:

$$-\frac{1}{A}\cdot\frac{d}{dz}(F_A)+r_A=0; \tag{15.14}$$

A reactor **energy balance** can also be developed based on similar assumptions leading to

$$-\frac{1}{A}\cdot\frac{d}{dz}\left(\sum_{i=1}^{N_C} Q\cdot C_i\cdot Cp_i\cdot T\right)+\sum_{j=1}^{N_R}(-\Delta H_{RXj})\cdot\Omega_j\cdot a_j\cdot r_j=0 \tag{15.15}$$

Note that this heat balance is for an adiabatic system and that additional terms for heat transfer from/to the surroundings would need to be added to account for cooling and heating.

A reactor **pressure balance** can also be provided, with the pressure drop assumed to be caused solely by the solids static head. The gas holdup can be calculated from separate functions that consider changes in gas velocity along the reactor height. The resulting pressure balance is

$$-\frac{dP}{dz}=[(1-\varepsilon)\cdot\rho_{solids}\cdot g] \tag{15.16}$$

The boundary condition for the pressure balance is

$$\text{At } z=0, P=(P)_{z=0}=P_f \tag{15.17}$$

15.4.3 Models Based on Standard Two-Phase Theory

15.4.3.1 Division of Flow and Calculation of Fluidized Bed Parameters

The models require division of the fluid flow among reactor phases. Most published models adopt the original "two-phase theory" of Toomey and Johnstone [14] (see Chapter 7) in one form or another. Moreover, all models require estimation of parameters that are functions of the fluidization hydrodynamics. Below we provide basic correlations required to set up a model and estimate basic fluidization parameters required to solve mole, energy, and pressure balances.

Flow Division Models typically assume that all or most of the gas in excess of that needed for minimum fluidization creates a low-density (also called L-phase, bubble, dilute, void, or dispersed) phase. Some widely separated particles may also be included in this phase. The remaining particles are assigned to a high-density (also identified as H-phase, dense, particulate, emulsion, or continuous) phase. Figure 15.5 shows a schematic of flow division. For bubbling beds, the two-phase theory of fluidization assumes that the "visible" bubble flow is given by

$$Q_{VB} = (U - U_{mf}) \cdot A \tag{15.18}$$

where Q_{VB} is the visible bubble flow, U is the superficial velocity, U_{mf} is the minimum fluidization velocity [15] (see Chapter 2), and A is the reactor cross-sectional area.

The intrinsic gas velocities of the phases are given by

$$U_{gas_\phi} = Q_\phi / (\psi_\phi \cdot \varepsilon_\phi \cdot A) \tag{15.19}$$

where for phase ϕ (high- or low-density phase), Q_ϕ is the phase volumetric flow rate, ψ_ϕ the phase volume fraction, and ε_ϕ the phase void fraction. The overall bed superficial gas velocity is

$$U = (Q_H + Q_L)/A \tag{15.20}$$

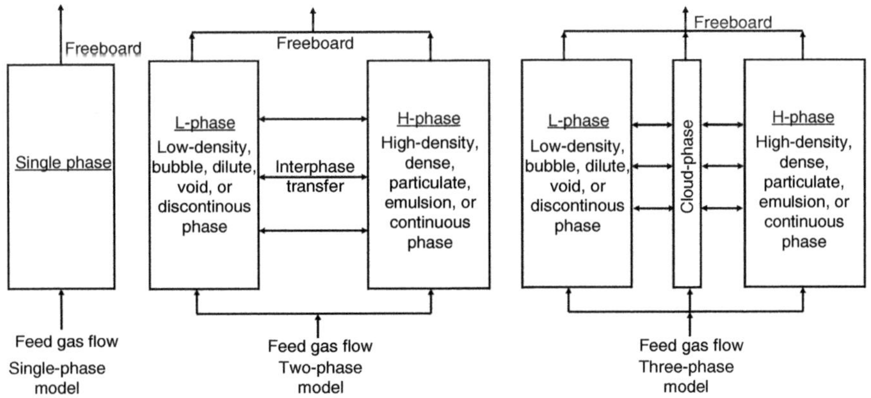

Figure 15.5 Schematic of flow division among phases.

Based on the ideal gas law, the total volumetric flow rate is given by

$$Q_\phi = \frac{R \cdot T_\phi}{P} \cdot \sum_{i=1}^{N_C} F_{i\phi} \tag{15.21}$$

where R is the gas constant, T_ϕ the temperature, and P the pressure. The most convenient way to account for the evolution of volumetric flow is to solve the conservation equations in terms of molar flows instead of species concentrations using the relation

$$F_{i\phi} = Q_\phi \cdot C_{i\phi} = \psi_\phi \cdot \varepsilon_\phi \cdot U_{gas_\phi} \cdot A \cdot C_{i\phi}$$

Calculation of Mass Transfer Coefficients A specific mass transfer coefficient should be calculated for each species because they are functions of molecular diffusion coefficients. There are different correlations for mass transfer, specific to different flow regimes. The following expressions are well established:

Bubbling regime	Turbulent regime	Fast fluidization regime
$(a_I \cdot k_{c_i}) = \left(\frac{U_{mf}}{3} + 2 \cdot \left(\frac{D_i \cdot \varepsilon_{mf} \cdot U_H}{\pi \cdot d_b}\right)^{1/2}\right) \cdot \frac{6}{d_b}$ [16]	$(a_I \cdot k_{c_i}) = 1.631 \cdot Sc_i^{0.37} \cdot U$ [17]	$(a_I \cdot k_{c_i}) = \left[\frac{4 \cdot D_i \cdot \varepsilon_H \cdot U}{\pi \cdot L_t}\right]^{\frac{1}{2}} \cdot \frac{2}{r_C}$ (15.22) [18]

where $a_{I_{(H\to L)}} = 6/d_b$ and $Sc_i = \mu/(\rho_{gas} \cdot D_i)$.

Calculation of Bed Volume Fractions The volume fraction of each phase can be calculated by using the following correlations:

Bubbling and turbulent regimes	Fast fluidization regime
Often ψ_L is assumed to be constant as per the inlet conditions. An alternative expression is $\psi_L = (\varepsilon - \varepsilon_{mf})/(\varepsilon_{L0} - \varepsilon_{mf})$	$\psi_L = r_C^2/\left(\frac{D_t}{2}\right)^2$, where core radius is calculated by $r_C = \frac{D_t}{2} - \frac{D_t}{2} \cdot (1 - (1.34 - 1.3.(1 - \varepsilon_{fast})^{0.2} + (1 - \varepsilon_{fast})^{1.4}))^{0.5}$ (15.23), [19]

The voidage is important because it determines the extent of reaction in the L-phase (e.g. bubbles). Although the value of this parameter has never been resolved, many models in the literature assume $\varepsilon_{L0} = 1$, which is a poor approximation for systems with fast chemical reactions. This voidage is expected to be between 0.97 and 0.999 or even higher.

Voidage Profiles The bed voidage profile in the freeboard is calculated from

Freeboard voidage profile (for bubbling and turbulent beds)	
Freeboard (if existing)	$\varepsilon_f = \varepsilon^* + (\langle\varepsilon_T\rangle - \varepsilon^*) - e^{-\left(\frac{3}{U - U_{mf}}\right) \cdot z}$ (15.24), [20]

where ε^* and U_{mb} are obtained from correlations, [21, 22], respectively:

$$\varepsilon^* = 1 - 0.022(U - U_{mb})^{3.64} \text{ (SI units)} \tag{15.25}$$

$$U_{mb} = \frac{U_{mf} \cdot 2300 \cdot \rho_{gas}^{0.126} \cdot \mu_{gas}^{0.523} \cdot e^{(0.176 \cdot Fines)}}{d_p^{0.8} \cdot g^{0.934}(\rho_{solid} - \rho_{gas})^{0.934}} \text{ or } U_{mb} = U_{mf}$$

$$\text{for group B or D particles} \tag{15.26}$$

The overall voidage for the dense bed section is given by

Bubbling regime	Turbulent regime	Fast fluidization regime
$\varepsilon = 1 - \left(\dfrac{(1-\varepsilon_{mf})}{\left(1 + \dfrac{U - U_{mf}}{0.711\sqrt{g \cdot db}}\right)} \right)$ [23]	$\varepsilon = \dfrac{U+1}{U+2}$ [24]	$\varepsilon = \left[1 + \dfrac{G_{so}\Psi_{slip}}{\rho_p U}\right]^{-1}$ (15.27) [25]

For the L-phase:

Bubbling regime	Turbulent regime	Fast fluidization regime
$\varepsilon_L = \varepsilon_{L0}$	$\varepsilon_L = \varepsilon$	$\varepsilon_L = \varepsilon$ (15.28)

For the H-phase:

$$\varepsilon_H = (\varepsilon - \psi_L \cdot \varepsilon_L)/\psi_H \tag{15.29}$$

Orcutt's Model Orcutt's model is simple and can be useful for modelling bubbling bed reactors [26, 27]. It considers two phases at steady state under bubbling fluidization conditions. The model covers the entire dense bed and neglects any reactions in the freeboard. The low-density phase is modelled as a 1-D plug flow system, considering interphase mass transfer, but ignores any chemical reactions in the bubbles. The low-density phase mole balance is then given by

$$\beta U dC_{Ab} = k_q(C_{Ad} - C_{Ab})a_b\varepsilon_b dz \tag{15.30}$$

where subscript b refers to bubbles and d refers to the dense phase. The high-density phase is modelled as a lumped (well-mixed) system considering mass transfer and chemical reaction. The model equations are written for a single nth-order reaction, without specific considerations for catalyst activity and other factors. The high-density phase mole balance is then represented by

$$(1-\beta)U(C_{Ain} - C_{Ad}) + \int_0^H (C_{Ab} - C_{Ad})a_b\varepsilon_b dz = (1-\varepsilon_b)(1-\varepsilon_{mf})Hk_nC_{Ad}^n \tag{15.31}$$

Kunii and Levenspiel Model This model considers three phases under steady-state bubbling fluidization conditions [28]. The low-density phase is modelled as a 1-D plug flow system, with interphase mass transfer, plus chemical reaction in the bubbles. Subscript b refers to bubbles, c to clouds, and e to an "emulsion phase" (dense-phase less clouds and wakes). The low-density phase mole balance is

$$-U_b \frac{dC_{Ab}}{dz} = \gamma_b K_r C_{Ab} + k_{bc}(C_{Ab} - C_{Ac}) \tag{15.32}$$

An intermediate cloud-wake phase (between the b and e phases) is also considered. Its mole balance corresponds to a lumped system that takes into account mass transfer to the high- and low-density phases, as well as chemical reaction. The cloud-wake-phase mole balance is

$$k_{bc}(C_{Ac} - C_{Ab}) = \gamma_c K_r C_{Ac} + K_{ce}(C_{Ac} - C_{Ae}) \tag{15.33}$$

The high-density phase is modelled as a stagnant region with interphase mass transfer and chemical reaction. The equations are written for a single reaction, without specific considerations of catalyst activity or complex reaction networks. The high-density phase mole balance is

$$K_{ce}(C_{Ac} - C_{Ae}) = \gamma_e K_r C_{Ae} \tag{15.34}$$

The Kunii and Levenspiel model can be effective in modelling bubbling beds.

Grace Model This model for bubbling beds can be regarded as a simplified version of the Kunii and Levenspiel model, with the cloud-wake and emulsion phases combined into a dense phase [29]. The low-density phase is subject to 1-D plug flow (no axial dispersion). It considers interphase mass transfer, based on Sit and Grace [16], as well as some nth-order chemical reaction in the bubbles (due to catalyst particles dispersed in the bubbles). The low-density phase mole balance is then

$$U\frac{dC_{Ab}}{dz} + k_q a_b \varepsilon_b (C_{Ab} - C_{Ad}) + k_n \phi_b C_{Ab}^n = 0 \tag{15.35}$$

where subscript b refers to bubbles and d to the dense phase. The high-density phase mole balance considers a lumped system with mass transfer and a simple chemical reaction function:

$$k_q a_b \varepsilon_b (C_{Ad} - C_{Ab}) + k_n \phi_d C_{Ad}^n = 0 \tag{15.36}$$

The model assumes that the reactor is isothermal and radially uniform. It makes no allowance for the distributor region, freeboard, or change in volumetric flow, but it can be easily upgraded to take into account several complex phenomena using the same basic mole balance structure. This model is regarded as a good compromise between complexity and accuracy.

15.4.4 More Sophisticated Models

15.4.4.1 Comprehensive Reactor Modelling

Some features that may be considered in a comprehensive model include:

- **Model flexibility**: The mechanistic model can treat a system of N_ϕ phases, N_C components, and N_R reactions. The conservation equations can be written in any coordinate system with 1-D, 2-D, or 3-D geometries using vector operators.

- **Phase approach**: Mole and energy balances include both gas and solids based on two pseudo-phases, designated here as the H-phase and L-phase.
- **Convective transport**: The convective velocities of the phases account for changes in molar and gas volumetric flows. Changes with time, temperature, and pressure can also be treated.
- **Equimolar interphase mass transfer**: Mass transfer is calculated proportional to a concentration driving force, with a different mass transfer coefficient for each chemical species to account for each species having a different molecular diffusivity.
- **Balancing interphase cross-flow**: Convective bulk mass transport of fluid from/to each phase is considered, determined by fluidization conditions and changes in volumetric flow.
- **Catalytic reactions**: Intrinsic chemical kinetics should be used. Overall effectiveness factors, as well as catalyst deactivation functions, may also be included.
- **Flow regime variation**: The model can be based on bubbling, turbulent, and fast flow regimes in different regions, or weighting factors can provide probabilistic averaging of parameters.
- **Selective removal of species**: If relevant, a model can include membranes.
- **Feed distribution along the reactor**: This can be included when applicable.
- **Other features**: Dynamic behaviour, mass dispersion, and anisotropic heat dispersion can also be included when needed.

Figure 15.6 presents a schematic of comprehensive reactor models. These models are often difficult to solve numerically and may require simplifying assumptions to be practical. A detailed derivation of a comprehensive model is provided elsewhere [30].

15.4.4.2 Computational Fluid Dynamics (CFD) Models for Fluidized Bed Catalytic Reactors

Computational fluid dynamics (CFD) modelling has improved rapidly in recent years due to advances in computational processing, numerical techniques, and the availability of commercial and open-source software. CFD is based on the same modelling principles as other reactor models, since it uses conservation equations for mass and energy, depends on the availability of good intrinsic kinetics, and requires that physical properties be estimated. However, CFD predicts the fluid mechanics of the fluidized bed by solving momentum balances, while still using correlations for certain parameters. CFD has the ability to estimate hydrodynamic parameters that other models obtain from empirical correlations. CFD also has the capacity to perform time-dependent simulations in 2-D and 3-D. However, as for any modelling approach, the quality of the results depends on the quality of the assumptions and input parameters.

15.4.5 Model Verification and Validation

The academic and industrial literature contains a wealth of modelling and experimental data for laboratory- and pilot-scale fluidized bed catalytic reactors.

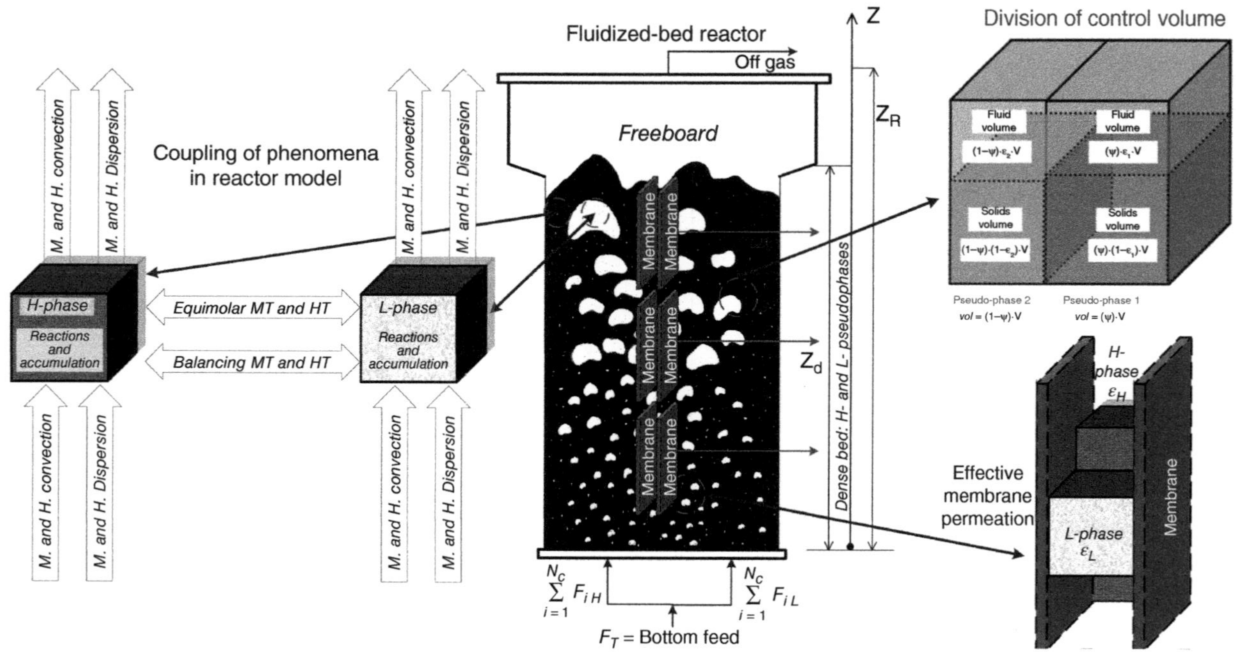

Figure 15.6 Schematic of comprehensive two-phase reactor models.

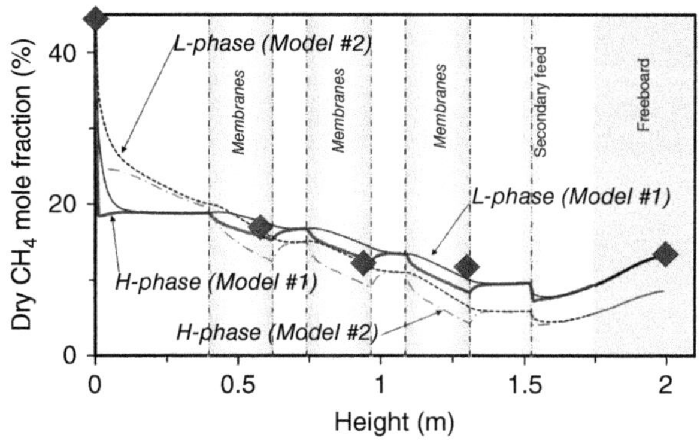

Figure 15.7 Example of a model assessment study for a fluidized bed catalytic reactor operating in the bubbling regime. Continuous lines show predictions of Model 1. Dashed lines show predictions of Model 2. The rhomboids provide experimental results based on gas sampling. Model 1 is seen to provide a better fit for the range of conditions studied.

However, it is difficult to put these studies in perspective to determine how well the models predict the behaviour of large-scale commercial units. The following should be considered to validate a model:

- Modellers should disclose what adjustable parameters they use. Best practices do not use adjustable parameters for base simulation cases, but may investigate the effect of adjusting parameters independently.
- It is essential to compare model predictions with experimental data (especially full-scale reactor data). Models should be used to simulate reactors in "as tested" conditions. Figure 15.7 shows an example of model assessment tests, where gas sampling data at different locations in a fluidized bed reactor under bubbling conditions are compared with a comprehensive model applied independently.
- It is important to test model assumptions by exploring their effect on predictions.

Some studies have compared model prediction with experimental reactor performance data, as discussed in references [31–38]. Some of the main findings of these studies are:

- Having excellent-quality intrinsic reaction kinetics is extremely important for all models.
- Interphase mass transfer considerations are most important for reactors with fast chemical kinetics and relatively slow interphase mass transfer.
- There is little or no improvement in model predictions when the number of phases is increased beyond two.
- 1-D models are sufficient unless there are internal baffles or significant radial variations.
- The freeboard region may have significant reverse reaction due to cooling. Simulation of the freeboard is important.
- Allowing for catalyst content in the low-density phase is important.

15.4.6 Recommendations for Programming and Numerical Solution of Reactor Models

A major part of reactor modelling is the solution of equations and analysis of numerical results. The following steps are recommended, once the model equations have been established:

(1) Before starting, write a program flowchart or logic diagram on paper. An example logic diagram is provided in Figure 15.8.
(2) Develop a plan for testing the validity of the model as early as possible. The best way to address possible problems is to test code elements and subroutines individually.
(3) Double check all units on paper. Proofread the code for typos. Check the code for logical/programming errors.
(4) To initialize the program, run the first pass without reactions, and then gradually increase the reaction rate constants until full rates are reached.
(5) Confirm that mass is conserved (total mass flow must be constant). Confirm that the moles of a major species are conserved (e.g. that total moles of carbon or sulphur are constant).
(6) Add comments throughout the code so that others can understand the program. This is important for technology transfer and a necessary practice in research.
(7) Insert original references in the code. This will allow you to understand it in the future.
(8) Check initial and boundary conditions.
(9) If there are variables undefined when equal to zero, increase the initial conditions by a very small value (e.g. $\sim 10^{-9}$), because numerical techniques may fail when encountering 0 values.
(10) You may sometimes need to reset/clear all matrices and vectors. It also sometimes helps to restart the solver.
(11) Print intermediate values from your code to confirm that they are reasonable.
(12) Simulate simple extreme cases, e.g. nonreactive, isothermal, and isobaric, to test the model.
(13) Simulate a system with well-known results or analytical solutions.
(14) Simulate the system of interest, checking against experiments and full-scale plant data.

15.5 Conclusions

Fluidized bed reactors are essential parts of many industrial processes, including some needed for the sustainability of mankind. It is therefore important to improve our understanding of complex phenomena taking place within the reactors and their peripherals to design and implement fluidized bed reactor processes, especially when new technology is developed.

Reactor modelling is necessary to design and scale up fluidized bed reactors. While reactor modelling has been an area of active research for many years and

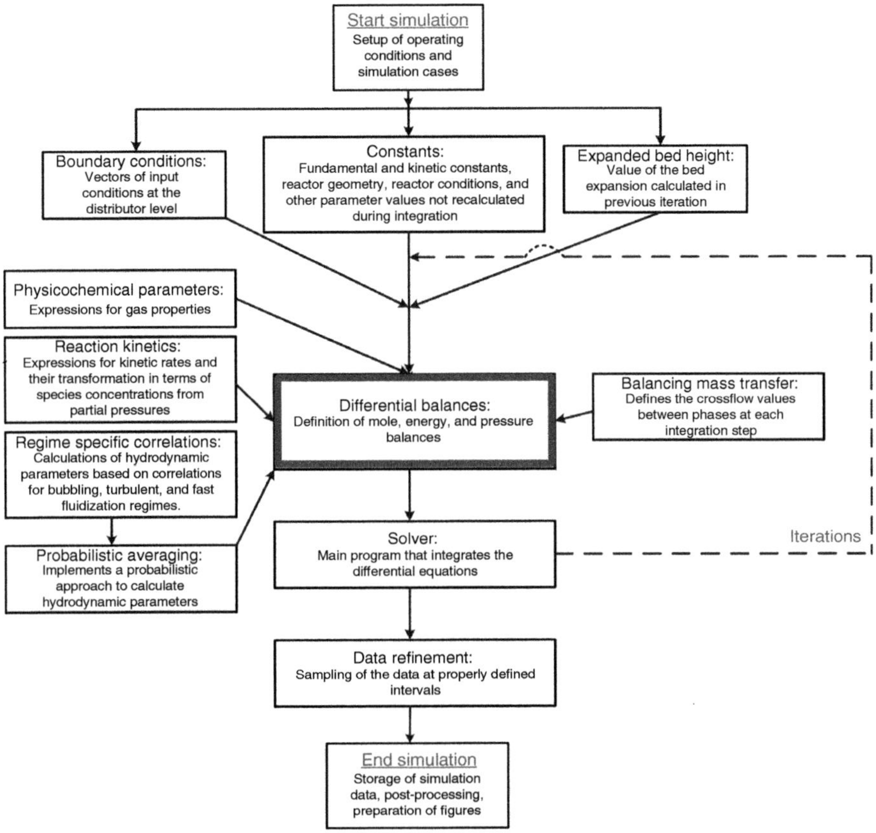

Figure 15.8 Example of a logic diagram for model programming and solution.

significant progress has been realized, there is no single model or set of correlations suitable for all systems. Therefore, it is important to explore the sensitivity of reactor model assumptions and simplifications to make sure that no systematic errors are introduced and also to review experimental, pilot plant, and industrial data to develop full-scale industrial designs.

Notations

A	cross-sectional area of reactor (m^2)
a_i	interphase area per unit volume (m^{-1})
a_j	catalyst activity for reaction j (–)
C_i	concentration of species i (mol/m^3)
Cp_i	specific heat of species i (J/kg K)
D_i	molecular diffusivity of species i (m^2/s)
D_t	reactor tube diameter (m)
F_i	molar flow rate (mol/s)
g	acceleration of gravity (m/s^2)
H	bed depth (m)

k_{bc}	bubble/cloud interphase mass transfer coefficient (m/s)
K_r	first-order reaction rate constant (s^{-1})
K_{ce}	cloud/emulsion mass transfer coefficient (m/s)
N_C, N_R	number of chemical species, reactions (–)
n	order of reaction (–)
P	pressure (kPa)
Q	volumetric gas flow (m^3/s)
Q_{VB}	gas flow due to translation of bubbles (m^3/s)
Q_H, Q_L	gas flow through H–L phase (m^3/s)
r_A	net rate of reaction (mol/m^3 s)
r_j, r_j'	rate of reaction j (mol/m^3 s or mol/kg s)
Sc	Schmidt number (–)
t	time (s)
T	temperature (K)
U	superficial gas velocity (m/s)
U_{mb}, U_{mf}	minimum bubbling, fluidization superficial velocity (m/s)
V	reactor volume
z	height coordinate above distributor (m)

Greek Letters

β	fraction of gas flow passing through the bubble phase (–)
ΔH_{RXj}	heat of reaction j
$\varepsilon, \varepsilon_{mf}$	voidage, voidage at minimum fluidization (–)
ε_b	fraction of bed volume occupied by bubbles (–)
γ_c, γ_e	fraction of cloud phase, emulsion phase occupied by particles (–)
μ	gas viscosity (Pa s)
ρ_{cat}	catalyst density (kg/m^3)
ρ_{gas}	gas density (kg/m^3)
υ_{ji}	stoichiometric constant of species i in reaction j (–)
Ω_j	overall effectiveness factor for reaction j (–)
ψ_ϕ	volume fraction occupied by phase ϕ, –

Subscripts

H, L	high-density phase, low-density phase
i, j	species, reaction

References

1 Grace, J.R. (1971). An evaluation of models for fluidized bed reactors. *AIChE Symp. Ser.* 67: 159–167.
2 Calderbank, P.H. and Toor, F.D. (1971). Fluidized beds as catalytic reactors. In: *Fluidization* (eds. J.F. Davidson and D. Harrison), 383–429. London: Academic Press.

3 Pyle, D.L. (1972). Fluidized bed reactors. Review. In: *Advances in Chemistry Series*, vol. 109 (ed. K.B. Bischoff), 106–130. Washington, DC: American Chemical Society.
4 Yates, J.G. (1975). Fluidised bed reactors. *Chem. Eng.* 303: 671–677.
5 Horio, M. and Wen, C.Y. (1977). An assessment of fluidized-bed modeling. *AlChE Symp. Ser.* 73: 9–21.
6 Grace, J.R. (1981). Fluidized bed reactor modeling: an overview. In: *Chemical Reactors*, ACS Symposium Series, vol. 168 (ed. H.C. Fogler), 3–18. Washington, D.C.: American Chemical Society.
7 Fane, A.G. and Wen, C.Y. (1982). Fluidized-bed reactors. In: *Handbook of Multiphase Systems* (ed. G. Hetsroni). New York: Hemisphere Publishing Corporation, Chapter 8.
8 Van Swaaij, W.P.M. (1985). Chemical reactors. In: *Fluidization*, 2e (eds. J.F. Davidson, R. Clift and D. Harrison). London: Academic Press.
9 Grace, J.R. (1986). Fluid beds as chemical reactors. In: *Gas Fluidization Technology* (ed. D. Geldart). Chichester, UK: Wiley.
10 Grace, J.R. (1986). Modeling and simulation of two-phase fluidized bed reactors. In: *Chemical Reactor Design and Technology*, NATO ASI Series, vol. 110 (ed. H.I. De Lasa). Dordrecht: Martinus Nijhoff Publishers.
11 Grace, J.R. and Lim, K.S. (1997). Reactor modelling for high-velocity fluidized beds. In: *Circulating Fluidized Beds* (eds. J.R. Grace, A.A. Avidan and T.M. Knowlton), 504–524. London: Chapman and Hall.
12 Ho, T.C. (2003). Modeling. In: *Handbook of Fluidization and Fluid Particle Systems* (ed. W.C. Yang), 239–255. New York: Marcel Dekker, Chapter 9.
13 Mahecha-Botero, A., Grace, J.R., Elnashaie, S.S.E.H., and Lim, C.J. (2009). Advances in modelling of fluidized-bed catalytic reactors: a comprehensive review. *Chem. Eng. Commun.* 196: 1375–1405.
14 Toomey, R.D. and Johnstone, H.F. (1952). Gaseous fluidization of solid particles. *Chem. Eng. Prog.* 48: 220–226.
15 Wen, C.Y. and Yu, Y.H. (1966). A generalized method for predicting the minimum fluidization velocity. *AlChE J.* 12: 610–612.
16 Sit, S.P. and Grace, J.R. (1981). Effect of bubble interaction on interphase mass transfer in gas fluidized beds. *Chem. Eng. Sci.* 36: 327–335.
17 Foka, J., Chaouki, J., and Guy, D.K. (1996). Gas phase hydrodynamics of gas-solid turbulent fluidized-bed reactors. *Chem. Eng. Sci.* 55: 713–723.
18 Pugsley, T.S., Patience, G.S., Berruti, F., and Chaouki, J. (1992). Modeling the catalytic oxidation of n-butane to maleic anhydride using a circulating fluidized bed reactor. *Ind. Eng. Chem. Res.* 31: 2652–2660.
19 Bi, H.T., Zhu, J.X., Qin, J., and Grace, J.R. (1996). Annular wall thickness in CFB risers. *Can. J. Chem. Eng.* 74: 811–814.
20 Kunii, D. and Levenspiel, O. (1991). *Fluidization Engineering*, 2e. Boston: Butterworth-Heinemann.
21 Morikawa, H., Bi, H.T., Lim, C.J., and Grace, J.R. (2001). Entrainment from pilot scale turbulent fluidized beds of FCC particles. In: *Fluidization X* (eds. M. Kwauk, J. Li and W.C. Yang), 181–188. New York: Engineering Foundation.

22 Abrahamsen, A.R. and Geldart, D. (1980). Behavior of gas-fluidized beds of fine powders. *Powder Technol.* 26: 35–46.

23 Clift, R. and Grace, J.R. (1985). Continuous bubbling and slugging. In: *Fluidization* (eds. J.F. Davidson, R. Clift and D. Harrison), 73–132. London: Academic Press.

24 King, D.F. (1989). Estimation of dense bed voidage in fast and slow fluidized beds of FCC catalyst. In: *Fluidization VI* (eds. J.R. Grace, L.W. Schemilt and M.A. Bergougnou), 1–8. New York: Engineering Foundation.

25 Patience, G.S., Chaouki, J., Berruti, F., and Wong, R. (1992). Scaling considerations for circulating fluidized bed risers. *Powder Technol.* 72: 31–37.

26 Orcutt, J.C., Davidson, J.F., and Pigford, R.L. (1962). Reaction time distributions in fluidized catalytic reactors. *Chem. Eng. Prog. Symp. Ser.* 58 (38): 1–15.

27 Davidson, J.F. and Harrison, D. (1963). *Fluidised Particles.* Cambridge: Cambridge University Press.

28 Kunii, D. and Levenspiel, O. (1969). *Fluidization Engineering*. New York: Wiley.

29 Grace, J.R. (1984). Generalized models for isothermal fluidized bed reactors. In: *Recent Advances in Engineering Analysis of Chemically Reacting Systems* (ed. L.K. Doraiswamy). New Deli: Wiley.

30 Mahecha-Botero, A., Grace, J.R., Lim, C.J. et al. (2009). Pure hydrogen generation in a fluidized bed membrane reactor: application of the generalized comprehensive reactor model. *Chem. Eng. Sci.* 64: 3826–3896.

31 Chavarie, C. and Grace, J.R. (1975). Performance analysis of a fluidized bed reactor. I. Visible flow behaviour. *Ind. Eng. Chem. Funds.* 14: 75–78.

32 Chavarie, C. and Grace, J.R. (1975). Performance analysis of a fluidized bed reactor. II. Observed reactor behavior compared with simple two-phase models. *Ind. Eng. Chem. Funds.* 14: 79–86.

33 Chavarie, C. and Grace, J.R. (1975). Performance analysis of a fluidized bed reactor. III. Modification and extension of conventional two-phase models. *Ind. Eng. Chem. Funds.* 14: 86–91.

34 Barreteau, D., Laguerie, C., and Angelino, H. (1978). An evaluation of some fluidized-bed reactor models for SO_2 sorption on copper oxide particles. In: *Fluidization* (eds. J.F. Davidson and D.L. Keairns), 297–302. Cambridge: Cambridge University Press.

35 Bolthrunis, C.O. (1989). An industrial perspective on fluid bed reactor models. *Chem. Eng. Prog.* 85: 51–54.

36 Mostoufi, N., Cui, H., and Chaouki, J. (2001). A comparison of two- and single-phase models for fluidized-bed reactors. *Ind. Eng. Chem. Res.* 40: 5526–5532.

37 Grace, J.R. and Taghipour, F. (2004). Verification and validation of CFD models and dynamic similarity for fluidized beds. *Powd. Technol.* 139: 99–110.

38 Mahecha-Botero, A., Boyd, T., Gulamhusein, A. et al. (2008). Pure hydrogen generation in a fluidized bed membrane reactor: experimental findings. *Chem. Eng. Sci.* 63: 2752–2762.

Problems

15.1 Conceptual questions:

- Develop a steady-state, two-phase one-dimensional mole balance for a fluidized bed reactor. List the major assumptions used.
- What are the main process variables required to characterize fluidized bed reactors?
- What parameter values are required to simulate a fluidized bed reactor?
- What is the difference between reactor design and reactor characterization?
- Derive and explain a mole balance for a plug flow reactor. How can you solve it?
- Write one typical expression to calculate reaction rate.
- Write the mathematical formula to convert a rate of reaction from a "per unit volume basis" to a "per unit mass of catalyst" basis.
- Depending on their independent variables, what type of models require algebraic equations? Which requires ODEs? And which requires PDEs?
- What are intrinsic reaction kinetics? Why are they important? What type of reactors can use them?
- Explain the effect of a catalyst on reaction rate, activation energy, and chemical equilibrium.
- What design documents and engineering steps are required to design and implement a reactor project at industrial scale?

15.2 You were hired as the new director of the reaction engineering division of a leading chemical corporation. The chemistry division has developed a promising catalyst to manufacture a valuable chemical "C." Your supervisor asks you to develop a research plan to implement this new catalyst by modelling a large-scale fluidized bed reactor using the new catalyst. The reaction network is $A \overset{r_1}{\rightarrow} 2B \overset{r_2}{\rightarrow} 4C$. A, B, and C are all gases at the conditions of interest. Both reactions are highly exothermic. The catalyst deactivates rapidly over time.

- Briefly comment on the general steps needed to implement the new catalyst.
- Propose a parameter estimation method to obtain intrinsic reaction kinetics for the required chemical reactions from laboratory tests.
- Write the required one-dimensional steady-state mole balances for a fluidized bed reactor based on a simple two-phase model.
- Write a steady-state energy balance and a pressure balance equation.
- What initial and boundary conditions would you use to solve the model?
- Propose expressions for the catalyst activity and effectiveness factor.
- Explain what strategies you would suggest to operate the reactor and to maximize the production of "C."

Identify and explain all assumptions that you consider to be necessary.

16

Fluidized Beds for Gas–Solid Reactions★

Jaber Shabanian[1] *and Jamal Chaouki*[2]

[1] *Natural Resources Canada, CanmetENERGY, 1 Haanel Drive, Ottawa, ON, Canada K1A 1M1*
[2] *Ecole Polytechnique, Chemical Engineering, Process Engineering Advanced Research Lab, C.P. 6079, succ. Centre-Ville, Montreal, QC, Canada H3C 3A7*

16.1 Introduction

This chapter focuses on non-catalytic gas–solid reactions in fluidized beds. Processes involving gas–solid reactions are of considerable importance in most industrialized countries. A few representative examples are thermal conversion (i.e. pyrolysis, gasification, and combustion) of solid fuels (coal, biomass, wastes), thermal treatment of mineral ores (e.g. roasting of sulfide ores, reduction of iron ores), regeneration of catalysts, and calcination reactions. As noted in Chapter 1, gas–solid fluidized beds have been widely adopted for these reactions owing to their advantages, such as ease of solids handling, enhanced rates of heat and mass transfer, reduced pressure drop, successful operation at different scales, and turndown flexibility [1].

Unlike solid-catalyzed gas-phase reactions, covered in Chapter 15, we consider in this chapter solids participating in non-catalytic gas–solid reactions. This results in continuous changes in the chemical composition and/or physical properties of the particles during the course of the reaction. Hence, the corresponding analysis tends to be more complicated than for catalytic gas-phase reactions [1, 2]. Gas–solid reactions may take place through one of the following routes [2]:

$$A_{(g)} + bB_{(s)} \rightarrow cC_{(g)} \tag{16.1}$$

$$A_{(g)} + bB_{(s)} \rightarrow cC_{(g)} + dD_{(s)} \tag{16.2,16.3a}$$

$$C_{(g)} + eE_{(s)} \rightarrow aA_{(g)} + fF_{(s)} \tag{16.3b}$$

The third route represents the solid–solid reactions proceeding through gaseous intermediates. Note that each of the above routes is intended to represent a dominant and thermodynamically feasible form of reaction among many reactions that could take place in the system under specific operating conditions. Hence, establishing the thermodynamic equilibrium relationship between the

★Natural Resources Canada, employer of Dr. Shabanian when he co-authored this chapter, has granted permission to Wiley-VCH to publish and reproduce this book contribution.

Essentials of Fluidization Technology, First Edition. Edited by John R. Grace, Xiaotao Bi, and Naoko Ellis.

reactants and products and calculating the heat effects of the reaction are important initial steps in designing a non-catalytic gas–solid reactive system. This immediately provides basic knowledge as to whether the reaction is thermodynamically feasible and whether precautions are required to handle the products and operate the system under different operating conditions, e.g. change in temperature, pressure, or gas composition [2]. Integrated thermochemical analysis software and database packages, such as FactSage [3] and Thermo-Calc [4], can provide great assistance in performing these calculations.

The performance of a fluidized bed reactor for gas–solid reactions with a continuous flow of solids greatly depends on three factors: (i) characteristic reaction time for single particles in the reactor environment, (ii) residence time distribution of particles in the bed, and (iii) spatial variation of gas composition in the reactor [2, 5]. The objective of this chapter is to combine these factors for the reliable prediction of reactor performance and/or successful design of a relevant fluidized bed reactor.

16.2 Gas–Solid Reactions for a Single Particle

The chemical interaction between a single particle and a moving gas stream is the smallest reactive unit typically considered in gas–solid reactor design. Figure 16.1 shows that the overall reaction process of such a reacting particle may involve several steps as follows:

(a) Mass transfer of gaseous reactant from the bulk of the gas stream to the external surface of the particle.
(b) Inward diffusion of gaseous reactant through the internal pores of the particle.
(c) Chemical reaction between the gaseous and solid reactants.
(d) Outward diffusion of gaseous product from the internal pores of the particle towards its external surface.
(e) Mass transfer of gaseous product from the external surface of the particle to the bulk gas stream.

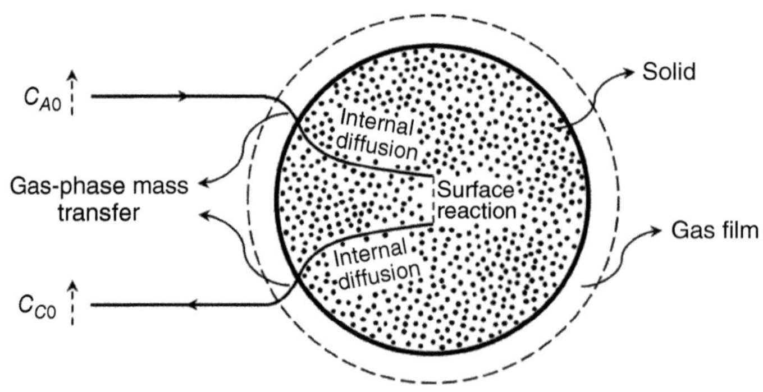

Figure 16.1 Steps in reaction of a single solid particle with a gas.

These steps often appear to occur in series with the help of some simplifying assumptions, but in reality this sequence is frequently not followed. Each of these steps could contribute to the overall rate of reaction, and their relative importance depends principally on the operating temperature, particle size, structural properties, and the nature of the reaction, reactants, and products. The relative importance may also change during the course of the reaction. Thus, it is critical to know which step or steps control the overall reaction rate in the reactor [1, 2, 5].

In general, steps related to mass transfer through the gas film surrounding the particle offer less resistance than the internal diffusion steps [1]. However, the external mass transfer resistance is critical in the early stages of the reaction when no porous product layer has yet formed on the particle. If a product layer has developed, then its internal diffusion resistance is usually more important than the external mass transfer [5]. Under circumstances where gas molecules can conveniently reach the reaction sites (e.g. small particle size and low temperature), the chemical reaction may be the rate-controlling step. However, if the chemical reaction step is very fast (e.g. at elevated temperatures), mass transfer may well be the rate-controlling step [2]. The overall reaction is very sensitive to temperature if it is controlled by the chemical reaction step, while it typically has a minor dependence on temperature when the process is mass transfer controlled. Also, smaller particle size drives the system towards chemical reaction control, whereas larger particle size can cause the rate-controlling step to shift to mass transfer [5].

With this background, it is necessary to select a reaction model that can adequately describe the complex interactions between the gas and solid phases. The model can assist in determining the characteristic reaction time of a single particle based on kinetic and particle structural properties. The reaction models presented in this chapter are classified based on the initial state of the particle, i.e. porous or non-porous, and the type of reaction. For the reaction models discussed here, we assume that (i) the reaction is irreversible and first order with respect to the gaseous reactant; (ii) the particles are spherical; and (iii) the reaction front retains its original shape during the course of the reaction, and (iv) the system is isothermal. For reversible reactions, the concentration of gaseous reactant A, C_A, can be replaced by $C_A - C_{A,\,equilib}$, and the development is identical. Although the models presented are for first-order kinetics to facilitate the treatment, Langmuir–Hinshelwood-type rate expressions provide a more realistic description of gas–solid reactive systems. Interested readers are referred to Szekely et al. [2] and Levenspiel [6] when these assumptions are not valid.

16.2.1 Reaction Models for Non-porous Particles

Depending on the structural changes that a non-porous particle may experience during the course of a gas–solid reaction, reactions of this type can be categorized into three groups [2]:

1. Shrinking particle reaction, where the particle shrinks during the reaction and may eventually disappear. Gaseous components and unreacted inert material (often very little of it) result from the reaction.

2. Reactions following the shrinking unreacted core model [7] for particles of unchanging overall size. An inert product layer forms and advances from the outer surface of the particle towards its core as the reaction proceeds. When the particle is completely reacted, it is composed wholly of product and is of the same size as the original particle.
3. Reactions with changing overall particle size (size may increase or decrease).

16.2.1.1 Shrinking Particle

Reactions of this type can be generally described by Eq. (16.1). Examples include combustion and gasification of coal, as well as fluorination and chlorination of metals [2]. We assume here that solid B is sufficiently pure that once the reaction is completed, negligible quantities of solid residue remain. Figure 16.2 illustrates the steps that a particle experiences during the reaction.

The overall reaction rate for this category can be controlled by chemical kinetics, external mass transfer, or their combined effects. A condensed version of relevant model equations is provided for each of these cases here.

Chemical Reaction Control The rate of conversion or disappearance of solid reactant B in this case is proportional to the area of the reaction front and can be represented [2, 5, 6] by

$$-\frac{1}{4\pi r_c^2 b}\frac{dN_B}{dt} = -\frac{\rho_s}{b}\frac{dr_c}{dt} = k_c C_{As} \tag{16.4}$$

where b is the stoichiometric coefficient of B, r_c is the radius of the reaction interface, N_B is the number of moles of B, t is time, ρ_s is the molar density of B, k_c is the reaction rate constant based on unit surface area of the solid, and C_{As} is the concentration of A at the reaction interface, i.e. the solid surface. Equation (16.4) can also be presented in dimensionless form [2] as

$$\frac{d\xi}{dt^*} = -1 \tag{16.5}$$

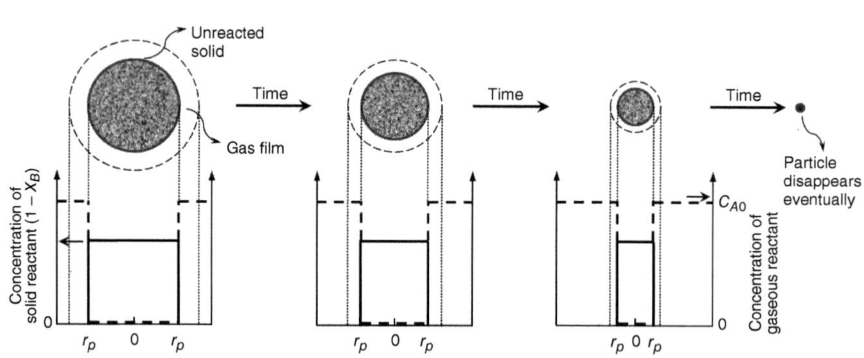

Figure 16.2 Steps in shrinking particle reaction. Solid and gaseous reactant concentration profiles are for chemical reaction control.

where

$$\xi = \left(\frac{A_p}{F_p V_p}\right) r_c \tag{16.6}$$

$$t^* = \frac{t}{\Theta} = \frac{bk_c C_{As}}{\rho_s r_p} t \tag{16.7}$$

Here ξ is the dimensionless position of the shrinking particle or reaction interface; A_p and V_p are the external surface area and volume of the particle, respectively; F_p is a particle shape factor (= 1, 2, or 3 for an infinite slab, long cylinder, and sphere, respectively); t^* is the dimensionless time; Θ is the characteristic reaction time, equal to the time required to completely convert an unreacted particle into product; and r_p is the particle radius. For a spherical particle, ξ is equal to (r_c/r_p). The initial condition for Eq. (16.5) is

$$\text{at } t^* = 0, \quad \xi = 1 \tag{16.8}$$

and the solution of Eq. (16.5) is

$$\begin{aligned} &\xi = 1 - t^* \ \text{ for } \ t^* < 1 \\ &\xi = 0 \ \text{ for } \ t^* \geq 1 \end{aligned} \tag{16.9}$$

For a single particle, the relation between the solid conversion X_B and ξ is expressed by

$$1 - X_B = \xi^{F_p} \tag{16.10}$$

Accordingly, the progress of the reaction in terms of X_B and ξ can be written as

$$t^* = 1 - \xi = 1 - (1 - X_B)^{\frac{1}{F_p}} = g_{F_p}(X_B) \tag{16.11}$$

where $g_{F_p}(X_B)$ is the conversion function and Θ is given by

$$\Theta = t|_{X_B=1} = \frac{\rho_s r_p}{bk_c C_{A0}} \tag{16.12}$$

C_{As} in the case of chemical reaction control is equal to C_{A0} the concentration of A in the bulk. Equation (16.12) shows that Θ is proportional to the particle radius when the chemical reaction resistance is the main resistance.

External Mass Transfer Control For very fast chemical reactions, the gaseous reactant A is quickly consumed at the reaction interface leading to $C_{As} = 0$, and external mass transfer controls the overall reaction rate. The rate of conversion of solid reactant can then be written [2, 6] as

$$-\frac{\rho_s}{b}\frac{dr_c}{dt} = k_D(C_{A0} - C_{As}) = k_D C_{A0} \tag{16.13}$$

where k_D is the mass transfer coefficient. For a spherical particle of fixed size, k_D can be calculated [8] as

$$Sh = \frac{2k_D r_p}{D_m} = 2.0 + 0.6 Re^{\frac{1}{2}} Sc^{\frac{1}{3}} \tag{16.14}$$

where Sh is the Sherwood number, D_m is the molecular diffusivity, Re is the Reynolds number, and Sc is the Schmidt number. For an individual particle reacting with the gas stream, where it shrinks as the reaction advances, k_D can be estimated from the following general expression [2]:

$$Sh = 2.0 + a' Re^{\frac{1}{2}} Sc^{\frac{1}{3}} \left(\frac{r_c}{r_p}\right)^{\frac{1}{2}} \tag{16.15}$$

where a' is a numerical coefficient. Note that this equation solely applies to single or widely dispersed particles surrounded by a gas stream. In a dense fluidized bed of coarse particles with unclouded bubbles, the overall bed mass transfer coefficient is identical to the single particle mass transfer coefficient as the fluidizing gas passes close to plug flow through the bed of solids. However, in a bed of fine particles, which contains bubbles with thin clouds and where most of the gas passes through the bed as bubbles, the overall bed mass transfer coefficient is less than the single particle mass transfer coefficient, owing to the additional mass transfer resistances from bubble to cloud to emulsion gas to particle for the reactant gas and, in the reverse direction, for the product gas [5]. Refer to Kunii and Levenspiel (K–L) [5] for further details. Rearranging Eq. (16.13) into a dimensionless form yields

$$\frac{d\xi}{dt^+} = -\frac{1}{2}\frac{\left(1 + a\xi^{\frac{1}{2}}\right)}{\xi} \tag{16.16}$$

where a is a constant and

$$a = a' Re^{\frac{1}{2}} Sc^{\frac{1}{3}} \tag{16.17}$$

$$t^+ = \frac{t}{\Theta} = \frac{2bD_m C_{A0}}{\rho_s r_p^2} t \tag{16.18}$$

where t^+ is the dimensionless time. The initial condition for Eq. (16.16) is

$$\text{at } t^+ = 0, \quad \xi = 1 \tag{16.19}$$

Integration of Eq. (16.16) with the initial condition defines the progress of reaction in terms of ξ as follows [2]:

$$t^+ = \frac{4}{3a}\left(1 - \xi^{\frac{3}{2}}\right) - \frac{2}{a^2}(1 - \xi) + \frac{4}{a^3}\left(1 - \xi^{\frac{1}{2}}\right) - \frac{4}{a^2}\ln\left(\frac{1 + a}{1 + a\xi^{\frac{1}{2}}}\right) = q(\xi) \tag{16.20}$$

The characteristic reaction time for external mass transfer control can be expressed as

$$\Theta = \frac{\rho_s r_p^2}{bD_m C_{A0}} \tag{16.21}$$

So Θ is proportional to the square of the particle radius or diameter in this case.

Combined External Mass Transfer and Chemical Reaction Control When the relative importance of the chemical reaction and external mass transfer is comparable, their contributions in defining the overall reaction rate must both be taken into account. As these steps act in series, by assuming pseudo-steady-state conditions, the overall rate of conversion of solid reactant is given [2] by

$$-\frac{\rho_s}{b}\frac{dr_c}{dt} = k_c C_{As} = k_D(C_{A0} - C_{As}) \tag{16.22}$$

By solving Eq. (16.22) for C_{As} and rearranging the resulting expression into dimensionless form, we have

$$\frac{dt^*}{d\xi} = -\left[1 + \frac{2\sigma_0^2\xi}{(1 + a\xi^{1/2})}\right] \tag{16.23}$$

Integrating Eq. (16.23) with the same initial condition as for Eq. (16.8), the progress of reaction is

$$t^* = (1 - \xi) + \sigma_0^2 q(\xi) \tag{16.24}$$

$$\sigma_0^2 = k_c r_p / 2D_m \tag{16.25}$$

Here σ_0^2 is a dimensionless parameter representing the relative importance of the chemical reaction vs external mass transfer. When σ_0^2 is large, the overall reaction is governed by the mass transfer resistance, but when σ_0^2 is small, the chemical reaction resistance is dominant [2]. Equation (16.24) reveals that the total time required to achieve any stage of conversion (or ξ) is the sum of the time needed if each resistance acted separately [2, 6]. This indicates that the mass transfer and reaction resistances are cumulative when they are acting in series. Therefore, the overall progress of reaction under the combined effects condition studied here can instead be approximated by Eqs. (16.11) and (16.12), with the reaction rate constant k_c replaced by an overall rate constant, $\overline{k}$ defined [1, 5] by

$$\overline{k} \cong \left(\frac{1}{k_c} + \frac{1}{k_D}\right)^{-1} \tag{16.26}$$

16.2.1.2 Shrinking Unreacted Core Model

In many gas–solid reaction systems of non-porous particles, a solid is produced, and the reaction can be represented by Eq. (16.2). If we assume that some of the stoichiometric coefficients are zero, examples of this type of reaction include the decomposition of metal compounds, oxidation of metals, reduction of metal oxides, and roasting of ores [2]. The most common reaction model to describe this kind of reaction is the shrinking unreacted core model [7], sometimes called the shell progressive model. As shown in Figure 16.3, this model describes the reaction as occurring at a sharp interface or within a narrow region between the unreacted core and a product layer. This product layer, which may or may not be porous, evolves from the outer surface of the particle towards its core as the reaction proceeds. Given that gaseous reactant needs to pass through this layer for

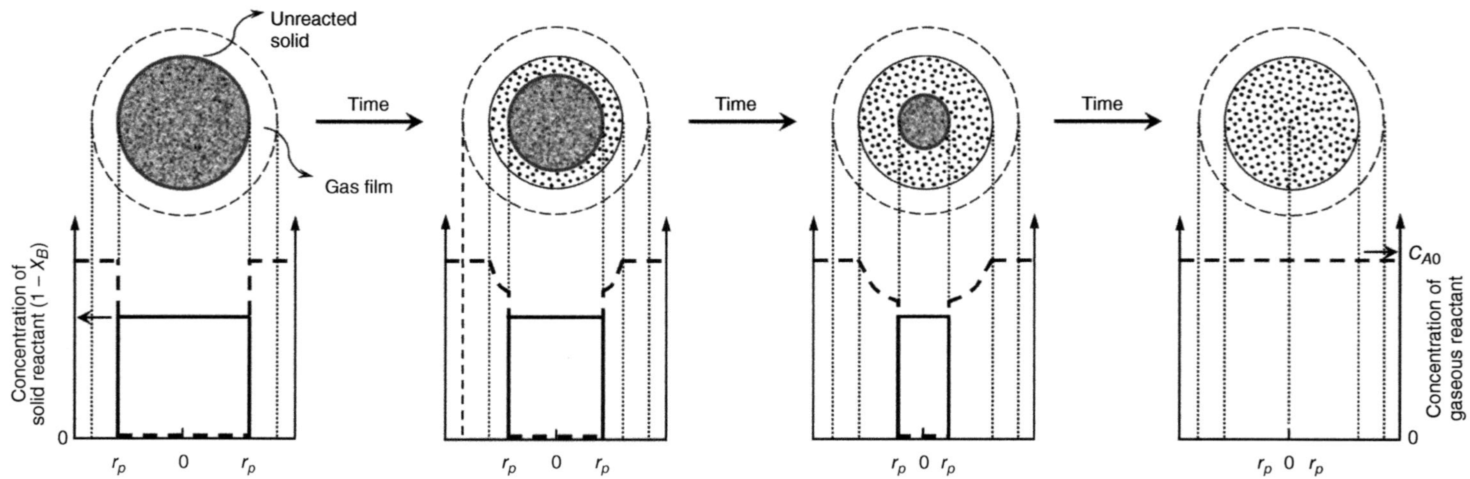

Figure 16.3 Steps in shrinking unreacted core particle reaction. Solid conversion and gas concentration profiles are for cases where the rate of reaction is controlled by internal diffusion.

the reaction to be completed, the effective diffusivity of gas through the porous solid or solid-state diffusion of the reactant plays an important role if the product layer is porous or non-porous, respectively. In both these situations, the model utilizes a pseudo-steady-state assumption, i.e. the migration rate of the reaction interface towards the particle core is markedly lower than the speed at which gas can diffuse to the interface [1, 2]. The rate-controlling step is then either a chemical reaction at the core interface, diffusion through the product layer, or their combined effects.

Chemical Reaction Control When the chemical reaction is rate controlling, the equations are identical to those provided in section "Chemical Reaction Control."

Internal Diffusion Control When the rate of the chemical reaction is high, the overall reaction rate can be controlled by the internal diffusion of gaseous reactant through the product layer. In order to develop the relevant conversion equations in the case of a porous solid product, we assume that the mass transfer within the product layer occurs either by equimolar counter-diffusion or is at a low concentration of a diffusing species. The rate of reaction of A at any instant is then equal to the rate of diffusion of this component through the product layer, given [2, 6] by

$$-4\pi r_c^2 D_e \frac{dC_A}{dr_c} = \text{constant} \tag{16.27}$$

where D_e is the effective diffusivity of A through the product layer and C_A is the concentration of A at any radius r_c. Integrating Eq. (16.27) across the product layer from r_p ($C_A = C_{A0}$) to r_c ($C_A = C_{As} = 0$) and correlating it with the consumption of solid reactant, we obtain

$$-4\pi r_c^2 D_e \frac{dC_A}{dr_c} = \frac{4\pi D_e C_{A0}}{\frac{1}{r_c} - \frac{1}{r_p}} = -\frac{1}{b} 4\pi r_c^2 \rho_s \frac{dr_c}{dt} \tag{16.28}$$

Integrating Eq. (16.28) for the initial condition of reacting solid

$$\text{at } t = 0, \quad r_c = r_p \tag{16.29}$$

we obtain

$$\frac{6bD_e C_{A0}}{\rho_s r_p^2} t = 1 - 3\left(\frac{r_c}{r_p}\right)^2 + 2\left(\frac{r_c}{r_p}\right)^3 \tag{16.30}$$

The characteristic reaction time under this condition can be expressed as

$$\Theta = \frac{\rho_s r_p^2}{6bD_e C_{A0}} \tag{16.31}$$

Rewriting Eq. (16.30) in terms of solid conversion, the progress of reaction is [2]

$$\widetilde{t} = \frac{t^*}{\sigma_s^2} = 1 - 3(1 - X_B)^{\frac{2}{3}} + 2(1 - X_B) = p_{F_p}(X_B) \tag{16.32}$$

where $\tilde{t}$ is the dimensionless time, $p_{F_p}(X_B)$ is the conversion function, and σ_s^2 is the shrinking core reaction modulus for the solid particle, defined as

$$\sigma_s^2 = \frac{k_c r_p}{6D_e} \tag{16.33}$$

The conversion equations reported here were obtained for a particle of unchanging overall size during the course of the reaction. Szekely et al. [2] showed that these equations are also valid for particles of changing size if $0.5 < \mathcal{r} < 2.0$, where $\mathcal{r}$ is the volume of solid product formed from unit volume of solid reactant. If the product layer is non-porous, solid-state diffusion determines the mass transfer across the product layer. Interested readers are advised to consult Szekely et al. [2] for the relevant equations.

Combined Internal Diffusion and Chemical Reaction Control If the chemical reaction and internal diffusion show comparable contributions in determining the overall reaction rate, the overall solid conversion expression must include both contributions. The overall rate of reaction can be obtained by either the progress rate equations for the chemical reaction step, i.e. Eqs. (16.4) and (16.11), or Eq. (16.32) for the internal diffusion step if C_{A0} is replaced by $(C_{A0} - C_{As})$. Accordingly, we have

$$\frac{dX_B}{dt} = \frac{bk_c C_{As}}{\rho_s r_p g'_{F_p}(X_B)} \tag{16.34}$$

and

$$\frac{dX_B}{dt} = \frac{6bD_e(C_{A0} - C_{As})}{\rho_s r_p{}^2 p'_{F_p}(X_B)} \tag{16.35}$$

Solving Eqs. (16.34) and (16.35) for C_{As} and representing the resulting expression in dimensionless form, we obtain

$$t^* = g_{F_p}(X_B) + \sigma_s^2 p_{F_p}(X_B) \tag{16.36}$$

The characteristic reaction time can then be given by

$$\Theta = \frac{\rho_s r_p}{bk_c C_{A0}}(1 + \sigma_s^2) \tag{16.37}$$

When $\sigma_s^2 < 0.1$, the chemical reaction step controls the overall reaction rate. However, the internal diffusion through the product layer is dominant when $\sigma_s^2 > 10$ [2]. An alternative approach similar to Eq. (16.26) can also be adopted here since the internal diffusion and chemical reaction steps take place in series. The overall rate constant $\overline{k}$ can be defined [1, 5, 6] as

$$\overline{k} \cong \left(\frac{1}{k_c} + \frac{r_p}{6D_e}\right)^{-1} \tag{16.38}$$

While the shrinking unreacted core model provides a simple representation of the reaction rate, it cannot adequately describe all non-catalytic gas–solid reactions for non-porous particles with the desired solid products. For instance, for reduction of very dense iron ore, the reaction rate is not well represented by the shrinking unreacted core model as the initial particle is partly converted at the primary stage of the reaction and is followed by further conversion of entrapped islets in the partly converted solid [9]. The crackling core model of Park and Levenspiel [10] can then be adopted instead.

16.2.2 Reaction Models for Porous Particles

Reactions involving porous particles are as important as those involving non-porous particles. Since porous particles may experience different changes in their structural properties, e.g. internal porosity, pore size distribution, and specific surface area, than non-porous particles during the course of the reaction, the simple shrinking unreacted core model is often not applicable to these systems [2]. Hence, relevant alternative reaction models have been developed. Reactions of porous particles can be categorized as (i) reactions that completely consume the particle and (ii) reactions for particles of unchanging overall size.

16.2.2.1 Reactions of Complete Consumption of the Particle

This type of reaction can generally be represented by Eq. (16.1). The combustion of porous graphite and the formation of carbonyls are examples of reactions of this type [2]. The overall rate of reaction can be controlled by either the chemical reaction, the combined effects of chemical reaction and internal diffusion, or external mass transfer. These scenarios occur when the intrinsic reactivity of the solid is low, at an intermediate level, and high, respectively. In the first scenario, the reaction takes place uniformly throughout the porous particle. In the latter cases, however, it occurs near the external surface of the particle, and the interior of particle remains unaffected by when it is exposed to the gaseous reactant [2].

Chemical Reaction Control A molecule of gaseous reactant can penetrate deeply into a porous particle prior to reacting with a solid when the chemical reaction is the rate-controlling step. As a result, the gaseous reactant can be evenly distributed throughout the particle and have a concentration identical to the bulk gas stream [2, 5]. In other words, the solid reactant homogeneously converts the porous particle, and the reaction follows the uniform reaction model [11]. For this case, the reaction rate can be expressed simply in terms of X_B, and the gaseous reactant concentration in the solid particle is obtained [5] by integrating

$$\frac{dX_B}{dt} = k_s C_{A0}(1 - X_B) \tag{16.39}$$

where k_s is the reaction rate constant based on unit volume of solid. The initial condition for Eq. (16.39) is

$$\text{at } t = 0, \quad X_B = 0 \tag{16.40}$$

Integration of Eq. (16.39) with the initial condition defines the conversion as a function of time [5] as

$$(1 - X_B) = \exp(-k_s C_{A0} t) \tag{16.41}$$

The characteristic reaction time for this condition can be obtained by monitoring X_B. A more complicated reaction model was proposed by Petersen [12] for this case, where he assumed that the reaction proceeded by enlarging the internal pores of the particle, while the overall size of the reacting particle remained unchanged until the particle was completely reacted.

Shrinking Porous Particle with Reaction Near the External Surface When the intrinsic reactivity of the solid in a porous particle is not low, gaseous reactant can be consumed before penetrating deeply into the particle. Thus, the reaction typically takes place in a narrow zone close to the external surface of the particle, and the reaction front moves towards the core, similar to the shrinking particle model for non-porous particles [2, 11]. When the intrinsic reactivity of the solid is at an intermediate level, the major difference between porous and non-porous particles is that the reaction occurs at a macroscopic sharp interface for non-porous particles, while the reaction zone is diffuse for porous particles [2]. Consequently, the expressions presented in section "Chemical Reaction Control" can be employed in this case also if discrepancies in the physical nature of these systems are considered. For this purpose, if the reaction rate is first order with respect to the gaseous reactant, k_c is replaced by k_c' defined as

$$k_c' = \sqrt{k_c S_v D_e} \tag{16.42}$$

where S_v is the surface area per unit volume of the particle. For a porous particle, ρ_s in Eq. (16.4) represents the molar density of the pore-free solid. If external mass transfer controls the overall reaction rate, a set of equations similar to those provided in section "External Mass Transfer Control" describes the progress of the reaction (see Szekely et al. [2] for details).

16.2.2.2 Reactions for Porous Particles of Unchanging Overall Size

There are many non-catalytic gas–solid reactions, such as roasting of sulfide ores and reduction of some metal oxides, in which a porous solid product is produced from the porous solid reactant. Reactions belonging to this category can be represented in the general form of Eq. (16.2). The chemical reaction step, internal diffusion through particle interstices, or their combined contributions may govern the overall rate of reaction. Although the overall size of the solid product may be larger or smaller than that of the solid reactant, the change is small, and the overall size of the particle can be assumed to remain invariant as the reaction proceeds [2]. In the reaction models presented in this section, a porous particle is considered to be composed of fine grains, identical to the assumption of the grain model [2, 13, 14]. It is assumed that grains are spherical ($F_g = 3$) and non-porous and that the reaction front retains its original shape within each grain as the reaction proceeds. Figure 16.4 shows a spherical porous particle made up of spherical grains with combined internal diffusion and chemical reaction control.

Chemical Reaction Control If diffusion of the gaseous reactant through particle interstices presents negligible resistance to the progress of the reaction, the concentration of the gaseous reactant is uniform throughout the particle. The gas–solid reaction then occurs homogeneously throughout the particle,

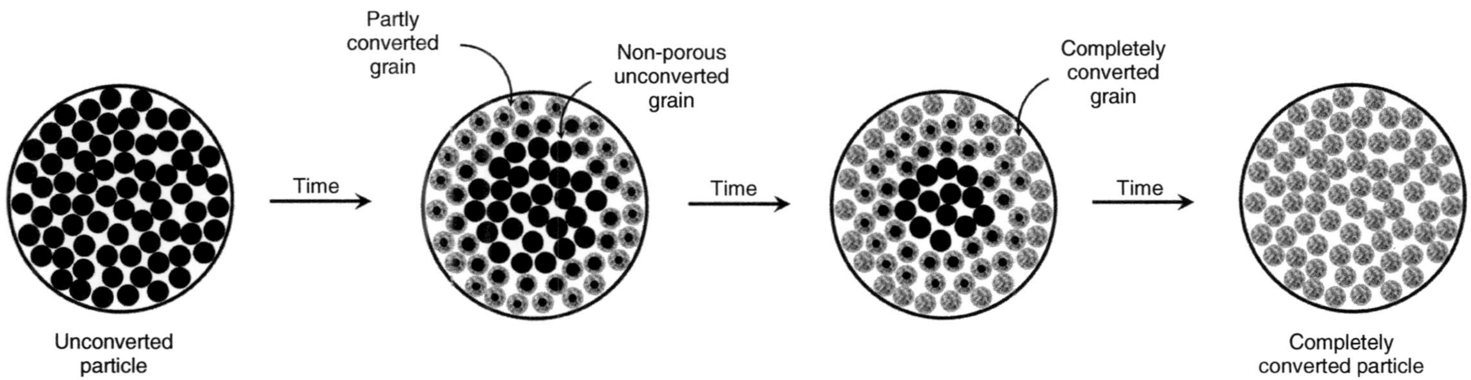

Figure 16.4 Schematic representation of grain model for gas–solid reaction with a porous particle.

and the reaction is that at the grain level. The overall rate of reaction is then determined by the rate of reaction for each grain in the particle structure, given [2] by

$$t^* = \frac{bk_c C_{A0}}{\rho_s r_g} t = g_{F_g}(X_B) + \sigma_g^2 p_{F_g}(X_B) \tag{16.43}$$

$$\sigma_g^2 = \frac{k_c r_g}{6D_g} \tag{16.44}$$

where σ_g^2 is the shrinking core reaction modulus for the grain and D_g is the effective diffusivity of A through the product layer within an individual grain. Accordingly, the characteristic reaction time can be written as

$$\Theta = \frac{\rho_s r_g}{bk_c C_{A0}}(1 + \sigma_g^2) \tag{16.45}$$

The simple reaction model presented in section "Chemical Reaction Control" can alternatively be adopted for this case.

Internal Diffusion Control When the diffusion of the gaseous reactant through the particle interstices acts as the rate-controlling step, the reaction takes place in a narrow zone between the completely reacted layer and the unreacted core. The concentration of reactant gas then drops from C_{A0} to a very small value at the reaction interface. The progress of the reaction in this case is very similar to that predicted by the shrinking unreacted core model with internal diffusion control for a non-porous particle. Hence, the analysis of section "Internal Diffusion Control" can be adapted here when considering differences in the physical nature of porous and non-porous systems. The rate of the reaction progress can be expressed [2] by

$$\tilde{t} = \frac{6bD_e C_{A0}}{(1 - \varepsilon_p)\rho_s r_p^2} t = p_{F_p}(X_B) \tag{16.46}$$

where ε_p represents the volume fraction occupied by the internal porosity and inerts in the porous particle. The characteristic reaction time under this condition can be expressed as

$$\Theta = \frac{(1 - \varepsilon_p)\rho_s r_p^2}{6bD_e C_{A0}} \tag{16.47}$$

Combined Internal Diffusion and Chemical Reaction Control When the overall gas–solid reaction rate of a porous particle is controlled by comparable chemical reaction and internal diffusion resistances, the reaction occurs in a diffuse region between the unreacted core and the completely reacted layer, as shown in Figure 16.4. Unlike the shrinking unreacted core model for non-porous particles under similar overall reaction control mode (discussed in section "Combined Internal Diffusion and Chemical Reaction Control"), the chemical reaction and internal diffusion steps occur in parallel for a porous particle. Sohn and Szekely [14] presented a reaction model for this case based on the grain model. By writing the mass balances for the solid and gas phases on a porous particle considering the aforementioned rate-controlling condition, they obtained a set of partial differential equations that should be solved based on the initial and boundary conditions and relevant information about the shape of a porous particle and its constituent grains. The resulting set of equations has analytical solutions under limited conditions, e.g. if the porous particle is an infinite slab made of flat grains. For other combinations of porous particle and grain shapes, the sets of equations must be solved numerically. An approximate solution for a spherical porous particle made of spherical grains was obtained based on numerical solution [14] as

$$t^* = \frac{bk_c C_{A0}}{\rho_s r_g} t \cong g_{F_g}(X_B) + \widetilde{\sigma}^2 p_{F_p}(X_B) \tag{16.48}$$

where

$$\widetilde{\sigma}^2 = \frac{(1 - \varepsilon_p) k_c r_p^2}{6 D_g r_g} \tag{16.49}$$

is the generalized gas–solid reaction modulus. Hence, the characteristic reaction time is

$$\Theta = \frac{\rho_s r_g}{bk_c C_{A0}}(1 + \widetilde{\sigma}^2) \tag{16.50}$$

When $\widetilde{\sigma}^2 < 0.1$, the chemical reaction step represents the principal resistance for the overall reaction rate, whereas the internal diffusion is the rate-controlling step when $\widetilde{\sigma}^2 > 10$ [2, 14]. If the diffusional resistance in the product layer around the individual grain is comparable with other resistances in the reactive system, the approximate solution and the corresponding Θ [2] are, respectively,

$$t^* \cong g_{F_g}(X_B) + \sigma_g^2 p_{F_g}(X_B) + \widetilde{\sigma}^2 p_{F_p}(X_B) \tag{16.51}$$

$$\Theta = \frac{\rho_s r_g}{bk_c C_{A0}}(1 + \sigma_g^2 + \widetilde{\sigma}^2) \tag{16.52}$$

16.2.3 Reaction Models for Solid–Solid Reactions Proceeding Through Gaseous Intermediates

In addition to the straightforward reaction routes described by Eqs. (16.1) and (16.2) discussed in Section 16.2.1, gas–solid reactions can form a component for more complex reactive systems, e.g. solid–solid reactions proceeding through gaseous intermediates, as represented by Eq. (16.3a). Examples of these reactions include metal oxide reduction by carbon at atmospheric pressure, segregation roasting, and lime pellet concentrate roasting [2]. One can adopt the treatment methods presented in Sections 16.2.1 and 16.2.2 to tackle these systems. Interested readers can obtain more details from Szekely et al. [2].

16.3 Reactions of Solid Particles Alone

In addition to the previously mentioned reactions, where a gaseous reactant participates in the reaction, gas–solid fluidized beds are also recognized as appropriate processing units for the sole reaction of solids. The high heat transfer rate and uniform temperature in these beds are the main driving factors for these applications. Examples of these reactions include carbonization of coal and biomass and calcination of limestone. While larger particles experiencing thermal decomposition may follow the shrinking unreacted core model, the overall rate of reaction can be controlled by heat conduction through the product layer to the heat-absorbing decomposition front. The decomposition reaction can, however, represent the controlling resistance for fine particles [5].

The second group of reactions in this class involves solid–solid reactions. This could be either in the form of variable composition particles, where solid–solid reactions take place between various components in the structure of a solid particle or when the solid reactants are distributed in different grains. Reactions that occur during the thermal processing of minerals to extract rare earth elements and during the production of cement, semiconductor, and electrochemical materials are examples of this type of reaction. Despite their important industrial roles, solid-state kinetics have not been as thoroughly investigated as their counterparts in fluid-phase reactions. Although many of these reactions occur in a network, single-step reaction is normally assumed, and relevant models derived from the gas–solid literature are typically adopted [15]. Heat transfer and diffusion, as well as reaction kinetics, affect solid–solid reactions [16]. If solid reactants are distributed in particulate solids, they need to contact one another, forming the initial surface where reaction occurs. These steps are absent in variable composition particles, as all the solid reactants are already embedded in a particle. However, in both cases, at least one of the reactants must diffuse through a growing product shell [17], within the ambit of the shrinking unreacted core model. The proposed kinetic models describing the mechanisms of these reactions are based on the fact that the product growth (overall reaction rate) is controlled by either diffusion of reactants through a continuous product layer, phase interface reactions, kinetic equations based on the order of reactions, or nuclei growth [17]. Readers are advised to see Tamhankar and Doraiswamy [17] and Khawam and Flanagan [18] for further details on solid-state kinetics.

If gas–solid fluidized beds are adopted for solid–solid mixed-powder reactions, since increasing the number of contacts between the solid reactants is essential, fine/ultrafine powders showing typical Geldart group C behaviour at ambient conditions [19] are likely to be employed. Alternatively, solid reactants can be pelletized prior to injection into a fluidized bed. For the former case, as fluidization of fine/ultrafine powders is challenging, a number of assisting methods have been proposed in the literature. A comprehensive review on the fluidization characteristics of these powders and a summary of the various assisting methods and their corresponding impact on enhancing their fluidization quality have been provided by Shabanian et al. [20].

16.4 Conversion of Particles Bathed by Uniform Gas Composition in a Dense Gas–Solid Fluidized Bed

If the concentration of gaseous reactant bathing single unchanging size particles in a dense gas–solid fluidized bed reactor is fixed, the overall performance of

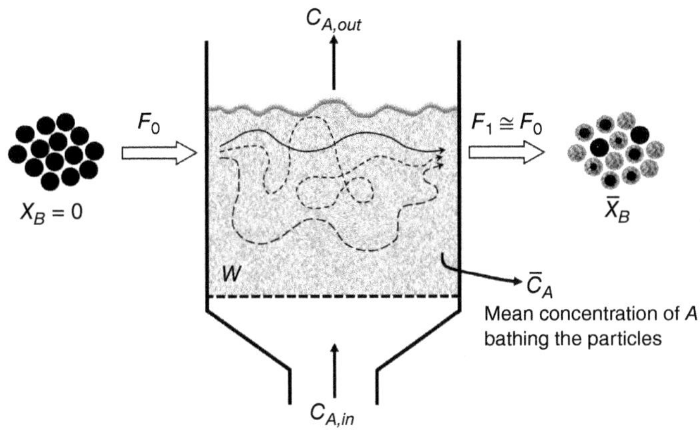

Figure 16.5 Schematic representation of fluidized bed reactor treating particles of identical size employing shrinking unreacted core model.

the reactor is determined by the chemical kinetics of the single particle and the residence time distribution of particles in the bed. Assuming that particles are bathed by the same mean composition of gaseous reactant in the bed is a reasonable approximation when the reaction is slow and the concentration of gaseous reactant remains nearly constant while passing through the bed [5]. A dense gas–solid fluidized bed is a good example of a reactor with a completely mixed flow of solids [6]. Therefore, as shown in Figure 16.5, individual particles may follow different trajectories before leaving the reactor, leading to different particle residence times in the bed. To account for this in a bed of single unchanging size particles and without particle entrainment, a mean conversion $\overline{X}_B$ of solids leaving the reactor needs to be calculated as follows [5, 6]:

$$\begin{pmatrix}\text{Mean value for the}\\ \text{fraction of}\\ \text{unconverted solid } B\\ \text{in the exit solid stream}\end{pmatrix} = \sum_{\substack{\text{particles}\\ \text{of all ages}}} \begin{pmatrix}\text{Fraction of}\\ \text{uncoverted solid } B\\ \text{in particles remaining}\\ \text{in the reactor for}\\ \text{interval } t \text{ to } t+dt\end{pmatrix} \times \begin{pmatrix}\text{Fraction of exit}\\ \text{solid stream that}\\ \text{has stayed for this}\\ \text{time interval in the}\\ \text{reactor}\end{pmatrix} \tag{16.53}$$

$$1-\overline{X}_B = \int_0^{\infty} (1-X_B)_{\substack{\text{single}\\ \text{particle}}} E(t)dt \tag{16.54}$$

where $\boldsymbol{E}(t)$ is the exit age distribution of solids. In a single fluidized bed reactor, particles are well mixed. $\boldsymbol{E}(t)$ is then given [6] by

$$\boldsymbol{E}(t) = \frac{e^{-t/\tau}}{\tau} \tag{16.55}$$

where τ is the mean residence time of solids in the reactor that can be estimated from the bed inventory mass W and the mass flow rate of the solids entering the reactor F_0 [5], i.e.

$$\tau = \frac{W}{F_0} \tag{16.56}$$

Since particles staying longer than Θ do not contribute to $1 - \overline{X}_B$, the upper limit of integration in Eq. (16.54) is Θ rather than ∞ [5].

Equations (16.54) and (16.55) can be employed with various conversion equations from Section 16.2 to provide the overall conversion expressions for a gas–solid fluidized bed reactor under various conditions. For instance, for a shrinking unreacted core particle with the rate of reaction controlled by chemical reaction, the overall conversion can be obtained by substituting Eqs. (16.11) and (16.55) into Eq. (16.54) so that

$$1 - \overline{X}_B = \int_0^{\Theta} \left(1 - \frac{t}{\Theta}\right)^3 \frac{e^{-t/\tau}}{\tau} dt \tag{16.57}$$

Integrating this expression by parts gives

$$1 - \overline{X}_B = 1 - 3\left(\frac{\tau}{\Theta}\right) + 6\left(\frac{\tau}{\Theta}\right)^2 - 6\left(\frac{\tau}{\Theta}\right)^3 \left(1 - e^{-\frac{\Theta}{\tau}}\right) \tag{16.58}$$

For large $\frac{\tau}{\Theta}$ or very high conversion, Eq. (16.58) can be written [5, 6] in equivalent expanded form as

$$1 - \overline{X}_B = \frac{1}{4}\left(\frac{\Theta}{\tau}\right) - \frac{1}{20}\left(\frac{\Theta}{\tau}\right)^2 + \frac{1}{120}\left(\frac{\Theta}{\tau}\right)^3 - \dots \tag{16.59}$$

As particles can be assumed to be perfectly mixed in a dense gas–solid fluidized bed reactor, a considerable portion of the fed solids have brief exit ages. Consequently, to achieve a high conversion of solids in a single-stage fluidized bed reactor, a long mean solids residence time is required. To overcome this hurdle, multi-staging by either countercurrent or crosscurrent flow is recommended [2, 5], with the crossflow arrangement usually preferred due to its relative simplicity [5].

If we assume that the complete conversion of a particle only depends on its total residence time in multiple fluidized bed reactors in series, one can employ Eq. (16.54) with a pertinent $\boldsymbol{E}(t)$ to estimate the overall conversion of solids leaving the system [5]. $\boldsymbol{E}(t)$ for N identical-sized well-mixed stages in series is given [6] by

$$\boldsymbol{E}(t) = \frac{1}{(N-1)!}\left(\frac{t}{\tau_i}\right)^{N-1} \frac{e^{-t/\tau_i}}{\tau_i} \tag{16.60}$$

where the mean residence of solids in each stage, τ_i, can be expressed as

$$\tau_i = \frac{W_i}{F_0} \tag{16.61}$$

Here W_i is the bed inventory in each stage. Substituting Eqs. (16.11) and (16.60) into Eq. (16.54) provides the overall conversion for an equi-sized multi-stage fluidized bed reactor system for shrinking unreacted core particles under chemical reaction control conditions [5]:

$$1-\overline{X}_B=\sum_{k=0}^{k=N-1}\frac{(N-k+2)!}{(N-m-1)!k!}\left(\frac{\Theta}{\tau_i}\right)^{k-3}e^{-\Theta/\tau_i}+\sum_{k=0}^{k=3}\frac{(N+k-1)3!}{(N-1)!k!(3-k)!}\left(-\frac{\tau_i}{\Theta}\right)^{k} \tag{16.62}$$

A simplified form of this expression for large $\frac{\tau_i}{\Theta}$ or a very high conversion is

$$N=2,\quad 1-\overline{X}_B=\frac{1}{20}\left(\frac{\Theta}{\tau_i}\right)^2-\frac{1}{60}\left(\frac{\Theta}{\tau_i}\right)^3+\frac{1}{280}\left(\frac{\Theta}{\tau_i}\right)^4-\dots \tag{16.63}$$

$$N=3,\quad 1-\overline{X}_B=\frac{1}{120}\left(\frac{\Theta}{\tau_i}\right)^3-\frac{1}{280}\left(\frac{\Theta}{\tau_i}\right)^4+\dots \tag{16.64}$$

If the mass of a reacting particle reaches zero during the reaction, its shrinking behaviour can be represented by the core of a shrinking unreacted core model, with the rate controlled by chemical reaction. The mean conversion of solids leaving a single-stage dense gas–solid fluidized bed reactor under this condition can be obtained directly from the mass flow rate of the solids exiting the reactor F_1 as follows [5]:

$$1-\overline{X}_B=\frac{F_1}{F_0}=1-3\left(\frac{\tau}{\Theta}\right)+6\left(\frac{\tau}{\Theta}\right)^2-6\left(\frac{\tau}{\Theta}\right)^3(1-e^{-\Theta/\tau}) \tag{16.65}$$

If the inlet solid stream consists of a size distribution of solids bathed by a gaseous reactant of the same mean composition in the reactor, one has to account for the size distribution, in addition to accounting for the residence time distribution of particles and the governing chemical kinetics of single particles, to estimate the overall conversion of solids leaving the fluidized bed reactor. See Refs. [21] and [5] for detailed information regarding this situation.

16.5 Conversion of Both Solids and Gas

The previous section dealt with gas–solid reaction systems, where particles are bathed by a gaseous reactant whose composition is uniform over the entire reactor. A more general case includes simultaneous variations in the conversion of both solids and gas in the reactor. For instance, any deviations in the rate of the inlet solid stream from its nominal value yield a change in the concentration of the gaseous reactant in the bed. This can alter the progress of reaction for solid particles in the reactor. Hence, for a proper evaluation of the reactor performance, both changes in the conversion of solids and gas must be considered concurrently. The problem addressed in this section illustrates the procedure to be followed in predicting the performance of a gas–solid fluidized bed reactor. A three-step calculation procedure outlined by K–L [5] can be adopted for this purpose.

Step 1 – Evaluate the gas conversion: Assume that the rate of disappearance of gaseous reactant follows a first-order reaction and that its progress can be expressed by the mean first-order reaction rate constant $\overline{K}_r$ and $\overline{C}_A$, the mean concentration of A bathing the particles in the bed. Depending on the fluidization behaviour of solids, pertinent performance equations of the fluidized bed reactor can be adopted to describe the progress of the disappearance of gaseous reactant. If the bed operates in the bubbling fluidization regime, "bubbling bed models" [5, 22–24] (see Chapter 15) for particles showing typical Geldart groups A, B, and D behaviour at ambient conditions can be employed. If the reactor operates in another fluidization regime (turbulent, fast fluidization, or pneumatic conveying regime), one needs to choose a model relevant to that case, as outlined in Chapter 15. We focus on the bubbling flow regime here.

Step 2 – Evaluate the solid conversion: The conversion expression of solid reactant can be obtained by one of the reaction models presented in Section 16.2, assuming that the solids are bathed by $\overline{C}_A$.

Step 3 – Match the overall material balances: In the last step, the total conversions of gaseous reactant A and solid reactant B must be related based on the stoichiometry of the reaction.

While the procedure presented here is applicable to solids of constant and changing sizes, various combinations of reactor flow models and kinetic models can be utilized to represent the progress of the disappearance of gaseous reactant and the progress of the solid-phase reaction. For instance, for an unchanging size solid reactant, four possible extremes may be encountered [5]:

Case I: Fine particles reacting in a bed of fine particles (e.g. calcination of alumina powders).

Case II: Fine particles reacting in a bed of coarse particles (e.g. reaction of fine alkali/alkaline earth metal-based sorbents with polluting gases, e.g. SO_2, in fluidized bed combustors; combustion or gasification of pulverized char in a bed of large ash particles).

Case III: Coarse particles reacting in a bed of fine particles (e.g. reduction of iron ore).

Case IV: Coarse particles reacting in a bed of coarse particles (e.g. combustion or gasification of large coal, biomass, or waste particles in a bed of coarse inert bed material or agglomerated ash).

In the following, we present the calculation procedure for cases I and II. In these derivations, we assume that the bed is composed of particles of uniform size and that it operates under isothermal conditions in the bubbling fluidization flow regime with negligible solids elutriation.

16.5.1 Reactor Performance Calculation for a Bed of Fine Particles (Case I)

The three-step procedure for this case is as follows.

Step 1: If the gas–solid fluidized bed reactor is composed of fine particles that behave like Geldart group A powders at ambient conditions, for the assumptions mentioned above, the K–L fast bubbling bed model [5, 22–24] can be adopted to describe the disappearance of gaseous reactant. This model postulates that the rate of disappearance of gaseous reactant A follows a first-order irreversible reaction [23, 24], i.e.

$$-\frac{1}{V_s}\frac{dN_A}{dt} = \overline{K}_r C_A \tag{16.66}$$

where V_s is the volume of solids in the fluidized bed reactor and N_A is the number of moles of A. $\overline{K}_r$ considers the reaction of A with the solids at various levels of conversion in the bed. According to the K–L bubbling bed model, the concentration of A leaving the reactor, $C_{A,\,out}$, can be given [23, 24] by

$$C_{A,out} = C_{A,in} \exp\left(-K_f \frac{f_b H_f}{U}\right) \tag{16.67}$$

where $C_{A,in}$ is the concentration of A entering the reactor, f_b is the bubble fraction, H_f is the height of fluidized bed, U is the superficial gas velocity, and K_f is the overall effective rate constant given by

$$K_f = \left[\gamma_b \overline{K}_r + \cfrac{1}{\cfrac{1}{K_{bc}} + \cfrac{1}{\gamma_c \overline{K}_r + \cfrac{1}{\cfrac{1}{K_{ce}} + \cfrac{1}{\gamma_e \overline{K}_r}}}}\right] \tag{16.68}$$

where γ_b, γ_c, and γ_e are the volumes of solids dispersed in bubbles, cloud–wake regions, and emulsion phase, respectively, divided by the volume of bubbles. From experimental measurements, γ_b can be taken to be 0.005, while γ_e and γ_c are given [24] by

$$\gamma_c = (1 - \varepsilon_{mf})\left[\cfrac{3}{\cfrac{\varepsilon_{mf} u_{br}}{U_{mf}} - 1} + f_w\right] \tag{16.69}$$

$$\gamma_e = \frac{(1 - \varepsilon_{mf})(1 - f_b)}{f_b} - \gamma_c - \gamma_b \tag{16.70}$$

Here ε_{mf} is the bed voidage at minimum fluidization, U_{mf} is the minimum fluidization velocity, and f_w is the volume fraction of cloud–wake region per bubble volume, which can be estimated as discussed in Chapter 7. u_{br} is the rise velocity of an isolated bubble relative to the emulsion phase [25]:

$$u_{br} = 0.711\sqrt{g d_b} \tag{16.71}$$

where g is the gravity acceleration and d_b is the volume-equivalent bubble diameter. In a bed of fine particles with typical Geldart group A behaviour at ambient conditions, bubbles quickly grow to their equilibrium size, a few centimetres, and stay at that size due to the equilibrium between bubble coalescence and splitting [5, 26]. Hence, it is rational to assume that d_b can be estimated as the equilibrium bubble diameter d_{be} proposed by Horio and Nonaka [27] as

$$d_{be} = \frac{D}{4}\left(-\gamma_m + \left(\gamma_m^2 + \frac{4d_{bm}}{D}\right)^{0.5}\right)^2 \tag{16.72}$$

for $60 \leq d_p \leq 450$ μm, $0.08 \leq D \leq 1.3$ m, and $0 \leq U \leq 0.48$ m/s. γ_m is a parameter of the model, D is the column diameter, and d_{bm} is the maximum bubble diameter from total coalescence of bubbles, expressed [27] as

$$\gamma_m = 7.22 \times 10^{-3} \frac{\sqrt{D/g}}{U_{mf}^{1.2}} \tag{16.73}$$

$$d_{bm} = 2.59g^{-0.2}((U - u_e)A_t)^{0.4} \tag{16.74}$$

Here u_e is the superficial gas velocity in the emulsion phase, and A_t is the bed cross-sectional area. K_{bc} and K_{ce} in Eq. (16.68) represent the coefficients of gas interchange between the bubble and cloud–wake region and between the cloud–wake region and emulsion phase, respectively. They are given [24] by

$$K_{bc} = 4.5\left(\frac{U_{mf}}{d_b}\right) + 5.85\left(\frac{D_m^{0.5} g^{0.25}}{d_b^{1.25}}\right) \tag{16.75}$$

$$K_{ce} = 6.77\left(\frac{D_m \varepsilon_{mf} u_{br}}{d_b^3}\right)^{0.5} \tag{16.76}$$

As industrial fine particle gas–solid fluidized bed reactors are usually operated at many multiples of U_{mf}, i.e. $U \gg U_{mf}$, f_b in a bed of these particles can then be estimated [5] as

$$f_b \cong \frac{U}{u_b} \tag{16.77}$$

where u_b is the rise velocity of bubble through the bed relative to the column wall. The K–L bubbling bed model for a bed of fine particles was established based on the hydrodynamic features of these particles that form very fast-rising thin cloud bubbles with u_{br} considerably larger than the upward velocity of gas at minimum fluidizing condition u_f, i.e. $u_{br} > 5u_f \cong 5U_{mf}/\varepsilon_{mf}$. Owing to this bubble characteristic, downward flow of solids in the emulsion phase of large beds of fine particles is so fast that it can develop gulf-stream circulation of solids, causing gas back-mixing in the emulsion phase. This can, subsequently, result in a deviation between the observed u_b and that estimated by the correlation developed by Davidson and Harrison [25] on the basis of simple two-phase flow theory. Hilligardt and Werther [28] proposed the following correlation to estimate u_b in a bed ($D \leq 1.0$ m) of fine particles while accounting for the column diameter:

$$u_b = 0.8(U - U_{mf}) + 3.2\,D^{\frac{1}{3}}\,u_{br} \tag{16.78}$$

If we assume that the bubble phase is approximately devoid of particles, H_f can be calculated by writing a mass balance for the bed of solids [5]:

$$H_f = \frac{H_{mf}(1 - \varepsilon_{mf})}{(1 - f_b)(1 - \varepsilon_e)} \tag{16.79}$$

where ε_e is the emulsion phase voidage and H_{mf} is the bed height at minimum fluidization. $\overline{C}_A$ can then be evaluated [5] as

$$\begin{aligned}\overline{C}_A &= \frac{1}{H_f}\int_0^{H_f} C_A(z)dz = \frac{1}{H_f}\int_0^{H_f} C_{A,in}\exp\left(-K_f\frac{f_b H_f}{U}\right)dz \\ &= \frac{(C_{A,in} - C_{A,out})U}{\overline{K}_r H_{mf}(1 - \varepsilon_{mf})}\end{aligned} \tag{16.80}$$

Step 2: If the solid reaction can be described by the shrinking unreacted core model with chemical reaction control, Eq. (16.58) presents the overall solid conversion in a single-stage fluidized bed reactor. Θ in the $\overline{C}_A$ relationship can be expressed as

$$\Theta = \frac{\rho_s d_p}{2bk_c\overline{C}_A} \tag{16.81}$$

where d_p is the particle diameter.

Step 3: If the gas–solid reaction is of the general form of Eq. (16.2), for the last step of the calculation, we have [5]

$$b\begin{pmatrix}\text{Disapperance of}\\ \text{gaseous reactant}\\ A\left(\frac{\text{mol}}{\text{s}}\right)\end{pmatrix} = \begin{pmatrix}\text{Disappearance}\\ \text{of solid reactant}\\ B\left(\frac{\text{mol}}{\text{s}}\right)\end{pmatrix} \tag{16.82}$$

$$b(UA_t(C_{A,in} - C_{A,out})) = \left(\frac{F_0}{M_B}\right)\overline{X}_B \tag{16.83}$$

where M_B is the molecular weight of solid reactant B.

16.5.2 Reactor Performance Calculation for a Bed of Coarse Particles (Case II)

Step 1: When the bed is mostly composed of coarse particles, the K–L slow bubbling bed model can represent the progress of the disappearance of gaseous reactant in a bubbling fluidized bed reactor. This is due to the fact that a bed of powders with typical Geldart group D behaviour at ambient conditions contains cloudless, slow-rising bubbles, i.e. $u_b < u_f \cong U_{mf}/\varepsilon_{mf}$. Adopting an identical assumption to that corresponding to Eq. (16.66), the model predicts [24]

$$C_{A,out} = C_{A,in}\exp\left(-\overline{K}_r\frac{U_{mf}(1 - f_b)H_{mf}(1 - \varepsilon_{mf})}{U^2}\right) \tag{16.84}$$

where f_b for these powders can be calculated [24] as

$$f_b = \frac{U - U_{mf}}{u_b + 2U_{mf}} \tag{16.85}$$

In addition, u_b for particles behaving similarly to Geldart group D powders at ambient conditions is given by Hilligardt and Werther [28] for $0.1 \leq D \leq 1.0$ m by

$$u_b = \psi(U - U_{mf}) + 0.87u_{br} \tag{16.86}$$

$$\text{with } \psi = \begin{cases} 0.26 & \text{for } \frac{h}{D} < 0.55 \\ 0.35\sqrt{\frac{h}{D}} & \text{for } 0.55 \leq \frac{h}{D} \leq 8 \\ 1 & \text{for } \frac{h}{D} > 8 \end{cases} \tag{16.87}$$

Accordingly, $\overline{C}_A$ can be obtained [5] as

$$\overline{C}_A = \frac{1}{H_f}\int_0^{H_f} C_{A,in} \exp\left(-\overline{K}_r \frac{U_{mf}(1-f_b)H_{mf}(1-\varepsilon_{mf})}{U^2}\right) dz$$
$$= \frac{(C_{A,in} - C_{A,out})U^2}{\overline{K}_r U_{mf}(1-f_b)H_{mf}(1-\varepsilon_{mf})} \tag{16.88}$$

Step 2: As fine particles are reacting with the gaseous reactant, adopting the same assumption as in step 2 of the previous case, Eqs. (16.58) and (16.81) apply here, too.

Step 3: Equation (16.83) applies here.

For cases III and IV, the same performance equations as presented in Sections 16.5.1 and 16.5.2 can be employed, while step 2 of those calculations should be adjusted based on a proper kinetic model for the reaction of coarse particles with the gaseous reactant.

16.6 Thermal Conversion of Solid Fuels in Fluidized Bed Reactors

The thermal conversion of solid fuels includes pyrolysis, gasification, and combustion processes. Pyrolysis is the thermal decomposition of the chemical bonds of a targeted carbonaceous material due to supplying heat at elevated temperatures in an oxygen-free environment. It produces gas and liquid products, as well as a solid residue that is richer in carbon. Gasification operates at higher temperatures than pyrolysis and converts the carbonaceous materials into syngas (a mixture of hydrogen and carbon monoxide, as its main product), with less oxygen present than needed for complete combustion. Carbon dioxide, methane, and steam are also gaseous products of this process, while tar, composed of high molecular weight aromatics, is produced as a by-product. During combustion, complete oxidization of carbonaceous materials occurs in an oxygen-rich environment to principally produce heat, carbon dioxide, and steam.

Solid fuel combustion is both a ubiquitous and complex non-catalytic gas–solid reaction. Combustion of solid fuels in a bed of inert particles involves multiple gas–gas and gas–solid reactions. Gas–gas reactions are mainly between the volatiles from the solid fuel interacting with other gaseous molecules, particularly oxygen, in the fluidizing gas. Gas–solid reactions are principally due either to the reaction between the remaining char with oxygen in the fluidizing gas or air-polluting gases produced during the combustion process with solid sorbents. In addition, oxidation and reduction reactions that take place for the oxygen carrier (OC) in oxygen carrier aided combustion (OCAC) or chemical looping processes [29, 30] represent other gas–solid reactions that occur during solid fuel combustion.

For successful design/performance evaluation of solid fuel fluidized bed combustors, knowledge of the order of magnitude of three characteristic times, related to devolatilization of solid particles, dispersion of solid particles in the bed, and reaction of remaining char, is essential. These characteristic times strongly depend on particle size. For instance, for combustion of large coal particles in a bed of large inert particles, the times related to the devolatilization, mixing, and reaction of the remaining solids are of order 100, 10, and 1000 seconds, respectively. However, they are respectively of order 1, 10, and 100 seconds when fine particles are combusted in a bed of large particles [5]. In addition, due to a shortage of conventional resources and stringent environmental constraints, and to support the growing needs of society, high-rank coal is being replaced by other fuel sources and blends of solid fuels, such as biomass, low-rank coal, urban and industrial wastes, and petroleum coke. Since these solid fuels have various origins, their physical, chemical, and reactive properties differ. Thus, proper characterizations are essential. If solid combustion is attempted with the aid of an OC, knowledge of the OC's oxidation and reduction characteristic times is also critical.

Due to the presence of fuel-bonded contaminants and those generated during combustion, air-blown coal power plants have long been recognized as major sources of air pollution through the emission of air pollutants, such as particulate matter, CO_2, SO_x, NO_x, Hg, HCl, trace elements (e.g. heavy metals), dioxins, furans, and aromatics. Carbon dioxide from fossil fuel combustion is the most important greenhouse gas contributing to climate change. Hence, mitigating this emission has received great attention, and several technologies are being widely investigated to enable carbon capture, utilization, and storage. Oxy-fuel and chemical looping combustion (CLC) technologies are among the most promising combustion technologies for power plants and industrial applications with inherent CO_2 capture [31, 32]. Oxy-fuel combustion is based on the removal of nitrogen from the oxidizer to burn the fuel with an oxygen-enriched gas, e.g. oxygen diluted with recycled CO_2. Recycling the CO_2-enriched flue gas (i) lowers the combustion temperature and (ii) acts as a heat carrier. By manipulating the recycle ratio, one can achieve heat transfer characteristics similar to air-blown combustors, making retrofits of existing coal-fired power plants possible [31]. Under these conditions, a flue gas is primarily composed of CO_2 and water, as well as small concentrations of impurities, like oxygen. The CO_2 is captured from the flue gas by condensing the water. A further purification step, e.g. catalytic deoxidation, provides a high-purity CO_2 stream for utilization/sequestration while meeting pipeline specifications.

In chemical looping, direct contact between air and fuel is prevented; instead, oxygen is transferred from the air to the fuel by a solid OC, normally a metal oxide. The OC experiences a redox cycle, where it is reduced in a fuel reactor to supply oxygen and re-oxidized again by air in an air reactor [32]. A general schematic of the CLC process using solid fuels, together with the principal reactions taking place in the air and fuel reactors, is presented in Figure 16.6a. Critical characteristics of a suitable OC for a CLC process include an appropriate oxygen transport capacity, high reactivity during multiple redox cycles, high selectivity towards CO_2 and H_2O, low attrition, high chemical stability and mechanical strength, enhanced resistance to agglomeration, low production costs, and minimal environmental impacts.

Depending on the oxygen release properties of the OC in the fuel reactor, the CLC process can be accomplished through in situ gasification chemical looping combustion (iG-CLC), chemical looping with oxygen uncoupling (CLOU), and chemical looping assisted by oxygen uncoupling (CLaOU). In the iG-CLC process, which has attracted great interest owing to the application of low-cost OCs, the OC does not generate gaseous oxygen in the fuel reactor environment. However, as depicted in Figure 16.6b, it reacts with the gas products of the pyrolysis and gasification of solid fuel in the fuel reactor (reactions R1–R3 in Figure 16.6a), following reaction R4, to produce CO_2 and H_2O. The water–gas shift reaction (R5 in Figure 16.6a) assists in modifying the gas composition in the fuel reactor. Almost pure CO_2 is obtained once the steam is condensed after polishing the flue gas. Gasification of char with H_2O and CO_2 is inherently slow, leading to a low overall reaction rate in the fuel reactor. This causes some char particles to be partially converted and transferred to the air reactor with the reduced OC. This, in turn, results in conventional combustion of char in the air reactor, decreasing the CO_2 capture efficiency. Installation of a carbon stripper at the exit of the fuel reactor towards the air reactor can assist in minimizing this undesirable phenomenon. In addition, CLOU (Figure 16.6c) can help resolve this problem as the OC releases its oxygen into the fuel reactor, making it possible to combust the solid fuel via gas-phase oxygen, as in normal combustion. Therefore, CLOU does not require steam gasification, slow gasification is avoided, and the CO_2 recirculated to the fuel reactor is sufficient to act as the fluidizing agent. Despite this advantage, the CLOU process suffers from the cost of synthetic OC. Research studies on developing low-cost OCs with oxygen uncoupling capability have resulted in some combined oxide OCs (e.g. Mn with other metals, such as Ca, Mg, Cu, Fe, or Si) that have not shown oxygen release capacities as high as copper oxide, the most efficient OC for the CLOU process. However, the CLC process with these combined oxides primarily operates as iG-CLC with an additional gaseous oxygen contribution. This results in the third avenue for the CLC process, CLaOU [32].

Chemical looping technology can also be adopted for hydrogen production when an appropriate OC is selected with complete control over its oxidation state during the process. In this process, known as chemical looping reforming (CLR), the air-to-fuel ratio is maintained low to avoid full oxidation of fuel to CO_2 and H_2O. Hence, a nitrogen-free gas stream concentrated in H_2 and CO is obtained from the fuel reactor [29] (see Adánez et al. [29, 32] for more information on the CLC and CLR processes).

Co-combustion of coal and biomass or municipal solid waste (MSW) is another technique to control greenhouse gas emissions from coal power plants. The partial substitution of coal by biomass can reduce the net generated CO_2

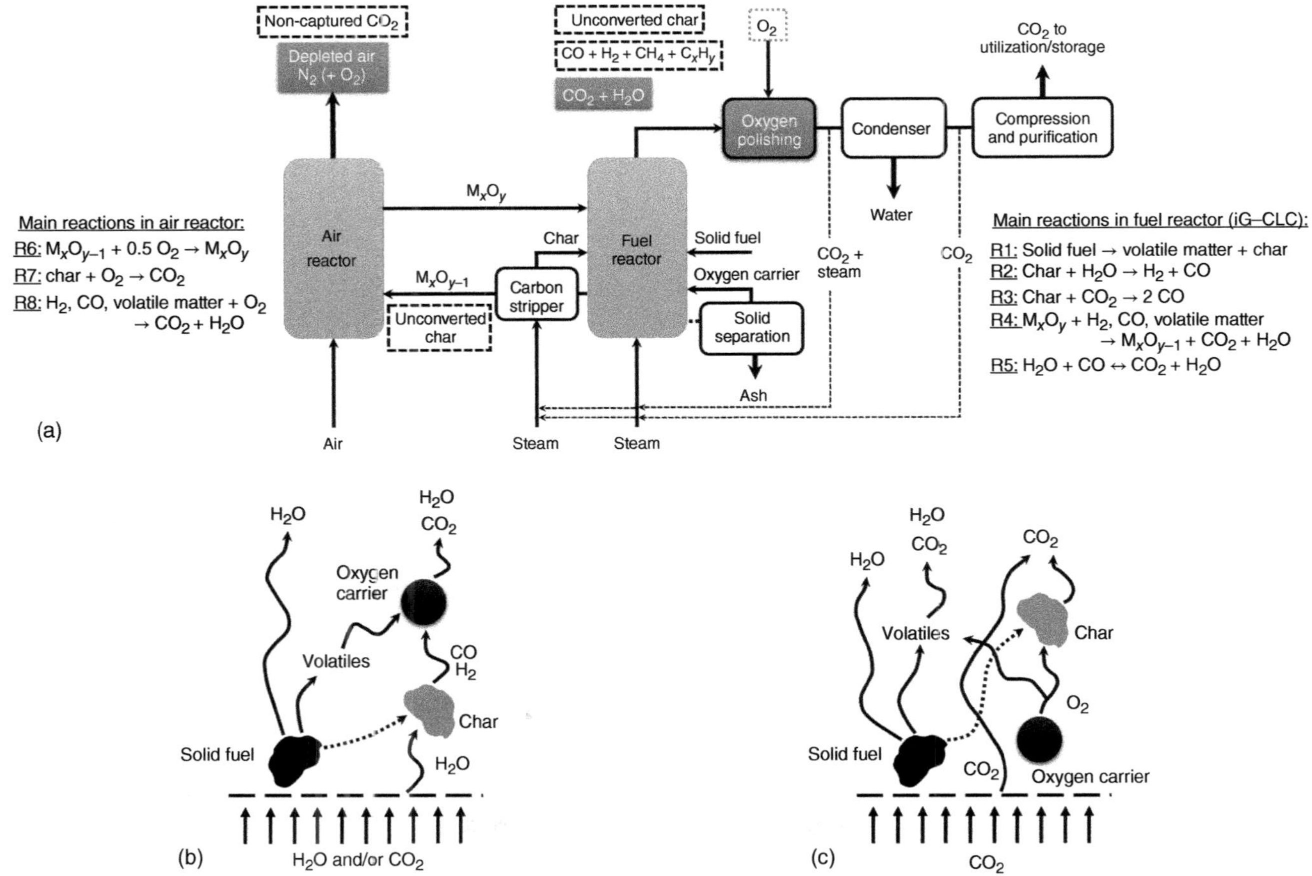

Figure 16.6 Schematic representation of (a) CLC, (b) iG-CLC, and (c) CLOU processes for solid fuel combustion. Source: Adapted from Adánez et al. 2018 [32].

as biomass combustion has the potential to be CO_2 neutral, particularly for agricultural residues and energy plants. CO_2 released during combustion in these plants is removed from the atmosphere during their growth by photosynthesis [33]. Partial substitution of coal by MSW is also beneficial in reducing SO_x and NO_x emissions since MSW generally contains negligible sulfur and less nitrogen than coal. The presence of renewable fractions in MSW can also assist in decreasing fossil fuel CO_2 generated by coal power plants. This strategy also helps limit landfill requirements of MSW and resulting emissions of methane, a more powerful greenhouse gas than CO_2 in terms of global warming. In addition to reducing emissions by the partial substitution of coal by MSW, alkali and alkaline earth sorbents can be added to the coal-waste solid fuel, further contributing to air emission control. This offers better emission control performance than direct sorbent injection as sorbents experience longer residence times in the reactor environment when bound to larger waste particles [34]. Despite these advantages, co-combustion of coal and biomass/MSW can cause defluidization of fluidized bed combustors and deposition issues due to the presence of alkali and alkaline earth elements in the solid fuel mixture [33, 34].

Combustion performance in thermal power plants is highly influenced by the contacting of fuel and air; achieving good contacting is most challenging when burning solid fuels. Insufficient contact between fuel and air leads to reducing regions in an otherwise oxidizing bed and, hence, undesirable outcomes, e.g. emissions of carbon monoxide, unburnt hydrocarbons, and char particles in the exhaust stream. Excess air (exceeding stoichiometric requirements) is typically employed to mitigate this problem. This strategy can, however, decrease thermal efficiency and increase fuel consumption, boiler and environmental footprints, and heat and power generation expenditures [30]. An alternative strategy is to have a multitude of fuel and oxidant injection points in a large-scale combustor. However, this negatively influences construction and operation expenditures, as well as reliability [35]. OCAC, a combustion concept inspired by the CLC process, is recommended to minimize local regions with insufficient oxygen. Through this approach, the inert bed material is partially or completely replaced by an OC material, similar to that adopted in CLC. The active bed material is oxidized in oxygen-rich regions of the combustion chamber and reduced in fuel-rich regions. Numerous advantages under both air and oxy-fired conditions achieved by this strategy are detailed by Rydén et al. [30] and Hughes et al. [35].

16.7 Final Remarks

1. Necessary steps in designing and developing commercial-scale fluidized beds for non-catalytic gas–solid reactions involve thermodynamic evaluation, kinetic study, and defining the operating regime/conditions when fluidization provides the best contacting mode to accomplish and scale-up the process. Thermodynamic evaluation is the first step to determine whether or not the reaction is thermodynamically feasible and what precautions need to be taken into account. Great caution must be exercised in collecting representative

kinetic data for the solid reactant in the reactor environment. Pilot-scale studies are required to provide information on the handling characteristics of the material (softening, agglomeration, attrition, etc.) [2] and hydrodynamic and reaction characteristics of the fluidized bed reactor.

2. The reaction models discussed in this chapter assume that no structural changes occur to the particle during the reaction. However, changes are inevitable in many cases; e.g. sintering at elevated temperatures can increase the diffusional resistance and cause agglomeration. While the former decreases the reactor performance, the latter may lead to complete defluidization and plant shutdown. With lower quality feedstocks and more stringent environmental regulations, chemical processes may be required to operate under extreme operating conditions, where agglomeration and defluidization are more likely. Thus, defluidization conditions must be identified in advance, and relevant counteractions must be implemented.
3. Fluidized beds are attractive for solids handling and providing bed temperature uniformity, a requisite for highly exothermic or endothermic reactive systems. However, there are two main drawbacks:
   i. When solids are the products of gas–solid reactions, fluidized bed reactors often operate in the dense (bubbling or turbulent) fluidization regime, where particles are nearly perfectly mixed. Hence, high solid conversions require long residence times owing to the wide residence time distributions of solids in the bed. This can be mitigated through the multi-staging of solids [2, 5].
   ii. Fluidized beds are not efficient in handling particles subject to cohesive interparticle forces, possibly leading to agglomeration [2].
4. Application of the computational fluid dynamics (CFD)–discrete element method (DEM) as a computational tool to simulate gas–solid flows has grown rapidly in recent decades. It can assist in decreasing costs associated with optimization, safety and scale-up, also eliminating the need to entirely rely on difficult and expensive experiments. Although CFD–DEM simulations have been widely employed by the academic community, their computational overhead, particularly for large multiphase systems, has created a significant barrier to widespread adoption by industry. Another challenge with the CFD–DEM simulations is that interparticle forces have not been fully addressed in calculations. Hence, agglomeration and deagglomeration of particles have not yet been adequately captured. Integrating non-catalytic gas–solid reactions into the CFD–DEM simulation offers valuable information about the operation, performance, design, and scale-up of fluidized bed reactors handling such reactions. Ongoing research with computational and hardware improvements will spur the adoption of CFD–DEM by industry [36].

Solved Problems

16.1 Spherical non-porous graphite particles of size $r_p = 500$ μm are burnt in a stagnant environment of 10% oxygen at 1000 °C and 10^5 Pa total pressure. The stoichiometry of the reaction is simply $C + O_2 \rightarrow CO_2$.

Assuming that the reaction is irreversible and first order with respect to oxygen, (a) calculate the time needed for complete conversion of the particles. (b) Repeat the calculations for particles of size $r_p = 50\ \mu m$.
Data: $\rho_p = 2600\ kg/m^3$, $k_c = 0.2\ m/s$, $D_m = 10^{-4}\ m^2/s$

Solution

(a) We initially calculate the concentration of gaseous reactant in bulk and molar density of pure solid reactant:

$$C_{A0} = \frac{P_{A0}}{RT} = \frac{y_{A0}P_T}{RT} = \frac{0.1 \times 10^5}{8.314 \times 1273} = 0.945\ mol/m^3$$

$$\rho_s = \frac{\rho_p}{M_B} = \frac{2600}{0.012} = 216\,666\ mol/m^3$$

From Eqs. (16.14) and (16.26), in a stagnant atmosphere,

$$\bar{k} \cong \left(\frac{1}{k_c} + \frac{1}{k_D}\right)^{-1} = \left(\frac{1}{k_c} + \frac{r_p}{D_m}\right)^{-1} = \left(\frac{1}{0.2} + \frac{0.0005}{10^{-4}}\right)^{-1} = 0.10\frac{m}{s}$$

This indicates that chemical reaction and mass transfer resistances are similar for these conditions. We reach the same conclusion from Eq. (16.25). From the stoichiometry, we have $b = 1$. Thus, from Eq. (16.12),

$$\Theta = \frac{\rho_s r_p}{b\bar{k}C_{A0}} = \frac{216\,666 \times 0.0005}{1 \times 0.10 \times 0.945} = 1146.4\ s$$

(b) For particles of $r_p = 50\ \mu m$, Eqs. (16.14) and (16.26) in a stagnant atmosphere provide

$$\bar{k} \cong \left(\frac{1}{k_c} + \frac{r_p}{D_m}\right)^{-1} = \left(\frac{1}{0.2} + \frac{0.000\,05}{10^{-4}}\right)^{-1} = 0.18\ m/s$$

This indicates that the chemical reaction step acts as the rate-controlling step for this condition. From Eq. (16.12), we have

$$\Theta = \frac{\rho_s r_p}{b\bar{k}C_{A0}} = \frac{216\,666 \times 0.000\,05}{1 \times 0.18 \times 0.945} = 63.7\ s$$

16.2 Experimental results for the conversion of spherical non-porous particles of uniform size with a reacting gas in a lab-scale fluidized bed reactor indicate that for the following reaction stoichiometry,

$$A_{(g)} + B_{(s)} \rightarrow C_{(g)} + D_{(s)}$$

the solid reactant experiences a shrinking unreacted core chemical reaction, while the reaction step is the rate-controlling step. (a) Design a commercial unit for 99% conversion of 5.0 ton/h of solid feed to three identical dense gas–solid fluidized bed reactors in series. (b) Illustrate how much bed weight reduction can be achieved by multi-staging compared with a single-stage fluidized bed reactor.

Data: $d_p = 500\ \mu m$, $\rho_p = 2600\ kg/m^3$, $\Theta = 1.0\ h$, $\varepsilon_{mf} = 0.45\ (-)$, $H_{mf} = 1.25D$

Solution

(a) From Eq. (16.64) we have

$$1 - \overline{X}_B = 1 - 0.99 = \frac{1}{120}\left(\frac{1.0}{\tau_i}\right)^3 - \frac{1}{280}\left(\frac{1.0}{\tau_i}\right)^4 \ (-)$$

from which the mean residence time of solids in each stage can be calculated as

$$\tau_i \cong 0.5\ h$$

From Eq. (16.61), the bed inventory in each stage is

$$W_i = F_0\tau_i = 5 \times 0.5 = 2.5\ ton$$

Accordingly, the required dimensions for each stage are

$$2.5\ ton = \frac{\pi}{4}D^2(1.25D) \times 2600 \times (1 - 0.45)/1000$$

$$D \cong 1.2\ m,\quad H_{mf} \cong 1.5\ m$$

(b) If the same process is carried out in a single gas–solid fluidized bed reactor, from Eq. (16.59), we have

$$1 - \overline{X}_B = 1 - 0.99 = \frac{1}{4}\left(\frac{1.0}{\tau}\right) - \frac{1}{20}\left(\frac{1.0}{\tau}\right)^2 + \frac{1}{120}\left(\frac{1.0}{\tau}\right)^3 \ (-)$$

from which the mean residence time of solids in the fluidized bed reactor is

$$\tau \cong 24.8\ h$$

From Eq. (16.56), the required bed inventory is

$$W = F_1\tau = 5 \times 24.8 = 124.0\ ton$$

The extent of bed weight reduction by multi-staging is then

$$\left(\frac{\begin{array}{l}\text{Total bed weight for single bed}\\ -\text{Total bed weight for three equal}\\ -\text{Sized beds in series}\end{array}}{\text{Total bed weight for single bed}}\right) \times 100$$

$$= \frac{124.0 - (3 \times 2.5)}{124.0} \times 100 = 94\%$$

Therefore, three equi-sized fluidized bed reactors in series can decrease the total bed weight to 6% of that of a single reactor, resulting in a major improvement in both capital and operating costs of the process.

16.3 Suppose a zinc oxide production company plans to retrofit a fluidized bed reactor of 6.5 m ID previously employed for calcination of alumina powders to roast irregular-shaped zinc blend particles, fluidized by air at 960 °C and 1 atm. The stoichiometry of the reaction is

$$\frac{2}{3}\text{ZnS} + \text{O}_2 \rightarrow \frac{2}{3}\text{ZnO} + \frac{2}{3}\text{SO}_2$$

To remove the exothermic heat of the reaction and decrease the gulf-stream circulation of fine solids in the bed, 564 vertical heat exchanger tubes of 0.1 m outer diameter, d_i, are placed in the reactor in a square pitch layout with a tube pitch, p_t, of 0.25 m. The company aims for a production rate of 110 ton/d. The reactor operates at $U = 0.45$ m/s. With the help of data provided below and adopting the three-step procedure, determine a proper value for H_{mf} to achieve a mean conversion of at least 99% for the solids. Suppose that the solid product is discharged through an overflow tube. The reactor operates at $H_{mf} \leq 1.2$ m to minimize solids entrainment.

Data: d_p = 60 μm, ρ_p = 4100 kg/m³, U_{mf} = 0.0018 m/s, ε_{mf} = 0.5 (−), D_m = 2.5 × 10⁻⁴ m²/s, γ_b = 0.005 (−), $M_B = M_{\text{ZnS}}$ = 0.097 44 kg/mol, M_{ZnO} = 0.081 39 kg/mol

Assume that the solid reaction follows the shrinking unreacted core model with chemical reaction control and that $k_c = 0.015$ m/s.

Solution

The desirable production rate can be converted to the mass flow rate of feed solids as follows:

$$F_0 = 110\,000 \times \frac{0.097\,44}{0.081\,39} \times \frac{1}{24 \times 3600} = 1.524 \text{ kg/s}$$

To start the calculations, consider H_{mf} = 1.0 m. We then follow the three-step procedure to evaluate the conversions of solids and gas and to determine whether or not the desired solid conversion will be achieved.

Step 1 – Evaluate the gas conversion: According to the Geldart powder classification [19], the powders for this process are close to the group A/B boundary. Let us assume that a bubbling bed of these fine particles contains fast-rising bubbles with thin clouds at an equilibrium size (to be verified below). Also, we postulate that cohesive interparticle forces vary little with the operating temperature; thus, the hydrodynamic correlations developed based on the experimental data at near ambient conditions are adopted for the high-temperature conditions [37] encountered here. Moreover, since $U \gg U_{mf}$ and the bed is composed of fine particles, $u_e \cong U_{mf}$.

As heat exchanger tubes are embedded in the reactor, an equivalent hydraulic bed diameter d_{te} should be calculated first (see Chapter 18).

$$d_{te} = 4\frac{p_t^2 - \frac{\pi}{4}d_i^2}{\pi d_i} = 4\frac{0.25^2 - \frac{\pi}{4} \times 0.1^2}{\pi \times 0.1} = 0.696 \text{ m}$$

We can then employ d_{te} in place of D to calculate the bubble size and velocity. From Eqs. (16.72) to (16.74),

$$\gamma_m = 7.22 \times 10^{-3}\frac{\sqrt{0.696/9.81}}{0.0018^{1.2}} = 3.782\,(-)$$

$$d_{bm} = 2.59 \times 9.81^{-0.2}\left((0.45 - 0.0018)\frac{\pi}{4} \times 0.696^2\right)^{0.4} = 0.808 \text{ m}$$

$$d_b \cong d_{be} = \frac{0.696}{4}\left(-3.782 + \left(3.782^2 + \frac{4 \times 0.808}{0.696}\right)^{0.5}\right)^2 = 0.057 \text{ m}$$

From Eq. (16.71),

$$u_{br} = 0.711\sqrt{9.81 \times 0.057} = 0.532 \text{ m/s}$$

$$\text{So that } \frac{u_{br}}{u_f} = \frac{u_{br}}{U_{mf}/\varepsilon_{mf}} = \frac{0.532}{0.0018/0.5} = 147.77\,(-)$$

This ratio indicates that the condition of applying the K–L fast bubbling bed model, i.e. $u_{br} > 5u_f \cong 5U_{mf}/\varepsilon_{mf}$, is well satisfied. Thus, we employ this model to evaluate the performance of the reactor for the disappearance of gaseous reactant. From Eqs. (16.77) and (16.78), we have

$$u_b = 0.8(0.45 - 0.0018) + 3.2 \times 0.696^{\frac{1}{3}} \times 0.532 = 1.867 \text{ m/s}$$

$$f_b = \frac{0.45}{1.867} = 0.241\,(-)$$

f_w for irregular particles 60 μm in size can be obtained from Rowe and Partridge [38] or Chapter 7 as 0.21. From Eqs. (16.69) and (16.70),

$$\gamma_c = (1 - 0.5)\left[\frac{3}{\frac{0.5 \times 0.532}{0.0018} - 1} + 0.21\right] = 0.115\,(-)$$

$$\gamma_e = \frac{(1 - 0.5)(1 - 0.241)}{0.241} - 0.115 - 0.005 = 1.455\,(-)$$

From Eqs. (16.75) and (16.76),

$$K_{bc} = 4.5\left(\frac{0.0018}{0.057}\right) + 5.85\left(\frac{(2.5 \times 10^{-4})^{0.5}(9.81)^{0.25}}{(0.057)^{1.25}}\right) = 6.020 \text{ s}^{-1}$$

$$K_{ce} = 6.77\left(\frac{2.5 \times 10^{-4} \times 0.5 \times 0.532}{0.057^3}\right)^{0.5} = 4.057 \text{ s}^{-1}$$

From the above calculations and Eq. (16.68),

$$K_f = \left[0.005\overline{K}_r + \frac{1}{\dfrac{1}{6.020} + \dfrac{1}{0.115\overline{K}_r + \dfrac{1}{\dfrac{1}{4.057} + \dfrac{1}{1.455\overline{K}_r}}}}\right] \text{s}^{-1} \tag{i}$$

From the ideal gas law,

$$C_{A,in} = \frac{P_{A,in}}{RT} = \frac{y_{A,in}P_T}{RT} = \frac{0.21 \times 10325}{8.314 \times 1233} = 2.076 \text{ mol } O_2/\text{m}^3$$

From Eqs. (16.79) and (16.67), and assuming $\varepsilon_e \cong \varepsilon_{mf}$,

$$H_f = \frac{1.0 \times (1 - 0.5)}{(1 - 0.241)(1 - 0.5)} = 1.32 \text{ m}$$

$$\frac{f_b H_f}{U} = \frac{0.241 \times 1.32}{0.45} = 0.707 \text{ s}$$

$$C_{A,out} = C_{A,in} \exp\left(-K_f \frac{f_b H_f}{U}\right) = 2.076 \exp(-0.707K_f) \text{ mol } O_2/\text{m}^3 \tag{ii}$$

From Eq. (16.80),

$$\overline{C}_A = \frac{(2.076 - C_{A,out}) \times 0.45}{\overline{K}_r \times 1.0 \times (1 - 0.5)} = \frac{0.9 \times (2.076 - C_{A,out})}{\overline{K}_r} \text{ mol } O_2/\text{m}^3 \tag{iii}$$

Step 2 – Evaluate the solid conversion: First, we need to determine ρ_s and then Θ from Eq. (16.81).

$$\rho_s = \frac{4100 \; kg/m^3}{0.097\,44 \frac{kg}{mol} ZnS} = 42\,077 \text{ mol ZnS/m}^3 \text{ solid}$$

$$\Theta = \frac{42\,077 \times 60 \times 10^{-6}}{2 \times \frac{2}{3} \times 0.015 \times \overline{C}_A} = \frac{126.23}{\overline{C}_A} \text{ s}$$

The mean residence time of solids in the reactor can be obtained by Eq. (16.56) and the calculations:

$$A_{te} = \frac{\pi}{4}D^2 - N_t \frac{\pi}{4} d_i^2 = \frac{\pi}{4} \times 6.5^2 - 564 \times \frac{\pi}{4} \times 0.1^2 = 28.739 \text{ m}^2$$

$$W = H_{mf}(1 - \varepsilon_{mf})A_{te}\rho_p = 1.0 \times (1 - 0.5) \times 28.739 \times 4100 = 58\,915 \text{ kg}$$

$$\tau = \frac{W}{F_0} = \frac{58\,915}{1.524} = 38\,658 \text{ s}$$

The mean conversion of a solid reactant can next be estimated by Eq. (16.58):

$$\overline{X}_B = 3\left(\frac{\tau}{\Theta}\right) - 6\left(\frac{\tau}{\Theta}\right)^2 + 6\left(\frac{\tau}{\Theta}\right)^3\left(1 - e^{-\frac{\Theta}{\tau}}\right)\ (-) \quad \text{(iv)}$$

where

$$\frac{\tau}{\Theta} = \frac{38658}{126.23/\overline{C}_A} = 306.25\overline{C}_A\ (-) \quad \text{(v)}$$

Step 3 – Match the overall material balances: From the above calculations and Eq. (16.83), the following condition should be satisfied:

$$\frac{2}{3}(0.45 \times 28.739(2.076 - C_{A,out})) = \left(\frac{1.524}{0.097\,44}\right)\overline{X}_B\ (-)$$

$$\overline{X}_B = 0.5512(2.076 - C_{A,out})\ (-) \quad \text{(vi)}$$

We proceed through a trial and error procedure with Eqs. (i)–(vi) to find an appropriate value for $\overline{K}_r$ to satisfy Eq. (vi). From this procedure, we obtain

$$\overline{K}_r = 15.20\ \text{s}^{-1}, C_{A,out} = 0.2757\ \text{mol O}_2/\text{m}^3,$$

$$\overline{C}_A = 0.1066\ \text{mol O}_2/\text{m}^3, \frac{\overline{C}_A}{C_{A,in}} = 0.0513\ (-)$$

$$K_f = 2.859\ \text{s}^{-1}, \Theta = 1184\ \text{s}, \overline{X}_B = 0.9924\ (-)$$

The conversion of gaseous reactant X_A under this condition can be estimated as

$$X_A = \frac{C_{A,in} - C_{A,out}}{C_{A,in}} = \frac{2.076 - 0.2757}{2.076} = 0.8672\ (-)$$

These results show that the high conversion ($\overline{X}_B > 0.99$) for solids can be reached by $H_{mf} = 1.0$ m. In order to select an optimized H_{mf} value, we need to repeat the calculations for a few bed heights around this value. Table 16P.1 summarizes the results for different values of H_{mf}.

These results indicate that we can achieve the desired conversion of solids at $H_{mf} \geq 1.0$ m. By selecting $H_{mf} = 1.0$ m, we meet all conditions.

Table P16.1 Results of Solved Problem 16.3.

H_{mf} (m)	$\overline{K}_r$ (s^{-1})	K_f (s^{-1})	τ (s)	Θ (s)	X_A (–)	$\frac{\overline{C}_A}{C_{A,in}}$ (–)	$\overline{X}_B$ (–)
0.7	47.05	3.882	27 061	2608	0.8532	0.0233	0.9764
0.8	31.88	3.479	30 926	2004	0.8599	0.0303	0.9840
0.9	21.86	3.143	34 792	1538	0.8643	0.0395	0.9890
1.0	15.20	2.859	38 658	1184	0.8672	0.0513	0.9924
1.1	10.80	2.618	42 524	923	0.8691	0.0658	0.9946
1.2	7.93	2.411	46 389	738	0.8704	0.0823	0.9960

Acknowledgments

The authors are grateful to Profs. John Grace and Xiaotao Bi, Dr. Robin Hughes, and Mr. Adrián Carrillo Garcia for their constructive comments.

Notations

Abbreviations

CFD	computational fluid dynamics
CLaOU	chemical looping assisted by oxygen uncoupling
CLC	chemical looping combustion
CLOU	chemical looping with oxygen uncoupling
CLR	chemical looping reforming
DEM	discrete element method
iG-CLC	in situ gasification chemical looping combustion
K–L	Kunii and Levenspiel
MSW	municipal solid waste
OC	oxygen carrier
OCAC	oxygen carrier aided combustion

Symbols

a	constant defined in Eqs. (16.16), (16.17), and (16.20) (–)
a'	numerical coefficient defined in Eqs. ((16.15)) and (16.17) (–)
A	gaseous reactant
A_p	external surface area of particle (m^2)
A_t	bed cross-sectional area (m^2)
A_{te}	bed cross-sectional area excluding area occupied by internal tubes (m^2)
b	stoichiometric coefficient of B (–)
B	solid reactant
C_A	concentration of A at any radius r_c (mol/m^3)
$\overline{C}_A$	mean concentration of A bathing particles (mol/m^3)
C_{As}	concentration of A at reaction interface (mol/m^3)
C_{A0}	concentration of A in bulk (mol/m^3)
$C_{A,\,in}$	concentration of A entering reactor (mol/m^3)
$C_{A,\,out}$	concentration of A leaving reactor (mol/m^3)
C_{C0}	concentration of gaseous product in bulk (mol/m^3)
d_b	volume-equivalent bubble diameter (m)
d_{be}	equilibrium bubble diameter (m)
d_{bm}	maximum bubble diameter from total coalescence of bubbles (m)
d_i	outer diameter of tube inserted into fluidized bed (m)
d_p	particle diameter (m)
d_{te}	equivalent hydraulic bed diameter (m)
D	column diameter (m)
D_e	effective diffusivity of A through product layer (m^2/s)

D_g	effective diffusivity of A through product layer around an individual grain (m^2/s)
D_m	molecular diffusivity (m^2/s)
$\boldsymbol{E}(t)$	exit age distribution of solids (s^{-1})
f_b	bubble fraction in a fluidized bed (–)
f_w	volume fraction of wake region per bubble volume (–)
F_g	grain shape factor (–)
F_p	particle shape factor (–)
F_0	mass flow rate of solids entering the reactor (kg/s)
F_1	mass flow rate of solids exiting the reactor (kg/s)
g	gravity acceleration (m/s^2)
$g_{F_p}(X_B)$	conversion function defined in Eq. (16.11) (–)
H_f	height of fluidized bed (m)
H_{mf}	bed height at minimum fluidization (m)
k_c	reaction rate constant based on unit surface area of solid (m^3 gas/m^2 solid s)
k_D	mass transfer coefficient (m/s)
k_s	reaction rate constant based on unit volume of solid (m^3 gas/mol A s)
$\overline{k}$	overall rate constant (m/s)
k_c'	reaction rate constant based on unit surface of solid for porous particle (m/s or m^3 gas/m^2 solid s)
K_{bc}	coefficient of gas interchange between bubble and cloud–wake region (s^{-1})
K_{ce}	coefficient of gas interchange between cloud–wake region and emulsion phase (s^{-1})
K_f	overall effective rate constant for a first-order chemical reaction in a fluidized bed reactor (s^{-1})
$\overline{K}_r$	mean first-order gaseous reaction rate constant (s^{-1} or m^3 gas/m^3 solid s)
M_B	molecular weight of B (kg/mol)
N	number of stages in a multi-stage system (–)
N_A	number of moles of A (mol)
N_B	number of moles of B (mol)
N_t	number of tubes (–)
$p_{F_p}(X_B)$	conversion function defined in Eq. (16.32) (–)
p_t	tube pitch in a tube bundle (m)
$P_{A,in}$	inlet partial pressure of A (Pa)
P_{A0}	partial pressure of A in bulk gas (Pa)
P_T	system total pressure (Pa)
$q(\xi)$	function defined in Eq. (16.20) (–)
r	radial coordinate (m)
$\boldsymbol{r}$	volume of solid product formed from unit volume of solid reactant (–)
r_c	radius of reaction interface (m)
r_p	original position of external surface/particle radius (m)
R	universal gas constant, (8.314 (J/mol K))
Re	Reynolds number $= \frac{2\rho_g U r_p}{\mu_g}$ (–)

Sc	Schmidt number (–)
Sh	Sherwood number given by Eq. (16.14) (–)
S_v	surface area per unit volume of particle (m^2 solid/m^3 solid)
t	time (s)
t^+	dimensionless time defined in Eq. (16.18) (–)
t^*	dimensionless time defined in Eqs. (16.7), (16.43), and (16.48) (–)
$\tilde{t}$	dimensionless time defined in Eqs. (16.32) and (16.46) (–)
T	temperature (°C or K)
u_b	rise velocity of bubble through bed relative to column wall (m/s)
u_{br}	rise velocity of an isolated bubble relative to emulsion phase (m/s)
u_e	superficial gas velocity in emulsion phase (m/s)
u_f	upward velocity of gas at minimum fluidizing condition (m/s)
U	superficial gas velocity (m/s)
U_{mf}	minimum fluidization velocity (m/s)
V_p	volume of particle (m^3)
V_s	volume of solids in fluidized bed reactor (m^3)
W	bed inventory mass (kg)
W_i	bed inventory in each stage (kg)
X_A	conversion of A (–)
X_B	solid conversion (–)
$\overline{X}_B$	mean solid conversion (–)
$y_{A,in}$	mole fraction of A in the inlet gas stream to the reactor (–)
y_{A0}	mole fraction of A in bulk gas (–)

Greek Letters

γ_b	volume of solids dispersed in bubbles divided by volume of bubbles (–)
γ_c	volume of solids dispersed in cloud and wake regions by volume of bubbles (–)
γ_e	volume of solids dispersed in emulsion phase divided by volume of bubbles (–)
γ_m	parameter in Horio and Nonaka [27] correlation, Eq. (16.72) (–)
ε_e	emulsion phase voidage (–)
ε_{mf}	bed voidage at minimum fluidization (–)
ε_p	volume fraction occupied by internal porosity and inerts in porous particle (–)
Θ	characteristic reaction time (s)
ξ	dimensionless position of shrinking particle or reaction interface (–)
μ_g	gas viscosity (Pa s)
ρ_g	gas density (kg/m^3)
ρ_p	particle density (kg/m^3)
ρ_s	molar density of B (mol/m^3)
σ_g^2	shrinking core reaction modulus for grain (–)
σ_s^2	shrinking core reaction modulus for solid particle (–)
σ_0^2	dimensionless parameter defined in Eq. (16.25) (–)
$\tilde{\sigma}^2$	generalized gas–solid reaction modulus (–)
τ	mean residence time of solids in reactor (s)

τ_i mean residence time of solids in each stage (s)
ψ parameter in Hilligardt and Werther correlation, Eq. (16.86) (–)

References

1 Grace, J.R. (1986). Fluid beds as chemical reactors. In: *Gas Fluidization Technology* (ed. D. Geldart), 285–339. Chichester: Wiley.
2 Szekely, J., Evans, J.W., and Sohn, H.Y. (eds.) (1976). *Gas–Solid Reactions*. New York: Academic Press.
3 Bale, C.W., Bélisle, E., Chartrand, P. et al. (2016). FactSage thermochemical software and databases, 2010–2016. *Calphad* 54: 35–53.
4 Andersson, J.O., Helander, T., Höglund, L. et al. (2002). Thermo-Calc & DICTRA, computational tools for materials science. *Calphad* 26 (2): 273–312.
5 Kunii, D. and Levenspiel, O. (eds.) (1991). *Fluidization Engineering*. Boston: Butterworth-Heinemann.
6 Levenspiel, O. (ed.) (1999). *Chemical Reaction Engineering*, 3e. New York: Wiley.
7 Yagi, S. and Kunii, D. (eds.) (1955). Studies on combustion of carbon particles in flames and fluidized beds. In: *Proceedings of the Fifth International Symposium on Combustion*, 231–244. New York: Reinhold.
8 Ranz, W.E. and Marshall, W.R. Jr., (1952). Evaporation from drops. *Chem. Eng. Prog.* 48 (3): 173–180.
9 Szekely, J. and Themelis, N.J. (eds.) (1971). *Rate Phenomena in Applied Metallurgy*. New York: Wiley.
10 Park, J.Y. and Levenspiel, O. (1975). The crackling core model for the reaction of solid particles. *Chem. Eng. Sci.* 30 (10): 1207–1214.
11 Walker, P.L. Jr., Rusinko, F., and Austin, L.G. (1959). Gas reaction of carbon. *Adv. Catal.* 11: 133–221.
12 Petersen, E.E. (1975). Reaction of porous solids. *AIChE J.* 3 (4): 443–448.
13 Szekely, J. and Evans, J.W. (1971). Studies in gas–solid reactions: Part I. A structural model for the reaction of porous oxides with a reducing gas. *Metall. Trans.* 2 (6): 1691–1698.
14 Sohn, H.Y. and Szekely, J. (1972). A structural model for gas–solid reactions with a moving boundary – III: a general dimensionless representation of the irreversible reaction between a porous solid and a reactant gas. *Chem. Eng. Sci.* 27 (4): 763–778.
15 Suresh, A.K. and Ghoroi, C. (2009). Solid–solid reactions in series: a modeling and experimental study. *AIChE J.* 55 (9): 2399–2413.
16 Murthy, J.S.N., Surendar Reddy, V., and Sankarshana, T. (2011). Solid–solid reaction in a fluidized bed. *Asia-Pac. J. Chem. Eng.* 6 (2): 244–256.
17 Tamhankar, S.S. and Doraiswamy, L.K. (1979). Analysis of solid–solid reactions: a review. *AIChE J.* 25 (4): 561–582.
18 Khawam, A. and Flanagan, D.R. (2006). Solid-state kinetic models: basics and mathematical fundamentals. *J. Phys. Chem. B* 110 (35): 17315–17328.
19 Geldart, D. (1973). Types of gas fluidization. *Powder Technol.* 7 (5): 285–292.

20 Shabanian, J., Jafari, R., and Chaouki, J. (2012). Fluidization of ultrafine powders. *Int. Rev. Chem. Eng.* 4 (1): 16–50.

21 Levenspiel, O., Kunii, D., and Fitzgerald, T. (1968). The processing of solids of changing size in bubbling fluidized beds. *Powder Technol.* 2 (2): 87–96.

22 Kunii, D. and Levenspiel, O. (1968). Bubbling bed model. Model for flow of gas through a fluidized bed. *Ind. Eng. Chem. Fundam.* 7 (3): 446–452.

23 Kunii, D. and Levenspiel, O. (1968). Bubbling bed model for kinetic processes in fluidized beds. Gas–solid mass and heat transfer and catalytic reactions. *Ind. Eng. Chem. Process Des. Dev.* 7 (4): 481–492.

24 Kunii, D. and Levenspiel, O. (1990, 1990). Fluidized reactor models. *Ind. Eng. Chem. Res.* 29 (7): 1226–1234.

25 Davidson, J.F. and Harrison, D. (1963). *Fluidised Particles*. Cambridge: Cambridge University Press.

26 Hilligardt, K. and Werther, J. (1987). Influence of temperature and properties of solids on the size and growth of bubbles in gas fluidized beds. *Chem. Eng. Technol.* 10 (1): 272–280.

27 Horio, M. and Nonaka, A. (1987). A generalized bubble diameter correlation for gas–solid fluidized beds. *AIChE J.* 33 (11): 1865–1872.

28 Hilligardt, K. and Werther, J. (1986). Local bubble gas hold-up and expansion of gas/solid fluidized beds. *German Chem. Eng.* 9 (4): 215–221.

29 Adanez, J., Abad, A., Garcia-Labiano, F. et al. (2012). Progress in chemical-looping combustion and reforming technologies. *Prog. Energy Combust. Sci.* 38 (2): 215–282.

30 Rydén, M., Hanning, M., Corcoran, A., and Lind, F. (2016). Oxygen carrier aided combustion (OCAC) of wood chips in a semi-commercial circulating fluidized bed boiler using manganese ore as bed material. *Appl. Sci.* 6 (11): 347.

31 Toftegaard, M.B., Brix, J., Jensen, P.A. et al. (2010). Oxy-fuel combustion of solid fuels. *Prog. Energy Combust. Sci.* 36 (5): 581–625.

32 Adánez, J., Abad, A., Mendiara, T. et al. (2018). Chemical looping combustion of solid fuels. *Prog. Energy Combust. Sci.* 65: 6–66.

33 Werther, J., Saenger, M., Hartge, E.U. et al. (2000). Combustion of agricultural residues. *Prog. Energy Combust. Sci.* 26 (1): 1–27.

34 Vekemans, O. and Chaouki, J. (2016). Municipal solid waste cofiring in coal power plants: combustion performance, Open access peer-reviewed chapter. In: *Developments in Combustion Technology* (eds. K.G. Kyprianidis and J. Skvaril), 117–141. IntechOpen.

35 Hughes, R.W., Lu, D.Y., and Symonds, R.T. (2017). Improvement of oxy-FBC using oxygen carriers: concept and combustion performance. *Energy Fuels* 31 (9): 10101–10115.

36 Cocco, R., William, D.F., Liu, P., and Hrenya, C.M. (2017). CFD–DEM: modeling the small to understand the large. *Chem. Eng. Prog.* 113 (9): 38–45.

37 Shabanian, J. and Chaouki, J. (2017). Effects of temperature, pressure, and interparticle forces on the hydrodynamics of a gas–solid fluidized bed. *Chem. Eng. J.* 313: 580–590.

38 Rowe, P.N. and Partridge, B.A. (1965). An X-ray study of bubbles in fluidized beds. *Trans. Inst. Chem. Eng.* 43: S116–S134.

39 Wen, C.Y. and Yu, Y.H. (1966). Mechanics of fluidization. *Chem. Eng. Prog. Symp. Ser.* 62 (62): 100–111.
40 Chitester, D.C., Kornosky, R.M., Fan, L.-S., and Danko, J.P. (1984). Characteristics of fluidization at high pressure. *Chem. Eng. Sci.* 39 (2): 253–261.

Problems

16.1 For the process of Solved Problem 16.3, suppose that the company plans to increase the production rate to 200 ton/d. For this purpose, the process engineering group recommends operating the existing reactor at identical conditions as before at $H_{mf} = 0.8$ m and $U = 0.50$ m/s. In addition, the exit solid stream of the reactor will be sent through an overflow tube to another bubbling fluidized bed reactor of 4.3 m ID equipped with 247 vertical heat exchanger tubes with the same d_i, p_t, and pitch layout as the first reactor. The second reactor will operate under similar operating conditions as the first. Through the three-step procedure, determine an optimized value of H_{mf} for the second reactor to ensure that solids reach a very high conversion ($\overline{X}_B > 0.99$) when exiting it. The second reactor should operate at $0.5 \leq H_{mf} \leq 1.2$ m to adequately remove the exothermic heat of the reaction and minimize solids entrainment. What conclusion can be made when comparing the results with those of Solved Problem 16.3?
Data: $\rho_{p,\,ZnO} = 5600$ kg/m^3
Use the Wen and Yu [39] correlation to calculate the U_{mf} of powders in the second reactor. For additional information, check the data provided for Solved Problem 16.3.

16.2 For the process of Solved Problem 16.3, suppose that the process engineering group recommends operating the reactor with coarse zinc blend particles rather than fine ones, at identical conditions as before. They believe that this will decrease the excessive operating costs of ore grinding. With the help of the data provided below,

(a) For $H_{mf} = 1.0$ m, determine whether or not the reactor can produce 110 ton/d of products, i.e. highly converted solids ($\overline{X}_B > 0.99$).
(b) If the solids entrainment from the reactor is minimized by $H_{mf} = 1.5$ m when processing coarse particles, determine the maximum daily production rate for highly converted solids ($\overline{X}_B > 0.99$).
(c) If a similar two-stage process as described in Problem 16.1 is adopted for coarse particles, determine the maximum daily production rate for highly converted solids ($\overline{X}_B > 0.99$) if H_{mf} is limited to 0.7–1.5 m in each reactor. The lower and upper bounds of H_{mf} are selected to ascertain the complete removal of the exothermic heat of reaction and the minimum solids entrainment, respectively. To reach an optimized design, attempt to distribute the total required bed inventory between reactors to satisfy all conditions with the minimum total bed inventory. The operating conditions of both reactors are identical, as described

previously, and consider the data provided below. Assume that the heat exchanger tubes are capable of completely removing the exothermic heat of reaction in all scenarios.

Data: $d_p = 700\ \mu m$, $U = 1.0$ m/s, $D_e = 8 \times 10^{-4}\ m^2/s$

Note with the large particle size, the contribution of internal diffusion should be considered along with the chemical reaction step in determining the overall reaction rate of solids, following the shrinking unreacted core model. Also, assume that the coarse particles form slow-rising cloudless bubbles when in the bubbling flow regime. Assume also that the heat exchanger tubes can stabilize the bubble diameter in both units at 0.2 m. Moreover, since the coarse particles are expected to show typical behaviour of powders close to the Geldart B/D boundary at ambient conditions, the Chitester et al. [40] correlation can be employed to estimate U_{mf} of powders in a high-temperature atmospheric bed, while gas properties at ambient temperature can be adopted for this purpose [37]. Furthermore, assume that cohesive interparticle forces are negligible for these coarse particles up to high operating temperatures. For additional information, check the data provided for Solved Problem 16.3 and Problem 16.1.

17

Scale-Up of Fluidized Beds

Naoko Ellis[1] *and Andrés Mahecha-Botero*[2]

[1] *Department of Chemical and Biological Engineering, University of British Columbia, 2360 East Mall, Vancouver, Canada V6T 1Z3*
[2] *NORAM Engineering and Constructors Ltd., 200 Granville Street, Suite 1800, Vancouver, Canada V6C 1S4*

17.1 Challenges of Scale

Laboratory-scale trials typically focus on proving the concept of a new process and do not necessarily consider its suitability using real feedstocks, full-scale equipment design, or process economics. For this reason, it is important to utilize scale-up techniques to design equipment and systems that are best suited for large-scale implementation and long-term operation.

The key goal of chemical reactor scale-up is "to design a pilot or industrial reactor able to replicate through a standard methodology the results obtained in the laboratory" [1]. The challenges include having confidence in the technical understanding of the phenomena across scales and the ability to assume the risk of the business [1]. In general, this is echoed in industry. For example, fluidized bed drying have been successfully scaled up owing to the confidence in single-particle drying kinetics applicable to the process. However, for catalytic fluidized bed reactors, there is still a deep chasm between industry and academics as the economic risk is extremely high. Some aspects that concern industry in this area include solids handling, attrition and wear, performance of mechanical parts, and fluidized bed hydrodynamics.

How to design a pilot-scale unit that addresses risks is another key question. In more general terms, pilot-scale experiments should test the technologies and equipment used in an industrial scale (not necessarily the same ones employed in the laboratory), confirm suitability of materials of construction, verify the reliability of equipment designs, evaluate product specifications, identify process conditions suitable for the operation of interconnected systems, review the safety systems used for the process, confirm process operability, and set up automation needs to be evaluated depending on such topics as hydrodynamics, residence times, diffusion effects, and heat exchange.

The design of fluidized beds is a difficult undertaking requiring experience and understanding of the phenomena involved. Laboratory and pilot plant

Essentials of Fluidization Technology, First Edition. Edited by John R. Grace, Xiaotao Bi, and Naoko Ellis.

units are usually required to reduce technical risk as the process is developed to large industrial scale. Failure to understand the principles involved has led to some spectacular scale-up failures in the past, e.g. in scaling up a fluid bed Fischer–Tropsch synthesis reactor in the 1950s [2]. Moreover, the safety and operability of innovative systems should be confirmed prior to the construction of large industrial plants.

It is useful to distinguish the types of pilot plant data that may be required for scale-up purposes:

- Chemical reaction considerations for scale-up:
  - How is the reaction system affected?
  - How are chemical performance metrics such as reactant conversion, selectivity, and yield affected?
  - To what extent do the existing reactor models and correlations require adjustment to represent the full-scale system?
  - What are the effects on gas-phase catalytic reactions and gas–solid reactions?
  - Is there an opportunity to optimize the operating conditions of the system to favour the performance of the reaction system?
- Physical scale-up considerations:
  - Equipment design issues
  - Operability issues
  - Materials of construction considerations
  - Solids feeding
  - Agglomeration and/or attrition of solids
  - Distributor-related design
  - Cyclone performance
  - Solids entrainment
  - Mechanical design of process vessels
- Extended operation considerations:
  - Confirmation of stability of catalyst over long-term operation
  - Effect of changes in particle size distribution, surface area, and pore structure over long-term operation (e.g. due to attrition or agglomeration)
  - Confirmation of reliability of equipment
  - Evaluation of the effect of impurities on equipment and product quality
  - Effect of trace species in the catalyst, including deactivation considerations
  - Impact on system reliability by off-spec feed and impurities, including variability in moisture content, feed composition, and dust loadings
  - Evaluation of fouling, erosion, and corrosion rates
  - Evaluation of reliability of instrumentation and control systems over long-term operation
  - Identification of equipment maintenance requirements
  - Operation of plant outside of normal conditions, including start-up, ramp-up, process upsets, turndown, and shutdown

Pilot plants may be required to address one, many, or all of the above purposes. Figure 17.1 shows a schematic diagram [3] demonstrating the interaction

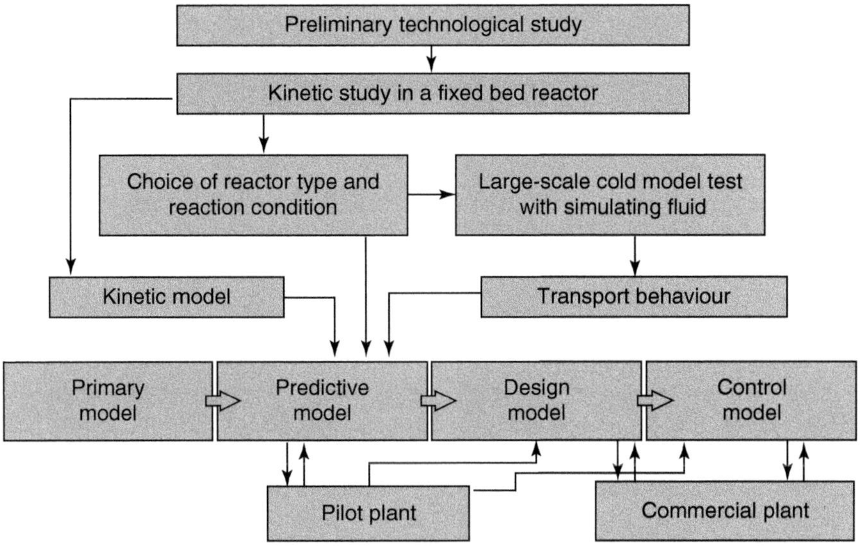

Figure 17.1 General development scheme of fluidized bed process. Source: Adapted from Chen et al. 1984 [3].

between kinetic and physical factors and the typical stages involved in a large-scale fluid bed process development.

17.2 Historical Lessons

Some earlier successes in scale-up of catalytic fluidized bed reactors were outlined by Kunii and Levenspiel [4]. Failures of fluidized bed scale-up are more challenging to list as they are often not well documented for a variety of reasons including marketing, pride, and confidentiality considerations. One well-documented failure relates to a Fischer–Tropsch reactor scale-up in Brownsville, Texas in the 1950s, where an ambitious scale-up from a 0.305 m diameter column to a 5 m diameter commercial-scale reactor resulted in unexpectedly poor performance. This was attributed to a reduced residence time of gas, leading to poor gas-phase contact and conversion [5]. It also coincided with a drop in the price of liquid hydrocarbon products, which led to shelving the unit [6].

In the mid-1950s, Sasol used data of Kellogg, based on 0.04 and 0.10 m diameter pilot plant units to scale up a commercial circulating fluidized bed (CFB) unit, leading to two commercial waves of Kellogg–Sasol CFB reactor development through 1962, followed by Badger–Sasol CFB reactor development from 1975 to 1980 [7].

The interesting part lies in Sasol re-examining the Brownsville, Texas, case in the 1980s, leading to the development of Sasol Advanced Synthol (SAS) technology. The motivation for this new development was based on the lower capital cost of a fixed (dense) fluidized bed (FFB) system being more compact in design and its self-supporting structure while also having a lower operating cost for less than 50% pressure drop across the FFB compared with CFB operation [8].

Based on the tenuous "quality of fluidization," Sasol studied 0.05, 0.14, and 0.64 m diameter tubes, 5–6 m in height, to optimize the particle size and density of Geldart group A and C catalyst particles at gas superficial velocities up to 0.6 m/s. In this extensive study assessing the quality of fluidization, bed voidage was used to compare their data with those of Volk et al. [9]. The transitional Geldart A–C powder type exhibited sufficient quality of fluidization to provide confidence in the construction of a demonstration reactor [8]. A pilot-scale unit of 0.9 m diameter served to resolve operational problems related to cyclone and distributor design, and by 1989, a 5.0 m diameter commercial SAS unit was successfully operated in Sasolburg, South Africa. This commercial unit was significant in establishing the critical catalyst bed density and voidage to avoid uncontrollable bed expansion. The accumulation of fine material in the catalyst bed through free carbon production via the Boudouard reaction affected the control over bed expansion, as the unit was operated in the transition from turbulent to fast fluidization flow regimes. The understanding of the operation based on catalyst flow characteristics and conversion from the 5.0 m diameter commercial unit led to successful scale-up to an 8.0 m diameter reactor [10].

17.3 Influence of Scale on Hydrodynamics

As gas-fluidized beds are scaled, achieving uniform flow across the unit may become difficult. With multiple feed points or multiple cyclones, preferential flow and uneven distribution are inevitable, given the complexity of multiphase systems. For large-scale units, multiple parallel cyclones can result in uneven flow distribution, and some cyclones become fouled over time, requiring costly shutdowns to clean. Feed distribution and solids circulation may behave quite differently in large units, requiring completely different mechanical design to achieve adequate performance.

Another major factor during scale-up is the change in external surface area per unit volume. Since smaller units have much higher surface area per unit volume than larger units, their heat loss per unit volume is much higher. This has a significant effect on the design of heat exchangers or heaters for removal or addition of heat from/to a unit. Furthermore, data obtained in small laboratory units may not accurately represent the process, since chemical reaction equilibrium may be shifted in regions such as the freeboard where significant cooling could take place.

In the following paragraphs, the effects of hydrodynamics in different fluidization flow regimes are summarized.

17.3.1 Bubbling Fluidization

As shown in Chapter 7, bubble diameter, bubble rise velocity, and bed expansion are all interrelated. The change in column diameter has different effects on bubble sizes, depending on the Geldart powder classification (see Chapter 2). Group A powders tend to have maximum bubble sizes, so increasing the column diameter may not affect bubble growth beyond a certain size. On the other hand, for group B particles, larger column diameter and deeper beds allow bubbles to grow much larger, influencing the rise velocity and bed expansion, thus potentially changing the interphase mass transfer and reaction rates. Furthermore, axial and radial mixing may change due to the change in bubble size with scale.

Hydrodynamic parameters such as bed voidage or bubble rise velocity generally vary with changes in column diameter. In many cases, hydrodynamic parameters reach asymptotic levels with increasing column diameter, thus ultimately becoming insensitive to scale. For this reason, cold models should be designed to target near the minimum column size at which the hydrodynamic parameters like bubble size reach their limits. For example, for the case shown in Figure 17.2, iron catalyst was used to monitor the bed voidage for varying column diameters. Increasing the column diameter resulted in a decrease in bed voidage to varying degrees. The difference can be attributed to the particle size distribution, as mentioned in Section 17.2. In the Sasol case, it was reported that higher fines content assisted in providing a higher dense-phase voidage [8].

Frictional wall effects are known to influence the bubble diameter resulting from bubble coalescence and growth, especially in smaller columns. In one study [11], the bubble rise velocity was compared using Geldart group A particles in 0.1, 0.19, and 0.38 m diameter columns, as shown in Figure 17.3. The results indicate that between columns of diameter 0.19 and 0.38 m, there was still an effect of scale on bubble velocity, reflecting variation in bubble size. As summarized in a review [12], there are contradictory reports on the effect of column size on bubble sizes, so further study is needed in this area to make sure general conclusions are not drawn from case-specific data.

For Geldart group B particles, a diameter of 0.152 m was considered adequate [13] for comparing axial solids holdup in columns of 0.05, 0.078, and 0.152 m diameter.

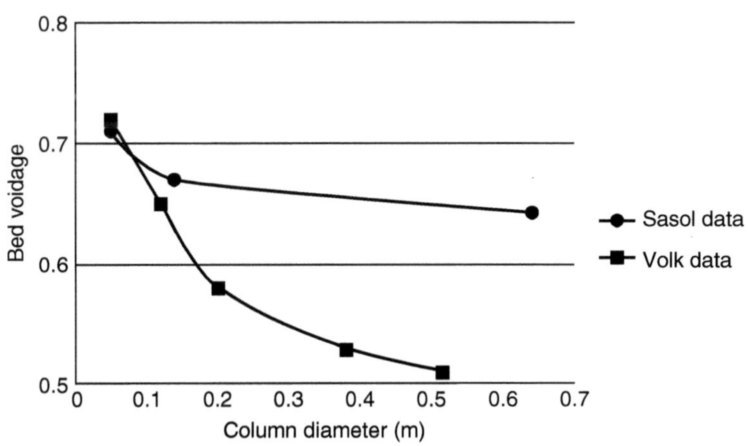

Figure 17.2 Effect of column diameter on voidage for Sasol Advanced Synthol cold model study. Source: Adapted from Volk et al.1962 [9].

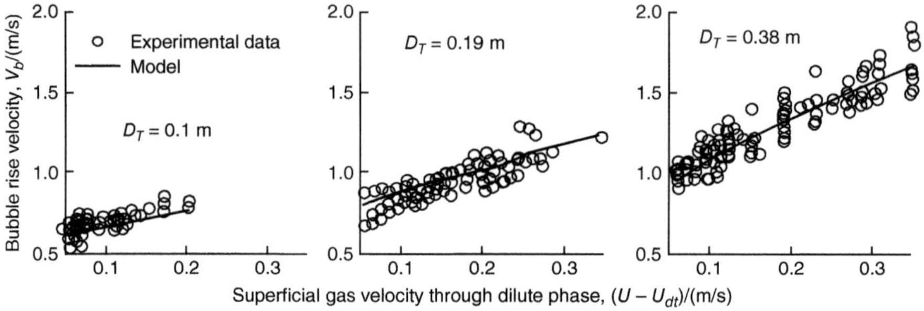

Figure 17.3 Effect of column diameter and superficial gas velocity on bubble rise velocity. Source: Adapted from Krishna and van Baten et al. 2001 [11]. Reproduced with permission of Elsevier.

Gas bypassing or channelling has been documented with Geldart A particles in deep beds [14]. In the process of scaling up, one common strategy is to maintain similar bed depth between a pilot unit and a demonstration plant, while the column diameter is significantly increased. In such a case, the pilot unit may exhibit preferential pathways for gas, while the rest of the bed may be defluidized. Clearly this is undesirable, with consequences of reduced interphase mass transfer. In a similar vein of poor gas–solid interaction, Geldart B particles may exhibit slugging in deep and narrow columns. In this case, a change in flow regime may occur as a result of scale-up. In small units with high bed height, bubble sizes can be larger than about 60% of the column diameter (see Chapter 8), resulting in slug flow. Any performance conclusions based on slug flow experiments will cause problems as the unit is scaled up. However, when the system is scaled up, bubbles may not reach this threshold or may start to break down, leading to transition to the turbulent fluidization flow regime. In either case, this will affect the interphase mass transfer, influencing the conversion and yield for catalytic reactions. Understanding the hydrodynamics underlying reactor performance is vital, especially when scaling chemical reactors.

17.3.2 Turbulent Fluidization Flow Regime

As covered in Chapters 4 and 9, transition to the turbulent fluidization flow regime occurs when bubbles reach a maximum size and start to break down. It is marked by a critical superficial velocity, U_c, at which pressure fluctuations reach a maximum. The effect of scale on U_c is shown in Figure 17.4, which depicts the scale of a column beyond which transition to the turbulent fluidization flow regime becomes insensitive to further increases in reactor scale [15, 39].

Radial voidage profiles comparing results from columns of diameter 0.076, 0.15, 0.29, and 0.61 m with Geldart group A particles have shown less sensitivity to column diameter compared to the effect of U [16].

Transport disengagement height (TDH) and entrainment are both affected by scale, affecting the design of the reactor and solids separation units. Prediction of entrainment rate through correlations often varies as entrainment is affected not only by the column diameter but also by the exit geometry (including

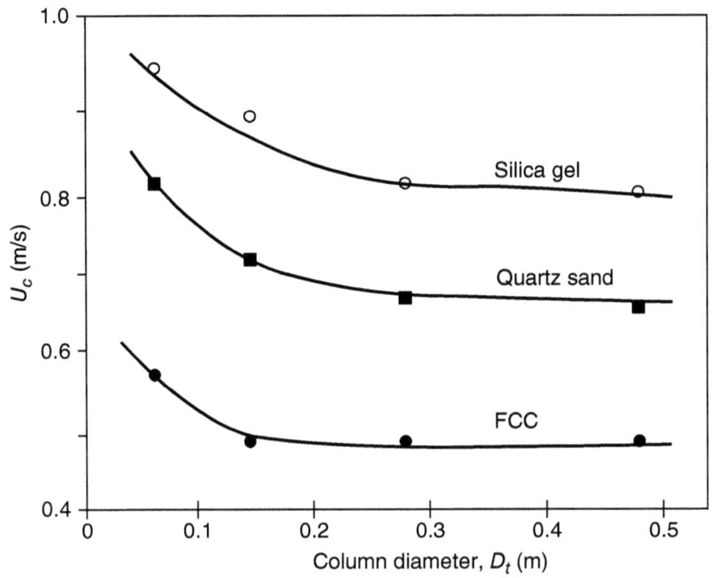

Figure 17.4 Effect of column diameter on transition velocity, U_c. Source: Adapted from Bi et al. 2000 [15, 39].

cross-sectional transitions to the freeboard) and particle size distribution. One study [17] reported higher total entrainment rate for a 0.152 m diameter compared with that of 0.076 m, while another [18] indicated no dependence of entrainment from columns of diameter 0.1, 0.3, and 0.9 m.

TDH has been reported [19] for various column diameters spanning from bubbling fluidization to the turbulent fluidization flow regime using Geldart group A particles, as shown in Figure 10.2 (chapter on entrainment). As summarized by Bi et al. [15], the TDH increased rapidly with increasing column diameter and U. Caution is required, especially when using correlations for entrainment and TDH, as the variabilities of predictions are huge (see Chapter 10).

17.3.3 Fast Fluidization

CFBs commonly operate in the fast fluidization flow regime. A comprehensive summary of commercial units and general considerations for scale-up was published by Matsen [20]. In general, small units of up to 0.3 m in diameter are used in laboratories, while commercial units may be as large as 10–20 m in cross section. A summary of hydrodynamic data from large-scale CFB units is available [21].

Slip factor, the ratio of gas velocity to average particle velocity expressed by

$$\varphi = \frac{U}{\varepsilon v} \tag{17.1}$$

where $v = \frac{G_s}{\rho_p(1-\varepsilon)}$ for seven columns with diameters from 0.051 to 0.18 m, was used to develop the correlation [22]

$$\varphi = 1 + \frac{5.6}{Fr} + 0.47Fr_t^{0.41} \tag{17.2}$$

where $Fr = \frac{U}{\sqrt{gD}}$ and $Fr_t = \frac{v_t}{\sqrt{gD}}$.

This can be used to estimate the gas and solid residence times, blower capacity, and standpipe design.

17.4 Approaches to Scale-Up

Large pilot plants are costly to fabricate and operate. Thus, designers need to decide the scale of pilot plants and the number of development steps required to scale up a new process. It is also a challenge to understand how scale affects hydrodynamics and performance, especially for operations at elevated temperatures and pressures. This section covers some of the approaches for scale-up.

17.4.1 Framing Questions

Mabrouk et al. [13] posed questions related to scale effects on fluidized bed hydrodynamics:

- Which laws govern the adaptation of process parameters in lab-scale measurements to those of the full-scale plant?
- Is it possible to achieve complete similarity between lab-scale and full-scale processes?
- Is one physical model sufficient?
- How small can the physical model be?

The answer to these questions is likely to depend on the process in question, how much previous knowledge the designer has of the process, the quality of the data used in the analysis, how critical the reactor application is, and the risk tolerance of the specific industry or company.

17.4.2 General Approaches

To develop a new process, there are several scale-up steps. General relationships between the kinetic and physical factors and models involved in commercializing a fluidized bed process are shown in Figure 17.1 [3]. Typical steps include the following:

- Hydrodynamic studies are initially carried out in a lab-scale unit. Fluidizability, particle attrition, and other prominent hydrodynamic characteristics are studied in small units in cold non-reacting units or reacting units as needed. These units are typically only operated for as long as needed to obtain performance data, generally less than ~100 hours of total on-stream time.
- Pilot-scale units (normally 200–600 mm diameter) are used to investigate particular uncertain aspects of the technology. The pilot may be a geometrically similar model or part of a system to test such characteristics as voidage distribution, pressure profiles, and mixing. Data from pilot plants can confirm the reliability of models with respect to scale and operating conditions. Pilot plants are operated to confirm several aspects of the design and are commonly operated for 1000 to a few thousand hours of total on-stream time.

- A demonstration plant of 800–1500 mm in diameter and of height similar to a commercial unit may be commissioned to further confirm the extent of reactions, feed system, and long-term operation. Further details on the steps and various models are summarized elsewhere [14]. Demonstration units are operated continuously long term while meeting the specifications of commercial products.
- More steps can be considered, depending on the results and performance of the systems of intermediate scale.

17.4.3 Dimensional Similitude (Scaling Models)

Scaling of equipment can be achieved through geometric similarity and matching of a sufficient number obtained from the Buckingham pi theorem to provide dynamic similitude [23, 24]. Scaling laws based on this approach have been applied to fluidized beds by Glicksman et al. [23] and Horio et al. [24]. For a system with negligible interparticle forces, Glicksman [25] proposed that the following dimensionless groups be matched:

$$\frac{U^2}{gd_p}, \frac{\rho}{\rho_p}, \frac{\rho d_p U}{\mu}, \frac{\rho(\rho_p - \rho)gd_p^3}{\mu^2}, \frac{G_s}{\rho U}, \frac{L}{D}, \text{bed geometry, dimensionless particle size distribution (PSD)} \quad (17.3)$$

As shown in Table 17.1, the conditions estimated by this approach give reasonable operating conditions for an atmospheric fluidized bed combustor.

Table 17.1 Atmospheric combustor modelled by a bed fluidized at ambient conditions.

Given	Commercial bed	Scaled model (dimensionless number used)
Temperature (°C)	850	25
Gas viscosity (10^{-5} kg/m s)	4.45	1.81
Density (kg/m^3)	0.314	1.20
Derived from scaling laws		
Particle density	ρ_{pc}	$3.82\rho_{pc}$ (from $\frac{\rho}{\rho_p}$ matching)
Bed diameter, length, etc.	D_c	$0.225D_c$ (from $\frac{L}{D}$ matching)
Particle diameter	d_{pc}	$0.225d_{pc}$ (from $\frac{\rho(\rho_p - \rho)gd_p^3}{\mu^2}$ matching)
Superficial velocity	U_{oc}	$0.47U_{oc}$ (from $\frac{U^2}{gd_p}$ matching)
Volumetric solids flux	$(G_s/\rho_p)_c$	$0.47(G_s/\rho_p)_c$ (from $\frac{G_s}{\rho_p U}$ matching)
Time	t_c	$0.47t_c$ (from $\frac{U^2}{gd_p}$ matching)
Frequency (turnover, etc.)	f_c	$2.13f_c$ (from $1/t_c$)

Source: Adapted from Glicksman 1984 [25].

Over the years, various experiments have been conducted to test the validity of full and simplified scaling approaches [26]. However, some aspects of fluidized bed systems that may be critical for scaling are not considered by the fluid dynamic scaling model approach. For example, interparticle interactions are not captured in the dimensionless groups included in Eq. (17.3). Interparticle forces such as van der Waals forces, capillary forces, electrostatics, etc. may be critical in dealing with Geldart group C powders and at times also for group A powders, for example, affecting particle agglomeration. In addition, particle collisions are not represented in the model. Through computational fluid dynamics (CFD) simulations, the effect of interparticle forces and collisions (e.g. via coefficient of restitution) can be included, emphasizing the significance of these mechanisms [27, 28]. Furthermore, the simplified model omitting the ratio of particle diameter to column diameter has been shown to not achieve hydrodynamic similitude [28].

In reviewing the scaling of fluidized bed combustors, Leckner et al. [29] reported reasonable success in hydrodynamic scaling between a small-scale solids test rig and a large boiler, with the challenge being finding particles of size, shape, and density that satisfy the criteria. In addition, combustion scaling was included to reflect combustion chemistry. Fluid dynamic similarity was limited in the horizontal direction, while the residence time for combustion conditions was better matched. Furthermore, heat transfer was only implicitly dealt with in the assumption of a bed temperature.

Rüdisüli et al. [12] examined the scaling law for bubbling fluidized bed reactors through hydrodynamic and reactive similarities. Chemical conversion and selectivity depend on the interactions of mass transfer, kinetics, and hydrodynamics. Kinetically limited reactions are less affected by the change in mass transfer due to scale-up. It is important to consider the rate-determining step in the process for chemical reactors. Various methods to test the validity of scaling approaches include comparison of pressure fluctuations analyzed through power spectral density and probability density functions and attractor comparison, based on fluidized beds exhibiting chaotic behaviour. Extensive discussions on early successes and shortcoming of dimensional similitude conclude with a cautionary note when interpreting the results, as scaling is difficult and risky [12].

Knowlton et al. [30] caution against using scaling laws for scaling of new processes without considering other reaction effects and attrition. Applying the scaling law while ensuring the same flow regime and Geldart particle groups in scaled units may lead to better success. They further suggest using scaling laws for modifying existing units, such as examining the addition of internals.

Matching all dimensionless numbers for testing is in practice difficult. Engineers often need to compromise to be able to find a scaled system that can be used for experimental testing. Furthermore, the suitability and assumptions of the model system must be verified.

17.4.4 Other Models

Various other models are available for characterizing fluidized bed reactors, as discussed in Chapter 15 and summarized by Detamore et al. [28]. More recently, the detailed behaviour of fluidized beds is being well investigated with the development of CFD codes and increasing computational power. Simulation of bubbling fluidized beds of 0.1–6 m in diameter based on an Eulerian–Eulerian model indicated a strong influence of column diameter on hydrodynamics [11]. A review on discrete particle modelling with consideration to multilevel modelling of fluidized beds [27] indicates that it is possible to include gas–particle and particle–particle interactions in simulations. As indicated in Chapter 6, CFD is becoming a powerful tool, assisting design and scale-up of fluidized beds.

17.5 Practical Considerations

Heterogeneous chemical and catalytic reactions are more difficult to scale up than drying or other physical processes. Cases where the flow regime changes with scale pose additional uncertainties. Pilot-scale units can play a key role in reducing uncertainties associated with scale-up.

17.5.1 Purpose of Pilot-Scale Units

The idea of a pilot-scale unit is to allow testing of predictive and design models against realistic conditions and scale, as depicted in Figure 17.1. The objectives of a fluidized bed pilot plant are likely to include examining some of the following:

- Effect of temperature and pressure
- Effect of feed composition
- Effect of catalyst configuration
- Effect of residence time, fluid velocity, and production rate
- Particle stickiness, agglomerating characteristics, and attrition
- Distributor design
- Baffle design, agitation, or other aids to fluidization
- TDH, placement of cyclones, cyclone design, and other collection devices
- Solids feeding and withdrawal
- Addition of liquid into a gas fluidized bed
- Heat transfer
- Erosion and corrosion problems
- Check the validity of reactor models
- Start-up and shutdown procedures
- Suitability of control strategy, control systems, and interlocks
- Ability to meet process specifications for long-term operation

Furthermore, auxiliary components including material feeding system, solids return, nonmechanical valves, and pressure balances between units are often examined in pilot-scale units. Other operational reasons may include establishing [20]:

- Solids handling techniques
- Equipment reliability
- Operating procedures

- A process to manufacture products for further testing or mechanical components
- Operator training
- Examination of operability constraints

17.5.2 Pilot-Scale Units

Some documented cases for pilot-scale unit operations and lessons learned are documented below. Some information has been published on many pilot-scale unit design and operations; however, it is not easy to capture the significance of each unit in the context of developing a whole process unless a narrative accompanies the experiences.

17.5.2.1 Biomass Combined Heat and Power (CHP) Güssing Case

The overall concept of this new process was to supply the heat required for endothermic biomass gasification by bed material (heat carrier) circulating between a combustor and a gasifier, as shown in Figure 17.5. This is a well-documented project, summarized in numerous publications (e.g. [31]). As shown in Figure 17.6, developmental stages included (i) 10 kW fuel equivalent unit (1993), (ii) 100 kW fuel equivalent unit (1995), (iii) a 100 kW fuel hot unit (1999), (iv) a 0.5 MW_{th} pilot plant (2012), and (v) an 8 MW_{th} demonstration plant (2002) leading to further development and scaled plants in Sweden and the United States. The years of starting fabrication are noted in brackets, indicating how long it takes to debug and scale up a new process. The initial design, (i), was an internally CFB; when scaled to (ii), it became an externally circulating unit due to cost and ease of construction and mechanical design considerations. Note that for both the 10 and 100 kW hot units, there were corresponding cold units in order to study the hydrodynamics. Unit (iii) was situated at TU Vienna, and various tests were run, including on the effect of temperature (between 780 and 900 °C) on product gas composition. Further tests on gas cooling and gas cleaning systems were conducted to apply the knowledge gained to the demonstration plant. Unit (iv) was similar in design to unit (iii); however,

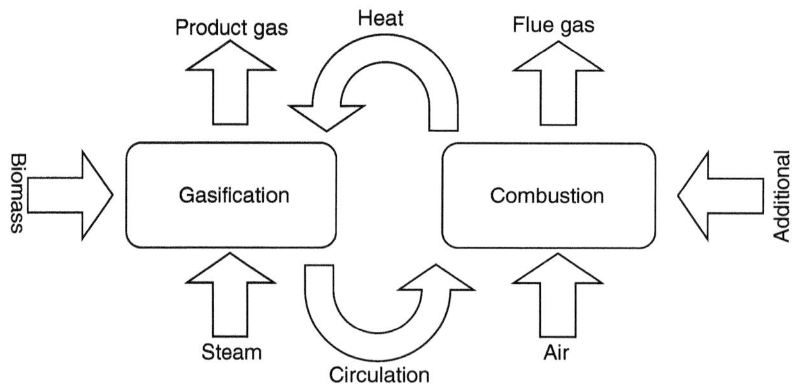

Figure 17.5 Conceptual diagram of the Biomass CHP Plant Güssing.

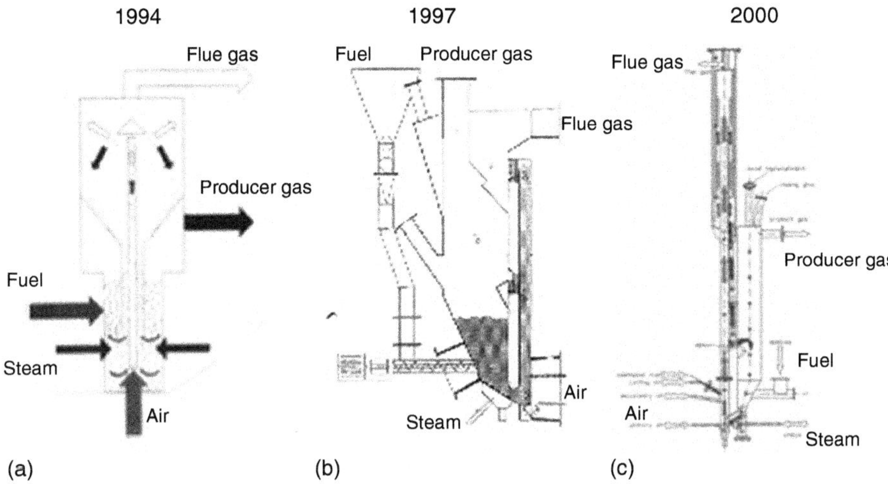

Figure 17.6 Schematic diagrams of pilot plants at TU Vienna depicting design development stages. (a) 10 kW fuel input pilot plant built in 1993; (b) 100 kW fuel input pilot plant built in 1995; (c) 100 kW fuel input pilot plant built in 2000.

knowledge from the cold flow model, hot test rig (unit iii), and simulation were all taken into consideration for scale-up.

As highlighted in this example, not everything needs to go through each scale-up phase, though the ultimate hot unit will require all the auxiliary components that a demonstration unit will have.

17.5.2.2 Dual Gasifier with CO_2 Capture

The second example comes from the University of British Columbia where a sorbent-enhanced biomass gasification was scaled up in collaboration with Highbury Energy Inc. As shown in Figure 17.7, exothermic carbonation is

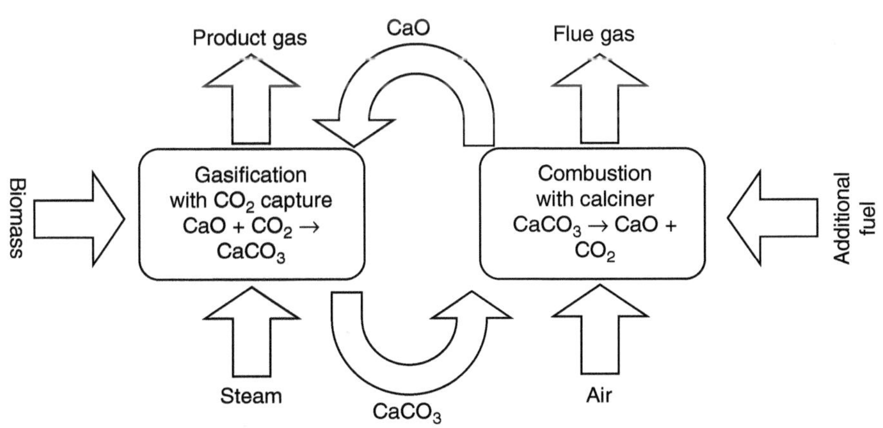

Figure 17.7 Conceptual diagram of the sorbent-enhanced biomass gasification at the University of British Columbia.

coupled with endothermic biomass gasification in the gasifier; in the calciner, endothermic calcination is coupled with exothermic combustion of added fuel. Removing CO_2 in the gasifier shifts the equilibrium to produce more product hydrogen.

The concept was initiated many years ago through thermogravimetric analysis (TGA) and using the unique dual environment TGA of sorbent study [32]. Testing was carried out in hot units of (i) 1.5 kW fuel bubbling fluidized bed gasifier, (ii) 7.5 kW fuel bubbling fluidized bed gasifier, and (iii) 75 kW fuel dual-bed CFB [33], shown in Figure 17.8.

Some of the challenges encountered were:

- Measurement of solids circulation rate
- Plugging of pressure transducer ports
- Standpipe flow
- Segregation of feed and bed particles

In hindsight, this project would have benefited from having a cold flow model where one could visualize the solids segregation and understand the flow pattern better. Due to frequent start-up and shutdown, leaks developed in the flanges over time. This can be avoided by using welded joints. Thermal expansion issues prevailed at the start and during steady-state conditions. A thermal stress analysis model provided excellent guidance in locating high stress points and for modifying piping around start-up burners [34]. Furthermore, using a CFD cold model at the design stage to examine impingement of solids leading to erosion problems, especially at the entrance to the riser, might have alleviated later problems.

It also became challenging to run the pilot plant due to training of operators, which included two research associates, seven graduate students/postdoctoral fellows, and three undergraduate/summer students. In general, it is difficult to

Figure 17.8 Photograph of the sorbent-enhanced biomass gasification at the University of British Columbia.

operate a pilot-scale unit in academia; however, the benefit of being able to contribute to the fundamental understanding of the phenomena is enticing. Pilot plants can assist in building confidence in scale-up to the demonstration unit and report on things that did and did not work.

17.5.2.3 Calcium Looping Technologies

The last example involves looping technologies in CFBs or dual fluidized bed systems. Calcium looping process is a post-combustion CO_2 capture technology for power generation plants. Several processes using calcium-based sorbents have been developed over recent decades. A comprehensive review of bench- and pilot-scale calcium looping systems [35] revealed available research capacity, and process modelling has assessed progress toward commercialization. As part of the EU-funded "CaOling" project (www.caoling.eu), a 1.7 MW_{th} pilot plant in La Pereda, Spain, was tested for post-combustion Ca looping in a CFB. They reported a total of 380 hours in the CO_2 capture mode [36]. Leading up to this pilot scale, there were three units (with capacities ranging from 10 to 75 kW_{th}) compared for this technology [37]. Collaboration between facilities in Spain, Germany, and Canada resulted in testing at various scales and at different locations, reducing the risk, while building confidence in the process by operating the pilot plant.

In another series of studies, process modelling of a calcium looping system was investigated by operating 10 and 200 kW_{th} units and finally a 1 MW_{th} unit [38]. Variation of key parameters in process modelling allowed scenario testing, helping to optimize the process. Hence, process modelling at different scales has become a powerful tool in scaling fluidized bed processes.

17.6 Scale-Up and Industrial Considerations of Fluidized Bed Catalytic Reactors

17.6.1 Challenges of Scale-Up of Fluidized Bed Catalytic Reactors

Scale-up benefits from knowledge from smaller units and reactor models to design larger reactors for industrial use. Unfortunately, there are risks associated with scale-up, and the assumptions valid at the laboratory scale (reactor diameter of ~50 mm) and pilot scale (reactor diameter up to ~0.5 m) may not be valid at full scale. Some important considerations for process scale-up are:

- **Demonstration plant tests**: To reduce technical risks, normal process development involves reactor modelling and the experience with new processes at laboratory, pilot, and demonstration scales before building a full-scale unit. "Demo-scale" systems can answer questions about fluidized bed challenges such as variability, bypassing, nonideal behaviour, gas mixing, feed gas distribution, and practical aspects of implementation. If there is limited time to conduct all these steps, it is recommended not to "skip steps," but instead consider the development of a development program that conducts the work above, with some tasks in parallel to expedite the work. For instance, the modelling can be carried out in parallel with the experimental work. Also the design, installation, and operation of the reactors at laboratory, pilot, and demonstration plant scales can be completed, with some steps in parallel to improve the schedule.

- **Emissions**: A design that may be appropriate at the pilot scale may have emissions that, when scaled, are too high for full-scale units. To comply with environmental regulations, reactors may require downstream units such as wet or dry scrubbers, absorption–desorption technologies, gas recirculation, catalytic oxidation, absorption systems, filters, electrostatic precipitators, or other tail-gas pollution abatement systems. This may cause technical and economic challenges due to requirements in purchased scrubbing chemicals, increased pressure drop, and disposal of scrubbing by-products.
- **Minimum, normal, and maximum plant rate**: The full-scale reactor must consider the process conditions at which it may be operated, including turndown and future expansion.
- **Emissions during start-up and process upset conditions**: To achieve ideal performance of the reactor and its catalyst, it is required to preheat the system to the required process conditions. Preheating and stabilization of the plant may take several days, depending on the size of the plant. During this period the emissions are often higher than at steady state. To address this issue, it is often necessary to modify plant operating procedures, make equipment upgrades, make feedstock changes, and sometimes utilize additional unit operations to minimize emissions during start-up. These issues are typically not considered for pilot plants.
- **Energy efficiency and integration**: Laboratory- and pilot-scale units normally focus on proving the reactor concept and are typically not designed for optimal energy use or energy recovery. This issue becomes extremely important at the industrial scale. For instance, if the reactions are exothermic, the design of a full-scale plant should be coupled with a heat removal system (e.g. a high-pressure steam system) that recovers energy for use elsewhere in the plant or for production of electricity in a turbogenerator set. Industrial reactors with endothermic reactions require a source of heat, which may be provided by burning hydrocarbons in situ or by indirect heat transfer from other systems. The integration of energy recovery and delivery systems is an important task that can improve the economics of industrial processes. To complete the design, the main process systems provide "master" process flow diagrams (PFDs), while ancillary systems, such as superheated steam systems and cooling water systems, provide "slave" PFDs.
- **Autothermal limits and heat losses**: Since the surface area of small reactors is very large for a given reactor volume or energy content, heat losses are likely to be the dominant terms in reactor heat balances. Laboratory units typically require continuous heating, even if the reactions are exothermic (e.g. by using electrical heaters). In contrast, heat losses in a large industrial reactor typically account for less than 2% of the total energy generated in the plant. The autothermal limit of a reactor (i.e. operating conditions where no additional heat is needed for steady-state operation) varies as the system is scaled up.

- **Process gas dew point considerations and fouling issues**: The process gas dew point determines the temperature limit before condensation of species in process gas begins. It is a function of the concentration of condensable gases, as well as system pressure. In industrial applications, it is important to consider the possibility of condensation to prevent corrosion and fouling.
- **Operating conditions**: The optimum operating conditions can be identified by reactor modelling or pilot plant operation and data analysis. The industrial application should operate under pressure and temperature conditions that are practical at large scale. For instance, high temperatures may reduce equipment reliability, and the use of high-pressure or pressure swings may be expensive. The total system pressure drop becomes more important at full scale since it directly affects the electrical costs required for the operation of blower and compressors.
- **Mechanical design**: Mechanical design typically becomes more complicated as the size of the equipment increases. For instance, the required vessel metal thickness increases, and thermal expansion issues become more severe for larger vessels. High-temperature applications may have long-term issues with metal creep and thermal stress that are not normally encountered in pilot plants. Conceptual designs used at the pilot scale may not be suitable at full scale, and new mechanical details have to be developed. For example, equipment with rectangular cross sections become prohibitively expensive at large scales due to wall thickness requirements. Moreover, the expectation for on-stream time and equipment longevity increases for industrial applications compared with pilot systems. Industrial plant equipment is typically designed to operate for 25+ years, and in many cases equipment is upgraded and maintained indefinitely with no fixed lifetime.
- **Materials of construction**: The reactor vessel must tolerate the process conditions and contact with process chemicals. It is important to select materials of construction that can withstand the peak reaction temperatures expected in the process and that is resistant to chemical corrosion over wide ranges of process species concentrations. Moreover, the reactor vessel and ancillaries must be tolerant to wear caused by the solid particles. High temperatures, corrosive chemicals, and mechanically erosive environments often occur in fluidized bed reactors, requiring construction in a variety of metal alloys, stainless steels, and brick/refractory-lined configurations.
- **Reactor economics**: An approach that may be convenient at small scale may not be economical at full industrial scale. In a pilot plant, it is often enough to prove a design concept, while in an industrial setting, the economics of the complete system has to be proven in the context of a minimum return on investment. Some reactor designs that prove to be a technical success might be economic failures.
- **Redundancy and reliability**: Full-scale systems should consider the consequences of equipment failure carefully. For this reason, it is often necessary to install redundant process equipment for blowers, pumps, critical systems, and safety systems. These increase the complexity and cost of the reactor system.
- **Industrial considerations**: Each plant has its technical requirements depending on the process industry and location. A full-scale reactor must comply

with those requirements, including the mechanical design and stamping by code (such as ASME Section VIII – Division 1), use of specific pressure relief strategies, and environmental emission standards (such as emission factors or guidelines of government agencies).

- **Safety**: It is of great importance to implement systems to mitigate safety risks. Applicable strategies include design reviews, hazards and operability studies (HAZOPs), relief studies, pressure tests, relief valves, venting to safe location, thermal insulation, minimizing time spent in the process area by operators, alarms and interlocks, redundancy of instrumentation, design by code, fabrication quality control, inspections and maintenance, and personal protection equipment.
- **Other practical issues**: When a new reactor design is implemented, there are a number of important practical issues such as process and mechanical engineering considerations, budgetary constraints, turnaround planning, technical risk, space availability, shipping size restrictions, permitting, logistic considerations, safety aspects, and the time required for fabrication and installation of major equipment. These considerations may play a major role in determining whether or not the new design is feasible.

17.6.2 Practical Recommendations for Industrial Implementation of Reactor Systems

Based on the steps and factors provided above, a complete reactor system can be designed. Similar considerations apply for laboratory-scale, pilot plant, and industrial systems, although the level of complexity varies widely. Typical design documents required to execute a fluidized bed reactor project include:

- **Block diagrams**: Block diagrams are sometimes useful to identify the main inputs and outputs of a system and to summarize consumption and production rates. However, block diagrams are only useful for high-level discussions, for small-scale lab experiments, but not for project execution.
- **Process flow diagrams (PFDs)**: These diagrams and stream tables contain the core process information. The total flow rate, species concentrations, temperature, and pressure are identified for each stream. A reactor model is used to calculate the inlet/outlet process stream of the fluidized bed, while simple calculation models for mass and energy balance are used to define the other process streams.
- **Piping and instrumentation diagrams (P&IDs)**: These drawings are required to integrate the reactor with other unit operations, its control system, and peripherals. The P&IDs require the following information:
  - Major equipment, including nozzles and internals
  - Piping, ducting, dampers, and valves with their sizing
  - Parameters to be measured (locally and in distributed control systems [DCS])
  - Parameters to control and major control loops
  - Identification inputs and outputs, as well as interlocks
  - Relief valves with set points

- **Equipment and mechanical documentation**: These documents and drawings are required to specify the complete system, including the reactor, any internals, and major unit operations:
  - Equipment mechanical drawings and detailed fabrication drawings
  - Equipment list with major dimensions and materials of construction
  - Process equipment datasheets and specification
  - Rotating equipment specification
  - Catalyst type and volumes
  - Equipment drawings
  - Specification of ducting, piping, expansion joints, and ancillaries
- **Process control documentation**: These documents are used to specify the control system for the reactor and its ancillaries. The following are required:
  - Control strategy and process control narratives
  - Control loop diagrams: for feedback, cascade, feed forward, and complex loops
  - Cause and effect matrix
  - Interlock list
  - Alarm and set point table
- **Instrumentation documentation**: Required to procure and install the instruments:
  - Instrument list
  - Instrument, process valves, relief valves, and dampers specification and/or datasheets
  - Instrument input/output lists
- **Layout and design documentation**:
  - System layout drawings (plan view, elevation, and/or 3-D model)
  - Ducting and piping drawings and isometrics
  - Other criteria and system drawings
- **Balance of plant documentation**: These include criteria documents and drawings suitable for the specifications of all other disciplines required for the construction and specification of the system, in addition to documents, drawings, and specifications for civil criteria, platforms, structural, electrical, wiring, motor controls, ancillaries, utilities, peripherals, cooling water, instrument air, insulation, and others.
- **Project life cycle support activities**: The engineering team in charge of the technology and design of a reactor system should be prepared to support the entire project cycle, as required by the application. Some of the activities and deliverables considered include:
  - **Basic engineering**: Including preliminary engineering drawings and calculations.
  - **Detailed engineering**: Including a final package of drawings suited for construction.
  - **Procurement support**: Often required to evaluate technical bids from suppliers of peripheral systems.
  - **Construction supervision support**: If required by end user.
  - **Installation supervision support**: If required by end user.

- **Final mechanical check-out and approval**: Required to ensure that the design intent of the reactor system is met by the system as built.
- **Pre-commissioning**: To evaluate all subsystems and ensure that they meet the design and safety specifications.
- **Operator training**: As required by end user.
- **Commissioning**: Functional checks on equipment and control systems, cold fluidization, and operation of systems in isolation to ensure they are ready to handle the chemicals and operating conditions required under normal operation.
- **Start-up**: Preheat reactor and ancillaries prior to introduction of chemicals; increase rate until the plant is stable and meeting performance requirements.
- **Performance test, troubleshooting, and support**: Evaluate performance indicators such as production capacity, efficiency, and emissions. Troubleshoot as needed.

Solved Problems

17.1 It is desired to build a cold model of width 1.0 m to study the hydrodynamics of a combustion unit of width 3.6 m that operates at 800 °C and 1 atm pressures and a superficial velocity range of 1.4–2.8 m/s with olivine sand of mean particle diameter 1.5 mm and density 3066 kg/m^3. Find the particle size, particle density, and superficial gas velocity required for the cold model.

Solution

In modelling of physical phenomena, it is possible to achieve full dynamic similarity if one assures the following: (i) perfect geometric similarity, i.e. all physical dimensions are scaled up or down according to a single scale ratio; and (ii) equality of the important dimensionless groups.

One can prove that similarity is obtained by writing the key governing physical laws as differential equations and then rendering the equations dimensionless. Condition (ii) above will assure that the dimensionless governing equations will be identical in the two cases, while condition (i) means that the boundary conditions are identical in dimensionless form. Hence the dimensionless solutions must be the same, and the results can be scaled.

A question of key importance that arises immediately is which dimensionless groups are important. If we choose d_p as the relevant length scale, then we may write

$$\text{Hydrodynamics} = f\,(d_p, U, \rho, g(\rho_p - \rho), \mu)$$

Note that it is implicitly assumed here that other factors causing forces on the particles (e.g. moisture or electrostatic charges) are absent in both cases.

With five dimensional qualities and three dimensions (mass, length, and time), we need two independent dimensionless groups, which we can choose as

$$\text{Hydrodynamics} = f\ (Re, Fr)$$

where $Re = \frac{\rho U d_p}{\mu}$ and $Fr = \frac{U^2 \rho}{d_p g(\rho_p - \rho)}$
Since both of these groups include U, it is more convenient to replace one of them by

$$Ar = \frac{g\rho(\rho_p - \rho)d_p^3}{\mu^2}$$

so scaling requires both geometric similarity and

$$\frac{g_1(\rho_{p1} - \rho_1)\rho_1 d_{p1}^3}{\mu_1^2} = \frac{g_2(\rho_{p2} - \rho_2)\rho_2 d_{p2}^3}{\mu_2^2} \tag{17.4}$$

Equality of Froude numbers requires

$$\frac{U_1^2 \rho_1}{d_{p1} g_1(\rho_{p1} - \rho_1)} = \frac{U_2^2 \rho_2}{d_{p2} g_2(\rho_{p2} - \rho_2)} \tag{17.5}$$

Now for the system to be modelled (designated by the subscript 1, i.e. olivine sand at 800 °C and 1 atm),

$$g_1 = 9.8\ \text{m/s}^2, d_{p1} = 1.5\ \text{mm}, \rho_{p1} = 3066\ \text{kg/m}^3,$$
$$\mu_1 = 4.3 \times 10^{-5}\ \text{Ns/m}^2, \text{and}\ \rho_1 = 0.33\ \text{kg/m}^3$$

In the cold model with scale factor of 3.6 m/1.0 m = 3.6

$$\text{Diameter of the particles}\ d_{p2} = \frac{1.5\ \text{mm}}{3.6} = 417\ \mu\text{m}$$

$$g_2 = g_1 = 9.8\ \text{m/s}^2$$

Also, since the cold model operates with air at atmospheric temperature and pressure, then $\rho_2 = 1.21\ \text{kg/m}^3$ and $\mu_2 = 1.82 \times 10^{-5}\ \text{Ns/m}^2$. This leaves the ρ_{p2} and U_2 to be specified.
From Eq. (17.4),

$$(\rho_{p2} - \rho_2) = (\rho_{p1} - \rho_1)\left(\frac{\mu_2}{\mu_1}\right)^2 \frac{\rho_1}{\rho_2} \frac{d_{p1}^3}{d_{p2}^3}$$

$$= 3065 \times \left(\frac{1.82}{4.3}\right)^2 \times \frac{0.33}{1.21} \times \left(\frac{1.5}{0.417}\right)^3 = 7000\ \text{kg/m}^3$$

$$\rho_{p2} \cong 7000\ \text{kg/m}^3$$

which is quite close to the density of steel. Therefore we choose shot steel of mean size about 417 μm.

Next the gas velocity is calculated using Eq. (17.5):

$$U_2 = U_1\sqrt{\frac{\rho_1}{\rho_2}\frac{d_{p2}}{d_{p1}}\frac{(\rho_{p2}-\rho_2)}{(\rho_{p1}-\rho_1)}} = U_1\sqrt{\frac{0.33}{1.21}\frac{0.417}{1.5}\frac{7000}{3065}} = 0.416U_1$$

Thus, the range of U_2 is 0.416×1.4 to 0.416×2.8 m/s = 0.582 to 1.16 m/s.

Notations

Ar	Archimedes number (–)
D	column diameter (m)
d_p	particle diameter (m)
f	frequency (s^{-1})
Fr	Froude number
Fr_t	Froude number based on terminal velocity
G_s	solids mass flux (kg/m^2 s)
g	acceleration of gravity (m/s^2)
L	length (m)
Re	Reynolds number (–)
U	superficial velocity (m/s)
U_c	transition velocity at which standard deviation of pressure fluctuations reaches a maximum (m/s)
v	average particle velocity (m/s)
v_t	particle terminal velocity (m/s)

Greek Letters

ε	void fraction (–)
μ	dynamic viscosity (Pa s)
ρ	density (kg/m^3)
ρ_p	particle density (kg/m^3)
φ	slip factor (Eq. (17.1))

Subscript

c	commercial (in Table 17.1)

References

1 Donati, G. and Paludetto, R. (1997). Scale up of chemical reactors. *Catal. Today* 34: 483–533.
2 Squires, A.M. (1982). Contributions towards a history of fluidization. In: *Proceedings of the Joint Meeting of Chemical Industry and Engineering Society of China and AIChE Conference*, 322–353. Beijing: AIChE.
3 Chen, G., Wang, Z., and Sun, G. (1984). On the methodology of process development of fluidized bed reactors. In: *Fluidization: Proceedings of Fourth*

International Conference on Fluidization (eds. D. Kunii and R. Toei), 583–590. New York: Engineering Foundation.

4 Kunii, D. and Levenspiel, O. (1991). *Fluidization Engineering*, 2e. Boston: Butterworth-Heinemann.

5 Maurer, S., Rüdisüli, M., Schildhauer, T. J., Biollaz, S., van Ommen, J. R. (2013). Scale-up of bubbling fluidized bed reactors with vertical internals: a new approach accounting for chemistry and hydrodynamics. *ECI Digital Archives, the 14th International Conference on Fluidization*. Noordwijkerhout, Netherlands.

6 Dry, M.E. (2002). The Fischer–Tropsch process: 1950–2000. *Catal. Today* 71 (3–4): 227–241.

7 Duvenhage, D.J. and Shingles, T. (2002). Synthol reactor technology development. *Catal. Today* 71: 301–305.

8 Silverman, R.W., Thompson, A.H., Steynberg, A. et al. (1986). Development of a dense phase fluidized bed Fischer–Tropsch reactor. In: *Fluidization V*, 441–448. Lyngby, Denmark: Engineering Foundation in New York.

9 Volk, W., Johnson, C.A., and Stotler, H.H. (1962). Effect of reactor internals on quality of fluidization. *Chem. Eng. Prog.* 58: 44–47.

10 Sookai, S., Langenhoven, P.L., and Shingles, T. (2001). Scale-up and commercial reactor fluidization related experience with synthol, gas to liquid fuel, dense phase fluidized bed reactors. In: *Fluidization X: Proceedings of the 10th Engineering Foundation Conference on Fluidization* (eds. M. Kwauk, J. Li and W.-C. Yang), 621–628. Beijing: United Engineering Foundation.

11 Krishna, R. and van Baten, J.M. (2001). Using CFD for scaling up gas-solid bubbling fluidized bed reactors with Geldart A powders. *Chem. Eng. J.* 82: 247–257.

12 Rüdisüli, M., Tilman, J., Schildhauer, S. et al. (2012). Scale-up of bubbling fluidized bed reactors – a review. *Powder Technol.* 217: 21–38.

13 Mabrouk, R., Radmanesh, R., Chaouki, J., and Guy, C. (2005). Scale effects on fluidized bed hydrodynamics. *Int. J. Chem. Reactor Eng.* 3 (1).

14 Knowlton, T. (2013). Fluidized bed reactor design and scale-up. In: *Fluidized Bed Technologies for Near-Zero Emission Combustion and Gasification* (ed. F. Scala), 481–523. Cambridge UK: Woodhead Publishing Limited, Elsevier.

15 Bi, H.T., Ellis, N., Abba, A.I., and Grace, J.R. (2000). A state-of-the-art review of gas–solid turbulent fluidization. *Chem. Eng. Sci.* 55 (21): 4789–4825.

16 Ellis, N., Bi, H.T., Lim, C.J., and Grace, J.R. (2004). Hydrodynamics of turbulent fluidized beds of different diameters. *Powder Technol.* 141 (1–2): 124–136.

17 Tasirin, S.M. and Geldart, D. (1998). Entrainment of FCC from fluidized beds – a new correlation for the elutriation rate constants $K_{i\infty}{}^{*}$. *Powder Technol.* 95 (3): 240–247.

18 Colakyan, M. and Levenspiel, O. (1984). Elutriation from fluidized beds. *Powder Technol.* 38: 223–232.

19 Zenz, F.A. and Othmer, D.F. (1960). *Fluidization and Fluid-Particle Systems*. New York: Reinhold.

20 Matsen, J.M. (1997). Design and scale-up of CFB catalytic reactors. In: *Circulating Fluidized Beds* (eds. J.R. Grace, A.A. Avidan and T.M. Knowlton), 489–503. Dordrecht: Springer.
21 Grace, J.R., Bi, X.T., and Golriz, M. (2003). Circulating fluidized beds. In: *Handbook of Fluidization and Fluid-Particle Systems*, Chemical Industries, vol. 2003, 485–544. New York: Marcel Dekker.
22 Patience, G.S., Chaouki, J., Berruti, F., and Wong, R. (1992). Scaling considerations for circulating fluidized bed risers. *Powder Technol.* 72 (1): 31–37.
23 Glicksman, L.R., Hyre, M.R., and Farrell, P.A. (1994). Dynamic similarity in fluidization. *Int. J. Multiphase Flow* 20: 331–386.
24 Horio, M., Ishii, H., Sawa, Y., and Muchi, I. (1986). A new similarity rule for fluidized bed scale up. *AlChE J.* 32: 1466–1482.
25 Glicksman, L.R. (1984). Scaling relationships for fluidized beds. *Chem. Eng. Sci.* 39: 1373–1379.
26 Glicksman, L.R. (2003). Fluidized bed scale-up. Chapter 13. In: *Handbook of Fluidization and Fluid-Particle Systems* (ed. W.-C. Yang), 343–378. New York: Marcel Dekker.
27 Deen, N.G., Annaland, M.V.S., Van der Hoef, M.A., and Kuipers, J.A.M. (2007). Review of discrete particle modeling of fluidized beds. *Chem. Eng. Sci.* 62 (1–2): 28–44.
28 Detamore, M.S., Swanson, M.A., Frender, K.R., and Hrenya, C.M. (2001). A kinetic-theory analysis of the scale-up of circulating fluidized beds. *Powder Technol.* 116 (2–3): 190–203.
29 Leckner, B., Szentannai, P., and Winter, F. (2011). Scale-up of fluidized-bed combustion – a review. *Fuel* 90 (10): 2951–2964.
30 Knowlton, T.M., Karri, S.B.R., and Issangya, A. (2005). Scale-up of fluidized-bed hydrodynamics. *Powder Technol.* 150 (2): 72–77.
31 Hermann, H., Reinhard, R., Klaus, B., Reinhard, K., Christian, A. (2002). Biomass CHP Plant Güssing – a success story. http://members.aon.at/biomasse/strassbourg.pdf (accessed 10 December 2018).
32 Sun, P., Lim, C.J., Grace, J.R., and Anthony, E.J. (2005). On sorbent performance for cyclic absorption of CO_2. *Greenhouse Gas Control Technol.* 7: 1801–1805.
33 Li, Y.H., Chen, Z., Watkinson, P. et al. (2018). A novel dual-bed for steam gasification of biomass. *Biomass Convers. Biorefin.* 8 (2): 357–367.
34 Ujash, S. (2015). Thermal analysis of a pilot dual fluidized system. University of British Columbia. BITS–UBC thesis report.
35 Hanak, D.P., Anthony, E.J., and Manovic, V. (2015). A review of developments in pilot-plant testing and modelling of calcium looping process for CO_2 capture from power generation systems. *Energy Environ. Sci.* 8 (8): 2199–2249.
36 Arias, B., Diego, M.E., Abanades, J.C. et al. (2013). Demonstration of steady state CO_2 capture in a 1.7 MW_{th} calcium looping pilot. *Int. J. Greenhouse Gas Control* 18: 237–245.
37 Rodríguez, N., Alonso, M., Abanades, J.C. et al. (2011). Comparison of experimental results from three dual fluidized bed test facilities capturing CO_2 with CaO. *Energy Procedia* 4: 393–401.

38 Ströhle, J., Junk, M., Kremer, J. et al. (2014). Carbonate looping experiments in a 1 MW_{th} pilot plant and model validation. *Fuel* 127: 13–22.
39 Cai, P. (1989). The transition of flow regime in dense phase gas-solid fluidized bed. PhD Thesis. Tsinghua University, Beijing 102249, China.

Problems

17.1 Fluid cracking catalysts (FCC) of mean particles size 65 μm and particle density of 1350 kg/m^3 is studied in a CFB of various sizes with air at ambient condition. Study A uses the column diameter of 0.152 m and varies the riser superficial gas velocities between 2 and 10 m/s. Plot the slip velocity predicted by correlation given by Patience et al. [22] against U. Comment on the trend. Study B uses four different column diameters (0.04, 0.15, 0.18, and 0.5 m) at G_s = 180 kg/m^2 s and U = 3 m/s. Calculate the voidage for each unit based on the correlation and comment on the trend.

17.2 A cold unit of diameter 1.02 m is used to study a fluidized bed combustion unit of 3.66 m diameter operating at 800 °C. If the combustion unit uses sand of 1.5 mm mean diameter and particle density of 3066 kg/m^3, determine whether the cold unit can be used as a scaled model for the study. Find a suitable type of particle use in the cold unit. If the hot unit is operated at U = 2.5 m/s, what would the gas velocity be in the cold unit? How would you ensure that both units operate in the same flow regime?

18

Baffles and Aids to Fluidization

Yongmin Zhang

State Key Laboratory of Heavy Oil Processing, China University of Petroleum, Fuxue Rd. 18th, Changping, Beijing 102249, P.R. China

18.1 Industrial Motivation

As summarized in earlier chapters, fluidized beds are widely used as reactors and contactors in many industries due to advantages with respect to heat transfer, temperature uniformity, and ability to transport solid particles. Despite their advantages and many applications in industry, fluidized beds have limitations. For low-velocity dense fluidized beds mostly operated in the bubbling and turbulent regimes, most of the gas passes through the solids as bubbles (or voids). When the bubbles are large, bubble rising velocity increases as well. Some gas in bubbles then has limited contact with solids. Therefore, bubble-emulsion interphase mass transfer is often the controlling step in many fluidized bed reactors, especially those with high reaction rates. Moreover, strong solids mixing is induced due to the vigorous bubble movement. As a result, solids are well mixed in fluidized beds, approaching continuous stirred tank reactor (CSTR) behaviour. The wide span of solids residence times tends to reduce reactor conversion and product selectivity.

To improve efficiency or operability of different fluidization systems, efforts have been made by both academic and industrial communities to develop various aiding methods. Among these, adding baffles is most often reported in the literature and quite widely practiced in industry. Baffles are mostly used in low-velocity dense fluidized beds. However, there are also reports of internals or baffles in high-velocity fluidized beds. Other aiding technologies have also been reported in the literature, including the introduction of electric or magnetic fields, pulsations, vibrations, or introducing glidants or other agents (e.g. antistatic agents).

Essentials of Fluidization Technology, First Edition. Edited by John R. Grace, Xiaotao Bi, and Naoko Ellis.

18.2 Baffles in Fluidized Beds

18.2.1 Clarification of Baffles in Low-Velocity Dense Fluidized Beds

In this section, the term "baffle" is understood to mean "internal" fixed surfaces, which include various types of baffles, tubes, and other specially designed structures inserted in fluidized beds. They are specially designed and arranged to improve the reactor performance by increasing the gas–solid contact efficiency and/or adjusting the residence time distributions of gas or solids. Other solid objects, e.g. cyclone diplegs, protruding nozzles, or probes, may also be present, but these objects usually have negligible effects on bed hydrodynamics and reactor performance. Thus, they are excluded from the discussion of this section. According to their shapes and arrangements, the baffles reported in the literature can be classified into three main categories: horizontal baffles, vertical baffles, and fixed packings.

Baffles in low-velocity dense fluidized beds, if installed properly, are expected to:

(a) Enhance gas–solid contact efficiency by decreasing average bubble size, by splitting bubbles, or restricting bubble coalescence
(b) Improve uniform lateral distribution of bubbles in the bed cross section
(c) Suppress axial gas–solid backmixing and reduce the span of residence time distributions of both gas and solids
(d) Reduce solids entrainment and the separation load of downstream cyclone separators or filters
(e) Reduce hydraulic bed diameter (free cross-sectional area divided by total contact perimeter) by increasing perimeter to reduce scale-up risk

However, it should be remembered that adding baffles may also result in some disadvantages including:

(a) Increase axial temperature gradients
(b) Decrease solids holdup and reduce solids residence time in the bed
(c) Increase equipment cost, increase maintenance, and introduce other possible engineering problems such as solids flooding and component failures
(d) Enhance solids segregation and appearance of defluidized zone in bed bottom due to more settlement of particles with larger size or higher density.

18.2.2 Geometric Characteristics of Baffles

18.2.2.1 Horizontal Baffles

Horizontal baffles are internal surfaces whose vertical depths are far less than their horizontal dimensions. Mesh grids, perforated plates, louver baffles, and horizontal tube bundles are typical horizontal baffles used in industrial fluidized bed reactors.

Mesh grids (Figure 18.1) are geometrically similar to wire meshes used in filters. In industrial fluidized beds, mesh grids are actually made of multiple crossing vertical slats. The mesh size is much larger than in wire meshes, usually several tens

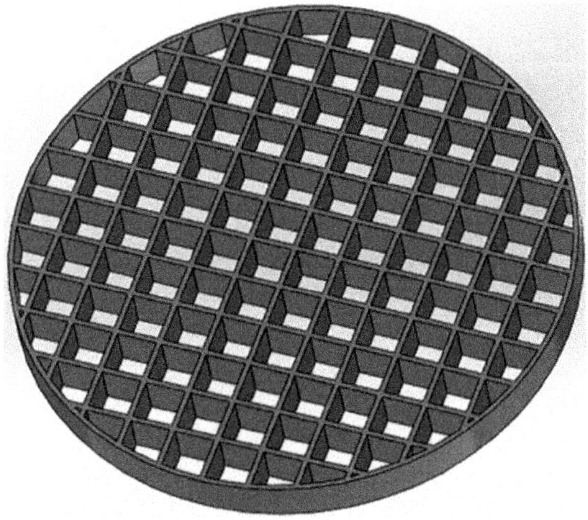

Figure 18.1 Mesh grid.

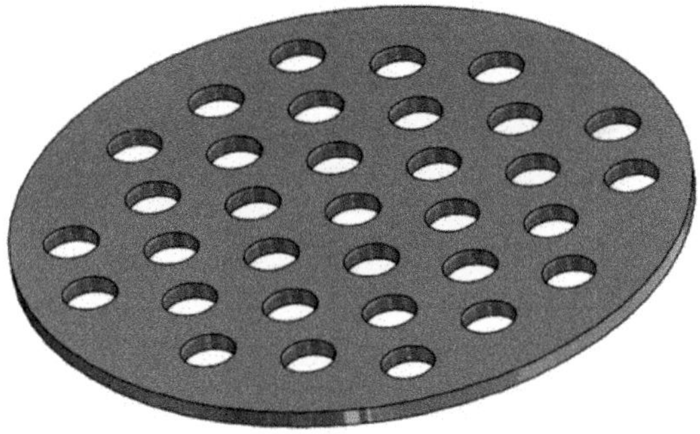

Figure 18.2 Perforated plate.

of millimetres or even greater than 100 mm. The grid thickness and slat thickness are also increased to several tens of millimetres and several millimetres to keep enough strength and stiffness to resist possible erosion or deformation during long periods of operation.

The perforated plate (Figure 18.2) is another type of horizontal baffle widely studied in low-velocity dense fluidized beds and applied in industry. Geometrically, it is similar to the perforated plate distributor in fluidized beds for gas distribution. The major difference is that the former has larger holes and higher open area ratios than the latter. The open area ratio is the sum of the areas of all holes divided by the cross-sectional area of the bed. The holes are usually several tens of millimetres in diameter. The range of open area ratios is 20–70%. The holes are usually distributed evenly across the surface of the plate, but a rarely used option is to have more holes towards the periphery in order to delay the natural tendency for development of patterns where more bubbles rise in the core, rather than near the outside.

Louver baffles are similar in geometry to louver shutters in house windows. As baffles in fluidized beds, they divert gas or solids by promoting horizontal movement. Figure 18.3a shows the simplest form of louver baffle. Usually, the slats are installed at the same pitch. When solids flow from top to bottom towards the louver baffle, solids will be diverted to provide horizontal movement, as shown by

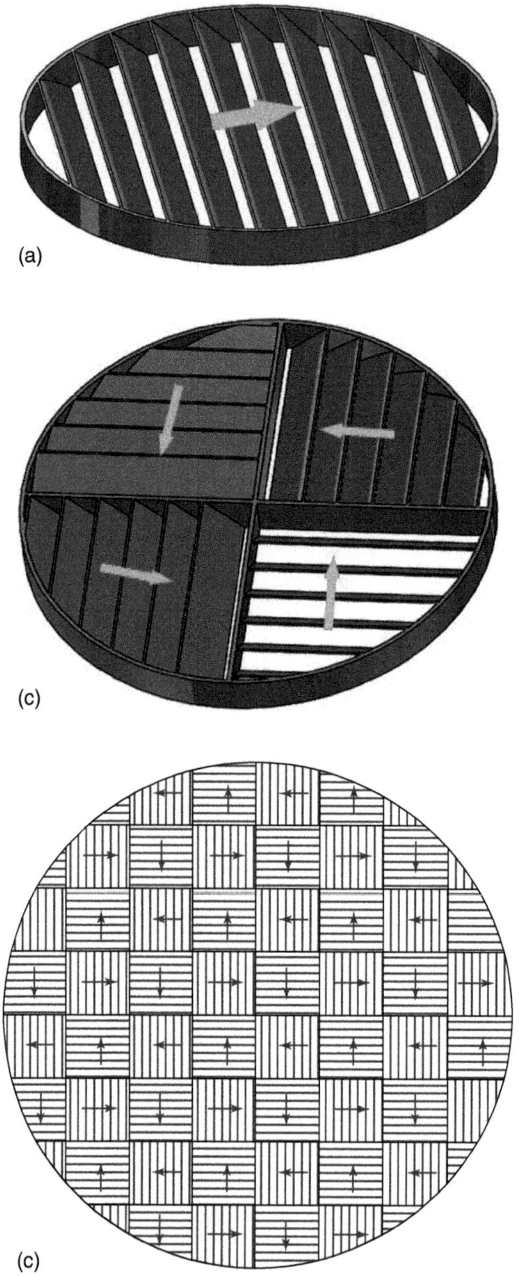

Figure 18.3 Different types of louver baffles. (a) louver baffle; (b) single-turn louver baffle; (c) multi-turn louver baffle.

the arrows. Louver baffles may have various shapes, such as the single-turn and multi-turn louver baffles shown in Figure 18.3b,c, respectively. In a single-turn louver baffle, the circular bed cross section is partitioned into four blocks. In each block, slats are installed at different orientations to divert flow in horizontal movements of different directions. In Figure 18.3b, solids flow towards the louver baffle is expect to travel as shown by the four arrows in the four blocks, rotating in a counterclockwise direction. It can also be designed to provide clockwise rotation. In industrial fluidized beds of large diameter, louver baffle can also be designed in more complex shapes as shown in Figure 18.3c, where more blocks exist in a cross section and multiple solids rotation patterns exist due to the different slat orientations in different blocks.

Zhang [1] proposed a modified horizontal louver baffle, called a Crosser grid, as shown in Figure 18.4, which can be viewed as another type of louver baffle. In this case, there are multiple parallel partition plates separating the baffle into multiple parallel flow regions. Parallel inclined vanes are installed in all flow regions, but their orientations are opposite in adjacent pair regions. Gas and solids flowing through adjacent flow regions can form interlaced contacting above and below the baffle layer. The induced stronger turbulence is expected to be favourable to bubble splitting and improved gas–solid contact.

In many fluidized beds, immersed heat exchange tubes are used to remove heat, e.g. in exothermic processes such as combustion in fluidized bed boilers. If multiple horizontal heat transfer tubes are installed at a given level, the assembly can act like a horizontal baffle. Due to the wide applications of tube bundles by industry, they have been the subject of considerable academic research. In industry, horizontal tube bundles seldom function only as baffles without heat transfer.

18.2.2.2 Vertical Baffles

Vertical baffles include heat exchange tubes, semicircular tubes, and planar plates mounted vertically in fluidized beds where their horizontal dimensions are far less than their vertical heights. A typical example of fluidized bed with vertical baffles is an external catalyst cooler used in a fluid catalytic cracking (FCC) unit (Figure 18.5 [2]), where multiple vertical heating tubes are installed to be in contact with hot catalyst and to transport residual heat arising from coke combustion in a regenerator.

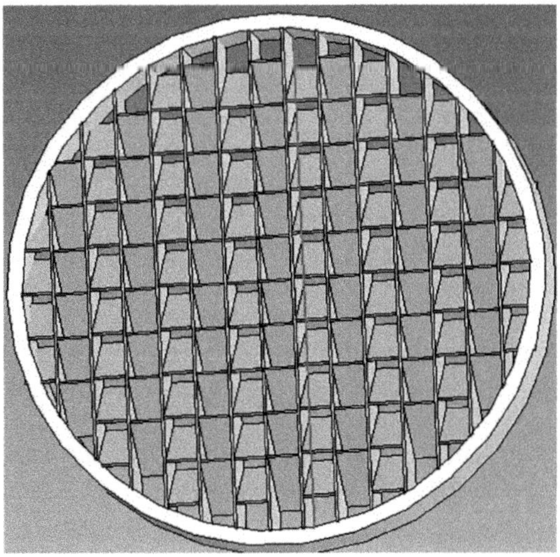

Figure 18.4 Crosser grid. Source: From Zhang 2009 [1].

Figure 18.5 FCC external catalyst cooler (KBR).

Addition of vertical baffles in dense fluidized beds is sometimes adopted to follow a scale-up criterion proposed by Volk et al. [3]. They found that if the hydraulic bed diameter of a large fluidized bed reactor is kept the same as the actual diameter of a small pilot reactor (diameter > 200 mm), by adding vertical rods, chemical conversion in the large baffled bed was equivalent to that in the small baffle-free bed. This is a simple, but oversimplified, criterion [4]. In practice, published reports of successful scale-up industrial fluidized bed reactors by other researchers following the Volk et al. [3] criterion are lacking.

18.2.2.3 Fixed Packings

Fixed packings in fluidized beds are similar to those used in gas–liquid packed towers. In both cases, the packings are intended to improve interphase mass transfer. However, packings in fluidized beds usually have larger internal passages for gas and solids flow through them to avoid solids bridging and defluidization. The dimensions of internal passages in fixed packings of industrial fluidized beds, e.g. pitch of adjacent slats or components, are mostly 50–200 mm, in contrast to the several millimetres or less in gas–liquid packed towers. Stronger structures are also required to resist deformation and erosion caused by solids impact. The packings combine the functions of both horizontal and vertical baffles, with highly efficient gas–solid contact, and they are easy to scale up. However, due to their complexity in structure, successful applications of packing in industrial fluidized beds are mostly confined to fine Geldart A particles with good flowability. Also the fixed packings make it difficult to remove all particles from the bed during shutdowns.

Pagoda-shape and ridge-shape internals proposed by Jin and his colleagues [5, 6] are two of the earliest reported fixed packings that have been successfully commercialized. They are mainly applied in fluidized bed reactors for chemical synthesis, with Geldart A particles and highly exothermic reactions. They are also sometimes referred to as "compound internals," as they combine features of both horizontal and vertical baffles. As shown in Figure 18.6, the pagoda-shape internal is actually a combination of vertical planar plates and inclined weir plates, with holes and saw-tooth notches in the V-shape beams. The beams are perpendicular in adjacent layers. The V-shape beams are attached in vertical tube bundles, which can both influence bed hydrodynamics (like vertical internals) and act as heat exchange tubes.

The KFBE packing of Koch-Glitsch [7], a company well known for structured packing technologies for gas–liquid towers, is a successful example of fixed packing in fluidized beds. As shown in Figure 18.7a, KFBE packing is composed of multiple pairs of inclined strips arranged in a crossing mode. The intersectional streams of gas and solids resulting from the guiding actions of the inclined strips can improve gas–solid contacting. The effectiveness of fixed packings in catalyst stripping has been widely recognized by petroleum companies worldwide. In addition to spent catalyst stripping in FCC units, similar packed strippers have also been applied in methanol-to-olefin (MTO) processes. Many petroleum companies have developed their own patented packed strippers,

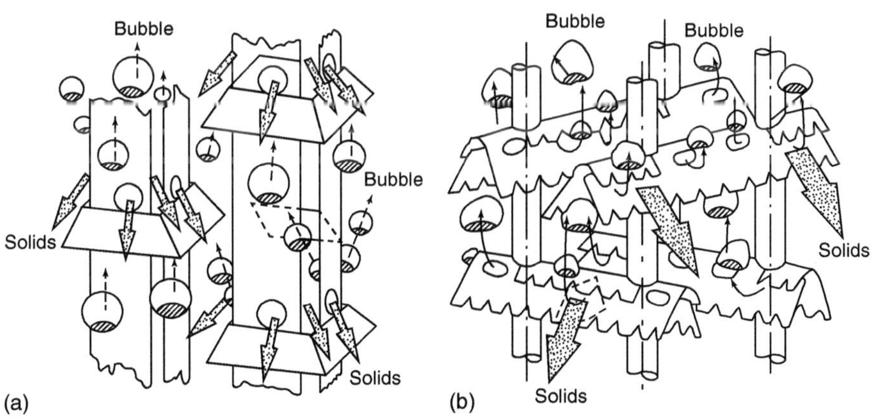

Figure 18.6 Pagoda-shape (a) and ridge-shape internals (b). Source: From Jin et al. 1982 [5] and Jin et al. 1986 [6].

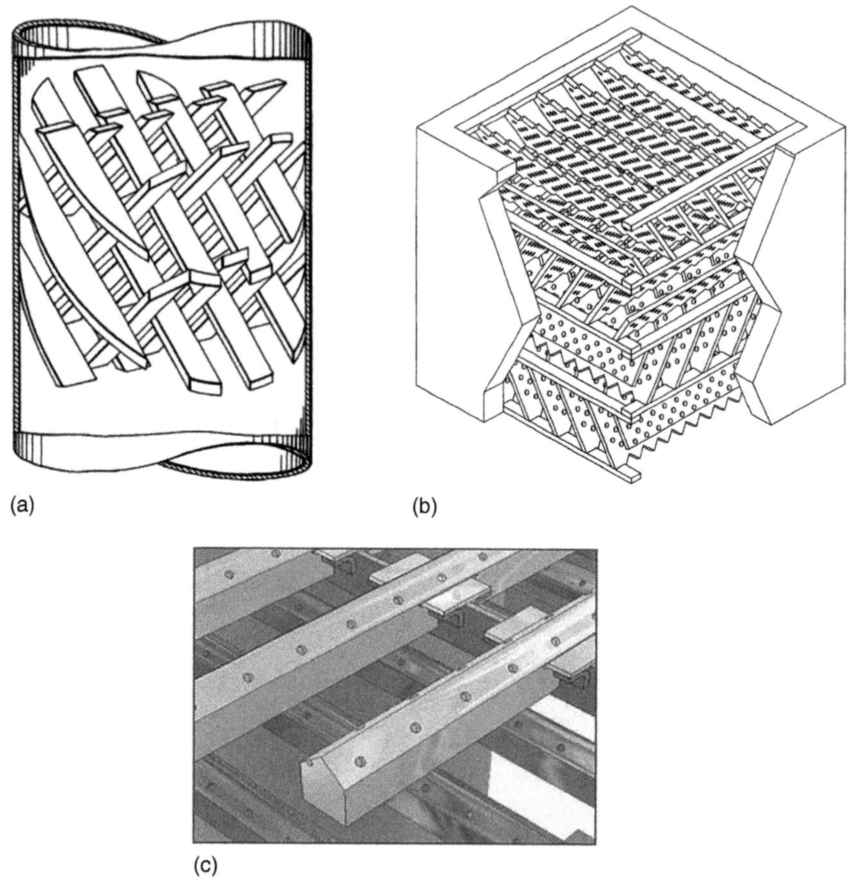

Figure 18.7 Various fixed packings used as catalyst stripper baffles. (a) KFBE baffle [7]; (b) MG baffle [8]; (c) Pentaflow baffle [9].

such as Lummus' MG stripper (Figure 18.7b) [8], Shell's PentaFlow stripper (Figure 18.7c) [9], and UOP's AF stripper [10].

Fixed packings have also been utilized in FCC regenerators, where coke deposited on FCC catalyst is burned off by fluidizing air to recover catalyst activity and provide heat needed for endothermic reactions. Insertion of fixed packing was first proposed by KBR [11, 12]. A section of fixed packing is inserted into a single-stage regenerator to realize better contact between air and coke on the catalyst. Due to the suppression of solids backmixing by the inserted fixed packing, the solids residence time distribution is narrowed to approach two CSTRs in series. The fraction of spent catalyst particles staying in the regenerator far from the mean residence time is thus significantly reduced. The fixed packing used by KBR [12] is shown in Figure 18.8. Multilayer Crosser

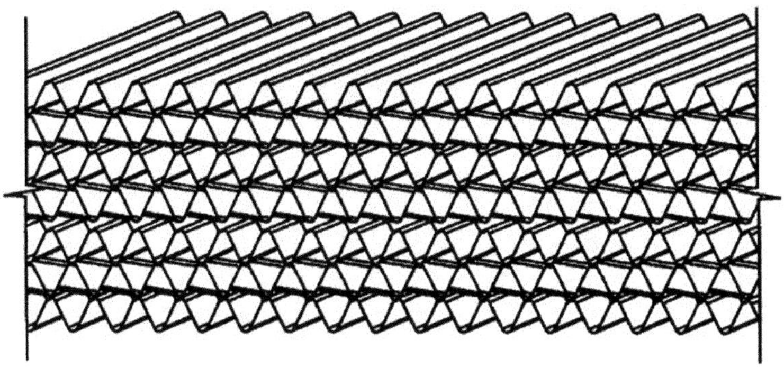

Figure 18.8 KBR's fixed packing for FCC regenerators. Source: From Miller et al. 1999 [11].

grids (Figure 18.4) can also be used in FCC regenerators to improve their performance [13].

18.2.3 Baffles in Low-Velocity Dense Fluidized Beds

18.2.3.1 Effect of Baffles on Bed Hydrodynamics

For the three categories of baffles covered above, horizontal baffles have been studied most, both by experimentation and by computational fluid dynamics (CFD), i.e. numerical simulation. When a horizontal baffle is inserted in a low-velocity bubbling or turbulent fluidized bed, a dilute zone appears beneath it. Figure 18.9 shows the gas–solid flow pattern in a fluidized bed with one layer of louver baffle. A trough appears in the axial profile of solids holdup near the position of the louver baffle, as shown in Figure 18.10 [1].

Figure 18.9 Diagram of dilute region immediately underneath a horizontal louver baffle in a bubbling fluidized bed of FCC particles. Source: From Zhang 2009 [1].

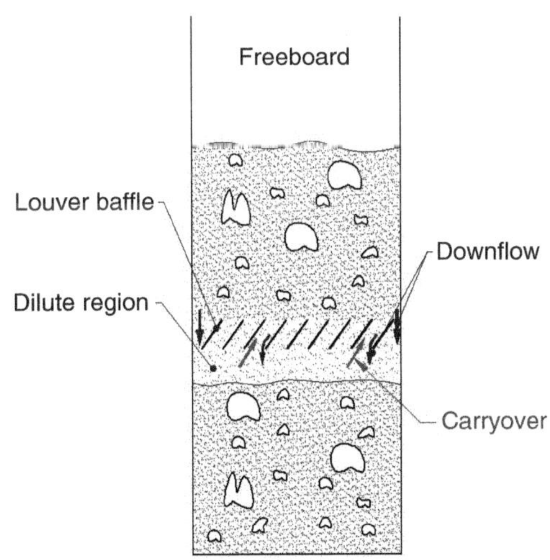

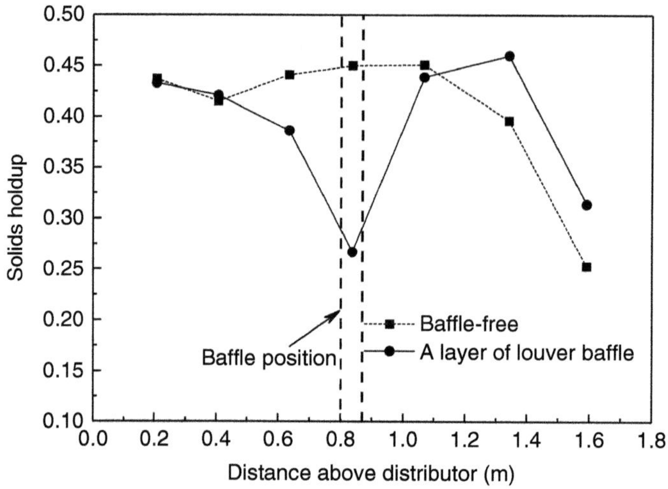

Figure 18.10 Effect of louver baffle on axial profile of solids holdup. FCC particles: $U = 0.371$ m/s, $H_0 = 1.28$ m; column cross section: 0.5 m (wide) × 0.03 m (thick); only one layer of louver baffle installed. Source: From Zhang 2009 [1].

With increasing superficial gas velocities, the solids holdup in the dilute freeboard zone is low above the surface of a conventional fluidized bed, like a "gas cushion." The change of gas cushion height is related to the solids flow rate across the baffle layer. As the gas velocity increases, downward solids flow is restrained. The upward flow rate of solids entrained by gas flow remains high, resulting in a decline of the bed level beneath the baffle layer. The upward solids flow rate decreases with decreasing height until it is equal to the downward solids flow rate. Therefore, the appearance of the dilute zone beneath a horizontal baffle is self-adjusting, helping the bed to maintain steady operation. In addition to being dependent on the gas velocity, the gas cushion height is also influenced by the structural geometry of the inserted baffle, e.g. its open area ratio, shape, and the size of passages for gas–solid flow. The open area ratio has the greatest influence. At the same gas velocity, higher gas cushion height appears below a baffle with lower open area ratio. A gas cushion that is too tall will cause a reduction in solids inventory in the reactor and hence in reactor performance. Most industrial fluidized bed reactors operate at relatively high gas velocities in the turbulent flow regime. In these beds, the baffle open area ratio must be limited to avoid reduction of reactor performance due to the appearance of large gas cushions. Horizontal baffles usually have lower open area ratios, so few layers are installed. Fixed packings can be viewed as multiple layers of horizontal baffle arranged compactly. To avoid exaggerated reduction of solids inventory, their open area ratios are usually controlled to be much higher than those for horizontal baffles.

Most baffles for low-velocity dense fluidized beds are intended to split bubbles. However, accurate measurement of bubble size in fluidized beds is difficult, especially in Geldart A particles, as bubbles continuously coalesce and break. Inserting baffles further increases the difficulty of bubble size measurement. Therefore, there are few reports comparing bubble sizes for baffle-free and baffled beds.

However, indirect evidence demonstrates the effectiveness of baffles in improving gas–solid contacting efficiency. One is the magnitude of pressure fluctuations in a baffled fluidized bed, which is usually much lower than for a corresponding baffle-free bed. The magnitude of pressure fluctuations in bubbling fluidized beds is closely related to bubble movement [14]. With proper processing of the pressure signals, the magnitude of pressure fluctuations can be related to the average bubble size in the bed.

Figure 18.11 plots results obtained by Zhang et al. [1, 5] in a small two-dimensional fluidized bed of FCC (Geldart A) particles. Multiple layers of compactly arranged louver baffles were used to investigate the effect of the number of baffle layers on pressure fluctuations. It is seen that the measured magnitudes of pressure fluctuations in the baffled beds are considerably lower in the baffle-free bed. The magnitudes can be further decreased by inserting more layers of louver baffles. An interesting phenomenon is that the clear reduction of magnitudes of pressure fluctuations only occurs in the bubbling flow regime ($U \leq 0.5$ m/s). Baffles have no clear effect on the magnitude of pressure fluctuations in the turbulent flow regime. The experimental results also demonstrate that baffle geometry, e.g. the inclination angle and pitch of vanes, also influences the magnitude of pressure fluctuations. The magnitudes of pressure fluctuations were utilized by Zhang [1] as a criterion for the optimization of baffle design.

There is also clear indirect evidence that baffles lead to higher bed expansion. Jin et al. [5, 16] compared the effect of different baffles on bed expansion in a two-dimensional fluidized bed. Figure 18.12 shows that both the bed with louver baffles and pagoda-shape baffles have higher bed expansions than for a baffle-free bed, with the louver baffle having a stronger influence than the pagoda-shape baffle. With increasing gas velocity, the difference in bed expansion between the baffle-free and bed with louver baffles clearly increases. The

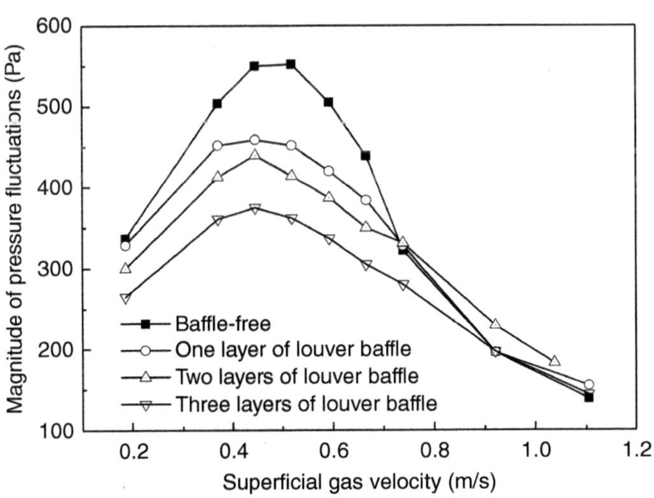

Figure 18.11 Effect of louver baffle on magnitudes of pressure fluctuations. FCC particles: $H_0 = 1.28$ m; column cross section: 0.5 m (wide) × 0.03 m (thick). Source: From Bi 2007 [14].

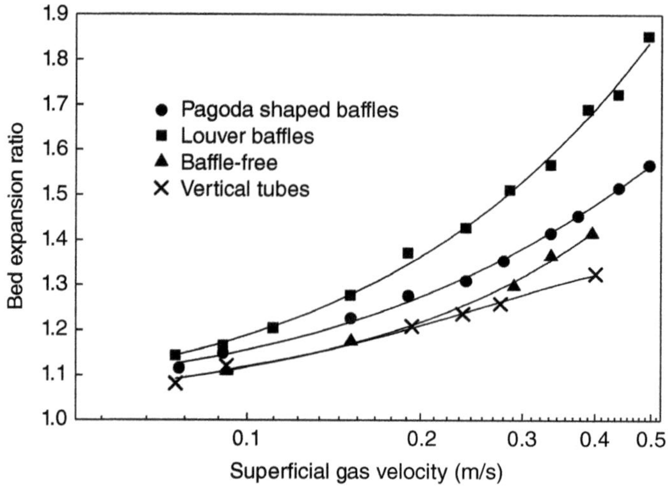

Figure 18.12 Effect of different baffles on bed expansion in a two-dimensional fluidized bed [5, 15]. Silica gel: $d_p = 208.5$ m, $\rho_p = 530$ kg/m^3; column cross section: 0.4 m (width) × 0.015 m (thickness). Source: From Jin et al. 1982 [16].

greater expansion of baffled beds is not only related to bubble breaking but also to the dilute "gas cushion" zone appearing below baffles. The lowest open area ratio of louver baffle leads to higher bed expansion. Also, the increased height of gas cushion at higher gas velocity results in a greater difference between the baffle-free and louver-baffled bed. Vertical tubes have almost no effect on bed expansion, consistent with inferior improvement in gas–solid contacting.

Cai et al. [17] found that adding baffles can decrease the onset velocity of the turbulent fluidization flow regime. Typical results are shown in Figure 18.13, where the peak magnitudes of pressure fluctuations, corresponding to the onset of the turbulent flow regime, appear earlier in the three baffled beds. The reason that baffles advance the transition to the turbulent flow regime is related to the promotion of bubble splitting. Among the three types of baffles, the pagoda-shape and ridge-shape internals demonstrated stronger reductions in the onset velocity than vertical tubes.

Another important function of baffles is to suppress axial mixing of both gas and solids. Clear evidence of the ability of baffles to suppress gas backmixing is provided by gas tracing, where tracer gas is injected at the top of a dense bed, then sampled beneath the injection level. Figure 18.14 compares typical axial gas dispersion coefficients, based on a one-dimensional steady mixing model in baffle-free and baffled beds of FCC particles [1]. It can be seen that adding louver baffles significantly reduced the axial gas dispersion coefficients. The mesh grid was found to be less effective in suppressing the gas backmixing. At low gas velocities, the measured axial gas dispersion coefficients were higher in the baffled bed with mesh grids than in the baffle-free bed. Differences in the configurations and net open area ratio of the Crosser and mesh grids account for the differences in gas backmixing suppression.

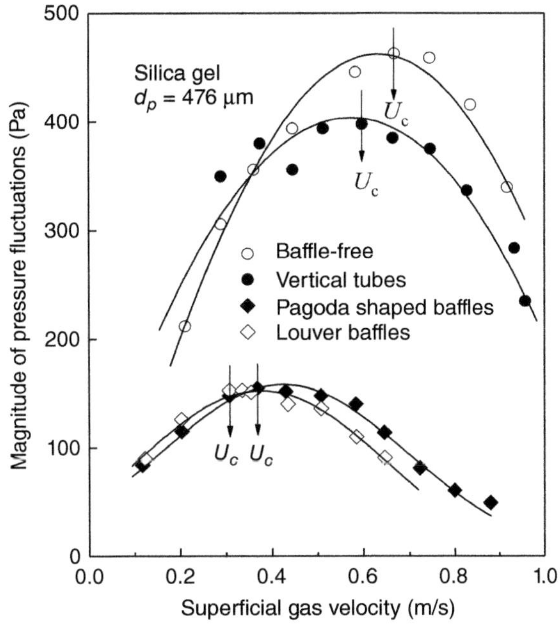

Figure 18.13 Effect of baffles on transition from bubbling to turbulent flow regime [16]. Silica gel: $d_p = 476$ m, $\rho_p = 834$ kg/m^3; column cross section: 0.28 m (width) × 0.28 m (thickness). Source: From Jin et al. 2003 [16].

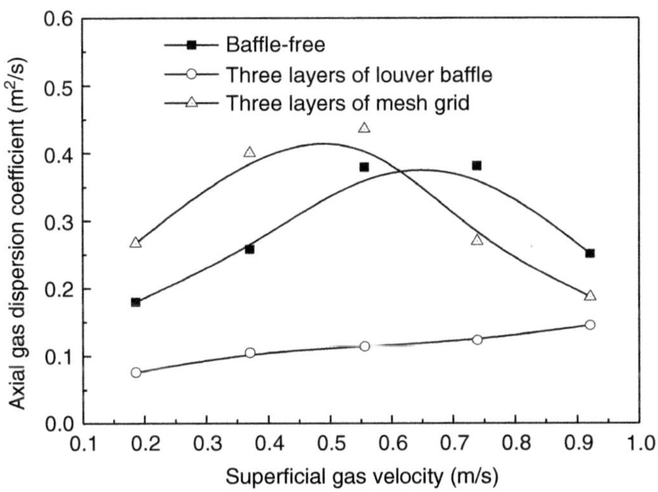

Figure 18.14 Effect of baffles on axial gas dispersion coefficients in fluidized beds. FCC particles: $H_0 = 1.28$ m; column cross section: 0.5 m (width) × 0.03 m (thickness). Source: From Zhang 2009 [1].

Gas mixing is strongly influenced by the solids flow pattern. In gas tracing experiments, it is usually found that locations where the concentration of measured tracer gas is high usually have high local downward solids flux. This is especially pronounced in fluidized beds of Geldart A particles. Similar to bubble size measurement, accurate measurement of solids mixing parameters in fluidized beds by solids tracing experiment is also difficult. Recently, direct evidence of the effect of baffles in suppressing solids backmixing was reported by Yang [18]. The transient vertical solids flux across the middle cross section of a louver baffle obtained by numerical simulation is plotted in Figure 18.15b. It is seen that both upward and downward solids circulation fluxes are near 10 kg/(m^2 s), nearly an order lower than the corresponding value in the baffle-free bed in Figure 18.15a. The ratio between the internal solids circulation flux in the baffled and free beds is consistent with the gas tracing experimental results, both qualitatively and quantitatively, further proving the dominant role of solids mixing on gas mixing in fluidized beds.

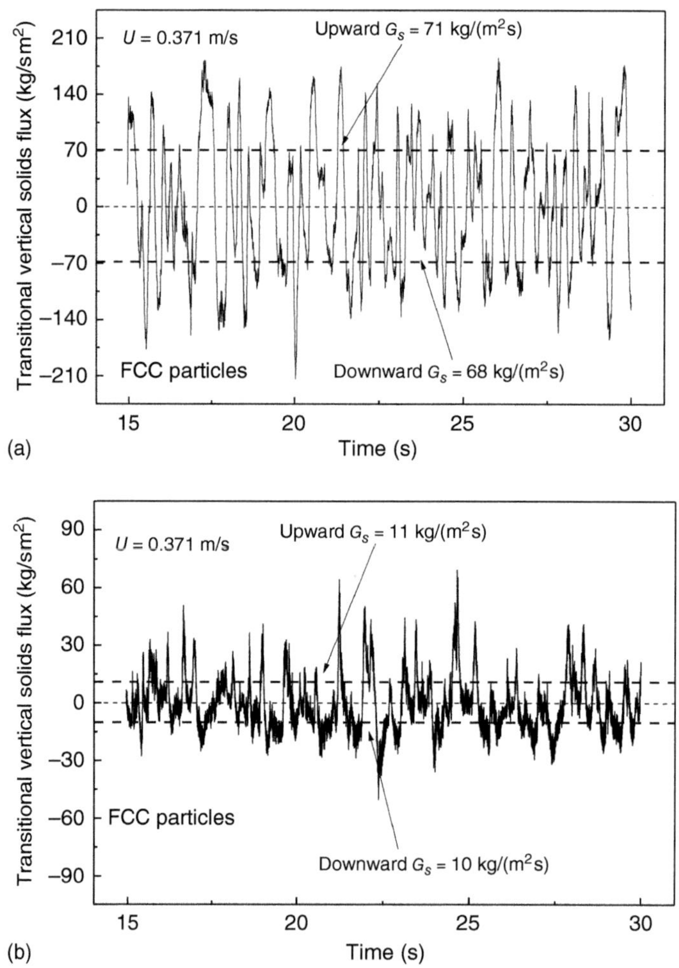

Figure 18.15 Comparison of simulated transient vertical solids flux in baffle-free and baffled beds. FCC particles: $U = 0.371$ m/s, $H_0 = 1.28$ m; column cross section: 0.5 m (width) × 0.03 m (thickness); only a single layer of louver baffle was installed. (a) baffle-free fluidized bed; (b) fluidized bed containing baffles. Source: (a) From Yang et al. [18].

Using X-ray imaging, van Dijk et al. [19] observed the interaction between rising bubbles and horizontal baffles in a fluidized bed. They found that particles in the wake are removed from a bubble when the bubble passes through a baffle layer, explaining the mechanism by which baffles suppress solids backmixing.

Based on previous studies of the effectiveness of various baffles, it can be summarized that baffles with inclined surfaces, e.g. horizontal louver baffles and fixed packings, are usually more effective in breaking bubbles and improving gas–solid contacting. Baffles containing vertical surfaces, e.g. mesh grid and most vertical baffles, are usually less effective in this respect, as indicated by the higher amplitudes of pressure fluctuations (Figure 18.13), lower bed expansion (Figure 18.12), and higher onset velocity of the turbulent flow regime (Figure 18.13). The effect of baffles on gas–solid mixing is primarily affected by their open area ratio. However, the shape and size of flow passages also have some influence. Vertical baffles are generally inferior to horizontal baffles with inclined surfaces and fixed packings in suppressing backmixing of both gas and solids.

Note that adding baffles has little influence on the reactor performance when a fluidized bed reactor is under chemical kinetics control. However, when a fluidized bed reactor is under the control of interphase mass transfer as in most fluidized beds of group A particles, then baffles, by controlling bubble size and improving interphase mass transfer, can significantly improve reactor performance [20].

18.2.3.2 Performance of Baffles in Industrial Fluidized Bed Reactors

Compared with results of laboratory experiments and numerical simulations, data from industrial fluidized bed reactors provide more persuasive proof of the value of baffles. Jin et al. [5, 6, 21] have used their pagoda-shape and ridge-shape internals in fluidized bed reactors for synthesis of phthalic anhydride, vinyl acetate, and acrylonitrile. After installing the pagoda-shape internals in an industrial fluidized bed reactor for synthesis of phthalic anhydride, the processing capacity was found to be increased by 80–100%, while the yield of phthalic anhydride increased by 4–6%.

Another example is the application of Koch-Glitsch's KFBE packing. In industrial FCC units, the hydrogen mass content in coke of spent catalyst is an indicator of the stripping efficiency, which is theoretically in the range of 5–9%. After replacing conventional disc–donut baffles with KFBE packing in an FCC unit, the hydrogen mass content in coke of spent catalyst decreased from 7% to 6%, even though the steam consumption decreased, resulting in a significant increase in stripping efficiency [22].

Crosser grids like those shown in Figure 18.4 have also been inserted in a single-stage FCC regenerator. [13] After the revamp, the maximum lateral temperature difference in the dense bed decreased from 15 to 3 °C, indicating stronger solids lateral mixing and better gas–solid contacting. Even with a 10.9% reduction of the main air flow rate, the carbon content in the regenerated catalyst decreased from 0.1% to 0.06%, resulting in better regeneration performance at lower cost.

18.2.3.3 Other Findings and Applications

In addition to the advantages of baffles covered above, baffles in low-velocity bubbling fluidized beds have been found to be capable of enhancing solids segregation. For fluidized beds of wide particle size distribution or particles of different densities, adding baffles assists larger or denser (jetsam) particles in sinking to the bed bottom and smaller or lighter (flotsam) particles in rising towards the top of the bed. Hence, some researchers have proposed that baffles be used to assist classification of solids [23–25]. However, for fluidized beds requiring uniformity of temperature or particle composition, adding baffles is likely to be inappropriate.

Baffles in low-velocity dense fluidized beds can also influence entrainment and elutriation of solids in the freeboard. For example, some researchers [26–29] have found that baffles immersed in the bed or in the splash zone can decrease particle carry-over in the freeboard, thus reducing the separation load of downstream cyclone separators or filters. According to these studies, baffles with inclined surfaces installed in the splash zone, i.e. just above the bed surface, are most effective in reducing particle carry-over.

18.2.4 Baffles or Inserts in High-Velocity Fast Fluidized Beds

A major distinct hydrodynamic feature of fast fluidized beds is their so called core–annulus flow structure, resulting in low and high solids holdups in the riser core and near the wall, respectively. To further improve the performance of these fast fluidized beds, researchers have investigated various baffles. The most studied geometry is the annular ring baffle (Figure 18.16a), which provides annular protrusions at the inner wall of a riser. Jiang et al. [30] further found that the conversion of the catalytic ozone decomposition reaction improved when ring baffles were present. Ring baffles have also been used in Shell's commercial FCC riser reactors [31]. Zheng et al. [32] found that radial profiles of solids holdup aremore uniformwith ring baffles installed inside a riser to disrupt the downward annular flow along the wall. Both gas and solid velocity profiles have

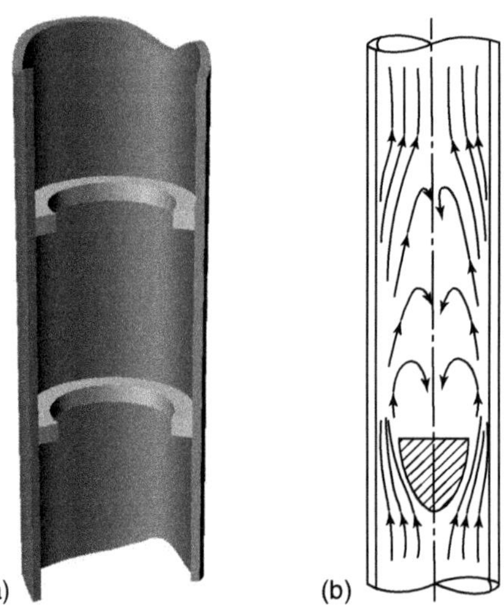

(a) (b)

Figure 18.16 Typical baffles for fast fluidized bed risers. (a) ring baffle; (b) bluff body [30, 31]. Source: (a) From Jiang et al. 1991 [30] and Chen 2006 [31].

been found to be more uniform after inserting ring baffles. Due to the reduced axial dispersion, better product selectivity was achieved. Another type of baffle tested in fast fluidized beds is the bluff body inserted by Gan et al. [33] (see Figure 18.16b). The objective of the ring baffles and bluff inserts is to change the flow pattern so that solids downflow near the wall is periodically disrupted, resulting both in stronger turbulence and more effective gas–solid contact efficiency. Moreover, the existence of the baffles or inserts can bring the gas and solids flow closer to desirable plug flow.

18.2.5 Design of Baffles for Industrial Fluidized Beds

Industrial fluidized beds are usually required to operate at high temperature and under highly erosive conditions for long periods (usually years). To guarantee long-term reliability of baffles, here are several suggestions for designers:

(a) Avoid flooding in fluidized beds with solids circulation through horizontal baffles.
Macro solids circulation exists in many industrial fluidized reactors, with solids continuously fed into the reactor and discharged simultaneously, e.g. to other vessels via transfer lines. The macro solids circulation flux with respect to the reactor cross section can range from several kg/(m^2 s) (e.g. in an FCC regenerator) to >100 kg/(m^2 s) (e.g. in an FCC stripper). Flooding may occur, especially in baffled fluidized beds equipped with horizontal baffles where gas is travelling upwards, counter-current to descending solids. This is similar to flooding in counter-current gas–liquid towers. During flooding, gas and solids gather below and above the flooding level, unable to maintain the contacting efficiency expected in a normal fluidized bed. The flooding is usually accompanied by solids bridging and gas bypassing.
For a given fluidized bed, each superficial gas velocity corresponds to a critical solids flux beyond which flooding occurs. The critical solids flooding flux is also influenced by the open area ratio of the horizontal baffle, the size and shape of open passages, solids properties, etc. The level with minimum flow area is usually where flooding is most likely to occur. Bi et al. [34] derived a mechanistic correlation to predict the critical solids flooding flux. It is suggested that a systematic calculation be done to check whether the range of macro solids circulation flux under normal operating conditions is safely below the critical solids flooding flux.

(b) Forces acting on internals under different operating conditions.
In industrial fluidized beds, baffles can over-deform or even be destroyed during long periods of operation. Once deformed, their structural properties and influence on the flow change. They may then depart from their optimal or acceptable performance. Fatigue due to the buffeting action of bubbles and surging of solids can lead to failure, serious accidents, and unscheduled shutdowns. To avoid these problems, the characteristics of forces acting on baffles need to be considered. Baffles in industrial fluidized beds may experience three different stress modes where the characteristic forces on baffles differ widely: (a) during steady fluidization operation, (b) when defluidized, and

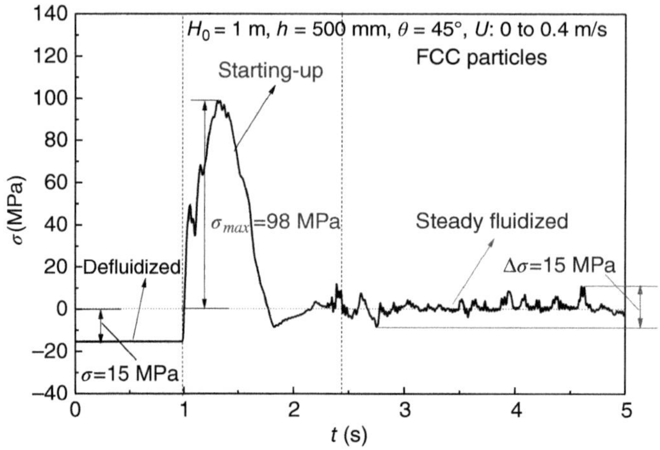

Figure 18.17 Typical transient stresses on an inclined slat under different operating conditions [35]. FCC particles: H_0 = 1.0 m; column cross section: 0.3 m (width) × 0.3 m (thickness); slat: 0.295 m (long) × 0.05 m (wide) × 0.008 m (thick), a single slat fixed at one end at 0.5 m above the bottom distributor. U switches from 0 to 0.4 m/s during start-up of the bed. Source: From Liu et al. [36].

(c) during start-up. Figure 18.17 shows typical forces acting on an inclined slat immersed in a fluidized bed of FCC particles [36].

In industrial fluidized bed reactors, mode (a) occupies the greatest duration, with baffles frequently subjected to impacts by particles due to bubble/void movement. According to Nagahashi et al. [35], when a rising bubble approaches and then passes a horizontal tube, the wake impact provides the strongest instantaneous forces. The periodic forces due to frequent rising bubbles may result in fatigue damage to baffles and their supporting structures. The magnitude of forces in the vertical direction is typically an order of magnitude greater than those in the horizontal direction. Bubble frequencies are typically of order 1–2 Hz.

The defluidized status may happen when the unit is shut down without discharging the solids in the bed or during unusual occurrences, such as power failures. During the defluidized period, the immersed baffle is subject to a downward force due to the weight of solids. The force can be estimated by the classic Jensen's formula [37]. Strength calculations are needed to avoid possible danger.

The start-up period may last only a small period in an industrial fluidized bed, but a strong force pulse that may act on baffles during this period requires special attention. Liu et al. [36] reported that when a bed is started up from a packed bed to a fluidized bed, an immersed horizontal slat will be subject to a strong force upward impulse whose peak magnitude may be several times higher than those encountered during steady fluidization. The difference can be seen clearly in Figure 18.17. This strong force impulse may damage or destroy horizontal baffles or support beams with large spans and horizontal projection areas. To avoid this during bed starting up, operators should ensure that the increase in gas flow rate occurs very slowly and begins with a low initial static bed level. Residual particles can be added after the bed reaches the steady operating flow rate. Moreover, it is suggested that designers use small-span vertical supporting beams, so their horizontal projected areas are small, reducing the forces acting on them.

(c) Anti-erosion measures
In industrial fluidized beds, baffles are continuously impacted by solid particles. Erosion is one of the most reported causes of damage and failure of baffles and other reactor components. According to Finnie [38], the erosion rate is usually proportional to the particle mass and the particle velocity raised to the power of 2–4. To restrain erosion on baffles, special attention should therefore be paid to regions with high solid velocities. In low-velocity dense fluidized beds, erosion is often not a serious problem. However, some local areas with high solid velocity near internal baffles may experience wear due to improper design, and baffle surfaces in these areas may be eroded during long periods of operation. An example is the erosive problem reported by Liang et al. [39] in FCC strippers. In fast fluidized beds, the average solid velocity is much higher. For the baffles and inserts in circulating fluidized bed (CFB) systems reported in Section 18.2.4, erosion and attrition problems need to be weighed against likely improvements in reactor performance. Anti-erosion measures are likely to be essential for these baffles. As erosion is a long-period process in industrial fluidized beds, experiments are difficult to carry out in laboratory. CFD simulation incorporating proper erosion models can be good tools in diagnosing and solving these erosion problems.

(d) Thermal stresses
Another common problem for baffles in industrial fluidized beds is caused by thermal stresses and improper structure design. For industrial fluidized beds, the temperature of baffles during installation is commonly much lower than during bed operation. The large temperature increase during operation results in significant thermal expansion of baffles, especially for those with high expansion coefficients. Without careful design, the resulting thermal stress can be high enough to cause deformation and high stresses. Therefore, careful consideration of thermal expansion is required during the design.
Note that different types of baffle have different functions and influences. Even for the same baffle, its function can also vary in fluidized beds of different particles, column dimensions, and operating conditions. Before designing a baffle for a specified industrial fluidization process, comprehensive understanding of the process, e.g. hydrodynamics, reaction kinetics, and solids properties, is very important.

18.3 Other Aids to Fluidization

18.3.1 Brief Introduction

In addition to baffles, attempts have been made to develop other aiding technologies to improve the performance of fluidization systems, such as using external electric or magnetic fields, pulsating gas feeding, external vibrations, and adding various glidants and other agents. Reducing average bubble size in low-velocity dense fluidized beds has been one of the major objectives. Another aim is to achieve smooth fluidization of particles that are otherwise difficult or impossible to fluidize, e.g. Geldart C particles and nanoparticles. Due to their small size and strong cohesive interparticle forces, abnormal fluidization behaviour, such as channelling, commonly exists in such beds.

18.3.2 Electrical Fields

When an electrical field is applied to a gas–solid fluidized bed, charge separation tends to occur, and particles become polarized. Strings of particles may then form in the direction of the electric field as shown in Figure 18.18. van Willigen et al. [40] simulated the effect of electric fields on particle behaviour by discrete element method (DEM) simulation.

To enhance performance by electric fields in a fluidized bed, van Willigen et al. [41] proposed alternating electric fields, where particles experience periodically oscillating attractive and repulsive interparticle forces. For insulating particles (e.g. glass beads or alumina catalyst particles) with a slightly conductive bulk and/or surface layer (e.g. by slightly humidifying the fluidized air), average bubble sizes in the bed can be significantly reduced, resulting in better gas–solid contact efficiency. For both fine Geldart A and B particles, frequencies less than 100 Hz were suggested, with an alternating electrical field [41].

18.3.3 Magnetic Fields

Fluidized beds can sometimes be enhanced by magnetic fields, similar to electrical fields. However, in order for magnetic fields to have an effect, the particles

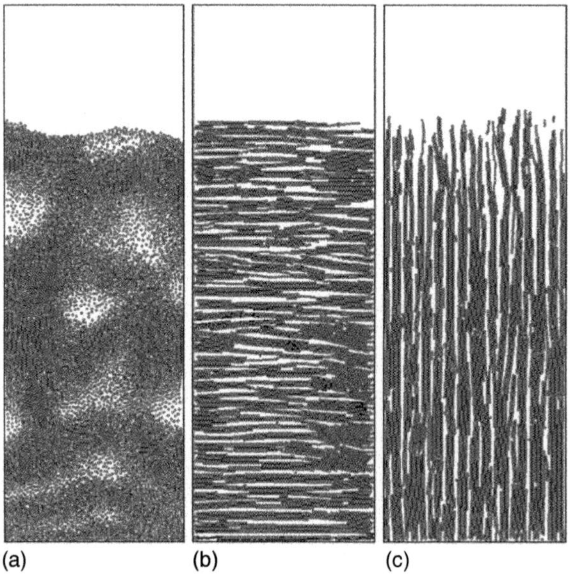

Figure 18.18 Particle behaviour in a freely bubbling bed: (a) without electric field, (b) after applying a horizontal electric field, and (c) after applying a vertical electric field. Monodisperse spheres: $d_p = 200\ \mu m$, $\rho_p = 2500\ kg/m^3$, relative dielectric constant = 7, 2D column: 42.3 mm (tall) × 14.2 mm (wide). Source: From van Willigen et al. [41].

must be magnetically susceptible. In a magnetic field, magnetic particles become polarized and rearrange themselves into particle strings.

Filippov [42] reported that a fluidized bed of magnetic particles can be stabilized if the applied magnetic field is controlled. Unlike a traditional fluidized bed, it is possible to fluidize without bubbles in a magnetically stabilized bed (MSB), resulting in homogeneous bed expansion and high gas–solid contact efficiency. In an MSB, bed expansion is not only influenced by gas velocity but also by the strength of the applied magnetic field. Thus, fine particles can be used without high pressure drop, a significant advantage over fixed beds. Also there is no particle entrainment, so solids separation devices used in conventional fluidized beds are unnecessary. Moreover, particles in an MSB are immobile, so gas and solids backmixing induced by bubble movement in conventional fluidized beds can be restrained. However, solids can also be easily transferred to other vessels as in conventional fluidized beds [43].

18.3.4 Pulsations and Vibrations

Geldart C particles and, to a lesser extent, Geldart A particles are subject to strong interparticle forces. When fluidizing these particles, channelling can occur due to agglomeration, leading to poor reactor performance. To assist in achieving stable fluidization and particle mobility, a number of aiding measures have been proposed.

One of the aiding measures is to pulse the fluidizing gas by (a) switching on and off a solenoid valve periodically, (b) changing the rotary speed of the valve block of a butterfly valve by a rotor, and (c) varying the piston speed or position of a reciprocating compressor, e.g. by an eccentric cam. The first of these tends to give a nearly rectangular waveform, whereas the other two measures result in nearly sinusoidal waveforms. In conventional fluidized beds, bubble size increases with increasing height due to coalescence. However, bubble size tends to decrease with increasing height in a pulsating fluidized bed, leading to better fluidization quality. Wang et al. [44] reported that resonance occurs in a pulsating bed near the natural frequency of the bed material. This resonance can effectively break large bubbles. For example, Liu et al. [45] used a pulsating bed to enhance the fluidization of wet particles of wide size distribution.

Another effective aiding measure is to vibrate the entire bed, thereby introducing external vibration energy into the fluidized particles. A vibrated fluidized bed (VFB) is realized by vibrating vertically the entire column. The magnitude and frequency can be changed, leading to different intensities of vibration energy. VFB are especially helpful for handling fine cohesive particles subject to agglomeration and for drying wet particles subject to liquid bridges.

18.3.5 Glidants and Antistatic Agents

For very fine particles, including nanoparticles, strong cohesion exists among the particles, making stable fluidization very difficult. However, some researchers (e.g. Kaliyaperumal et al. [46]) have found that adding some large particles that are easily fluidized (Geldart A or B) can facilitate smooth fluidization of the finer particles. These large particles serve as "glidants," aiding the flow ability of the fine particles. Some dielectric particles readily accumulate electrostatic charges during fluidization. As discussed in Chapter 13, strong electrostatic charging can not only lead to solids agglomeration and reduced fluidization quality but also endanger safe operation. To reduce the harmful effects of electrostatic charge accumulation, adding some conductive fine particles (Geldart type C) is often effective in reducing electrostatic charges [47]. Humidifying the fluidizing gas can also effectively restrain electrostatics in systems where the introduction of moisture can be tolerated.

18.4 Final Remarks

Space constraints have limited the content that can be included in this chapter. For readers needing more information on fluidized bed baffles, see reviews by Harrison and Grace [4] and Jin et al. [16].

Notations

d_p	particle diameter (m)
G	mass flux (kg/m^2 s)
H	bed height (m)
U	superficial gas velocity (m/s)
U_c	onset velocity of turbulent regime (m/s)

Greek Letters

$\Delta\sigma$	fluctuation range of stress (Pa)
θ	angle
ρ	density (kg/m^3)
σ	stress (Pa)
σ_{max}	maximum stress (Pa)
μ	viscosity (Pa s)

Subscripts

0	static bed
p	particle
s	solids

References

1 Zhang, Y. (2009). Hydrodynamic and mixing properties of a novel baffled fluidized bed. Doctoral Dissertation. China University of Petroleum, Beijing, China.

2 Pillai, R. and Niccum, P.K. (2011). FCC catalyst coolers open window to increased propylene. *Grace Davison FCC Conference*. Munich, Germany.

3 Volk, W., Johnson, C.A., and Stotler, H.H. (1962). Effect of reactor internals on quality of fluidization. *Chem. Eng. Prog.* 58: 44–47.
4 Harrison, D. and Grace, J.R. (1971). Fluidized beds with internal baffles. In: *Fluidization* (eds. J.F. Davidson and D. Harrison), 599–626. London: Academic Press.
5 Jin, Y., Yu, Z., Zhang, L. et al. (1982). Pagoda-shaped internal baffles for fluidized bed reactors. *Int. Chem. Eng.* 22 (2): 269–279.
6 Jin, Y., Yu, Z., Zhang, L. et al. (1986). Ridge type internal baffle for fluidized bed reactor. *Pet. Technol.* 15 (5): 269–277.
7 Rall, R.R. and Wichita, K.S. (2001). Apparatus for contacting of gases and solids in fluidized beds. USP 6224833, filed 10 December 1999.
8 Marchant, P.E., Soni, D.S., and Castagnos L. (2007). Apparatus for countercurrent contacting of gas and solids. USP 7179427, filed 25 November 2002.
9 Dries, H., Muller, F., Willbourne, P., et al. (2005). Consider using new technology to improve FCC unit reliability. *Hydrocarbon Processing* (February), pp. 69–74.
10 Hedrick, B.W., Xu, Z., Palmas, P., et al. (2008). Stripping apparatus and process. USP 7332132, filed 19 March 2004.
11 Miller, R.B., Yang, Y., and Johnson, T.E. (1999). RegenMax™ technology, staged combustion in a single regenerator. *NPRA*, San Antonio, USA.
12 Miller, R.B., Katy, T.X., and Yang, Y. (2003). Staged catalyst regeneration in a baffled fluidized bed. USP 6503460, filed 17 March 1999.
13 Li, X., Zhang, Y., Yu, Q. et al. (2013). Revamping for reinforced regeneration of a 500,000 TPY residue FCC unit. *Petroleum Refinery Eng.* 43 (4): 10–13.
14 Bi, H.T. (2007). A critical review of the complex pressure fluctuation phenomenon in gas-solids fluidized beds. *Chem. Eng. Sci.* 62: 3473–3493.
15 Zhang, Y., Grace, J.R., Bi, X. et al. (2009). Effect of louver baffles on hydrodynamics and gas mixing in a fluidized bed of FCC particles. *Chem. Eng. Sci.* 64 (14): 3270–3281.
16 Jin, Y., Wei, F., and Wang, Y. (2003). Effects of internal tubes and baffles. In: *Handbook of Fluidization and Fluid Particle Systems* (ed. W.-C. Yang), 171–199. New York: Marcel Dekker, Inc., Chapter 7.
17 Cai, P., Jin, Y., Yu, Z., and Wang, Z. (1987). A criterion for transition from bubbling to turbulent fluidization. *J. Chem. Industry Eng. (China)* 2 (2): 234–244.
18 Yang, Z., Zhang, Y., and Zhang, H. (2019). CPFD simulation on effects of louver baffle in a dense fluidized bed of Geldart A particles. *Adv. Powder Technol.* 30: 2712–2725.
19 van Dijk, J.J., Hoffmann, A.C., Cheesman, D., and Yates, J.G. (1998). The influence of horizontal internal baffles on the flow pattern in dense fluidized beds by X-ray investigation. *Powder Technol.* 98 (3): 273–278.
20 Grace, J.R. (1974). Fluidization and its application to coal treatment and allied processes. *AIChE Symp. Ser.* 70 (141): 21–26.
21 Wang, Z., Jin, Y., Yao, W., and Yu, Z. (1995). Industrial experiment of a turbulent fluidized bed reactor for synthesis of vinyl acetate. *Chem. React. Eng. Technol..* 11 (1): 50–55.

22 Rall, R. R. and DeMulder, B. (2000). New internal for maximizing performance of FCC catalyst stripers. *Twelfth Refining Seminar*, San Francisco, California, USA.
23 Bosma, J.C. and Hoffmann, A.C. (2003). On the capacity of continuous powder classification in a gas-fluidized bed with horizontal sieve-like baffles. *Powder Technol.* 134: 1–15.
24 Chyang, C.S., Wu, K.T., and Ma, T.T. (2002). Particle segregation in a screen baffle packed fluidized bed. *Powder Technol.* 126: 59–64.
25 Zhang, Y., Wang, H., Chen, L., and Lu, C. (2012). Systematic investigation of particle segregation in binary fluidized beds with and without multilayer horizontal baffles. *Ind. Eng. Chem. Res.* 51 (13): 5022–5036.
26 Lewis, W.K., Gilliland, E.R., and Lang, P.M. (1962). Entrainment from fluidized beds. *AlChE Symp. Ser.* 58: 65–78.
27 Tweddle, T.A., Capes, C.E., and Osberg, G.L. (1970). Effect of screen packing on entrainment from fluidized beds. *Ind. Eng. Chem. Proc. Des. Dev.* 9 (1): 85–88.
28 George, S.E. and Grace, J.R. (1981). Entrainment of particles from a pilot scale fluidized bed. *Can. J. Chem. Eng.* 59: 279–284.
29 Kong, G. (1993). Research of entrainment rate of fluidized bed with baffle plate. *J Shengyang Institute Chem. Technol.* 7 (4): 254–262.
30 Jiang, P., Bi, H.T., Jean, R.H., and Fan, L.S. (1991). Baffle effects on performance of catalytic circulating fluidized bed reactor. *AlChE J.* 37: 1392–1340.
31 Chen, Y.-M. (2006). Recent advances in FCC technology. *Powder Technol.* 163: 2–8.
32 Zheng, C., Tung, Y., Li, H., and Kwauk, M. (1992). Characteristics of fast fluidized beds with internals. In: *Fluidization VII* (eds. O.E. Potter and D.J. Nicklin), 275–284. New York: Engineering Foundation.
33 Gan, N., Jiang, D.Z., Bai, D.R. et al. (1990). Concentration profiles in fast-fluidized bed with bluff-body. *J. Chem. Eng. Chin. Univ.* 3 (4): 273–277.
34 Bi, H.T., Cui, H., Grace, J. et al. (2004). Flooding of gas-solids countercurrent flow in fluidized beds. *Ind. Eng. Chem. Res.* 43: 5611–5619.
35 Nagahashi, Y., Asako, Y., Lim, K.S., and Grace, J.R. (1998). Dynamic forces on a horizontal tube due to passing bubbles in fluidized beds. *Powder Technol.* 98: 177–182.
36 Liu, D., Zhang, S., Zhang, Y., and Grace, J.R. (2017). Forces on an immersed horizontal slat during starting up a fluidized bed. *Chem. Eng. Sci.* 173C: 402–410.
37 Janssen, H.A. (1895). Versuch uber getreidedruck in silozallen. *Zeitschrift des Vereines Deutscher Ingenieure* 39 (35): 1045–1049.
38 Finnie, I. (1960). Erosion of surfaces by solid particles. *Wear* 3: 87–103.
39 Liang, Y., Zhang, Y., and Lu, C. (2015). CPFD simulation on wear mechanisms in disc-donut FCC strippers. *Powder Technol.* 279: 269–281.
40 van Willigen, F.K., Demirbas, B., Deen, N.G. et al. (2008). Discrete particle simulations of an electric-field enhanced fluidized bed. *Powder Technol.* 183: 196–206.

41 van Willigen, F.K., van Ommen, J.R., Turnhout, J.R., and van den Bleek, C. (2003). Bubble size reduction in a fluidized bed by electric fields. *Int. J. Chem. Reactor Eng.* 1: A21.

42 Filippov, M.V. (1960). The effect of a magnetic field on a ferromagnetic particle suspension bed. *Prik. Magnitogidrodin. Trudy Instituta fiziologii, Akademiia nauk. Latvia SSR (USSR)* 12: 215–236.

43 Rosensweig, R.E. (1995). Process concepts using field-stabilized two-phase fluidized flow. *J. Electrostat.* 34: 163–187.

44 Wang, Y., Wang, T., Yang, Y., and Jin, Y. (2002). Resonance characteristics of a vibrated fluidized bed with a high bed hold-up. *Powder Technol.* 127: 196–202.

45 Liu, Y., Ohara, H., and Tsutsumi, A. (2017). Pulsation-assisted fluidized bed for the fluidization of easily agglomerated particles with wide size distributions. *Powder Technol.* 316: 388–399.

46 Kaliyaperumal, S., Barghi, S., Briens, L. et al. (2011). Fluidization of nano and sub-micron powders using mechanical vibration. *Particuology* 9: 279–287.

47 Park, A.-H., Bi, H., and Grace, J.R. (2002). Reduction of electrostatic charges in gas–solid fluidized beds. *Chem. Eng. Sci.* 57: 153–162.

Problem

18.1 A horizontal baffle (e.g. the Crosser grid shown in Figure 18.4) is to be installed in an FCC regenerator of inner diameter (ID) 8 m, whose working temperature is 700 °C. A typical regenerator wall consists of three layers, i.e. steel wall, heat insulation lining, and wear-resisting lining from outside to inside as shown in Figure 18P.1. To avoid dew point corrosion, the steel wall is required to be 200 °C during unit operation. What steel should be selected for the horizontal baffle? How should the horizontal baffle layer be supported and fixed to ensure safe operation for years? Consider the suggestions in Section 18.2.5 to find a preliminary design scheme.

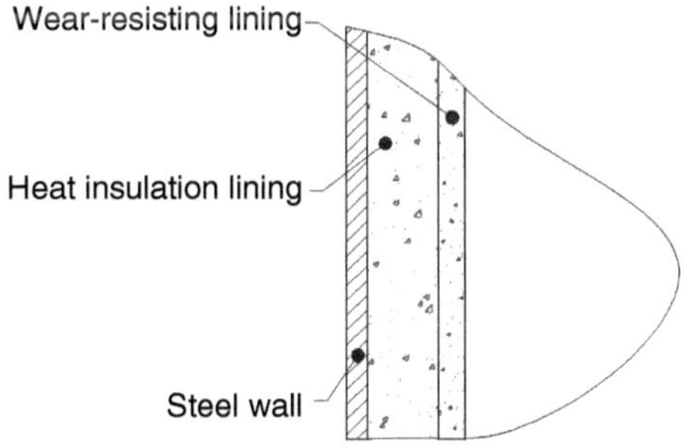

Figure 18P.1 Schematic of an industrial FCC regenerator wall.

19

Jets in Fluidized Beds

Cedric Briens[1] and Jennifer McMillan[2]

[1] *Western University, Institute for Chemicals and Fuels from Alternative Resources, Department of Chemical and Biochemical Engineering, Thompson Engineering Building, London, Ontario, Canada N6A 5B9*
[2] *Syncrude Canada Limited, Research and Development, 9421-17 Avenue NW, Edmonton, Alberta, Canada T6N 1H4*

19.1 Introduction

In the particulate, bubbling, and turbulent fluidization flow regimes, jets can form when gas is blown into the bed, forming a cavity where gas occupies more than 80% of the volume. Jets are produced:

- At the gas distributors through which fluidizing gas is introduced. The gas velocity through the distributor holes typically ranges from 10 to 80 m/s, resulting in subsonic flow.
- When gas is introduced at sonic or supersonic velocities to promote grinding of bed particles.
- When gas is used to inject particles into a fluidized bed.
- When gas is used to spray liquid into a fluidized bed.

19.2 Jets at Gas Distributors

The main purpose of a gas distributor is to uniformly distribute the fluidizing gas into the bed, so the same amount of gas enters each section of the bed. See also Chapter 5 for further information on gas distributors. There are two general types of gas distributors:

- Perforated plates are metal plates in which holes are drilled in a regular pattern to distribute the fluidizing gas as uniformly as possible into the bed, as shown in Figure 19.1. The design is simple and robust.
  The plate must be designed to withstand the entire bed weight. In industrial applications, plates may thus be quite thick. Their thickness can be reduced by using convex or concave designs that improve their mechanical performance; in this case, either the upper regions must have smaller holes or a smaller concentration of holes to avoid a larger gas flow through these regions, unless uneven flow is desired.

Essentials of Fluidization Technology, First Edition. Edited by John R. Grace, Xiaotao Bi, and Naoko Ellis.

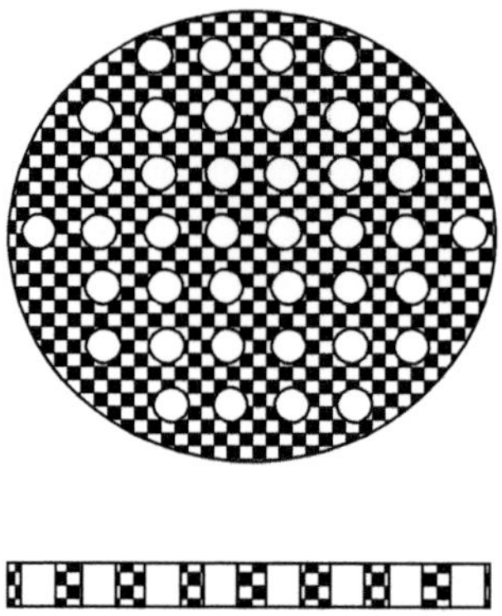

Figure 19.1 Perforated plate.

A drawback of a perforated plate is that the gas is first introduced in a space below the plate, called a "windbox" or plenum chamber (Figure 19.2). This increases the height and thus the cost of the fluidization column.

- Spargers, sometimes called pipe grids, are metal tubes in which holes are drilled in a regular pattern to distribute the fluidizing gas as uniformly as possible. Figure 19.3 shows a typical sparger tube configuration where the fluidization gas first enters through a main header, is split between a series of smaller tubes, and then enters the bed through holes in the tubes. These holes are usually located on the lower part of the tubes so that, if any solids enter the tubes, gravity will help them drain. The holes are usually inclined, as shown in Figure 19.4, alternating from one side to the other.

Distributor plates are usually cheaper to manufacture than spargers but require an expensive windbox. They are also simpler to design than spargers, mechanically stronger, and less likely to warp. They are also easier to clean: for example, during major maintenance shutdowns in large reactors, workers walk on top of the plate and use metal rods to clear the distributor holes of deposits. Plates can be designed to nearly completely avoid the formation of defluidized zones, which is not possible with spargers. Plates are also easier to permanently adapt to new

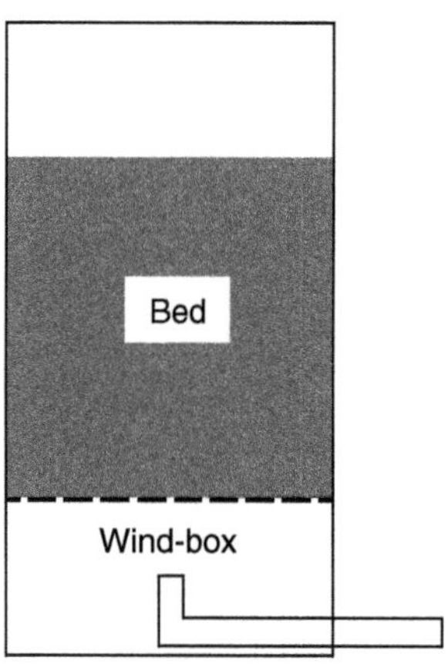

Figure 19.2 Windbox.

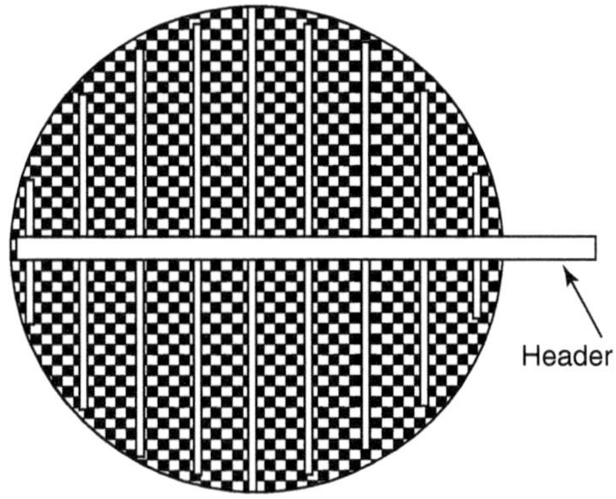

Figure 19.3 Sparger.

Figure 19.4 Cross-sectional view of sparger pipe.

conditions by plugging some holes, drilling more holes, or over-drilling existing holes.

On the other hand, spargers can accommodate a larger turndown ratio by using more than one header and shutting off flow to some of the headers when operating at lower gas flow rates. Plates require more expensive seals at high pressure, as the area to be sealed is much larger. Solids weeping is more of an issue with plates. Spargers can be used for the segregated injection of several feeds (e.g. in acrylonitrile reactors, one sparger is used to inject oxygen and another to inject a mixture of ammonia and propylene).

Gas distributors are often the cause of operating problems with industrial fluidized beds. The issues are:

- Nonuniform gas distribution
- Defluidized zones
- Erosion of internals such as heat exchanger tubes, cyclone diplegs, vessel wall, or the distributor itself
- Particle attrition caused by grid jets
- Mass, heat transfer, and reaction within the jets

19.2.1 Criterion for Uniform Gas Distribution

In a fluidized bed, there are pressure fluctuations, and the instantaneous pressure on one side of the distributor may differ from that on the other side. This will lead to a higher flow of gas through the distributor on the low-pressure side (Figure 19.5). Pressure fluctuations result from various causes such as waves at the bed surface, random coalescence of bubbles, intermittent solids feeding, and intermittent solids discharge.

A poor bubble distribution tends to be self-perpetuating: the hydrostatic pressure drop is smaller on the side that contains more bubbles, leading to a higher

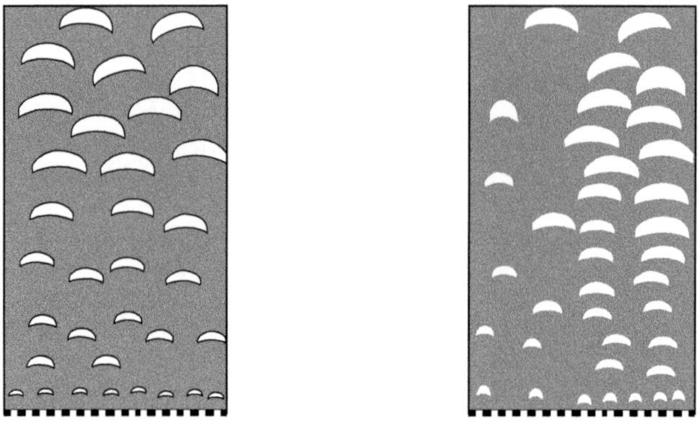

Figure 19.5 Bubble flow patterns with a good and a poor distributor.

gas flow through the grid holes on that side. The best way to limit the impact of pressure fluctuations on distributor gas flow is to install a high pressure drop distributor.

The following derivation shows how a high grid pressure drop to bed pressure drop ratio minimizes the impact of bed pressure drop fluctuations on one side of the bed:

- As shown below, the grid pressure drop is roughly proportional to the hole gas velocity squared:

$$\Delta P_{grid} \propto U_h^2 \propto U_g^2 \tag{19.1}$$

- Taking the logarithm and differentiating gives

$$\frac{d(\Delta P_{grid})}{\Delta P_{grid}} = 2\frac{dU_g}{U_g} \tag{19.2}$$

- Pressure transmits very quickly through a pure gas and the relatively low solids concentration gases in the windbox and freeboard (see Figure 19.2). As a result, the pressure is the same on all sides in both the windbox and freeboard, and the sum of the grid and bed pressure drops is the same on all sides of the column:

$$\Delta P_{grid} + \Delta P_{bed} = \text{constant}$$

and

$$d(\Delta P_{grid}) = -d(\Delta P_{bed}) \tag{19.3}$$

- Combining Eqs. (19.2) and (19.3) gives

$$\frac{dU_g}{U_g} = \frac{1}{2}\frac{d(\Delta P_{grid})}{\Delta P_{grid}} = -\frac{1}{2}\frac{d(\Delta P_{bed})}{\Delta P_{grid}} \tag{19.4}$$

- Or

$$\frac{dU_g}{U_g} = -\frac{1}{2}\frac{1}{\left(\frac{\Delta P_{grid}}{\Delta P_{bed}}\right)}\frac{d(\Delta P_{bed})}{\Delta P_{bed}} \tag{19.5}$$

- Therefore, the impact of a relative change in bed pressure drop on one side of the column on the gas flow rate on the same side is inversely proportional to the grid pressure drop to bed pressure drop ratio.

According to Kunii and Levenspiel [1], the ***dry*** grid pressure drop (i.e. pressure drop with no solids present) must be such that it is:

1. Larger than 10% of the bed pressure drop (higher values are better). With shallow beds, the grid pressure drop should be larger than 30% of the bed pressure drop.
2. Larger than 3400 Pa.
3. Larger than 100 times the pressure drop due to gas expansion from the inlet line into the vessel.

Karri and Knowlton [2] proposed criteria similar to those proposed by Kunii and Levenspiel [1]. The ***dry*** grid pressure drop must be such that it is:

1. Larger than 30% of the bed pressure drop for upwardly or laterally directed gas introduction into the bed, or 10% for downwardly directed gas introduction (e.g. sparger with downward-directed holes).
2. Larger than 2500 Pa.

Poor gas distribution is difficult to detect with pressure measurements. With exothermic or endothermic processes, thermocouples may indicate maldistribution.

19.2.1.1 "Dry" Distributor Pressure Drop

McAllister et al. [3] provided an accurate graphical correlation for the "dry" distributor pressure drop. For quicker estimates, one can use the following correlations:

- For ambient conditions, use the correlation [4]

$$\Delta P_{grid} = \frac{1}{2\ \rho_f}\ (\rho_f\ U_h)^2\ \left[1.4139 + \frac{1}{2}\ \frac{h}{d_h}\ Re_h^{-1/4}\right] \quad \text{where}$$

$$Re_h = \frac{\rho_f\ d_h\ U_h}{\mu_f} \tag{19.6}$$

- For other conditions, the following correlation [5] is recommended [6]:

$$\Delta P_{grid} = \frac{1.49}{2\rho_f}\ (\rho_f\ U_h)^2\ \left(\frac{d_h}{h}\right)^{0.26} \tag{19.7}$$

19.2.1.2 Actual Distributor Pressure Drop

The actual pressure drop is often larger than the "dry" grid pressure drop, mainly due to intermittent solids leakage into the distributor holes and from defluidized zones [7]. In practice, assume a distributor pressure drop equal to double the "dry" pressure drop when designing gas compressors.

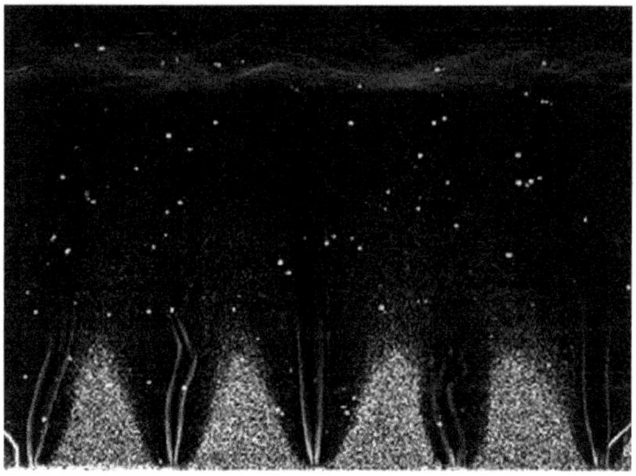

Figure 19.6 PIV measurements in a 2-D fluidized bed.

19.2.2 Defluidized Zones

Figure 19.6 shows particle image velocimetry (PIV) measurements in a 2-D fluidized bed [8]. Stagnant defluidized zones are formed between the holes of the perforated plate. Figure 19.7 shows similar results obtained in a more realistic 3-D bed using magnetic resonance imaging (MRI) measurements [9]. Similar results have been obtained in another MRI study [10]. Defluidized zones are also formed with sparger distributors. Defluidized zones must be avoided if the reaction is exothermic, and either the solids or the gas can be degraded by heat or long residence times (e.g. solids melt or the gas cokes). However, in some cases, defluidized zones may be used to advantage. For example, some calciners use a settled layer of solids to protect the gas distributor from the high bed temperatures.

There are three main causes of defluidized zones:

1. **Low distributor pressure drop**: If the distributor pressure drop is too low, all the distributor holes will not be cleared of solids at start-up. When the extra pressure drop needed to clear the plugged holes exceeds the extra hydrostatic pressure drop of fluidized solids inside the plugged holes, holes remain

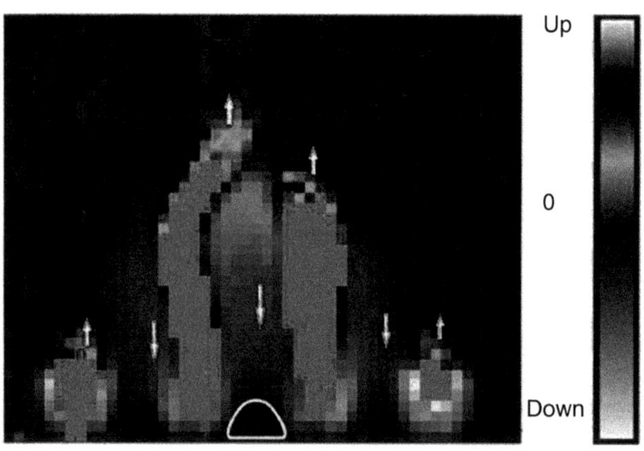

Figure 19.7 MRI measurements in a 3-D fluidized bed.

plugged [11]. The number of plugged holes can be roughly estimated from correlations [12].

2. **Sticky or agglomerating solids**: If solids form agglomerates, the agglomerates may grow to such a size or such a concentration that gas bubbles can no longer carry the agglomerates to the bed surface. Deposition of agglomerates on the distributor usually leads to major problems. However, some reactors, such as slagging gasifiers, use the settling of agglomerates to separate ash-rich agglomerates from the rest of the bed. If the solids are sticky, special attention should be given to shutdown and start-up procedures to avoid plugging the grid holes.
3. **Distributor design**: If there is significant space between distributor holes, defluidized zones will form between holes (Figure 19.6). Similarly, if there is a zone without holes near the vessel wall, to avoid erosion of the wall by the gas jets, a defluidized zone will form near the wall. With sparger distributors, if the holes are directed upwards, or if the gas jets issuing from downward holes do not penetrate deep enough into the bed, a defluidized layer will form at the bottom of the bed. To avoid defluidized zones, special perforated plate designs avoid flat surfaces between grid holes: use assemblages of pyramids that slope down towards the grid holes, as shown in Figure 19.8.

Some distributors eliminate or greatly reduce defluidized zones by creating swirling flow. An example is the punched plate design shown in Figure 19.9 [13].

Figure 19.8 Design to avoid defluidized zones (pyramidal additions in red).

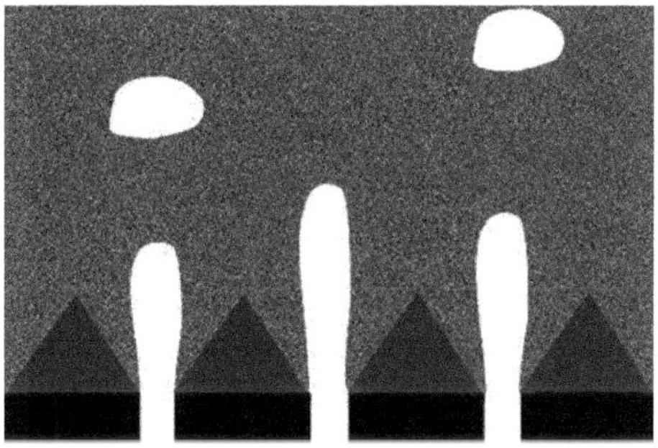

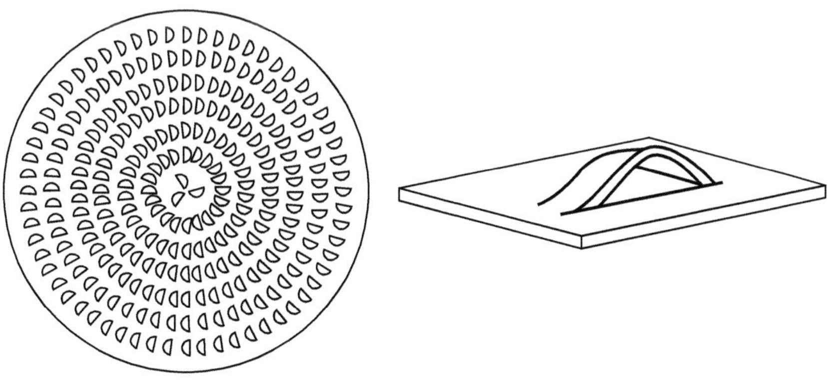

Figure 19.9 Punched plate distributor with details of one hole.

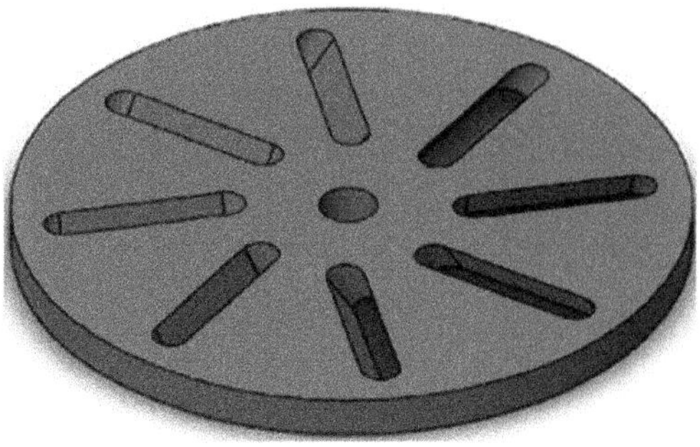

Figure 19.10 Slotted distributor (slots at a 45° angle).

Swirling flow is also used to prevent the permanent settling on the distributor of large particles, such as granules in granulation equipment. Swirling of the bed can also be achieved with slotted distributors such as shown in Figure 19.10 [14]. See Chapter 5 for more on distributors.

19.2.3 Erosion of Internals

Gas distributors operate in either the bubbling regime, where gas bubbles form directly at the distributor, or the jetting regime, where pulsating jets from the distributor periodically release gas bubbles from their tip. A low gas velocity through the distributor hole, below about 30 m/s, is associated with the bubbling flow regime, while higher velocities are associated with jetting [15].

Figure 19.11 shows the expansion cycle of a typical vertical, upward pulsating jet. The jet starts at its minimum penetration depth, L_{min}. The jet then expands gradually and reaches its maximum penetration depth, L_{max}. A gas bubble, with a volume-equivalent diameter equal to about $L_{max}/2$, detaches from the jet tip, with its top a distance L_B above the distributor hole. Most jet penetration correlations predict the maximum penetration depth, L_{max}. This pulsating behaviour has been confirmed by MRI measurements, as shown in Figure 19.12 [16]. Similar pulsating behaviour occurs for horizontal jets [17].

The gas jets issuing from the distributor holes entrain particles at high velocity, quickly eroding any surface that they strike (sandblasting). Figure 19.7 shows that particles descend between gas jets, enter jets near their base, and are accelerated to a high velocity within the dilute-phase gas jet. The grey represents regions where particle velocities could not be measured because their concentration was too low or their velocity too high.

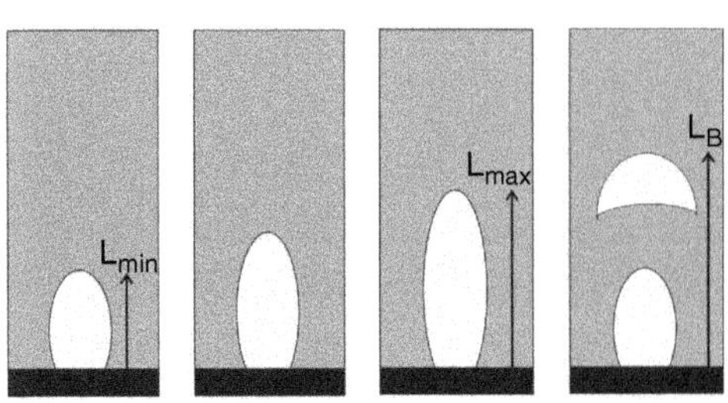

Figure 19.11 Expansion cycle of a pulsating vertical gas jet.

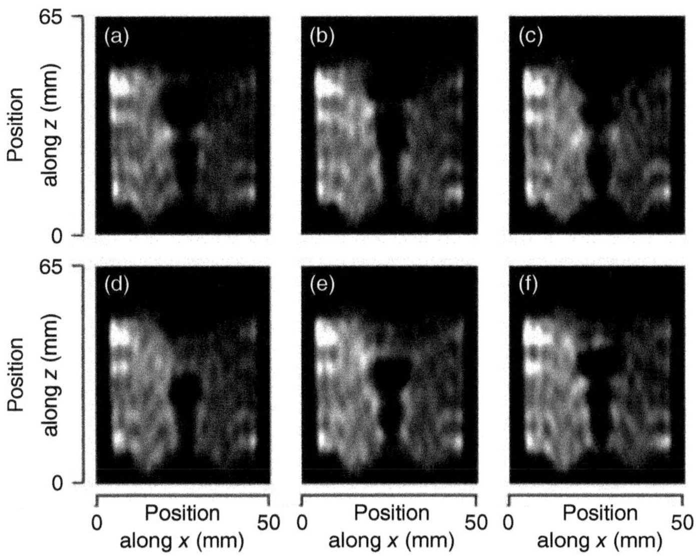

Figure 19.12 Magnetic resonance imaging of a pulsating vertical jet. Images taken at different times of jet expansion-contraction cycle.

If jets contact any solid surface, entrained particles will cause erosion. It is thus important to avoid contact between jets and internals such as heat exchanger tubes and cyclone diplegs. Also avoid contact between the jets and vessel wall or supports. No internal should, therefore, be located at a distance of less than L_{max} from the distributor hole. To be on the safe side, one may use L_B, as there is significant erosion of internals between L_{max} and L_B, although it is much less than the erosion on internals closer than L_{max} from the distributor. For more on erosion/wear, see Chapter 13.

Several correlations are available to predict jet penetration depth. Gas jets are not stable, but oscillate in both vertical and horizontal directions. For a given hole velocity and diameter, vertical upward-directed jets penetrate approximately twice as far as horizontal jets and three times as far as downward jets [2].

19.2.3.1 Penetration of Upward Vertical Jets

Correlations have been reviewed [18]. A popular correlation [19] gives

$$\frac{L_j}{d_h} = 5.2 \left[\frac{\rho_f \; d_h}{\rho_p \; d_p}\right]^{0.3} \left\{1.3 \left[\frac{U_h^2}{g \; d_h}\right]^{0.2} - 1\right\} \tag{19.8}$$

More recently, all data published to date, including high pressure and temperature data, have been used to derive the following correlation [18] for subsonic jets:

$$\frac{L_j}{d_h} = 8.37 \left[\frac{\rho_f \; U_h^2}{\rho_{bed} \; g d_j}\right]^{0.30} \left[\frac{U_g}{U_{mb}}\right]^{-0.30} \tag{19.9}$$

19.2.3.2 Penetration of Horizontal, Inclined, and Downward Vertical Jets

A horizontal jet bends upwards with part of its jet expansion in the vertical direction. The jet penetration increases by only ~4% when going from a horizontal to a vertically downward jet [20]. Since this is much less than the accuracy of predicted penetrations, most correlations do not consider the effect of nozzle inclination.

Benjelloun et al. [21] proposed:

$$\frac{L_j}{d_h} = 5.52 \left[\frac{\rho_f\ U_h^2}{(\rho_p - \rho_f)\ g\ d_h}\right]^{0.27} \tag{19.10}$$

More recently, all data published to date, including high pressure and temperature data, have been used to derive the following correlation [18] for subsonic jets:

$$\frac{L_j}{d_p} = 93.8 \left[\frac{\rho_f\ U_h^2}{\rho_p\ g\ d_p}\right]^{0.30} \left[\frac{g\, d_p^3\, \rho_f\, (\rho_p - \rho_f)}{\mu_f^2}\right]^{-0.25} \left[\frac{U_g}{U_{mb}}\right]^{0.05} \tag{19.11}$$

19.2.3.3 Angle of Upward Jets

Measurements [22] indicated that the jet half-angle varied from 5° to 45°, depending on the gas velocity, temperature, pressure, and solids properties. The correlation from Merry [19] for jet half-angle did not work. Hensler et al. [23] found a jet half-angle of 16° with positron emission particle tracking (PEPT) measurements. A more recent correlation [10] provides the jet half-angle in degrees:

$$\theta = 2.86 \left(\frac{d_h^2}{A_h}\right)^{-0.41} \left(\frac{U_h^2}{g\, d_h}\right)^{-0.059} \left(\frac{U_{mf}^2}{g\, d_h}\right)^{0.33} \tag{19.12}$$

19.2.3.4 Merging and Coalescence

If the pitch between adjacent distributor holes is not large enough, jets issuing from adjacent holes are likely to merge. Figure 19.13 shows the merging of two jets issuing horizontally in opposite directions from two holes of the same tuyere distributor [17].

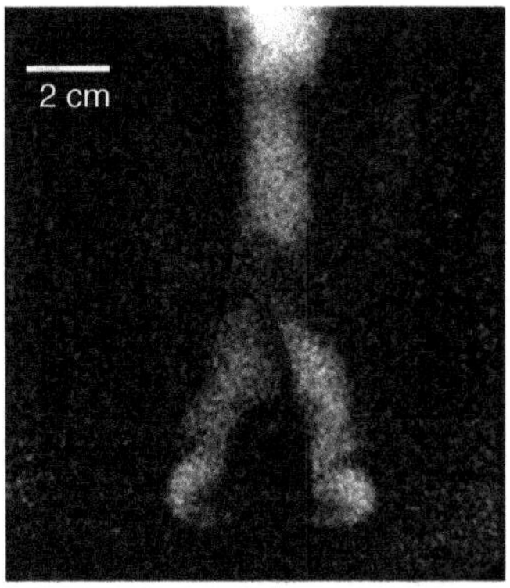

Figure 19.13 X-ray picture of the merging of two jets.

When jets merge at the distributor, this has a significant impact on jet penetration [17, 24].

19.3 Mass Transfer, Heat Transfer, and Reaction in Distributor Jets

In many catalytic fluidized bed processes, a large part of the reaction occurs in the grid region [25] due to high rates of heat and mass transfer in the grid jets. When a cold gas is injected into a bed of hot particles, it nearly reaches the bed temperature by the end of the jet cavity [26]. The high heat transfer rate is due to entrainment of particles, which enter the jet cavity near its base and accelerate to high speed, as shown in Figure 19.7. The high relative velocity between the particles and gas ensures a high heat transfer coefficient.

The high rates of mass transfer between jet gas and bed particles are due to the entrainment of particles into the gas jet, as with heat transfer [27]. Similar behaviour has been observed with downward gas jets [28]. Emulsion gas is also entrained into the jet cavity, entering with the emulsion particles near the base of the jet cavity [29]. Gas entrainment increases with increasing fluidization velocity [29].

Some processes inject downward-inclined gas jets into a fluidized bed. Computational fluid dynamics (CFD) modelling and experiments have shown that good combustion and gasification can be achieved whatever the inclination [30].

19.4 Particle Attrition and Tribocharging at Distributor Holes

In many fluidized beds, particle attrition mainly occurs in the distributor jets and at the cyclone inlet. Attrition associated with gas bubbles, far from the distributor, is usually much smaller than attrition caused by the distributor jets [31]. As shown in Figure 19.7 [9], particles are picked up near the base of distributor jets, accelerated to high velocities and slammed against slow-moving emulsion solids just beyond the jet tip. This occurs with upward, inclined, and downward jets [32].

Figure 19.14 shows the two attrition processes that occur when a particle hits another particle or a solid surface at high speed. With the abrasion process, small fines are removed from the particle surface. With the fragmentation process, the particle shatters into several fragments of similar size.

Some researchers have found that the particle attrition rate is proportional to grid jets kinetic energy ($\rho_f \cdot U_h^2$) [33]. This means that if the orifice velocity is kept constant, increasing the pressure will increase the attrition rate. Increasing the temperature would theoretically reduce the attrition rate, but, in practice, there may be a strong additional effect of thermal shocks.

Attrition rates in upward and horizontal jets are similar, but attrition is 5–15 times higher with downward jets [34]. An empirical correlation has been proposed to predict the attrition rate caused by gas jets in fluidized beds [35].

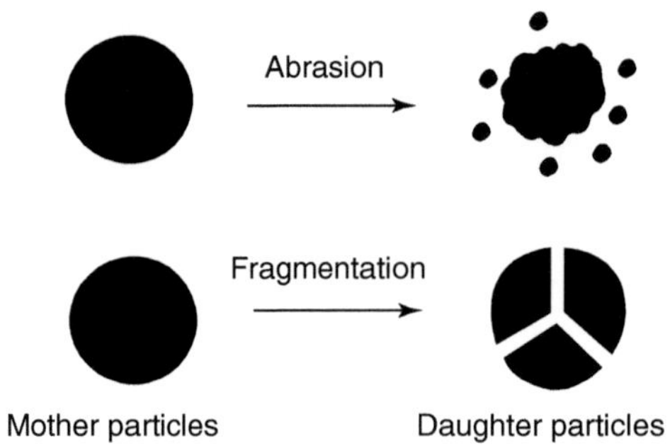

Figure 19.14 Attrition processes.

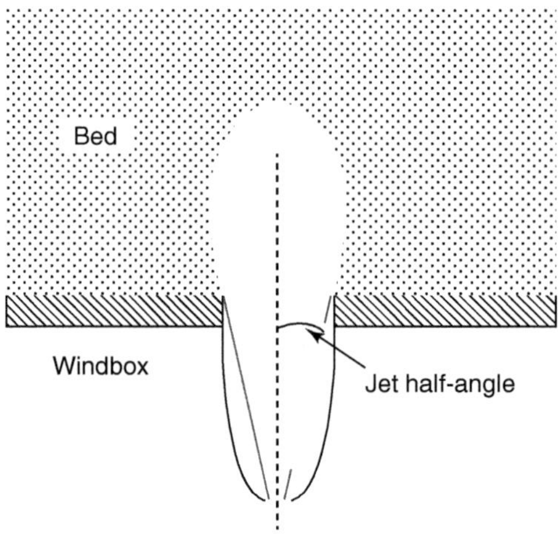

Figure 19.15 Shrouded jet design.

With soft (e.g. polyethylene or polypropylene) particles, attrition is not a major issue, but tribocharging occurs when particles strike each other at high speed [36]. Special designs have been developed to minimize attrition and tribocharging in the distributor jets. They attempt to prevent solids from reaching the base of the jet. A simple example is shown in Figure 19.15. The jet half-angle is set to the half-angle of a jet expanding into a pure gas. A reasonable value for the jet half-angle is 5.5° [2]. Figure 19.15 represents a limiting case where the jet "touches" the shroud wall at its end; in practice, a longer shroud is used to be on the safe side. However, unless special precautions are taken, this design will promote formation of defluidized zones between the distributor holes. A similar shrouded design has been proposed [37]. A similar design can be used for the holes of sparger distributors.

An ASTM standard jet cup attrition test measures the powder attritability with small amounts of solids [31]. Modified designs have been proposed to provide attrition data even more relevant to fluidized bed processes [38–40]. A possible issue with these tests is that they do not take thermal effects into account [39]. See Chapter 13 for more on attrition.

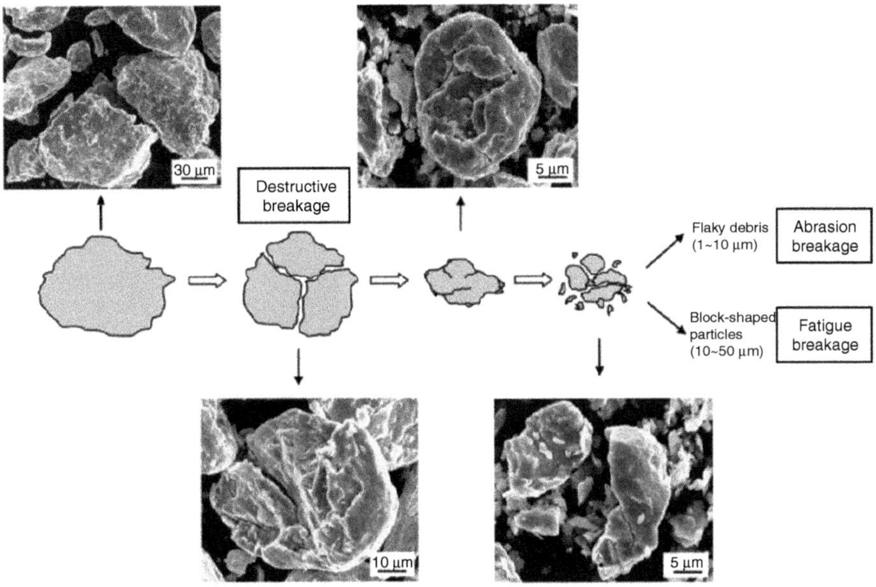

Figure 19.16 Jet milling mechanisms.

19.5 Jets Formed in Fluidized Bed Grinding

High-velocity gas jets can be introduced into a fluidized bed to grind particles. The fluidized bed can also be used as a classifier to avoid grinding the particles below a set size: the fluidization velocity is then set to elutriate the fine particles (see Chapter 10).

19.5.1 Mechanisms

Figure 19.16 shows the breakage mechanisms of a powder during jet milling in a fluidized bed [41]. High-velocity collisions of coarse particles result in destructive breakage, in which cracks propagate through weaker regions within the particles. Smaller particles are more resistant to breakage since they have fewer flaws. Fatigue breakage requires repeated impacts to propagate microcracks within the particles. Abrasion breakage smoothes particles by trimming edges. An alternate nomenclature considers only two mechanisms: abrasion, which results in a mixture of large smooth particles and fines, and fragmentation, which results in fragments of similar sizes (Figure 19.14).

Figure 19.7 shows how, in a distributor jet, particles are entrained from bed regions near the base of the jet and accelerated to high speed within the jet. Achieving particle breakage requires high particle velocities, typically with sonic jets. Figure 19.17 shows how particles are entrained and accelerated in the first part of the jet in a fluidized bed with a high-velocity gas nozzle [42].

Figure 19.18 shows PIV measurements of particle velocity along the centre line of an attrition jet using a sonic jet of air (whose critical pressure ratio was

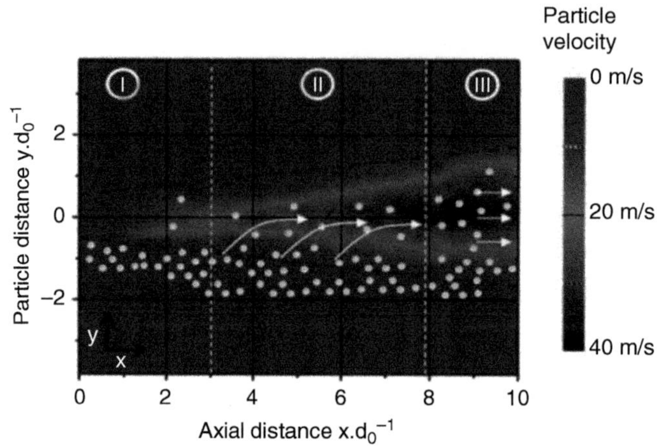

Figure 19.17 Particle velocities within the first part of the jet formed by gas issuing from a nozzle at 250 m/s.

1.26) [42]. Particles that enter the jet near its base, just downstream of the nozzle tip, are first accelerated by high-velocity gas. As the jet expands, the gas velocity decreases and the particle velocity peaks at ~43 m/s. Because the particles are glass spheres of Sauter mean diameter 92 μm, they have significant inertia, and their velocity is still nearly 30 m/s at the jet tip. The particles within the jet therefore hit the dense bed particles at the jet tip with a high velocity, causing significant grinding.

With smaller particles, inertial effects are reduced and the particle velocity at the jet tip is reduced. This is why a limit to fineness is observed with jet grinding, i.e. particles below a certain size are not ground [43].

Figure 19.18 shows that the particle velocity at the jet tip is significantly lower than the maximum particle velocity within the jet. Two methods can intensify

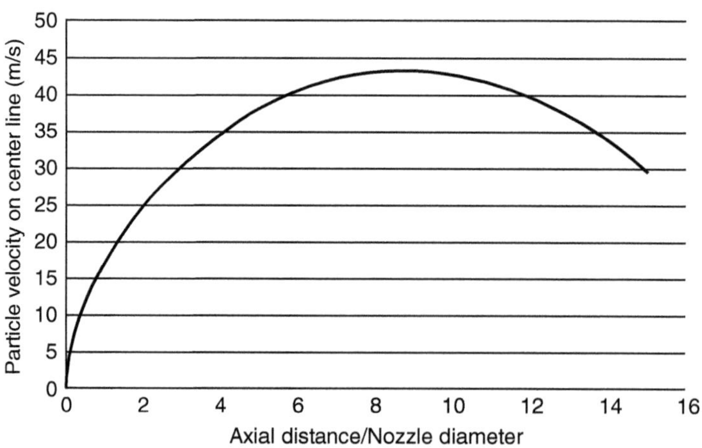

Figure 19.18 Particle velocity on jet centreline (glass spheres with a Sauter mean diameter of 92 μm; air at room temperature).

particle collisions: a solid target can be placed near the position of maximum particle velocity, or opposing jets promote head-on particle collisions.

19.6 Applications

Jet grinding is typically used to avoid contamination to/from grinding surfaces, to prevent rapid erosion of the grinding surfaces with abrasive particles, to prevent thermal degradation of the product, and when grinding needs to be integrated with other processes. An advantage of jet grinding is that there is no contamination from steel or ceramic grinding surfaces, making it especially attractive for applications such as silicon grinding [44], carbon nanotubes [45], titanium–aluminum alloys [41], and tungsten powder [46]. Sometimes, contamination of the grinding surfaces should be avoided, as their safe disposal would be an issue. An example is the reprocessing of spent nuclear fuel [47].

The grinding of particles is inefficient and most of the spent energy goes to heat. In jet grinders, however, the superior heat transfer capability of fluidized beds helps evacuate the heat of grinding. Jet grinding is, therefore, attractive when thermal degradation of ground material is an issue, as in pharmaceutical [48] or nutraceutical [49] applications.

Conventional grinders suffer from materials constraints at low or high temperatures. Jet grinding is, thus, used for applications such as the cryogenic grinding of scrap tire rubber [50], green sand reclamation [51], and particle size control in fluid cokers [52].

19.7 Jet Penetration

Two approaches have been used to predict the penetration of the jets formed with sonic attrition jets. The first approach uses an empirical modification of a correlation [21] for the penetration of subsonic jets [53]. For example, Eq. (19.13) uses different values of the parameters α and β for different solids (e.g. $\alpha = 0.15$ and $\beta = 0.664$ for sand and $\alpha = 0.38$ and $\beta = 0.54$ for petroleum coke) [53]. CFD can also be used to predict jet penetration:

$$\frac{L_j}{d_h} = \beta \left[\frac{\rho_f \; U_h^2}{(\rho_p - \rho_f) \; g \; d_h} \right]^{\alpha} \tag{19.13}$$

19.8 Solids Entrainment into Jets

Most of the solids entrained into a jet enter the jet cavity just downstream of the nozzle tip [54]. From Bernoulli's equation, the high gas velocity in this part of the jet cavity is associated with a low pressure, causing particles from the dense bed to be sucked into the jet. Gas is also sucked into the jet; with high-velocity jets used for grinding, so much gas is entrained from the bed that defluidization occurs locally near the nozzle tip [54].

The entrainment efficiency, defined as the mass of solids entrained into the jet per unit mass of attrition gas, is much larger for smaller nozzles: reducing the nozzle diameter by half increases the entrainment efficiency by a factor of 5 [55].

19.9 Nozzle Design

To improve the process efficiency, the injection nozzles for attrition gas are typically convergent–divergent Laval-type nozzles [56], as shown in Figure 19.19. Experiments have shown that the nozzle geometry has a large impact on its attrition performance [56, 57]. If the half-angle of the expansion section (α in Figure 19.19) is too large, too much gas momentum is lost in shocks, while if the half-angle is too small, there is excessive boundary layer friction [56]. Simulations of jet grinding have confirmed that a perfectly expanded nozzle, with exit pressure equal to the local bed pressure, performs better than under- or over-expanded nozzle designs [58]. In practice, the optimum half-angle is usually between 3.5° and 5°.

The jet grinding efficiency is often expressed as the ratio of the new particle surface to the mass of attrition gas used. Increasing the pressure upstream of the attrition nozzle increases the grinding efficiency [55]. Nozzle internals can help produce jets with larger interfacial area with the fluidized bed dense phase, enhancing solids entrainment from the bed into the jet. This increases the grinding efficiency [59].

Increasing the nozzle diameter has two effects. First, as shown in Section 3.3, the jet penetration increases, providing more time for particles to accelerate, so they can reach a higher velocity, enhancing grinding. Second, as shown above, the mass of solids entrained into the jet per unit mass of attrition gas is reduced, which has a detrimental impact on grinding. In practice, the first effect predominates, so it is usually beneficial to use fewer larger nozzles [55]. Twin-jet nozzles can increase the grinding efficiency by combining the advantages of small and large nozzles [55].

19.9.1 Nozzle Inclination

Vertical downward jets provide more effective jet grinding than horizontal jets [60]. This is because downward jets densify the bed region where the particles accelerated in the jet impact. A jet with an inclination of 60° with the horizontal direction is even more effective as it entrains more particles than a vertical jet [60].

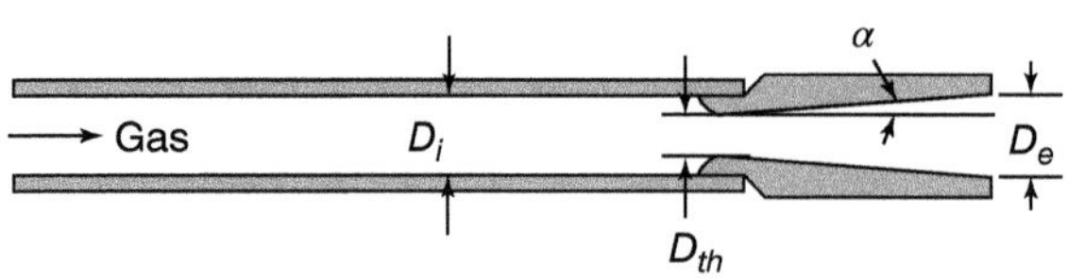

Figure 19.19 Typical attrition nozzle.

19.9.2 Impact of Bed Hydrodynamics

Jet grinding can be enhanced by modifying the local fluidized bed hydrodynamics. Three bed regions can be modified: the region near the base of the jet, the region around the middle part of the jet, and the region near the jet tip.

As shown in Section 2.2, so much gas is entrained from the bed into the jet that the bed region near the base of the jet may defluidize. Defluidization should be avoided as it greatly reduces particle entrainment into the jet. A possible solution is to ensure that the nozzle tip is in a region of high gas velocity [61]. However, this also means that a significant fraction of the jet energy is wasted to entrain gas instead of particles. Significant improvements in particle entrainment and, hence, grinding efficiency can be obtained by adding a shroud of appropriate geometry around the base of the jet: an ideal shroud greatly reduces gas entrainment without hindering particle entrainment [43]. In some cases, a shroud can increase the grinding efficiency by an order of magnitude [43].

The region around the middle part of the jet can also be modified to enhance the grinding efficiency. As shown in Figure 19.18, as the particles travel down the jet, their velocity increases and then peaks because the gas velocity decreases due to the jet expansion. The gas velocity can be kept high for longer by using a cylindrical tube around the jet cavity to constrain the jet and prevent its expansion [62]. This tube should only be around the middle part of jet, so it does not restrict solids entrainment into the jet, near the base of the jet. In some cases, such a tube can increase the grinding efficiency by 1 order of magnitude [43]. However, this method may not be practical with abrasive particles, as the tube would be eroded quickly. Even better results can be obtained if it is used in combination with a shroud [43].

Finally, it is best to locate the jet tip in a low fluidization region [61]. This ensures that the particles accelerated by the jet undergo impact with a denser bed region.

19.9.3 Opposing Jets

The concept underlying the use of opposing jets is to promote more violent collisions between particles with jets coming from opposite directions. Using two horizontal opposing jets did not much improve the grinding over the results obtained with a single jet [63]. This is probably because the jets could deflect slightly, which would prevent true head-on collisions of the particles.

Commercial opposing jet grinding systems use three opposing jets. Such a system is inherently self-centring and jets cannot avoid each other [64]. Figure 19.20 shows the solids concentration in the horizontal plane of a fluidized bed with three opposing jets [65].

19.10 Jet-Target Attrition

Several grinding systems insert a solid target in the path of the jet so that particles impact on the target. Using a target can increase the production of fines by a factor of 2 or 3 [63]. Flat targets are more effective than cup-shaped targets [59].

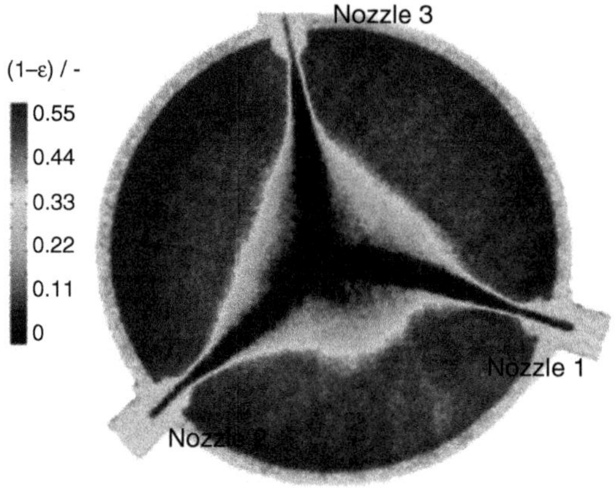

Figure 19.20 Solids distribution in a fluidized bed with three opposing jets (ε is the voidage).

Jet-target systems are typically used with relatively soft particles. With abrasive particles, the target would erode quickly, and the product would be contaminated with target material.

The distance between the nozzle tip and the target can be adjusted to modulate the grinding intensity. Locating the target in the jet region with the highest particle velocities (Figure 19.18) provides fragmentation (Figure 19.14), while locating it at longer or shorter distances reduces the intensity of the collisions of the particles with the target and promotes abrasion (Figure 19.14) [63]. For example, a fluidized bed for the commercial production of carbon nanotubes uses an attrition jet integrated with the gas distributor with a target; the target is adjusted so that carbon nanotubes growing from the surface of catalyst particles are broken off but the catalyst particles are not fragmented [45].

A spiral jet mill retains the advantages of a jet-target system while minimizing the erosion of a solid target. Four jets are used to create a vortex within the fluidized bed, which promotes the formation of a dense layer of particles near the wall [66]. Particles accelerated in each jet impact at high speed on this dense layer of particles.

19.10.1 Prediction of Attrition Rates

Different approaches have been used to predict attrition rates. Empirical correlations predict the rate at which new particle area is generated. Particle breakage models predict not only the attrition rate but also the size distribution of the ground particles. Finally, more fundamental models focus on particle dynamics.

While there are correlations to predict the attrition rate from subsonic jets [67], few correlations exist to predict the attrition rate from the sonic jets that are typically used in jet grinding [56, 57]. An example is the following correlation that predicts the new particle surface created per unit mass of attrition gas [57]:

$$\eta = 7.81 \times 10^{-7}\, \alpha\beta\, d_h^{1.131}\, U_{hj}^{0.55}\, (\rho_j U_{hj}^2)^{1.635} \left(\frac{U_g - U_{mf}}{U_{mf}} \right)^{0.494} \tag{19.14}$$

This correlation includes two empirical parameters. The first, α, accounts for the effect of the particle properties, which can vary with the bed temperature [68]. The second parameter, β, accounts for nozzle geometry [56, 57].

Several particle breakage models have been proposed to predict the size distribution of the ground particles. For example, a model for supersonic jet grinding was developed from three basic assumptions: particles are well mixed and all equally likely to enter the attrition jet, particle breaks result in two daughter particles with relative sizes set by an empirical symmetry coefficient, and fines below a critical size are not broken [43]. Particle breakage kinetics have also been modelled for both batch [69] and continuous grinding with opposed jets [70].

Some models have focused on particle dynamics. An Eulerian–Eulerian approach for the gaseous and particulate phases, coupled with the kinetic theory of granular flow and an attrition model based on particle properties and granular temperature, predicted the attrition performance of single supersonic jets [71]. CFD was used to model opposed jets grinding [72], while using CFD for the fluids and discrete element modeling (DEM) for the particles could model spiral jet grinding [73].

19.11 Jets Formed When Solids Are Fed into a Fluidized Bed

Gas can be used to inject solids into fluidized beds. The gas–solid stream then forms a jet cavity. Pulsed injection is used to achieve good penetration of solids into the bed while minimizing conveying gas. See also Chapter 11 for further information on solids feeding.

19.11.1 Mechanisms

The gas injected with the solids conveys the particles in the feeding line. It also accelerates the particles to a high speed, so they do not spend much time in sections of the line that are near the bed, which is essential in processes with hot beds. Finally, the gas helps create a jet cavity in the fluidized bed, so the injected solids reach the heart of the fluidized bed, far from the wall.

19.11.2 Applications

The main application of pulsed feeders is to inject highly reactive catalyst or prepolymer particles in gas-phase polyolefin reactors. Since the polymerization reactions are exothermic, the particles must be quickly dispersed throughout the fluidized bed to avoid the formation of hot spots, which would lead to agglomerate formation [74].

Pulsed feeders can also be used to inject particles in combustion [75], gasification, and pyrolysis [76] fluidized bed reactors. A benefit is that feed particles are injected into well-agitated regions of the bed, where good heat transfer from the hot bed particles ensures rapid heating and fast pyrolysis conditions [77]. Pulsed feeders are also useful when the feed material can melt at low temperatures and cannot be allowed to spend too much time in the hot regions of the feed line, near the hot bed [76].

19.11.3 Jet Penetration

The following empirical correlation predicts the jet penetration depth into the bed:

$$L_j = 2.27 \times 10^{-11}\ m_p^{-1.48}\ V_{inj}^{0.61}\ \rho_{bed}^{3.4}\ \text{(SI units)} \tag{19.15}$$

where m_p is the mass of the solids plug and V_{inj} the velocity at which it enters the bed.

19.11.4 Solids Entrainment

There is no information on the entrainment of bed solids into gas–solid jets, but, as a first approximation, one could use information on bed solids entrainment into gas–liquid jets.

19.11.5 Injection System Design

Two types of injection systems are used. The first system feeds solids intermittently into an injection line that is continuously flushed with gas [78]. The second system uses intermittent gas pulses into the conveying line, synchronized with the solids pulses [74, 79].

A simple model was developed to help design pulsed injection systems with a downward-inclined injection line [74]. It assumes that the solids plug is propelled by the high-pressure gas at its back, and takes into account the friction of the solids plug on the line wall, and the leakage of gas through the porous solids plug. As the plug mass is increased, gas leakage is reduced, but friction on the line wall is increased: there is an optimum plug mass.

A model was developed to design horizontal injection systems [79]. The empirical particle–wall friction coefficient was determined with a lab-size feeder, and the model was then used to design an industrial-size feeder, which performed as expected [79].

19.11.6 Nozzle Inclination

A horizontal injection system [79] does not perform as well as an inclined one [74]. With a horizontal injection line, the solids plug spreads as it travels along the line, and solid "stragglers" are left at the bottom of a horizontal line after each pulse [79]; this does not occur with an inclined line.

19.11.7 Impact of Bed Hydrodynamics

It is best to inject the solids near the bottom of the bed, where there is a high concentration and frequency of gas bubbles that pick up injected solids in their wake and disperse them to other bed regions [74, 75]. Increasing the superficial gas velocity greatly accelerates the dispersion of the injected solids [74].

19.12 Jets Formed When Liquid Is Sprayed into a Gas-Fluidized Bed

In some industrial processes, liquid is sprayed into fluidized beds with an atomization gas, which greatly helps distribute the liquid. After a brief sub-section on studies of pure liquid jets in fluidized beds, this section focuses on gas–liquid spray jets.

19.12.1 Pure Liquid Jets

Few studies have focused on the injection of a pure liquid into gas-fluidized beds, without atomization gas. Typically a liquid is used that vaporizes at the bed temperature and pressure [80, 81]. Liquid evaporates more quickly when the fluidization velocity is increased or the liquid flow rate is reduced [81]. High-velocity liquid jets penetrate deep into a fluidized bed but, compared with gas–liquid jets, form much smaller jet cavities, as their expansion angle is very small [82]. The formation of wet deposits at the bed wall suggests that the liquid distribution is rather poor compared to what can be achieved with gas atomization [80].

19.12.2 Mechanism for Gas–Liquid Jets

In most applications, atomization gas is used to spray liquid into a fluidized bed. It provides smaller droplets that are typically in the same size range as the bed particles and creates a jet cavity in the fluidized bed. Ariyapadi et al. [82] used X-rays to study the interactions of a small gas–liquid spray jet with a fluidized bed (Figure 19.21). Some bed particles are entrained into the jet cavity near the nozzle tip, where the pressure is lower, while others enter with the gas bubbles that rise into the jet. Entrained particles and high-velocity liquid droplets mix imperfectly as they move through the jet cavity and then strike dense fluidized bed particles, near the tip of the jet cavity. Liquid–solid agglomerates are periodically removed from the cavity by rising gas bubbles detaching from the jet tip. As the agglomerates move through the bed, they may break into smaller agglomerates or individual wet particles. Wet agglomerates, especially when freshly broken, may also grow by capturing dry particles [83]. Figure 19.21 shows liquid–solid agglomerates, made visible by a radio-opaque liquid. Other studies have found evidence

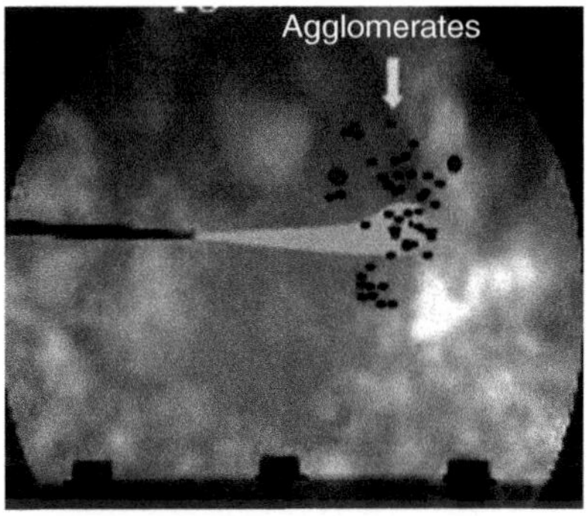

Figure 19.21 X-ray picture of jet cavity created by gas–liquid spray in a bubbling fluidized bed. The liquid was radio-opaque.

of agglomerates, even when the bed temperature is higher than the boiling point of the injected liquid [84].

19.12.3 Applications

Non-reacting hydrocarbon liquids such as *n*-butane are injected into fluidized bed polymerization reactors. The resulting evaporative cooling greatly increases the throughput of gas-phase polyethylene and polypropylene reactors [85, 86]. In this application, standard heat exchangers cannot be used to remove the heat of polymerization since they would become coated with an insulating layer of polymer.

Liquid formulations are sprayed into fluidized beds to promote granulation of fluidized particles or to apply a coating on the particle surfaces. Such processes are important in the food, fine chemicals, and pharmaceutical industries [87].

In fluid catalytic cracking (FCC), hydrocarbons are sprayed into a circulating bed of hot catalyst particles. The heat carried by the particles helps to vaporize the hydrocarbons, and their vapours are then cracked at the surface of the catalyst particles. Industrial experience and models have shown that improving feed atomization can greatly increase the yield of valuable products [88].

In Fluid Coking™ and Flexicoking™, heavy hydrocarbon residues are sprayed into a downflowing fluidized bed of hot coke particles. As the liquid contacts hot particles, it heats up and undergoes thermal cracking reactions that produce vapours of lighter, more valuable hydrocarbons [89]. Enhancing initial liquid–solid contact can improve the operability of the process and the yield of valuable products [90–92].

19.13 Jet Penetration

When vertical gas–liquid jets enter a fluidized bed, the jet cavity contracts as a large bubble detaches from its tip and then grows back as more gas and liquid flow into the cavity. The maximum jet penetration increases with increasing gas density and with increasing nozzle exit diameter [93]. Increasing the nozzle exit velocity also increases horizontal jet penetration [91]. The penetration of gas–liquid jets [94] can be predicted by adapting a correlation developed for pure gas jets [21] and by CFD modelling [95]. Jet penetration is affected by the spray nozzle design; generally, nozzle designs that reduce jet penetration increase the jet expansion angle [96]. The expansion angle of spray in the fluidized bed is about half its value in open air [86, 97].

With the assistance of high-frequency pulsations in gas–liquid flow into the nozzle, jet penetration can be expanded and shortened rapidly. This helps distribute the liquid over a larger number of fluidized particles [98, 99].

19.13.1 Solids Entrainment

Maximizing entrainment of fluidized particles into the jet cavity enhances the liquid distribution [100]. The flow rate of solid particles entrained from a fluidized bed into a spray jet cavity was measured with a new method that uses a partitioned fluidized bed [54, 101]. The momentum of the liquid–gas spray jet does not greatly affect the solids entrainment rate. Trebling the fluidizing gas velocity can double the solids entrainment rate [102, 103]. Under vigorous bubbling conditions, as in fluid cokers, most solids entering the jet cavity come from the wakes of bubbles that rise into the cavity [100, 104].

19.13.2 Injection System Design

The injection system design affects the behaviour of the spray jet in fluidized beds. Here we consider three aspects of the injection system design: the piping upstream of the nozzle, the actual spray nozzle, and attachments downstream of the nozzle. Conventional gas-atomized spray nozzles use either internal mixing or external mixing of liquid and atomization gas. For spraying in fluidized beds, internal mixing is preferred [84].

19.13.2.1 Upstream Piping Design

The ideal upstream piping design for most industrial applications effectively blends liquid and atomization gas streams and prevents large flow pulsations while having a reasonable pressure drop. Good premixing has been shown to have a beneficial impact on droplet size [105, 106] and liquid–solid contact [107, 108], and its impact on spray performance has been modelled [109]. Pulsations of gas–liquid flow within spray nozzles are caused by unstable flow upstream of the nozzle [110, 111] and can be detected from the pressure fluctuations at the premixer [110]. The two main types of flow in the upstream conduit are slug flow and bubbly flow [112]: slug flow creates low-frequency, high-amplitude pressure fluctuations, while bubbly flow causes high-frequency, low-amplitude pressure fluctuations [108]. The Taitel and Dukler [113] flow regime map is a reasonable predictor of nozzle stability for fully developed two-phase flow [114]. As shown below, many nozzles incorporate premixers near the nozzle tip so that the two-phase flow does not develop fully. Therefore, a properly designed premixer is a critical component of the upstream injection system to ensure that a stable gas–liquid mixture reaches the nozzle tip.

For applications such as fluid coking, nozzles must be "rodable," i.e. a metal rod can be pushed through to clear the nozzle of deposits without shutting down the fluidized bed. This requires dedicated rodable premixers. While a simple T design, with the liquid flowing straight and the gas entering at 90°, works well for small-scale nozzles, large nozzles require more sophisticated designs such as a Venturi or a bilateral flow conditioner (BFC) in order to effectively mix the gas and liquid. In the Venturi design, liquid flows through the Venturi, and the atomization gas enters the throat at a 90° angle [115, 116]. The BFC design combines the pressurized liquid with the atomizing gas as they enter from separate pipes at an angle of 30° to the axis of the nozzle and at about 90° from each other along the circumference in the direction of the nozzle tip [117]. While BFC premixers usually perform well, Venturi premixers perform well over wider ranges of liquid and gas flow rates.

The flow conduit is located between the gas–liquid premixer and the nozzle tip [115]. In general, the conduit is required to inject beyond the wall thickness, insulation, and any deposit accumulation on the internal wall, ensuring that the feed is always injected into a well-fluidized, unblocked area of the fluidized bed [118]. Roughening the conduit surface or adding flow restrictions may help reduce slugging but could affect the rodability of the system [119]. A helical insert may also be used [120].

19.13.2.2 Spray Nozzle Design

This section reviews laboratory nozzles, non-rodable commercial nozzles, and rodable commercial nozzles. The ranking of various nozzle types according to their performance in a fluidized bed depends on the atomization gas flow rate [121].

19.13.2.3 Laboratory Nozzles

A very simple nozzle that can be used in the laboratory is a straight cylindrical tube. If it is operated with a very high atomization gas-to-liquid ratio (GLR) of 50 wt% [112], this provides a nearly perfect dispersion of the liquid on the fluidized solids [122]. The larger expansion angle obtained with these nozzles increases solids entrainment and jet dispersion [123], benefiting liquid–solid contact. Rectangular or "flat" nozzles are also used in the laboratory, with a fan angle that provides wide spray coverage in the fluidized bed [124]. In commercial nozzles, the GLR typically ranges from 1 to 5 wt%, and the performance of scaled-down versions of commercial nozzles depends greatly on the GLR: with increasing GLR, the droplet size usually decreases in open air [112], and the liquid distribution on particles improves in fluidized beds [112].

19.13.2.4 Non-rodable Commercial Nozzles

This type of nozzle is used, for example, in FCC units, in which clogging risks must be minimized. The simplest design uses a capped cylindrical pipe with a single slot through the cap to generate flat sprays [125]. It can be modified by using two slots to generate two flat jets [126]. Recent designs have greatly improved the premixing of liquid and atomization gas within the nozzle body [127–131].

When clogging is less of an issue, nozzles of more complex geometry can be used. For example, some nozzles use internals to spin the gas–liquid mixture at high velocity just upstream of the nozzle tip: the gas–liquid mix continues to spin as it leaves the nozzle, forming either a full or a hollow cone spray [121].

19.13.2.5 Rodable Commercial Nozzles

Spray nozzles designed for fluid cokers must be rodable so that a straight rod can be forced through the nozzle. This feature is required because the heavy residual oils processed in fluid cokers have the propensity to form deposits within the nozzle, which can then be cleared with a rod, followed by a high-pressure water wash. The series of converging and diverging sections within the spray nozzle have a stabilizing effect, improving gas–liquid mixing within the nozzle body, and thus reducing droplet size, resulting in improved contact between the sprayed liquid and the bed particles. The convergent sections are incorporated into the design in order to accelerate the flow mixture and atomize the liquid into small droplets through elongation and shear stress mechanisms. The half-angles are optimal at 20° when the flow is most stable, with relatively low flow rates and minimal increase in feed pressure [115, 132]. Finally, the mixture is accelerated to reach sonic velocity conditions before ejecting through the nozzle tip. The high shear forces between the gas and liquid phase promote external atomization [95]. Increasing the atomization gas flow rate improves performance [133]. Decreasing the tip exit area is also beneficial as it increases the momentum flux and produces smaller droplets [108]. The high momentum of the exiting gas–liquid flow also helps create a low-pressure zone in the jet cavity downstream of the nozzle tip, entraining gas and particles from the dense fluidized bed into the jet cavity [54].

Minor modifications to the design of these nozzles can have a significant impact on their performance in a fluidized bed [121, 134, 135]. For example, eroded commercial spray nozzles may perform better than the undamaged, new spray nozzles [136]. Using a "cloverleaf" design at the tip of the nozzle can also enhance its performance [116].

19.13.2.6 Downstream Attachments

This section considers different attachments that can be placed downstream of a spray nozzle tip to enhance its performance in a fluidized bed.

19.13.2.7 Impact Attachments

The best impact attachment is a cone placed right in front of a spray nozzle exit, with its apex directed at the nozzle [108, 134]. This solid conical obstruction redirects the liquid spray issuing from the nozzle from the central core of the spray jet cavity to its peripheral region, where solids are concentrated. It is placed axially with the spray nozzle. The angle with which the conical attachment is designed produces a maximum jet dispersion angle [96]. A possible limitation of these attachments is that they are prone to be destroyed over time by either erosion from the fluidized solids or shear forces from the gas bubbles.

19.13.2.8 Shrouds

Attaching a shroud to a spray nozzle greatly affects the bed hydrodynamics around the nozzle tip [59]. With the shroud, the gas-liquid jet entrains a larger amount of solids while entraining a significantly lower volume of gas from the fluidized bed. The shrouded jet is narrower, is more stable, and penetrates less. It has been observed that while the solids move in intermittent waves towards the free jet, the solids move in a smooth continuous manner towards the shrouded jet at a faster rate [137]. The design of the shroud and the operating conditions should be such that the cross-sectional area of the spray at the nozzle exit is not larger than ¾ the area of the base. The gas velocity must be increased enough to help limit the amount of particles sticking to the internal walls of the shroud. Collection and material decomposition on the walls limit the useful life of the attachment [138].

19.13.2.9 Gas Jets

Instead of a solid shroud, one can also use a gas shroud. Two shroud configurations have been tested: the sonic shroud attachment consists of three small sonic gas jets surrounding the nozzle, whereas an annular shroud consists of an annular jet all around the nozzle [134]. The gas shrouds produce a dilute environment, where the gas–liquid mixture can expand into an effective solids entrainment zone creating a larger interfacial area, resulting in better contact between the solid and liquid phases [134]. Both configurations improve the nozzle performance [134, 139]. They also perform well when combined with an impact attachment or a draft tube [134].

19.13.2.10 Draft Tubes

A draft tube attachment is placed just downstream of the spray nozzle, co-axial with the nozzle. The draft tube enhances turbulent mixing within the jet cavity, promoting contact between entrained particles, which tend to concentrate in the peripheral region of the jet cavity, and the liquid droplets, which are concentrated in the central region of the jet cavity [95, 140]. It performs very well with any size of spray nozzle [90, 141, 142].

The shape of the draft tube attachment affects its performance. Three shapes have been tested: a straight cylindrical tube, a cylindrical tube with a rounded inlet, and a Venturi-shaped tube. Solids entrainment rates were maximized with the Venturi-shaped tube for gas-only jets and with the standard cylindrical draft tube shape for liquid–gas jets [143]. For each application, there are optimum draft tube dimensions and an optimum distance from the spray nozzle tip to the draft tube. While increasing the draft tube length promotes mixing within the tube, it also dissipates more jet momentum within the draft tube, reducing jet penetration and mixing downstream of the draft tube [123]. The draft tube diameter must be large enough to capture all the jet spray, but small enough to adequately confine the jet and promote mixing [54, 140]. As the distance between the nozzle tip and draft tube increases, more solids can be entrained into the jet cavity, but the spray expands more before reaching the draft tube, and the draft tube diameter must then be increased to confine the whole spray jet, which reduces the turbulence and mixing intensity within the draft tube [59, 144, 145].

Draft tubes are often associated with liquid spray in fluidized bed coating [146] to promote more even coating of particles by liquid [147]. Experiments and modelling have confirmed that a draft tube, often called a "Wurster tube," greatly reduces the formation of liquid–solid agglomerates [87].

19.13.3 Nozzle Inclination

Spray nozzles tend to be horizontal in fluid cokers, vertical in fluidized beds for agglomeration or coating, and inclined upwards in FCC risers. Modelling suggests that, in FCC risers, the feed could be injected at a downward angle to enhance initial contact between liquid and particles and reduce wall regions where fouling could occur [148].

19.13.4 Interactions Between Spray Jets

Industrial FCC risers use multiple spray jets that interact [149]. These interactions affect initial liquid–solid contact and riser hydrodynamics [150]. In fluid cokers, liquid is sprayed through several banks spread over the coker height. Each bank includes several nozzles spraying from the wall into the bed. Spray jets from the same bank do not interact. Interactions between spray jets from different banks would be detrimental to the liquid distribution, as jets from lower banks would capture gas bubbles and transfer their gas towards the central core of the bed, starving upper bank jets from gas bubbles [151]. Experiments suggested that proper positioning of the various spray nozzles could be used to minimize the detrimental impacts of the lower nozzle banks on the upper nozzle banks [151]. A detailed study of various spray nozzle configurations showed how current nozzle bank configurations could be modified to improve liquid distribution on the fluidized coke particles by taking advantage of beneficial interactions [151].

19.13.5 Impact of Bed Hydrodynamics

Increasing the fluidization velocity greatly improves the liquid distribution on fluidized particles. This effect has been observed for a wide variety of nozzles, liquids, and particles [152–156].

The local bed hydrodynamics during both the initial liquid injection and the subsequent drying greatly affect agglomerate properties and the amount of liquid trapped in agglomerates [157]. In a fluidized bed with a high gas velocity (>20 times the minimum fluidization velocity) where the nozzles are located, the solids entering the spray jet cavity come primarily from the wake of gas bubbles entering the cavity [100, 151]. The size distribution of particles trapped in agglomerates is the same as the size distribution of the bed particles expected if they came from the bubble wakes [158].

Liquid distribution can be improved by modifying the local bed hydrodynamics by either injecting additional gas just below the spray jet cavity [100] or redirecting gas flow with internals [159] such as baffles already used to modify solids flow patterns in fluid cokers [160, 161].

Vaporization of the injected liquid greatly affects bed hydrodynamics and can enhance the liquid distribution [80, 152]. This is why, when there are several spray nozzles as in most industrial processes, the location and orientation of the spray nozzles is important, even if there are no direct interactions between the spray jets [150]. Gas jets can also be used to induce rotation of the fluidized bed, thus improving the liquid distribution [119].

Solved Problems

19.1 Gas Distributor Design

A fluidized bed has an inner diameter of 4 m. The expanded fluidized bed height is 6 m. The pressure above the bed surface is 2×106 Pa, i.e. 20 atm.

In the freeboard, the gas has $\rho_g = 9\,\text{kg/m}^3$, $\mu_g = 2.1\times10^{-4}\,\text{kg/(m s)}$. As a first approximation, you may assume that the gas properties do not vary significantly throughout the column. The particles have the following properties: $\rho_p = 950\,\text{kg/m}^3$; Sauter mean diameter: $d_{psm} = 390\,\mu\text{m}$; sphericity = 1. Assume a bed voidage at minimum fluidization of $\varepsilon_{mf} = 0.42$. The superficial gas velocity may be assumed constant over the bed height and equal to 0.60 m/s. The bed mass is 35 583 kg. Design a perforated gas distributor plate. For mechanical reasons, its thickness should be 20 mm. The distributor has 4 mm diameter holes. To protect the column wall, no part of any hole can be within 35 mm of the column inner wall.

1. What should the grid pressure drop be?
2. What is the required number of holes?
3. What is the centre-to-centre pitch using a square pitch pattern?
4. What is the jet penetration using Merry's correlation (Eq. (19.8))?

Solution

1. Since this is not a shallow bed, the dry grid pressure drop must be at least:
   - 10% of bed pressure drop (higher values are better)
   - 3400 Pa
   - 100 × (pressure drop due to gas expansion from inlet line into the vessel)

   $$\text{Bed pressure drop: } \Delta P_{bed} = \frac{W_{bed}}{A} = \frac{M_s g}{\frac{\pi}{4}D^2} = 27\ 778\ \text{Pa}$$

   $$10\%\ \text{of}\ \Delta P_{bed} = 2778\ \text{Pa} < 3400\ \text{Pa}$$

   $$\text{Grid pressure drop: } \Delta P_{grid} = 3400\ \text{Pa}$$

2. Number of holes required: From Eq. (19.8)

$$\Delta P_{grid} = \frac{1.49}{2\rho_f}(\rho_f\ V_h)^2\left(\frac{d_h}{h}\right)^{0.26} = 0.745\rho_f V_h^2\left(\frac{d_h}{h}\right)^{0.26}$$

$$V_h = \sqrt{\left(\frac{h}{d_h}\right)^{0.26}\frac{\Delta P_{grid}}{0.745\rho_f}}$$

Mass balance on fluidization gas:

$$\rho_f\left(\frac{\pi}{4}D^2\right)V_g = \rho_f\left(N_h\frac{\pi}{4}d_h^2\right)V_h$$

$$N_h = \frac{V_g}{V_h}\left(\frac{D}{d_h}\right)^2$$

$$V_h = \sqrt{\left(\frac{h}{d_h}\right)^{0.26}\frac{\Delta P_{grid}}{0.745\ \rho_f}}$$

$h = 20\ \text{mm};\ d_h = 4\ \text{mm};\ \rho_f = 9\ \text{kg/m}^3;\ \Delta P_{grid} = 3400\ \text{Pa}$

$$\Rightarrow V_h = 27.76 \text{ m/s}$$

$$N_h = \frac{V_g}{V_h}\left(\frac{D}{d_h}\right)^2$$

$$D = 4 \text{ m}; \; V_g = 0.6 \text{ m/s}$$

$$\Rightarrow N_h = 21\ 614$$

3. Centre-to-centre pitch:
   x_w is the minimum distance between edge of holes and column wall. The minimum distance between the wall and centre of the holes is

   $$\left(x_w + \frac{d_h}{2}\right)$$

   The hole centres are located within a circle of diameter:

   $$(D - 2x_w - d_h)$$

   Column cross-sectional area irrigated by 1 hole:

   $$a_{i1} = \frac{\frac{\pi}{4}(D - 2x_w - d_h)^2}{N_h}$$

   Pitch is centre-to-centre distance p between adjacent holes. With a square pitch, each hole irrigates a cross-sectional area of p^2:

   $$p = \sqrt{a_{i1}} = (D - 2x_w - d_h)\sqrt{\frac{\pi}{4N_h}}$$

   $$D = 4 \text{ m}; \; d_h = 4 \text{ mm}; \; x_w = 35 \text{ mm}; \; N_h = 21\ 614$$

   $$\Rightarrow \; p = 23.7 \text{ mm}$$

4. Jet penetration:

   $$\text{Merry's correlation: } L_j = 5.2\ d_h \left[\frac{\rho_f\ d_h}{\rho_p\ d_p}\right]^{0.3} \left\{1.3 \left[\frac{V_h^2}{g d_h}\right]^{0.2} - 1\right\}$$

   $$\rho_f = 9 \text{ kg/m}^3; \; \rho_p = 950 \text{ kg/m}^3$$

   $$d_h = 4 \text{ mm}; \; d_p = 390\ \mu\text{m}$$

   $$V_h = 27.76 \text{ m/s}$$

   $$\Rightarrow L_j = 87 \text{ mm}$$

19.2 Penetration of a Gas–Liquid Jet

Heavy oil is sprayed continuously into a fluidized bed of petroleum coke particles. As with gas–solid jets, the tip of the jet cavity should be in region with a high concentration of gas bubbles.

Conditions:

Liquid and steam atomization gas are preheated to 350 °C. Liquid density is 950 kg/m^3.
Reactor pressure is 150 kPa (absolute), giving a steam density of 0.52 kg/m^3 at the nozzle temperature and bed pressure. The nozzle tip diameter is 15 mm.
The atomization gas-to-liquid mass ratio is 1 wt%; i.e. $GLR = 0.01$ wt/wt.
Ariyapadi's nozzle geometry parameter is $C_g = 2.5$.
Liquid flow rate can range from 2 to 10 kg/s.

Solution

Detailed calculation for 6 kg/s. All units are SI units.

1. Assume a mean voidage ε' of 0.5 to start.
2. As suggested by Diechsel and Winter [162], the following relationships can be used to obtain $\varepsilon(\eta)$.
   At the nozzle periphery:
   $$\varepsilon(\eta = 1) = 1 - (1 - \varepsilon')^{\frac{\varepsilon'}{3}} = 0.109$$
   At the nozzle centre:
   $$\varepsilon(\eta = 0) = 1 - (1 - \varepsilon')^{3.1} = 0.883$$
   At all other locations:
   $$\varepsilon(\eta) = \varepsilon(\eta = 0) - [\varepsilon(\eta = 0) - \varepsilon(\eta = 1)]\eta^n$$
   The index (n) can be calculated as follows:
   $$n = \frac{1.015}{(1.015 - \varepsilon')} = 1.971$$
3. One of the other unknowns is the pressure at the exit of the nozzle. This means that the gas density (ρ_g) at the nozzle tip is also unknown. To resolve this problem, a second iterative procedure (inner loop) can be adopted based on the assumption that the gas density is constant over the entire nozzle exit cross section.
4. Assume a gas density, $\rho_g = 2.0$.
5. Now, using the equation that relates the speed of sound in a gas–liquid mixture to that in a pure gas Ariyapadi et al. [94], the local two-phase sonic velocity can be calculated:
   $$U_{mix}(\eta) = \sqrt{\frac{\gamma RT}{m_W}} \times \left(\frac{\rho_g}{\varepsilon(\eta)\,\rho_{mix}(\eta)}\right)^{0.5}$$
   where the local mixture density is given by $\rho_{mix}(\eta) = \rho_g\ \varepsilon(\eta) + \rho_L\ [1 - \varepsilon(\eta)]$.

η	0.05	0.15	0.25	0.35	0.45	0.55	0.65	0.75	0.85	0.95
ρ_{mix}	115	130	160	205	265	339	427	529	645	776
U_{mix}	0.141	0.133	0.122	0.111	0.102	0.0957	0.0921	0.0923	0.0982	0.118

The gas superficial mass velocity at the nozzle tip can be represented by

$$G_{g,guess} = 2\rho_g \times \int_0^1 U_{mix}\varepsilon(\eta)(\eta)\,\eta\,d\eta = 63.8$$

6. Since the actual gas superficial mass velocity at the nozzle tip ($G_g = G_L \times GLR$) is already known, by minimizing $\Rightarrow (G_{g,guess} - G_g)^2$, an optimum value of ρ_g can be obtained for the assumed ε'. After iterations, we obtain 4.1709.
7. Now, the mean slip factor (S) can be calculated using Eq. (19.8) by numerical integration. We find $S = 0.4819$.
8. Using this S value, a new void fraction can be calculated from the following relation:

$$\varepsilon'_{new} = \frac{1}{\left(\frac{\rho_g}{\rho_L}(ALR)S + 1\right)} = 0.825$$

9. The final step involves minimizing $\Rightarrow (\varepsilon'_{new} - \varepsilon')^2$ to get the optimum value of ε' and S. It should be noted that by using this approach, the calculated ε' accounts for the gas–liquid slip correction, and thereby eliminates the homogeneous assumption ($U_g = U_L$), which tends to overestimate the jet momentum. We obtained $\varepsilon' = 0.7716$.

$$L_{jet} = \frac{5.5}{g^{0.27}} \frac{1}{(\rho_p - \rho_g)^{0.27}} \left(\frac{G_L^2(1 + ALR)}{\rho_L(1 - \varepsilon')}\right)^{0.27} d^{0.73}\, C_g = 2.63\text{ m}$$

Notations

A_h	area associated with each distributor hole $(0.25 \cdot \pi \cdot D_c{}^2/N_h)$ (m^2)
d_h	grid hole diameter (m)
d_p	particle diameter (or Sauter mean diameter for multisize particles) (m)
D_c	inner column diameter (m)
g	gravity constant (m/s^2)
h	grid thickness or nozzle length (m)
L_j	jet penetration length (m)
m_p	mass of plug (kg)
N_h	number of distributor holes (–)
P_{bed}	pressure in the bed at the orifice level (Pa)
P_j	pressure within the orifice (Pa)
ΔP_{grid}	grid pressure drop (Pa)
Re_h	hole Reynolds number: $Re_h = \rho_f \cdot d_h \cdot U_h / \mu_f$

U superficial gas velocity (m/s)
U_h gas velocity through the grid holes (at bed pressure) (m/s)
U_{hj} gas velocity through the grid holes (at pressure P_j) (m/s)
U_{mb} minimum bubbling velocity (m/s)
U_{mf} minimum fluidization velocity (m/s)
V_{inj} injection velocity (m/s)

Greek Letters

ε bed voidage (−)
μ_f fluid viscosity (kg m/s)
η grinding efficiency (m^2/kg)
ρ_{bed} bed density (kg/m^3)
ρ_f fluid density (at bed pressure) (kg/m^3)
ρ_j fluid density at pressure, P_j (kg/m^3)
ρ_p particle density (kg/m^3)
θ jet expansion half-angle (°)

References

1 Kunii, D. and Levenspiel, O. (2013). *Fluidization Engineering*. Amsterdam, Netherlands: Elsevier.
2 Karri, R. and Knowlton, T. (1999). Gas distributor and plenum design in fluidized beds. In: *Fluidization Solids Handling and Processing* (ed. W.C. Yang), 209–235. Westwood, NJ: Noyes Publications.
3 McAllister, R.A., McGinnis, P.H., and Plank, C.A. (1958). Perforated-plate performance. *Chem. Eng. Sci.* 9: 25–35.
4 Briens, C.L., Tyagi, A.K., and Bergougnou, M.A. (1988). Pressure drop through multiorifice gas distributors in fluidized bed columns. *Can. J. Chem. Eng.* 66: 740–748.
5 Qureshi, A.E. and Creasy, D.E. (1979). Fluidised bed gas distributors. *Powder Technol.* 22: 113–119.
6 Sánchez-Prieto, J., Soria-Verdugo, A., Briongos, J.V., and Santana, D. (2014). The effect of temperature on the distributor design in bubbling fluidized beds. *Powder Technol.* 261: 176–184.
7 Behie, L.A., Voegelin, B.E., and Bergougnou, M.A. (1978). The design of fluid bed grids using the orifice equation. *Can. J. Chem. Eng.* 56: 404–405.
8 Agarwal, G., Lattimer, B., Ekkad, S., and Vandsburger, U. (2012). Experimental study on solid circulation in a multiple jet fluidized bed. *AlChE J.* 58: 3003–3015.
9 Pore, M., Chandrasekera, T.C., Holland, D.J. et al. (2012). Magnetic resonance studies of jets in a gas–solid fluidised bed. *Particuology* 10: 161–169.
10 Rees, A.C., Davidson, J.F., Dennis, J.S. et al. (2006). The nature of the flow just above the perforated plate distributor of a gas-fluidised bed as imaged using magnetic resonance. *Chem. Eng. Sci.* 61: 6002–6015.

11 Serviant, G.A., Bergougnou, M.A., and Baker, C.G.J. (1972). Further studies on grid leakage in gas fluidized beds. *Can. J. Chem. Eng.* 50: 690–694.
12 Fakhimi, S., Sohrabi, S., and Harrison, D. (1983). Entrance effects at a multi-orifice distributor in gas-fluidised beds. *Can. J. Chem. Eng.* 61: 364–369.
13 Wormsbecker, M., Pugsley, T., Van Ommen, J.R. et al. (2009). Effect of distributor design on the bottom zone hydrodynamics in a fluidized bed dryer using 1-D X-ray densitometry imaging. *Ind. Eng. Chem. Res.* 48: 7004–7015.
14 Yudin, A.S.M., Anuar, S., and Oumer, A.N. (2016). Improvement on particulate mixing through inclined slotted swirling distributor in a fluidized bed: an experimental study. *Adv. Powder Technol.* 27: 2102–2111.
15 Sauriol, P., Cui, H., and Chaouki, J. (2012). Gas–solid structure in the vicinity of a sparger nozzle in a fluidized bed. *Powder Technol.* 228: 131–140.
16 Müller, C.R., Holland, D.J., Davidson, J.F. et al. (2009). Geometrical and hydrodynamical study of gas jets in packed and fluidized beds using magnetic resonance. *Can. J. Chem. Eng.* 87: 517–525.
17 Panariello, L., Materazzi, M., Solimene, R. et al. (2017). X-ray imaging of horizontal jets in gas fluidised bed nozzles. *Chem. Eng. Sci.* 164: 53–62.
18 Sauriol, P., Cui, H., and Chaouki, J. (2013). Gas jet penetration lengths from upward and downward nozzles in dense gas–solid fluidized beds. *Powder Technol.* 235: 42–54.
19 Merry, J.M.D. (1975). Penetration of vertical jets into fluidized-beds. *AIChE J.* 21: 507–510.
20 Hong, R., Li, H., Li, H., and Wang, Y. (1997). Studies on the inclined jet penetration length in a gas-solid fluidized bed. *Powder Technol.* 92: 205–212.
21 Benjelloun, F., Liegeois, R., and Vanderschuren, J. (1995). Penetration length of horizontal gas jets into atmospheric fluidized beds. In: *Proceedings of Fluidization-VIII* (eds. J.-F. Large and C. Laguerie), 239–246. New York: Engineering Foundation.
22 Vaccaro, S. (1997). Analysis of the variables controlling gas jet expansion angles in fluidized beds. *Powder Technol.* 92: 213–222.
23 Hensler, T., Tupy, M., Strer, T. et al. (2015). Particle tracking in fluidized beds with secondary gas injection. *Procedia Eng.*: 850–857.
24 Zhang, K., Pei, P., Brandani, S. et al. (2012). CFD simulation of flow pattern and jet penetration depth in gas-fluidized beds with single and double jets. *Chem. Eng. Sci.* 68: 108–119.
25 Floarea, O. and Straja, S. (1989). A non-isothermal grid model for catalytic fluidized-bed reactors. *Chem. Eng. Sci.* 44: 641–648.
26 Fan, L.T., Song, J.C., and Yutani, T.N. (1984). Transient axial solid and gas temperature profiles in the grid zone of a gas-solid fluidized bed. *Ind. Eng. Chem. Fund.* 23: 170–174.
27 Freychet, N., Briens, C.L., Bergougnou, M.A., and Large, J.F. (1989). A new approach to jet phenomena gas entrainment and recirculation in a bidimensional spouted fluidized bed. *Can. J. Chem. Eng.* 67: 191–199.
28 Shen, Z., Briens, C.L., Bergougnou, M.A., and Kwauk, M. (1991). Jet presence probability in a two-dimensional fluidized bed. *Powder Technol.* 67: 93–101.

29 Gbordzoe, E.A.M., Littman, H., and Bergougnou, M.A. (1988). Gas-solid contacting in a two-dimensional fluidized bed containing mobile internals: non-reactive tracer studies. *Can. J. Chem. Eng.* 66: 158–162.
30 Nemoda, S., Mladenović, M., Belošević, S. et al. (2009). Numerical model of gaseous fuel jet injection into a fluidized furnace. *Int. J. Heat Mass Trans.* 52: 3427–3438.
31 Wu, D., Wu, F., and Gu, Z. (2016). Catalyst attrition in an ASTM fluidized bed. *Catal. Today* 264: 70–74.
32 Xuereb, C., Laguerie, C., and Baron, T. (1991). Behavior of horizontal or inclined continuous jets gas-injected into a fluidized-bed. 1. Morphology of the jets. *Powder Technol.* 67: 43–56.
33 Sishtla, C., Findlay, J., Chan, I., and Knowlton, T. (1989). The effect of temperature and gas velocity on fines generation in non-reactive fluidized beds of coal char. In: *Fluidization VI* (eds. J.R. Grace, L.W. Shemilt and M.A. Bergougnou), 581–588. New York, NY: Engineering Foundation.
34 Werther, J. and Xi, W. (1993). Jet attrition of catalyst particles in gas fluidized beds. *Powder Technol.* 76: 39–46.
35 Zhang, H., Degrève, J., Baeyens, J., and Wu, S.-Y. (2016). Powder attrition in gas fluidized beds. *Powder Technol.* 287: 1–11.
36 Fotovat, F., Bi, X.T., and Grace, J.R. (2018). A perspective on electrostatics in gas-solid fluidized beds: challenges and future research needs. *Powder Technol.* 329: 65–75.
37 Lim, J.H., Lee, Y., Shin, J.H. et al. (2014). Hydrodynamic characteristics of gas–solid fluidized beds with shroud nozzle distributors for hydrochlorination of metallurgical-grade silicon. *Powder Technol.* 266: 312–320.
38 Cocco, R., Arrington, Y., Hays, R. et al. (2010). Jet cup attrition testing. *Powder Technol.* 200: 224–233.
39 Amblard, B., Bertholin, S., Bobin, C., and Gauthier, T. (2015). Development of an attrition evaluation method using a jet cup rig. *Powder Technol.* 274: 455–465.
40 Rydén, M., Moldenhauer, P., Lindqvist, S. et al. (2014). Measuring attrition resistance of oxygen carrier particles for chemical looping combustion with a customized jet cup. *Powder Technol.* 256: 75–86.
41 Lu, X., Liu, C.-C., Zhu, L.-p., and Qu, X.-h. (2014). Influence of process parameters on the characteristics of TiAl alloyed powders by fluidized bed jet milling. *Powder Technol.* 254: 235–240.
42 Koeninger, B., Hensler, T., Schug, S. et al. (2017). Horizontal secondary gas injection in fluidized beds: solid concentration and velocity in multiphase jets. *Powder Technol.* 316: 49–58.
43 McMillan, J., Briens, C., Berruti, F., and Chan, E. (2007). Particle attrition mechanism with a sonic gas jet injected into a fluidized bed. *Chem. Eng. Sci.* 62: 3809–3820.
44 Hsu, G., Rohatgi, N., and Houseman, J. (1987). Silicon particle growth in a fluidized-bed reactor. *AlChE J.* 33: 784–791.
45 Baddour, C.E., Briens, C.L., Bordere, S. et al. (2009). The fluidized bed jet grinding of carbon nanotubes with a nozzle/target configuration. *Powder Technol.* 190: 372–384.

46 Li, R., Qin, M., Liu, C. et al. (2017). Particle size distribution control and related properties improvements of tungsten powders by fluidized bed jet milling. *Adv. Powder Technol.* 28: 1603–1610.

47 Wang, J., Liu, B., Shao, Y. et al. (2014). The reprocessing of reactor core materials. *Nucl. Eng. Des.* 271: 162–165.

48 Lim, R.T.Y., Hoong, A.Y.J., Ng, W.K., and Tan, R.B.H. (2015). Anomalous size evolution of partially amorphized pharmaceutical particles during post-milling storage. *Powder Technol.* 286: 1–8.

49 Zhong, C., Zu, Y., Zhao, X. et al. (2016). Effect of superfine grinding on physicochemical and antioxidant properties of pomegranate peel. *Int. J. Food Sci. Technol.* 51: 212–221.

50 Xu, X., Li, X., Liu, F. et al. (2017). Batch grinding kinetics of scrap tire rubber particles in a fluidized-bed jet mill. *Powder Technol.* 305: 389–395.

51 Cruz, N., Briens, C., and Berruti, F. (2009). Green sand reclamation using a fluidized bed with an attrition nozzle. *Resour. Conserv. Recycl.* 54: 45–52.

52 Dunlop, D., Griffin, L., and Moser, J. (1958). Particle size control in fluid coking. *Chem. Eng. Prog.* 54: 39–43.

53 Li, F., Briens, C., Berruti, F., and McMillan, J. (2013). Modelling of penetration of horizontal supersonic nozzles in high temperature fluidised beds. *Can. J. Chem. Eng.* 91: 1115–1130.

54 Briens, C., Berruti, F., Felli, V., and Chan, E. (2008). Solids entrainment into gas, liquid, and gas–liquid spray jets in fluidized beds. *Powder Technol.* 184: 52–57.

55 Li, F. (2011). Particle attrition with supersonic nozzles in a high temperature fluidized bed. Western University, University of Western Ontario.

56 Cruz, N., Briens, C., and Berruti, F. (2010). Supersonic attrition nozzles in gas–solid fluidized beds. *Chem. Eng. Process. Process Intensif.* 49: 225–234.

57 McMillan, J., Briens, C., Berruti, F., and Chan, E. (2007). High velocity attrition nozzles in fluidized beds. *Powder Technol.* 175: 133–141.

58 Pougatch, K., Salcudean, M., and McMillan, J. (2011). Nozzle design influence on particle attrition by a supersonic steam jet. *Powder Technol.* 209: 35–45.

59 McMillan, J. (2007). Characterization of the interactions between high velocity jets and fluidized particles. PhD thesis. Western University.

60 Zhang, Q., Jamaleddine, T.J., Briens, C. et al. (2013). Effect of nozzle inclination on jet attrition. *Powder Technol.* 241: 236–243.

61 Zhang, Q., Jamaleddine, T.J., Briens, C. et al. (2012). Jet attrition in a fluidized bed. Part II: effect of fluidized bed hydrodynamics. *Powder Technol.* 229: 170–177.

62 Smith, L.S. and Mastalski, H.T. (1992) Throughput efficiency enhancement of fluidized bed jet mill. US Patent, Xerox Corp, US5133504A, 27 November 1990.

63 Tasirin, S.M. and Geldart, D. (1999). Experimental investigation on fluidized bed jet grinding. *Powder Technol.* 105: 337–341.

64 Koeninger, B., Hensler, T., Romeis, S. et al. (2018). Dynamics of fine grinding in a fluidized bed opposed jet mill. *Powder Technol.* 327: 346–357.

65 Koeninger, B., Koegl, T., Hensler, T. et al. (2018). Solid distribution in fluidized and fixed beds with horizontal high speed gas jets. *Powder Technol.* 336: 57–69.

66 Baddour, C.E., Briens, C.L., Bordere, S. et al. (2008). An investigation of carbon nanotube jet grinding. *Chem. Eng. Proc.* 47: 2195–2202.

67 Ghadiri, M., Cleaver, J., Tuponogov, V., and Werther, J. (1994). Attrition of FCC powder in the jetting region of a fluidized bed. *Powder Technol.* 80: 175–178.

68 Li, F., Briens, C., Berruti, F., and McMillan, J. (2012). Particle attrition with supersonic nozzles in a fluidized bed at high temperature. *Powder Technol.* 228: 385–394.

69 Berthiaux, H. and Dodds, J. (1999). Modelling fine grinding in a fluidized bed opposed jet mill. Part I: batch grinding kinetics. *Powder Technol.* 106: 78–87.

70 Berthiaux, H., Chiron, C., and Dodds, J. (1999). Modelling fine grinding in a fluidized bed opposed jet mill. Part II: continuous grinding. *Powder Technol.* 106: 88–97.

71 Pougatch, K., Salcudean, M., and McMillan, J. (2010). Simulation of particle attrition by supersonic gas jets in fluidized beds. *Chem. Eng. Sci.* 65: 4829–4843.

72 Rajeswari, M.S.R., Azizli, K.A.M., Hashim, S.F.S. et al. (2011). CFD simulation and experimental analysis of flow dynamics and grinding performance of opposed fluidized bed air jet mill. *Int. J. Miner. Process.* 98: 94–105.

73 Brosh, T., Kalman, H., Levy, A. et al. (2014). DEM–CFD simulation of particle comminution in jet-mill. *Powder Technol.* 257: 104–112.

74 Guedon, M.O., Baron, T., Briens, C.L., and Knowlton, T.M. (1994). Intermittent injection of prepolymer in a pressurized fluidized bed. *Powder Technol.* 78: 25–32.

75 Hartman, M., Svoboda, K., Pohořelý, M., and Trnka, O. (2005). Combustion of dried sewage sludge in a fluidized-bed reactor. *Ind. Eng. Chem. Res.* 44: 3432–3441.

76 Tumbalam Gooty, A., Li, D., Berruti, F., and Briens, C. (2014). Kraft-lignin pyrolysis and fractional condensation of its bio-oil vapors. *J. Anal. Appl. Pyrolysis* 106: 33–40.

77 Tumbalam Gooty, A., Li, D., Briens, C., and Berruti, F. (2014). Fractional condensation of bio-oil vapors produced from birch bark pyrolysis. *Sep. Purif. Technol.* 124: 81–88.

78 Pohořelý, M., Svoboda, K., and Hartman, M. (2004). Feeding small quantities of particulate solids. *Powder Technol.* 142: 1–6.

79 Berruti, F.M. and Briens, C.L. (2013). Novel intermittent solid slug feeder for fast pyrolysis reactors: fundamentals and modeling. *Powder Technol.* 247: 95–105.

80 Fan, L.-S., Lau, R., Zhu, C. et al. (2001). Evaporative liquid jets in gas–liquid–solid flow system. *Chem. Eng. Sci.* 56: 5871–5891.

81 Yang, Y., Sun, J., Huang, Z. et al. (2018). Dynamic and steady-state characterization of the liquid spray zone in an externally heated gas-solid fluidized bed. *Ind. Eng. Chem. Res.* 57: 2988–3001.

82 Ariyapadi, S., Holdsworth, D.W., Norley, C.J. et al. (2003). Digital X-ray imaging technique to study the horizontal injection of gas-liquid jets into fluidized beds. *Int. J. Chem. Reactor Eng.* 1: A 56.

83 Weber, S.C. (2009). Agglomerate stability in fluidized beds. Western University.

84 Bruhns, S. and Werther, J. (2005). An investigation of the mechanism of liquid injection into fluidized beds. *AIChE J.* 51: 766–775.

85 Alizadeh, A. and McKenna, T.F.L. (2013). Condensed mode cooling in ethylene polymerisation: droplet evaporation. *Macromol. Symp.* 333: 242–247.

86 Newton, D., Fiorentino, M., and Smith, G. (2001). The application of X-ray imaging to the developments of fluidized bed processes. *Powder Technol.* 120: 70–75.

87 Fries, L., Antonyuk, S., Heinrich, S., and Palzer, S. (2011). DEM-CFD modeling of a fluidized bed spray granulator. *Chem. Eng. Sci.* 66: 2340–2355.

88 Gupta, A. and Subba Rao, D. (2003). Effect of feed atomization on FCC performance: simulation of entire unit. *Chem. Eng. Sci.* 58: 4567–4579.

89 Gray, M.R., Zhang, Z., McCaffrey, W.C. et al. (2003). Measurement of adhesive forces during coking of Athabasca vacuum residue. *Ind. Eng. Chem. Res.* 42: 3549–3554.

90 House, P., Saberian, M., Briens, C. et al. (2004). Injection of a liquid spray into a fluidized bed: particle-liquid mixing and impact on fluid Coker yields. *Ind. Eng. Chem. Res.* 43: 5663–5669.

91 Knapper, B.A., Gray, M.R., Chan, E.W., and Mikula, R. (2002). Measurement of efficiency of distribution of liquid feed in a gas-solid fluidized bed reactor. *Int. J. Chem. Reactor Eng.* 1.

92 Stanlick, C., Berruti, F., and Briens, C. (2017). Effect of mixing and vapor residence time on thermal cracking of bitumen in a mechanically fluidized reactor. *Fuel* 200: 481–487.

93 Briens, C., Li, F., Berruti, F., and McMillan, J. (2010). Penetration of high velocity horizontal gas jets into a fluidized bed at high temperature. In: *Fluidization XIII*, ECI Symposium Series. New York: Engineering Conferences International.

94 Ariyapadi, S., Berruti, F., Briens, C. et al. (2004). Horizontal penetration of gas-liquid spray jets in gas-solid fluidized beds. *Int. J. Chem. Reactor Eng.* 2 (1).

95 Pougatch, K., Salcudean, M., and McMillan, J. (2012). Three-dimensional numerical modelling of interactions between a gas-liquid jet and a fluidized bed. *Chem. Eng. Sci.* 68: 258–277.

96 Pougatch, K. and Salcudean, M. (2011). Computational investigation of liquid spray dispersion modification by conical nozzle attachments. *J. Fluids Eng. Trans. ASME* 133: 031301.

97 Dawe, M., Briens, C., and Berruti, F. (2008). Study of horizontal sonic gas jets in gas–solid fluidized beds. *Can. J. Chem. Eng.* 86: 506–513.

98 Leach, A., Sabouni, R., Berruti, F., and Briens, C. (2013). Use of pulsations to enhance the distribution of liquid injected into fluidized particles with commercial-scale nozzles. *AIChE J.* 59: 719–728.

99 Sabouni, R., Briens, C., and Berruti, F. (2013). Enhancement of the liquid feed distribution in gas solid fluidised beds by non-mechanically induced nozzle pulsations. *Can. J. Chem. Eng.* 91: 1161–1167.

100 Mohagheghi Dar Ranji, M. (2014). Impact of local bed hydrodynamics on jet-bed interaction. PhD thesis. Western University.

101 Felli, V. (2002). Solids entrainment from a fluidized bed into a gas–liquid jet: effect of the draft tube. Dissertation. The University of Western Ontario.

102 Hulet, C., Briens, C., Berruti, F. et al. (2002). Entrainment and stability of a horizontal gas-liquid jet in a fluidized bed. *Int. J. Chem. Reactor Eng.* 1.

103 Cocco, R., McMillan, J., Hays, R., and Karri, S. (2013). Liquid injection into fluidized beds. *The 14th International Conference on Fluidization, ECI Symposium Series*, The Netherlands.

104 Mohagheghi, M., Berruti, F., and Briens, C. (2016) Impact of local fluidized bed hydrodynamics on interactions between particles and gas-liquid sprays. Fluidization XV. *ECI Proceedings*, Montebello, Canada.

105 Gomez, J., Fleck, B., Olfert, J., and McMillan, J. (2011). Influence of two-phase feed bubble size on effervescent atomization in a horizontal nozzle assembly. *Atomization Spray.* 21: 249–261.

106 Zaremba, M., Weiß, L., Malý, M. et al. (2017). Low-pressure twin-fluid atomization: effect of mixing process on spray formation. *Int. J. Multiphase Flow* 89: 277–289.

107 Portoghese, F., Ferrante, L., Berruti, F. et al. (2010). Effect of the injection-nozzle geometry on the interaction between a gas-liquid jet and a gas-solid fluidized bed. *Chem. Eng. Process.* 49: 605–615.

108 House, P. (2007). Interaction of gas-liquid jets with gas-solid fluidized beds: effect on liquid-solid contact and impact on fluid coker operation. PhD thesis, Western University.

109 Pougatch, K., Salcudean, M., and McMillan, J. (2014). Influence of mixture non-uniformity on the performance of an effervescent nozzle. *Fuel* 116: 601–607.

110 Ariyapadi, S., Berruti, F., Briens, C. et al. (2005). Stability of horizontal gas-liquid sprays in open-air and in a gas-solid fluidized bed. *Powder Technol.* 155: 161–174.

111 Hong, M., Fleck, B.A., and Nobes, D.S. (2014). Unsteadiness of the internal flow in an effervescent atomizer nozzle. *Exp. Fluids* 55: 1855–1859.

112 Portoghese, F., Ferrante, L., Berruti, F. et al. (2008). Effect of injection-nozzle operating parameters on the interaction between a gas-liquid jet and a gas-solid fluidized bed. *Powder Technol.* 184: 1–10.

113 Taitel, Y. and Dukler, A. (1976). A model for predicting flow regime transitions in horizontal and near horizontal gas-liquid flow. *AlChE J.* 22: 47–55.

114 Maldonado, S., Fleck, B.A., Heidrick, T. et al. (2008). Development of an experimental method to evaluate the stability of gas-liquid sprays. *Atomization and Sprays* 18.

115 Chan, E., Knapper, B., Mueller, E. et al. (2015). Fluid injection nozzle for fluid bed reactors. Google Patents, WO2016089821A1, filed 1 December 2015.

116 Reid, K.L., McMillan, J., Pougatch, K., and Salcudean, M.E. (2017). Fluid injection nozzle for fluid bed reactors. Google Patents, US973836B2, filed 29 July 2015.

117 McCracken, T.W., Bennett, A.J., Jonasson, K.A. et al. (2006). Mixing arrangement for atomizing nozzle in multi-phase flow. US Patent 7,140,558 B2.

118 Briens, L., Book, G., Albion, K. et al. (2011). Evaluation of the spray stability on liquid injection in gas-solid fluidized beds by passive vibrometric methods. *Can. J. Chem. Eng.* 89: 1217–1227.

119 Maroy, P., Loutaty, R., and Patureaux, T. (1994). Process and apparatus for contacting a hydrocarbon feedstock with hot solid particles in a tubular reactor with a rising fluidized bed. US patent US5348644A, filed 13 November 1990.

120 Keon, L.E. (1992). FCC process using feed atomization nozzle. US Patent 5108583, filed 20 October 1989.

121 Leach, A., Chaplin, G., Briens, C., and Berruti, F. (2009). Comparison of the performance of liquid-gas injection nozzles in a gas-solid fluidized bed. *Chem. Eng. Process. Process Intensif.* 48: 780–788.

122 Farkhondehkavaki, M., Soleimani, M., Latifi, M. et al. (2014). Characterization of moisture distribution in a fluidized bed. *Measurement* 47: 150–160.

123 Briens, C., Dawe, M., and Berruti, F. (2009). Effect of a draft tube on gas–liquid jet boundaries in a gas–solid fluidized bed. *Chem. Eng. Process. Process Intensif.* 48: 871–877.

124 Qureshi, M.M.R. and Zhu, C. (2006). Crossflow evaporating sprays in gas–solid flows: effect of aspect ratio of rectangular nozzles. *Powder Technol.* 166: 60–71.

125 Kolb, N., Yuan, E., Anderson, L.R., and Jackson, G. (1994). Fluidized catalytic cracking feed nozzle. US Patent 5,306,418, filed 2 April 1992.

126 Haruch, J. (1997). Enhanced efficiency nozzle for use in fluidized catalytic cracking. US Patent 5,673,859, filed 13 December 1994.

127 Chen, Y.-M. (2011). Feed nozzle assembly. US Patent 7,992,805 B2, filed 12 November 2004.

128 LaCroix, M., Gbordzoe, E.A., and Santner, C.R. (2018). Catalytic cracking spray nozzle with internal liquid particle dispersion ring. US Patent 9925508B2, filed 3 October 2014.

129 Quiroz-Pérez, E., Vázquez-Román, R., Castillo-Borja, F., and Hernández-Barajas, R. (2016). A CFD model for the FCC feed injection system. *Fuel* 186: 100–111.

130 Holtan, T.P. and Muldowney, G.P. (2000). Atomizing feed nozzle. US Patent 6142457A, filed 30 January 1998.

131 Williatte, C., Sigaud, J.-B., Patureaux, T., and Loutaty, R. (1991). Device for injection of a hydrocarbon feedstock into a catalytic cracking reactor. US patent 5037616, filed 21 June 1988.

132 Base, T., Chan, E., Kennett, R., and Emberley, D. (1999). Nozzle for atomizing liquid in two phase flow, Canadian Oil Sands. US 6,003,789, filed 15 December 1997.

133 Ali ZirGachian, M., Soleimani, M., Briens, C., and Berruti, F. (2013). Electric conductance method for the assessment of liquid–gas injection into a large gas–solid fluidized bed. *Measurement* 46: 893–903.

134 House, P.K., Briens, C.L., Berruti, F., and Chan, E. (2008). Effect of spray nozzle design on liquid-solid contact in fluidized beds. *Powder Technol.* 186: 89–98.

135 Farkhondehkavaki, M. (2012). Developing novel methods to characterize liquid dispersion. PhD thesis. Western University, London, ON.

136 Prociw, N. (2014). Effect of nozzle geometry on jet bed interaction: experiments with commercial scale nozzles and eroded nozzles. MESc thesis. Western University.

137 Hulet, C., McMillan, J., Briens, C. et al. (2007). Visualization of the effect of a shroud on entrainment of fluidized solids into a gas jet. *Int. J. Chem. Reactor Eng.* 5 (1).

138 Mckinley, J.B. and Derrig, M.J. (1957). Fluid contacting process involving fluidized particles. US Patent 2786801, filed 4 April 1952.

139 Saha, M. (2012). Simultaneous particle agglomeration and attrition in a high temperature fluidized bed. Western University.

140 Ariyapadi, S., McMillan, J., Zhou, D. et al. (2005). Modeling the mixing of a gas-liquid spray jet injected in a gas-solid fluidized bed: the effect of the draft tube. *Chem. Eng. Sci.* 60: 5738–5750.

141 McMillan, J., Zhou, D., Ariyapadi, S. et al. (2005). Characterization of the contact between liquid spray droplets and particles in a fluidized bed. *Ind. Eng. Chem. Res.* 44: 4931–4939.

142 ZirGachian, M.A., Briens, C., Berruti, F., and McMillan, J. (2018). Impact of a draft tube on industrial-scale fluid Coker™ spray jets in fluidized beds. *Can. J. Chem. Eng.* 96: 317–322.

143 Hulet, C. (2006). Interaction phenomena of submerged jets and fluidized solids. PhD thesis. Western University.

144 Ariyapadi, S., Berruti, F., Briens, C. et al. (2003). Modeling the injection of gas-liquid jets into fluidized beds of fine particles. *Can. J. Chem. Eng.* 81: 891–899.

145 Hulet, C., Briens, C., Berruti, F., and Chan, E.W. (2008). Effect of a shroud on entrainment into a submerged jet within a fluidized bed. *Chem. Eng. Process.* 47: 1435–1450.

146 Teunou, E. and Poncelet, D. (2002). Batch and continuous fluid bed coating–review and state of the art. *J. Food Eng.* 53: 325–340.

147 Jones, D.M. (1995) Fluidized bed with spray nozzle shielding. US Patent 5437889A, filed 9 August 1993.

148 Chen, S., Fan, Y., Yan, Z. et al. (2016). CFD optimization of feedstock injection angle in a FCC riser. *Chem. Eng. Sci.* 153: 58–74.

149 Patel, R., Wang, D., Zhu, C., and Ho, T.C. (2013). Effect of injection zone cracking on fluid catalytic cracking. *AIChE J.* 59: 1226–1235.

150 Pelissari, D.C., Alvarez-Castro, H.C., Mori, M., and Martignoni, W. (2016). Study of feedstock injection to improve catalyst homogenization in the riser of a FCC. *Braz. J. Chem. Eng.* 33: 559–566.

151 Elkolaly, H. (2015). Effect of spray jet interactions on the liquid distribution in a fluidized bed. MESc thesis. Western University.
152 Ahmadi Motlagh, A.H., Grace, J.R., Briens, C. et al. (2017). Experimental analysis of volatile liquid injection into a fluidized bed. *Particuology* 34: 39–47.
153 Copan, J., Clarke, N., and Berruti, F. (2001). The interaction of single-and two-phase jets and fluidized beds. In: *Fluidization X* (eds. M. Kwauk and J. Li), 77–84. New York: Engineering Foundation.
154 Mohagheghi, M., Hamidi, M., Briens, C. et al. (2014). The effects of liquid properties and bed hydrodynamics on the distribution of liquid on solid fluidized particles in a cold-model fluidized bed. *Powder Technol.* 256: 5–12.
155 Morales, C.B., Jamaleddine, T.J., Berruti, F. et al. (2016). Low-temperature experimental model of liquid injection and reaction in a fluidized bed. *Can. J. Chem. Eng.* 94: 886–895.
156 Li, L. (2016). Effect of local bed hydrodynamics on the distribution of liquid in a fluidized bed. MESc thesis. Western University.
157 Li, B., Sanchez Careaga, F., Briens, C., and Berruti, F. (2016). Effect of local hydrodynamics on the distribution of liquid sprayed into a fluidized bed. *Fluidization XV, ECI Proceedings*, Montebello, Canada.
158 Reyes, L.A.P. (2015) Effect of temperature and successive sprays on liquid distribution in fluidized beds. MESc thesis. Western University.
159 Jahanmiri, M. (2017). Use of a baffle to enhance distribution of a liquid sprayed into a gas-solid fluidized bed. MESc thesis. Western University.
160 Issangya, A.S., Reddy Karri, S.B., and Knowlton, T.M. (2008, 2008). Effect of baffles on jet streaming in deep fluidized beds of group A particles. In: *Proceedings 9th International Conference on Circulating Fluidized Beds* (ed. J. Werther), 105–110. Hamburg, Germany: TuTech Innovation GmbH.
161 Wyatt, J.T., Jones, E.N., Chen, A.U. et al. (2013). Circulating fluid bed reactor with improved circulation. US Patent 8,435,452 B2, filed 17 February 2011. ExxonMobil Research and Engineering Company.
162 Deichsel, M. and Winter, E.R.F. (1990). Adiabatic two-phase pipe flow of air-water mixtures under critical flow conditions. *Int. J. Multiphase Flow.* 16: 391–406.

Problem

19.1 Penetration of a Gas–Solid Jet

The objective is to inject pulses of 0.05 kg of prepolymer in a gas-phase polyolefin reactor. To achieve rapid dispersion of the injected prepolymer, the tip of the injection jet should reach a fluidized bed region with a high gas bubble density.

The normal design procedure would go through the following steps:

(1) Select an injection height.
(2) Predict the radial profile of the gas bubbles concentration at that height, using information.

(3) Select an injection velocity of the solids plug as it enters the bed, so that the solids plug reaches a bed region with a high gas bubbles concentration.
(4) Design an injection system that will deliver the required injection velocity, using Ref. [74].

To help with step 3, this example will show how the jet penetration varies with the injection velocity of the prepolymer plug as it enters the bed. Once the appropriate injection velocity has been selected, the injection system would have to be designed to deliver that injection velocity.

Conditions:

Fluidized bed density (kg/m^3): 200
(*Note*: the relatively low bed density value, as the bed particles are porous).

20

Downer Reactors

Changning Wu[1] *and Yi Cheng*[2]

[1] *Clean Energy Institute, Southern University of Science and Technology, Shenzhen, Guangdong, 518055, P.R. China*
[2] *Department of Chemical Engineering, Tsinghua University, Beijing 100084, P.R. China*

20.1 Downer Reactor: Conception and Characteristics

Fluidization is a process in which particles are driven to a dynamic fluid-like state by means of an upward or downward flow of gas (or liquid). This chapter is mainly devoted to consideration of fluidized bed reactors where the fluidizing fluid is a gas and the gas and the particles both move downwards co-currently in a vertical pipe or column.

Typical types of gas–solid two-phase flows are illustrated in Figure 20.1, with four quadrants obtained by placing fluid flow direction and solid flow direction ("+" and "−" mean upwards and downwards, respectively) on the F and S axes. The top-right quadrant represents a gas–solid co-current upflow reactor with different fluidization regimes (e.g. bubbling, turbulent, and fast fluidization); the bottom-right quadrant represents gas–solid counter-current contact; the bottom-left quadrant represents gas–solid co-current downflow, corresponding to the downer reactor. Though this chapter only considers gas–solid systems, conceptually the physical scenario in the top-left quadrant is possible for a liquid–solid fluidized bed if the density of the solid particles is less than that of the liquid.

Figure 20.2 illustrates the essential concept of (a) a downer reactor and (b) a practical downer in a circulating fluidized bed system. Feed (gas) meets solid particles at the top of a vertical column, and the mixture flows downwards in the direction of gravity. Products are separated from the solid particles at the bottom of the reactor and quenched in most cases to terminate further reactions. If necessary, the solid particles (typically catalyst particles) need to be regenerated and circulated back to the top of the reactor [1].

In comparison with riser reactors, downer reactors have some key advantages based on the fact that particles flow co-currently downwards with the gas phase

Essentials of Fluidization Technology, First Edition. Edited by John R. Grace, Xiaotao Bi, and Naoko Ellis.

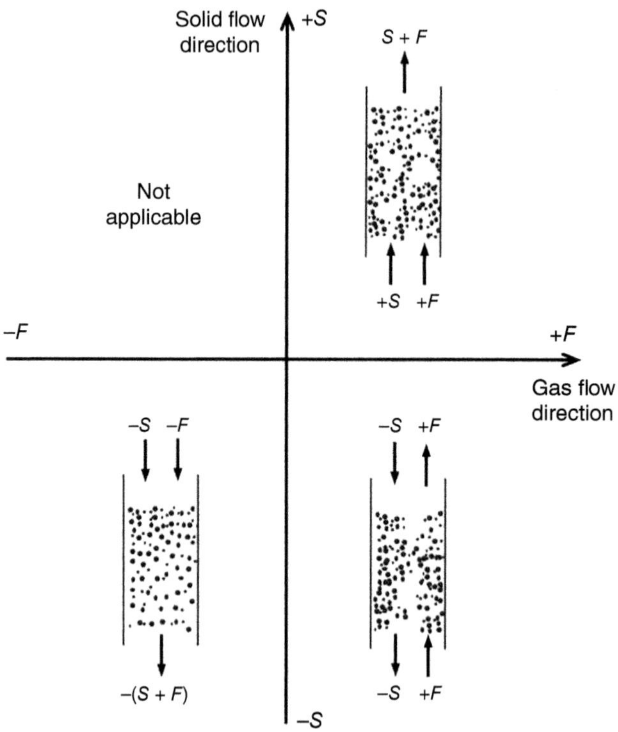

Figure 20.1 Four quadrants of gas–solid two-phase flows.

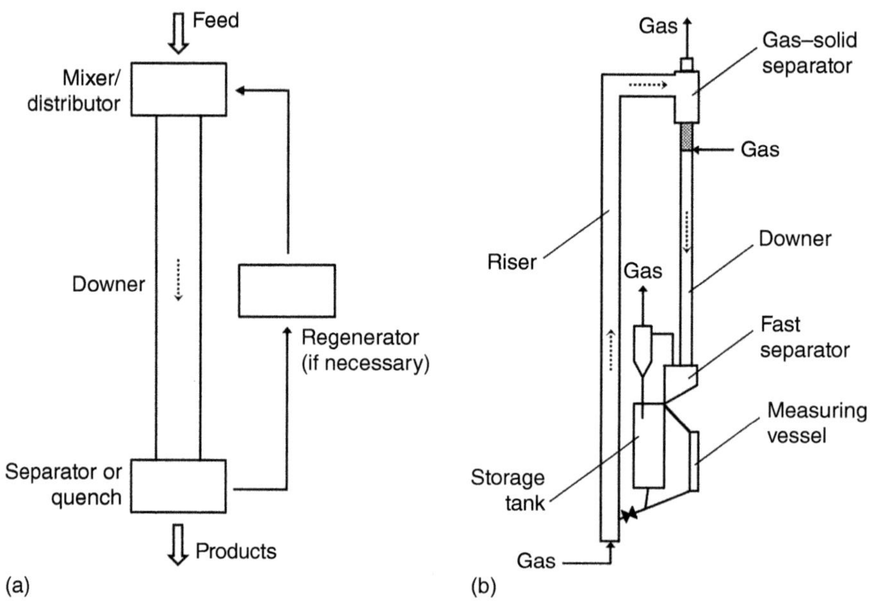

Figure 20.2 Diagrams of downer reactors: (a) essential concept design and (b) a practical downer in a circulating fluidized bed system.

in the direction of the gravitational force, rather than against it, as in conventional fluidized bed and riser reactors. Key advantages include the following:

(1) There is no minimum gas velocity, such as minimum fluidization velocity and minimum transport velocity below which conventional and circulating fluidized beds cannot operate; as a result, the solid/gas loading ratio can be much higher.
(2) Both axial and radial gas–solid flow structures are much more uniform.
(3) Short contact time reactions can be carried out in downers under near plug flow conditions owing to the reduced axial dispersion and more uniform gas–solid contact time. Accordingly, better product selectivity can be achieved.

Downers are therefore novel multiphase reactors with great potential for high-severity processes, such as high-temperature, ultra-short contact time consecutive reactions (e.g. milliseconds to seconds) with intermediates as the desired products. From a reaction engineering viewpoint, such processes put strong demands on the solids feeding and gas–solid mixing at the downer inlet, as well as the fast separator at the outlet. Thus, the performance of a downer reactor is very sensitive to the structure design, although the concept design shown in Figure 20.2 looks rather simple. On the other hand, multiphase flows in downers exhibit unique features, which challenge the predictability of modelling and simulation for gas–solid flows.

20.2 Hydrodynamics, Mixing, and Heat Transfer of Gas–Solid Flow in Downers

20.2.1 Basic Hydrodynamic Behaviour

Solid particles and gas are fed from the top of downer reactors. The particles are generally given an initial velocity, whereas the gas is immediately given a certain velocity downwards (somewhat in excess of the superficial gas velocity, U_g). The particles are then accelerated by the gas-particle drag force (F_D) and gravitational force (F_G). Neglecting the gas-particle force along the horizontal direction and interaction among particles, the particles continue to travel downwards, with a force balance on a single particle in the downflowing gas stream given by

$$\rho_p V_p \frac{d\overline{u}_p}{dt} = \rho_p V_p g + \frac{1}{2} C_D \rho_g |\overline{u}_g - \overline{u}_p| (\overline{u}_g - \overline{u}_p) \frac{\pi d_p^2}{4} - \rho_g V_p g \tag{20.1}$$

where ρ_g is the gas density, ρ_p is the particle density, d_p is average diameter of the particles, V_p is average volume of the particles, $\overline{u}_g$ is cross-sectional average gas velocity, $\overline{u}_p$ is cross-sectional average particle velocity, and C_D is a drag coefficient. The three terms on the right-hand side arise due to gravity, drag, and buoyancy forces, respectively. Depending on the sign of gas-particle slip velocity, $\overline{u}_{sl} = \overline{u}_g - \overline{u}_p$, the drag force acts in the same direction or opposite to gravity.

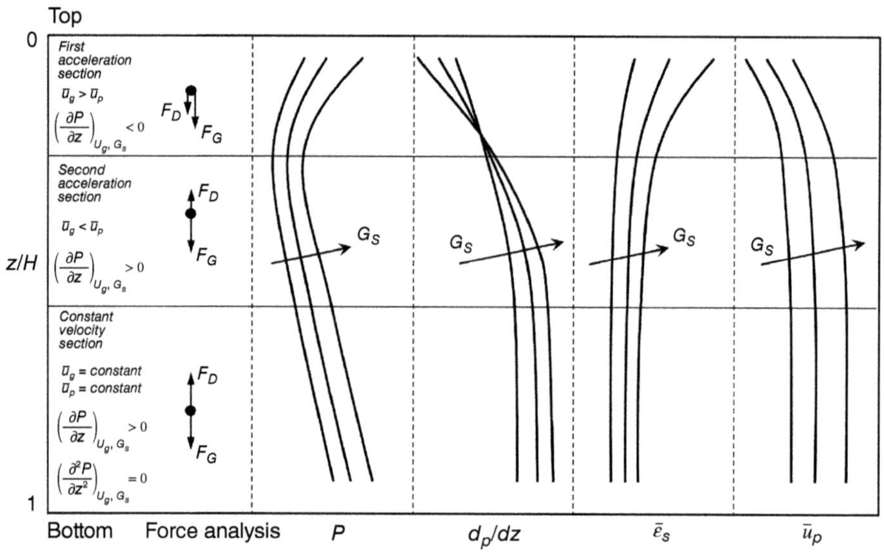

Figure 20.3 Basic hydrodynamic behaviour in downers with force balance analysis.

The particles undergo three stages of hydrodynamic behaviour as illustrated in Figure 20.3 and summarized below [1–3]:

(1) First, an acceleration section extending from the top of a downer reactor to the level where the particle velocity is equal to the gas velocity. In this section, drag has the same direction as gravity and varies from a positive value to zero, while $\overline{u}_g$ is always greater than $\overline{u}_p$.
(2) Second acceleration section: From the position where the gas–solid drag force is 0 to a lower level where drag balances $(\rho_p - \rho_g)V_p g$ (the buoyancy force is negligible compared with gravity and is not shown in Figure 20.3). In this section, drag, whose direction is opposite to gravity, varies from zero to $(\rho_p - \rho_g)V_p g$, while $\overline{u}_g$ is always lower than $\overline{u}_p$.
(3) Constant velocity section occupying the remainder of the downer after the second acceleration section, excluding any exit effect. In this section, the particles descend faster than the gas, but $\overline{u}_p$ remains constant due to a net force of zero, while $\overline{u}_g$ also remains constant due to the constant gas-particle slip velocity.

The axial profiles of gas pressure (P) and gas pressure gradient at a given U_g and different solids flux (G_s) are plotted in Figure 20.3, as measured experimentally by Wang et al. [2] and discussed by Zhu et al. [3].

The gas pressure in the flow direction drops substantially to compensate for the energy loss due to drag on particles and the friction between the wall and the gas–solid suspension in the first acceleration section, i.e.

$$\left(\frac{\partial P}{\partial z}\right)_{U_g,G_s} < 0 \quad (20.2)$$

where z is distance between the downer inlet and the position of particles observed.

Beyond the position of zero drag, the gas pressure increases gradually as the particles are further accelerated by gravity and impose a downward drag force on the gas in the second acceleration section, where

$$\left(\frac{\partial P}{\partial z}\right)_{U_g,G_s} > 0 \tag{20.3}$$

The gas pressure increases continuously and linearly as the particles exert a constant downward drag force on the gas equal to the gravity force, after the particles enter into the constant velocity section, where

$$\left(\frac{\partial P}{\partial z}\right)_{U_g,G_s} > 0, \left(\frac{\partial^2 P}{\partial z^2}\right)_{U_g,G_s} = 0 \tag{20.4}$$

As discussed above, the cross-sectional averaged particle velocity increases from the top of the downer to the bottom of the second acceleration section and then remains constant, showing the same variation trends at different solids flux, G_s. The corresponding profiles of cross-sectional averaged solid volume fraction ($\overline{\varepsilon}_s$) illustrated in Figure 20.3 could be generated using the following continuity equation of the solid phase:

$$\overline{\varepsilon}_s = \frac{G_s}{\rho_p \overline{u}_p} \tag{20.5}$$

Equation (20.1) can be solved to obtain the variation of particle velocity at any time since the drag coefficient can be determined empirically and the cross-sectional average gas velocity can be expressed as

$$\overline{u}_g = \frac{U_g}{\overline{\varepsilon}} = \frac{U_g}{1 - \overline{\varepsilon}_s} \tag{20.6}$$

where $\overline{\varepsilon}$ is the cross-sectional average voidage.

Lehner and Wirth [4] successfully measured the development of local solid volume fraction along a downer from the inlet to the developed region near the exit by a novel X-ray computed tomography (CT) technique as shown in Figure 20.4. A typical distribution of a dense ring of solids near the wall was clearly disclosed, in good agreement with experimental work by the Fluidization Lab of Tsinghua University (FLOTU).

The radial profiles of hydrodynamic parameters, such as the solid volume fraction and velocities, differ greatly in a riser and a downer [1]. As summarized in Figure 20.5, a downer has a rather uniform radial flow structure compared with a riser. The dense ring distribution of solids in the downer is unique, with its physical mechanism unsolved.

20.2.2 Mixing Behaviour of Solids in Downers

Solids mixing in reactors can be determined by measuring residence time distributions (RTDs) of solids using a phosphorescent tracer, which becomes

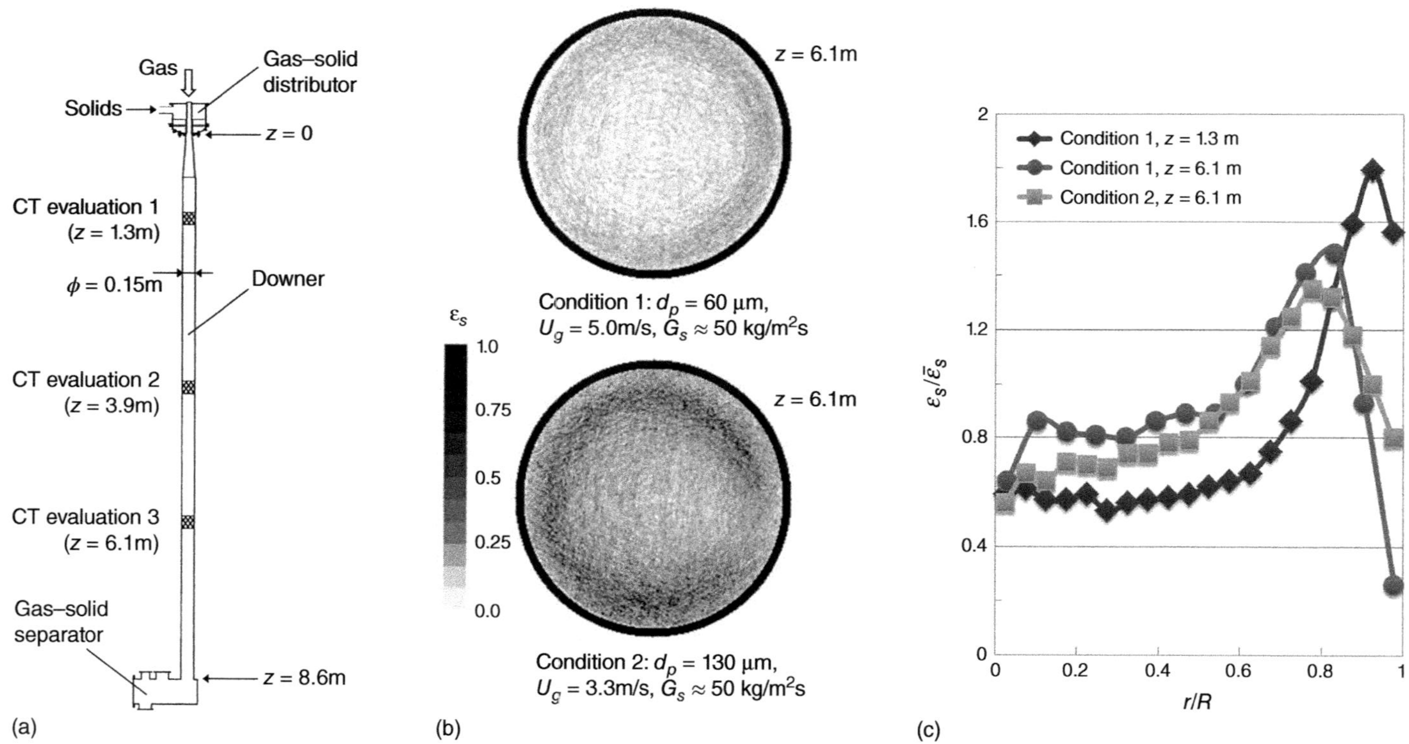

Figure 20.4 Solid volume fraction at different axial positions in a downer: (a) schematic of the downer investigated by X-ray CT technique, (b) cross-sectional distribution of solid volume fraction measured at height of 6.1 m under two conditions, and (c) normalized radial profiles of solid volume fraction at height of 1.3 and 6.1 m. Source: Lehner and Wirth 1999 [4]. Reproduced with permission of Elsevier.

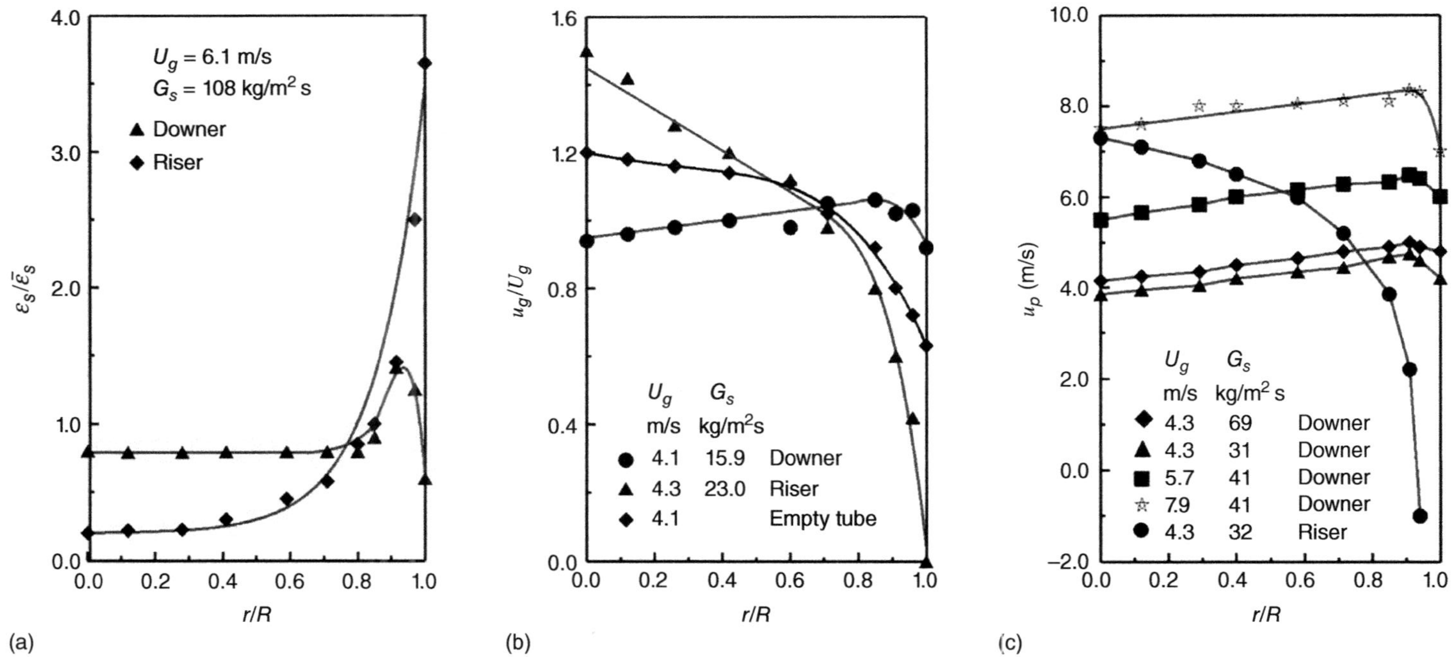

Figure 20.5 Typical radial profiles of solid volume fraction, gas velocity, and particle velocity in riser and downer reactors of the same 140 mm ID.

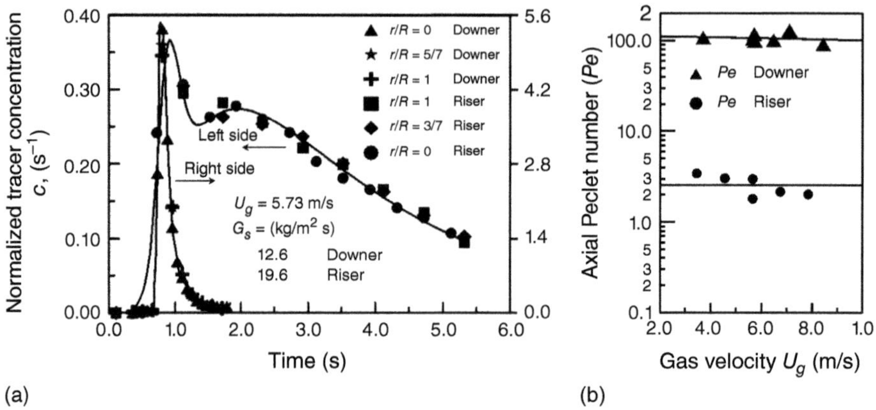

Figure 20.6 Typical solids RTDs and Péclet numbers (*Pe*) in riser and downer reactors of the same 140 mm ID. Source: Wei and Zhu 1996 [5]. Reproduced with permission of Elsevier.

fluorescent for a short period of time, with a known decay constant, after illumination [5]. Figure 20.6 shows typical solids RTDs as represented by the profiles of normalized tracer concentration in a riser and a downer. The narrower RTD profile illustrates the near plug flow behaviour in the downer, whereas the bimodal RTD profile with a long tail demonstrates relatively strong backmixing of solids in the riser, with corresponding Péclet numbers, *Pe*, of ~100 and 2 for the two reactors, respectively. The flow direction clearly has a remarkable influence on the axial solids mixing in the two vertical gas–solid suspension systems. In comparison with solids mixing, gas mixing in the axial direction in the downer is very weak, so the gas phase can generally be considered to be in plug flow.

20.2.3 Heat Transfer in Downers

Heat transfer is governed by the hydrodynamics in multiphase flows. Suspension density is a dominant factor influencing heat transfer since particle convection is mainly dependent on the local bed density and solid velocity near heat transfer surfaces. The heat transfer coefficient between a suspended surface and the gas–particle flow suspension in a downer of 9.3 m height using a miniature cylindrical heat transfer probe was measured by Ma and Zhu [6]. Their results demonstrated that bed suspension density is the most influential factor. The radial profiles of heat transfer coefficient are consistent with that of solid volume fraction (see Figure 20.7). The average heat transfer coefficient decreases with decreasing solids circulation rate due to the decreased solid volume fraction. The suspension-to-wall heat transfer coefficient is generally in the range of 100–150 W/m^2 K, not substantially less than coefficients reported for risers (see Chapter 14).

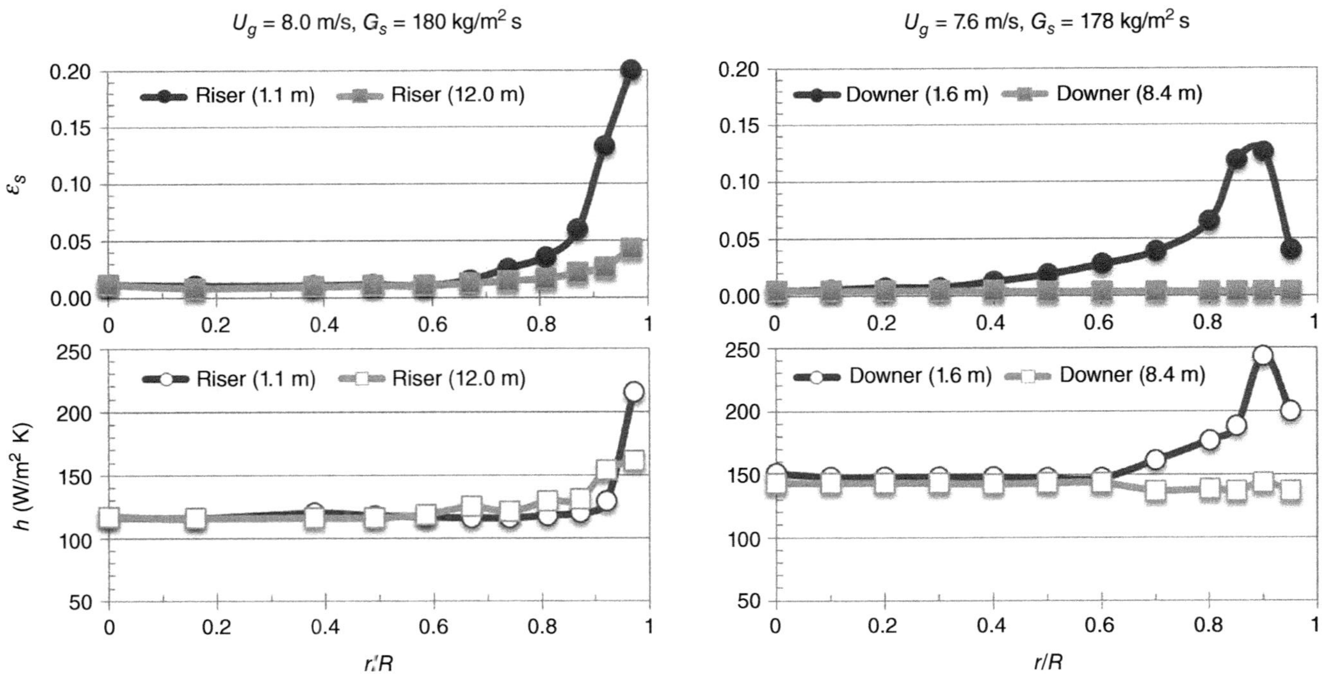

Figure 20.7 Typical radial profiles of heat transfer coefficient and solid volume fraction in riser and downer reactors with the same 100 mm ID, at different distance to reactor inlet. Source: Ma and Zhu 2001 [6]. Reproduced with permission of John Wiley and Sons.

20.3 Modelling of Hydrodynamics and Reacting Flows in Downers

Theoretical modelling is helpful for scale-up, reactor design, and operation of downers. Different hydrodynamic models have been proposed, varying from simple one-phase to three-phase models and one-dimensional (1-D) to three-dimensional (3-D) space. These can be classified into three categories:

(a) Reaction engineering model (generally 1-D or 2-D), normally without consideration of the complex nature of multiphase flow.
(b) Eulerian–Eulerian computational fluid dynamics (CFD) model (see Chapter 6), where the solid and gas phases are treated as overlapping continua.
(c) Eulerian–Lagrangian CFD model (again see Chapter 6), where the path of each individual particle is calculated based on Newton's second law.

For solid-catalyzed processes, the solid (particle) phase consists of a heterogeneous catalyst for gas-phase reactions. The dispersed motion of the particle phase may lead to different pathways, and therefore residence times, along which the catalyst activity changes due to the micro-environment (e.g. temperature and concentration fields). In gas–solid non-catalytic processes, the particles act as a reactant material and/or heat carrier. For such cases, tracking each particle's history in a multiphase flow reactor is crucial to predict the macroscale reactor. From a theoretical point of view in modelling multiphase reactions, it is important to choose the most appropriate modelling approach.

Most of the literature on catalytic processes in downer reactors focuses on the fluid catalytic cracking (FCC) process, which converts low-value heavy hydrocarbons into a series of more valuable products such as gasoline and light olefinic compounds. The parallel–serial reaction kinetics are shown in Figure 20.8. Mainly due to coke deposition, the FCC catalyst activity drops rapidly, e.g. by 50% within about one second, and the catalyst particles must be regenerated by continuous combustion of the coke with air. The deactivation of the catalyst particles due to coke deposition is commonly included in FCC kinetic models. The catalyst activity in terms of time depends on the history of each particle's motion and the history of coke formation on each particle.

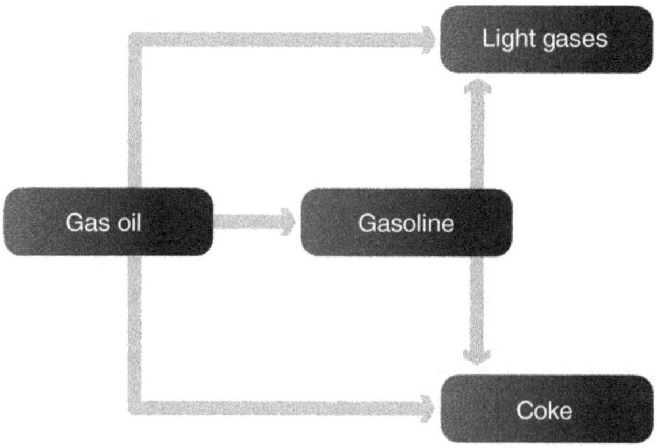

Figure 20.8 Simplified reaction scheme of FCC process: four-lump reaction kinetics.

20.3.1 Reaction Engineering Model

In a typical 1-D reaction engineering model for FCC reactors, the axial distributions of velocity, temperature, and volume fraction for each phase and concentration of individual chemical species are predicted via 1-D mass, energy, and chemical species balances (see the example reported by Bolkan-Kenny et al. [7]) by considering the conservation of materials and momentum and neglecting radial nonuniformity of flow structures. The distinct solid backmixing behaviour in risers and downers is often characterized by axial Péclet numbers, e.g. <10 for risers and >100 for downers. Based on this type of model, the overall performance of the reactors can be predicted.

A semi-empirical approach to improve the accuracy of the reaction engineering model is to refine the mass, energy, and chemical species balances by considering radial nonuniformity of flow structures. Two-dimensional (2-D) mass, energy, and chemical species balance equations are then solved to obtain axial and radial profiles of the two-phase reacting flow in FCC reactor, taking into account the changes of gas densities due to cracking reactions, and hence also the gas velocities along the reactor length. The 2-D flow field is determined by axial distributions of hydrodynamic characteristics and radial correlations of these characteristics. The axial distributions are then solved by a 1-D model, based on the conservation of materials and momentum, the same as for the 1-D reaction engineering model mentioned above. The empirical correlations describing the radial profiles of local hydrodynamic characteristics are from the published literature.

20.3.2 Eulerian–Eulerian Model

The Eulerian–Eulerian model is based on the conservation equations of mass, momentum, energy, and species for each phase of the fluid flow (two phase: gas and catalyst; three phase: gas, catalyst, and liquid droplets). These models are more suitable for theoretical studies of the physics of multiphase flow phenomena inside the reactors.

In the Eulerian–Eulerian scheme, a *gas turbulence–solid turbulence* model (k–ε–Θ–k_p) was developed, taking into account the fast and dense gas–solid flow in downers [8]. The model comprises a k–ε turbulence model for the gas phase, a k_p turbulence model, and a kinetic theory description of solid stresses characterized by granular temperature (Θ) for the solid phase. Due to the comprehensive consideration of gas–solid turbulent behaviour and the collisions between particles, the k–ε–Θ–k_p model is suitable for describing the gas–solid pipe flow over wide operating ranges and different flow sections along the downer. For the gas–solid flows simulated by the k–ε–Θ–k_p model, the predicted radial profiles of solid volume fraction and particle velocity in a downer of 92 mm in inner diameter (ID) gave very good agreement with experimental data for the solids jet process (see Figure 20.9a) [9]. The solid volume fraction in the developed region of a downer of 418 mm ID was also in reasonably good agreement with experimental data (see Figure 20.9b) [10]. Scale-up of the fully developed region of downers was also well captured by the

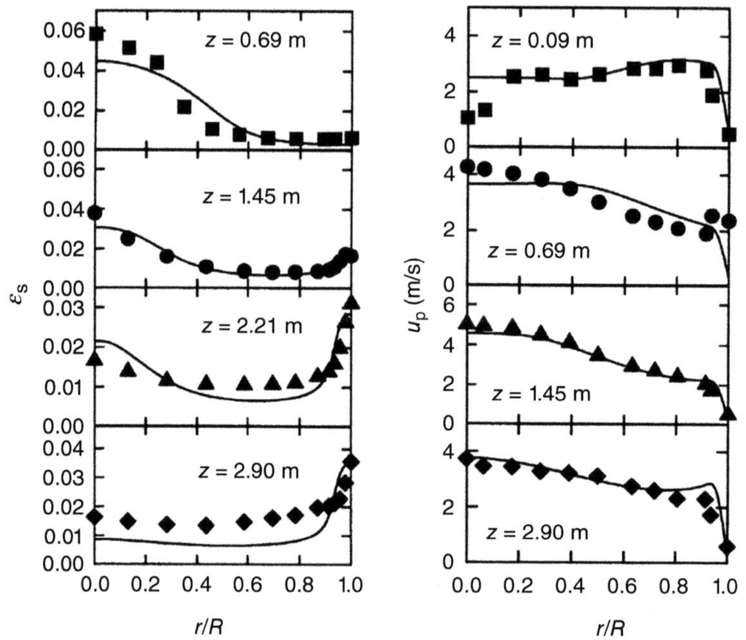

(a)

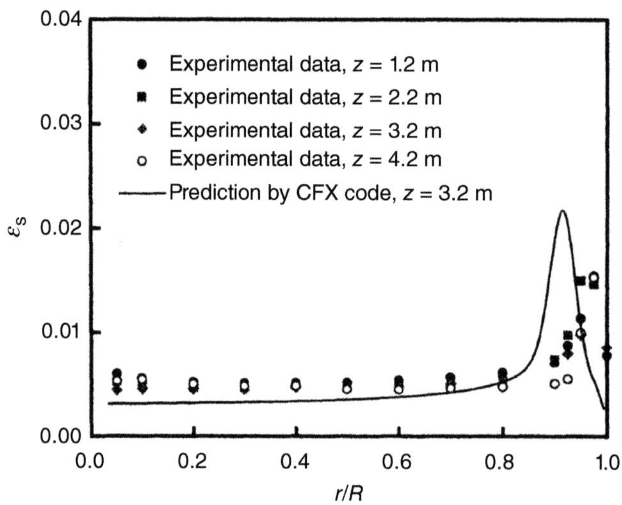

(b)

Figure 20.9 Comparison of model predictions and experimental data in downers: (a) radial profiles of solid volume fraction and particle velocity in the entrance region of a downer (92 mm in ID, 4 m in height, $U_g = 2.8$ m/s, $G_s = 47$ kg/m^2 s) and (b) radial profiles of solid volume fraction in developed region of a downer (418 mm in ID, 6 m in height, $U_g = 4.2$ m/s, $G_s = 44$ kg/m^2 s).

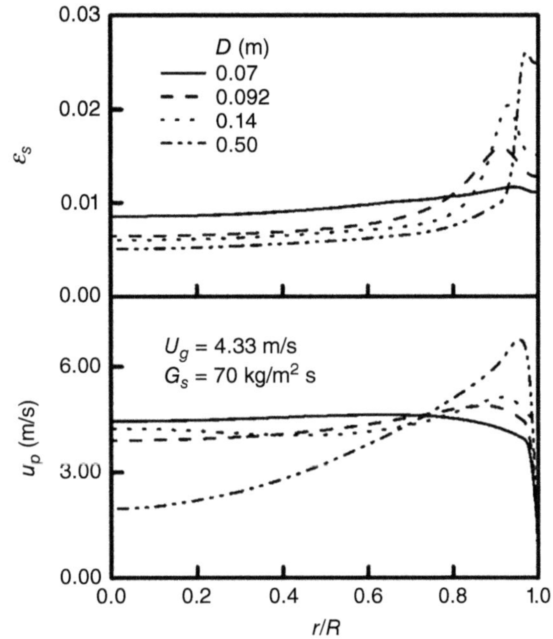

Figure 20.10 CFD predictions of scale-up effect of downers.

model. As shown in Figure 20.10, in smaller downers, e.g. <0.1 m ID, the radial solids distribution is relatively flat; the dense ring flow structure only appears in medium-size downers, such as 0.14 m ID; in much larger downers (e.g. >0.50 m ID), the solid volume fraction may show a monotonous trend along the radial direction, with a uniform core region and a very thin dense region at the wall.

20.3.3 Eulerian–Lagrangian Model

Unlike the Eulerian–Eulerian method, the Eulerian–Lagrangian model has clearer physics by considering the particle–particle and particle–fluid contact at particle scale using Newton's laws. A general modelling scheme is illustrated in Figure 20.11 for describing the chemically reacting flow problems with a continuous gas phase and a discrete particle phase.

In most fluidized beds, the interaction between particles must be considered to capture the major features of the underlying multiphase transport phenomena. The discrete element method (DEM), also called the soft-sphere model, is frequently applied in simulating gas–solid flows in fluidized beds. A three-equation linear dash-pot model for DEM was developed [11] taking full account of the particle–particle interactions at the microscale, including the normal contact force, tangential contact force (including static sliding friction force and dynamic sliding friction force), and the moment (including the moment generated by the tangential force, static rolling friction force, and rolling friction force). The gas–solid flows in a 2-D downer of width 100 mm and height 10 m (to ensure that the two-phase flow becomes fully developed) were simulated using the CFD–DEM method. The predicted macroscale flow structure agreed well with

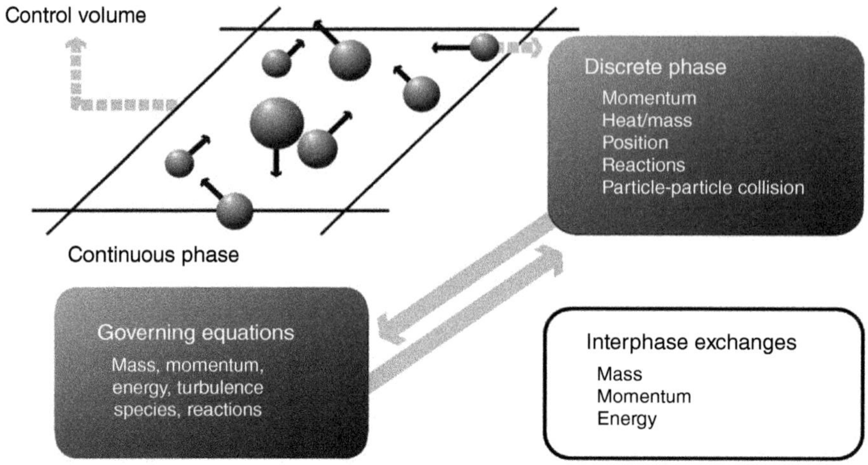

Figure 20.11 Schematic of Eulerian–Lagrangian model for simulation of reacting flows.

the experiments. Distinct clustering at the mesoscale was revealed throughout the downer. Figure 20.12 shows the flow development characterized by the particle distribution. In the 3–4 m section, particles start to form clusters. From the instantaneous state of the bed predicted by simulation, the clusters show anomalous behaviour and are often in the form of flocs. Compared with the Eulerian–Eulerian model, the CFD–DEM approach is better able to explain the basic physics of the gas–solid flow in downers, e.g. the unique flow structure, scale-up effect, and the influence of the wall and particle properties.

A novel numerical scheme, named computational particle fluid dynamics (CPFD), incorporating the multiphase particle-in-cell (MP-PIC) method was proposed [12] to describe the solid phase. This calculates the fluid phase using a Eulerian computational grid and the solid phase using Lagrangian computational particles, linking the two phases with a particle stress gradient term added to the equation of motion of the particles. In CPFD, a "numerical particle" is identified as an ensemble of particles with similar properties such as species, size, and density. The computational time is greatly reduced by providing a numerical approximation for the solid phase, similar to the numerical control volume provided for the fluid phase. The CPFD approach has been shown to be able to describe particle motions and particle distributions, also providing snapshots along the downer [13].

Under the Eulerian–Eulerian modelling scheme, the inherent assumption that the particle phase is a continuous medium imposes constraints to deep understanding of the FCC reaction system. One of the key constraints is that the history of the catalyst particles does not exist, making it difficult to estimate the catalyst activity. The inherent assumptions, i.e. neglecting the coke transportation by catalyst particles, obviously conflict with the nature of the gas–solid reacting flow with deactivated catalyst in the process (e.g. radial motion and concentration

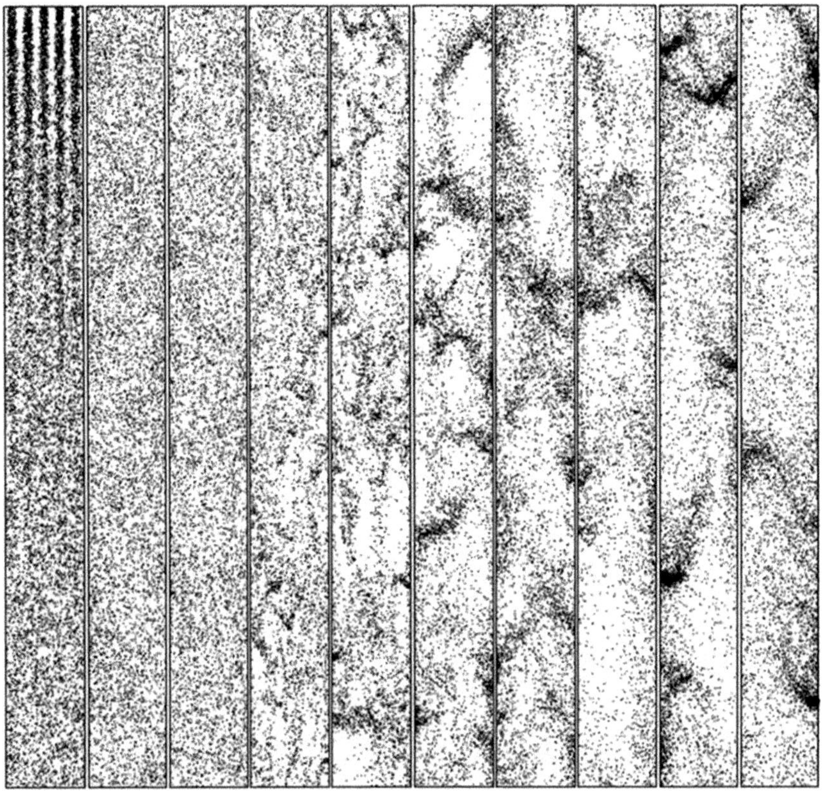

Figure 20.12 CFD–DEM simulation of hydrodynamics in downer.

gradient of the particles). The FCC reacting flow cannot be physically described by the Eulerian–Eulerian modelling scheme.

A Eulerian–Lagrangian scheme-based model was developed [14] to simulate the complex gas–solid reacting flows in FCC processes accommodated in 2-D risers or downer reactors (2 m in height and 50 mm in width). Considering solid-catalyzed gas-phase reactions, the model particularly incorporated the descriptions for heat transfer between particles and between gas and particles, instantaneous catalyst deactivation, and lumped kinetics in the gas phase for the FCC process, together with the governing equations for the hydrodynamics. By tracking the trajectory of each particle, the catalyst activity could be calculated from the residence time of the particle. Figure 20.13a shows the spatial distribution of particle residence time at certain time in the middle section of a riser operated at a catalyst-to-oil (CTO) ratio of 15. The residence time of the catalyst particles upstream (e.g. in the height of 0.98 m, near the left wall) is even larger than the one downstream (e.g. in the 1.32 m tall region near the centre), with a lower activity of 0.564 vs. 0.754 (see Figure 20.13b). This result can be explained

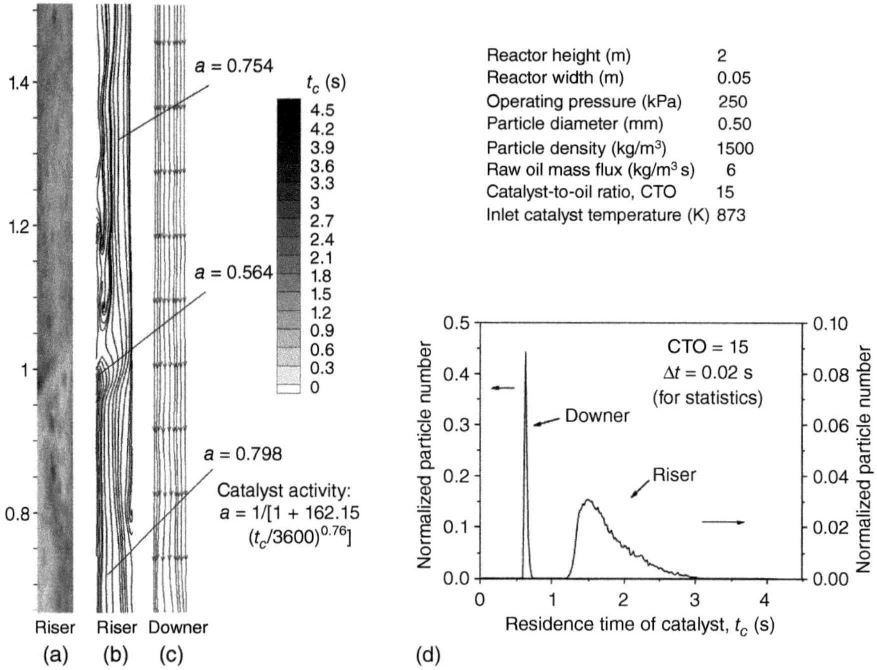

Figure 20.13 Comparison of reacting flow simulation results between riser and downer: (a) cell-averaged particle residence time in riser, (b) gas flow streamlines in riser, (c) gas flow streamlines in downer, and (d) predicted particle residence time distributions (RTDs).

by the formation and slow movement of clusters in the riser. The eddy at a height of 0.98 m corresponds to a cluster in the riser, while no large eddies are observed in the downer according to Figure 20.13c. Once particles are added to a cluster, they stay longer inside the reactor, resulting in the long tail of the RTD shown in Figure 20.13d.

20.4 Design and Applications of Downer Reactors

20.4.1 Inlet Design

The ultra-short contact times between phases in the whole downer impose stringent demands on the solids feeding and rapid gas–solid mixing at the downer inlet. In other words, the solids must be *rapidly* dispersed in the gas flow while having intimate contact with the gas. Otherwise, the non-ideal distribution of highly concentrated gas–solid flows in the initial stage would cause fast reaction rates at local positions, leading to worsening of the product distribution.

In principle, the inlet design of a downer should provide uniform distribution of phases, quick acceleration of solids, and excellent control of gas–solid mixing to make best use of the advantages of the near plug flow in downers. Table 20.1 summarizes representative inlet geometries, which can be classified into three main categories: inlet design with premixed gas–solid flows, inlet design with pre-distributed solids, and inlet design with jet nozzle(s).

Table 20.1 Representative inlet design for downers.

Categories	Inlet design	Key features
Inlet design with premixed gas–solid flows	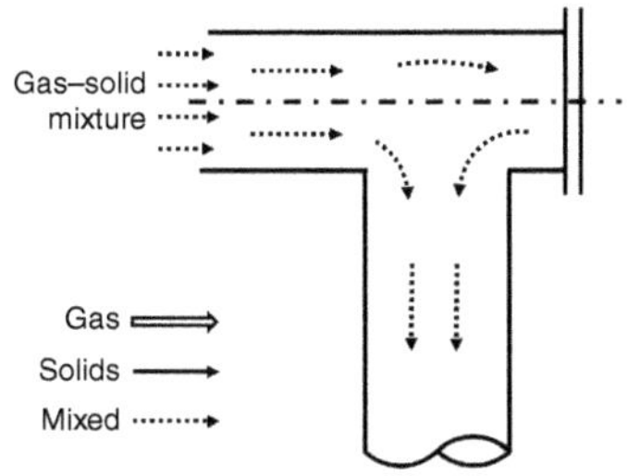	• Gas–solid mixture enters the downer directly through a 90° bend, or a smooth connection • Solids may not be well distributed
	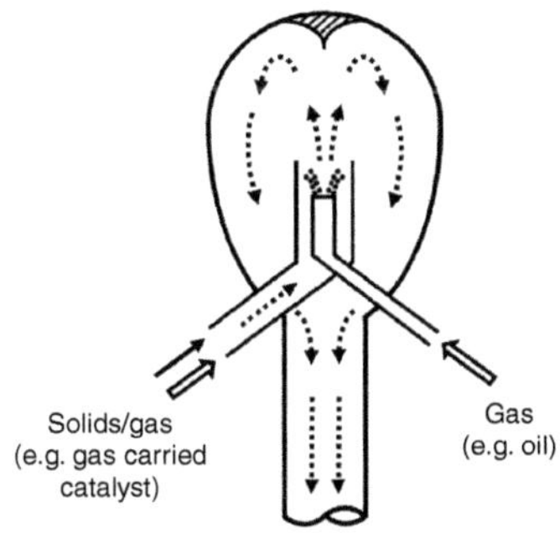	• Gas and solids move upwards, respectively, in two nested feeding pipes and then mix with each other in the top region containing the feeding pipes
Inlet design with pre-distributed solids	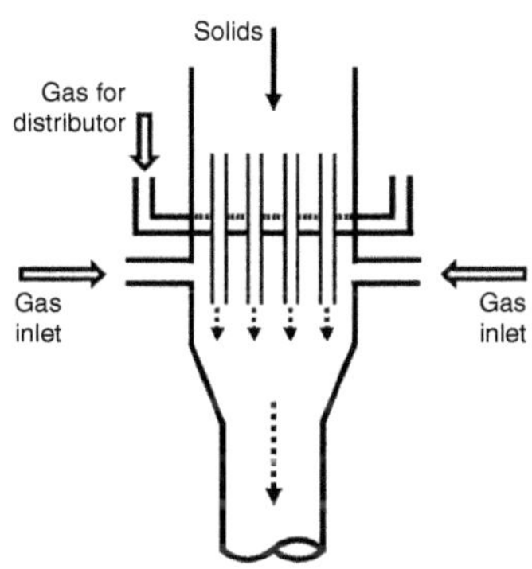	• Solids are fluidized above the downer inlet and flow through several tubes into the downer, while gas is introduced through the ring slots • Solid flow rate can be adjusted flexibly
	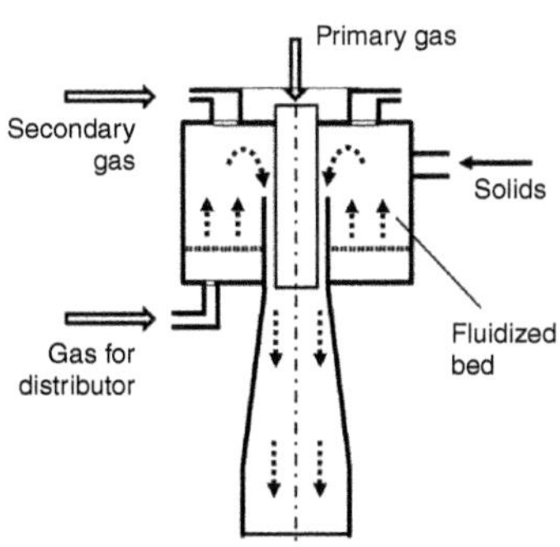	• The inlet structure consists of two concentric pipes in the centre. Gas is sucked through the inner tube for primary air, while solids are fed into the annular gap surrounding the primary air tube

(Continued)

Table 20.1 (Continued)

<table>
<tr><th>Categories</th><th>Inlet design</th><th>Key features</th></tr>
<tr><td>Inlet design with jet nozzle(s)</td><td>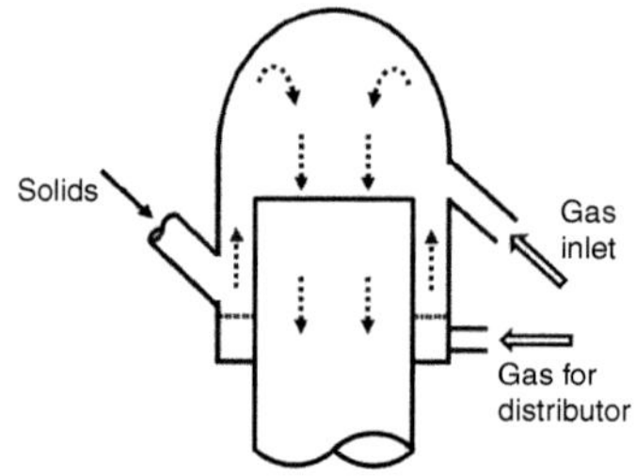
</td><td>• The inlet structure consists of an annular riser and a downer, connected smoothly at the top, by choosing an annular riser to obtain a more uniform flow
• Feedstock can be injected at different positions along the riser or at the top of the downer</td></tr>
<tr><td></td><td>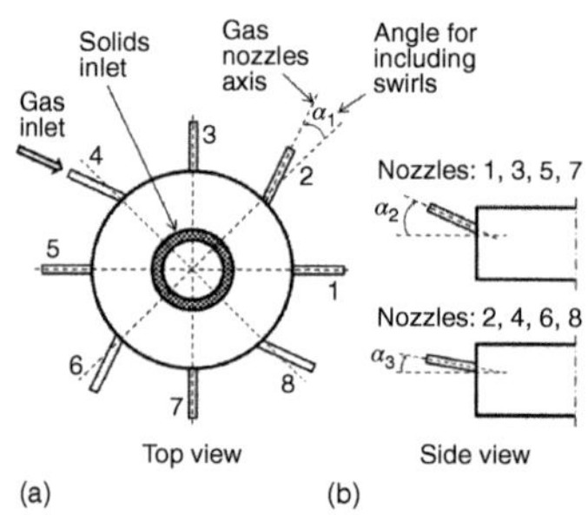
</td><td>• A mixing chamber equipped with eight jet nozzles, through which the gas enters the mixing chamber can be oriented independently
• Nozzles can be angled in the horizontal plane to induce swirl and inclined to contact the solids jets at different heights</td></tr>
<tr><td></td><td>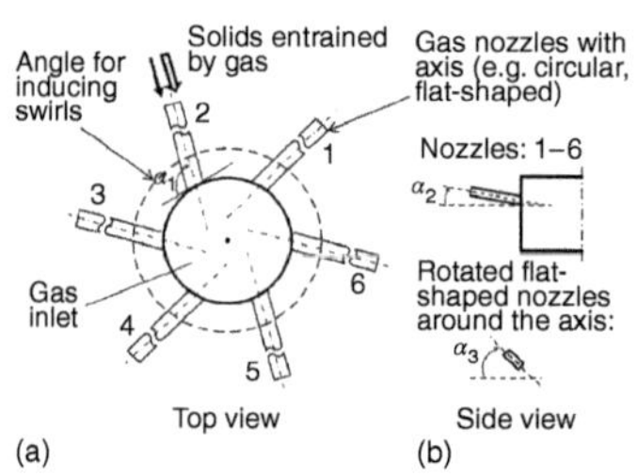
</td><td>• A mixing chamber equipped with six jet nozzles, through which the solids enter the chamber entrained by gas; the reacting gas enters the chamber vertically from the top of the reactor
• Nozzles can be angled in the horizontal plane to induce a swirl and inclined to contact the gas jet at different heights</td></tr>
<tr><td></td><td>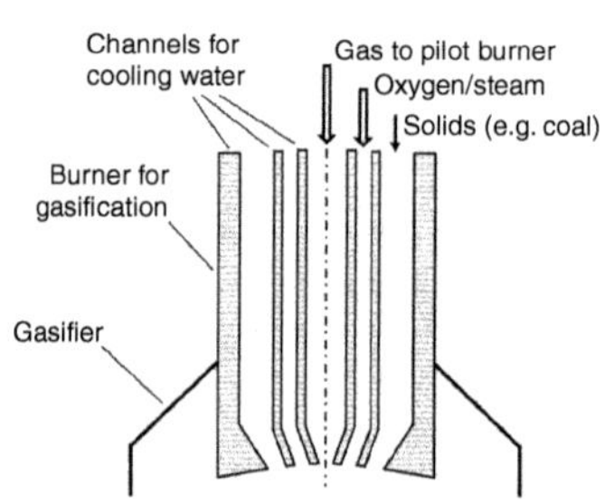
</td><td>• A burner for gasification typically consists of three concentric nozzles with the tip of the central nozzle recessed
• The central nozzle conveys fuel gas to pilot the burner while the two concentric nozzles supply oxygen/steam and solids (e.g. coal particles), respectively. The solids may be fed into the concentric nozzle by a carrier gas or liquid</td></tr>
</table>

20.4.2 Fast Separation of Gas and Solids at Downer Exit

The gas–solid separator at the downer outlet plays a similarly significant role to the downer inlet. In a typical gas–solid heterogeneous catalytic process, the short, uniform contact times between catalysts and gas require rapid separation of gas and solids at the outlet to avoid further unwanted reactions, e.g. excessive coke formation and over-cracking in FCC process. An alternative could be to cool the entire products flow very quickly (i.e. quenching) to stop the reaction almost instantaneously. Considering that the residence time in a downer is only a fraction of a second, a cyclone with a solids residence time of the order of one to two seconds would be unacceptable.

Representative designs of gas–solid fast separator for downers are illustrated in Figure 20.14. For the separator proposed in early literature (see Figure 20.14a) with a one-quarter turn, a separation time of 30 ms and an efficiency of 98% were claimed. For another early proposed gas–solid separator (see Figure 20.14b) with a severely size-constrained outlet in the bottom region and a gas outlet in the top region, a solids separation efficiency >96–99% from the main stream within 0.05 to 0.3 s was reported. The third typical fast separator design (Figure 20.14c), within which the solids leave the separator after one-quarter circular motion

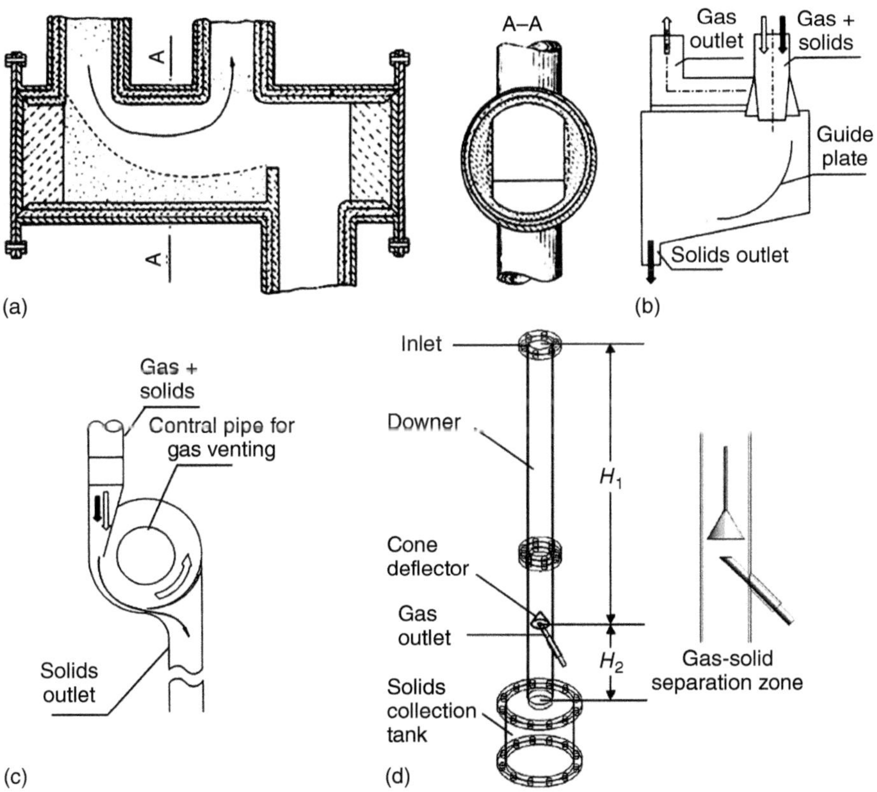

Figure 20.14 Representative outlet designs for downers.

through the bottom gap structure in the wall and gas, keeps spirally revolving around the central pipe for some revolutions until it is collected into the central pipe for backward gas venting, making the volume of the separator small, but the separation efficiency is very high, e.g. 98%. The last fast separator, shown in Figure 20.14d, mainly consists of flow deflector blades (i.e. swirl vanes) and a gas outlet pipe [15] to decrease the product vapor residence time and to reduce the severity of vapor over-cracking compared to other fast separation methods. Recovery of glass beads or FCC catalyst particles in air in cold tests was found to be very high, with a minimum measured separation efficiency of 98.5%.

20.4.3 High-Density Downer

Frequently, the reported solid volume fraction in the developed region of downers is <1%, while researchers prefer high-density operation with solids flux >200 kg/m^2 s or solid volume fraction >5% for enhanced interphase heat transfer and catalytic reaction. Specially designed downer systems have been reported to accomplish the high-density operations by Zhu's group at Western University (Canada) and the Institute of Process Engineering at the Chinese Academy of Sciences and FLOTU (both in China). The effective approaches to accomplish high-density operations mainly include pre-accelerating solid particles before the solids are fed into the downer inlet to increase the solids flux, utilizing a relatively low gas velocity to keep the terminal velocity of solid particles relatively low, and utilizing the full height of downcomer to obtain a high solids circulation driving force.

20.4.4 Downer–Riser Coupled Reactors

It has been observed that the quality of the product mixture from an FCC unit (e.g. selectivity to intermediate products such as gasoline and light olefins) can be improved if at least one of the products is cracked further. Based on the fact that the utility of simply recycling one/more of the cracked products back to the (single) riser reactor is limited due to unwanted secondary cracking under the same conditions, several special FCC processes utilizing different multiple reaction zones have been patented or tested on a commercial scale to achieve higher selectivity to gasoline and light olefins. These include a flexible dual-riser FCC process, a two-stage riser FCC process, and a modified FCC process for clean gasoline and propylene.

To take full advantage of riser and downer for the FCC process, there are two patented FCC techniques using downer-based coupled reactors. As the illustrated coupled riser-to-downer reactor (RTDR) in Figure 20.15a, feed oil first enters the riser, facilitating initial phase contact and good conversion of feed oil, ensured by the high suspension density of catalysts in the riser. Then the reaction mixture enters the downer, where favourable RTDs, near plug flow, and uniform flow structure are guaranteed. Accordingly, over-cracking is limited and selectivity is improved in the downer. This coupled reactor design also makes the operation very flexible. For example, the feed positions can be adjusted either along the riser or along the downer, and feeds can differ at different feeding locations.

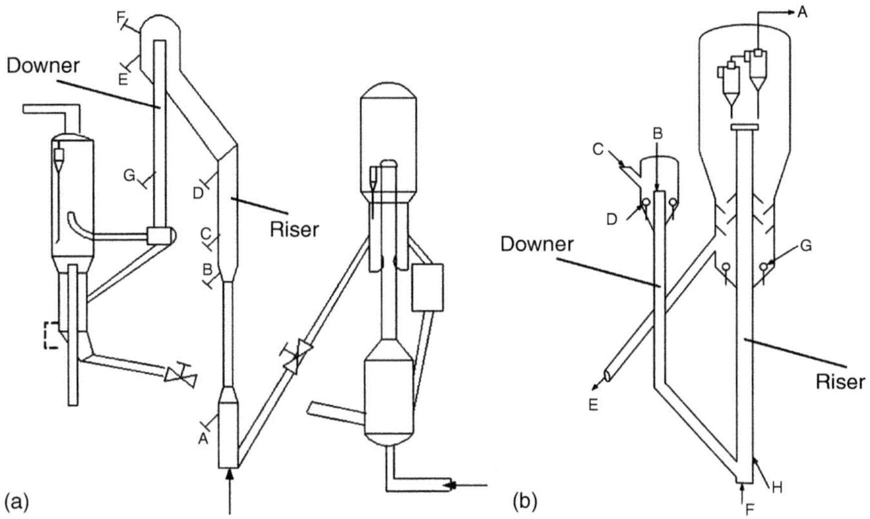

Figure 20.15 Schematic diagrams of coupled downer-to-riser reactors (DTRR) for FCC processes.

The RTDR design was demonstrated by FLOTU in the Ji'nan refinery (SINOPEC, China) at a throughput of 150 000 tonnes per annum (residue oil). As a comparison, for the illustrated coupled downer-to-rise reactor (DTRR) in Figure 20.15b, the feed oil first enters a downer reactor, undergoing high-severity operation (e.g. high temperature and/or high catalyst loading), but a short contact time between phases, where thermal cracking and catalytic cracking take place simultaneously. After a certain conversion of feed oil, the reacting flow transfers into a riser for further cracking reactions of the heavy components adsorbed on the catalysts. In the riser, a lower temperature and a longer residence time are favourable, especially to promote hydrogen shift reactions, reducing the olefin content of the gasoline product.

20.4.5 Application Case 1: FCC

Major contributions on hot downer experiments are from FLOTU and joint research from the Petroleum Energy Center (PEC) in Tokyo and King Fahd University of Petroleum and Minerals (KFUPM) in Dhahran. Common interest was focused on the integration of refining and petrochemicals for greater economic benefits from the process.

KFUPM's research on comparison of downer- and riser-based FCC processes at high-severity conditions in pilot plant experiments was summarized by Abul-Hamayel [16]. Features of the high-severity fluid catalytic cracking (HS-FCC) process include high reaction temperatures (>550 °C), downflow reactor, short contact time, and high CTO ratio. Depending on the operating mode, the HS-FCC doubles the amount of light olefins and, in another mode, provides three times more light olefins, accompanied by minimal loss of gasoline. The production of propylene is 2.1–3.6 times higher than in the conventional

Table 20.2 Comparison of DCC process in commercial riser and hot downer.

Operating conditions			
	Units	**Commercial riser**	**Downer experiment**
Feeding temperature	°C	380	350
Reaction temperature	°C	565	606
Operating pressure	MPa	0.08	0.008
Catalyst/oil ratio	kg/kg	10.87–11.80	37.54
Residence time	s	3.0–4.0	0.74
Material balance			
	Name	**Commercial riser**	**Downer experiment**
Feed (wt%)		100.00	100.00
Products (wt%)	1. Dry gas	11.03	4.26
	H_2S	0.20	Not tested
	H_2	0.29	0.13
	CH_4	3.39	0.91
	C_2H_6	2.14	0.36
	C_2H_4	5.01	2.86
	2. Liquefied petroleum gas (LPG)	39.77	42.97
	C_3H_8	3.27	0.89
	C_3H_6	19.54	23.25
	C_4H_{10}	2.84	3.46
	C_4H_8	14.12	15.37
	3. Gasoline	21.47	28.77
	4. Light gas oil (LGO)	19.23	17.82
	5. Coke	8.50	6.52
	6. Loss	0	3.67
	Sum	100.00	100.00

Source: Reprinted with permission from Deng et al. 2002 [17]. Copyright 2002, American Chemical Society.

FCC process. HS-FCC gasoline has a high octane number and contains more heavy fractions. The content of olefins in the HS-FCC gasoline dropped by 50–85 wt%, depending on the operating severity. By increasing the CTO ratio, the effects of operating at high reaction temperature (thermal cracking) are minimized.

Deep catalytic cracking (DCC) in a downer reactor (4.5 m height and 13 mm ID) was studied by Deng et al. [17]. The results showed that a downer can significantly improve the selectivity of desired products compared with a commercial riser (see Table 20.2) for the same feed and catalyst: the yields of propylene and gasoline increase by 3.71 and 7.30 wt%, respectively, while the yields of the dry gas and coke were reduced by 6.77 and 1.98 wt%.

20.4.6 Application Case 2: Gasification

Gasification processes convert fossil fuel-based or organic carbonaceous materials into carbon monoxide, hydrogen, and carbon dioxide by reacting feedstock (e.g. coal and/or biomass) at high temperatures (>700 °C) with a controlled amount of oxygen and/or steam. A typical gasification stages include pyrolysis, partial oxidation, and hydrogenation. The resulting gas mixture is called syngas and is itself a fuel. The energy derived from the syngas is considered to be renewable energy if the gasified feedstock was obtained from biomass.

Reactors for coal/biomass gasification can be grouped into three categories: moving bed gasifiers, fluidized bed gasifiers, and entrained flow gasifiers. Entrained flow gasifiers often operate with fuel and oxidants (e.g. oxygen and steam) in co-current downward flow, which serve as another kind of downer reactor. The residence time of solid particles in these processes is a few seconds. The feed is ground and wet-milled to a mean size of about 100 μm to allow good transport in the gas and a high gasification reaction rate. This type of gasifier operates at high temperature and pressure, resulting in rapid reactions at high throughput. The high temperatures also reduce the concentrations of tar, oil, and other liquids in the product gas.

Many kinds of entrained flow gasifiers have been developed for commercial use, such as the GE Texaco gasifier, Shell gasifier, Siemens gasifier, East China University of Science and Technology (ECUST) gasifier, and Tsinghua oxygen-staged entrained flow (OSEF) gasifier. The fine feed can be fed into a gasifier in either a dry or pumpable slurry form. The slurry feed is a simpler, but more reliable, although it introduces water into the reactor, with heat required to evaporate the water.

In a typical GE Texaco process for coal gasification, a coal/water slurry (60–65 wt%) and oxygen (>95%) were fed downwards into a single-stage, refractory-lined entrained flow reactor to undergo reaction at up to 1400 °C and 4 MPa, with a carbon conversion efficiency up to 99%. The slurry is pumped to a custom-designed injector mounted at the top of the gasifier. The raw syngas leaving the gasifier can be cooled by a radiant heat exchanger, a direct quench system, or a combination of the two. GE Texaco gasifiers have been operated commercially for about 50 years using a wide variety of feedstocks such as natural gas, heavy oil, coal, and petroleum coke.

20.4.7 Application Case 3: Coal Pyrolysis in Plasma

Coal pyrolysis to acetylene in thermal plasma provides a direct route to make chemicals, especially acetylene (C_2H_2) from coal resources [18]. When coal is heated to temperatures above 1800 K within milliseconds, a gaseous mixture (volatiles) will be released and soon converted to final products, with C_2H_2 as the principal hydrocarbon component and H_2, C_2H_4, CO, and CH_4 as by-products. A downer reactor with the highest severity operation and ultra-short solid residence time would promote coal pyrolysis in thermal plasma. The rapid coal pyrolysis process essentially consists of several steps: (i) formation of hydrogen plasma in a plasma generator, (ii) fast heating of coal powders in milliseconds to release volatile matter in coal in a downer reactor, and (iii) fast quenching of the reacting stream to prevent acetylene from decomposing into soot and hydrogen. Since the complex reaction process is operated at very high-severity conditions, high gas velocity, and residence time in milliseconds, the reactor design must meet great challenges such as the nozzle design for coal injection into the hot hydrogen stream and severe coking problems, endangering the continuous operation.

Rapid pyrolysis of coal in thermal plasmas has gained worldwide attention. One of the most promising pilot plant tests was the Avco arc-coal process using a prototype of 1 MW plasma reactor in the early 1980s, with an acetylene yield of about 35% and specific energy consumption at 10.5 kWh/kg C_2H_2 for water quenching conditions. Although further attempts to scale up and optimize were terminated for lack of commercial sponsorship, the coal pyrolysis process was proven to be technologically viable and economically favourable. Encouraging progress has been made in recent years in China [1, 19]. With collaboration of Tsinghua University and Xinjiang Tianye (Group) Co. Ltd., a major polyvinyl chloride (PVC) producer in China has made breakthroughs on a 2 MW industrial platform (e.g. 10.5 kWh/kg C_2H_2). The whole reactor units were then scaled up to the 5 MW level. So far, the whole process involving the 5 MW plasma generator, reactor, and separation units have been operated continuously and stably for more than 75 h per run.

Table 20.3 Summary of key features of riser and downer reactors.

	Riser	Downer
Flow direction	Co-current upwards	Co-current downwards
Hydrodynamics	Nonuniform Core–annular radial distribution	Uniform Core–annular, but less
Mixing/solids residence time	Relatively strong solids backmixing Bimodal RTD profile, generally with a long tail	Much less backmixing Narrow RTD
Heat transfer	Local heat transfer coefficient generally in the range of 100–150 W/m^2 K	Not significantly lower than that in risers
Scale-up	Already scaled up to commercial scale with unavoidable backflow due to clusters and streamers along the wall	Already scaled up to 600 mm in diameter in a commercial demonstration unit with much less scale-up effect
Applications	Commercial fast fluidization processes, such as, FCC, Fischer–Tropsch synthesis, combustion of coal/biomass/wastes/off-gases, gasification of coal and biomass, calcination, catalyst regeneration, flue gas dry scrubbing [20]	Ultra-short contact processes, such as high-severity FCC, gasification of coal and biomass, coal pyrolysis in plasma, catalytic pyrolysis and cracking of hydrocarbons, oxidation of $TiCl_4$

20.5 Conclusions and Outlook

Downers have favourable flow structures and operate essentially at plug flow. There are unique advantages for applications in ultra-short contact time processes. Downer reactors have been utilized for some attractive applications in energy and chemical industry, including oil refining, coal/biomass gasification, and coal pyrolysis. Key features of downer reactors are summarized in Table 20.3, together with a comparison with riser reactors.

Solved Problem

20.1 In a circular downer reactor of 10 m in length and 0.14 m in ID, spherical solid particles of density 1545 kg/m^3 and mean diameter 59 μm are fed at the downer inlet, with an initial solid volume fraction of 0.4, superficial gas velocity 6.14 m/s, gas density 1.29 kg/m^3, gas viscosity 1.8e – 5 kg/m s, and solids circulation rate of 100 kg/m^2 s. Assume that radial profiles of gas and solids flow over the cross section of the downer are flat. For the case where particle interactions are not considered, the drag coefficient, C_D, as a function of particle Reynolds number, $Re_p = \rho_g|\overline{u}_g - \overline{u}_p|\, d_p/\mu$ given by

$$C_D = \begin{cases} 24/Re_p, Re_p < 0.3 \\ 24(1 + 0.15Re_p^{0.687})/Re_p, 0.3 \leq Re_p < 500 \\ 0.44, Re_p \geq 500 \end{cases}$$

(a) Calculate the particle velocity and solid volume fraction at the downer outlet, where the particles have already reached the constant velocity section.

(b) Calculate the distance to downer inlet for completion of particle acceleration, and compare with the measured result, 6.0 m found experimentally.

Solution

(a) Applying Eq. (20.1) to the downer reactor with constant particle velocity in the fully developed flow region, we obtain the force balance equation:

$$\rho_p V_p g + \frac{1}{2} C_D \rho_g |\overline{u}_g - \overline{u}_p|(\overline{u}_g - \overline{u}_p)\frac{\pi d_p^2}{4} - \rho_g V_p g = 0$$

The first step is to calculate the dimensionless group $C_D Re_p^2$:

$$C_D Re_p^2 = \frac{4}{3}\frac{\rho_g(\rho_p - \rho_g) g d_p^3}{\mu^2}$$
$$= \frac{4}{3}\frac{1.29 \times (1545 - 1.29) \times 9.8 \times (5.9 \times 10^{-5})^3}{(1.8 \times 10^{-5})^2}$$
$$= 16.4942$$

Based on $Re_p = \frac{16.4942}{C_D Re_p} = \frac{16.4942}{24} > 0.3$ and $Re_p = \sqrt{\frac{16.4942}{0.44}} = 6.1226 <$ 500, we find
$C_D Re_p^2 = \frac{24}{Re_p}(1 + 0.15 Re_p^{0.687}) Re_p^2 = 16.4942$, i.e. $Re_p = \frac{16.4942}{24(1+0.15 Re_p^{0.687})}$
The above equation can be solved by iterating, with a starting value of 100. After six iterations, we obtain the following solution with better than 0.0001 accuracy:

$$Re_p = 0.6202$$

The second step is to calculate terminal particle velocity $\overline{v}_t$ based on $Re_p = \frac{\rho_g \left(\overline{v}_t - \frac{U_g}{1-\overline{\varepsilon}_s}\right) d_p}{\mu}$ and $\overline{\varepsilon}_s = \frac{G_s}{\rho_p \overline{v}_t}$, with the rearranged equation as follows:

$$v_t - \frac{U_g}{1 - \frac{G_s}{\rho_p \overline{u}_t}} = \frac{\mu Re_p}{\rho_g d_p}, \text{ i.e. } \overline{uv}_t = \frac{6.14}{1 - \frac{100}{1545\overline{u}_t}} + \frac{(1.8 \times 10^{-5}) \times 0.6202}{1.29 \times (5.9 \times 10^{-5})}$$

$$= \frac{6.14}{1 - \frac{100}{1545\overline{u}_t}} + 0.1467$$

The above equation can be solved by iteration with a starting value of 7. After three iterations, we find the following solution with accuracy better than 0.0001:

$$\overline{u}_t = 6.3499, \text{with } \overline{\varepsilon}_s = \frac{G_s}{\rho_p \overline{v}_t} = \frac{100}{1545 \times 6.3499} = 0.0102$$

Hence, particle velocity and solid volume fraction at the downer outlet are 6.35 m/s and 0.0102, respectively.

(b) Applying Eq. (20.1) to the downer reactor for calculating particle velocity at any time and distance to the inlet (z), we can rewrite Eq. (20.1) as a system of ordinary differential equations (ODEs) as follows:

$$\begin{cases} \rho_p V_p \frac{d\overline{u}_p}{dt} = \rho_p V_p g + \frac{1}{2} C_D \rho_g |\overline{u}_g - \overline{u}_p| (\overline{u}_g - \overline{u}_p) \frac{\pi d_p^2}{4} - \rho_g V_p g, \overline{u}_p|_{t=0} \\ \quad = \frac{G_s}{\rho_p (\overline{\varepsilon}_s|_{t=0})} \\ \frac{dz}{dt} = \overline{u}_p, z|_{t=0} = 0 \end{cases}$$

The above ODEs can be solved numerically by using function ode45 in Matlab:

[t, y] = ode45(@myfun, tspan, y_ini)

The time interval, tspan, is given as [0, 5] in units of seconds. y_ini represents the initial value of vector $(h, \overline{u}_p)$, and myfun represents coded function of the ODEs. The returned matrix y has 485 rows and 2 columns, containing the first column vector as different distances to the inlet and the second one for particle velocity.

The particle velocity increases rapidly in the inlet region in the downer and reaches a constant value at $h = 0.7$ m. The calculated distance to downer inlet for completion of particle acceleration is much shorter than the 6.0 m measured result, which means that particle interactions in the downer reactor cannot be neglected.

Notations

A	cross-sectional area (m^2)
a	catalyst activity (–)
c	normalized tracer concentration (s^{-1})
C_D	drag coefficient (–)
D	column diameter (m)
d_p	particle diameter (m)
F	axis represents fluid flow direction (–)
F	force (N)
G	mass flux (kg/m^2 s)
g	acceleration of gravity (m/s^2)
H	bed height (m)
h	heat transfer coefficient (W/m^2 K)
ID	inner diameter (–)
k	turbulent kinetic energy (m^2/s^2)
P	pressure (Pa)
Pe	Péclet number (–)
Re	Reynolds number (–)
R	column radius (m)
r	radial coordinate (m)
S	axis represents solid flow direction (–)
t	time (s)
t_c	catalyst residence time (s)
U	superficial velocity (m/s)
u	local velocity (m/s)
V	volume (m^3)
z	distance to reactor inlet (m)

Greek Letters

ε	phase volume fraction (–)
ε	turbulent energy dissipation rate (m^2/s^3)
Θ	granular temperature (m^2/s^3)
μ	dynamic viscosity (kg/m s)
ρ	density (kg/m^3)

Subscripts

C	catalyst
D	drag force
G	gravity
g	gas

p	(solid) particle
s	solid
sl	slip
t	terminal settling

Superscripts

$\bar{}$	spatial average

Abbreviations

1-D	one-dimensional
2-D	two-dimensional
3-D	three-dimensional
Avco	Avco Corporation
CFD	computational fluid dynamics
CPFD	computational particle fluid dynamics
CT	computed tomography
CTO	catalyst-to-oil
DCC	deep catalytic cracking
DEM	discrete element method
DTRR	downer-to-rise reactor
ECUST	East China University of Science and Technology
FCC	fluid catalytic cracking
FLOTU	Fluidization Lab of Tsinghua University
GE	General Electric Company
HS-FCC	high-severity fluid catalytic cracking process
KFUPM	King Fahd University of Petroleum and Minerals in Dhahran
MP-PIC	multiphase particle-in-cell method
OSEF	oxygen-staged entrained flow
PEC	Petroleum Energy Center in Tokyo
PVC	polyvinyl chloride
RTD(s)	residence time distribution(s)
RTDR	riser-to-downer reactor

References

1 Cheng, Y., Wu, C.N., Zhu, J.X. et al. (2008). Downer reactor: from fundamental study to industrial application. *Powder Technol.* 183 (3): 364–384.

2 Wang, Z.W., Bai, D., and Jin, Y. (1992). Hydrodynamics of cocurrent downflow circulating fluidized bed (CDCFB). *Powder Technol.* 70 (3): 271–275.

3 Zhu, J.X., Jin, Y., Yu, Z.Q. et al. (1995). Cocurrent downflow circulating fluidized bed (downer) reactors – a state of the art review. *Can. J. Chem. Eng.* 73 (5): 662–677.

4 Lehner, P. and Wirth, K.E. (1999). Characterization of the flow pattern in a downer reactor. *Chem. Eng. Sci.* 54 (22): 5471–5483.

5 Wei, F. and Zhu, J.X. (1996). Effect of flow direction on axial solid dispersion in gas solids cocurrent upflow and downflow systems. *Chem. Eng. J.* 64 (3): 345–352.
6 Ma, Y.L. and Zhu, J. (2001). Heat transfer in the downer and the riser of a circulating fluidized bed – a comparative study. *Chem. Eng. Technol.* 24 (1): 85–90.
7 Bolkan-Kenny, Y.G., Pugsley, T.S., and Berruti, F. (1994). Computer simulation of the performance of fluid catalytic cracking risers and downers. *Ind. Eng. Chem. Res.* 33 (12): 3043–3052.
8 Cheng, Y., Guo, Y.C., Wei, F. et al. (1999). Modeling the hydrodynamics of downer reactors based on kinetic theory. *Chem. Eng. Sci.* 54 (13–14): 2019–2027.
9 Cheng, Y., Wei, F., Guo, Y.C., and Jin, Y. (2001). CFD simulation of hydrodynamics in the entrance region of a downer. *Chem. Eng. Sci.* 56 (4): 1687–1696.
10 Zheng, Y., Cheng, Y., Wei, F., and Jin, Y. (2002). CFD simulation of hydrodynamics in downer reactors. *Chem. Eng. Comm.* 189 (12): 1598–1610.
11 Zhao, Y.Z., Cheng, Y., Ding, Y.L., and Jin, Y. (2007). Understanding the hydrodynamics in a 2D downer by CFD–DEM simulation. In: *Proceedings of 12th International Conference on Fluidization* (eds. F. Berruti, X.T. Bi and D. Puglsey), 855–862. Canada: Vancouver.
12 Snider, D.M. (2007). Three fundamental granular flow experiments and CPFD predictions. *Powder Technol.* 176 (1): 36–46.
13 Abbasi, A., Islam, M.A., Ege, P.E., and de Lasa, H.I. (2013). CPFD flow pattern simulation in downer reactors. *AIChE J.* 59 (5): 1635–1647.
14 Wu, C.N., Cheng, Y., Ding, Y.L., and Jin, Y. (2010). CFD–DEM simulation of gas-solid reacting flows in fluid catalytic cracking (FCC) process. *Chem. Eng. Sci.* 65 (1): 542–549.
15 Yu, X., Makkawi, Y., Ocone, R. et al. (2014). A CFD study of biomass pyrolysis in a downer reactor equipped with a novel gas-solid separator – I: hydrodynamic performance. *Fuel Process. Technol.* 126: 366–382.
16 Abul-Hamayel, M.A. (2004). Comparison of downer and riser based fluid catalytic cracking process at high severity condition: a pilot plant study. *Petrol. Sci. Technol.* 22 (5–6): 475–490.
17 Deng, R.S., Wei, F., Jin, Y. et al. (2002). Experimental study of the deep catalytic cracking process in a downer reactor. *Ind. Eng. Chem. Res.* 41 (24): 6015–6019.
18 Bond, R.L., Galbraith, I.F., Ladner, W.R., and McConnell, G.I.T. (1963). Production of acetylene from coal, using a plasma jet. *Nature* 200 (4913): 1313–1314.
19 Yan, B.H., Xu, P.C., Guo, C.Y. et al. (2012). Experimental study on coal pyrolysis to acetylene in thermal plasma reactors. *Chem. Eng. J.* 207–208: 109–116.
20 Grace, J.R., Bi, H., and Golriz, M. (2003). Circulating fluidized beds, Chapter 19. In: *Handbook of Fluidization and Fluid-Particle Systems* (ed. W.-C. Yang). New York: CRC.

Problems

20.1 Take the same conditions as listed in Solved Problem 20.1, and consider that particle interactions can be described by adding a correction factor for drag coefficient, $\frac{14.1(1+2.78\ \rho_g U_g/G_s)}{U_g/\sqrt{gd_p}}$.

(a) Calculate the particle velocity and solid volume fraction at the downer outlet, where the particles have already reached the constant velocity section.

(b) Calculate the distance to downer inlet for completion of particle acceleration.

Solution

(a) Applying the force balance equation of particles listed in the Solution for Solved Problem 20.1 and the correction for gas-particle drag coefficient, $\frac{14.1(1+2.78\ \rho_g U_g/G_s)}{U_g/\sqrt{gd_p}} = 0.0674$, we can recalculate the dimensionless group $C_D Re_p^2$:

$$C_D Re_p^2 = \frac{16.4942}{0.0674} = 244.7$$

Based on the facts $Re_p = \frac{244.7211}{24} > 0.3$ and $Re_p = \sqrt{\frac{244.7}{0.44}} = 23.5836 <$ 500, we get
$C_D Re_p^2 = \frac{24}{Re_p}(1 + 0.15 Re_p^{0.687}) Re_p^2 = 244.7$, i.e. $Re_p = \frac{244.7}{24(1+0.15 Re_p^{0.687})}$
The above equation can be solved by using iteration with a starting value of 100. After 10 iterations, we obtain the following solution with accuracy better than 0.0001.

$$Re_p = 6.5881$$

The next step is to calculate the terminal particle velocity $\overline{v}_t$ as follows:

$$\overline{u}_t - \frac{U_g}{1 - \frac{G_s}{\rho_p \overline{u}_t}} = \frac{\mu Re_p}{\rho_g d_p}, \text{ i.e. } \overline{u}_t = \frac{6.14}{1 - \frac{100}{1545\overline{u}_t}} + \frac{(1.8 \times 10^{-5}) \times 6.5881}{1.29 \times (5.9 \times 10^{-5})}$$

$$= \frac{6.14}{1 - \frac{100}{1545\overline{u}_t}} + 1.5881$$

The above equation can be solved by using iteration with a starting value of 7. After three iterations, we can obtain the following solution with accuracy better than 0.0001.

$$\overline{u}_t = 7.7796, \text{with } \overline{\varepsilon}_s = \frac{G_s}{\rho_p \overline{u}_t} = \frac{100}{1545 \times 7.7796} = 0.0083$$

Hence, particle velocity and solid volume fraction at the downer outlet are 7.78 m/s and 0.0083, respectively.

(b) Applying the same numerical algorithm for solving ODEs listed in Solution for Solved Problem 20.1, with the correction factor for gas-particle drag coefficient, $\frac{14.1(1+2.78\,\rho_g U_g/G_s)}{U_g/\sqrt{gd_p}} = 0.0674$, we can recalculate the distance to the downer inlet for completion of particle acceleration, i.e. 6.9 m. This predicted distance is close to the measured one, i.e. 6.0 m.

21

Spouted (and Spout-Fluid) Beds

Norman Epstein

University of British Columbia, Department of Chemical and Biological Engineering, 2360 East Mall, Vancouver, Canada V6T 1Z3

21.1 Introduction

The wet summer of 1952 in western Canada prompted the development of a compact high-capacity wheat dryer in the laboratories of the National Research Council of Canada [1]. The aim was to improve on both existing batch dryers, with their stagnant beds of grain, and existing continuous driers, with their absence of vigorous grain mixing. Air fluidization of wheat particles (3.2 mm × 6.4 mm) was attempted first, but resulted in unacceptable slugging. Progressive attempts to remedy this problem by reducing the area of the gas inlet failed, until it was reduced to the size of a single small orifice. The circulation pattern of the solids and distribution pattern of the gas differed completely from those of a fluidized bed. The detailed development story of this "new" fluid-solids contacting technique, labelled a *spouted bed*, has been told by Gishler [2].

Figure 21.1 illustrates a spouted bed schematically. Fluid, usually a gas, is injected vertically through a centrally located small opening at the base of a conical, cylindrical, or conical–cylindrical (as in Figure 21.1) column containing relatively coarse solid particles, e.g. $d_p \gtrsim 1$ mm. If the fluid injection rate is high enough, the resulting jet causes a stream of particles to rise rapidly through a hollowed central core, or *spout*, within the bed of solids. These particles, after rising to a height above the surface of the surrounding packed bed, or *annulus*, rain back as a *fountain* onto the annulus, where they slowly move downwards and, to some extent, inwards, as a loosely packed bed. Fluid from the spout leaks into the annulus and percolates upwards through the moving packed solids there. These solids are re-entrained into the spout over the entire dense bed height. The overall system thereby includes a centrally located dilute-phase co-current upward transport region and a surrounding dense-phase downwardly moving packed bed through which fluid percolates counter-currently. A systematic cyclic pattern of solids movement is thus established, with effective contact between fluid and solids and with unique hydrodynamics [3].

The spouted bed flow regime, which occurs over a limited range of fluid velocity, is bracketed by fixed packed bed (i.e. static bed) operation at lower velocities,

Essentials of Fluidization Technology, First Edition. Edited by John R. Grace, Xiaotao Bi, and Naoko Ellis.

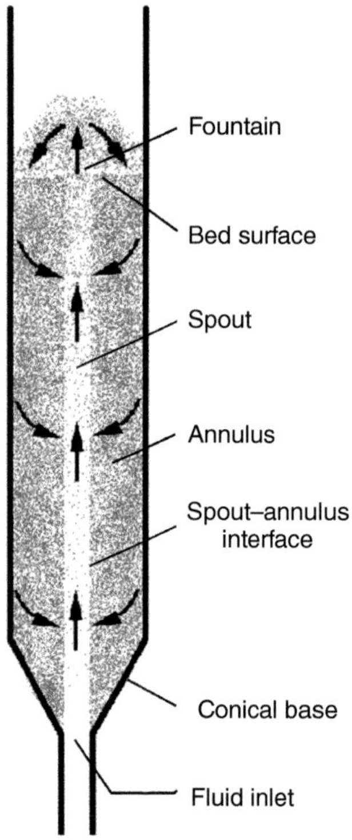

Figure 21.1 Schematic diagram of a spouted bed. Arrows indicate direction of solids movement.

by bubbling fluidization at higher velocities and higher bed depths, and by slugging fluidization at still higher velocities at all bed depths. An example of a flow regime map, represented by a plot of initial static bed depth vs. superficial fluid velocity (based on internal cylinder diameter) is shown in Figure 21.2. The demarcation line obtained by decreasing the fluid velocity until the spout collapses to give a loose-packed static bed represents the superficial *minimum spouting velocity*, U_{ms}, at various bed depths. The horizontal transition line separating spouting and bubbling represents the *maximum spoutable bed height*, H_m, for the given system.

21.2 Hydrodynamics

21.2.1 Constraints on Fluid Inlet Diameter and Cone Angle

Above some critical value of the inlet nozzle-to-column diameter ratio, D_i/D, there is no spouting regime. Instead, the bed changes directly from the fixed to the aggregatively fluidized state with increasing fluid velocity. This critical value is well approximated at $H = H_m$ by $(U_{mf}/v_t)^{1/2}$ or $\varepsilon_{mf}^{n/2}$ [6]. The same critical value can be safely applied at $H < H_m$. Since for smooth uniform-size spheres, $\varepsilon_{mf} = 0.415$, and the Richardson–Zaki [7] index n (see Chapter 3) increases from values as low as 2.4 for coarse spheres ($d_p \gtrsim 1$ mm) to values as high as 4.8 for finer particles ($d_p \ll 1$ mm), $(D_i/D)_{crit}$, decreases from 0.35 for gas spouting of coarse particles

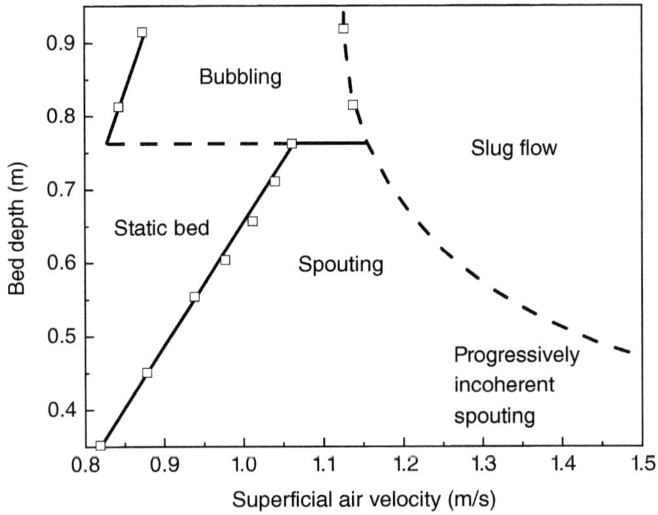

Figure 21.2 Flow regime map for wheat particles (prolate spheroids: 3.2 mm × 6.4 mm, $\rho_p = 1376\ \text{kg/m}^3$), $D = 152.4$ mm, $D_i = 12.5$ mm. Fluid is air. Note that the bubbling zone is accessible by raising the bed depth in the spouting zone or by lowering the gas velocity in the slugging region at $H > H_m$, but not by increasing the bed depth in the static bed zone. Source: Epstein and Grace 1997 [4] and Mather and Gishler 1955 [5].

to 0.1 for fine particles [8, chapter 3]. Increasing particle sphericity increases this critical ratio due to the accompanying increase of ε_{mf}.

Another critical diameter ratio is D_i/d_p, which must not exceed about 25 [9] if stable non-pulsatile spouting is to be achieved. The included angle of the *conical base* is a less critical parameter and needs only to exceed about 35° for stable spouting and to be smaller than $\pi - 2\alpha$, where α is the angle of repose of the solids, if dead solids at the bottom are to be avoided. The latter condition is well satisfied by $\theta = 90°$.

21.2.2 Minimum Spouting Velocity

For cylindrical columns up to about 0.5 m in diameter, with or without a conical base and with static bed heights exceeding both D and the cone height, the empirical but dimensionally consistent Mathur–Gishler [3, 5] equation,

$$U_{ms} = \left(\frac{d_p}{D}\right)\left(\frac{D_i}{D}\right)^{1/3}\sqrt{\frac{2gH(\rho_p - \rho_f)}{\rho_f}} \tag{21.1}$$

continues to be the most reliable predictor available for the superficial velocity at minimum spouting, U_{ms}, within the column for a wide variety of solids, bed dimensions, nozzle diameters, and both ambient air (within ±10%) and water (within ±15%) as fluids. Here d_p is the arithmetic average of bracketing screen apertures for closely sized near-spherical particles and as the Sauter mean diameter for mixed sizes, using the equi-volume sphere diameter for nonspherical particles. An exception is the case of particles such as prolate spheroids (e.g.

wheat) that align themselves vertically in the spout, for which prediction by Eq. (21.1) is best if d_p is taken as the horizontally projected diameter, i.e. the smaller of the two principal dimensions. The rest of this chapter is devoted to gas, rather than liquid, spouting, as explained in Section 21.5.

For spouting with air temperatures up to 580 °C, Eq. (21.1) was found to somewhat under-predict U_{ms}, primarily because the effect on U_{ms} of gas density ρ_f, which decreases with increasing temperature, is experimentally larger than shown by Eq. (21.1) [10]. The countervailing effect of gas viscosity, which is missing in Eq. (21.1) and which increases with increasing temperature, is insufficient to nullify the under-prediction. Modified dimensionless empirical equations applicable to these high temperatures are available elsewhere [8, chapter 3; 10].

For bed diameters exceeding 0.5 m, Eq. (21.1) increasingly underestimates U_{ms}. A rough working approximation for such diameters is that U_{ms} is 2D times the value given by Eq. (21.1), with D in metres [8, chapter 3].

For conical beds, including those in conical–cylindrical columns in which the static bed height H barely exceeds that of the conical base, and for flat-based columns in which $H \overline{<} D$, U_{ms} is no longer proportional to $H^{1/2}$ but approaches direct proportionality with H [8, chapter 5].

21.2.3 Maximum Spoutable Bed Height

Three distinct mechanisms that could cause conical–cylindrical or flat-base cylindrical spouted beds to become unstable beyond a certain initial bed height are (i) choking of the spout as particles from the annulus accumulate below the fountain, (ii) collapse of the increasingly incoherent spout–annulus interface, and (iii) fluidization of the upper annulus by the increasing annulus fluid flow at the upper bed levels [3]. It is the third mechanism that has yielded verifiable results.

Based on the third mechanism, the maximum value of U_{ms}, denoted by U_m, is obtained by substituting $H = H_m$ in Eq. (21.1) rewritten for that specific condition as

$$U_m = \left(\frac{d_p}{D}\right)\left(\frac{D_i}{D}\right)^{1/3}\sqrt{\frac{2gH_m(\rho_p - \rho_f)}{\rho_f}} \tag{21.2}$$

Empirically, U_m/U_{mf} can vary from 0.9 to 1.5, with a value of 1.1 found in spouting a large variety of solid materials with ambient air [11] and air temperatures up to 580 °C. If one utilizes the Wen and Yu [12] correlation, after Ergun [13], for the minimum fluidization velocity from Eq. (2.17), the result is

$$H_m = \frac{D^2}{d_p}\left(\frac{D}{D_i}\right)^{2/3}\frac{568}{Ar}(\sqrt{1 + 35.9 \times 10^{-6}Ar} - 1)^2(U_m/U_{mf})^2 \tag{21.3}$$

where $Ar = d_p{}^3(\rho_p - \rho_f)g\rho_f/\mu^2$.

For conical spouted beds, the same mechanism, viz fluidization of the upper annulus, is nonexistent, since any tendency for the increased volumetric flow fluid at the top of the annulus to increase the fluid velocity is counteracted by the increasing bed cross-sectional area with increasing height. Slugging may be caused by one or both of the other two mechanisms, choking of the spout or collapse of the spout wall, but this has not received rigorous quantification [8, chapter 5].

21.2.4 Fluid Flow in Annulus

Based on a force balance on a differential thickness dz of the annulus and on the assumption that Darcy's law applies to the vertical component of fluid flow through the downward moving loose-packed annulus, with the boundary condition that the annulus solids are incipiently fluidized at $z = H_m$, Mamuro and Hattori [14] derived the following equation for the superficial fluid velocity U_{az} in the annulus at height z in a cylindrical column:

$$\frac{U_{az}}{U_{mf}} = 1 - \left(1 - \frac{z}{H_m}\right)^3 \tag{21.4}$$

Although Eq. (21.4) was originally derived only for $H = H_m$, Grbavčić et al. [15] showed that for a given vessel geometry, solid material, and spouting fluid, U_{az} at any level z is independent of total bed height H. Thus, provided H_m is known, Eq. (21.4) can be used to predict U_{az} for all bed heights below H_m when operating at $U \simeq U_{ms}$. It has also been shown that, despite the fact that fluid velocities in the annulus are typically far in excess of those applicable to Darcy's law, U_a by Eq. (21.4) deviates by no more than 2% from U_{az} derived by the same force balance applied to the opposite extreme of inviscid flow [8, chapter 3].

Practical operations are usually carried out with U between $1.1U_{ms}$ and $1.5U_{ms}$. This results in a measurable decline in gas flow through the annulus, caused by increases in spout diameter and in annular solids downflow. Thus, by increasing U from U_{ms} to $1.1U_{ms}$, U_{aH_m} at $z = H = H_m$ was found to decrease from U_{mf} to $0.9U_{mf}$ [16]. Eq. (21.4) must then be modified by substituting $0.9U_{mf}$ for U_{mf}. Typical gas streamlines in the annulus are shown in Figure 21.3.

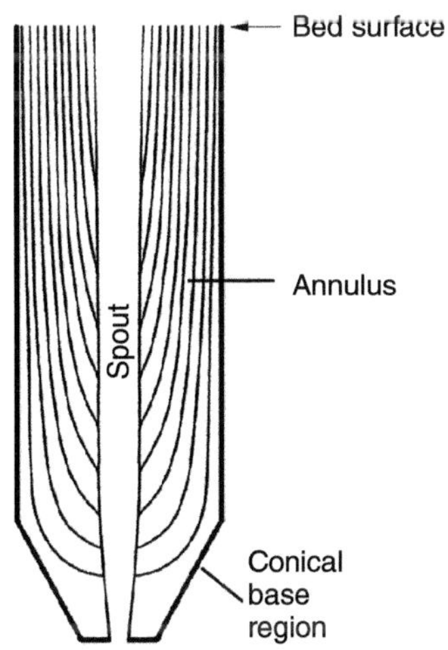

Figure 21.3 Calculated and tracer-observed gas streamlines for a 0.24 m diameter × 0.72 m deep bed of polystyrene pellets (d_p = 2.93 mm, D_i/D = 0.12), spouted at $U = 1.1U_{ms}$. Source: From Lim and Mathur 1976 [17] and Epstein and Grace 1997 [4].

21.2.5 Fluid Flow in Spout

Assuming a circular fluid inlet diameter D_i, the spout cross section at any level z is also circular, but on the way up, as observed on the transparent flat face of a semi-cylindrical half column, it follows various expansion and/or contraction contours, the shapes of which depend on D, d_p, and D_i [4]. At height z, continuity dictates that for a cylindrical column,

$$UD^2 = U_{sz}D_{sz}^2 + U_{az}(D^2 - D_{sz}^2) \tag{21.5}$$

where U_{az} is given by Eq. (21.4), or its possible modification if $U > U_{ms}$. For $z = H$, knowing D_{sH}, the spout superficial velocity U_{sH} entering the fountain can be determined by Eq. (21.5). Usually D_{sH} is approximately twice D_i [8, chapter 3].

21.2.6 Pressure Drop

Figure 21.4 shows typical plots of $-\Delta P$ vs. U for the transition from flow through a static packed bed to a spouted bed. Before arriving at the fully spouting condition labelled D, the bed passes through a peak pressure drop, $-\Delta P_m$, associated with the energy required by the fluid jet to disrupt the packing [8, chapter 2]. Unlike the curve for increasing gas flow, the decreasing gas flow trajectory is independent of the initial bed voidage, and point C' on this trajectory defines the minimum spouting velocity, U_{ms}. The pressure drop at this point remains constant as U is increased above U_{ms}. It is possible to avoid the pressure drop peak entirely by inserting vertically a pipe of inside diameter somewhat greater than that of the fluid inlet, over the inlet, next introducing the required solids and then the required fluid flow, before removing the pipe vertically.

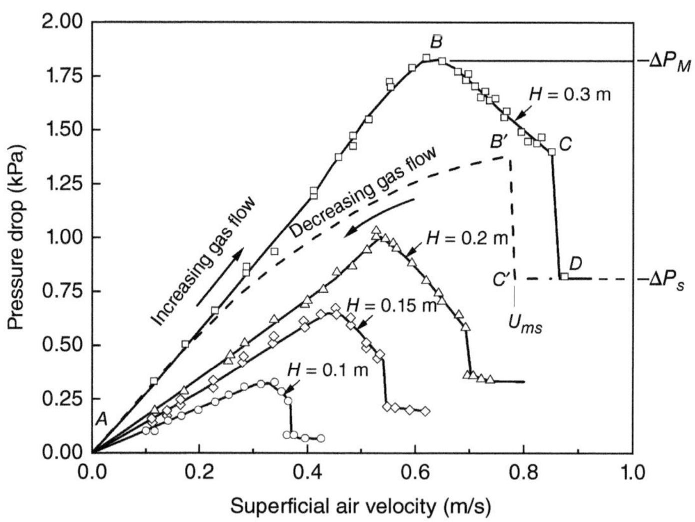

Figure 21.4 Typical pressure drop vs. flow rate curves for the onset of wheat spouting. $D = 152.4$ mm, $D_i = 12.7$ mm, $\theta = 60°$. Source: From Madonna et al. 1961 [18] and Mathur and Epstein 1974 [3].

Darcy's law, already applied in the derivation of Eq. (21.4), is used in conjunction with that equation to derive an expression for pressure drop across a spouted bed. Thus, for a vertical differential thickness dz of the annulus, the pressure drop $-dP$ is given by

$$-dP = KU_{az}dz = KU_{mf}\left[1-\left(1-\frac{z}{H_m}\right)^3\right]dz \tag{21.6}$$

where K is the Darcy or viscous flow constant. Integrating between the limits $z = 0, P = P_0$ and $z = H, P = P_H$, the result is

$$\frac{-\Delta P_{sp}}{-\Delta P_{fl}} = \frac{P_o - P_H}{KU_{mf}H} = \frac{3H}{2H_m} - \left(\frac{H}{H_m}\right)^2 + \frac{1}{4}\left(\frac{H}{H_m}\right)^3 \tag{21.7}$$

For a bed with $H = H_m$, the pressure drop for spouting is 75% of that for fluidization according to Eq. (21.7). The corresponding ratio for inviscid flow is 0.643 [8, chapter 3]. For $H = 0.5H_m$, $-\Delta P_{sp}/-\Delta P_{fl} = 0.531$, but for beds that are shallow, particularly with conical bases, the neglect of radial velocity gradients makes 0.75 a safer figure to use to estimate power consumption.

21.2.7 Behaviour of Solid Particles

The gas streamlines shown in Figure 21.3, when reversed in direction, represent quite accurately the streamlines for the flow of solids in the annulus region. Starting from $z = H$, where the fountain overflow meets the annulus, the particles descend in the loose-packed annulus of the cylindrical column, decreasing in their downward velocity linearly with decreasing z, until just above the height of the conical base. This occurs because the annular bed, as it moves downwards, is being progressively depleted of particles entering, and being entrained by, the spout.

At any cross section of the spout, the radially symmetric upward particle velocities vary approximately parabolically from zero at the spout–annulus interface to a maximum at the centre. The radial average particle velocity increases from zero at $z = 0$ to a maximum at $z \ll H$ and then gradually decreases with increasing z in the freeboard until it reaches zero again at the top of the fountain.

The voidage ε_a in the annulus at $U = U_{ms}$ is equal to ε_{mf}, slightly higher if U exceeds U_{ms} significantly, ε_{mf} being the void volume fraction of a random loose-packed bed (see Chapter 2). For uniform-size smooth spheres, ε_{mf} can be taken as 0.42, with its value increasing as the sphericity decreases. The voidage in the spout varies almost linearly with z from unity at $z = 0$ to ε_{mf} in the core of the fountain.

Because of solids cross-flow from the annulus into the spout over the entire annulus height and because of the shower effect of the fountain, a spouted bed is an excellent mixer when a single solid species is used. For a continuously fed spouted bed, assuming that the solids feed and discharge ports are located at opposite sides to avoid short circuiting, that the cone included angle is small enough to prevent dead solids at the base, and that the mean residence time of the solids exceeds a few minutes, almost perfect mixing of the solids occurs [3]. When more than one species of solids is used, e.g. particles of different sizes and/or densities, considerable segregation occurs, and special measures must be taken to eliminate it [4; 8, chapter 8].

21.3 Heat and Mass Transfer

Heat can be delivered to the cold particles of a spouted bed from (i) hot spouting fluid, or (ii) a heated column wall, or (iii) a heated surface immersed in the bed. Although a variety of empirical equations giving lumped heat transfer coefficients are available for each of these three situations [8, chapter 9], each presumably for a specific range of design and operating conditions, (i) and (iii) are best represented by a *two-region model* that treats the spout and annulus separately [3, 4]. In item (i), thermal equilibrium between the fluid and particles is readily achieved in the annulus, but not the spout. In all three cases, because spouted bed particles are usually relatively large (>1 mm) and non-metallic, intra-particle temperature gradients and heat transfer must also be accounted for.

For wall-to-bed (or vice versa) heat transfer, the theoretical two-dimensional Higbie penetration model [3] applied to the thermal boundary layer at the wall results in

$$h_w = 1.128\left[\frac{v_w \rho_b c_{pp} k_b}{(H-z)}\right]^{1/2} \tag{21.8}$$

where the heat transfer surface extends over a length of $H - z$. Equation (21.8) over-predicts h_w somewhat because it neglects the higher voidage at the wall than in the bulk of the moving packed bed annulus [3].

Particle-to-fluid mass transfer, which is especially important in spouted-bed particle drying [8, chapter 11], has, like its heat transfer analogue, also given rise to several empirical equations, in this case for lumped external mass transfer coefficients. These coefficients can be applied to the constant rate drying period, during which external mass transfer controls the drying rate. As it turns out, however, drying of materials, such as agricultural products and fertilizer granules, for which hot air-spouted beds have proven to be most effective, is usually carried out over ranges of particle moisture content within the falling rate period, so that internal diffusion controls the drying rate. Assuming thermal equilibrium between air and well-mixed solids and knowledge of the internal moisture diffusivity of the particles, drying rates of cereal grains have been accurately predicted for both batch and continuous operation [19].

21.4 Chemical Reaction

A two-region model, mentioned above, is essential in the analysis of chemical reaction in a spouted bed, whether the solids act as a catalyst [8, chapter 19] or as a reactant [8, chapter 15]. Since fluid elements in the central spout travel more quickly and with a much higher voidage, than in the annulus, it is important to distinguish between the two regions, reaction being more favourable in the annulus, where the fluid is in more intimate contact with the solids, than in the spout.

The simplest representation of the two regions undergoing chemical reaction is via a *one-dimensional model*, in which radial gradients are ignored within each region. Steady-state mass flow balances are written for a differential height dz of each region. These two equations each contain a term that keeps track of the upwardly changing flow rate and targeted chemical species composition, another involving the relevant chemical kinetics, and a third that describes the almost negligible inter-region diffusional mass transfer. The annulus equation also contains a term arising from the net outflow of fluid from the spout into the annulus. Plug flow of fluid is assumed in reach region and chemical-kinetic control of the targeted chemical species generation rate. Flow within the annulus is given by Eq. (21.4) and within the spout by Eq. (21.5).

A more sophisticated approach is the *streamtube model*. Vertical plug flow through the spout is again assumed, but flow in the annulus is treated as occurring in the streamtubes bounded by the streamlines such as those shown in Figure 21.3. Streamlines situated so that each streamtube carries an equal proportion of the fluid flow can be established by a vectorized form of the Ergun [13] equation for flow through a loose-packed bed. Plug flow is assumed in each streamtube and bulk flow from spout to annulus, but inter-region diffusional mass transfer is ignored.

The above two models were tested in experiments involving the decomposition of ozone in a 155 mm diameter spouted bed of catalyst pellets [20]. Conversions were calculated from the numerical solution of both models and compared with experimental results. Not surprisingly, the two models were in excellent agreement with each other for a first-order reaction, since theoretically they predict identical overall conversions [21], but they were also in excellent agreement with the experimental conversions. However, the streamtube model gives more accurate representation of the flow patterns and concentration profiles within the reactor, especially for columns of large diameter [4].

21.5 Spouting vs. Fluidization

In processes such as hot air drying of agricultural products, spouted beds accomplish for relatively coarse particles what gas fluidized beds achieve for finer particles. The cut-off between fine and coarse is an equi-volume sphere diameter of 0.5–1.0 mm. While it is possible to spout finer particles efficiently in a miniature spouted bed, the holdup and capacity of such a bed would be too small for practical purposes.

Apart from achieving favourable gas–solids contact, spouting is better than fluidization at agitating coarse particles with a gas causing substantive solid–solid contacts. This additional feature of a spouted bed is particularly applicable to mechanical operations such as comminution of solids and de-husking of grains.

Also, the systematic cyclic movement of the particles in a spouted bed and the almost perfect particle mixing result found by determining residence time distributions are an important feature of spouting of direct benefit for such processes as granulation, particle coating, and solids blending.

Spouting can be obtained with either a gas or a liquid as the jet fluid and even by gas–liquid combinations [8, chapter 20]. However, as a substitute for liquid fluidization, liquid spouting does not offer the same advantages as gas spouting, since, even with coarse particles, liquid fluidization is usually of good quality, without “bubbling,” slugging, or excessive aggregation. The systematic cyclic movement of the particles is retained by liquid-spouted beds, but apparently has not yet been turned to practical advantage. The fluid referred to in this chapter is therefore almost always a gas, usually air.

Table 21.1 from Cui and Grace [22] summarizes key differences between gas-fluidized and gas-spouted beds, including that shown in Eq. (21.7). The absence of a distributor and its associated solids-plugging problems is a distinct advantage of spouting over fluidization.

21.6 Spout-Fluid Beds

If, in addition to supplying a central spouting jet, auxiliary fluid is also supplied to the annulus region of a spouted bed through a flat horizontal distributor below a cylindrical column or a conical distributor for a conical–cylindrical column, the result is a *spout-fluid bed* [8, chapter 6]. Five flow regimes have been identified with progressive increases in fluid flow [23]:

1. **Fixed packed bed**: $U < U_{ms}$, $U + U_{aux} < U_{mf}$, where U and U_{aux} are the superficial spouting and superficial auxiliary fluid velocities, respectively.
2. **Spouting with aeration**: $U \geq U_{ms}$, $U + U_{aux} < U_{mf}$. Normal spouting continues, but additional fluid to the annulus enhances heat and mass transfer between the fluid and particles.
3. **Spout-fluidization:** $U \geq U_{ms}$, $U + U_{aux} > U_{mf}$. The bed fluidizes, but spout breakthrough continues unsteadily.
4. **Jet-in-fluidized bed**: Further increase in U_{aux} results in submergence of the spout jet in a bubbling fluidized bed.
5. **Slugging**: Still further increase in U_{aux} results in formation of slugs (see Chapter 8) occupying most of the bed cross-sectional area.

Although the total gas flow required for spout-fluidization of a given bed of solids exceeds that for either spouting or fluidization on its own, the additional gas serves to fluidize the annulus, thereby improving the particle circulation and intermixing, as well as the effectiveness of gas–solid contact (resulting in favourable heat transfer between gas and particles, wall, and submerged objects). Also, the maximum spoutable bed depth can be exceeded, restrictions with respect to particle size are eliminated [3], and it is more likely to be able to process cohesive or “sticky” particles. Consequently spout-fluid beds have been increasingly favoured for some applications.

Table 21.1 Significant differences between gas-spouted beds and gas-fluidized beds [21].

Aspect	Fluidized bed	Spouted bed
Mean particle size	~0.03–3 mm; usually <1 mm	~0.6–6 mm; usually >1 mm
Particle size distribution	Usually broad	Usually narrow
Pressure drop/height within the bed	96–100% of that needed to support the particles, i.e. of $\sim(\rho_p - \rho)g(1 - \varepsilon)$	Less than ~75% of that needed to support the weight of the particles, i.e. $<0.75 \times (\rho_p - \rho)g(1 - \varepsilon)$
Pressure drop across entry orifices	Usually 30–50% of that across the bed	As small as possible consistent with satisfying the other constraints, e.g. orifice diameter <25 mean particle diameter
Axial gradient of pressure	Virtually independent of height in the column	Varies with height
Temperature gradient	Very uniform temperature over entire fluidized bed region	Significant temperature gradients both axially and radially
Column geometry	Usually cylindrical columns	Usually diverging conical base with or without cylindrical portion above
Orifice diameter	No restriction	Must not exceed 25 mean particle diameters
Orifice number density and configuration	Large number of orifices (typically >100 per square meter); many geometries; no constriction needed at the entrance	Usually a single orifice. For multiple spouting, smaller number density of orifices (typically no more than ~10–50 per square meter); it is helpful to have a constriction at the entrance
Orifice orientation	Most often horizontal, downward facing, or oblique with downward component, but may face upwards	Always upward-facing
Orifice feed system	Windbox (sometimes call plenum chamber) feeds all orifices	Separate gas supply and control of each orifice
Gas motion	Less ordered; depends on flow regime and specific geometry	Outwards from the spout into the dense phase, except just above inlet

(Continued)

Table 21.1 (Continued)

Aspect	Fluidized bed	Spouted bed
Particle motion	Complex flow regimes and particle motion; region surrounding the gas entry orifices is usually fluidized with few particle–particle contacts	Systemic circulation patterns, up the spout and slowly downwards in the annulus. Annulus is in moving packed bed flow with substantial particle–particle contacts
Particle segregation	Very little segregation providing that particles are well fluidized	Significant segregation according to both size and density of particles
Attrition	Normally little except in cyclone or jet regions	Significant in spout region and in fountain above
Bed depth	Broad range of depths, e.g. from 0.1 to 20 m dense bed depth	More limited range of depths, usually between 0.2 and 2.0 m
Superficial gas velocity	Broad range, typically $(U - U_{mf}) = 0.2–10$ m/s	More limited range, typically $1.1–1.8U_{ms}$
Orifice velocity variation	Some variation with time across each individual orifice as bubbles form and leave the orifice or jet	Essentially steady, as flow is controlled to each hole from an upstream valve
Internals	Common in fluidized bed processes	Rare in spouted bed processes, except for some use of draft tubes and heat transfer coils

Source: Cui and Grace 2008 [22]. Reproduced with permission of Elsevier.

21.7 Non-conventional Spouted Beds

Conical, cylindrical, and especially conical–cylindrical spouted beds, described in earlier sections of this chapter, are commonly referred to as conventional spouted beds (CSBs). Many modifications and extensions of CSBs, in addition to the spout-fluid bed, have been recorded in the literature [3, 4, 8]. A few are briefly described here.

Since the H/D ratio must exceed unity for stable spouting [3], the large bed heights required for large column diameters give rise to excessive pressure drops. Also, the long times spent by particles in the annulus of a tall bed is a distinct disadvantage for processes such as particle drying, where excessive time in contact with the hot gas could thermally damage the particles. Consequently, spouted bed diameters exceeding 1 m are rare in practice, an exception being polymer blenders of approximately 3 m diameter [3]. Scale-up without these disadvantages can, however, be accomplished by a *multiple spouted bed*, shown schematically in Figure 21.5. Though the spouting fluid may come from a single manifold, the flow to each inlet nozzle must be controlled separately. Figure 21.5 does not show the dead stagnant solids that accumulate at the bottom of a flat base, as opposed to their absence in cone-based multi-spouting [8, chapter 17; 25].

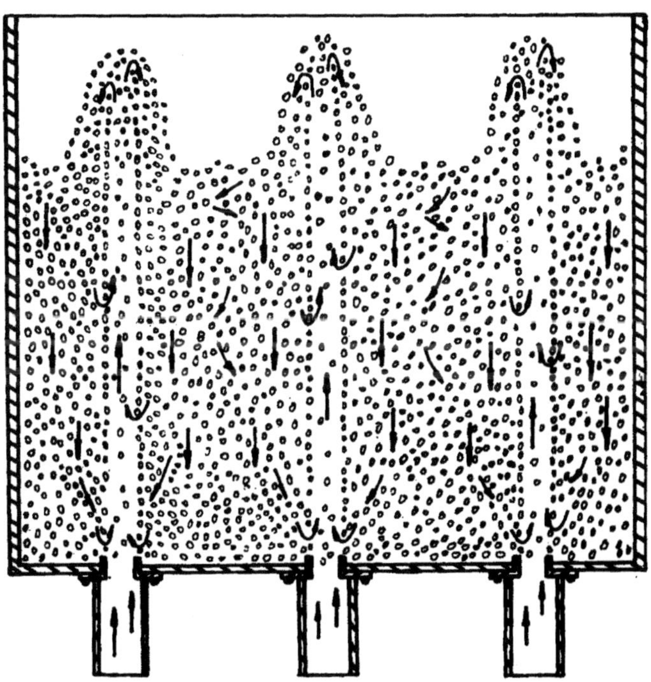

Figure 21.5 Schematic diagram of a multiple spouted bed. Source: From https://pascal-francis.inist.fr/vibad/index.php?action=getRecordDetail&idt=PASCAL7688006771. Licensed under CC BY 4.0 [24]. Reproduced with permission from Engineers Australia.

An alternative for carrying out multiple spouting is a multi-compartment *slot-rectangular spouted bed* (SRSB), a double compartment version that is illustrated in Figure 21.6. Originated in Leningrad [27] and applied to grain drying by Mujumdar [28] and coworkers, who described a SRSB as a two-dimensional spouted bed, three-dimensional effects exist [8, chapter 17] even when the slot length spans the entire thickness L of the column with its divergent base of height H_d. A separation panel of height H_{sp} limits, but does not entirely eliminate, cross-flow of adjacent solids. Capacity can be increased by adding additional compartments, thus increasing the column width W, and/or by adding one or more adjacent but independently operating compartmentalized columns in front, thus increasing the total multi-column thickness L.

In contrast to multiple spouting in parallel chambers, *multistage spouting* involves spouting in series, with outlet gas from one bed jetted into another

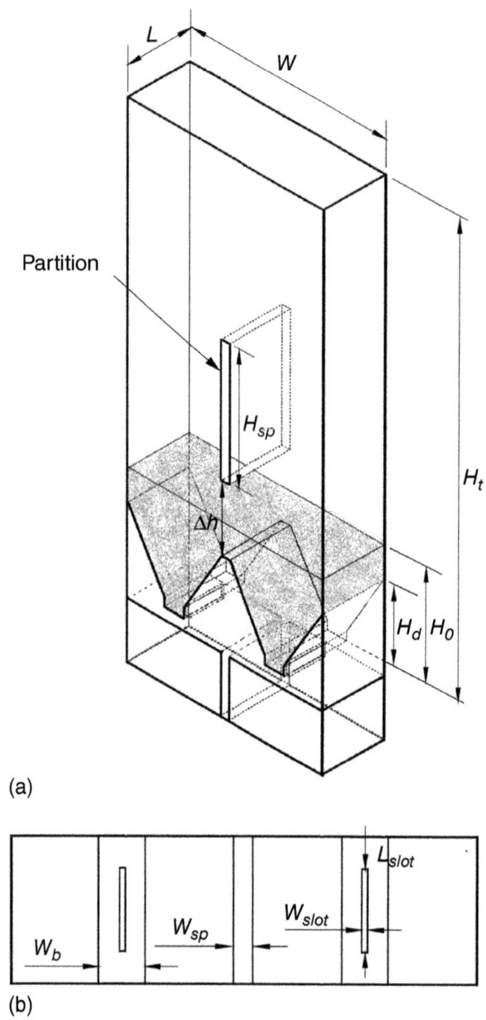

Figure 21.6 (a) Schematic of a two-compartment slot-rectangular spouted bed. (b) Top view. Source: Wang et al. 2016 [26]. Reproduced with permission of Elsevier.

bed sitting on top of it. Solids from the top can travel downwards from one bed to another by internal downcomers, an early example being a four-stage unit in which particulate solids were heated by compressed hot air [29]. Another example is a two-stage unit in which the same or different combustible solid fuels were separately fed to and spent versions removed from each stage. The spouting gas to the bottom stage, mainly air and steam, gave rise to combustion there, producing an oxygen-depleted hot flue gas that triggered pyrolysis in the upper stage [8, chapter 15; 30].

Cross-flow of both fluid and solids between the spout and annulus can be eliminated over most of the height of both a spouted and spout-fluid bed by inserting in the spout region, starting at least $10d_p$ above the fluid inlet nozzle, an open *draft tube* with walls that are impermeable to both phases [8, chapter 7]. The draft tube diameter is usually chosen to be about $2D_i$. The draft tube is aligned vertically, with its axis collinear with the axis of the column (see Figure 21.7). As a result, the spouted bed can now function at depths $>H_m$ and handle smaller particles. There are also large reductions in U_{ms} and solids circulation rate, as well as less solids mixing [3]. These changes are advantageous for granulation and particle coating [8, chapter 13], where plug flow of solids increases the product uniformity. If a draft tube is permeable to fluid, but impervious to the solids, it is well suited to applications in which the time spent in the annulus by all the particles is maximized without curtailing annular fluid flow [3, 4]. To perform the same functions as a draft tube in a SRSB, draft plates are added, and the slot length is extended to span the full thickness of the column.

In a *top-sealed spouting* vessel with a fluid outlet at the bottom [3] or part way up, fluid is forced to travel downwards though the annulus, resulting in a narrower residence time distribution of the fluid, especially if a draft tube is installed

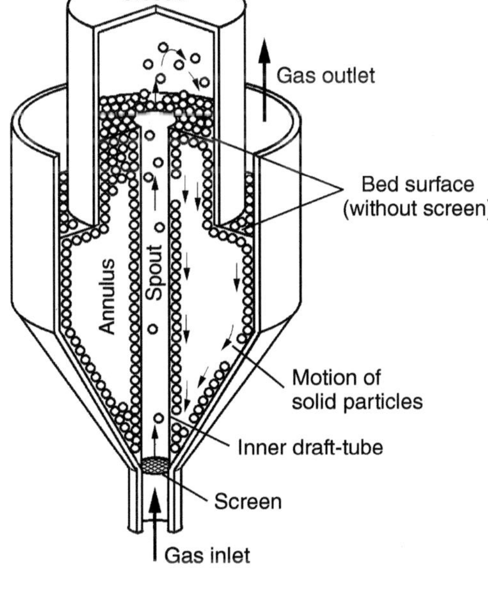

Figure 21.7 Schematic diagram of a top-sealed spouted bed with draft tube and side outlet. Source: Hattori et al. 1984 [31]. Reproduced with permission of Japan Science and Technology Agency.

(Figure 21.7). The combination of side outlet and draft tube also allows operation with smaller particles, and each of these three factors – side outlet, draft tube, small particles – contributes to increased conversions when a gaseous chemical reaction is catalyzed by the particles [32]. The position of the side outlet is subject to optimization [33].

If a shallow static bed of solids ($H/D_i = 2$–5, $\varepsilon \approx 0.4$, $d_p = 1$–5 mm) in a conical or conical–cylindrical column with $\theta = 30$–$45°$ is first conventionally spouted at, say, $U/U_{ms} = 1.5$ and the gas velocity is then gradually increased, the bed will first undergo a long transition region of progressively incoherent spouting during which the annulus voidage increases, until at $U/U_{ms} = 7$–8, ε finally exceeds 0.9, and there is no longer any distinction in this respect between the spout and the annulus [8, chapter 5; 34]. This *dilute-phase spouting* regime has been mis-named a "jet-spouted bed" [35], a term that persists in the literature. It is a misnomer because it incorrectly implies that a CSB, unlike this unconventional spouted bed, is not actuated by a jet. Because of the much lower solids holdups and therefore solids residence times, as well as high gas-particle relative velocities and correspondingly higher heat transfer coefficients, dilute-phase spouted beds have been reported to give superior performance in drying paste, slurries, and solutions [8, chapter 12] of heat-sensitive materials, especially bio-products [36], on inert solids.

Three-phase spouting [8, chapter 20] may be performed with gas and liquid flowing either co-currently or counter-currently. In a co-current gas–liquid spouted bed, important in granulation and particle coating [8, chapter 13], gas is used to atomize the liquid feedstock through the inside of the flow entry nozzle. Installing a draft tube generates higher gas holdups [37], while the introduction of additional liquid through a conical distributor gives rise to a spout-fluid bed with vigorous gas–liquid–solid contacting of interest in catalytic heavy hydrocarbon processing [38]. In counter-current gas–liquid spouting, applicable to gas absorption, dust removal, and stripping, low-density solid spheres are spouted by an upward flow of gas and irrigated by downward flow of liquid [39]. A relatively high-pressure drop can be reduced by using a counter-current gas–liquid spout-fluid bed, in which a portion of the gas is introduced via a centrally located nozzle and the rest through a surrounding gas distributor [40].

21.8 Applications

As applied industrially, spouted (and spout-fluid) beds always operate continuously with respect to the fluid flow but in some cases operate batchwise and in others continuously with respect to the solids, with particles fed from above by gravity or entrained into the bed by the spouting gas. Though CSBs were invented to dry wheat [1], the first commercial spouted beds were applied to drying peas, lentils, and flax [3, 4]. Since that time, the number of applications of gas-spouted and spout-fluid beds has increased greatly to include coating of fertilizer granules, gas cleaning, blending of polymer chips, comminution, and decarbonization. Each of these applications of spouted and spout-fluid beds depends on one or more of six features [4]:

1. Effective gas-particle contact, supported by extensive solids mixing, thereby accomplishing for coarse particles ($d_p = 1 - 10$ mm) what a fluidized bed does for fine particles ($d_p \lesssim 1$ mm).
2. Higher gas velocities and correspondingly lower gas residence times than for low voidage fluidized beds of fine solids.
3. Systematic cyclic movement of solids that enhances the reasonably good solids mixing of even a single cycle.
4. Solids attrition and de-agglomeration caused by high-velocity inter-particle collisions in the spout.
5. Absence of a gas-distributor plate, in contrast to a fixed or fluidized bed.
6. Counter-current heat transfer between ascending gas and descending annular solids.

Note that, unlike item 1, items 2–6 are inapplicable to fluidized beds, even with fine particles.

21.9 Multiphase Computational Fluid Dynamics

The last three decades have seen intense development of computational fluid dynamics (CFD) internationally. Modelling of spouted [8, chapter 4] and spout-fluid [8, chapter 6] beds has advanced accordingly. The necessarily limited range of most empirical equations and the necessarily simplified assumptions underlying most theoretical derivations are largely overcome by CFD. Spouted beds also lend themselves well to testing of CFD codes, given their smaller number of larger particles and more regular flow patterns than fluidized beds. One can therefore expect this approach to spouted and spout-fluid beds to continue to grow, though it will never fully replace the need for physical three-dimensional experiments.

Solved Problem

21.1 A flat-based vertical cylindrical column of diameter 0.3 m contains a 0.5 m high loose-packed bed (voidage 0.42) of 2.0 mm diameter glass spheres (specific gravity 2.5). Compare the superficial velocity for minimum spouting with a 15 mm diameter centrally located orifice with the minimum fluidization velocity of the same bed with a multi-orifice distributor at its base. The fluid in both cases is ambient air at 20 °C.

Solution

From Eq. (21.1),

$$U_{ms} = \left\langle \frac{0.002}{0.3} \right\rangle \left\langle \frac{0.015}{0.3} \right\rangle^{1/3} \sqrt{\frac{2(9.8)(0.5)(2500 - 1.2)}{1.2}} = 0.35 \text{ m/s}$$

From Eq. (2.18),

$$Ar = \frac{1.2(9.8)(2500 - 1.2)0.002^3}{0.000\,018^2} = 7.26 \times 10^5$$

and from Eq. (2.17),

$$U_{mf} = \frac{0.000\,018}{1.2 \times 0.002}(\sqrt{33.7^2 + 0.0408 \times 7.26 \times 10^5} - 33.7) = 1.06\ \text{m/s}$$

In this case, U_{mf} is approximately three times U_{ms}. Note that U_{mf} is independent of the bed height, whereas U_{ms} is dependent on H, increasing as the bed is made deeper.

Notations

Ar	Archimedes number (–)
c_{pp}	specific heat capacity of solid particles (J/kg K)
d_p	particle diameter (m)
D	diameter of cylindrical, or cylindrical extension of, column (m)
D_i	fluid inlet orifice diameter (m)
D_s	spout diameter at any level (m)
D_{sz}	spout diameter at height z (m)
D_{sH}	spout diameter at entrance to fountain (m)
g	acceleration of gravity (m/s^2)
H	height of loose-packed static bed; height of annulus, measured from fluid inlet orifice (m)
H_d	height of SRSB divergent base (m)
H_m	maximum spoutable value of H (m)
H_{sp}	separation panel height (m)
H_t	column height (m)
h_w	surface mean heat transfer coefficient between wall and bed (W/m^2 K)
K	dimensional constant in Eq. (21.6) (Pa s/m^2)
k_b	effective thermal conductivity of loose-packed bed (W/m K)
L	thickness of SRSB (m)
L_{slot}	slot length (m)
n	Richardson–Zaki [7] index (–)
Δh	distance between the bottom of separation panel and top of the diversion base (m)
ΔP_m	peak pressure drop (Pa)
ΔP_s	spouting pressure drop (Pa)
ΔP_{sp}	pressure drop for spouting (Pa)
ΔP_{fl}	pressure drop for fluidization (Pa)
U	superficial fluid velocity based on cross-sectional area of cylindrical part of column (m/s)
U_a	superficial fluid velocity in annulus at any level (m/s)
U_{aHm}	superficial fluid velocity in annulus at the maximum spoutable bed height (m/s)

U_m	U_{ms} at maximum spoutable bed height (m/s)
U_{mf}	minimum fluidization value of U (m/s)
U_{ms}	minimum spouting value of U (m/s)
U_s	superficial fluid velocity in spout at any level (m/s)
v_t	terminal settling velocity of isolated particle in spouting fluid (m/s)
v_w	downward particle velocity at column wall (m/s)
W	width of SRSB (m)
W_b	base width (m)
W_{slot}	slot width (m)
W_{sp}	separation panel width (m)
z	vertical distance from fluid inlet orifice (m)

Greek Letters

α	angle of repose of the solids (°)
ε	voidage (–)
ε_a	annulus voidage (–)
μ	viscosity (kg/m s)
ρ_b	bulk density (kg/m^3)
ρ_f	fluid density (kg/m^3)
ρ_p	particle density (kg/m^3)
θ	cone included angle (°)

Subscripts

aux	auxiliary
b	bulk
crit	critical
fl	fluidization
H	at height H above fluid inlet orifice
0	at zero level with respect to the fluid inlet orifice
sp	spouting (except as used in H_{sp})
z	at height z above fluid inlet orifice

Abbreviations

CSB	conventional spouted bed
SRSB	slot-rectangular spouted bed

References

1 Mathur, K.B. and Gishler, P.E. (1955). A study of the application of the spouted bed technique to wheat drying. *J. Appl. Chem.* 25: 624–636.

2 Gishler, P.E. (1983). The spouted bed technique – discovery and early studies at N.R.C. *Can. J. Chem. Eng.* 61: 267–268.

3 Mathur, K.B. and Epstein, N. (1974). *Spouted Beds*. New York: Academic Press.

4 Epstein, N. and Grace, J.R. (1997). Spouting of particulate solids. In: *Handbook of Powder Science and Technology*, 2e (eds. M. Fayed and L. Otter), 532–567. New York: Chapman and Hall.

5 Mather, K.B. and Gishler, P.E. (1955). A technique of contacting gases with coarse solid particles. *AIChE J.* 1: 157–164.

6 Littman, H., Morgan, M.H. III,, Vucović, D.V. et al. (1979). Prediction of the maximum spoutable height and the average spout inlet tube diameter. *Can. J. Chem. Eng.* 57: 684–687.

7 Richardson, J.F. and Zaki, W.N. (1954). Sedimentation and fluidization. *Trans. Inst. Chem. Eng. A* 32: 35–53.

8 Epstein, N. and Grace, J.R. (eds.) (2011). *Spouted and Spout-Fluid Beds*. Cambridge: Cambridge University Press.

9 Chandnani, P.P. and Epstein, N. (1986). Spoutability and spout destabilization of fine particles with a gas. In: *Fluidization V* (eds. K. Ostergaard and A. Sorensen), 233–240. New York: Engineering Foundation.

10 Li, Y., Lim, C.J., and Epstein, N. (1996). Aerodynamic aspects of spouted beds at temperatures up to 580°C. *J. Serb. Chem. Soc.* 61: 253–266.

11 McNab, G.S. and Bridgwater, J. (1977). Spouted beds – estimation of spouting pressure drop and the particle size for deepest bed. *Proceedings of the European Congress on Particle Technology*, Numenburg.

12 Wen, C.Y. and Yu, Y.H. (1966). A generalized method for predicting the minimum fluidization velocity. *AIChE J.* 12: 610–612.

13 Ergun, S. (1952). Fluid flow through packed columns. *Chem. Eng. Prog.* 48 (2): 89–94.

14 Mamuro, T. and Hattori, H. (1968). Flow pattern of fluid in spouted beds. *J. Chem. Eng. Jpn.* 1: 1–5.

15 Grbavčić, Z.B., Vuković, D.V., Zdanski, F.K., and Littman, H. (1976). Fluid flow pattern, minimum spouting velocity and pressure drop in spouted beds. *Can. J. Chem. Eng.* 54: 33–42.

16 Epstein, N., Lim, C.J., and Mathur, K.B. (1978). Data and models for flow distribution and pressure drop in spouted beds. *Can. J. Chem. Eng.* 56: 436–447.

17 Lim, C.J. and Mathur, K.B. (1976). A flow model for gas movement in spouted beds. *AIChE J.* 32: 674–680.

18 Madonna, L.A., Lama, R.F., and Brisson, W.L. (1961). Solids air jets. *Br. Chem. Eng.* 6: 524–528.

19 Zahed, A.H. and Epstein, N. (1992). Batch and continuous spouted bed drying of cereal grains: the thermal equilibrium model. *Can. J. Chem. Eng.* 70: 945–953.

20 Piccinini, N., Grace, J.R., and Mathur, K.B. (1979). Vapour phase chemical reaction in spouted beds: verification of theory. *Chem. Eng. Sci.* 34: 1257–1263.

21 Brereton, C.M.H. and Grace, J.R. (1984). A note on comparison of spouted bed reactor models. *Chem. Eng. Sci.* 39: 1315–1317.

22 Cui, H. and Grace, J.R. (2008). Spouting of biomass particles: a review. *Bioresour. Technol.* 99: 4008–4020.

23 Zhang, J. and Tang, F. (2006). Prediction of flow regimes in spout-fluidized beds. *China Particuol* 4: 189–193.

24 Foong, S.K., Barton, R.K., and Ratcliffe, J.S. (1975). Characteristics of multiple spouted beds. *Mech. Chem. Eng. Trans.* 1 (2): 7–12.

25 Gong, X., Hu, G., and Li, Y. (2006). Hydrodynamic characteristics of a novel annular spouted bed with multiple air nozzles. *Ind. Eng. Chem. Res.* 45: 4830–4836.

26 Wang, Z., Saidi, M., Lim, C.J. et al. (2016). Comparison of DEM simulation and experiments in a dual-column slot-rectangular spouted bed with a suspended partition. *Chem. Eng. J* 290: 63–73.

27 Romankov, P.G. and Rashkovskaya, N.B. (1968). *Drying in a Suspended State*, 2end ed. (in Russian). Leningrad: Chemistry Publishing House.

28 Mujumdar, A.S. (1984). Spouted bed technology – a brief review. In: *Drying 84* (ed. A.S. Mujumdar), 224–239. New York: Hemisphere.

29 Malek, M.A. and Walsh, T.H. (1966). The treatment of coal for coking by the spouted bed process. *Rep. No. FMP 66/54-SP*. Department of Mines and Technical Surveys, Ottawa, Canada.

30 Rovero, G. and Watkinson, A.P. (1990). A two-stage spouted bed process for autothermal pyrolysis or retorting. *Fuel Proc. Technol.* 26: 221–238.

31 Hattori, H., Kobayashi, A., Aiba, I., and Koda, T. (1984). Modification of the gas outlet structure on the spouted bed with inner draft tube. *J. Chem. Eng. Jpn.* 17 (1): 102–103.

32 Hattori, H. and Takeda, K. (1976). Modified spouted beds with the gas outlet located in the side wall surrounding the annular dense bed. *J. Fac. Text. Sci. Technol. Shinshu Univ. Ser B Eng.* 12: 1–13.

33 Brereton, C.M.H., Epstein, N., and Grace, J.R. (1996). Side-outlet spouted bed with draft tube. *Can. J. Chem. Eng.* 74: 542–546.

34 Barrazo, M.A.S., Duarte, C.R., Epstein, N. et al. (2010). Experimental and computational fluid dynamics study of dense-phase, transition region, and dilute-phase spouting. *Ind. Eng. Chem. Res.* 49: 5102–5109.

35 Markowski, A. and Kaminski, W. (1983). Hydrodynamic characteristics of jet-spouted beds. *Can. J. Chem. Eng.* 61: 377–381.

36 Markowski, A.S. (1993). Quality interaction in a jet spouted bed dryer for bio-products. *Drying Technol.* 11: 369–387.

37 Fan, L.S., Hwang, S.J., and Matsuura, A. (1984). Hydrodynamic behaviour of a draft tube gas–liquid–solid spouted bed. *Chem. Eng. Sci.* 39: 1677–1688.

38 Kono, H. (1980). A new concept for three phase fluidized beds. *Hydrocarbon Process.*: 123–129.

39 Vuković, D.V., Zdanski, F.K., Vunjak, G.V., and Grbavčić, Z.B. (1974). Bed expansion and liquid holdup in a three phase spouted bed contactor. *Can. J. Chem. Eng.* 52: 180–184.

40 Vuković, D.V. and Vunjak-Novaković, V. (1978). The three-phase spout-fluid bed – a novel gas-liquid contacting system. *6th CHISA Conference*, Prague, Paper C3.11.

22

Three-Phase (Gas–Liquid–Solid) Fluidization

Dominic Pjontek[1], Adam Donaldson[2], and Arturo Macchi

[1] *Western University, Department of Chemical and Biochemical Engineering, Thompson Engineering Building, London, Ontario, Canada N6A 5B9*
[2] *Dalhousie University, Department of Chemical Engineering, PO Box 15000, Halifax, Nova Scotia, Canada B3H 4R2*
[3] *University of Ottawa, Centre for Catalysis Research and Innovation, Department of Chemical and Biological Engineering, Ottawa, Canada K1N 6N5*

22.1 Introduction

Cases where relatively small amounts of liquid are present locally, which occurs when injecting liquid jets into gas-fluidized beds, are treated in Chapter 19. This chapter describes systems where liquid plays a major role, in addition to gas, in driving the flow. Building on previous chapters, the operating modes, applications, and general design and modelling of gas–liquid–solid (GLS) fluidized beds are summarized in this chapter. The works of Fan [1], Nigam and Schumpe [2], Duduković et al. [3], Fan and Yang [4], and Buwa et al. [5] are drawn upon to illustrate the application of transport phenomena models to characterize reactor performance for the mode of operation where both gas and liquid flow upwards and liquid is the continuous wetting phase.

22.1.1 General Description and Classification

GLS fluidization occurs when a gas is injected into a liquid-fluidized bed or when a substantial flow of liquid is injected into a gas-fluidized bed (Figure 22.1). In applications where liquid is the continuous wetting phase (Figure 22.1a), flow of the liquid relative to the gas depends on the particle density: liquid usually flows upwards (co-currently) with the gas when the solid particle density is greater than the liquid density and downwards (counter-currently) to the gas when the solid particles are less dense than the liquid. Reactors are typically considered to be composed of four distinct regions: plenum, distributor, bulk bed, and freeboard. Liquid, often with gas, is introduced into the reactor via the plenum. Gas bubbles form, detach, and develop via coalescence and breakup to reach their equilibrium size/shape in the distributor region of the fluidized bed. The bulk region of the bed usually constitutes the main portion where the variance of volume-average transport phenomena (e.g. interphase mass transfer coefficient)

Essentials of Fluidization Technology, First Edition. Edited by John R. Grace, Xiaotao Bi, and Naoko Ellis.

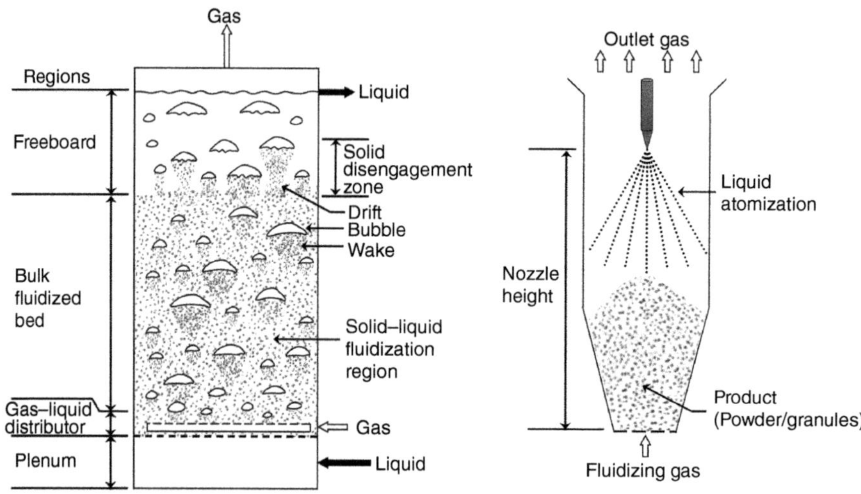

(a) (b)

Figure 22.1 Three-phase fluidization examples: (a) Co-current gas and liquid flow, liquid as continuous phase. Source: Fan 1989 [1]. Reproduced with permission of Elsevier (b) Evaporative liquid jets in a gas-fluidized bed, gas as continuous phase.

is lower. The freeboard is situated above the fluidized bed and consists primarily of gas and liquid, since typically only small amounts of particles are entrained in the wakes of rising bubbles. Furthermore, a three-phase fluidized bed differs from a "slurry bubble column" in that it is limited to a net liquid flow and significant relative velocity between the liquid and solids (with particle diameters typically greater than 0.5 mm). With gas as the continuous phase, Figure 22.1b portrays a fluidized bed granulator in which liquid is atomized onto particles that receive successive coatings (and may agglomerate) until they are large enough to classify towards the bottom of the vessel and be preferentially removed. Note that the bed material remains partially mixed, so the exit contains some smaller particles that are separated and recycled to the granulator.

A complete classification scheme of three-phase contactors, with their associated relative advantages and disadvantages, depending on the continuous phase and net phase flow direction, can be found in Fan [1]. The selection of a desired phase contacting pattern is a function of many factors but usually considers an analysis of at least three: (i) the relative transport and reaction rates to determine, and then maximize, the rate of the limiting step; (ii) rate of heat removal from, or addition to, the reactor; and (iii) rate of solids make-up, with further details in Section 22.2.

22.1.2 Applications

Applications of three-phase fluidization occur across several industrial sectors and have been presented in Refs. [1–5]. With liquid as the continuous phase, the catalytic hydrogenation of heavy oils is one of the more successful commercial applications. Figure 22.2a presents an LC-FINING® unit, often referred to as an

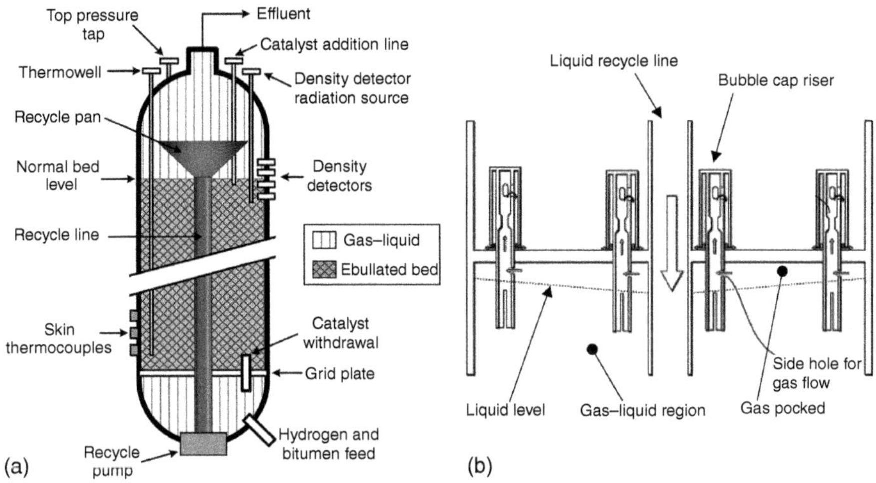

Figure 22.2 (a) LC-FINING unit schematic. Source: From McKnight et al. 2003 [6]. Reproduced with permission of Wiley (b) Grid schematic with plenum and fluidized bed respectively below and above. Source: From US Patent 2014/0147352 A1.

"ebullated" bed. Heavy oil and hydrogen are heated separately, then mixed, and introduced into the reactor plenum through a perforated ring. The fresh feed, combined with the internally recycled liquid products, passes through a bubble cap grid (Figure 22.2b) to ensure even distribution of the phases into the fluidized bed of cylindrical catalyst particles. Internal liquid recycle provides the necessary fluid velocity to maintain a specific catalyst bed height, ensure temperature uniformity, and reach the desired product yield. γ-Ray density detectors allow the bed height to be determined, while thermocouples monitor the bed temperature. Online catalyst addition and removal are carried out to maintain the desired activity in the face of catalyst fouling from carbon, sulphur, and heavy metal deposits (e.g. vanadium from bitumen).

Heavy oil hydroprocessing reactors operate at elevated pressures (5.5–21 MPa) and temperatures (300–440 °C) with multicomponent liquids, subject to foaming [4]. By increasing gas density, these elevated pressures enhance the bubble breakup rate, resulting in smaller bubbles ($d_b < 1$ mm), lower bubble rise velocities, and greater gas holdups. Naturally occurring surfactants can further reduce the rise velocity ($u_b < 0.1$ m/s) of these small bubbles by creating surface tension gradients and immobilizing the interface. These two synergistic effects increase the gas holdup ($\varepsilon_g > 0.3$) in the hydroprocessor, which in turn decreases the liquid residence time, resulting in lower product space–time yields. The reactor geometry, in particular its entrance (gas–liquid distribution system) and exit (liquid recycle pan), also affects the reactor performance at these operating conditions. An even spatial distribution of the fluids is essential for proper hydrogenation and to limit the formation of coke that, by settling on the grid and detrimentally affecting the distribution, can become self-propagating. The gas-phase separation efficiency of the recycle pan is also critical, as any gas entrained with the recycled liquid increases the gas flowrate and holdup at the expense of liquid holdup and product yield.

With gas as the continuous phase, there are several commercial applications of evaporative liquid jets injected into a gas-fluidized bed, such as particle granulation and coating in the food and pharmaceutical industries. Analogously, thermal and catalytic cracking of liquid hydrocarbons as well as polymerization in the condensation mode are other important commercial applications [7, 8]. The injection nozzle(s) should be designed to produce liquid droplets that uniformly coat individual particles with a liquid film thickness that minimizes liquid bridging between particles, which can lead to particle clustering, reduced fluid–particle contact area, and even defluidization. Details regarding the design of such reactors are covered in Chapter 19.

22.2 Reactor Design and Scale-up

The production of a given chemical starts with a conceptual design of the overall process. Knowing the process chemistry, physics, and resulting sequence of unit operations, idealized steady-state process simulations enable preliminary assessment of technical and economic viability, as well as environmental impact (e.g. via life cycle analysis). Socioeconomic impacts should also be explored as policymakers greatly influence technical development and vice versa (e.g. governments' strategies towards climate change and reduction of greenhouse gas emissions). This analysis often depends on reactor design and scale-up assumptions, requiring a good understanding of fluidization fundamentals, transport phenomena, and reaction engineering when the system in question involves multiple phases.

22.2.1 Reactor Design

The design procedure for a three-phase fluidized bed reactor is philosophically similar to that for a two-phase system (see Chapter 15), which itself is an extension of idealized single-phase plug flow reactor (PFR) and continuous stirred tank reactor (CSTR) models covered in undergraduate chemical reaction engineering courses [9]. The performance (i.e. yield) of a multiphase reactor, as a function of its operating conditions and phase physical properties, is therefore obtained via species mass balances, as well as energy and momentum balances to predict the concentration, temperature, and pressure profiles. Following Duduković et al. [3], the universal balance equation applied to any conserved property (e.g. species mass) for a particular phase (i) in any multiphase reactor is

$$\begin{aligned}&(\text{Rate of input by phase } i) - (\text{Rate of output by phase } i)\\&\quad + (\text{Net rate of interphase transport into phase } i)\\&\quad = (\text{Rate of accumulation in phase } i)\\&\quad - (\text{Net rate of generation in phase } i) \qquad (22.1)\end{aligned}$$

The left side of the balance equation contains information/sub-models of the reactor phase holdup and flow pattern (i.e. mixing) and contacting pattern (i.e.

interphase mass transfer), whereas the generation term on the right accounts for chemical reactions and their kinetics (i.e. rate expressions). The sophistication of the resulting reactor model depends on the size and location of the control volume to which Eq. (22.1) is applied (see Figure 22.3), which in turn depends on the accuracy of available information regarding the rate-limiting step. The reactor total volume, over which Eq. (22.1) is integrated, is generally divided into regions with distinctive transport features, such as those identified in Figure 22.1a. Compartmentalization is discussed in Section 22.3, and a worked example for a three-phase fluidized bed hydrocracker is presented in Solved Problem 22.1. Sections 22.4 and 22.5 present mathematical expressions for the fluid dynamic and interphase mass transport terms in Eq. (22.1) for the bulk fluidized bed region.

When designing multiphase reactors, it is important to understand and quantify transport–reaction interactions on a particle or single eddy scale, interphase transport rates on particle and reactor scales, flow and contacting patterns of each phase, and their responses to changes in reactor scale and operating conditions [3]. As a result, there is continuing development of simulation tools that integrate detailed reaction chemistry models, energy minimization multi-scale approaches [11], and computational fluid dynamic (CFD) models capturing meso- and macroscale structures. Some models provide overall flow features and then rely on empirical transport and reaction kinetic relations to describe reactor performance, while others attempt to describe the phenomena on various scales from first principles, with moderate success [10, 12]. While available commercial CFD codes have made it possible to simulate gross flow patterns in large reactors, understanding the phenomena on various length and time scales still requires experimental validation and innovative nonintrusive

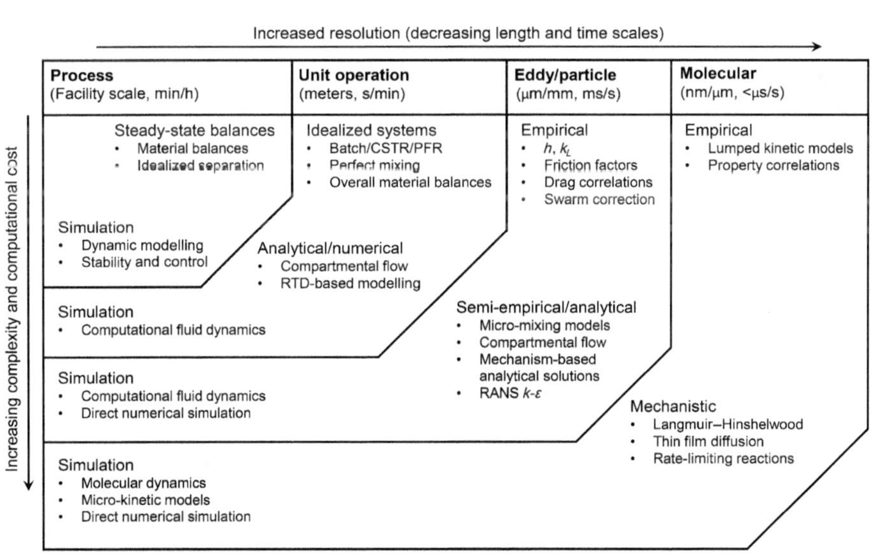

Figure 22.3 Length and time scales in multiphase reactor modelling. Source: Adapted from Dudukovic and Mills 2014 [10].

measuring techniques capable of monitoring the dynamics of bubbles and solids. A review of measurement techniques in gas–liquid and GLS systems is available from Boyer et al. [13].

22.2.2 Reactor Scale-up

If a multiphase reactor model is developed using robust sub-models at the appropriate length and time scales, then the Damköhler numbers and resulting yield predictions should be scalable. The general design cycle is further depicted in Figure 22.4, where the intrinsic reaction kinetics are coupled (via the conservation equation, Eq. (22.1)) with transport phenomena models (fluid dynamics, heat and mass transfer) for a reactor model iteratively tested under reactive conditions in a pilot plant. The reactor model and pilot operating experience are then applied to the design and optimization of the commercial size reactor.

The challenge is to obtain transport phenomena models that can be applied with confidence, since transport parameters can vary with reactor scale, whereas intrinsic reaction kinetics do not. One method to fundamentally investigate fluid dynamics is via dimensional analysis and similitude (see Chapter 17), but this has its difficulties, as the physics need to be well understood. Also, failing to include a key factor (refer to Table 22.1) leads to misleading and inaccurate results. Similitude can be achieved by rendering dimensionless the governing conservation equations and boundary conditions, by means of force balances or the Buckingham pi theorem. Once the dimensionless groups have been formed, the results obtained in experiments yield empirical data that describe the phenomena studied for all systems that are geometrically similar and have the same values of key dimensionless groups (i.e. dynamic similitude).

For example, to model three-phase fluidized bed holdups, Larachi et al. [15] considered 14 of the factors listed in Table 22.1, which yielded 11 (a prohibitively high number) dimensionless groups. As a result, cross-correlation coefficients of the liquid gravity force, liquid viscous force, capillary force, and gas, liquid, and solid inertial forces were examined to generate an optimal assortment of six dimensionless groups that describe the gas holdup, for example:

$$\varepsilon_g = f(Fr_g, Ca_g, \chi_{gL}, Mo/Eo_p, d_p/d_c, \text{binary bubble coalescence index}) \tag{22.2}$$

22.3 Compartmental Flow Models

Transport phenomena can vary significantly between the entrance, bulk, and exit regions of a multiphase reactor. Some reactors may be described by a single simplified geometry if variability in these features is low, but often a more flexible and robust reactor model is needed. In these cases, the total vessel volume may be divided into sub-regions, each individually modelled and collated via boundary conditions. From a reaction engineering view, this is similar to describing a PFR as a series of CSTR, while, at the extreme, one would approach

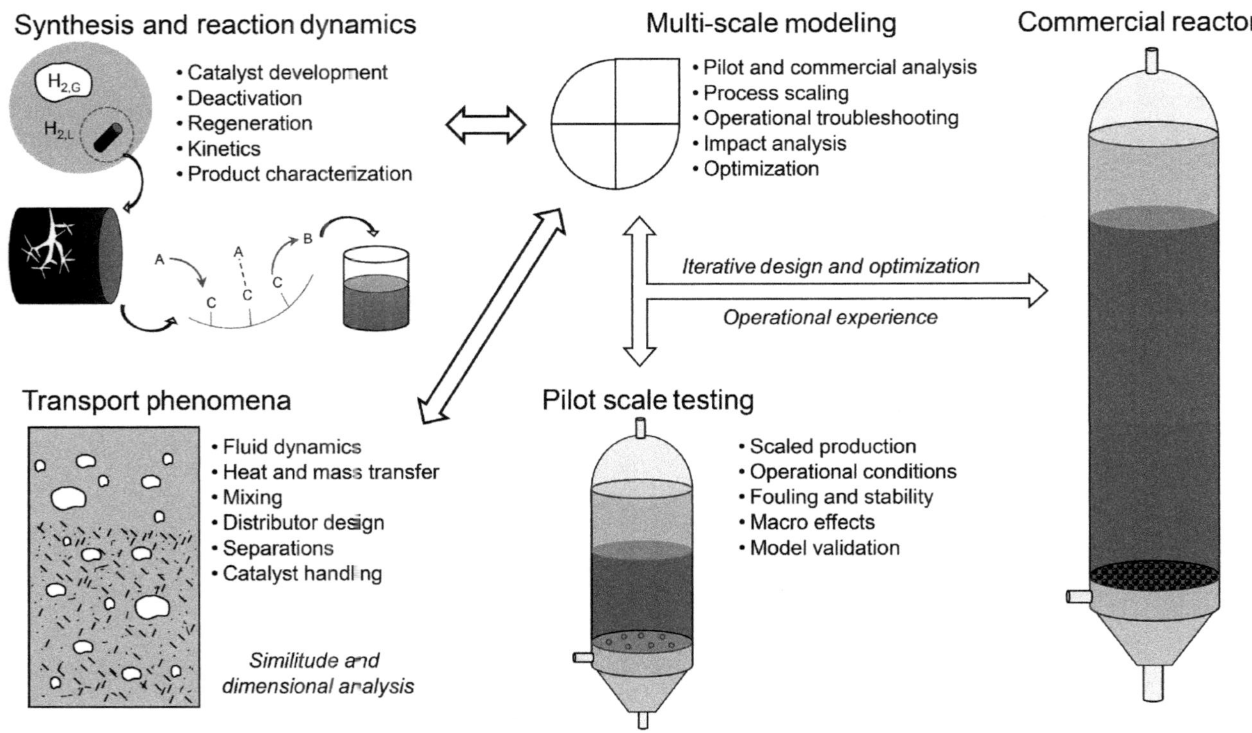

Figure 22.4 Schematic diagram of fluidized reactor development. Source: Adapted from Jiang et al. 2003 [14].

Table 22.1 Parameters that affect bed transport phenomena.

Phases	• **Solid particles**: average size, size distribution, average density, density distribution, sphericity, wettability, porosity, thermal conductivity, heat capacity • **Liquid**: density, surface tension, rheology, foaming characteristics, thermal and electrical conductivity, volatility • **Gas**: density, viscosity, solubility, diffusivity
Operating conditions	Gas and liquid superficial velocities, temperature, pressure
Reactor geometry	• **Column**: cross-sectional shape, diameter, angle of inclination, bed length (aspect ratio) • **Internals**: position, length, and diameter of a draft tube; presence and geometry of baffles or fixed packing • **Entrance (fluids distributor)**: single- or dual-fluid injection nozzles • **Exit (phase separator)**: number of separation stages, liquid and/or solids recycling, gas demister, foam destruction device

CFDs with mass and energy balances for all cells. Within three-phase fluidized systems, four regions are described in the following sections, as identified in Figure 22.1a.

22.3.1 Plenum and Fluid Distributor

The plenum chamber, which may contain packing, is a calming region intended to flatten the radial profile of liquid velocity as it enters the fluidized bed. In Figure 22.1a, the gas is independently introduced into the liquid-fluidized bed via a sparger situated just above the liquid distributor. As a result, the design of distributors for two-phase fluidized beds (Chapters 3 and 5) and bubble columns [2, 16] holds, where the goal is to optimize mechanical energy consumption, while evenly distributing the fluids and preventing solids from plugging the orifices. For liquid–solid fluidized beds, the distributor fractional open area is typically such that its pressure loss is around 20–30% of the bed frictional pressure loss. In order to achieve a steady gas jet (as opposed to periodic bubbling that is subject to liquid weeping into the gas manifold), the fractional open area and orifice diameter for the gas sparger are selected to achieve $We_{or} = d_{or}\rho_g U_{or}^2/\sigma > 2$ with $d_{or} > 1$ mm for liquids with viscosities similar to water [17]. Moreover, the pressure loss in the manifold channels leading to the orifice must be orders of magnitude lower than through the orifice to achieve nearly equal gas flow through all orifices.

Other fluid distribution systems combine the gas and liquid before injection into the fluidized bed. This generally leads to forming smaller bubbles due to a greater energy dissipation rate than for single-phase gas flow. LC-FINING units have a two-phase distribution system (Figure 22.2b) since direct injection of hydrogen into the catalytic reactor can create hot spots and associated operational problems. The gaseous and liquid hydrocarbons are thus premixed with hydrogen in the plenum chamber, forming a gas pocket beneath the grid. The size of the pocket grows until the pressure loss of the gas entering through the side slits on the bubble cap riser is equal to the pressure loss of the liquid entering the riser from the bottom centre, ensuring that fluids enter the bubble cap at the same pressure. The plenum gas–liquid interface under ideal conditions is horizontal, resulting in uniform fluid distribution into the catalyst bed above. Selecting the fractional opening and resulting two-phase flow pressure loss in the bubble cap riser is complex [16] but must again ensure that dynamic pressure fluctuations in the bed do not affect the position of the gas–liquid interface in the plenum. The two-phase flow pressure loss is typically related to the pressure loss for the liquid alone via a frictional pressure loss multiplier that considers the multiphase flow as pseudo-homogeneous, with no slip between the phases, or as separate phases with various flow regimes (e.g. Lockhart–Martinelli approach) [18].

22.3.2 Fluidized Bed

The fluidized bed region can be further divided into entrance and bulk bed regions (see Figure 22.1a), depending on the relative volumes and transport features in each region. The gas–liquid distribution system greatly influences the bubble mean size and size distribution when coalescence rates are slow, such as at high pressure, in the presence of surfactants and/or with relatively large particles, several mm in diameter (see Section 22.4.1.2). On the other hand, equilibrium bubble size can be quickly reached in systems with rapid bubble coalescence, to such a degree that distributors have minimal influence, especially for tall reactors with $H/d_c > 5$ [19].

Available correlations for fluid dynamic and heat and mass transport parameters in the three-phase fluidized bed region rarely consider the effect of distributor design and are thus volume averaged over the entire bed. Caution is therefore essential to ensure that the influence of system geometry is considered, in addition to the range of phase physical properties and operating conditions. Table 22.1 presents a list of the relevant variables, some of which are discussed in the subsequent sections; more details can be found in Refs. [1–5]. Note that the impacts of temperature and pressure are captured via changes in phase physical properties. However, very few data are available in the literature at elevated pressures and temperatures, as well as for multicomponent liquids subject to foaming. Pressures up to ~6 MPa have been found to significantly impact the rates of bubble coalescence and breakup, lowering the maximum stable bubble size [4]. Many industrial grade liquids contain impurities (e.g. alcohols, organic acids, electrolytes) that accumulate at the gas–liquid interface, usually hindering bubble coalescence, and leading to frothing and foaming in the extreme. The sorption of surfactants and resulting equilibrium interfacial condition requires some time to establish. Surface activity in bubbling liquids should thus be considered a dynamic phenomenon, primarily affecting rising spherical and ellipsoidal bubbles ($Re_b < 1$; $Re_b > 1$ and $Eo_b < 40$), where surface tension influences bubble shape [20].

22.3.3 Freeboard

The co-current gas–liquid upflow in the freeboard can be modelled as a bubble column since particle carry-over is generally negligible, resulting in a relatively short particle transport disengagement zone (refer to Section 22.4.2.3). The fluidized bed thus "acts" as another gas–liquid distribution system for the freeboard. There are several models to estimate bubble properties and resulting transport parameters in gas–liquid systems [2, 16, 20]. For example, an air bubble rising in tap water at atmospheric conditions ($Mo = 2.6 \times 10^{-11}$) is considered spherical below a diameter of ~1.0 mm ($Eo_b = 0.14$, $Re_b = 193$), where surface and/or viscous forces dominate, and spherical-cap above a volume-equivalent spherical diameter of ~17 mm ($Eo_b = 40$, $Re_b = 5308$), where inertial forces dominate. Between these sizes, bubbles are considered ellipsoidal with the resulting shape, rise velocity, and motion influenced by all three types of forces.

Note that the reactor exit design can have a significant impact on bed and freeboard fluid dynamics, especially when phases are recycled. For example, poor and/or unpredictable gas–liquid separation performance within the gravity-based recycle pan of the LC-FINING unit may negatively affect product space–time yields and throughput capacity due to increased gas holdup. This form of gas–liquid separation and its bubble cut diameter are strongly dependent on the average and distribution of liquid residence time in the recycle pan.

22.4 Fluid Dynamics in Three-Phase Fluidized Beds

22.4.1 Flow Regimes

22.4.1.1 Minimum Fluidization

The minimum liquid fluidization velocity (U_{Lmf}) can be obtained from the intersection of the fluidized and fixed bed dynamic pressure drop slopes when the liquid velocity is reduced for a constant gas velocity (see examples in Figure 22.5a). The dynamic pressure drop ($-\Delta P_{dyn}$) is defined as the total pressure drop corrected for the hydrostatic pressure of liquid (i.e. $-\Delta P_{dyn} = -\Delta P_{Total} - g\rho_L \Delta z$). Briens et al. [22] suggest that this experimental method yields the transition from a fixed bed to an intermediate fluidized state or agitated bed, where particle movement is primarily due to gas bubbles passing through the bed.

The liquid velocity necessary to fluidize the bed decreases with increasing gas velocity (Figure 22.5b). However, the extent of this decrease depends on the particle diameter, particle shape, and liquid physical properties. Note that wall effects may become important for smaller column diameters when $d_c/d_p < 30$ and for $d_c < \sim 0.1$ m since slugging can occur. While the minimum liquid fluidization velocity in a liquid–solid fluidized bed is minimally affected by pressure, there is an impact when gas is introduced, especially at higher gas velocities where bubbles are well dispersed, leading to high gas holdups.

Several empirical and semi-empirical correlations have been proposed to predict U_{Lmf} see Ref. [23]. The gas-perturbed liquid model described below is relatively successful for larger particles, typically above 1 mm in diameter, since it

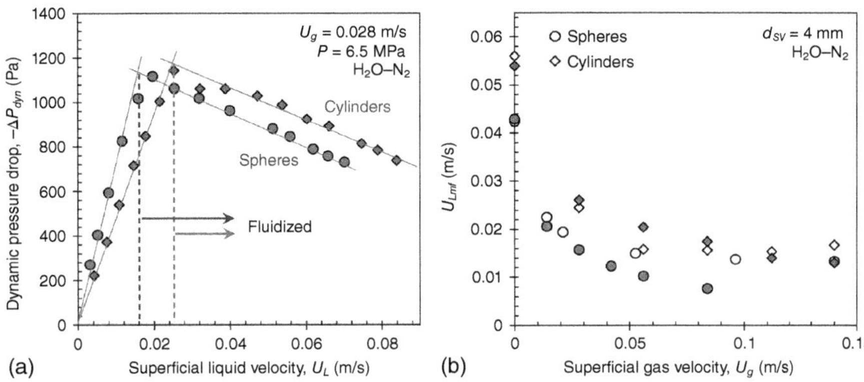

Figure 22.5 (a) Bed dynamic pressure drop as function of U_L at constant U_g. (b) U_{Lmf} as a function of U_g for pressures of 0.1 MPa (hollow symbols) and 6.5 MPa (filled symbols). Measurements used nitrogen, water, and glass spheres or aluminum cylinders with d_{sv} = 4 mm. Source: Adapted from Pjontek and Macchi 2014 [21].

assumes that full support of the solids is provided by the liquid, whose interstitial velocity is increased because of the volume occupied by gas [24]. The force balance on the particles then reduces to that for a liquid–solid fluidized bed (see Chapter 3) when no gas is present.

$$Re_{Lmf} = \frac{d_p U_{Lmf} \rho_L}{\mu_L} = \sqrt{\left[150\frac{1-\varepsilon_{mf}}{3.5\phi}\right]^2 + \varepsilon_{mf}^3 (1-\alpha_{mf})^3 \frac{Ar'_L}{1.75}} - 150\frac{1-\varepsilon_{mf}}{3.5\phi} \tag{22.3}$$

A better fit is obtained when accounting for the gas–liquid mixture in the buoyancy [25], modifying the Archimedes number ($Ar_L{}'$), and when approximating the experimentally observed decrease in bed porosity at minimum fluidization (ε_{mf}) with the addition of gas, resulting in

$$Ar'_L = \frac{\rho_L(\rho_p - (\rho_g \alpha_{mf} + \rho_L(1-\alpha_{mf})))g d_v^3}{\mu_L^2} \tag{22.4}$$

$$\varepsilon_{mf} = \varepsilon_{mf}|_{U_{g=0}} \left(1 - 0.34\left(1 - \frac{U_{Lmf}}{U_{Lmf}|_{U_{g=0}}}\right) + 0.22\left(1 - \frac{U_{Lmf}}{U_{Lmf}|_{U_{g=0}}}\right)\right) \tag{22.5}$$

The gas holdup on a solids-free (α_{mf}) basis at minimum fluidization is estimated from the empirical relation [26]:

$$\alpha_{mf} = \frac{0.16 U_g}{\varepsilon_{mf}(U_g + U_{Lmf})} \tag{22.6}$$

The bed porosity at minimum fluidization and $U_g = 0$ can be estimated based on the following approximation [27]:

$$\varepsilon_{mf}|_{U_{g=0}} \approx \frac{0.415}{\sqrt[3]{\phi}} \tag{22.7}$$

22.4.1.2 Bubbling Regimes

At lower gas flowrates, spherical or ellipsoidal bubbles of relatively uniform size may be produced with little interaction between them. The flow regime is referred to as "dispersed or homogeneous bubble flow," and an increase of gas flowrate increases the bubble population more than the bubble size. However, this flow regime cannot be sustained indefinitely. Beyond a certain gas flowrate, the bubble population becomes sufficiently great that bubbles start to coalesce, producing larger spherical-cap bubbles that rise faster than their smaller counterparts. The resulting wider bubble size distribution induces relatively strong liquid circulation patterns and mixing. This flow regime is referred to as "coalesced or heterogeneous bubble flow." Further increases in gas velocity (e.g. above 0.1 m/s) result in "slug flow" in columns of diameter <0.1 m, while operation at high liquid velocities yields the "transport flow" regime, encountered in three-phase circulating beds [28]. These flow regimes are illustrated in Figure 22.6; however, the bridging and annular flow regimes are not typically observed for three-phase flow.

Bubble breakup can occur via bubble–particle collision and bubble instability (e.g. Kelvin–Helmholtz), limiting growth to a maximum size [17]. Bubble growth via coalescence occurs in three steps: bubbles approach and form a thin film of liquid between each other, and the liquid film then drains due to gravity and capillary forces and finally ruptures upon reaching a critical thickness. Bubble collisions can be induced by liquid velocity fluctuations due to turbulence and by the wake of preceding larger bubbles, while the duration of liquid film thinning is considered rate limiting [17]. As a result, the probability of bubble coalescence is reduced, and transition from dispersed to coalesced bubble flow is delayed

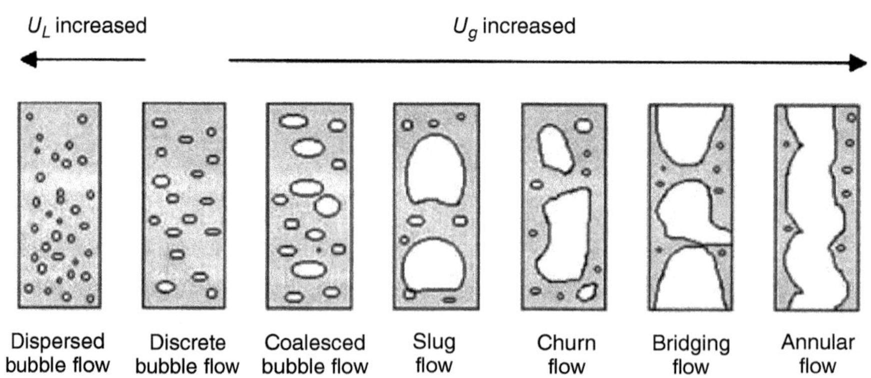

Figure 22.6 Schematic diagram of flow regimes observed for an air-water-1.5 mm glass beads three-phase fluidized bed. Source: From Zhang et al. 1997 [29]. Reproduced with permission of Elsevier.

with an increase in pressure and improved cross-sectional distribution of the gas, which lower the bubble collision frequency and maximum stable bubble size (and its wake), as well as with the presence of surfactants that create surface tension gradients, slowing the liquid film drainage rate, i.e. *Gibbs–Marangoni* effect [17, 20]. The impact of particles on the bubbling regime is complex and dependent on the relative particle and bubble sizes. Solid particles increase the apparent viscosity of the medium surrounding the bubble, which can favour bubble coalescence if the particle inertia is insufficient to break bubbles. The boundary between smaller and larger particles is, for example, around 3 mm in diameter for an air–tap water–glass sphere system at atmospheric conditions [2]. Therefore, three-phase systems with $d_p = 4$ mm will achieve dispersed and then coalesced bubble flow with increasing gas velocity, whereas solely coalesced or dispersed and coalesced bubble flow will be achieved, depending on the liquid flow rate, for $d_p = 1.5$ mm (See experimental data in Problem 22.2).

The dispersed-to-coalesced bubble flow transition velocity is an important reactor design parameter that can be estimated experimentally via analysis of pressure fluctuation signals or identified by a marked drop in the rate of gas holdup increase with increasing superficial velocity [19]; no sufficiently generalized model or map currently exists.

22.4.2 Phase Holdups

22.4.2.1 Modelling: Global (Bed Volume Averaged)

Global phase holdups can be determined experimentally by measuring the dynamic pressure drop throughout the bed and freeboard regions. The bed height (H) is then estimated from the intersection of the bed and freeboard dynamic pressure profiles, obtained by linear regression, as illustrated in Figure 22.7. A linear pressure gradient is indicative of constant axial phase holdups. The solid particle holdup (ε_p) is calculated knowing M_p, the mass of

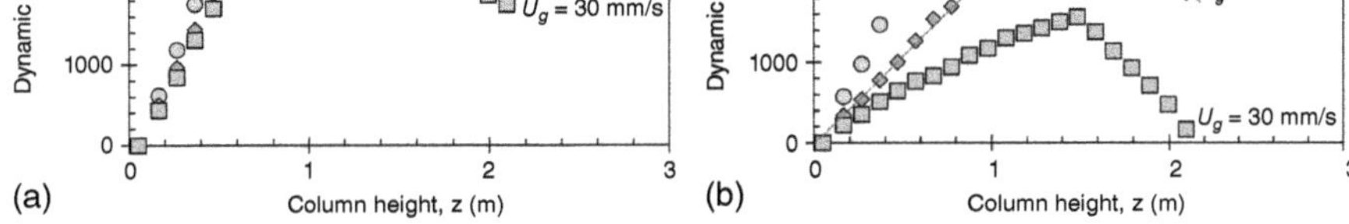

Figure 22.7 Gas–liquid–solid fluidized bed dynamic pressure profiles in an air-aqueous ethanol-spent LC-FINING catalyst system for (a) U_L = 15 mm/s and (b) U_L = 36 mm/s. Source: Adapted from Parisien et al. 2017 [30].

solids in the fluidized bed:

$$\varepsilon_p = \frac{4M_p}{\pi d_c^2 H \rho_p} \tag{22.8}$$

Neglecting frictional drag on the wall and accelerations of the phases in the vertical direction, the gas holdup in the bed region is calculated via the bed region dynamic pressure profile:

$$\varepsilon_g = \frac{(\Delta P_{dyn}/\Delta z)g^{-1} + (\rho_p - \rho_L)\varepsilon_p}{\rho_L - \rho_g} \tag{22.9}$$

The bed region liquid holdup (ε_L) is calculated knowing that the sum of phase holdups must be unity. Gas holdups in the freeboard region are also estimated based on the dynamic pressure profile above the bed with ε_p set to zero.

The overall impact of operating conditions, within a given flow regime, is qualitatively summarized by Table 22.2. Note that an increase in liquid velocity can improve "distributor quality" when gas and liquid are jointly injected. The impact of gas velocity is complex as the introduction of gas to a liquid–solid fluidized bed can lead to bed expansion or contraction and then expansion with further increase in gas flow rate (see Section 22.4.2.2).

Larachi et al. [15] correlated global phase holdups using a combination of artificial neural networks and dimensional analysis. Correlations were developed using a database containing 55 Newtonian liquids (20 500 data) and 19 non-Newtonian liquids (2500 data), 5 gases (including H_2), and 110 particles differing in shape and size, with volume-equivalent diameters up to 15 mm, 17 column diameters, and operation at elevated pressure and temperature (up to 15.6 MPa and 363 K). They provide a user-friendly Microsoft Excel™ spreadsheet that allows the phase holdups to be computed (see https://www.gch.ulaval.ca/flarachi/). The gas holdup dependency was already presented in Eq. (22.2); below

Table 22.2 Qualitative effect of changes of operating conditions and physical properties on bed global phase holdups.

Increase of	ε_p	ε_g	ε_L
U_L	−	=	+
U_g	+ or −	+	−
d_p	+	?	?
ρ_p	+	−	?
ρ_L	−	−	?
μ_L	−	−	+
ρ_g	−	+	−
Coalescence inhibition	−	+	−
Distributor quality	−	+	=

+, increase; −, decrease; =, constant; ?, unclear.

Source: Duduković et al. 2002 [3]. Reproduced with permission of Taylor and Francis.

are the resulting dimensionless groups for the liquid holdup and bed porosity. Note that only two of the three holdup equations are independent since the sum of the phase holdup fractions is unity:

$$\varepsilon_L = f(U_L/u_{p,t}, Ca_{gL}, Mo, 1 - \rho_L/\rho_p, d_p/d_c) \tag{22.10}$$

$$\varepsilon = 1 - \varepsilon_p = f(Fr_L, Re_{Lg}, St_{Lg}, U_L/u_{p,t}, Eo_p, d_p/d_c) \tag{22.11}$$

The correlations developed by Han et al. [31] are also acceptable for estimating global phase holdups at atmospheric conditions for liquids without surface-active solutes that significantly affect bubble coalescence (e.g. liquids that do not froth/foam). Note that for the particle holdup, the liquid–solid fluidized bed behaviour (expansion or contraction) at the introduction of gas is an a priori:

$$\varepsilon_L = \left(\frac{U_L}{u_{p,t}}\right)^{1/n} (1 - 0.374 Fr_g^{0.176} We_c^{-0.173}) \tag{22.12}$$

For $0.054 \leq U_L/u_{p,t} \leq 0.899$, $0.000\,039\,5 \leq Fr_g \leq 2.551$, $0.000\,918 \leq We_c \leq 5.013$

When the liquid–solid fluidized bed expands at gas introduction,

$$\varepsilon_p = 1 - \left(\frac{U_L}{u_{p,t}}\right)^{1/n} (1 + 0.123 Fr_g^{0.347} We_c^{0.037}) \tag{22.13}$$

For $0.029 \leq U_L/u_{p,t} \leq 0.481$, $0.000\,017\,3 \leq Fr_g \leq 3.788$, $0.008\,95 \leq We_c \leq 5.479$

When the liquid–solid fluidized bed contracts at gas introduction,

$$\varepsilon_p = 1 - \left(\frac{U_L}{u_{p,t}}\right)^{1/n} \left[0.359 Fr_g^{0.552} We_c^{0.124} + \exp\left(-\left(\frac{U_L}{u_{p,t}}\right)^{0.305} Fr_g^{0.5}\right)\right] \tag{22.14}$$

For $0.024 \leq U_L/u_{p,t} \leq 0.899$, $0.000\,092\,8 \leq Fr_g \leq 2.551$, $0.000\,065\,9 \leq We_c \leq 1.529$

Alternatively, the semi-empirical k–x generalized "bubble wake model" [32] can be used to estimate the bed porosity. Analogous to the bubbling bed model in gas–solid fluidization, the three-phase fluidized bed is then divided into three regions: gas bubbles, liquid–solid wakes behind rising bubbles, and a gas-free liquid–solid fluidized bed (see illustration in Figure 22.1a). The model is based on three assumptions: (i) wakes rise at the same linear velocity as the bubbles, and thus the entrained liquid volume is proportional to the gas volume ($k = \varepsilon_w/\varepsilon_g$), (ii) the solids holdup in the wake is proportional to that in the liquid–solid fluidized bed ($x = \varepsilon_{p,w}/\varepsilon_{p,LS}$), and (iii) the liquid holdup and relative liquid–solid fluidized bed velocity follow the Richardson–Zaki model [33]. After material balances, the following bed porosity equation is then obtained:

$$\varepsilon = 1 - \varepsilon_p = \left(\frac{U_L - U_g k(1-x)}{u_{p,t}(1 - (1+k)\varepsilon_g)}\right)^{1/n} (1 - (1 + k(1-x))\varepsilon_g) + (1 + k(1-x))\varepsilon_g \tag{22.15}$$

The challenge in using this model is accurately choosing the parameters k and x, as well as acquiring the gas holdup from an independent source. Models for k and x have been summarized by Larachi et al. [15], with x usually assumed to be zero (i.e. no solids in the wake). Epstein [32] recommends the following models where k is determined by considering that the bubble wake has the sphere-completing volume to a spherical-cap bubble and the solids partition coefficient x is then extracted from measured holdups and Eq. (22.15). More details on bubble wake dynamics are presented by Fan and Tsuchiya [34]:

$$k = \left(0.61 + \frac{0.037}{\varepsilon_g + 0.013}\right)\varepsilon^3 \tag{22.16}$$

$$x = 1 - 0.877y \text{ for } 0 \leq y \leq 1.14 \tag{22.17}$$

$$x = 0 \text{ for } y > 1.14 \tag{22.18}$$

$$y = u_{p,t}\left(\frac{U_g}{\varepsilon_g} - \frac{U_L}{\varepsilon_L}\right)^{-1} \tag{22.19}$$

22.4.2.2 Bed Contraction vs Expansion

Introducing a small gas flow into a liquid–solid fluidized bed can lead to either bed contraction or expansion. The rise velocity of gas bubbles is greater than the mean velocity of the liquid, and the wake that rises with the bubbles may present a much lower solids concentration than the bulk of the bed. The presence of gas thus has a twofold effect, with the net impact on the bed movement depending on the bubble size and rise velocity as well as the volume and solid content of the bubble wakes. First, the volume occupied by the gas decreases the volume available for the liquid and solid, which increases the liquid interstitial velocity and drag on the particles, causing increases in bed height and porosity. Second, the liquid in the bubble wake may not participate in the fluidization, resulting in less drag on the particles, lower bed height, and reduced porosity. The k–x generalized bubble wake model can predict the bed behaviour when gas is introduced. When

$$\psi = \left(\frac{n}{n-1} + k\right)\frac{U_L}{\varepsilon_L} + \frac{xk(U_g/\varepsilon_g)}{n-1} - \left[(1+k)U_L + \frac{k}{n-1}\left(\frac{U_g}{\varepsilon_g} - \frac{U_L}{\varepsilon_L}\right)\right] + xk\left(U_L - \frac{U_L}{\varepsilon_L}\right)\left(\frac{n}{n-1}\right) \tag{22.20}$$

is less than zero, addition of gas leads to contraction; if, on the other hand, $\psi > 1$, the bed expands when a little gas is added.

22.4.2.3 Particle Entrainment

For liquids of low to moderate viscosity (i.e. similar to water), the interface between the bulk bed and freeboard regions is typically more distinct for relatively large/heavy particles (e.g. $d_p > 3$ mm; $\rho_p > 2500$ kg/m^3) than for small/light particle systems (e.g. $d_p < 1$ mm; $\rho_p < 1700$ kg/m^3). As in gas fluidization (see Figure 22.1a and Chapter 10), particles can be entrained from the bed surface into the freeboard via the bubble wake (primarily smaller particles) and drift (primarily larger, less dense particles). Particles in the drift are confined near the bed surface and tend to settle in the freeboard due to gravity, whereas particles in the bubble wake travel further upwards and are progressively released via vortex shedding from the wake. Furthermore, depending on the apparent rheology of the bed, the bubbles may split in the freeboard due to the lower effective liquid viscosity than in the bed. The transport disengagement height (H_m) can thus be modelled based on wake formation and shedding:

$$H_m = u_b N_m / f_c \tag{22.21}$$

where u_b is the bubble rise velocity; f_c the vortex shedding frequency, which is approximated by the bubble rocking frequency; and N_m is the number of shedding/rocking cycles before all particles are released from the wake. Models for all three parameters, as well as step-by-step visualization of the particle entrainment/de-entrainment mechanism, have been provided by Miyahara et al. [35].

22.5 Phase Mixing, Mass Transfer, and Heat Transfer

Trends and models to characterize phase mixing patterns, surface-to-bed heat transfer, and interphase mass transfer rates are summarized here, with greater details available in Refs. [1–5]. Bubble dynamics (size, shape, rise velocity, and motion) again play a crucial role on the flow regime and its associated transport rates, so the resulting correlations are often limited to the experimental conditions studied.

22.5.1 Phase Mixing

If dispersed plug flow is selected as the liquid flow model for the material balance in Eq. (22.1), then the liquid axial dispersion coefficient (E_{zL}) spatially averaged over the entire bed volume can be predicted [36] by

$$Pe_{zL} = \frac{d_p U_L}{E_{zL}} = c_1 \left(\frac{U_L}{U_L + U_g} \right)^{1.03} \left(\frac{d_p}{d_c} \right)^{1.66} \tag{22.22}$$

Here, axial liquid mixing is primarily due to wake formation and shedding behind rising bubbles. The dispersion coefficient is thus typically lower in the dispersed than for the coalesced bubble flow regime. For coalesced bubble flow, $c_1 = 11.96$, and the equation applies for $0.004 \le d_p/d_c \le 0.024$ and $0.250 \le U_L/(U_L + U_g) \le 0.857$. For dispersed bubble flow, $c_1 = 20.72$, and the equation is intended for $0.004 \le d_p/d_c \le 0.012$ and $0.143 \le U_L/(U_L + U_g) \le 0.857$.

The applicability of the dispersion model for the gas phase is questionable when a relatively large bubble size distribution exists (i.e. coalesced bubble flow regime). For dispersed bubble flow and narrow bubble size distributions, the gas is usually assumed to be in plug flow.

Particle mixing in the bed follows a similar mechanism as for entrainment into the freeboard: Larger bubbles capture particles in their wake and periodically release them via vortex shedding, whereas smaller bubbles displace particles via their drift. The impact of bubble wake and drift dynamics within the bed increases as particles become smaller or less dense. Solids axial gradients can also occur when there is a significant particle size and/or density distribution. A segregation or sedimentation–dispersion model, such as for slurry bubble columns, is then used to model the solids axial concentration profile [2]. Consider the dynamic pressure drop axial profiles presented in Figure 22.7. The spent catalyst particles are considered to be relatively heavy, but they have a wide density distribution. This leads to axial particle holdup gradients when the bubbles are sufficiently large (U_g = 30 mm/s) for effective wake transport, leading to nonlinear curves in Figure 22.7a and the associated image in Figure 22.8, right. Since the liquid and gas are co-fed below the distributor, a greater liquid flow rate increases the energy dissipation rate across the distributor, resulting in smaller bubbles [16]. The impact of both wake and drift dynamics is then insufficient to measure a significant phase holdup axial gradient, as evidenced by the near-linear pressure profile (Figure 22.7b) and a sharper interface between the bed and freeboard (Figure 22.8, left).

22.5.2 Surface-to-Bed Heat Transfer

The surface-to-bed heat transfer coefficient (h) is used to estimate the rate of heat removal from, or addition to, the reactor, affecting the temperature profile and thus reaction rates. Equation (22.23) extends a heat transfer model for liquid–solid fluidization, Eq. (22.24), for a given particle holdup. It is applicable for both atmospheric and elevated pressures, via the gas holdup term, and

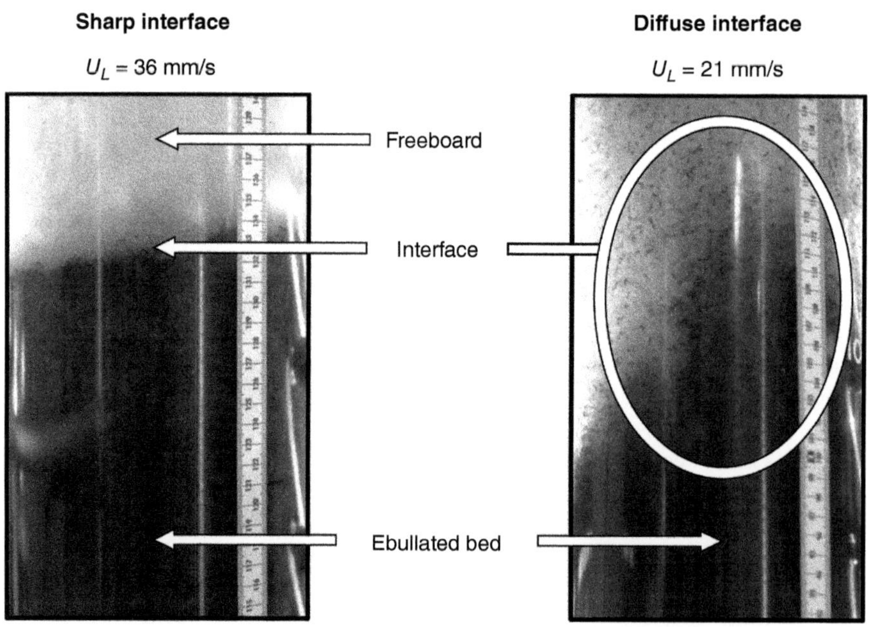

Figure 22.8 Impact of particle–bubble interaction on sharpness of bed–freeboard interface at U_g = 15 mm/s for gas and liquid co-fed below distributor. Source: Adapted from Parisien et al. 2017 [30].

is based on experimental data of Luo et al. [37]:

$$h_{GLS} = h_{LS}\varepsilon_g^{0.45}\left(\frac{0.396}{U_g^{0.45}} + \frac{0.677}{u_{p,t}}\right) \text{ (SI units)} \tag{22.23}$$

$$Nu_{LS} = \frac{d_p h_{LS}}{\kappa_L} = 0.67\left(\frac{d_p \rho_L U_L}{\mu_L}\right)^{0.62}\left(\frac{C_{pL}\mu_L}{k_L}\right)^{0.33}\frac{\varepsilon_p^{0.38}}{1-\varepsilon_p} \tag{22.24}$$

The heat transfer coefficient is more dependent on the liquid, than on the gas physical, and thermal properties, and it is also strongly influenced by the phase velocities. A maximum h_{GLS} is reached when increasing the liquid superficial velocity due the increased bed porosity, while the introduction of gas increases h relative to a liquid–solid fluidized bed of the same particle holdup. Table 22.3 summarizes the impact of various operating conditions.

22.5.3 Interphase Gas–Liquid and Liquid–Solid Mass Transfer

Solid-catalyzed gas–liquid reactions can be expressed as

$$A(g \xrightarrow{\text{absorption}} L) + B(L) \xrightarrow{\text{catalyst site}} \text{products} \tag{22.25}$$

Species transfer from the bubble to the liquid (gas–liquid mass transfer). Liquid and dissolved gas species then transfer to the solids surface (liquid–solid mass transfer) and finally through the particle pores (intra-particle diffusion) to the reaction site where adsorption and reaction ultimately occur. Reaction products make their way back in reverse order. The relative rates of intra-particle diffusion and surface reaction are characterized by the Thiele modulus, from which an internal effectiveness factor ranging from 0 (diffusion limited) to 1 (reaction

Table 22.3 Qualitative effect of varying operating conditions and physical and thermal properties on three-phase heat and mass transfer coefficients.

Increase of	h	$k_L a_b$
U_L	+; −	+
U_g	+	+
d_p	+	+
d_c	+	?
C_{pL}	+	
κ_L	+	
ρ_p	+	+
ρ_L	?	?
μ_L	−	−
σ	?	−
Coalescence inhibition	−	+
Distributor quality	?	+

+, increase; −, decrease; ?, unclear.
Source: Duduković et al. 2002 [3]. Reproduced with permission of Taylor and Francis.

limited) can be calculated (see Ref. [9] for details). Figure 22.9 illustrates this interphase material transfer via concentration gradients, with equilibrium assumed at the interfaces. Note that the system pressure is uniform among all phases. Temperature differences across the phase boundaries and within the solid particle occur due to energy transfer from gas absorption and reaction, although they are typically minor because of the relatively large heat capacity of the continuous liquid phase [9].

22.5.3.1 Gas–Liquid Mass Transfer

For gases that are sparingly soluble, usually true, the resistance to mass transfer resides in the liquid phase. The liquid-side mass transfer coefficient (k_L) accounts for the liquid flow field surrounding rising gas bubbles, while the interfacial area (a_b) is the important bubble dimension. Gas–liquid mass transfer correlations commonly combine these two quantities into a volumetric mass transfer coefficient ($k_L a_b$) that is again typically averaged over the entire bed volume and dependent on the accuracy of the liquid mixing model selected. Schumpe et al. [38] developed a correlation at atmospheric conditions using Newtonian and non-Newtonian aqueous solutions, which is suited to biological processes such as fermentation:

$$k_L a_b / D_{i,L}^{0.5} = 2988\, U_g^{0.44} U_L^{0.42} \mu_{eff}^{-0.34} u_{p,t}^{0.71} \text{ (SI units)} \tag{22.26}$$

and applies for $0.017 \leq U_g \leq 0.118$ m/s, $0.03 \leq U_L \leq 0.16$ m/s, $0.001 \leq \mu_{eff} \leq 0.119$ Pa s, and $0.08 \leq u_{p,t} \leq 0.60$ m/s.

As previously mentioned, hydroprocessing operating conditions often involve surface-active species and elevated pressures, leading to the formation of bubbles smaller than 1 mm in diameter with relatively narrow size distributions. Since published three-phase fluidized bed mass transfer correlations do not cover this fluid dynamic range, it is best to estimate $k_L a_b$ with correlations developed for bubble columns and slurry bubble columns that achieve similar gas–liquid interfacial areas, such as for Fischer–Tropsch synthesis [39, 40].

22.5.3.2 Liquid–Solid Mass Transfer

Liquid–solid mass transfer coefficients (k_s) in GLS fluidized beds can be correlated by application of Kolmogorov's theory of local isotropic turbulence. For example, a correlation that relates the increase in energy input by the gas to the increment in k_s relative to the corresponding liquid–solid (two-phase) fluidized bed [41] is

$$\frac{Sh_{GLS} - Sh_{LS}}{Sh_{LS}} = 0.237 \left(\frac{U_g^3 \rho_L^5}{g \mu_L (\rho_p - \rho_L)^4} \right)^{0.144} \varepsilon_g^{0.170} \tag{22.27}$$

$$\text{with } Sh_{LS} = 0.34 (Ar_L Sc)^{\frac{1}{3}} \tag{22.28}$$

which applies for $6.4 \leq Re_L \leq 2530$, $0 \leq Re_g \leq 120$, $8500 \leq Ar_L \leq 23.4 \times 10^6$, $860 \leq Sc \leq 19\,900$, and $0.27 \leq (\rho_p/\rho_L - 1) \leq 7.0$.

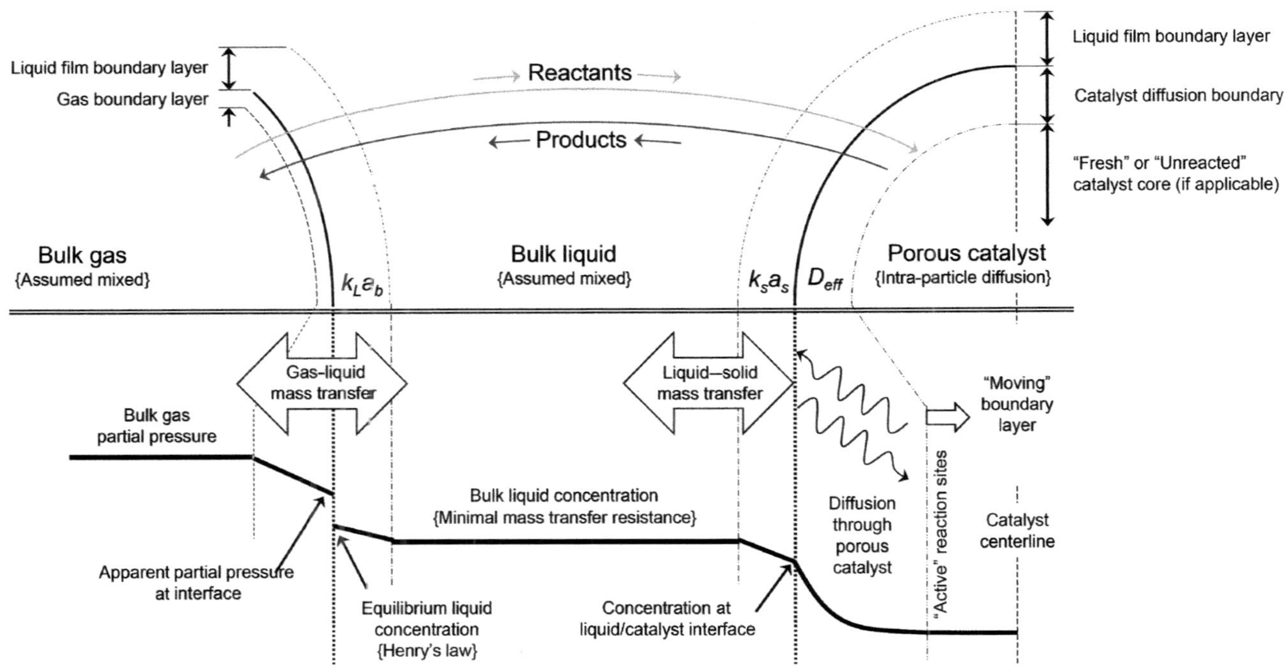

Figure 22.9 Species concentration gradients across phase boundaries for a solid-catalyzed gas–liquid reaction. Source: Adapted from Levenspiel 1999 [9].

22.6 Summary

The performance of three-phase fluidized bed systems can be successfully modelled by understanding the regions of interest (plenum, distributor, bulk bed, and freeboard), relevant scales (time, length), and impact of physical properties and operating conditions on fluid dynamics (incipient fluidization, flow regimes, phase holdups), transport correlations (heat and mass transfer), and reaction rates. Current best practices involve compartmentalizing the regions of interest and developing scale-appropriate conservation equations. These equations can then be simplified and solved through understanding of the rate-limiting steps and phenomena of interest, with the impact of assumptions assessed through sensitivity analysis to determine how these assumptions affect the outcome. This last step continues to be of great importance, since reliable models describing transport phenomena in these systems have often not been derived under industry-specific conditions and should be experimentally tested through pilot-scale trials if critical to the success of an application. Solved Problem 22.1 provides an example of how this approach is applied to the analysis of the catalytic conversion of heavy oil in a commercial three-phase fluidized bed hydrocracker.

Solved Problems

22.1 The design and sizing of a gas–liquid–solid fluidized bed reactor is demonstrated in this example based on the reaction kinetics and transport phenomena of an ebullated bed hydrocracker. Figure 22.2a shows four regions that can be considered for the hydrocracker reactor model: (i) plenum including the bubble cap grid, (ii) ebullated bed, (iii) freeboard including the gas–liquid separator, and (iv) liquid recycle line [42]. The complexity and interdependence of these regions must be considered to calculate the ebullated bed transport parameters and overall reactor performance. For example, at selected overall fluid throughputs for the hydrocracker, initial estimates of the liquid recycle ratio and gas–liquid separation efficiency are required to determine the recycled gas and liquid flow rates through the ebullated bed; however, the resulting total gas and liquid flow rates exiting the ebullated bed also influence the phase separation efficiency in the freeboard. This demonstrates the control volume interdependence due to the internal fluid recirculation; for the hydrocracker nevertheless, such an iterative procedure is typically required when designing an ebullated bed reactor (e.g. varying the diameter) to obtain desirable transport properties (e.g. phase holdups, bed inventory).

Solution

Figure 22P.1 recommends a procedure to size and design a gas–liquid–solid fluidized bed reactor, including an iterative compartmentalized model to obtain converged transport parameters in the ebullated bed region. This worked example focuses on simplified heavy oil conversion

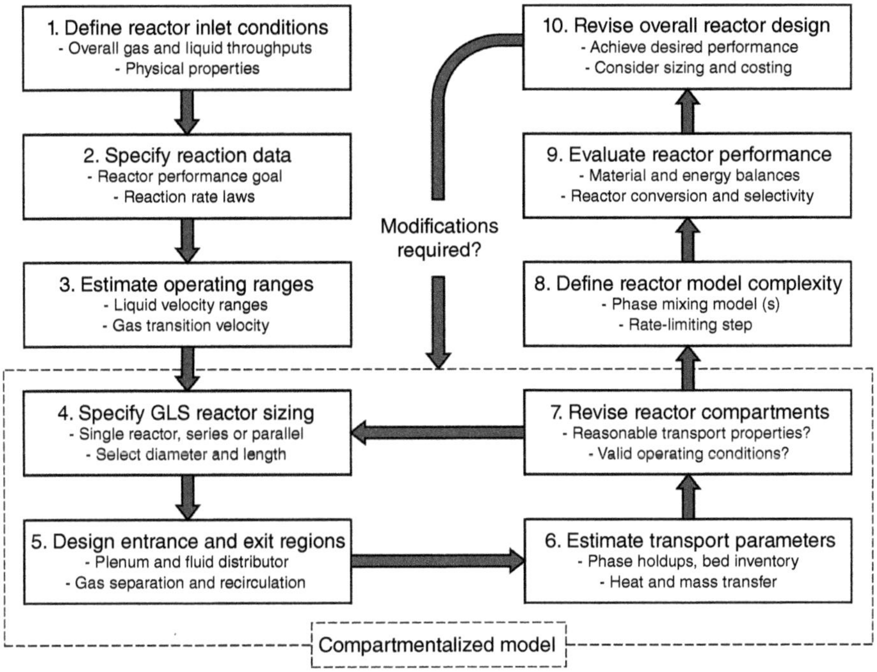

Figure 22P.1 Gas–liquid–solid fluidized bed reactor sizing and design procedure.

reactions and includes assumptions based on prior work to estimate a converged compartmentalized model (i.e. resolved control volume boundary conditions), streamlining the overall reactor performance calculation.

Define reactor inlet conditions: Gas and liquid throughputs and phase physical properties for an industrial scale gas–liquid–solid fluidized bed hydrocracker can be estimated from the literature. Table 22P.1 provides approximate ranges and selected values for this example.

Specify reaction data: Heavy oil hydrocracking models often use lumped kinetics based on boiling point ranges. This example, focusing on heavy oil conversion, uses the reaction model proposed by Parulekar and Shah [44], which accounts for heavy oil (A) conversion to light oil (B) and volatile organics (C) (Table 22P.2).

Estimate operating ranges: The operating ranges for the gas and liquid superficial velocities must be identified in the ebullated bed region. The liquid superficial velocity must be greater than the minimum liquid fluidization velocity and significantly lower than the particle terminal velocity (i.e. the presence of gas increases the interstitial liquid velocity in the ebullated bed). The minimum liquid fluidization velocity is estimated using the correlation of Zhang et al. [24], based on a first estimate without gas flow ($\alpha_{mf} = 0$). The particle terminal velocity for cylindrical particles is estimated using the correlation of Haider and Levenspiel [46].

Table 22P.1 Operating conditions and phase physical properties for worked example.

	Parameters	Ranges	Example values	References
Operating conditions	Temperature (°C)	400–450	400[a]	[6, 43]
	Pressure (MPa)	7–20	15	[6, 43]
	Liquid throughput (bbl/d)	20 000–70 000	30 000	[43]
	Gas throughput (m^3(STP)/d)	1 650 000	1 650 000	[6]
Solid particle properties	Particle cylinder length (mm)	1.5–6.0	5	[43]
	Particle cylinder diameter (mm)	0.9–1.3	1.2	[6, 43]
	Wet particle density (kg/m^3)	1300–2000	2000	[30, 44]
Liquid properties	Liquid density (kg/m^3)	700–900	800	[6, 44]
	Liquid viscosity (Pa s)	0.001–0.009	0.005	[44, 45]
	Interfacial tension (N/m)	0.008	0.008	[44]
Gas properties	Gas density (kg/m^3)	40–50	45	[6, 45]

[a] Operating temperature selected based on applicable range for reaction kinetic data.

Table 22P.2 Heavy oil conversion reactions and kinetic parameters.

Reaction	Rates of reaction (r_j)	Units for r_j	Frequency factor (A_j)	Activation energy (E_j)
(1) $A_{(L)} \rightarrow B_{(L)}$	$r_1 = k_1 C_A C_{slurry}$	$\frac{\text{kg A}}{\text{m}^3\ \text{liquid} \cdot \text{s}}$	$A_1 = 20\ \frac{\text{m}^3\ \text{slurry}}{\text{kg cat} \cdot \text{s}}$	$E_1 = 82.8\ \frac{\text{kJ}}{\text{mol}}$
(2) $A_{(L)} \rightarrow C_{(g)}$	$r_2 = k_2 C_A C_{H_2} C_{slurry}$	$\frac{\text{gmol C}}{\text{m}^3\ \text{liquid} \cdot \text{s}}$	$A_2 = 10\ \frac{\text{m}^3\text{gas} \cdot \text{m}^3\text{slurry}}{\text{kg cat} \cdot \text{kg A} \cdot \text{s}}$	$E_2 = 116.0\ \frac{\text{kJ}}{\text{mol}}$

Here C_A is the heavy oil concentration in the liquid (kg A/m^3 liquid), C_{H_2} is the hydrogen concentration in the gas (gmol H_2/m^3 gas), C_{slurry} is the catalyst concentration in the liquid–solid slurry (kg cat/m^3 slurry), and the kinetic constants are determined using the Arrhenius equation ($k_j = A_j e^{-E_j/RT}$). The liquid fed to the reactor is assumed to consist entirely of heavy oil ($x_{A0} = 1$). The gas fed to the reactor is assumed to be pure hydrogen ($P_{H_2} = P_{inlet}$). Note that the rate laws and component concentrations are dependent on the overall phase holdups.

Source: Parulekar and Shah 1980 [44]. Reproduced with permission of Elsevier.

Hydrocrackers typically operate in the dispersed bubble flow regime to reduce gas holdup, thus increasing the liquid holdup (i.e. higher liquid residence time). The transition velocity from dispersed to coalesced bubble flow therefore provides an upper limit to the gas velocity, which is estimated based on the correlation proposed by Wilkinson et al. [47] for bubble columns, since no more appropriate model currently exists for three-phase fluidized beds.

Table 22P.3 summarizes the results for the liquid and gas operating ranges. The liquid superficial velocity should thus be maintained in the range $0.035\ \text{m/s} < U_L \ll 0.165\ \text{m/s}$ to ensure fluidization with negligible particle carry-over. The gas superficial velocity should be significantly below 0.39 m/s, which is an elevated estimate for high-pressure conditions, to ensure operation in the dispersed/homogeneous bubble flow regime.

Specify GLS reactor sizing: A single reactor is first assumed, which can be revised later based on the predicted reactor performance. The column diameter is then fixed as an ebullated bed sizing parameter due to its influence on the gas and liquid superficial velocities. The ebullated bed height can then be adjusted to obtain the desired heavy oil conversion. Experiments described in McKnight et al. [6] provide an ebullated bed inner

Table 22P.3 Minimum fluidization velocity, particle terminal velocity, and gas transition velocity.

Parameter	Symbol	Equation	Value	Units
Volume-equivalent diameter	d_v	$(1.5d_p^2L_P)^{1/3}$	2.2	mm
Archimedes number	Ar_L	$\rho_L d_v^3(\rho_p - \rho_L)g/\mu_L^2$	4.1×10^5	—
Bed voidage at minimum fluidization	ε_{mf}	Equation (22.7)	0.46	—
Sphericity	ϕ	$d_v^2/(0.5d_p^2 + d_pL_P)$	0.73	
Reynolds number at minimum fluidization	Re_{Lmf}	Equation (22.3)	123	—
Minimum liquid fluidization velocity	U_{Lmf}	$Re_{Lmf}\mu_L\rho_L^{-1}d_p^{-1}$	0.035	m/s
Reynolds number at terminal velocity	$Re_{p,t}$	$Ar_L^{1/3}\left(\frac{18}{Ar_L^{2/3}} + \frac{2.335-1.744\phi}{Ar_L^{1/6}}\right)^{-1}$	582	—
Particle terminal velocity	$u_{p,t}$	$Re_{p,t}\mu_L\rho_L^{-1}d_p^{-1}$	0.165	m/s
Small bubble rise velocity	u_{small}	$2.25\left(\frac{\sigma}{\mu_L}\right)\left(\frac{\sigma^3\rho_L}{g\mu_L^4}\right)^{-0.273}\left(\frac{\rho_L}{\rho_g}\right)^{0.03}$	0.15	m/s
Transition velocity	$U_{g,\,trans}$	$0.5\,u_{small}\exp(-193\,\rho_g^{-0.61}\mu_L^{0.5}\sigma^{0.11})$	0.39	m/s

diameter estimate of 3 m for the selected 30 000 bbl/d liquid throughput. The impact of the cross-sectional surface area occupied by an internal liquid recycle line is neglected for the volume and superficial velocity calculations.

Design entrance and exit regions: As described in Section 22.3, the plenum chamber and fluid distributor design affect the gas and liquid dispersion in the ebullated bed. The gas–liquid separation in the freeboard for liquid recirculation also affects the ebullated bed gas and liquid holdups and thus the reactor performance. The following assumptions are adopted for this example to define the boundary conditions between the reactor control volumes:

- **Plenum chamber and fluid distribution**: Recirculating liquid and feed flow are perfectly mixed; gas and liquid flows are one-dimensional (i.e. distributed evenly across the column cross-sectional area).
- **Freeboard gas–liquid separation**: 10% of the total gas flow rate exiting the ebullated bed is entrained in the liquid recycle line.
- **Liquid recirculation**: 85% of the liquid flow rate exiting the ebullated bed is recycled; therefore, the recycle flow is 5.7 times larger than the liquid fed to the reactor. The liquid flow pattern can thus be considered to be perfectly mixed.

It should be noted that the constant gas entrainment and liquid recycle estimates are simplifications based on a converged compartmentalized model. Gas and liquid superficial velocities in the ebullated bed region, required to estimate transport parameters, can thus be calculated based on the feed flow rates and assumed entrance and exit performances (Table 22P.4).

Estimate transport parameters: Most ebullated bed correlations in the literature are based on aqueous systems at atmospheric pressure, complicating the estimation of transport parameters for elevated pressure and temperature conditions. The relatively high liquid recirculation, effective fluid distribution at the grid, and phase physical properties at hydrocracking conditions lead to dispersed bubble flow at relatively high gas holdups, which does not correspond to most available correlations. Phase holdups are thus estimated using the correlations proposed by Pjontek et al. [45], as they were developed under analogous operating conditions. The particle holdup is estimated using a modified version of the well-known Richardson–Zaki model [33]; the gas holdup is calculated

Table 22P.4 Gas and liquid superficial velocities in the ebullated bed region.

	Reactor throughput (m^3/s)	Recycled flow rate (m^3/s)	Superficial velocity (m/s)
Liquid	0.055	0.313	0.052
Gas	0.318	0.035	0.050

Table 22P.5 Ebullated bed and freeboard region phase holdups.

Parameter	Symbol	Equation	Value	Units
Index for bed expansion	n	$\frac{4.8-n}{n-2.4} = 0.043Ar_L^{0.57}\left[1 - 1.24\left(\frac{d_v}{d_c}\right)^{0.27}\right]$	2.45	—
Wall effect correction factor	k_{wall}	$k_{wall} = 1 - 1.33 \cdot \frac{d_v}{d_c}$	0.999	—
Freeboard gas holdup	$\varepsilon_{g,GL}$	$\frac{\varepsilon_{g,GL}}{1-\varepsilon_{g,GL}} = 0.79\left(\frac{u_g}{u_L}\right)^{1.12}$	0.45	—
Ebullated bed particle holdup	ε_p	$1 - \varepsilon_p = \left(\frac{u_L}{k_{wall}u_{p,t}}\right)^{1/n}\left(1 + 0.22\left(\frac{u_g}{u_L}\right)^{0.92}\right)$	0.24	—
Ebullated bed gas holdup	ε_g	$\varepsilon_g = \varepsilon_{g,GL} \cdot (1 - \varepsilon_p)$	0.34	—
Ebullated bed liquid holdup	ε_L	$\varepsilon_L = 1 - \varepsilon_p - \varepsilon_g$	0.42	—

by assuming that the ebullated bed region gas holdup on a particle-free basis is comparable with the freeboard gas holdup; and the liquid holdup is determined by difference, i.e. by 1 minus gas holdup minus solids holdup (Table 22P.5).

The interphase mass and energy transport phenomena are based on the following observations in prior work [43, 48, 49]:

1. The reaction network kinetics [44] are based on the hydrogen gas-phase concentration; thus, the rate constants embed the mass transport resistances from the bubbles to the reaction sites in the catalyst particle. In order to compare transport and reaction rates, reaction 2 can be rendered pseudo-first order with respect to C_A in view of the excess of hydrogen, yielding an apparent rate constant of ~0.0003 s^{-1}. Under hydrocracking conditions, though, gas–liquid and liquid–solid interphase mass transport rates are typically orders of magnitude greater than the reaction itself. For example, the volumetric gas–liquid mass transfer coefficient ($k_L a_b$) and volumetric liquid–solid mass transfer coefficient ($k_S a_S$) both range from 0.1 to 1 s^{-1}, based on approximate bubble sizes and prior work [2]. Moreover, the catalyst pore structure is such that intra-particle diffusion does not significantly affect the reaction rate; thus the internal effectiveness factor (η) is ~1 [50].
2. Temperature gradients between phases are considered negligible, and the high liquid recirculation results in approximately isothermal conditions throughout the reactive volume.

Revise reactor compartments: It is recommended at this step in the design procedure to re-evaluate the reactor sizing and configuration based on their impact on the operating conditions and transport parameters. As previously stated, the interdependency of the reactor model control volumes (i.e. plenum, ebullated bed, gas–liquid separation, recycle line) has been simplified in this example. Below are some suggested considerations for gas–liquid–solid fluidized bed reactors:

- Column diameter modifications to adjust the gas and liquid superficial velocities.
- Liquid recirculation rate to ensure operation between minimum fluidization and well below particle terminal settling velocity.
- Modification of the gas–liquid separator to reduce gas entrainment in the recycle line.

Define reactor model: Material balances for the gas, liquid, and solid phases are presented. The equations assume no variation in the radial and angular directions and can be applied to any control volume (i.e. compartment) within the overall reactor. Following Eq. (22.1), a hypothetical component "*i*" is defined, which is present in the gas phase, absorbs in the liquid phase, is transported to the catalyst particle surface, and diffuses into the catalyst pores where the chemical reaction occurs:

$$\text{Gas phase}: \frac{\partial}{\partial z}\left(\varepsilon_g E_{zg,GLS}\frac{\partial C_{gi}}{\partial z}\right) - \frac{\partial (U_g C_{gi})}{\partial z} - k_L a_b\left(\frac{P_i}{H_i} - C_{Li}\right) = \frac{\partial (\varepsilon_g C_{gi})}{\partial t}$$

$$\text{Liquid phase}: \frac{\partial}{\partial z}\left(\varepsilon_L E_{zL,GLS}\frac{\partial C_{Li}}{\partial z}\right) - \frac{\partial (U_L C_{Li})}{\partial z} + k_L a_b\left(\frac{P_i}{H_i} - C_{Li}\right) - k_S a_S(C_{Li} - C_{Si}) = \frac{\partial (\varepsilon_L C_{Li})}{\partial t}$$

$$\text{Solid phase}: k_S a_S(C_{Li} - C_{Si}) = \sum_{j=1}^{n} \eta_j r_j$$

The above model accounts for material convection, axial dispersion, interphase mass transfer, and catalyst effectiveness factor. For control volumes without particles (i.e. a_S and $C_{Si} = 0$), the model reduces to include only the gas–liquid terms. Important characteristics of the ebullated bed hydrocracker allow for simplifications to the material balances. First, the heavy oil conversion reactions only occur in the presence of the solid catalyst. Since the focus is to size the reactor, the remainder of this worked example determines the ebullated bed control volume required to obtain the desired conversion (i.e. design and sizing of the gas–liquid distribution at the entrance, gas–liquid separation in the freeboard, and recycle line are not performed in this example). Furthermore, the internal liquid recirculation rate of 85% causes the heavy oil concentration in the ebullated bed to be highly dependent on the concentration exiting the recycle line, rather than the reactor feed concentration. The previous flow behaviour results in well-mixed conditions for the liquid phase in the ebullated bed, which will be approximated as equivalent to a continuous stirred tank reactor. The following assumptions are used to further simplify the model:

(1) Gas phase is described by the ideal gas law.

(2) Hydrogen is greatly in excess and minor production of volatiles, maintaining constant hydrogen partial pressure and axial gas and liquid superficial velocities.
(3) Interphase and intra-particle transport rates are significantly greater than reaction rates.
(4) Isothermal operation such that an energy balance is only required to estimate the rate of energy transfer between the bed and heat exchanges surfaces (not performed here).
(5) Isobaric operation since the pressure loss across the fluidized bed is negligible relative to the absolute pressure.

The ebullated bed model for this example then simplifies to the following steady-state mass balance for the heavy oil component:

$$F_{A0} - F_A + (-r_1' - r_2')V = 0$$

where F_{A0} and F_A are the inlet and outlet heavy oil mass flowrates (kg A/s), r_1' and r_2' are modified reaction rates based on the ebullated bed region volume (kg/m^3 s), and V is the ebullated bed volume (m^3). The above equation can be rewritten as a function of the liquid throughput (Q_L), column cross-sectional area (A_C), ebullated bed height (H), and exiting heavy oil liquid concentration (C_A):

$$C_A - C_{A0} + (r_1' + r_2')A_C H/Q_L = 0$$

Evaluate reactor performance: The simplified material balance using the CSTR approximation determines the relationship between ebullated bed height and heavy oil conversion. Reactor performance for the selected ebullated bed configuration and operating conditions is calculated as follows (Table 22P.6 and Figure 22P.2):

Table 22P.6 Reactions rates and heavy oil conversion estimates for H = 10 m.

Parameter	Symbol	Equation	Value	Units
Heavy oil concentration in the liquid	C_{A0}	$C_{A0} = x_{A0}\rho_L$	800	kg A/m^3 liquid
Catalyst concentration in the slurry	C_{slurry}	$C_{slurry} = \rho_{slurry}\varepsilon_p/(1-\varepsilon_g)$	731	kg cat/m^3 slurry
Hydrogen concentration in the gas	C_{H_2}	$C_{H_2} = P_{H_2}/RT$	2713	gmol H_2/m^3 gas
Modified reaction rates	r_1'	$r_1' = k_1 C_A C_{slurry}\varepsilon_L$	1.24	kg/m^3 s
Modified reaction rates	r_2'	$r_2' = k_2 C_A C_{H_2} C_{slurry}\varepsilon_L M_{W,C}$	0.17	kg/m^3 s
Heavy oil concentration	C_A	$C_A - C_{A0} + (r_1' + r_2')A_C H/Q_L = 0$	212	kg A/m^3 liquid
Heavy oil conversion	X_A	$X_A = (C_{A0} - C_A)/C_{A0}$	73.5	%

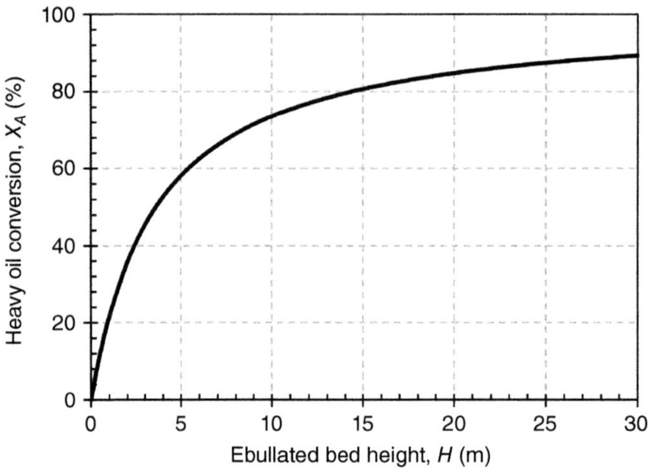

Figure 22P.2 Heavy oil conversion as a function of ebullated bed height.

Revise overall reactor design: Following the ebullated bed conversion calculations, crucial design decisions can be re-evaluated by considering their impact on the overall reactor sizing, performance, and costing criteria. The reactor modelling complexity can also be revisited to estimate the sensitivity of simplifying assumptions. It should now be apparent that, to obtain a proper design, gas–liquid–solid fluidized bed reactors require an iterative approach, potentially with directed experiments to determine unknown reaction and/or transport parameters as well as entrance and exit region performance criteria.

Notations

a_i	bubble or particle surface area per unit reactor volume (m^{-1})
A_c	column cross-sectional area (m^2)
A_j	frequency factor of reaction j in Arrhenius equation (s^{-1} if first order reaction)
Ar_L	Archimedes number $= g\rho_L(\rho_p - \rho_L)d_v^3/\mu_L^2$ (–)
Ar'_L	Archimedes, mixed fluid density $= g\rho_L(\rho_p - (\rho_g\alpha_{mf} + \rho_L(1-\alpha_{mf})))d_v^3/\mu_L^2$ (–)
c_1	constant in Eq. (22.22) (–)
Ca_g	gas-phase capillary number $= U_g\mu_g/\sigma$ (–)
Ca_{gL}	gas–liquid phase capillary number $= (U_g + U_L)\mu_g/\sigma$ (–)
C_i	concentration of species i ($kmol/m^3$)
C_p	specific heat capacity (J/(kg K))
d_i	diameter according to subscript i (m)
$D_{i,j}$	diffusivity of species i in phase j (m^2/s)
E_j	activation energy of reaction j (J/mol)
Eo_i	bubble or particle Eötvös number $= gd_v^2\lvert\rho_i - \rho_L\rvert/\sigma$ (–)

E_{zL}	liquid axial dispersion coefficient (m^2/s)
f_c	vortex shedding frequency (s^{-1})
F_i	mass flow rates of species i (kg/s)
Fr_g	gas-phase Froude number $= U_g^2/(gd_v)$ (–)
Fr_L	liquid-phase Froude number $= U_L^2/(gd_v)$ (–)
g	acceleration of gravity (m/s^2)
h	surface-to-bed heat transfer coefficient ($W/(m^2\,K)$)
H	bed height (m)
H_i	Henry's constant of species i ((Pa m^3)/mol)
H_m	transport disengagement height (m)
i, j	indices (–)
k	ratio of wake volume to bubble volume (–)
k_j	kinetic rate constant of reaction j (s^{-1} if first order)
k_L	liquid-side gas–liquid mass transfer coefficient (m/s)
k_s	liquid–solid mass transfer coefficient (m/s)
k_{wall}	wall effect correction factor (–)
L	length (m)
Mo	Morton number $= g(\rho_L - \rho_g)\mu_L^4/(\rho_L^2\sigma^3)$ (–)
M_p	mass of solids particles (kg)
M_w	molecular weight (kg/mol)
n	Richardson–Zaki index (–)
N_m	number of vortex shedding/rocking cycles (–)
Nu_i	Nusselt number in multiphase system $i = h_i d_p/\kappa_L$ (–)
P	pressure (Pa)
Pe_{zL}	liquid axial Péclet number $= d_p U_L/E_{zL}$ (–)
Q	volumetric flowrate (m^3/s)
r_i	rate of reaction i (mol/m^3 s)
R	ideal gas constant (J/(mol K))
Re_b	bubble rise Reynolds number $= d_b u_b \rho_L/\mu_L$ (–)
Re_{Lg}	liquid–gas phases Reynolds number $= d_p(U_L + U_g)\rho_L/\mu_L$ (–)
Re_{Lmf}	liquid-phase Reynolds number at minimum fluidization $= d_p U_{Lmf}\rho_L/\mu_L$ (–)
$Re_{p,t}$	single particle terminal settling Reynolds number $= d_p u_{p,t}\rho_L/\mu_L$ (–)
Sc	Schmidt number $= \mu_L/(\rho_L D_{i,j})$ (–)
Sh_i	Sherwood number in multiphase system $i = d_p k_S/D_{i,j}$ (–)
St_{Lg}	liquid–gas phase Stokes number $= \mu_L(U_L + U_g)/(g\rho_L d_v^2)$ (–)
t	time (s)
T	temperature (K)
u	local velocity (m/s)
U	superficial velocity (m/s)
V	volume (m^3)
We_c	Weber number based on column $= d_c\rho_L U_L^2/\sigma$ (–)
We_{or}	Weber number based on orifice $= d_{or}\rho_g U_{or}^2/\sigma$ (–)
x	ratio of solids holdup in wake to that in liquid–solid fluidized region (–)
x_i	mass fraction of species i (–)

X_i	chemical conversion of species i (–)
y	symbol defined in Eq. (22.19) (–)
z	Cartesian coordinate in vertical direction (m)

Greek Letters

α	solids-free gas holdup $= \varepsilon_g/(\varepsilon_L + \varepsilon_g)$ (–)
Δ	difference (–)
ε	bed porosity $= 1 - \varepsilon_p$ (–)
ε_i	holdup of phase i (–)
η	internal effectiveness factor (–)
κ_L	liquid thermal conductivity (W/(m K))
μ	dynamic viscosity (Pa s)
ρ	density (kg/m^3)
σ	gas–liquid interfacial tension (N/m)
ϕ	sphericity (–)
χ_{gL}	gas–liquid phase inertial Lockhart–Martinelli ratio $= \rho_g U_g^2/(\rho_L U_L^2)$ (–)
ψ	parameter in Eq. (22.20) (–)

Subscripts

0	initial
1, 2	reaction 1, 2 in Solved Problem 22.1
A	heavy oil in Solved Problem 22.1
b	bubble
c	column
dyn	dynamic
eff	effective, associated with liquid–solid mixture
g	gas
GL	gas–liquid freeboard
GLS	gas–liquid–solid fluidized bed
L	liquid
LS	liquid–solid fluidized bed
mf	minimum fluidization
p	particle
or	orifice
S	solid
small	small bubbles in a large/small binary bubble population
sv	spherical surface–volume ratio equivalent
t	terminal rise or settling
trans	dispersed-to-coalesced bubble flow transition
v	spherical volume equivalent
w	wake

References

1 Fan, L.-S. (1989). *Gas–Liquid–Solid Fluidization Engineering*. Boston: Butterworths.
2 Nigam, K.D. and Schumpe, A. (1996). *Three-Phase Sparged Reactors*. Netherlands: Gordon and Breach Publishers.
3 Duduković, M.P., Larachi, F., and Mills, P.L. (2002). Multiphase catalytic reactors: a perspective on current knowledge and future trends. *Catal. Rev.* 44 (1): 123–246.
4 Fan, L.-S. and Yang, G. (2003). Gas–liquid–solid three-phase fluidization, Chapter 27. In: *Handbook of Fluidization and Fluid-Particle Systems* (ed. W.-C. Yang). New York: Marcel Dekker Inc.
5 Buwa, V.V., Roy, S., and Ranade, V.V. (2016). Three-phase slurry reactors. In: *Multiphase Catalytic Reactors – Theory, Design, Manufacturing, and Applications* (eds. Z.I. Önsan and A.K. Avci), 132–151. Hoboken: Wiley Inc.
6 McKnight, C.A., Hackman, L.P., Grace, J.R. et al. (2003). Fluid dynamic studies in support of an industrial three-phase fluidized bed hydroprocessor. *Can. J. Chem. Eng.* 81 (3–4): 338–350.
7 Gray, M.R. (2002). Fundamentals of bitumen coking processes analogous to granulations: a critical review. *Can. J. Chem. Eng.* 80 (3): 393–401.
8 Fan, L.-S., Lau, R., Zhu, C. et al. (2001). Evaporative liquid jets in gas–liquid–solid flow system. *Chem. Eng. Sci.* 56 (21–22): 5871–5891.
9 Levenspiel, O. (1999). *Chemical Reaction Engineering*. New York: Wiley.
10 Dudukovic, M.P. and Mills, P.L. (2014). Challenges in reaction engineering practice of heterogeneous catalytic systems. *Adv. Chem. Eng.* 45: 1–40.
11 Jin, G. (2006). Multi-scale modeling of gas–liquid–solid three-phase fluidized beds using the EMMS method. *Chem. Eng. J.* 117 (1): 1–11.
12 Powell, J.B. (2017). Application of multiphase reaction engineering and process intensification to the challenges of sustainable future energy and chemicals. *Chem. Eng. Sci.* 157: 15–25.
13 Boyer, C., Duquenne, A.-M.M., and Wild, G. (2002). Measuring techniques in gas–liquid and gas–liquid–solid reactors. *Chem. Eng. Sci.* 57 (16): 3185–3215.
14 Jiang, P., Wei, F., and Fan, L.-S. (2003). General approaches to reactor design, Chapter 12. In: *Handbook of Fluidization and Fluid-Particle Systems* (ed. W.-C. Yang). New York: Marcel Dekker Inc.
15 Larachi, F., Belfares, L., Iliuta, I., and Grandjean, B. (2001). Three-phase fluidization macroscopic hydrodynamics revisited. *Ind. Eng. Chem. Res.* 40 (3): 993–1008.
16 Deen, N., Mudde, R.F., Kuipers, J.A.M. et al. (2010). Bubble columns. In: *Ullmann's Encyclopedia of Industrial Chemistry*, vol. 38 (ed. Wiley-VCH), 382–391. Weinheim, Germany: Wiley-VCH.
17 Yang, G.Q., Du, B., and Fan, L.S. (2007). Bubble formation and dynamics in gas–liquid–solid fluidization – a review. *Chem. Eng. Sci.* 62 (1–2): 2–27.
18 Zeghloul, A., Azzi, A., Hasan, A., and Azzopardi, B.J. (2018). Behavior and pressure drop of an upwardly two-phase flow through multi-hole orifices. *Proc. Inst. Mech. Eng., Part C* 232 (18): 3281–3299.

19 Besagni, G., Inzoli, F., and Ziegenhein, T. (2018). Two-phase bubble columns: a comprehensive review. *Chem. Eng.* 2 (2): 13.

20 Clift, R., Grace, J., and Weber, M. (1978). *Bubbles, Drops, and Particles*. New York: Academic Press.

21 Pjontek, D. and Macchi, A. (2014). Hydrodynamic comparison of spherical and cylindrical particles in a gas–liquid–solid fluidized bed at elevated pressure and high gas holdup conditions. *Powder Technol.* 253: 657–676.

22 Briens, L.A., Briens, C.L., Margaritis, A., and Hay, J. (1997). Minimum liquid fluidization velocity in gas-liquid-solid fluidized beds. *AIChE J.* 43 (5): 1180–1189.

23 Larachi, F., Iliuta, I., Rival, O., and Grandjean, B.P.A. (2000). Prediction of minimum fluidization velocity in three-phase fluidized-bed reactors. *Ind. Eng. Chem. Res.* 39 (2): 563–572.

24 Zhang, J.P., Epstein, N., and Grace, J.R. (1998). Minimum fluidization velocities for gas-liquid-solid three-phase systems. *Powder Technol.* 100 (2–3): 113–118.

25 Lee, D., Epstein, N., and Grace, J.R. (2003). The variable-voidage effect on modeling the minimum liquid fluidization velocity of gas-liquid fluidized beds. *J. Chem. Eng. Japan* 36 (9): 1111–1115.

26 Yang, X.L., Euzen, J.P., and Wild, G. (1993). Study of liquid retention in fixed-bed reactors with upward flow of gas and liquid. *Int. Chem. Eng.* 33 (1): 72–84.

27 Wen, C.Y. and Yu, Y.H. (1966). A generalized method for predicting the minimum fluidization velocity. *AIChE J.* 12 (3): 610–612.

28 Atta, A., Razzak, S.A., Nigam, K.D.P., and Zhu, J.-X. (2009). (Gas)–liquid–solid circulating fluidized bed reactors: characteristics and applications. *Ind. Eng. Chem. Res.* 48 (17): 7876–7892.

29 Zhang, J.-P., Grace, J.R., Epstein, N., and Lim, K.S. (1997). Flow regime identification in gas-liquid flow and three-phase fluidized beds. *Chem. Eng. Sci.* 52 (21–22): 3979–3992.

30 Parisien, V., Pjontek, D., McKnight, C.A. et al. (2017). Impact of catalyst density distribution on the fluid dynamics of an ebullated bed operating at high gas holdup conditions. *Chem. Eng. Sci.* 170: 491–500.

31 Han, J.H., Wild, G., and Kim, S.D. (1990). Phase holdup characteristics in three phase fluidized beds. *Chem. Eng. J.* 43 (2): 67–73.

32 Epstein, N. (1981). Three-phase fluidization: some knowledge gaps. *Can. J. Chem. Eng.* 5: 649–657.

33 Khan, A.R. and Richardson, J.F. (1989). Fluid–particle interactions and flow characteristics of fluidized beds and settling suspensions of spherical particles. *Chem. Eng. Commun.* 78 (1): 111–130.

34 Fan, L.-S. and Tsuchiya, K. (1990). *Bubble Wake Dynamics in Liquids and Liquid–Solid Suspensions*. Stoneham: Butterworth-Heinemann.

35 Miyahara, T., Tsuchiya, K., and Fan, L.-S. (1989). Mechanism of particle entrainment in a gas–liquid–solid fluidized bed. *AIChE J.* 35 (7): 1195–1198.

36 Kim, S.D. and Kim, C.H. (1983). Axial dispersion characteristics of three phase fluidized beds. *J. Chem. Eng. Japan* 16 (3): 172–178.

37 Luo, X., Jiang, P., and Fan, L.-S. (1997). High-pressure three-phase fluidization: hydrodynamics and heat transfer. *AIChE J.* 43 (10): 2432–2445.
38 Schumpe, A., Deckwer, W.-D., and Nigam, K.D.P. (1989). Gas–liquid mass transfer in three-phase fluidized beds with viscous pseudoplastic liquids. *Can. J. Chem. Eng.* 67 (5): 873–877.
39 Basha, O.M., Sehabiague, L., Abdel-Wahab, A., and Morsi, B.I. (2015). Fischer–Tropsch synthesis in slurry bubble column reactors: experimental investigations and modeling – a review. *Int. J. Chem. Reactor Eng.* 13 (3): 201–288.
40 Vik, C.B., Solsvik, J., Hillestad, M., and Jakobsen, H.A. (2018). Interfacial mass transfer limitations of the Fischer–Tropsch synthesis operated in a slurry bubble column reactor at industrial conditions. *Chem. Eng. Sci.* 192: 1138–1156.
41 Arters, D.C. and Fan, L.-S. (1990). Experimental methods and correlation of solid–liquid mass transfer in fluidized beds. *Chem. Eng. Sci.* 45 (4): 965–975.
42 Lane, C.D., Macchi, A., Pjontek, D. et al. Modelling the fluid dynamics of a commercial ebullated bed hydroprocessor. *Ind. Eng. Chem. Res.* 58 (39): 18387–18396.
43 Martínez, J., Sánchez, J.L., Ancheyta, J., and Ruiz, R.S. (2010). A review of process aspects and modeling of ebullated bed reactors for hydrocracking of heavy oils. *Catal. Rev.* 52 (1): 60–105.
44 Parulekar, S.J. and Shah, Y.T. (1980). Steady-state behaviour of gas—liquid—solid fluidized-bed reactors. *Chem. Eng. J.* 20 (1): 21–33.
45 Pjontek, D., Mcknight, C.A., Wiens, J., and Macchi, A. (2015). Ebullated bed fluid dynamics relevant to industrial hydroprocessing. *Chem. Eng. Sci.* 126 (under review): 730–744.
46 Haider, A. and Levenspiel, O. (1989). Drag coefficient and terminal velocity of spherical and nonspherical particles. *Powder Technol.* 58 (1): 63–70.
47 Wilkinson, P.M., Spek, A.P., and van Dierendonck, L.L. (1992). Design parameters estimation for scale-up of high-pressure bubble columns. *AIChE J.* 38 (4): 544–554.
48 Eccles, R.M. (1993). Residue hydroprocessing using ebullated-bed reactors. *Fuel Process. Technol.* 35 (1–2): 21–38.
49 Schweitzer, J.-M.M. and Kressmann, S. (2004). Ebullated bed reactor modeling for residue conversion. *Chem. Eng. Sci.* 59 (22–23): 5637–5645.
50 Cheng, Z.-M., Huang, Z.-B., Yang, T. et al. (2014). Modeling on scale-up of an ebullated-bed reactor for the hydroprocessing of vacuum residuum. *Catal. Today* 220–222: 228–236.

Problems

22.1 **U_{Lmf} in a gas–liquid–solid fluidized bed**: Determine the minimum liquid fluidization velocity based on the dynamic pressure drops measured while varying the liquid superficial velocity at a constant gas flow rate. Experimental data were collected for glass spheres ($\rho_p = 2500\,\text{kg/m}^3$) in air and water at atmospheric temperature and pressure [21]. Discuss the impact of gas addition and particle size, and compare with predictions from Eq. (22.3).

1.5 mm diameter glass spheres				4 mm diameter glass spheres			
$U_g = 0$ m/s		$U_g = 0.056$ m/s		$U_g = 0$ m/s		$U_g = 0.052$ m/s	
U_L (m/s)	$-\Delta P_{dyn}$ (Pa)	U_L (m/s)	$-\Delta P_{dyn}$ (Pa)	U_L (m/s)	$-\Delta P_{dyn}$ (Pa)	U_L (m/s)	$-\Delta P_{dyn}$ (Pa)
0.037	945	0.032	866	0.068	1020	0.066	585
0.034	985	0.029	925	0.063	1066	0.060	677
0.030	1031	0.022	916	0.056	1121	0.056	736
0.028	1057	0.019	899	0.050	1169	0.049	863
0.027	1072	0.017	876	0.041	1178	0.046	920
0.024	1117	0.015	865	0.039	1074	0.039	1001
0.022	1154	0.014	869	0.035	929	0.035	1031
0.019	1195	0.012	880	0.034	841	0.029	1096
0.016	1220	0.010	864	0.028	612	0.022	1148
0.014	1066	0.008	817	0.023	462	0.017	1162
0.011	896	0.004	834	0.019	354	0.013	1043
0.009	680	0.003	646	0.016	285	0.012	920
0.006	448	0.002	510	0.011	163	0.007	550
0.002	229	0.001	339	0.007	109	0.003	201

22.2 Phase holdups in a gas–liquid–solid fluidized bed: The following figures provide the gas, liquid, and solid phase holdups in the bed and freeboard for (a) 1.5 mm and (b) 4.0 mm diameter glass spheres ($\rho_p = 2500\,\text{kg/m}^3$) in air and water at atmospheric temperature and pressure [21].

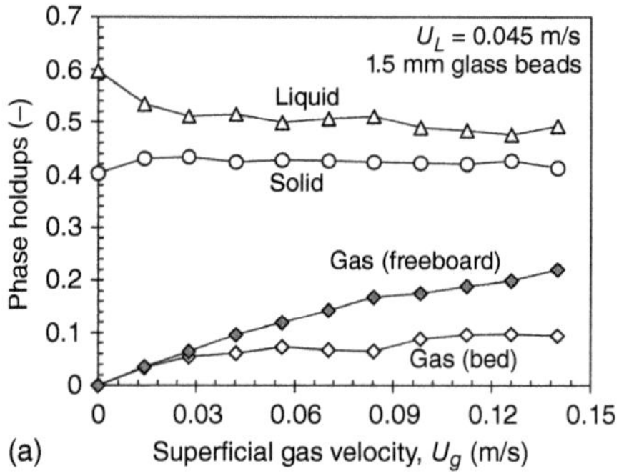

(a)

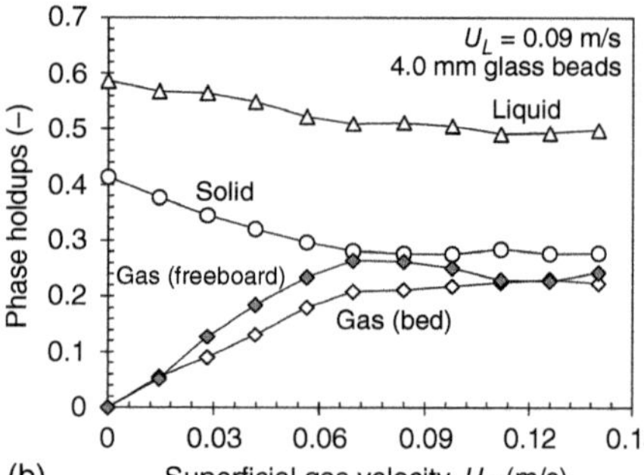

(b)

a. Discuss the effects of gas velocity and particle size on the phase holdups and the transition velocity from dispersed to coalesced bubble flow.
b. Compare phase holdups in the bed to predictions from the Larachi et al. [15] and Han et al. [31] models.
c. Compare and discuss gas holdups obtained in the freeboard vs. those obtained in the bed.
d. Observe and discuss the liquid–solid fluidized bed behaviour at the introduction of gas for 1.5 and 4 mm glass spheres. Discuss the effects of particle size.

Index

a

abrasion 275, 276, 279, 467, 469, 474
absolute pressure transducers 87
acoustic emission measurements 92
  defluidized zones 462–464
  internals erosion 464–467
agglomerates 62, 190, 252, 273, 274, 463, 475, 477, 478, 483, 554
agglomeration 2, 87, 94, 98, 173, 190, 192, 194, 270, 273–274, 388, 391, 406, 414, 451, 482
angle of internal friction 18, 67, 82, 85–86
angle of repose 18, 82, 84, 85, 533
angle of upward jets 466
annulus 534
  fluid flow in 535
anti-static agents 451–452
apparent reaction kinetics 341, 342
Archimedes number 21, 22, 26, 34, 40, 169, 278, 304, 563
attrition 2, 4, 274–278
  jet-target 473–474
  modeling 276–278
  nozzle 472
  particle 467–468
  rate 278, 474–475
automatic solids flow devices 219–222
automatic solids recirculation systems 212–213
autothermal limits 420
axial gas dispersion coefficient 168–170, 256, 442, 443
axial solids dispersion coefficient 171–173

b

backmixing coefficient 169–170
baffles
  advantages of 446
  on bed expansion 442
  on bed hydrodynamics 439–445
  catalyst stripper 438
  for fast fluidized bed risers 446
  fixed packings 436–439
  in fluidized beds 432
  gas dispersion coefficients 443
  geometric characteristics of 432–439
  high-velocity fast fluidized beds 446–447
  horizontal 432–435
  industrial fluidized beds 445
    design of 447–449
  louver 432, 433, 435, 441, 442
  low-velocity dense fluidized beds 432, 439–446
  vertical 435–436
bed contraction *vs.* expansion 568
bed defluidization 274
bed diameter 192, 394, 534, 543
bed dynamic pressure drop 562, 563
bed expansion 2, 48, 119, 139, 142–143, 159, 173, 408, 442
bed hydrodynamics 439–445
  impact of 476
bed pressure drop fluctuations 460
bed-to-surface heat transfer 47–48

bed-to-surface heat transfer (*contd.*)
correlations for 302–304
fast and circulating fluidized beds 316–318
fluidized bed and immersed surfaces 304–314
gas convection component 298–299
maximum heat transfer coefficient 299–302
methods of augmenting 314
particle convective component 294–298
radiative 314–315
bed transport 560
bed voidage profile 350–351
bed volume fractions 350
bench-scale testing 271
biomass combined heat and power (CHP) 416–417
bivariate time series
cross-correlation coefficients 96–97
cross-spectral density function 97
joint probability density function 96
boundary conditions 122–123, 133, 348, 377, 535, 558, 578
bubble caps 79, 555, 561
bubbles *see also* voids
bed expansion 142–143
breakup 137, 564
bursting 184
cloud formation 137–138
coalescence 139
definition of 2
diameter 167
effect on mixing 143–144
flow of gas by translation of 139–140
flow patterns 460
in fluidized beds 134–135
form 133
in freely bubbling bed 142
gas-fluidized beds 146
gas mixing 145–146
hydrodynamic properties of 135–138
interactions 139
inter-phase mass transfer 137–138, 320
in liquids 134–135
maximum stable size 137
mean diameter 140–142
particle segregation 144–145
radial non-uniformity of 143–144
rising velocity 135
single 135
solids mixing 144–145
turnover time 144–145
volume fraction 142
wakes 135–137, 567
bubbling
fluidized bed with tube bundle 123–124
freely bubbling beds 129–146
onset of bubbling fluidization 55–57
turbulent fluidized beds 190–191
bulk density
and flowability 16–18
voidage 16–18
buoyancy force 43, 281, 501, 502

c

calcium looping technologies 419
carryover, see entrainment
catalyst stripper baffles 438
catalytic fluidized-bed reactor
catalytic process 334
CFD models 353
components 76–78
comprehensive reactor modelling 352–353
vs. gas-solid reactors 336
mass/mole and energy balances 345
model verification and validation 353–355
programming and numerical solution 356
reaction rate expressions 345–348
requirements 335
scale-up of 419–422
single-phase models 346–348
standard two-phase theory 349–352
CFD–DPM 110
limitations of 119

of small-scale circulating fluidized bed 124–125
chemical looping assisted by oxygen uncoupling (CLaOU) 388
chemical looping combustion (CLC) 387, 388
chemical looping processes 387
chemical looping reforming (CLR) 388
chemical looping technology 388
chemical looping with oxygen uncoupling (CLOU) 388
chemical reactor scale-up 405
choking 2, 63, 534, 535
circulating fluidized bed (CFB)
advantages 241
axial and radial distributions (profiles) 243–246
basic parameters 241–243
characteristics of 240–241
CFD–DPM Simulation of 124–125
circulating turbulent fluidized bed 249–250
definition of 1–2, 239–240
fast fluidization and 62–64
flow pattern transitions in 64
gas and solids mixing 256–258
fast heat transfer 316–318
hydrodynamics 261–262
loop configuration 218
micro-flow structure 250–256
pilot plant units 407
radial profiles of solids distribution 246–248
reactor performance of 258–260
solids distribution 246–248
solids holdup/voidage 243–246
circulating fluidized bed combustors (CFBC) 212, 245
circulating turbulent fluidized bed (CTFB) 241, 249–250
cloud formation 137–138, 157
cloud-wake phase 352
clusters 2, 190, 251–253, 256
coal pyrolysis 521–523
coarse-grained particle (CGP) 110
coefficient of restitution 15, 414
coherent frequency 97
column diameter effect 172–173, 409–411
compartmental flow models
fluidized bed 561
freeboard 562
plenum and fluid distributor 560–561
comprehensive reactor modeling 336, 352–353
comprehensive two-phase reactor models 354
computational fluid dynamics (CFD)
boundary conditions 122–123
development of 415
discrete element method (DPM) 115–119
for fluidized-bed catalytic reactors 353
gas–solid interaction 119–122
two-fluid model (TFM) 110–115
conical spouted beds 534
contact angle, of liquid 12
contact electrification 269
continuity equations 111, 503
continuous stirred tank reactor (CSTR) 347, 431, 432, 556, 580
controlled solids recirculation systems 213–215
conventional spouted beds (CSBs) 543
core–annulus flow structure 64, 165–167, 247, 251, 257, 446
Coulomb's law 269
cross-correlation function 96, 97, 243
crosser grid 435, 439, 445
cross-spectral density function 96, 97
cyclone diplegs 206, 212, 213, 219–222, 227, 432, 459, 465
cyclone recirculation system 213
cyclones 222–223
dimensions and design 228–229
dipleg 206, 219, 222
flow patterns 225, 226
gas outlet tube (GOT) 228
inlet design 227
inlet gas velocity for 228
internal *vs.* external 227

cyclones (*contd.*)
in parallel 226–227
pressure drop 224
reverse-flow 225, 226
in series 226
solid loadings effect 227–228
types 223–225
uniflow 223
vortex length of 228

d

Darcy's law 535, 537
data analysis
bivariate time series 96–97
frequency analysis 95–96
other signal analyses 98
deep bubbling beds 144
defluidization 94, 210, 274, 390, 436, 471, 473, 556
defluidized zones
distributor design 463
low distributor pressure drop 462–463
sticky/agglomerating solids 463
demo-scale systems 419
dense bed height 159, 192, 531
dense phase
definition of 2
interphase 320
solids holdup 167
density, of liquid 12
dielectric constant 16
differential pressure transducers 19, 88, 155
diffusion coefficient 113, 145, 347, 350
dilute phase 2, 61, 62, 90, 92, 163, 222, 244, 251, 257, 464, 531
dilute-phase spouting 546
dilute transport zone 183
dimensional similitude 413–414
direct numerical simulation (DNS) 110, 120
discharge coefficient 80, 82, 106
discrete particle method (DPM)
governing equations 116–119
limitations of CFD–DPM 119
distributor 2
design 48, 79, 123, 408, 415, 463, 561
plates 39, 79–82, 84, 88, 105, 272, 274, 275, 278, 458
pressure drop 461–462, 465
three-phase system 560–561
actual distributor pressure drop 461–462, 465
downer-based coupled reactors 519
downer reactor
CFD–DEM simulation 512, 513
coal pyrolysis 521–522
conception and characteristics 499–501
DCC process 520
definition of 2
downer-riser coupled 518, 519
Eulerian–Eulerian model 509–511
Eulerian–Lagrangian model 511–514
fast separation of gas and solids 517–518
FCC 519–520
features of 522
with force balance analysis 502
gasification 521
heat transfer in 506–507
high-density 518
hydrodynamic 501–503
inlet design 514–516
mixing 503–506
outlet design 517
reaction engineering model 509
scale-up effect of 511
downer–riser coupled reactor 518–519
downward vertical jets 466
draft tube attachment 482
drag force 14, 40, 43, 111, 117, 119–121, 138, 181, 184, 193, 217, 253, 281, 501
drag models 111, 119–121, 125
dual gasifier 417–419

e

ebullated bed 555, 578–580, 582
sizing 577

electrical conductivity 16
electrical fields 450
electrostatic charge 194–195
  Faraday cage 272
  measurement and prediction of 271–272
  mitigation techniques 272–273
electrostatic force 55, 195, 270, 273, 282
electrostatic precipitators 94, 229, 230, 420
elutriation
  definition of 2, 184
  rate constant 62, 185–186, 190, 191
endothermic reaction 340, 420, 439
energy balance 347, 348
  equation 112
  for gas phase 111
  mass/mole balance 345
energy minimization multiscale (EMMS) model 121, 557
entrainment 181–196
  ejection of particles into the freeboard 184–185
  reducing 195
  of solid particles from fluidized beds
    dense bed height 192
    electrostatic charges 194–195
    geometry and shape of freeboard 192
    internals 192–193
    particles properties 191, 192
    pressure and temperature 193–194
  transport disengagement height (TDH) 182–183, 185–190
  from turbulent fluidized beds 190–191
equilibrium reactor simulation 341
equimolar inter-phase mass transfer 353
erosion, *see also* wear
  anti-erosion measures 449
Eulerian–Eulerian computational fluid dynamics (CFD) model 508
Eulerian–Eulerian (E–E) models 109, 272, 415, 475, 509–513
Eulerian–Lagrangian (E–L) models 109, 115, 119, 272, 508, 511–514
exothermic reactions 293, 317, 340, 437
external mass transfer 341, 365–369, 373, 374, 538

f

fabric filters 229
fast fluidization 411–412
  and circulating fluidized bed 62–64
fast Fourier transform (FFT) 95
feeding system 78, 81–83
fines content effect 173
Fischer–Tropsch synthesis 406, 407, 572
fixed-(dense)-fluidized bed (FFB) system 407
fixed packings 432, 436–440, 445
flowability 16–18, 84
  characterization of 84–86
  particles or powders 76
flow regimes
  bubbling regime 564–565
  minimum liquid fluidization velocity 562–564
fluctuation, pressure 59
fluid catalytic cracking (FCC) 3, 25, 63, 140, 174, 183, 239, 245, 274, 435, 478, 508
fluid flow
  in annulus 535
  in spout 536
fluidizability 17, 18, 75
  quality of fluidization 84–87
fluidized bed combustor 1, 273, 291–293, 387, 390, 413, 414
fluidized bed operation
  shutdown 94
  start-up 94
  steady-state operation 94
fluidized bed reactors 112
  for gas–solid reactions 364

fluidized bed reactors (*contd.*)
  design 334
  modelling
    mass/mole and energy balances 345
    model development 337–338
    reaction rate expressions 345–348
    structure and reaction 338–342
    two-phase theory 349–352
fluidized bed systems
  acoustic emission measurements 92
  configuration and design
    auxiliary components 83–84
    column material 78–79
    distributor or orifice plate 79–81
    feeding system 81–83
    hopper design 82
    plenum chamber 81
  gas/solids sampling 92–93
  isolated location 76
  mixing and residence time distribution 93
  non-invasive measurements 90–91
  optical probes 89–90
  pressure measurement 87–89
  scale-up 75, 76
  secondary injection of fluid 84
  solids circulation flux 92
  thermal sensors 89
  visualization measurements 91–92
fluidized catalytic cracking (FCC) process 203, 204
fossil fuel combustion 387
freeboard
  definition of 2, 562
  diameter 192
  ejection of particles into 184, 185
  gas-liquid separation 578
  geometry and shape of 192
  voidage profile 350–351
freely bubbling bed 93, 139–146, 450
  rising velocity of bubbles in 142
frictional force 14, 280, 282
frictional model 114–115
Froude numbers 425

g
gas backmixing 169, 170, 173, 246, 249, 256, 442, 443
gas cushion 439, 440, 442
gas dispersion coefficients 168–171, 173, 256, 442, 443
gas distributors
  operating problems 459
  perforated plates 457, 458
  spargers 458
gaseous reactant 241, 364–367, 369, 371, 373, 374, 376–379, 381–383, 385, 386
gas fluidization
  circulating fluidized bed 62–64
  fast fluidization 62–64
  gas–solid fluidized beds 64–65
  gas–solid vertical transport 65–68
  Geldart powder classification for 24–27
  onset of bubbling fluidization 55–57
  onset of fluidization 55
  onset of slugging fluidization 57–58
  onset of turbulent fluidization 58–61
  pressure and temperature 68–69
  termination of turbulent fluidization 62
gas-fluidized beds 131, 541
  advantages 3
  applications 5, 9
  bubbles formation 133
  disadvantages 3, 4
  effects of hydrodynamics 408
  factors influencing bubbles in 146
  gas properties 11
gasification 417–419
gas jets 463–465, 467, 469, 478, 481–483, 560
gas–liquid jets
  applications 478
  mechanism for 477–478
gas–liquid mass transfer 571, 572, 579
gas–liquid–solid fluidization 552–582
  applications 554–556
  compartmental flow models 558–562

description and classification 553–554
reactor design 556–558
reactor scale-up 558
gas mixing 145–146, 168–171, 256, 257, 443, 445
gas outlet tube (GOT) 212, 223, 225, 228
gas-perturbed liquid model 562
gas–solid conversion 381–386
gas–solid fluidized beds 55, 64, 65, 378, 380, 382, 384
application 475–476
bed hydrodynamics, impact of 476
injection system design 476
jet penetration 476
flow regime diagram for 64–65
mechanisms 475
nozzle inclination 476
solids entrainment 476
gas–solid interaction
drag force 119–121
heat transfer 121–122
gas–solid jets entrainment 476
gas–solid reactions 363
fluidized bed reactor for 364
for single particle 364–377
gas/solid sampling 92–93
gas–solid two-phase flows 499, 500
gas–solid vertical transport 65–68
gas-spouted beds 540, 541
gas streamlines 535, 537
gas-to-liquid ratio (GLR) 480
gas transition velocity 577
gas turbulence–solid turbulence model 509
Geldart powder classification 24–27, 57, 86, 191, 408
Geldart powder groups 2, 11, 24, 27
Gidaspow drag model 113, 120
glidants 431, 449, 451–452
governing equations
discrete particle method 115–119
two-fluid model (TFM) 110–115
granular-bed filters 230
gravity force 116, 281, 503, 558
grid
crosser 435, 439, 445
definition of 2

h

hard-sphere model 116
Hausner ratio 17, 84
heat and mass transfer
bed-to-surface transfer 47–48
interphase transfer 46–47
heat exchanger 80, 278, 280, 317, 330, 408, 478
heat exchange tubes 435, 437
heat losses 347, 408, 420
heat transfer
augmenting methods 314
bed-to-surface 294–302, 314
correlations, for fluidized beds 302–304
distributor jets 467
downer reactor 506
effect of pressure 303
in fast and circulating fluidized beds 316–318
fluidized bed and immersed surfaces
correlations for vertical tubes 306–309
finned tubes and non-cylindrical tubes 312–313
horizontal tubes 313–314
Martin's correlation 309–312
Molerus and Wirth correlation 304
gas–solid interactions 121–122
gas–solid phases 111, 112
interphase 293–294
radiative 314–315
spouted beds 538
heat transfer coefficients
bed and vertical tubes 309
measurements of 305
heavy oil hydroprocessing reactors 555
Higbie penetration model 538
high-density downer 518
hopper design 82–83

horizontal baffles 432–435, 439, 440, 445, 447, 448
horizontal jet 80, 464–467, 472
hybrid standpipes 203, 214
hydrocrackers 575, 577, 580
hydrodynamic force 117, 187
hydrodynamics
  fluid inlet diameter and cone angle 532–533
  fluid flow in annulus 535
  fluid flow in spout 536
  influence of scale 408–412
  maximum spoutable bed height 534–535
  minimum spouting velocity 533–534
  pressure drop 536–537
  solid particles 537–538

i
injection system design 476, 479–482
  draft tube attachment 482
  gas jets 482
  impact attachment 481
  laboratory nozzles 480
  non-rodable commercial nozzles 480
  rodable commercial nozzles 480–481
  shrouds 481
  spray nozzle 480
  upstream piping 479–480
inlet gas velocity 226–228
in situ gasification chemical looping combustion (iG–CLC) 388
instrumentation documentation 423
internal porosity 13–14, 373, 376
internals, *see* baffles
internals erosion 464–465
interphase mass transfer 112, 137–138, 318, 355
  bubble to dense-phase 320
  bubble-emulsion 431
  equimolar 353
interphase transfer 46–47
intrinsic reaction kinetics 341, 342, 558
inverse fluidization 39

j
jet grinding 471
jet-in-fluidized-bed 540
jet milling mechanisms 469
jet penetration 471, 478
  bed hydrodynamics, impact of 483
  depth 476
  injection system design 479–482
  lengths 80
  spray jets, interactions between 483
  nozzle inclination 482
  solids entrainment 478–479
jets
  formed in fluidized bed grinding 469–471
  nozzle design 472
  solids entrained into 471–472
jets coalescence 466–467
jets, gas distributors 457–467
  actual distributor pressure drop 461–464
  merging and coalescence 466–467
  uniform 459–461
  upward vertical jets 465
jet-spouted bed 546
jet-target attrition 473–474
joint probability density function 96

k
kinetic energy 277
kinetic granular theory 112–114, 123
Kunii and Levenspiel (K–L) bubbling bed model 351–352, 382–384

l
laminar flow 158, 304
Langmuir–Hinshelwood type rate expressions 365
laser Doppler velocimetry (LDV) 90 256
laser sheet methods 90
layer inversion 41–43
  condition 42

velocity 46
voidage 43–46
Le Chatelier's principle 339, 340
liquid
properties 12
recirculation 578
liquid fluidization 33–48
field of existence 33–35
layer inversion 41–46
superficial velocity 37–40
liquid-fluidized bed 35, 36, 40, 46
distributor design 48
heat and mass transfer 46–48
homogeneous expansion 35, 36
particle segregation and mixing 40–41
voidage 37–40
liquid–solid mass transfer 571–572
local void size 166
lockhopper 232
log-normal distribution 16
loop seal 2, 219, 222, 239
louver baffles 432–435, 440–442
low-density circulating fluidized beds (LDCFB) 245, 258
lumped model, single-phase 346

m

macro solids circulation 447
magnetic fields 146, 450–451
mass transfer
bubble to dense phase interphase 320
in distributor jets 467
particle and fluid 318–319
spouted beds 538
mass transfer coefficient 350, 367, 368
bed-to-surface 47
interphase 138, 320, 553
gas–liquid 572
liquid 572
mass/mole balance 345, 346, 352
Mathur–Gishler equation 533
maximum heat transfer coefficient 299–304, 309, 312
maximum spoutable bed height 532, 534–535
mean particle diameter 16, 18, 25, 77, 136, 169, 278, 424
membrane reactor 2
membrane walls 2, 317
membrane water-wall assembly 317
mesh grids 432, 433, 442, 443, 445
methanol-to-olefin (MTO) 5, 438
micro flow structure 250–256
minimum bubbling velocity 24, 26, 55, 131, 206
minimum fluidization velocity 11, 18–25, 27, 34, 39, 55, 83, 104, 131, 147, 190, 205, 299, 314, 321, 330, 349, 383, 501, 534, 547, 577
minimum spouting velocity 532–534
mixing time distribution 93
model balance 347
model development, reactor 337–338
molecular diffusivity 169, 347, 368
multiphase computational fluid dynamics 547
multiphase particle-in-cell (MP-PIC) methods 110, 119, 123, 512
multiphase reactor modeling 557, 558
multiple spouted bed 543
multistage spouting 544

n

Navier–Stokes equations 120
Newtonian fluid 112
Newton's equations 116, 120
Newton's law 15, 511
non-conventional spouted beds 543–546
non-invasive measurements 87, 90–91
non-mechanical solids flow devices
automatic solids flow devices 219–222
control mode of operation 216–218
non-porous particles, reaction models for 365–373
shrinking particle reaction 365–366

non-porous particles, reaction models for (*contd.*)
  shrinking unreacted core particle reaction 369–373
non-rodable commercial nozzles 480
nozzle design
  bed hydrodynamics, impact of 473
  inclination 472
  laboratory nozzle 480
  opposing jets 473
Nusselt number 122, 293, 298, 310, 318
Nyquist sampling theorem 87

o
opposing jets 473, 474
optical probes 89–90, 94, 96, 107, 243
Orcutt's model 351
ordinary differential equations (ODEs) 337, 524
orifice plate 79, 321
overflow fluidized-bed dipleg 222
overflow fluidized standpipe 207–208, 212, 219
oxy-fuel 387
oxygen carrier (OC) 387
oxygen carrier aided combustion (OCAC) 387, 390

p
packed-bed flow 20, 206
packet theory 294–297, 301
pagoda-shape internals 437, 442, 4445
parallel cyclones 226–228, 408
partial differential equations (PDEs) 337, 377
particle aggregation processes 253, 255
particle attrition 274–276
  modeling 276–278
  tribocharging at distributor holes 467–468
particle-bubble interaction 570
particle clusters 35, 193, 252, 256
particle density
  definition of 86
  nonporous particles 13
particle entrainment 62, 568–569
particle fluidizability 18, 86, 196
particle mixing 5, 40–41, 139, 144, 171, 173, 569
particle segregation 20, 40–41, 144–145, 542
  model 43, 45, 49
particle shape 12–13, 17, 27, 562
particle sheets 253
particle size 12, 26
  critical 191
  distribution 3, 16, 23, 24, 27, 173, 187, 275, 413, 541
  transfer for 296
  mean 68, 86, 541
particle streamers/strands 253, 255
particle swarms 253, 255
particle terminal velocity 38–40, 62, 172, 257, 575, 577
particle velocity
  jet in fluidized bed 469, 470
  on jet centreline 469–470
parvoids 35
Peclet number 145, 169, 506
perforated plates 79, 433, 457, 458
phase holdups 565–569, 579
  estimation of 578
  in gas–liquid–solid fluidized bed 588–589
phase mixing 569–570
phosphorus particle tracer technique 258
pierced sheet grids 79
piezometric pressure 34
pilot scale unit 405, 408, 412
  biomass combined heat and power (CHP) 416–417
  calcium looping technologies 419
  dual gasifier 417–419
  purpose of 415–416
piping and instrumentation diagrams (P&IDs) 422
piston test 85
plenum chamber 3, 78, 81, 84, 88, 96, 458, 560, 561, 578
plug flow reactor (PFR) 347, 361, 556

porous particles 373
particle density 86
reactions of complete consumption 373–374
of unchanging overall size, reactions 374–377
positron emission particle tracking (PEPT) 91, 466
Prandtl number 122, 318
pressure drop 536–537
actual distributor 461
cyclone 223, 224
dry distributor 461
fluidized underflow standpipes 210
in overflow fluidized standpipe 207, 208
vs. superficial velocity method 18–20
underflow packed bed standpipe 208
pressure fluctuations
in bubbling fluidized beds 441
in fluidized beds 95, 459
standard deviation of 20, 56, 59–60, 163
pressure gradient force 111, 117
pressure measurement 87–89, 95
pressurized vessels 78
primary cyclone diplegs 212, 219, 220, 232
primary diplegs 219–220
process control documentation 423
process flow diagrams (PFDs) 420, 422
pulsating fluidized bed 451
pulsating vertical gas jet
expansion cycle 464
MRI measurements 464, 465
punched plate distributor 463
pure liquid jets 477
pycnometer 86

r

radial gas dispersion coefficient 170, 171, 173
radiative heat transfer 299, 310, 314–315
radioactive particle tracking (RPT) 91
rate-controlling step 365, 371, 373, 376, 377
Rayleigh–Taylor instability 137
reaction engineering model 508, 509
reaction equilibrium 339–340
reaction kinetics 340–342, 344, 378
apparent 342
intrinsic 342, 558
reaction rates 345–348
reactor compartments 579
reactor configurations 334
reactor design 582
fluidized bed catalytic reactors 334
gas–liquid–solid (GLS), fluidized beds 556–558
gas–solid 364
reactor economics 421
reactor energy balance 347, 348
reactor geometry 262, 555, 560
reactor model 580
reactor performance
coarse particles 385–386
fine particles 382–385
of circulating fluidized beds 258–260
evaluation of 581
reactor pressure balance 348
rectangular fluidized bed 321
relative humidity 17, 82, 273
residence time distribution (RTD) 93, 168, 169, 514
reverse-flow cyclones 223–226
Reynolds number 21, 22, 34, 38, 40, 46, 80, 120, 133, 169, 172, 368
Richardson–Zaki equation 37–40
riser
bubble cap 561
CFB 63, 219, 239, 243, 249, 253, 255, 256, 258, 260, 270
definition of 3, 63
vs. downer 514, 520, 522
FCC 63, 214, 245, 258, 482, 483
HDCFB 247
particle clusters in 252
two-stage 518

rise velocity 57, 166, 167
 of gas bubbles 568
rodable commercial nozzles 480–481
Rosin–Rammler distribution 16
rotameters 83, 84
rotary valves 231
round-nose slugs 57, 153–154, 158

S

saturation carrying capacity 62
Sauter mean 12, 16, 25, 44, 77, 278, 533
scale-up
 dimensional similitude 413–414
 framing questions 412
 hydrodynamic studies 412
 pilot-scale units 412
Schaeffer's model 115
Schmidt number 169, 318, 368
scrubbers 229
secondary cyclone dipleg 212, 213, 219, 220, 236
secondary diplegs 212, 213, 219, 220, 236
segregation 3, 40–41, 144–145
separation devices
 electrostatic precipitators 230
 fabric filters 229
 granular-bed filters 230
 scrubbers 229
 U-beam separator 230
series cyclones 226
Sherwood number 318, 368
shrinking particle
 chemical reaction control 366–367
 combined chemical reaction 369
 external mass transfer control 367–368
 external surface 374
 steps in 366–370
shrinking unreacted core model
 chemical reaction control 371
 combined chemical reaction 372–373
 internal diffusion control 371–372
shrouded jet 468, 481
shrouds 80, 481, 482
single bubbles, rising velocity of 135
single-phase models 346–348
single slugs
 cloud formation 157
 properties 156–157
slot-rectangular spouted bed (SRSB) 544, 545
slotted distributor 464
slug
 definition of 156
 frequency 159
 rising velocity 158
 single (gas) 156–157
 spacing and length 158, 159
slug flow 153–160
 continuous, hydrodynamics of 158–159
 experimental identification of 155–156
 gas in slugging beds 159–160
 in fluidized beds 155
 in liquids 155
 mixing of solids 159–160
 round-nosed 153–154
 square-nosed 154–155
 transition to 156
 types of 153–155
 uniformity and symmetry of 159
slugging bed
 as chemical reactors 160
 mixing of solids and gas 159–160
slugging fluidization, onset of 57–58
soft-sphere model 116, 117, 511
solid flow control devices/feeders 230–232
solid fuel combustion 386, 387, 389
solid particles 3
 and bed transport 560
 behaviour of 537–538
 heat capacity of 116
 parameters affecting entrainment of 191–195
 reactions of 377–378
 shrinking core reaction modulus 372
 single 364
solids
 circulation flux 92, 241

definition of 3
distribution 246–248
entrainment 478–479
holdup/voidage 243–246
mixing 144–145, 171–172
return systems 203–206
sampling 92–93
solid–solid reactions 363, 377, 378
spargers 79, 458, 459
pipe 211, 459
tube configuration 458
uses 459
sphericity
definition of 13
for nonspherical particles 15
splash zone 182
spouted beds 531–548
applications 546–547
chemical reaction 538–539
fluidized bed 541–542
heat and mass transfer 538
hydrodynamics 532–538
non-conventional 543–546
slot-rectangular 544
spout-fluid beds 540, 543, 545, 546
spouting *vs.* fluidization 539–540
spray nozzle design 478, 480
square-nosed slugs 154, 155, 157
standpipes 203, 206
automatic solid recirculation systems 212–213
configurations 207
controlled solid recirculation systems 213–215
function 215–216
gas-solids flow in 214
hybrid 203
for nonmechanical solids flow devices 216–222
overflow fluidized 207–208, 219
in recirculating solid systems 212
relative gas velocity flow 205
types of 206–212
underflow fluidized 210–212
underflow packed bed 208–209
stoichiometric coefficients 345, 366, 369
Stokes law 22
Stokes terminal velocity 15
streamtube model 539
subsonic jets 465, 466, 471, 474
superficial gas velocity 241, 257
superficial velocity
definition of 3
gas 4, 25, 408, 577
liquid 47, 571, 575, 577, 587
vs. pressure drop 18–20
spout 536
surface roughness 14, 17
surface tension, of liquid 12
surface-to-bed heat transfer 339, 569–571

t

tangential force 116, 117, 511
teetering 1
terminal settling velocity 14–15, 31, 39, 49, 58, 62, 66, 67, 119, 181, 184, 191, 193, 194, 580
thermal conductivity
effective 295, 297, 298, 302
of gas phase 122
probes 89
thermal conversion 363
of solid fuels 386–390
thermal flux probes 89
thermal properties, particle 16, 47
thermal response time 89
thermal sensors 89
thermal stresses 275, 418, 421, 449
thermoconductivity detector 93
thermocouples 89, 93, 278, 461, 555
thermodynamic equilibrium 339, 363
three-phase fluidization 554
three-phase spouting 546
tracer-observed gas streamlines 535
transition velocity 59, 64, 69, 163, 173, 411, 565, 577
transport disengagement height (TDH) 62, 139, 182–183, 185–190, 313, 410, 569

tribocharging 270–272
  in distributor jets 467–468
triboelectric charging 269, 270, 272, 273
turbulent fluidization
  flow regime 410–411, 442, 457
  onset of 58–61, 95
  termination of 62
  transition from bubbling to 164
turbulent fluidized beds 165, 166, 172
  axial and radial voidage distribution 165–166
  column diameter, effect of 172–173
  fines content, effect of 173
  gas mixing 168–171
  local void size and rise velocity 166
  solids mixing 171–172
  void phase and dense phase solids holdup 167
  void phase volume fraction 167
turnover time 144–145
two-fluid model (TFM) 110
  frictional model 114–115
  governing equations 110–112
  kinetic granular theory 112–114
  limitations of 115
  bubbling fluidized bed with tube bundle, simulation 123–124

u

U-beam separator 230
underflow fluidized standpipes 210–212, 214
underflow packed-bed standpipe 207–209, 217, 219
uniflow cyclones 223
uniform gas composition 378–381
uniform gas distribution 459–461
uniform reaction model 373
uniform voidage slip velocity 256
unsteady state gas convective heat transfer 295
upstream piping design 479–480
upward vertical jets 465

v

van der Waals forces 14, 55, 116, 181, 192, 270, 301
vertical baffles 432, 435–437, 445
viscosity, of liquid 12
visualization measurements 91–92
voidage
  bulk 16
  contour plot 124
  definition of 3
  distribution 165–166
  layer inversion 43–44
  at minimum fluidization 27–28, 147
  relationship 37, 40
  time-average 107, 108
voids 131
  fraction 14, 16, 111, 167, 168, 345
  phase volume fraction 167
von Mises yield conditions 115
vortex length, of cyclone 228

w

wastage, *see* wear
wear 4, 278–280, 464–465
wheat particles 531, 533
wheat spouting 536
windbox 3, 81, 83, 458, 460, 468, 541
Wurster tube 482

y

Young's modulus 117